RECENT ADVANCES IN EXPERIMENTAL MECHANICS
VOLUME 2

PROCEEDINGS OF THE 10TH INTERNATIONAL CONFERENCE ON EXPERIMENTAL MECHANICS / LISBON / PORTUGAL / 18-22 JULY 1994

Recent Advances in Experimental Mechanics

J.F. SILVA GOMES
Faculdade de Engenharia, Universidade do Porto, Portugal

F.B. BRANCO
I.S.T., Universidade Técnica de Lisboa, Portugal

F. MARTINS DE BRITO, J. GIL SARAIVA & M. LURDES EUSÉBIO
Laboratório Nacional de Engenharia Civil, Lisboa, Portugal

J. SOUSA CIRNE
Faculdade de Ciencias e Tecnologia, Universidade de Coimbra, Portugal

A. CORREIA DA CRUZ
Instituto de Soldadura e Qualidade, Lisboa, Portugal

VOLUME 2

A.A. BALKEMA / ROTTERDAM / BROOKFIELD / 1994

This edition of the proceedings is sponsored by the Calouste Gulbenkian Foundation and Junta Nacional de Investigação Científica e Tecnológica.

The texts of the various papers in this volume were set individually by typists under the supervision of each of the authors concerned.

Published by
A.A. Balkema, P.O. Box 1675, 3000 BR Rotterdam, Netherlands
A.A. Balkema Publishers, Old Post Road, Brookfield, VT 05036, USA

For the complete set of two volumes, ISBN 90 5410 395 7
For volume 1, ISBN 90 5410 396 5
For volume 2, ISBN 90 5410 397 3
© 1994 A.A. Balkema, Rotterdam
Printed in the Netherlands

Recent Advances in Experimental Mechanics, Silva Gomes et al. (eds) © 1994 Balkema, Rotterdam, ISBN 90 5410 395 7

Table of contents

7 Quality control and testing of materials and components

8 *Fracture mechanics and fatigue*

9 Biomechanics

5 Residual stresses

Recent Advances in Experimental Mechanics, Silva Gomes et al. (eds) © 1994 Balkema, Rotterdam, ISBN 90 5410 395 7

Automatic analysis of residual stresses in rails using grating interferometry

Małgorzata Kujawińska, Leszek Sałbut & Artur Olszak
Optical Engineering Division, Warsaw University of Technology, Poland

Colin Forno
National Physical Laboratory, UK

ABSTRACT

The method of automatic analysis of residual stress distribution in railway rail section using grating interferometry is presented. An annealing process for stress relieving connected with the application of high temperature resistant specimen grating is described. The experimental results of strain distributions: $\varepsilon_x, \varepsilon_y, \gamma_{xy}$, compared with the results obtained by hybrid FEM for a model rail are presented.

1 INTRODUCTION

Rail manufacturing processes and the action of train load in service, as well as the thermal stresses create residual stresses in railroad rails. These stresses, acting in conjunction with stresses produced by live loads, can threaten the safety of train operation by increasing the hazard of creation and propagation of fatigue cracks and the posibility of violent brittle breakages. To avoid these effects it is necessary to determine the level of residual stresses developed during the manufacturing process and under various operating conditions.

Several analytical and experimental methods have been described for estimating the magnitude of residual stresses in rails. The most frequently used model for prediction of residual stress in rails was developed by Orkisz et al. (1990). This formulation relies on the notion of a shakedown state established by the highest load to which the rail is subjected. Efficient implementation of this model is much less expensive than the recently used experimental techniques however it cannot take into account the influence of all manufacturing and exploitation condition.

As a result several experimental methods using strain gauges (Groom 1983), ultrasonic testing, X-ray diffraction, neutron diffraction (Webster et al. 1992) and grating interferometry (Czarnek 1992) have been proposed. High accuracy, high resolution, full-field monitoring and relatively low costs indicate that grating interferometry should be considered as the attractive technique for residual stress analysis. Several studies have been carried out on the rails in which the residual stresses were released by cutting a rail slice with a reflection grating fixed on its surface (Czarnek 1992, Sałbut et al. 1993). However, the cutting process performed on a high-frequency (1200 1/mm) grating easily causes its damage

and may introduce additional stresses in the sample tested. Also, the cutting process decreases the full-field analysis advantages of grating interferometry. A much better solution is to use an annealing process for the residual stresses relieving. This approach has been implemented for in-plane residual stresses measurement in a slice of railway rail.

2 OBJECT OF EXPERIMENT AND ITS ANALYTICAL MODEL

For verification of analytical and experimental methods of residual stress analysis in railway rails a controlled laboratory loading process is required. It can be realised by rolling the special wheel along the top of the testing rail (see Fig. 1a) with know contact force.

The part of new UIC 60 rail (made in Huta Katowice steel works) was rolled on EMS-60 testing machine (Świderski et al. 1992) in Central Research Institute of the Polish State Railways, Warsaw. The rolling process was

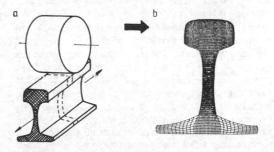

Fig. 1. Scheme of the rolling process(a) and the geometrical model of a rail slice (b).

performed with the load 150 kN, at a reciprocation frequency of 1 Hz and the number of cycles applied was 5×10^5. After the process, a 10 mm thick vertical slice was cut off from the central part of the rail.

The strain and stress distributions in the rail were calculated (Orkisz 1990) and later simplified for the in-plane strain and stress components in the rail slice by Magiera et al. (1992). Calculations were performed on a geometrical model of the rail which includes 2480 elements and 2768 nodes in the rail cross-section (see Fig. 1b). For experimental testing of the residual stresses and strains in railways rail, the grating interferometry method for in-plane displacement measurement was applied.

3 HIGH TEMPERATURE GRATING TECHNOLOGY

The annealing process for stress relieving in connection with the grating interferometry method can be applied if a high temperature resistant grating is attached on the specimen. Herein the process developed out at the National Physical Laboratory, UK, has been adapted (Kearney and Forno 1993, Forno 1994).
This method consist of the following steps:

a) Surface preparation
A smooth surface on the specimen is as a prerequisite for this sort of application of grating interferometry and as the pitch of the printed grating is 0.84 μm, the finish has to be mirror-like. In practice, this means polishing the specimen with the inevitable introduction of stresses close to the surface. These can be minimised through careful preparation which, in this case involved an initial slow grinding followed by a slow lap polish. In this way, the polish-induced stresses are within first 12 micrometers of the surface.

b) Coating with photoresist
Following the cleaning process, the samples are coated with photoresist. Spinning the resist onto fast rotating samples has proved to be the most effective means of producing a uniformly thin layer.

c) Grating preparation
A grating structure is printed into the photoresist by a method similar to the two-beam interference technique used in grating interferometry for interrogating the specimen grating. Instead of the usual He-Ne (633 nm) illumination, a blue wavelength laser such as the He-Cd (442 nm) is employed. The photoresist is sensitive to this wavelength. To generate the two-beam interference pattern, the angle between the overlapping beams is reduced to 30.8°, corresponding to a projected grating frequency of 1200 lines per mm. Next, before developing the resist, a stencil is placed on the top and a second exposure is made. The stencil consists of a photographic plate bearing a fine transparent line array which has a much coarser pitch than the grating (ap. 12 l/mm is suitable). It is important that the exposure is made long enough, so that after development the lines through the resist penetrate through to the component's surface. For the production of a crossed grating in the resist, the sample is exposed a second time, with a rotation of 90° between the two. Typical exposure times are in the order of 5 minutes total.

d Coating with heat resistant materials
Purpose of the resist grating is to form a mold for a subsequent thin layer of a heat resistant metal (here Nichrome was used). Metal is applied by plating, or vacuum deposition. Providing the layer is sufficiently thick, it will retain its molded shape when the resist is eliminated during the heating process. At temperatures above 300°, the resist will decompose and eventually vaporise through any cracks or holes in the metal film, leaving the molded grating attached to the substrate. Although the majority of the metal is attached to resist, in the regions of the course grid where the component is bare, the metal is "keyed" onto the component and acts as a fixture for the grating.

4 THE EXPERIMENTAL ARRANGEMENT AND METHODOLOGY

4.1 Experimental system
System for the measurement of x,y-displacements and strain calculation is shown in Fig. 2 (Sałbut et al.1993). It is based on the grating conjugated diffraction order interference (Post 1987, Patorski 1993). He-Ne laser beam, after passing through the pinhole system PS, is collimated by objective OC (F'=1000 mm, φ=180 mm. Two parts A and B of this beam illuminate symmetrically the reflection type diffraction grating fixed to the specimen. First diffraction order beams interfere and produce a fringe pattern with the information about in-plane displacements.

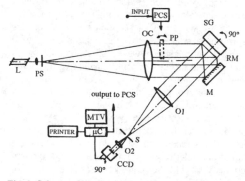

Fig. 2. Scheme of the grating interferometry system. L-He-Ne laser, PS-pinhole system, OC-collimator objective; 01, 02-imaging objectives; M-mirror; SG-sample with diffraction grating; RM-rotary mount; S-diaphragm; PSC-phase shift controller; PP-tilting plate

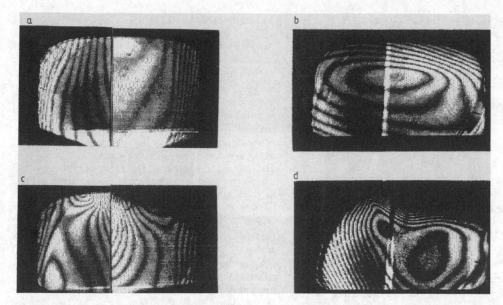

Fig. 3. The interferograms representing the in-plane displacement fields on a two-dimensional vertical slice of rail. The initial interferograms representing u_0 (a) and v_0 (b) in-plane displacement fields and the relevant interferograms after the residual stress relieving by annealing: u_1 (c) and v_1 (d).

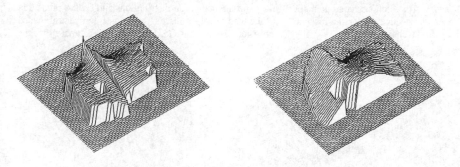

Fig. 4. The 3-D plots of the calibrated displacement fields u (a) and v (b).

Interferogram is collected by CCD camera using imaging objectives O1 and O2 and video signals from CCD are digitised using VFG-512 frame grabber in an IBM or compatible personal computer. The computer program DZF for automated fringe pattern analysis (copyright by IKPPiO-PW) based on the discrete temporal phase shifting method is used (Kujawińska 1992). Phase shifting is introduced by tilting plane parallel plate PP. Using the five-intensity algorithm, the phase ϕ is calculated from:

$$\phi(x,y) = a\tan\left[2(I_4 - I_2)/(I_1 - 2I_3 + I_5)\right] \quad (1)$$

where: I_1 to I_5 are the intensity values obtained for phase shifts 0, $\pi/2$, π, $3\pi/2$ and 2π.

After the unwrapping process the u displacement field can be calculated according to

$$u(x,y) = (d/4\pi)\phi(x,y) \quad (2)$$

where: d is the grating period (0.834 μm in our system). Rotary mounts of the specimen (RM) and CCD camera enable the measurement of u and v displacements for the cross-type diffraction grating used. After obtaining the u and v displacement fields, the strains can be calculated by numerical differentiation, i.e.,

$$\varepsilon = \frac{\partial u}{\partial x}, \ \varepsilon = \frac{\partial v}{\partial y}, \ \gamma_{xy} = \frac{\partial u}{\partial y} + \frac{\partial v}{\partial x} \quad (3)$$

4.2 The methodology of measurement

After preparing the sample in the way described in Sections 2 and 3, the measurement process is carried out as follows:

1 Mounting and adjusting the sample on the rotary mount RM

2 Registration of the interferograms: for u-direction, and after rotating the mount and CCD camera by 90° for v-direction.

3. Calculation of the u_0 and v_0 in-plane displacement fields

4 Annealing of the sample (in our case the sample was annealed in 600°C for three hours).

5 Mounting the sample on the rotary mount in the same position as in point 1. This operation is possible due to the three point support system used.

6 Registration and calculation of the u_1 and v_1 displacement fields.

7 calculation of the displacement values $u = u_1 - u_0$ and $v = v_1 - v_0$. In this step the initial distortion of the specimen grating and interferometer imperfections are removed.

8 Calculation of the strains: $\varepsilon_x, \varepsilon_y$ and γ_{xy}.

The results of the measurement can be presented in the form of 3-D plots, x and y cross-sections, contour maps and data tables and can be used directly for the residual stress determination by hybrid techniques .

5 EXPERIMENTAL RESULTS

According to the methodology of the measurement given in Sections 4.2, the initial fringe patterns representing u_0- and v_0- displacement fields were recorded (Fig. 3a) and analysed. After the annealing process the rails were repositioned in the interferometer and the final fringe patterns of u_1- and v_1-displacement (Fig.3b) were analysed. All u displacement maps suffer the lack of data in the central area due to the restricted field of view during exposure of the y-direction grating.

The resultant displacement fields (u, v) were obtained by the subtraction of the relevant maps and are shown in Fig. 4. These displacements are then applied for calculations of the residual strains relieved by annealing. Figures 5 a-c present the 3-D maps of $\varepsilon_x, \varepsilon_y, \gamma_{xy}$ with the region where the fringe patterns had the sufficient quality for the proper analysis.

Contour strain maps within the boundary of the rail are presented in Fig. 5 d-f respectively. Theoretically expected strain maps calculated for the centrally loaded slice of the rail (Magiera et al 1992) are shown in Fig. 6 a-f. Qualitative and quantitative comparison of the theoretical and experimental strain distributions indicates the difference in the theoretical and actual position of the applied load. Correcting for the shift of the strain peaks in the experimental strain distributions and on the prediction of the theoretical results for the case of non central loaded rail slice (Perlan and Gordon 1992) we estimate that the

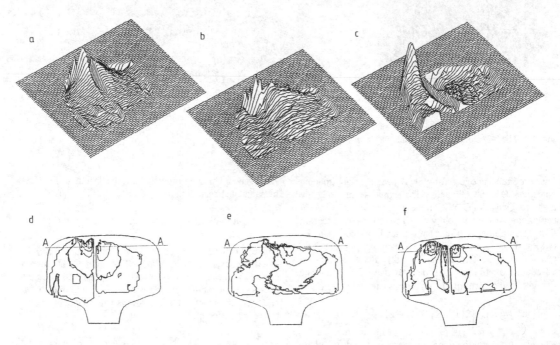

Fig. 5. The strain distributions calculated from u and v displacement fields: ε_x, ε_y and γ_{xy} given in form of 3-D plots (a-c) and the contour maps (d-f) respectively

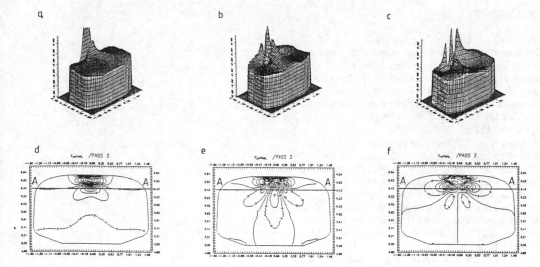

Fig. 6. The theoretical strain distributions calculated for the centrally loaded rail slice. The maps of ε_{xt} (d), ε_{yt} (e) and γ_{xyt} (f).

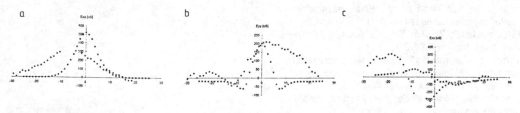

Fig. 7. The comparison of horizontal cross-sections (A-A) of the relevant theoretical and experimental strain maps of ε_x (a), ε_y (b) and γ_{xy} (c), respectively. ✱ ✱ ✱ -experimental; ■ ■ ■ -theoretical;

wheel load was applied 7 mm off-axis from the vertical centre line. Fig. 7 a-c show the horizontal cross-sections taken through the maximum at relevant theoretical and experimental strain maps. The magnitudes of ε_x and ε_y strain are in good agreement, although the experimental peaks are wider to the theoretical ones due to the asymmetric character of the strain distributions in non centrally loaded rails. This is also the reason of the increased values of the experimental shear strain γ_{xy} as shown in Fig. 7c.

Degree of agreement between the theoretical and experimental results of the in-plane components of the residual strain depends on:
- credibility of the theoretical model (Orkisz et al. 1990, Perlman and Gordon 1992)
- correctness of the assumption that the residual stress state is invariant in the longitudinal direction of rail
- accuracy of the displacement measurement using the automatic fringe pattern analysis (estimated at 40 nm)
- systematic errors caused by the initial distribution of the specimen grating and the interferometer imperfections.

(corrected by the subtraction of the initial displacement fields u_0, v_0).

6 CONCLUSIONS

Grating interferometry system for automatic analysis of residual stresses in railway rails has been presented. Several modifications of the interferometer set-up, technology of the high-temperature resistant grating and the analysis software were described. The initial results of the strain distribution determination and its comparison with the theoretical results proves the suitability of the experimental approach.

In future, the presented system will be applied for:
- determining the full stress field $\left(\sigma_x, \sigma_y, \sigma_z\right)$ by analysing obliquely cut samples,
- developing the hybrid technique for the full residual stress determination.

This project was supported by the National Scientific Research Council under the grant No. 295/P4/92/03.

7 REFERENCES

Magiera J., M.Hołowiński, J.Krok, J.Orkisz 1992 "Determination of residual stress distribution in plane, vertical slice of UIC-60 railway rail on the base of numerical solution of 3-D elastic-plastic problem" *Internal Report, Cracow University of Technology*

Groom J.J. 1983 "Determination of residual stresses in rails" *Battelle Columbus Laboratories, Columbus, OH, report no. DOT/FRA/ORD-83-05*

Orkisz J.et al. O.Orringer, M.Hołowiński, M.Pazdanowski, W.Cecot 1990 "Discrete analysis of actual residual stresses resulting from cyclic Loadings", *Computers & Structures" 35(4), 397-412.*

Perlman A.B., J.E.Gordon "Application of the constrained minimisation method to the prediction of residual stresses in actual rail sections", *Residual Stress in Rails, vol. II,151-177, Kluwer Academic Publ. 1992*

Czarnek R., J.Lee, S.-Y. Lin 1992 "Moiré interferometry and its potential for application to residual stresses measurement in rails", *Residual Stress in Rails, vol.I, 153-167, Klawer Academic Publ. 1992*

Sałbut L., M.Kujawińska, J.Kapkowski 1993 "Automatic analysis of residual stresses in rails in the modified moiré interferometry system", *Proc. SPIE 2004*

Świderski Z., A.Wójtowicz 1992 "Plans and progress of controlled experiments on rail residual stress using the EMS-60 machine", *Residual Stress in Rails, Vol I, 57-66, Kluwer Academic Publ. 1992*

Kearney A., C.Forno 1993 "High temperature resistant gratings for moiré interferometry", *Experimental Techniques, November/December 9-12 1993*

Forno C. 1994 "Moiré interferometry gratings for high temperature applications", *Proc. SPIE Conf. 'INTERFEROMETRY'94' Warsaw 1994*

Post D. 1987 "Moiré interferometry" in *Handbook of Experimental Mechanics, A.S.Kobayashi ed., Englewood Cliffs 1987*

Patorski K. 1993 "Handbook of the - *Moire Fringe Technique", Elsevier Science Publishers, Amsterdam 1993*

Kujawińska M. 1992 "Expert system for analysis of complicated fringe patterns", *Proc. SPIE 1755, 1992*

Recent Advances in Experimental Mechanics, Silva Gomes et al. (eds) © 1994 Balkema, Rotterdam, ISBN 90 5410 395 7

Electron beam moiré measurement of residual strains around Vickers impression in a tungsten carbide composite*/**

Ouk S. Lee
Mechanical Engineering Department, Inha University, Inchon, Korea

David T. Read
Materials Reliability Division, National Institute of Standards and Technology, Boulder, Colo., USA

ABSTRACT: An electron beam moiré method was employed to measure residual strain produced by the Vickers indenter in a WC-4.7 wt.% Co specimen. Line gratings, 57 μm wide by 45 μm high, with a pitch of 87 nm were written by electron beam lithography. An interior region of the grating was loaded by the Vickers indenter with 9.8 N for 30 s. The measured residual strains were fitted to the theoretical values estimated by two available models (Yoffe model, and Chiang, Marshall, and Evans (CME) model). Two models used experimental parameters such as the strength of field in the Yoffe model and the free surface correction factors in the CME model. The tangential surface residual strain (TSRS) were fitted well by two models except in the near vicinity of the Vickers impression. The ratio of radial surface residual strains to TSRS was measured as 2.4, while the Yoffe and CME models predicted 2.1, and 2.39 - 2.43 with Poisson's ratio $\nu = 0.22$, normalized plastic zone size $\beta = 2.52$, and free surface correction factors m and m_r = 0.125, and 0.235, respectively.

1 INTRODUCTION

The indentation technique has been used for many years to characterize fracture and deformation processes in hard materials such as carbides and glasses simply because of its simplicity (Marshall & Lawn 1979). Furthermore, it has been claimed that fracture toughness can be successfully estimated from Vickers indentation results.

Very recently, the residual stresses in soda-lime glasses measured using the deflection method and X-ray diffraction were found to agree with those obtained by the pointed-indentation technique (Chandrasekar & Chaudhri 1993) in which the theoretical background was adopted from two models (Yoffe 1982, Chiang, Marshall & Evans (CME) 1982). It is interesting to note that two models predict different stress distribution; i.e., the compressive (Yoffe) and tensile (CME) tangential surface stresses at the vicinity of the indent under the identical loading. However, the tangential surface residual stresses (TSRS) under full unloading are predicted as tensile by the two models. The Yoffe model does not include the plastic zone at the surface of specimen around the impression, while the CME model allowed large plastic zone below the indentation as well as the specimen surface around the indent. Furthermore, controversy on the detailed deformation mechanism in the region around various indenter impressions persists.

The main purpose of this paper was to measure nanoscale residual deformation around the Vickers impression in a hard material. The electron beam moiré (EBM) technique (in which the pitches of specimen line and reference gratings were 87 and 76 nm, respectively) was used to measure the residual strains after full unloading of the Vickers indenter. The measured residual strains were compared to those estimated using the Yoffe and the CME models.

2 THEORETICAL BACKGROUND

The pertinent equations for TSRS within elastic region around Vickers impression were derived based on the residual stress fields given by the Yoffe and the CME models, respectively, as the following:

$$TSRS_{Yoffe} = \frac{B(1+7\nu)}{Er^3} \qquad (1)$$

$$TSRS_{CME} = \frac{H(1-\nu)(1-2\nu)}{E(1-m)\Omega^2}$$

$$x \left[\frac{1}{1+3\ln\beta} \left(3\beta\ln\beta - \frac{1}{2}\beta + \frac{3}{2} - \frac{\beta^3}{4\Omega^2} \right) \right.$$

$$\left. - (\beta-1) - (1-m_r)\left(\frac{1}{\Omega^2} - 1 \right) \right] \qquad (2)$$

where B = a constant representing the strength of the field; r = radial distance from the center of ball indentation; E = elastic modulus; and ν = Poisson's ratio; H = hardness; m = free surface correction factor for peak load; m_r = free surface correction factor for residual stress; Ω = normalized radial distance from the center of ball indentation (= r/a); β = normalized plastic zone size (= b/a); and b = plastic zone size around indenting impression.

a = radius of ball indentation

$$= \bar{a} \times (\cot (\psi)/4.44)^{1/3} \qquad (3)$$

$\bar{a}$ = half diagonal of the Vickers impression; and ψ = half included angle between opposite faces of the indenter pyramid (= 68°).

3 EXPERIMENTAL

3.1 Specimen material and EBM grating

The material used in this study was tungsten carbide with 4.7 wt.% cobalt as a binder (WC4.7Co). The modulus of elasticity and Poisson's ratio are 620 GPa and 0.22, respectively. The nominal size of tungsten carbide skeleton is 1-5 μm. A typical microstructure taken by using scanning electron microscope (SEM) is shown in Fig. 1.

To free specimen surface from residual stresses, 0.1 mm in thickness (including 0.02 mm deep skin in which residual stresses by EDM procedure may exist) was carefully polished away with 0.25 μm diamond particles.

The second step was the application of 100 nm thin polymethylmethacrylate (PMMA) coating on the mirror surface specimen. This was achieved by spinning 2% volume PMMA (molecular weight of 950000 in chlorobenzene) at a speed of 2250 rpm for 30 s. The PMMA coating serves as an EB resist. The specimen was then baked on a hot plate for 90 minutes at 171°C. Some critical information on the writing of line gratings are as the following: time gap between baking and writing = 3 hours; writing duration = 220 s; probe current = 4 pA, vacuum of SEM = 4.8 x 10^-6 torr; accelerating voltage of SEM = 20KeV; and filament current of

Fig. 1. A typical micrograph of WC4.7Co taken by SEM.

Fig. 2 A typical image of line gratings in a 2000x pattern at a magnification of 60000x on SEM.

SEM = 255 μA. The general procedure for producing the line gratings should be referred to a recent paper (Lee & Read 1994). A typical image of line gratings in a 2000x pattern recorded on an SEM at a magnification of 60000x is shown in Fig. 2. The wandering in

Fig. 3 Linear mismatch EBM fringes by specimen line gratings (pitch = 87 nm) and scanning lines of EB (pitch = 76 nm) (grating area = 57 μm wide by 45 μm high).

line gratings appeared in Fig. 2 is due to wavy motion of the incident EB and random nature of electron backscattering.

3.2 EBM fringe patterns

In the EBM method, the scan pattern of the SEM acts as a reference grating to produce EBM fringes associated with specimen line gratings written on the thin PMMA resist. The reference pitch can be varied by adjusting either the magnification of SEM or the number of scan lines in the image. Fig. 3 shows a small grating area of 57 μm wide by 46 μm high covered with EBM fringe pattern in the WC4.7Co specimen. The EBM fringes shown in Fig. 3 were generated by the mismatch in pitches between specimen line gratings P_s = 87 nm and scanning lines of EB P_r = 76 nm.

4 RESULTS AND DISCUSSION

The surface residual deformations produced by Vickers indentation were recorded by using the EBM method. In the EBM method, we could modulate the sensitivity of measurement by adjusting the pitches of specimen line gratings and/or reference gratings. Very fine specimen line gratings as shown in Fig. 2 were needed since the actual surface residual deformation produced by the Vickers indentation was very small.

Fig. 4 shows EBM fringe patterns at magnification of 2300x on SEM, appeared around the Vickers indenting impression after removing the indenter. The indenting load and time were 9.8 N and 30 s, respectively. The change in EBM fringe pattern (comparing to the EBM linear mismatch fringes) at the near vicinity of the Vickers impression is clearly visible in Fig. 4. The half diagonal of the Vickers impression $\bar{a}$ and the radius of ball indentation a were 15, and 6.75 μm, respectively. Strain measurements were carried out at 6 positions along a radial line passing the center and an apex of the Vickers impression.

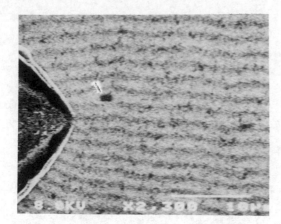

Fig. 4 EBM fringe patterns by the linear mismatch and residual deformation at 2300x on SEM.

The displacement (= N x p where N = fringe order number by residual displacement and linear mismatch EBM fringes; and p = pitch of reference gratings) along each of the vertical lines were fitted to polynomials of six degrees. The residual strains along vertical lines were estimated from the displacement values as:

$$e_{yyres} = p\frac{dN}{dy} - constant \qquad (3)$$

where constant = p x dN'/dy; and N' = fringe order number of linear mismatch EBM fringe patterns.

It was assumed that there was no residual strain at the position 1 (ψ_1 = r/a = 6.47) since the residual stresses decrease to zero very rapidly. The reference strain produced by the linear mismatch EBM fringe pattern at position 1 was estimated as 0.03987 by using a 2 term polynomial best fit procedure with 17 experimental measuring points. Fig. 5 (a) and (b) shows two dimensional isometric strain contours and three dimensional view of surface residual strain distributions measured using the EBM method around the Vickers impression in a WC4.7Co specimen.

The theoretical residual strains obtained using the Yoffe and CME models were compared to measured values. The CME model requires the plastic zone size to be known a

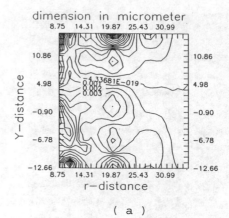

(a)

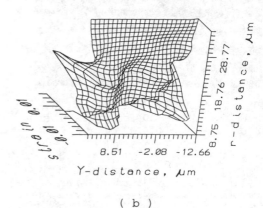

(b)

Fig. 5(a) 2-dimensional isometric contours; and (b) 3-dimensional view of surface residual strains measured using the EBM method.

priori around the indenting impression on the specimen surface since the free surface correction factors appeared in eq. (2) are given as normalized values by the plastic zone size. The apparent plastic zone size at near vicinity of the Vickers impression was assumed to be the same as that produced in PMMA coating. The size of apparent plastic zone b was measured as about 2 μm by identifying the permanently deformed region in PMMA coating as distinctly shown in Fig. 6. Parameters β (= b/a), m, and m_r appeared in eq. (2) were determined as 2.52, 0.125, and 0.235, respectively, by using Fig. 6 in the paper by CME.

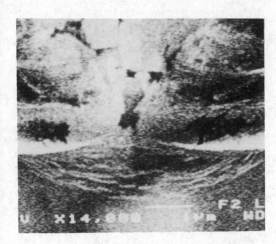

Fig. 6 Permanent deformation produced in PMMA coating around the Vickers impression.

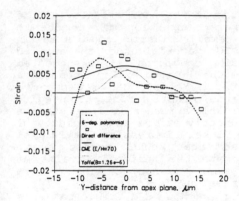

Fig. 7 Residual strain determined by displacement fittings, direct difference, and two models.

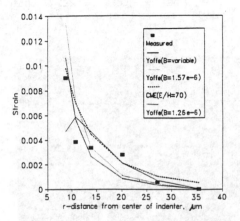

Fig. 8 TSRS determined by the EBM method and two models.

In eq. (1), B is a constant which represents the strength of the field. Yoffe chose $B = 0.06 \bar{p} a^3$ (where $\bar{p}$ = a uniform normal pressure; and a = radius of ball indentation) after a little computation as a fair average value considering the unavoidable uncertainties involved. B can be rewritten as the following since $\bar{p} = P/(\pi a^2)$:

$$B = 0.019 \ P \ x \ a \qquad (5)$$

where P = applied load.

Fig. 7 shows a typical residual strain distribution obtained by the EBM method, direct difference, and two models, along a y-line 22.09 μm ahead of the center of the Vickers impression. The measured residual strains at 6 positions on a radial line passing the center and an apex of the Vickers impression allowed us to estimate B as the following:

Position	1	2	3	4	5	6
r (μm)	35	27	20	14	11	9
B x 10^6	0	2	4	1.6	.8	1.1
Averaged B	1.57 x 10^{-6}					

The calculated B using eq. (5) (Yoffe 1982) with the Vickers indenting load of 9.8 N and a = 6.75 μm was 1.26 x 10^{-6}. This is considered to be in good agreement with the experimental results.

TSRS measured using EBM technique and calculated by two models along a radial line passing the center and an apex of the Vickers impression are shown in Fig. 8. We plot the Yoffe model three times: $B = 1.26 \ x \ 10^{-6}$; $B = 1.57 \ x \ 10^{-6}$; and B = variable. The variable B was obtained as B x 10^7 = -0.16 x r^2 + 6.71 x r + 41.18 by using a 3 term polynomial best fit procedure with six B. In the CME model, we used E/H = 70. CME gave no value for this parameter. Figure 8 shows that the measured TSRS agreed well with predictions by two models except at the near vicinity of the Vickers impression. The discrepancies at the near vicinity of the Vickers impression may be due to the small plastic zone size in WC4.7Co material under the applied loading condition. It is suggested from the limited experimental data that the Yoffe model may need to include plastic zone to predict residual stresses and/or strains accurately at the near vicinity of the Vickers impression. It is also noted that both models over-estimate the residual strains at the near vicinity of the Vickers impression.

Fig. 9 shows EBM fringe patterns around two apexes of the Vickers impression at a magnification of 1100x from which we can measure RSRS and TSRS simultaneously. From these EBM fringes, we measured the ratio of RSRS to TSRS (= REXY) and compared to theoretical predictions.

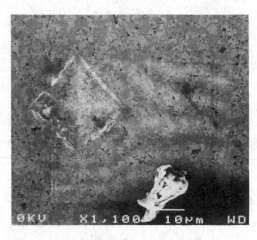

Fig. 9 EBM fringe patterns around two apexes of the Vickers impression at a magnification of 1100x on SEM.

The REXY were determined by the Yoffe and CME models to be 2.1, and 2.39 - 2.43, respectively, with ν = 0.22, β = 2.52, m = 0.125, and m_r = 0.235. We measured a value of 2.41 for REXY from Fig. 9, by using the distance between fringes in y direction at both apexes of the Vickers impression. Our experimental result agrees well with the predictions by the Yoffe and CME models.

5 CONCLUSION

An EBM method has been successfully applied to measure very small residual deformation of the order of a few nm in a WC4.7Co specimen. It was found that the Yoffe and CME models successfully predicted the TSRS produced by the Vickers indentation after full unloading except the near vicinity of the Vickers impression. It was also found that the measured ratio of RSRS to TSRS, 2.4, agreed well with 2.1 and 2.39 - 2.43 predicted by the Yoffe and the CME models, respectively.

ACKNOWLEDGMENT

The first author acknowledges the support of Korea Science and Engineering Foundation (Project number 92-23-00-13) during this investigation.

REFERENCES

Chandrasekar, S. & M.M.Chaudhri 1993. Indentation cracking in soda-lime glass and Ni-Zn ferrite under Knoop and conical indenters and residual stress measurements: Philosophical Magazine A 67: 1187-1218.

Chiang, S.S., D.B.Marshall & A.G. Evans 1982. The response of solids to elastic/plastic indentation, 1. Stresses and residual stresses: J. Appl. Phys. 53: 298-311.

Lee, O.S. & D.T.Read 1994. Electron beam moiré measurement of elastic-plastic strain transition at the tip of a sharp notch: will appear.

Marshall, D.B. & B.R.Lawn 1979. Residual stress effects in sharp contact cracking: Journal of Materials Science 14: 2001-2012.

Yoffe, E.H. 1882. Elastic stress fields caused by indenting brittle materials: Philosophical Magazine A 46: 617-628.

Recent Advances in Experimental Mechanics, Silva Gomes et al. (eds) © 1994 Balkema, Rotterdam, ISBN 90 5410 395 7

Local stress analysis in ULSI by microscopic Raman spectroscopy and computer simulation

H. Sakata, N. Saito, N. Ishitsuka & H. Miura
Mechanical Engineering Research Laboratory, Hitachi, Ltd, Japan

ABSTRACT : To investigate the residual stresses in semiconductor chips, we used a stress simulator called SIMUS to evaluate local stresses in silicon substrates with tungsten gate electrodes for ultralarge-scale integrated circuits and we compared the simulation results with the results of microscopic Raman measurements. The simulated values of stress in the silicon substrate at nanometer-order depths from the surface agreed with the measured values. When using microscopic Raman spectroscopy in combination with stress analysis, it is necessary to weight the near-surface stress values, especially for structures in which stress changes markedly with changes in depth.

1 INTRODUCTION

With the continuing move toward higher levels of integration in semiconductor devices, residual stresses in the silicon substrates are becoming a serious problem. These stresses, caused by lattice mismatch and differences between the thermal expansion coefficients of silicon and other film materials, degrade the electrical properties of Si substrates. They also cause mechanical problems during the large-scale integration (LSI) fabrication processes, problems such as cracking or peeling of the thin films and the creation of various defects in the silicon substrates. To evaluate these residual stresses and to ensure device reliability, there is an increasing need for techniques that can measure stresses in microareas. Hitachi has developed such a technique based on microscopic Raman spectroscopy (Sakata 1990, Miura 1990, Sakata 1990b, Sakata 1988).

In the work described in this paper, microscopic Raman spectroscopy (Sakata 1990, Sakata 1990b) was used to measure residual stresses in silicon substrates with tungsten gate electrodes for ultralarge-scale integrated circuits (ULSIs). Tungsten gate electrodes are of great interest for ULSI applications because of their low resistivity. The local stress distributions in a submicrometer area of a Si substrate, near the edge of a tungsten thin film, were obtained by the type of Raman spectrum analysis developed by Hitachi (Sakata 1990, Sakata 1990b). The stresses were also simulated on a supercomputer by using a simulator called SIMUS (Saito 1991) developed for finite-element stress analysis of electron devices with multilayer structures. The simulation results were compared with these of the Raman microprobe measurements.

2 ANALYSIS METHOD

2.1 *Object of analysis*

The object of analysis was a gate electrode structure (Fig. 1) fabricated in the following way. After a 20-nm-thick SiO_2 layer was formed by oxidizing a p-type (100) silicon

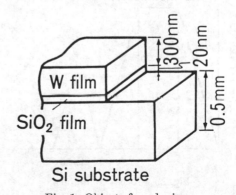

Fig. 1. Object of analysis.

substrate, a 300-nm-thick tungsten film was deposited by dc magnetron sputtering and patterned as a gate electrode (Sakata 1990, Sakata 1990b). The simulated process conditions are shown in Fig. 2 (temperature-time chart).

2.2 Method

The sample shown in Fig. 1 can be modeled for analysis as shown in Fig. 3. This analytical model uses the equivalent substrate thickness concept (Saito 1991). Rather than the entire 500-μm thickness of the silicon substrate, a 100-nm thick dummy layer inserted below a certain thickness (5 μm) of substrate was analyzed. In this way, the analysis region could be made very small.

The SIMUS stress simulator (Saito 1991) was used for the analysis. There are 2491 node points and 788 elements in the mesh diagram used in this work. The smallest element dimensions (near the edges) were 20 nm in both the x and z directions.

2.3 Boundary conditions

The analysis was done for generalized plane strain condition (tying* of the y direction displacement) using the boundary conditions shown in Fig. 3:

- displacement u = 0 at x = -1 μm
- tying* is applied for u direction displacement at x = 10 μm
- displacement w = 0 at z = -5.1 μm

[* Applied in such cases as transformation that preserves the linearity of the edge by boundary conditions that make the displacement of node points dependent on the displacement of other node points.]

2.4 Thin-film material constants

The material constants used in the analysis are listed in Table 1. The value given for the intrinsic stress of tungsten at the time of deposition (-1500 MPa, where the negative sign indicates compressive stress) was obtained during sputtering with an argon pressure of 0.133 Pa (Yamamoto 1987). However, because this data was obtained from the warping of the substrate at room temperature, the thermal stress component (approximately 20 MPa) was neglected and it was assumed that the warping is due only to intrinsic stress.

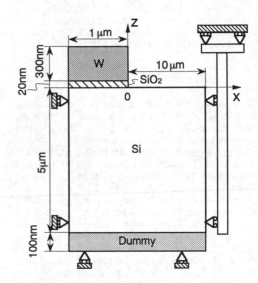

Fig. 3. Device structure with the boundary conditions for simulation.

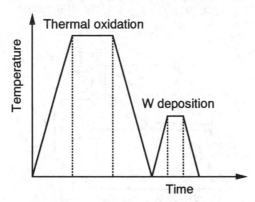

Fig. 2. Process conditions.

Table 1. Material constants for simulation.

Material	E (GPa)	ν	α(10^{-6}/°C)	Intrinsic Stress σ_I (MPa)
W	362	0.35	4.3	-1500
SiO2	70	0.17	0.6	0
S i	150	0.2	3.8	0
Dummy	180000	0.2	3.8	0

712

3 COMPARISON OF ANALYSIS RESULTS AND MICROSCOPIC RAMAN MEASUREMENTS

3.1 *Stress analysis results*

The analysis results for the stress distribution (here, σx) at various depths from the surface of the silicon substrate are shown in Fig. 4. We can see that the stress σx in the part of the silicon substrate from which the tungsten has been removed is a compressive stress that becomes larger nearer the surface of the silicon substrate and changes markedly as the distance from the tungsten edge increases. At depths beyond about 200 to 300 nm the edge has no observable effect on the stress concentration . In the part of the silicon substrate under the tungsten film, on the other hand, for depths of up to about 200 nm the stress σx at about 70 nm from the tungsten edge changes to a tensile stress. In addition, this tensile stress is larger nearer the substrate surface.

3.2 *Comparison with microscopic Raman measurements*

The stress measurements made by microscopic Raman spectroscopy are plotted in Fig. 5 with the distance from the tungsten edge as the horizontal axis and the compressive stress as the vertical axis. For comparison with these measured values,

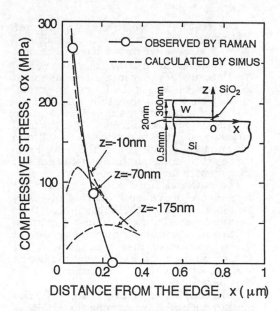

Fig. 5. Local stress distributions (comparison of Raman measurements and analysis results).

the stress analysis results from Section 3.1 (Fig. 4) are represented in Fig. 5 as broken lines. We can see good agreement between the measured stress distribution and the one simulated for a depth of about 10 nm.

The depth to which light from a 514.5-nm Ar+ laser will penetrate a silicon substrate in microscopic Raman spectroscopy is theoretically about 0.7 μm.

We conclude from these considerations and the agreement between the measured and calculated stress distributions in the region far nearer to the surface than the penetration depth of the laser beam (Fig. 5) that when the stress distribution in the depth direction changes markedly, as it does in this specimen, the values very near the silicon surface (on the nanometer order) should be given reasonable weighting. This suggests that the Raman technique can be used to measure the stress in a region much shallower than is penetrated by the laser beam.

3.3 *Identification of the stress measurement region*

In this section, we describe how the stress measurement region for the microscopic Raman spectroscopy is identified (i.e., the quantitative weights are determined).

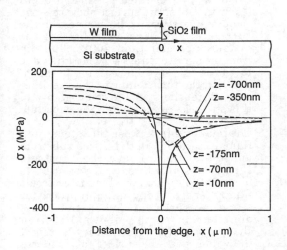

Fig. 4. Stress distributions at various depths from the surface of the silicon substrate.

The results of considering the Raman scattering intensity from a silicon substrate show that even when the laser light does penetrate to a depth of 0.7 μm, about 60% of the Raman scattering intensity (equivalent to the vertical axis of the Raman spectrum) comes from the region within 0.1 μm of the surface. Furthermore, the region between 0.1 and 0.2 μm accounts for 25% of the total intensity and the region between 0.2 and 0.3 μm accounts for 10% of the total. From there, the emission intensity drops off sharply with increasing depth.

The data described above quantify the dependence of Raman scattering intensity on distance below the substrate surface. Because of this strong relationship between depth and intensity it is necessary to weight the calculated values when the stress values change markedly very near the surface, as they do in the specimen under consideration here.

Consider the weighting of the measurements. If we assume that 100% of the observed scattering intensity is emitted from within a depth of 40 nm (that is, 53% is emitted from the region between the surface and the depth of 20 nm and 47% is emitted from the region between 20 and 40 nm), there is good agreement with the measured values (Fig. 6). Although it was stated in

Section 3.2 that the simulated stress distribution for a depth of about 10 nm from the silicon substrate surface agrees with the measured one (Fig. 5), consideration of the principle of microscopic Raman measurements suggests that it might not be possible to observe only the stress at the point 10 nm below the surface. Thus, the stress measurement region assumed here is considered appropriate.

4 CONCLUSIONS

We used a stress simulator called SIMUS, developed in our laboratories, to evaluate local stresses in the silicon substrate of a ULSI tungsten gate. Comparison of the simulation results with the results of microscopic Raman measurements leads to the following conclusions.

1) The SIMUS analysis results show that deposition of the tungsten layer generates a compressive stress in the exposed surface of the silicon substrate near the tungsten edge and that this compressive stress decreases sharply with increasing distance from the edge. These results are qualitatively consistent with measurements by microscopic Raman spectroscopy. Quantitatively, the measured values agree with the simulated stress distribution values for the region very near the substrate surface (i.e., within nanometers).

2) The results of analyzing the relationship between the generated stress and depth below the substrate surface show that the compressive stress is larger nearer the surface of the exposed substrate: at distances greater than 200 to 300 nm from the surface, stress concentration due to the tungsten edge cannot be seen. Beneath the tungsten layer, on the other hand, a tensile stress is generated and this stress is larger nearer the substrate surface.

3) The relationship between the intensity of Raman scattering in the silicon substrate and the distance below the substrate surface was quantified and was used to identify the region for which microscopic Raman spectroscopy is appropriate for stress measurement. When the stress distribution in the depth direction changes sharply, it is necessary to weight the values near the surface (at distances of nanometer order).

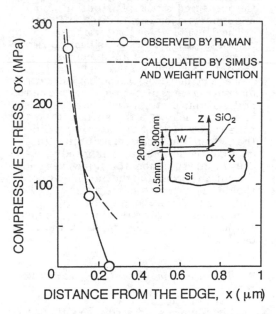

Fig. 6. Local stress distributions (comparison of Raman measurements and analysis results with weight function).

ACKNOWLEDGEMENTS

The authors wish to thank Dr. N. Okamoto, Dr. S. Sakata, Dr. T. Hatsuda and Mr. H. Ohta for helpful discussions, Mr. T. Furusawa for help with the calculation and Dr. N. Yamamoto for help with the sample preparation.

REFERENCES

Miura H., Sakata H. and Sakata S. 1990. Proc. of 9th Int. Conf. on Exp. Mech. 3: 1301-1306.

Saito N., Sakata S., Shimizu T. and Masuda H. 1991. Proc. of Int. Conf. on Compu. Eng. Sci.: 880-883.

Sakata H., Dresselhaus G., Dresselhaus M. S. and Endo M. 1988. *J. Appl. Phys.* 63: 2769-2712.

Sakata H., Hatsuda T. and Kawai S. 1990. Proc. of 9th Int. Conf. on Exp. Mech. 3: 1307-1313.

Sakata H., Hatsuda T. and Yamamoto N. 1990. Proc. of 12th Int. Conf. on Raman Spectroscopy. 822-823.

Yamamoto N., Iwata S. and Kume H. 1987. *IEEE Trans. on Electron Devices* ED-34, 3: 607-614.

Recent Advances in Experimental Mechanics, Silva Gomes et al. (eds) © 1994 Balkema, Rotterdam, ISBN 90 5410 395 7

Measurement of stress distribution during laser irradiation by using irreversible stress-induced phase transformation in tetragonal zirconia

T. Miyake, F. Asao & C. O-oka
Nagoya Municipal Industrial Research Institute, Japan

ABSTRACT: To acquire stress which is difficult to measure directly because of temporal and spatial limitations, measurement of phase transformation in tetragonal zirconia by Raman microprobe was employed, noticing some of the phase transformation is induced irreversibly by stress and Raman microprobe has high spatial resolution. Applying this idea to measurement of stress in a limited area caused by a transient phenomenon, spatial distribution of transformed particles corresponding to experienced stress during the process can be obtained in the suffered specimen after the process has finished. Actually the stress distribution in zirconia-toughened Al_2O_3 caused by irradiation of CO_2 laser, which attracts attention as an alternative cutting method of brittle materials, was obtained experimentally in micrometer order resolution, with correlating fraction of transformed particle to stress through an uniaxial calibration test.

1 INTRODUCTION

Advanced ceramics are going to be applied widely in such engineering fields where thermal, abrasive and chemical resistance characteristics are required. With increase of such ceramics use, the traditional machining process with diamond tools becomes insufficient because of its inflexibility and cost. Laser is one of alternative machining methods because of the advantage of removing brittle materials efficiently and flexibly.

But the heat affected layer in the laser machined work results in both strength and surface quality degradation. Especially the thermal stress due to temperature gradient may initialize cracking and be a main cause of strength degradation.

To evaluate the thermal stress, many numerical simulations have been performed. But many assumptions have to be made in the simulations for characteristic values of machined materials and boundary conditions, i.e. absorptivity of laser light, thermal emissivity, coefficient of heat transfer, and so on. Moreover those values have temperature dependency and also change with both surface conditions and its phase, like melted or not. Simulated results may vary with those assumptions, therefore experimental measurement of the thermal stress becomes very impor-

tant. However, to measure the stress is difficult due to both temporal and spatial limitations. The phenomena finish only in a few milliseconds within area of only some hundred micrometers in diameter.

In the previous work (Miyake 1993), we clarified the extent of surface layer affected by grinding, noticing phase transformation from tetragonal to monoclinic form, occurs beyond some critical stress in tetragonal zirconia polycrystal (TZP) ceramics. From experimental results by Swain (1985), residual strain was observed in TZP. This implied the stress-induced phase transformation is irreversible and monoclinic phase could be observed even after stress removal and closely related to the applied stress. By applying this idea to laser machining, the stress during laser irradiation could be frozen as monoclinic form like plastic deformation in metals. The fraction of monoclinic particles in machined works, i.e. the quantity of the irreversible transformation, could be used as an index which represents the maximum stress experienced during the process.

To measure monoclinic concentration in TZP containing ceramics, X-ray diffraction method is commonly used (Garvie 1972), but it is inadequate to determine the extent of the transformation on laser irradiated surface because of limited spacial resolution. On the other hand, the Raman microprobe has high ability to distinguish between monoclinic and tetragonal polymorphs in micro spatial resolution and is useful for investigating distribution of transformed particles (Dauskardt 1990).

In the present work, the transformation behavior of Ce-TZP toughened alumina (ZTA) on the laser irradiated surface and the cross section normal to the irradiated surface was investigated by Raman microprobe. With correlating transformed fraction to stress through an uniaxial compression test, the distribution of thermal stress suffered during laser irradiation was evaluated both in the radial direction and the direction of depth,

2 EXPERIMENTAL PROCEDURE

The materials used in the experiments were TZP containing 5 mol% CeO_2 and 1 mol% Y_2O_3, and its toughened Al_2O_3 (30 wt% ZrO_2) sintered at atmospheric pressure and obtained commercially (Hitachi Chemical Co., Ltd., Tokyo, Japan). The materials were cut and polished on the irradiation surface to plates of 4mm thick for laser irradiation test and also ground to rectangular bars of 10mm height and 4mm width for compression test. No monoclinic phase was indicated on the ground and polished surface by Raman microprobe analysis.

Laser irradiation tests were performed using CO_2-laser at an output intensity of 5W. In order to change the induced thermal stress, the irradiating duration was controlled between 8 and 33 ms by shutting the cw-mode laser beam. Spot diameter on specimen surface was enlarged to about 1mm by defocusing to avoid melting at the irradiated spot. To avoid making the phenomena complicated cooling gas was not employed.

Calibration tests were carried out for Ce-TZP in uniaxial compression at a crosshead speed 0.5mm/min, measuring strain by four strain gauges mounded in the axial direction. The monoclinic content was measured after loading to a certain level below failure and unloading.

The Raman spectra were acquired in a 180° backscattering geometry using a microscope system. A 100× objective lens was used to focus the incident light and the illuminated spot diameter on the sample surface was 1-2μm order. The 488 nm line from an argon ion laser at the laser output power of 300mW was used. The scatter light was collected by the microscope objective and dispersed with a triple monochromator (MR-1100, Japan spectroscopic Co., Ltd. Tokyo, Japan) and detected with an intensified photodiode array. The distribution of monoclinic phase was measured by mapping with 10μm step in the radial direction and 5μm step in the direction of depth.

The monoclinic concentration is calculated using peak intensities ratio of the monoclinic doublet (at 181 and 192 cm^{-1}) to the tetragonal bands (at 148 and 264 cm^{-1}), with an factor to coincide with result by X-ray diffraction (Clarke 1982).

3 RESULTS AND DISCUSSION

3.1 Monoclinic distribution on the laser irradiated surface and its normal section

In Raman microprobe measurement, laser light must penetrate to a certain depth of specimen before scattering, and then return to the surface. Therefor the intensity of the collected light for Raman measurement, scattered from any depth z, will be proportional to $e^{-2\alpha z}$, here α is the absorption coefficient. To acquire the penetration depth of ZTA experimentally, a wedge shape specimen as illustrated in figure 1 are prepared and the Raman spectra intensities obtained from various thickness were examined. In figure 1, obtained Raman signal intensities normalized by that from a specimen thick enough are plotted versus thickness. From this result, it is possible

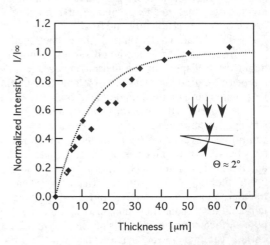

Figure 1. Raman signal intensities obtained from various penetration depth.

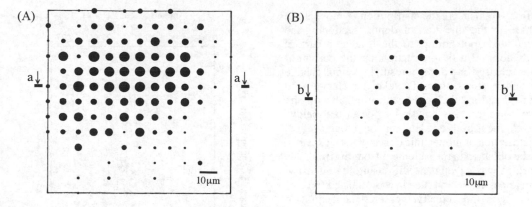

Figure 2. Distribution of monoclinic phase on laser irradiated surface, (A) laser irradiating duration ≈ 17ms; (B) laser irradiating duration ≈ 8ms. Location and area of dots indicate the location and the concentration of monoclinic phase, respectively.

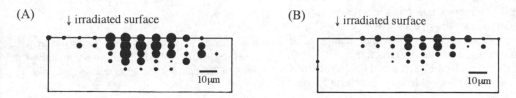

Figure 3. Distribution of monoclinic phase on cross section normal to the laser irradiated surface, (A) laser irradiating duration ≈ 17ms (a-a section in figure 2) ; (B) laser irradiating duration ≈ 8ms (b-b section in figure 2). Dots indicate the same in figure 2.

to know how the light penetrating to any depth contributes to the obtained Raman spectra. The information from less than 10μm depth contains over 50% of the acquired overall signal. Eliminating the effect of light penetration with this intensity-thickness curve, the observed monoclinic concentration was adjusted into that just on the surface.

Due to laser irradiation, the material in the vicinity of the irradiated spot will expand more than the surrounding material. This causes compressive stress within the irradiated spot and tensile stress at the surroundings. The compressive stress is supposed to be much higher than the tensile stress because the mass which is subjected to the compressive stress is much less than that of the tensile stress.

Figure 2 shows the distribution of monoclinic phase observed by Raman microprobe on the laser irradiated surface of ZTA, after adjusted the laser penetration effect in Raman microprobe measurement. The monoclinic phase distributed sym-

metrically with respect to the beam center as expected. But the monoclinic phase was observed only around the center of the irradiation beam, thereby knowing that the transformation to monoclinic phase caused only in the compressive stress field. The compressive stress exceeded the critical stress, beyond which the irreversible stress-induced transformation mechanism operated, but the tensile stress did not. This implied that the compressive stress much higher than the tensile stress and was in good agreement with the preceding consideration and the simulated result (Ito 1990). Moreover the monoclinic fraction well reflected the difference in the laser irradiating duration, that the longer duration caused the larger transformed region, agreed with simulated results that the high compressive stressed region increased with laser irradiating duration (Morita 1991).

Not only stress distribution on the irradiated surface but also that in the direction of depth is important information to clarify the laser induced stress. To measure the distribution of monoclinic phase in the direction of depth, sectioning the irradiated spot normal to the irradiated surface was done. But by the sectioning, the constraint was released and it might result in serious change with the stress state. To take the change into consideration, the measurement results on the cross section was adjusted as described below. From the observed data on the cross section, distribution of monoclinic concentration expected to be obtained in measurement from the irradiated surface was calculated with using the intensity-thickness curve and then it was adjusted to coincide with the actually measured concentration from the irradiated surface in maximum value.

The adjusted distribution of the monoclinic phase in the direction of depth are shown in figure 3. It showed same tendency in the radial direction as that on the irradiated surface. In the direction of depth the monoclinic phase decreased rapidly with depth and disappeared over about 20μm. The depth coincided with the compressive stressed region predicted by simulations (Ito 1990, Morita 1991). From these facts the adjustment is supposed to be reasonable and the adjusted results are reliable.

These experimental results suggest that the monoclinic concentration is closely related to the induced stress during laser irradiation and it will be possible to know the stress amplitude by correlating it with monoclinic concentration.

3.2 Uniaxial compressive calibration test

Many experimental results show that tetragonal zirconia shows inelastic deformation caused by the volumetric dilatation due to t-m phase transformation under the influence of applied stress (Swain 1985). Therefore residual strain would be related directly to monoclinic concentration after removal of load. But in the experimental results of bending, it was shown that the deviation from linear stress-strain relation differs between under tensile stress and compressive stress (Marshall 1986). Moreover Chen (1986) pointed so-called yield stress of Mg-PSZ (partially stabilized zirconia) in tension is less than a half of that in compression. Now in the laser irradiation the transformation was caused by compressive stress, so an compression test was employed to correlate the monoclinic concentration to the applied stress. Although material is supposed to be subjected to multi-axial compression in actual,

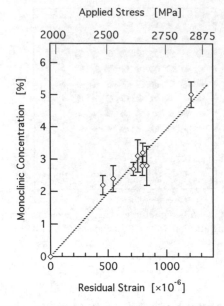

Figure 4. Variation of monoclinic concentration with residual strain after unloading in uniaxial compression. Relation with applied stress is represented by using top axis.

an uniaxial compression test was employed based on the experimental result (Marshall 1986) that irreversible response in four-point bending and biaxial flexure test was almost same in Mg-PSZ.

The monoclinic concentration after unloading, generated by irreversible transformation, increased with residual strain lineally (figure 4), from 0% at zero residual strain to 5% just prior to failure. From mechanics the macroscopic residual strain ε_r due to the dilatation component of the transformation strain is given by

$$\varepsilon_r = f\varepsilon_T/3 \qquad (1)$$

where f is the volumetric fraction of transformed particles and ε_T is the unconstrained strain (dilatation component). With $\varepsilon_T \approx 0.045$ (for PSZ) and $e_r \approx 1200 \times 10^{-6}$ (just prior to failure), equation 1 gives $f = 0.08$. This is a little larger than the experimental result $f \approx 5\%$. Same difference was obtained in other experiments for Mg-PSZ and it was explained by the contribution of shear strain to elongation (Marshall 1986). So it is able to say that the measured monoclinic concentration is reasonably correlated with the residual strain by the linear relation in figure 4.

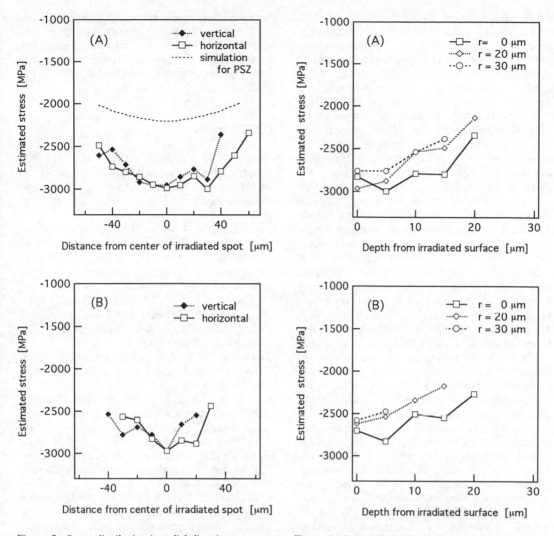

Figure 5. Stress distribution in radial directions perpendicular each other on laser irradiated surface for (A) irradiating duration ≈ 17ms: (B) irradiating duration ≈ 8ms.

Figure 6. Stress distribution with various radii on cross section normal to laser irradiated surface for (A) irradiating duration ≈ 17ms: (B) irradiating duration ≈ 8ms.

And then the relation between the residual strain and the applied stress is needed to acquire the induced stress during the laser irradiation directly from the monoclinic concentration. In uniaxial compression tests, variation of the measured residual strain was generated with bending component even under same load. So one-to-one correspond between the residual strain ε_r and the applied stress σ was made by approximating the result in good alignment, which contained less than 2.5% bending component, with a polynomial equation of $\sigma = \sigma_y + a\varepsilon_r^n$ (where σ_y is yield stress ≈

1855MPa, a and n are material constants. $a \approx$ 38.6 ; $n \approx 0.46$). The acquired result was also shown in figure 4, using the top axis with values obtained by the approximation. Using this relation, monoclinic concentration could be directly converted to the suffered stress in laser irradiation.

3.3 *Maximum stress distributions during laser irradiation*

The converted stress will represent the maximum

stress caused in the Ce-TZP particles of ZTA during heating and cooling in the laser irradiation process. Figure 5 shows the distribution of the converted stress in two directions intersecting perpendicularly each other at the center of the irradiated spot on the irradiated surface. And figure 6 shows stress distribution in the direction of depth on the cross section. The reproducibility of measurement of monoclinic content was ≈0.5%, thereby knowing the resolution of the stress converted from the monoclinic concentration was ≈45MPa. The stress distribution in the two directions reasonably agreed with each other.

The extent subjected over 2500MPa reflected the variation in the irradiating duration, but the maximum compressive stress reached almost 3000MPa regardless the variation in the irradiating duration contrary to the simulated result (Morita 1991) that compressive stress increases with irradiating duration. The measured maximum value is thought as the limitation in compression the material can bear and beyond that microcracks are generated. Actually the compressive strength of monolithic Ce-TZP is 2800MPa, almost same as the maximum and also microcracks were observed within the area of 80μm and 30μm in diameter on the 17 and 8 ms irradiated surface respectively. These were well coincided with the high stressed region in figure 5 and 6.

In figure 5, simulated circumferential stress on irradiated surface for PSZ (Ito 1990) the condition of CO_2-laser output intensity of 5W, duration of 5ms, and spot diameter of 200mm, was also shown. From comparison of the experimental results with the simulation, Ce-TZP particles may be subjected additional compressive stress due to CTE's (coefficient of thermal expansion) mismatch, because Ce-TZP's CTE is lager than that of Al_2O_3. For almost all zirconia containing ceramics, CTE of matrix is less than that of zirconia, so this idea works efficiently in acquiring compressive stress.

4 CONCLUSION

To acquire stress distribution with temporal and spatial difficulty in measurement, such an method was employed that measuring the stress-induced transformation in tetragonal zirconia by Raman microprobe, noticing that the transformation caused by stress irreversibly and Raman microprobe has micrometer order spatial resolution. This was applied practically to measurement of the thermal stress induced by CO_2 laser irradiation in ZTA and reasonable stress distribution was obtained. This idea worked well and is expected to be available for zirconia-containing ceramics in measurement of transient stress distribution in small area.

REFERENCES

Chen, I.-W. & P.E. Reyes Morel 1986. Implications of transformation plasticity in ZrO2-containing ceramics: I, shear and dilatation effects. *J. Am. Ceram. Soc.* 69:181-189.

Clarke ,D.R. & F. Adar 1982. Measurement of the crystallographically transformed zone produced by fracture in ceramics containing tetragonal zirconia. *J. Am. Ceram. Soc.* 65: 284-288

Dauskardt, R. H., W.C. Carter, D.K. Veirs & O. Ritchie 1990. Transient subcritical crack-growth behavior in transformed toughened ceramics. *Acta Metall.* 38: 2327-2336.

Garvie, R.C. & P.S. Nicholson 1972. Phase analysis in zirconia system. *J. Am. Ceram. Soc.* 55: 303-305.

Ito, M., K. Ueda & T. Sugita 1990. Fracture mechanics analysis of thermal shock cracking of ceramics in laser heating. *J. Jpn. Soc. Prec. Eng.* 56: 1487-1492.

Marshall, D.B. 1986. Strength characteristics of transformation-toughen zirconia. *J. Am. Ceram. Soc.* 69: 173-180.

Miyake ,T. ,F. Asao & C. O-oka 1993. Relation between flexural strength and phase transformation in ground tetragonal zirconia. *Science and technology of zirconia V*: 401-409. Lancaster: Technomic Publishing Co., Inc.

Morita, N., T. Watanabe & Y. Yoshida 1991. Crack-free processing of hot-pressed silicon nitride ceramics using pulsed YAG laser. *Trans. Jpn. Soc. Mech. Eng.* 57: 1031-1039.

Swain , M.V. 1985. Inelastic deformation of Mg-PSZ and its significance for strength-toughness relationships of zirconia-toughened ceramics. *Acta Metall.* 33: 2083-2091

Recent Advances in Experimental Mechanics, Silva Gomes et al. (eds) © 1994 Balkema, Rotterdam, ISBN 90 5410 395 7

Elastic waves and non-destructive ultrasonic method of determination of two-axial stresses

A.N.Guz'
Institute of Mechanics, Kiev, Ukraine

ABSTRACT: Non-destructive method of determination of residual stresses in solid bodies, presented in the report, is based on regularities of propagation of elastic waves in compressible (relatively rigid) materials with initial stresses. Main attention is directed also towards determination of two-axial stresses. This is a new aspect in non-destructive ultrasonic method. Measurements procedure is considered.

1 MAIN STATEMENTS

Theoretical basis for presented variant of ultrasound non-destructive method of determination of stresses in solid bodies are regularities of propagation of elastic waves in bodies with initial stresses. Complete presentation of the subject in monograph by A.N.Guz' (1986). Regularities are used of propagation of longitudinal and transverse (shear) waves in infinite body with initial stresses.

These special feature define the area of application of the method considered by following four statements.

1. The method is intended for materials of metals and alloys type and for measurement of stresses of magnitudes lower than the yield point. Consequently, the method allows to measure only "elastic" stresses. Many various sources of occurrence of "elastic" stresses are known: mechanical and thermal loading; residual stresses caused by conditions of mounting and erection of structures; residual stresses caused by electric welding and radiation and so on.

2. As to structures elements the method is intended for plane (plates) and curved (shells) structures elements fabricated from sheet materials, for flange and other elements of rolling profiles, for various structures elements of constant thickness or of slowly changing thickness. In Fig.1 part of element of structure is presented, fabricated from sheet material (h - sheet thickness; R - minimal dimension in plan of structure element considered).

It follows from the foregoing that for the method considered conditions should be satisfied

$$h << R \tag{1}$$

3. Method is intended for measurement of stresses which are constant along the thickness (membrane stresses). In accordance with Fig.1 following conditions are assumed to be satisfied (at assumed location of axes)

$$6_{11}^{\circ} = 0; \; 6_{22}^{\circ} = const; \; 6_{33}^{\circ} = const \tag{2}$$

Consequently, the method allows to measure only stresses which change insignificantly (little) along the thickness.

4. Method is intended for measurement of stresses which (in the plane of the sheet - in the plane $X_2 O X_3$ in Fig.1) are constant within the limits of dimensions of ultrasonic generator and of receiver. In Fig.1 the ultrasonoc generator and the receiver are shown

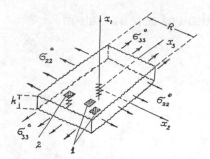

Figure 1. General scheme for
formulation of problems

as rectangles which are shaded.
Two variants of excitation and recording of waves are realized:
first variant - ultrasonic generator and receiver are located on
different face surfaces (variant 1
in Fig.1); second variant - ultrasonic generator and receiver are
united in one element and are located on one of face surfaces (variant 2 in Fig.1). The restriction
discussed is the consequence of
assumption of conditions (1) in
the theory presented by A.N.Guz'
(1986). Accordingly, the method is
intended for measurements of
stresses which change little in
sheet's plane within the limits of
dimensions of ultrasonic generator
and receiver. All theoretic results (validation of method) were
obtained within the framework of
linearized theory of propagation
of elastic waves in bodies with
initial stresses. Moreover, for
relatively rigid materials the linear approximation was used characterized by satisfaction of following inequalities

$$\sigma_{ij}^{o}\, \mu^{-1} << 1, \qquad (3)$$

where μ - shear modulus. Linear
approximation was assumed by para-
meter $\sigma_{ij}^{o}\, \mu^{-1}$. Foregoing statements determine the limits of applicability of the method discussed.

2 MAIN RELATIONS

Main acoustic relations were obtained in monograph by A.N.Guz'
(1986) at arbitrary homogeneous

initial stressed three-axial state
within the framework of the linearized theory of propagation of
elastic waves in bodies with initial stresses. Results were obtained with application to the theory
of finite (great) initial stresses
and to two variants of the theory
of small initial deformations for
"natural" (relative to the path
without accounting for initial deformation) and for "true" (relative to the path according for initial deformation) waves propagation speeds. "True" waves propagation speeds will be denoted by C
and "natural" waves propagation
speeds will be denoted by V . By
index "zero" will be denoted all
values related to the initial state. It should be remarked, that
stresses acting in the body and
measured by the method considered,
according to accepted terminology
also are initial ones.

For treatment of results, related to isotropic materials, as in
most publications of other authors, the elastic potential of
the Murnagan type is used in following form

$$\varPhi = \tfrac{1}{2}\lambda A_1^2 + \mu A_2 + \tfrac{a}{3}A_1^3 + b A_1 A_2 + \tfrac{c}{3}A_3 , \qquad (4)$$

where: λ , μ - Lame constants; a ,
b and c - constants of third or-
der; A_j (j =1,2,3) - algebraic in-
variants of Green deformations
tensor. In the case of isotropic
bodies without initial stresses -
speeds of propagation of longitudinal (dilatational waves) and of
transverse (shear waves) waves are
determined by relations

$$C_\ell^o = \sqrt{\tfrac{\lambda + 2\mu}{\rho}}; \quad C_s^o = \sqrt{\tfrac{\mu}{\rho}} \qquad (5)$$

For treatment of results related
to materials with insignificant
orthotropy of elastic properties
in natural (in the absence of initial stresses) state the elastic
potential is in the form

$$\varPhi = \tfrac{1}{2}E_{ijnm}\varepsilon_{ij}\varepsilon_{nm} + \tfrac{a}{3}A_1^3 + b A_1 A_2 + \tfrac{c}{3}A_3 , \qquad (6)$$

where in addition to foregoing notations are introduced: ε_{ij} - components of Green deformation ten-
sor, E_{ijnm} - elastic constants of
linear anisotropic body. In the
following it will be assumed that

axes of coordinate system chosen coincide with elastic- equivalent directions of orthotropic body; in this case, in addition to expressions (6) following expressions should be used

$$E_{1123} = E_{1113} = E_{1112} = E_{2223} = E_{2221}$$
$$= E_{3332} = E_{3331} = E_{3312} = E_{2213} = E_{2313}$$
$$= E_{2312} = E_{1312} = 0 \qquad (7)$$

Since the elastic potential (6) and (7) should describe the properties of material with insignificant orthotropy of elastic properties, in addition (6) and (7) the following relations should be assumed

$$E_{jjjj} = \langle \lambda \rangle + 2\langle \mu \rangle + \varepsilon_{jjjj}, \quad j = 1,2,3;$$
$$E_{ijji} = \langle \mu \rangle + \varepsilon_{ijji}, \quad E_{iijj} = \langle \lambda \rangle + \varepsilon_{iijj},$$
$$i, j = 1,2,3, \, i \neq j \qquad (8)$$

At notations (8), according to assumption made the conditions should also be satisfied

$$|\varepsilon_{ijnm}| < min\{\langle \lambda \rangle, \langle \mu \rangle\} \qquad (9)$$

Materials with elastic potential (6)-(9) in monograph by A.N.Guz' (1986) are named quasiisotropic materials. This class of materials includes polycrystalline materials in which insignificant orthotropy occurs as the result of technology of fabrication (for example, rolled products). This orthotropy of properties (although it is insignificant) should be accounted for, since its influence on waves propagation speeds, quantitatively, may be equal to effects of initial stresses.

In the first chapter monograph by A.N.Guz' (1986) on an example of sheet aluminium alloy AMG6 and of sheet steel 09G2S it is shown that with the use of (6)-(9) regularities of elastic waves propagation could be described in the case of occurrence of initial stresses. In the case (6)-(9) along the Ox_1 axes (Fig.1) three waves along each axis will already propagate (even when initial stresses are absent). For example, along the Ox_1 axis (Fig.1) when initial stresses are absent follo-

wing values are obtained: $C_{\ell x_1}^o$ - longitudinal wave velocity; $C_{s x_1 x_2}^o$ - velocity of transverse wave, propagating along the Ox_1 axis and polarized in the plane $x_1 O x_2$; $C_{s x_1 x_3}^o$ - velocity of transverse wave, propagating along the Ox_1 axis and polarized in the plane $x_1 O x_3$

$$\rho C_{\ell x_1}^{o2} = E_{1111} = \langle \lambda \rangle + 2\langle \mu \rangle + \varepsilon_{1111} \equiv \rho C_\ell^{o2} + \varepsilon_{1111};$$
$$\rho C_{s x_1 x_2}^{o2} = E_{1221} = \langle \mu \rangle + \varepsilon_{1221} \equiv \rho C_s^{o2} + \varepsilon_{1221};$$
$$\rho C_{s x_1 x_3}^{o2} = E_{1331} = \langle \mu \rangle + \varepsilon_{1331}$$
$$\equiv \rho C_s^{o2} + \varepsilon_{1331} \qquad (10)$$

It should be remarked that in the first chapter of second volume of monograph by A.N.Guz' (1986) values are presented of constants of the third order (a , β and C) for up to 39 various materials with application to six different variants of the theory (variants $И_к$, $И_{1M}$ and $И_{2M}$ - "true" propagation velocities within the framework of the theory of finite and of two variants of the small initial deformations theory; variants $E_к$, E_{1M} and E_{2M} - "natural" propagation velocities within the framework of the theory of finite and of two variants of the small initial deformations theory). Numerical values of different variants of the theory may differ from each other which is confirmed by results presented in the first chapter of second volume of monograph by A.N.Guz' (1986).

Using the foregoing approach, in Appendix of second volume of monograph by A.N.Guz' (1986) for all variants of the theory acoustic relations under biaxial stressed state are obtained (Fig.1). Accordingly, for isotropic materials with application to "true" velocities the expression is obtained

$$\sigma_{33}^o - \sigma_{22}^o = \left(\frac{C_{s x_1 x_3} - C_s^o}{C_s^o} - \frac{C_{s x_1 x_2} - C_s^o}{C_s^o} \right) A;$$
$$\sigma_{33}^o + \sigma_{22}^o = \left(\frac{C_{s x_1 x_3} C_s^o}{C_s^o} + \frac{C_{s x_1 x_2} - C_s^o}{C_s^o} \right) B \qquad (11)$$

For quasi-isotropic materials with application to "true" velocities the expression is obtained

$$\sigma_{33}^o - \sigma_{22}^o = \left(\frac{C_{s x_1 x_3} - C_{s x_1 x_3}}{C_s^o} - \right.$$

$$- \frac{C_{sx_1x_2} - C^o_{sx_1x_3}}{C^o_S}\Big) A;$$

$$\sigma^o_{33} + \sigma^o_{22} = \Big(\frac{C_{sx_1x_3} - C^o_{sx_1x_3}}{C^o_S} + \frac{C_{sx_1x_2} - C^o_{sx_1x_2}}{C^o_S} \Big) B. \tag{12}$$

In (11) and (12) and below with application to elastic body with initial stresses (Fig.1) notations are introduced: $C_{sx_1x_2}$ – velocity of transverse wave, propagating along the Ox_1-axis and polarized in the plane $x_1 O x_2$; $C_{sx_1x_3}$ – velocity of transverse wave, propagating along the Ox_1-axis and polarized in the plane $x_1 O x_3$. It should be remarked that values A and B in (11) and (12) are expressed by elastic constants of the second and third order and are of different form for various variants of the theory. As an example, we present expressions for determination of A and B with application to relations (12). In case of the theory of small initial deformations ($\mu_K \equiv \mu_{1M}$), according to second volume of monograph by A.N.Guz' (1986) we obtain

$$A = 8 <\mu>^2 (4<\mu>+c)^{-1};$$

$$B = 6K_o <\mu> [2b - (<\lambda> - 2<\mu>)(1 +$$

$$+ \tfrac{1}{4} <\mu>^{-1} c)]^{-1}; \quad K_o = <\lambda> + \tfrac{2}{3} <\mu> \tag{13}$$

For the second variant of the theory of small initial deformations second volume of monograph by A.N.Guz' (1986) we obtain

$$A = 8 <\mu>^2 c^{-1}; \quad B = 6K_o <\mu> [2(\theta - <\lambda>) -$$

$$- (<\lambda> - 2<\mu>) \tfrac{1}{4} <\mu>^{-1} c]^{-1} . \tag{14}$$

3 DESCRIPTION OF METHOD OF MEASUREMENTS

The expressions (11) and (12) for various theories relate, for example (11), the difference $(\sigma^o_{33} - \sigma^o_{22})$ an the sum $(\sigma^o_{33} + \sigma^o_{22})$ of principal stresses with the difference or the sum of relative changes of velocities $(C_{sx_1x_2}, C_{sx_1x_3}, C^o_S)$ waves which propagate along the Ox_1-axis (Fig.1) and which are polarized in mutually orthogonal planes $x_1 O x_2$

and $x_1 O x_3$ (Fig.1). In (11) and (12) values A and B are contained which are expressed by elastic constants of the second and the third order second volume of monograph by A.N.Guz' (1986); examples of such expressions are presented in (13) and (14) for specific theory. For this reason it becomes possible using know values (C^o_S – velocity of the shear wave in unloaded material; A and B are expressed by elastic constants of the second and the third order) and measured values ($C_{sx_1x_2}$ and $C_{sx_1x_3}$ – propagation velocities of shear waves in loaded material) by means of (11) and (12) to determine by non-destructive method the biaxial stresses state (in principal stresses). The method presented resembles the photoelastic method, however, it is of more general character. Indeed, in the photoelastic method only one relation is know for the difference of principal stresses – of the type of first expression (11). In photoelastic method the relation for the sum of principal stresses – of the second expression (11) type is lacking. In view of this using the photoelastic method only it is impossible to determine the biaxial stresses. For this reason in the photoelastic method additional procedures are used (measurements near the free border, numerical integration, measurement of thickness change and so on). In contrast, the method of acoustoelasticity, based on relation of the type of (11) and (12) allows, within the framework of this method, to determine the biaxial stresses. Another merit of the acoustoelastic method (as compared with photoelastic method) is the possibility to carry out non-destructive measurements not only on models but also on real structures. Main drawback of the acoustoelastic method is the necessity of carrying out measurements with high degree of precision, since operating stresses resulted in very insignificant changes in elastic waves propagation velocities.

It should be remarked also that in theacoustoelastic method the values A and B (11) and (12) may not be determined by elastic constants of the second and the third order using relations (13) and (14). Prescribing the magnitudes of σ_{33}° and σ_{22}° and using the relations (11) and (12) the values A and B for a given material may be determined. By this procedure the calibration of the instrument is carried out.

The present method of determination of values A and B is a prefereble procedure. The instrument for measurements was designed in the Institute of Electric Welding of the Academy of Sciences of the Ukraine named after E.O.Paton, chief engineer O.I.Gushcha (second volume of monograph by A.N.Guz' (1986)). The instrument is based on the method of impulse recirculation, blok-scheme of the instrument is presented in Fig.2. The precision instrument is intended for measurement of ultrasound velocity with relative precision of 10^{-5}. The master oscillator operates in self-exciting oscillatory regime and generates impulses which follow with period not less than 100 mks. These impulses energize, start the high-power generator 2, which excites ultrasonic impulses in specimen 4 with the help of acoustic transformer 3. Impulses excited in the specimen, after multiple reflections from opposite surfaces (along the OX_1-axis in Fig.1) circulate in the specimen and gradually attenuate. These oscillation excite in the transformer 3 electric signals which enter in the form of echo-signals 5 through key device 5 in the amplifier 6. The key device 5 separates the input of the amplifier 6 and the transformer 3 when high-voltage probing impuls enters – and unites them for reception of echo-signals. Amplified echo-signals in the form of row of impulses enter the input of the coincidence schene 8. In the second input of the coincidence scheme 8 the gating pulse enters formed in the output of regulated delay line 7. The delay line 7 is activated by probing impulse from generator 2. With the help of regulated delay line 7 the operator obtains the coincidence in time of the gating pulse with the wave chosed in the traced echo-signal.

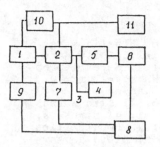

Figure 2. Block-scheme of the instrument

During this operation in the output of the coincidence scheme 8 the impulse is generated which enters through the delay line 9 in the input of the generator 1 and synchronizes its operation. Consequently, in the system a periodicity is established of movement of impulses which is determined by time of delay of echo-signals in the specimen. The frequency of movement of these impulses is measured by countingtype electronic frequency meter 11. The signal of the generator of probing impulses 2 enters also the discriminator 10, which at stationary frequency of recirculation generates the signal switching over the generator 1 in the waiting regime. In conclusion it may be stated that in operating of this device the measurements are made on one surface of the specimen or of structure element which corresponds to the case 1 in Fig.1. Measurements are made at the ultrasonic frequency of 5 MHz, the complete set of the instrument includes serial frequency meter r3-35. Acoustic transformers are fabricated from quartz plates, mounted on the specimen or on structure elements by electromagnet. In the location of measurement the surface of specimen or of structure element is prepared by polishing.

Results on the check of the ultrasonic method and on determination of two-axial residual stresses occurring in structural elements in electric welding are considered in second volume of monograph by A.N.Guz' (1986) and paper by A.N.Guz' (1990).

REFERNCES

Guz', A.N. 1986. Elastic waves in
  bodies with initial stresses.
  In 2 vol. Vol.1. General problems. Kiev: Naukova Dumka.
  276 pp. (in Russian).
Guz', A.N. 1986. Elastic waves in
  bodies with initial stresses.
  In 2 vol. Vol.2. Regularities
  of propagation. Kiev: Naukova
  Dumka. 536 pp. (In Russian).
Guz', A.N. 1990. Non-destructive
  ultrasound method of determination of biaxial stresses. Proceedings of the 9th International Conference on Experimental
  Mechanics. Vol.3, pp.1171-1179.
  Copenhagen: Aaby Truk.

Recent Advances in Experimental Mechanics, Silva Gomes et al. (eds) © 1994 Balkema, Rotterdam, ISBN 90 5410 395 7

Surface stresses measurement using ultrasonic techniques

D. Rojas González, A. Correia da Cruz & E. Dias Lopes
Structural Integrity Group, R&D Direction, ISQ OEIRAS, Portugal

P. Barros
NDT Group, Industrial Quality Direction, ISQ OEIRAS, Portugal

ABSTRACT: The linear change of ultrasonic velocity with the applied or material stress condition can be used for residual stresses measurements. The technique developed uses longitudinal "creeping" waves, propagating parallel to the material surface and is based on the measurement of the time-of-flight (TOF), change with the applied stresses. The materials acousto-elastic constant, gives the relationship between TOF and stress, and can be obtained by uniaxial tensile calibration. The acousto-elastic constant of stainless steels and low alloys steels were measured and the method applied to welded joints.

INTRODUCTION.

The measurement of residual stresses has win considerable interest due to the important role of that type of stresses on the catastrophic failure of structures or equipments.

In fact, they influence failure mechanisms like, brittle failure, stress corrosion, buckling, fatigue, etc..

The Hole drilling, and X-Ray diffraction techniques are the more common methods for residual stresses measurements. However the application of these traditional techniques present difficulties, related respectively to its semi-destructive character, interpretation of results and time consuming.

This situation, led to the development of other techniques and ultrasonic techniques proved to be a good alternative method.

Most of the available Ultrasonic testing, for residual stresses assessment are based on the measurement of sound velocity changes due to the material stress condition.

Longitudinal, Shear, Shear Polarized and Rayleigh waves or a combination of them have been used.

Measurement techniques using Rayleigh waves (sub-surface propagation) are very sensitive to surface condition or imperfection on the surface, like liquid drops or weld spatter.

Longitudinal waves are less sensitive than the Rayleigh waves to such factors. Longitudinal waves propagating parallel to the component surface are called "creeping" waves. The principle of creeping waves is shown on fig.1

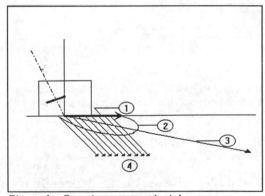

Figure 1.- Creeping waves principle.
 1.- Creeping Waves.
 2.- Compressional waves envelope.
 3.- Main compressional angle.
 4. Shear waves.

The fact, that creeping waves are less sensitive than Rayleigh waves to surface imperfections and condition, led to its use on the present method for surface residual stresses measurement.

METHOD PRINCIPLE

An ultrasonic method for measuring bolt-up and long term relaxation stresses in steel studs, by measuring the ultrasonic wave velocity change, was suggested by Pantermuehl et Al [1].

A similar method for residual stresses measurements, using the velocity change of creeping waves, was used by P. Barros [2].

Both methods are based on the relationship between stress (σ) and ultrasonic velocity change , (ΔV), that is expressed by equation 1.

$$\frac{\Delta V}{V} = \beta \sigma \qquad (1)$$

where $\Delta V/V$ is the ultrasonic wave velocity change, β is the material acousto-elastic constant and σ is the applied stress.

The ultrasonic velocity changes, can be evaluated by measuring, the time of sound travel across a given distance, i.e. the time-of-flight, (TOF), for that distance, and the previous equation can be written as follows:

$$\frac{\Delta V}{V} = \frac{V_f - V_0}{V_0} = \frac{t_0 - t_f}{t_f} = \beta \sigma \qquad (2)$$

where t_0 is the initial, time-of-flight corresponding to a stress free condition, and t_f is the time-of-flight of a component under load or with residual stresses.

That equation permits to establish a relationship between the TOF's and the existing stresses.

A good relationship was found by D. Rojas et Al [3], on SA387 Gr22 and HII steels.

The acousto-elastic constant β, can be experimentally determined by measuring changes in the time-of-flight for different applied stresses.

Then, the residual stresses can be calculated by expressing equations as follows:

$$\sigma_{Res} = \frac{1}{\beta}\left[\frac{t_0 - t_f}{t_f}\right] + \frac{S}{\beta} \qquad (3)$$

where S is the method application error or residual stresses existing on the supposed stress free calibration bar.

EXPERIMENTAL PROCEDURE

In order to measure the TOF, three probes were used, one emitter, (T), and two receiver, (R_1 and R_2), probes, (fig 2).

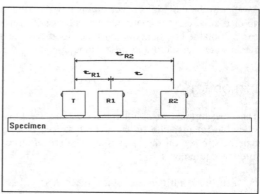

Figure 2.- Transducer arrangement for the time-of-flight.

This arrangement prevents the need of knowing the actual angle and probe delay, necessary if only two probes were used.

So the time-of-flight between probes R_1 and R_2 is given by

$$t = t_{R2} - t_{R1} \qquad (4)$$

where t_{R1} and t_{R2}, are the TOF between the emitter and probes R_1 and R_2 respectively.

An ultrasonic, Krautkramer USIP-12 equipment with a digital DTM-12 module was used. Measurements of t_{R1} and t_{R2} by independent gates and wave evaluation on RF mode, was made.

The probes are mounted on a probe holder device fig.3, in order to assure a fixed distance between probes as well as a good coupling to the test material surface.

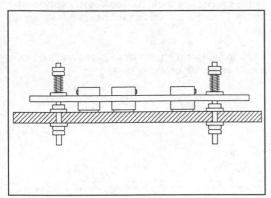

Figure 3.- *Probe holder device.*

The TOF´s data collected is processed and by linear regression the material acousto-elastic constant β,and method application error, S, are obtained. Fig. 4 to 6 present the data obtained by performing the calibration procedure on stainless steel, AISI 316 and low alloy steels SA387 Gr.11 and Gr.12 respectively.

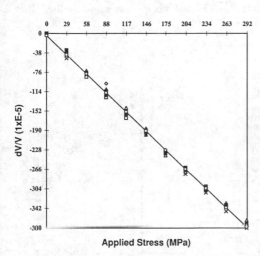

Figure 4.- *AISI 316 calibration results.*

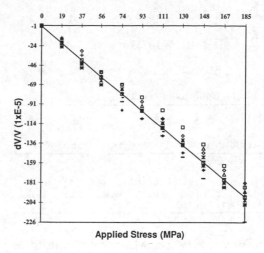

Figure 5.- *SA387 Gr.11 calibration results.*

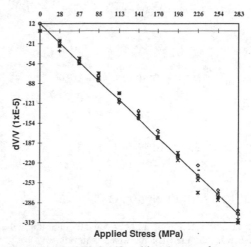

Figure 6.- *SA387 Gr.12 calibration results*

The following equations were obtained for the materials and table 1 presents the related data.

AISI 316

$$\frac{t_0 - t_f}{t_f} = (-1.284*10^{-5})*\sigma - 4.195*10^{-5} \quad (5)$$

SA387 Gr.11

$$\frac{t_0 - t_f}{t_f} = (-1.0603*10^{-5})*\sigma + 0.7568*10^{-5} \quad (6)$$

731

SA387 Gr.12

$$\frac{t_0 - t_f}{t_f} = (-1.118*10^{-5})*\sigma + 11.75*10^{-5} \quad (7)$$

Table 1.- *Calibration results.*

MATERIAL	β (*10⁻⁵)	sd (*10⁻⁵)	S (*10⁻⁵)	sd (*10⁻⁵)	R²
AISI 316	-1.284	0.011	-4.195	3.247	0.999
SA387 Gr.11	-1.060	0.122	-0.756	2.37	0.998
SA387 Gr.12	-1.118	0.021	11.746	6.47	0.996

Transformation of equations 5 to 6 permits the evaluation of residual stresses on components using the studied materials.

$$\sigma_{(316)} = -7.788*10^4*(\frac{t_0 - t_f}{t_f}) + 3.3 \quad (8)$$

$$\sigma_{(Gr.11)} = -9.431*10^4*(\frac{t_0 - t_f}{t_f}) + 0.714 \quad (9)$$

$$\sigma_{(Gr.12)} = -8.444*10^4*(\frac{t_0 - t_f}{t_f}) + 10.5 \quad (10)$$

RESULTS AND CONCLUSIONS.

A good linear relationship between TOF and applied stress was obtained for the three material.

The method is sensitive to the material surface condition, so the surface finish is an important parameter. An emery paper, 220 grade surface finish was used and is adequate.

Another important parameter is the coupling of probes to the material surface. The coupling pressure and the coupling media play and important role. In order to have a constant pressure the probe holder device used on calibration incorporates springs and Better results (less scatter), were obtained fig. 4. A similar device is being developed for field applications. Scatter improvement was also obtained by changing the coupling media from oil to a oil and grease mixture.

The improvement can be observed in fig 4 where the scatter is considerable less in comparison with fig. 5 and 6.

The application of this method to a butt weld joint , gave a transverse distribution of longitudinal residual stresses according to theory and other experimental results. The residual stresses distribution is shown on fig.7.

The application to a fillet weld joint also gave good results as shown on fig 8.

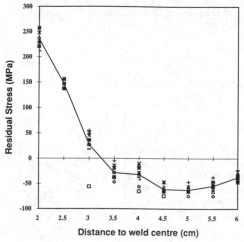

Figure 7.- Residual stresses on a butt welded joint (SA387 Gr.11).

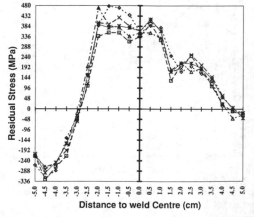

Figure 8.- Residual stresses on a fillet welded joint (SA387 Gr.12).

REFERENCES.

Joe Pantermuehl. P. and Birring, A.S. 1988
*Ultrasonic Procedure For Measuring Bolt-Up and
Long-Term Relaxation Stress in Steel Studs*
Materials Evaluation 46: 708-711

P. Barros 1991. *Medição de Tensões por Ultrasons*
Internal Report ISQ Oeiras Portugal.

D.Rojas, Dias Lopes, P. Barros, A. Correia da Cruz.
1992 *Medición de Tensiones Residuales Utilizando
Ondas Creeping.* Proceedings of the IX - IIW Latin
American Regional Welding: 1271-1292. Rio de
Janeiro, Brasil.

Creeping Wave Probes
RTD QUALITY SERVICES

Smith, P. H. 1987. *Practical Applications of Creeping Waves.* British Journal of NDT: 318-321.

Recent Advances in Experimental Mechanics, Silva Gomes et al. (eds) © 1994 Balkema, Rotterdam, ISBN 90 5410 395 7

Depth profiles of residual stress in Zn-Ni alloy electro-plating with different thickness using new method of X-ray stress analysis

T. Sasaki & Y. Hirose
Department of Materials Science and Engineering, Kanazawa University, Japan

ABSTRACT: The residual stree analysis was made on Zn−Ni alloy electroplating by the new procedure of the X−ray diffraction technique developed by the authors. The distribution of the residual stress in the thickness direction can be obtained non−destructively by the present method. Thickness of the plating examined was between 2 to 10 μ m. Non−linear $\sin^2 \phi$ diagrams were obtained from all specimens. It was found from the microscope observation on the cross section that the cracks in the plating emerged at the area where tensile stresses were occured.

1 INTRODUCTION

The Zn−Ni alloy electro−plating has a good ability for the corrosion proof at high temperature and has been used for automobiles [1] [2]. The prevention of the crack initiation in the plating layer is important for maintaining the ability of the parts used. The tensile residual stress occured in the layer is one of the main factors on this problem. From this point of view, the method of the X−ray residual stress measurement is the most powerful skill for the experimental stress analysis.

The X−ray stress measurement was made by the authors and they observed severly curved 2θ vs. $\sin^2 \phi$ plots from this plating [3] [4]. It was also found that the curvature in the plot was seriously affected by its thickness. A steep gradient of the residual stress parallel to the surface normal is thought to cause such X−ray diffraction data. It is important to make the state of the residual stresses clear from a practical point of view on the crack initiation.

The conventional $\sin^2 \phi$ method for the X−ray stress analysis is inadequate to the present material because of the strong non−linearity of the 2θ vs. $\sin^2 \phi$ plots [5] [6]. So the authors developed a new method of the X−ray stress analysis in order to analyze thin films with steep stress gradients. The authors' method includes the methods for both the stress calculation and the determination of the X−ray elastic constants [5] – [11]. By means of this method, one can obtain the depth profiles of the residual stress in a non−destructive way from X−ray data.

As a result of the application of the present method to X−ray diffraction data obtained from Zn−Ni alloy electro−plating layers with different thickness ranged from 2 μ m to 10 μ m, the residual stress distributions for each sample were found out respectively. Samples with the thickness of the plating less than 4 μ m have tensile stresses in the surface and compressive ones at the boundary to the substrate. Those with the thickness more than 6 μ m have opposite distributions of the residual stresses. Cracks were observed at the locations where the tensile stresses were occured.

2 PRINCIPLE OF X−RAY STRESS ANALYSIS FOR THIN FILMS WITH GRADIENT OF STRESS WITH RESPECT TO THICKNESS DIRECTION

The penetration depth of X−ray beams for thin films, expressed as T r, can be obtained from the following equations [12] [13].

$$T_r = \frac{\sin\theta_{oo}\cos\psi}{2\mu} f_{(t)} \qquad (\Psi-goniometer)$$

$$\frac{\sin^2\theta_{oo}-\sin^2\psi}{2\mu\sin\theta_{oo}\cos\psi} f_{(t)} \qquad (\Omega-goniometer) \qquad (1)$$

where θ is Bragg's angle, μ is a linear absorption coefficient, t is the thickness of the film and f(t) is a correction factor on the film thickness expressed as follow;

$$f_{(t)} = l n \left\{ \frac{1}{1-(1-exp(-t/T)R)} \right\} \qquad (2)$$

where T is a X−ray penetration depth for bulk materials which equals to T_f in eq(1), because $f(t) =1$ for $t \to \infty$. And R is a constant related to the ratio of the intensity of X−ray beams diffracted from whole material and that from the area between surface and depth T_f. Though the value of R is not clear at the present stage, we assume that $R=1-1/e$ in this paper.

The stress distribution inside is assumed to be a linear function of depth, z, from the surface expressed as follow;

$$\sigma_{ij}=\begin{bmatrix} \sigma_{110} & \sigma_{120} \\ \sigma_{120} & \sigma_{220} \end{bmatrix}+\begin{bmatrix} A_{11} & A_{12} \\ A_{12} & A_{22} \end{bmatrix}z \quad (3)$$

where σ_{ij0} (i,j=1,2) are stresses at the surface (z= 0) and A_{ij} (i,j=1,2) are stress gradients with respect to the thickness direction. The strains obtained by the X−ray diffraction technique are averaged over the volume sampled by the X−ray beams and are expressed as follow;

$$\langle\varepsilon_{\phi\psi}\rangle=\frac{\int_0^{T_f}\varepsilon_{\phi\psi}exp(-z/T)dz}{\int_0^{T_f}exp(-z/T)dz} \quad (4)$$

$$=(s_2/2)(\sigma_{110}cos^2\phi+\sigma_{120}sin2\phi$$
$$+\sigma_{220}sin^2\phi)sin^2\psi+s_1(\sigma_{110}+\sigma_{220})$$
$$+\{(s/2)(A_{11}cos^2\phi+A_{12}sin2\phi$$
$$+A_{22}sin^2\phi)sin^2\psi\}+s_1(A_{11}+A_{22})W_fT$$

where s_1 and s_2 are the X−ray elastic constants consisted of both Young's modulus, E, and Poisson's ratio, ν, as follow;

$$s_1=-\frac{\nu}{E} \quad , \quad s_2/2=\frac{1+\nu}{E} \quad (5)$$

The weight coefficient, W_f, in eq(4) is given by the next equation;

$$W_f=\frac{1-\{f_{(t)}+1\}exp\{-f_{(t)}\}}{1-exp\{-f_{(t)}\}} \quad (6)$$

Equation (4) is the basic equation of the present method of the X−ray stress analysis.

In order to solve all stress components, σ_{ij0} and A_{ij}, from X−ray diffraction data, the following parameters, $<c_i>$ (i=1,2,3), are introduced.

$$\langle c_1\rangle=\{\langle\varepsilon_{\phi\psi}\rangle_{\phi=0}+\langle\varepsilon_{\phi\psi}\rangle_{\phi=90}\}=u_1X_1+u_2X_2$$
$$\langle c_2\rangle=\{\langle\varepsilon_{\phi\psi}\rangle_{\phi=0}-\langle\varepsilon_{\phi\psi}\rangle_{\phi=90}\}=u_3X_3+u_4X_4 \quad (7)$$
$$\langle c_3\rangle=\{\langle\varepsilon_{\phi\psi}\rangle_{\phi=45}-\langle\varepsilon_{\phi\psi}\rangle_{\phi=135}\}=u_5X_3+u_6X_4$$

where u_m (m=1,2, $\cdot$ $\cdot$,6) are unknown coefficients

consisted of stress components expressed as follow;

$$\begin{array}{ll} u_1=\sigma_{110}+\sigma_{220} & , \quad u_4=A_{11}-A_{22} \\ u_2=A_{11}+A_{22} & , \quad u_5=2\sigma_{120} \\ u_3=\sigma_{110}-\sigma_{220} & , \quad u_6=2A_{12} \end{array} \quad (8)$$

Variables X_l (l=1,2,3,4) consists of material constants and experimental conditions;

$$\begin{array}{l} X_1=(s_2/2)sin^2\psi+2s_2 \\ X_2=X_1W_fT \\ X_3=(s_2/2)sin^2\psi \\ X_4=X_3W_fT \end{array} \quad (9)$$

Applying the least square fitting to eq(7), all the stress components in eq(3) can be obtained.

3 DETERMINATION OF X−RAY ELASTIC CONSTANTS FROM NON−LINEAR SIN $^2\phi$ DIAGRAMS

If the mechanical stress $\sigma_{110}{}^A$, is applied to the film in which there exists residual stress expressed as eq(3), we can get the next equation;

$$\langle\varepsilon_{\phi\psi}\rangle=\sigma_{110}{}^A\{(s_2/2)cos^2\phi sin^2\psi+s_1\}$$
$$+(s_2/2)(\sigma_{110}{}^Rcos^2\phi+\sigma_{120}{}^Rsin2\phi$$
$$+\sigma_{220}{}^Rsin^2\phi)sin^2\psi+s_1(\sigma_{110}{}^R+\sigma_{220}{}^R)$$
$$+\{(s_2/2)(A_{11}{}^Rcos^2\phi+A_{12}{}^Rsin2\phi$$
$$+A_{22}{}^Rsin^2\phi)sin^2\psi+s_1(A_{11}{}^R+A_{22}{}^R)W_fT$$

$$(10)$$

where A indicates the applied stress and R indicates the residual stress. We can get the following equation from the partial differentiation of the above euation with respect to the applied stress, $\sigma_{110}{}^A$.

$$\left\{\frac{\partial\langle\varepsilon_{\phi\psi}\rangle}{\partial\sigma_{110}}\right\}_{\phi=0}=\frac{s_2}{2}sin^2\psi+s_1 \quad (11)$$

So we can get the X−ray elastic constants of s_1 from the slope of the above function and s_2 from its intercept as follow;

$$\frac{s_2}{2}=\frac{\partial}{\partial sin^2\psi}\left\{\frac{\partial\langle\varepsilon_{\phi\psi}\rangle}{\partial\sigma_{110}{}^A}\right\}_{\phi=0}$$

$$s_1=\left\{\frac{\partial\langle\varepsilon_{\phi\psi}\rangle}{\partial\sigma_{110}}\right\}_{\phi=0,\psi=0} \quad (12)$$

This procedure is applicable to the material with any thickness and non−linear gradient of stress.

4 MATERIALS AND EXPERIMANTAL METHOD

The specimens used were electroplated with Zn−Ni alloy onto the steel with the thickness of 0.8 mm. The content of nickel in the plating layer was about 12 wt.% . Five types of the thickness of the plating layer such as 2, 4, 6, 8, 10 μ m were prepaired onto the steel sheets of the size of 180 mm long and 90 mm wide. Another specimen with the thickness of the plating of 5.4 μ m was also prepared for the determination of the X−ray elastic constants. The size of the substrate was 175 mm long and 15 mm wide.

Diffraction conditions were: Zn−Ni γ (552) reflection with Cr−K α radiation, psi−constant method, side−inclination method, parallel beam optics. The half value breadth method was used for the determination of the peak positions. A device of the four−point−bending was manufactured for apply-ing mecanical stresses to the specimen and used at the experiment on the X−ray elastic constants. Applied straines were monitored by means of the strain gage method. The mechanical value of Young's modulus was calculated sccording to the formula on the composite beams with data obtained by the strain gage method.

5 RESULTS AND DISCUSSION

Figure 1 shows the relation between the slopes of the function of $< \varepsilon_{\phi\phi} >$ vs. σ_{110}^{Λ} and $\sin^2 \phi$. We can see a linear relation in the figure, which agrees with eq(11). According to eq(12), the X−ray elastic constants were obtained as: $s_1 =10.3 \pm 0.17$ (TPa^{-1}) and $s_2 /2=2.02 \pm 0.60$ (TPa^{-1}). The mechanical value of Young's modulus of the plating layer was founf as $E_{mech} =118$ (GPa).

Figure 2 shows the 2 θ vs. $\sin^2 \phi$ diagrams obtained from the Zn−Ni alloy electroplating layer with different thickness such as 2, 4 and 10 μ m respectively. The solid lines in the figure show the results of the stress calculations by the present method. We can see that they agree with measured data. Figure 3 shows the distributions of the residual stress in the thickness direction calculated from the present method. It is found from the figure that the stress gradients in the thin plating layers (less than 4 μ m) are different from those in the thicker ones (more than 6 μ m).

Figure 4 shows the proposed model by the authors for the distribution of the residual stresses in the Zn−Ni alloy electroplating layers. The large compressive stress near the boundary between the plating layer and the substarte is thought to be occured due to the crystallographic unbalance, that is,

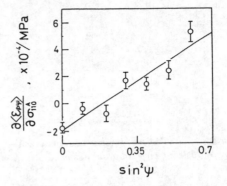

Fig.1 Determination of X−ray elastic constrants from Zn−Ni alloy erectroplating by the present method.

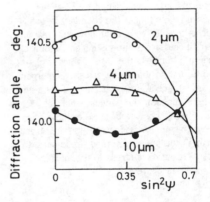

Fig.2 2 θ vs. $\sin^2 \phi$ diagram obtained from Zn−Ni alloy electroplating layers of the thickness of 2, 4, 10 μ m.

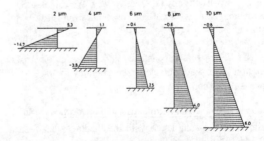

Fig.3 Distributions of residual stresses in Zn−Ni alloy e rectroplatings with different thickness obtained by the present method.

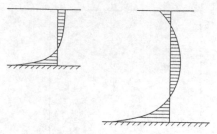

Fig.4 Models of distributions of residual stresses in Zn—Ni alloy erectroplating proposed by the authors.

the structure of Zn—Ni γ (552) phase is bcc with a lattice constant of a=0.89168 nm [14], and bcc with a=0.286 nm for the substrate (α —Fe). If the plating grows on the substrate with a similar crystallographic structure as of the substrate at the early stage, a huge compressive stress field as observed at the experiment can be built up at the boundary in the plating layer. A tensile stress field, however, may be built up near the surface of the plating as a result of the equilibrium of force. In the case of thicker platings more than 6 μ m, a compressive stress is emerged again near the surface as shown in figure 4. The infor matins on the strain near the boundary can not be reflected into the X—ray data because of the absorption of X—ray beams by the material. So the stress gradient through the present method would output the opposite distribution of stress to that of thin platings as shown in figure 4.

6 CONCLUSIONS

1. A new method was proposed for the stress analysis of thin films from which non—linear sin ^{2} ϕ diagrams were obtained. Applying the present method, we can obtain the distribution of stress in the thickness direction without any destructive way. We can also determine the X—ray elastic constants from these thin films.

2. The X—ray elastic constants of Zn—Ni alloy electroplating layer were obtained experimentaly by the present method.

3. The depth profiles of the residual stress in Zn—Ni alloy electroplating layers with the thickness between 2 and 10 μ m were found.

ACKNOWLEDGEMENTS

The authors are grateful to Dr. Michio Katayama of Kawasaki steel Co. for useful discussions and kindly supplying the samples.

REFERENCES

(1)K. Yamato, T. Honjo, T. Ichida, H. Ishitobi and M. Kawai, Technical Report of Kawasaki Steel, 16, 72 (1984).
(2)M. Kurachi and K. Fujiwara, Trans. JIM, 11, 311 (1970).
(3)K. Kyono, K. Yamato and M. Katayama, The 24th Symposium on X—Ray Studies on Mechanical Behaviors of Materials, Kyoto, July 23—24(1987), p. 31.
(4)T. Sasaki, Y. Yoshioka and M. Kuramoto, J. of the Japanese Society for Non—destructive Inspection, 32, 8, 614(1983).
(5)H. Doelle, J. Appl. Cryst., 12, 489(1979).
(6)I. C. Noyan and J. B. Cohen, Mat. Sci. and Eng., 75, (1985).
(7)T. Sasaki, M. Kuramoto and Y. Yoshioka, Advances in X—Ray Analysis 27, 121 (1984).
(8)Y. Yoshioka, T. Sasaki and M. Kuramoto, Advances in X—Ray Analysis 28, 255 (1985).
(9)T. Sasaki, M. Kuramoto and Y. Yoshioka, Advances in X—Ray Analysis 28, 265 (1985).
(10)T. Sasaki, M. Kuramoto and Y. Yoshioka, Proc. of the 32nd Japan Congress on Materials Research, 115(1989).
(11)T. Sasaki, M. Kuramoto and Y. Yoshioka, J. of the Japanese Society for Non—Destructive Inspection, 39, 8, 660(1990).
(12)B. D. Cullity, "Elements of X—ray Diffraction", Addison—Wesley, (1956).
(13)U. Wolfstieg, Haerterei—Tech. Mitt., 31, (1976).
(14)K.Tamura and A.Osawa, Science Repts. Tohoku Imp. Univ., 21, 344 (1932).

Recent Advances in Experimental Mechanics, Silva Gomes et al. (eds) © 1994 Balkema, Rotterdam, ISBN 90 5410 395 7

Characterisation of mechanical properties and residual stresses in hardened materials

A.C. Batista & A.M. Dias – *FCTUC, Universidade de Coimbra, Portugal*

P. Virmoux & G. Inglebert – *ISMCM, Saint-Ouen, France*

T. Hassine – *GLCS, Montreuil, France*

J.C. Le Flour – *Direction des Etudes Matériaux, Renault S.A., Boulogne-Billancourt, France*

J.L. Lebrun – *LM3, UA CNRS1219, ENSAM, Paris, France*

ABSTRACT: Various mechanical tests such as four points monotonic bending and spherical indentation were performed on laboratory samples and industrial gears from an automotive gearbox, submitted to surface treatments of carbonitriding and shot-peening. Metallurgical investigation and specific stresses and strains measurements during or after tests, coupled with a numerical elasto-plastic and fatigue analysis were then used to identify the hardened layers behaviour of the samples. Our measurements and a numerical optimisation performed on all our tests give us consistent information on the behaviour law. It allows a prediction of the life and the residual stresses evolution under the actual loading of a chosen gear.

1 INTRODUCTION

Mechanical parts, such as gears, are submitted in motion to high contact loads which lead to an important gradient of stresses, generating a phenomenon of contact plasticity (Johnson 1985, Inglebert 1991). This phenomenon is characterised by the appearance of plastic deformations and residual stress changes. If the parts work under strongly loaded contact situations, we can also observe the appearance of a progressive degradation of their surface, in the form of irreversible prejudicial marks, like pits or spalls.

Those parts are often submitted to surface hardening treatments such as carburizing or nitriding, which generate compressive residual stresses in the surface layers and modify the mechanical properties of those layers when compared to the bulk material. Variations of the elastic behaviour law are most often assumed negligible, but microstructural properties and plastic and fatigue behaviour laws are severely modified. The knowledge of the behaviour of the surface layers is important, because it allows to predict the evolution under service of the whole treated parts.

Due to the particular nature of these hardened layers, which do not exist freely and are submitted to residual stresses from their elaboration process, their mechanical properties cannot be handled in the classical ways. Inverse methods have to be used.

Moreover, the quite brittle behaviour in tensile conditions makes it very difficult to obtain precisely the plastic properties from tensile tests.

In this paper we attempt to obtain the mechanical properties of the surface layers by mechanical tests of monotonic four points bending, or by spherical indentation tests coupled with a numerical elasto-plastic and fatigue analysis implemented in a finite element program.

2 ELASTO-PLASTIC ANALYSIS METHOD

The elasto-plastic calculation program has been developed on a PC microcomputer from the finite element code ACORD2D. It uses the simplified analysis of inelastic structures developed by Zarka and Inglebert (1989), which is based on:

1. An analytic calculation of contact stresses in elasticity calculated from Hertz theory at each mesh point, from contact parameters such as the geometry of the contact and the maximal contact pressure.

2. An estimation of the elasto-plastic evolution using only purely elastic calculations and local projections. This enables the evaluation of the residual stress state after the contact.

The code has an initialisation module and a multiaxial high-cycle fatigue module too. The first one allows us to introduce the initial state of residual stresses and plastic deformations in the calculations.

The second one uses the calculated stresses after contact to evaluate the life of the modelled part, using a multiaxial high-cycle fatigue criterion such as Crossland, Sines, Dang Van, etc...

The program works both on axial-symmetry or plane strain, and uses isoparametric 3-node triangular elements in the mesh definition. To describe the elasto-plastic behaviour of the materials, a linear kinematic hardening law has been implemented (figure 1). With such a law only two properties are sufficient to characterise the plastic behaviour:

1. The initial yield strength, σ_y.
2. The linear kinematic hardening modulus, h.

The modulus h allows us to link the elastic Young modulus E and the plastic tangent modulus E_T through the formula:

$$\frac{1}{E_T} = \frac{1}{E} + \frac{1}{h} \tag{1}$$

A more detailed description of the elasto-plastic calculation method can be found in a recent publication (Virmoux 1993).

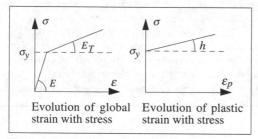

Evolution of global strain with stress

Evolution of plastic strain with stress

Figure 1. Linear kinematic hardening law.

Table 1. Chemical composition of the 27CrMo4 steel.

Chemical element	Weight %
C	0.24 - 0.31
Mn	0.60 - 0.85
Cr	0.95 - 1.25
Mo	0.20 - 0.30
Si_{max}	0.40
Al_{max}	0.050
Ni_{max}	0.30
Cu_{max}	0.30
P_{max}	0.025
S	0.020 - 0.040

3 MATERIALS

The material studied was the steel 27CrMo4 normally used in motor car gearboxes. The chemical composition of the steel is indicated in table 1. The parts were studied after two different types of surface treatments: (i) carbonitriding followed by quenching in oil, and (ii) carbonitriding followed by quenching and a shot-peening treatment. From this point we will refer these samples (ii) with their last surface treatment as the shot-peened ones.

4 HERTZIAN CONTACT FATIGUE TESTS

Contact fatigue tests on real parts of an automotive gearbox were carried out on a gearbox simulator developed by Renault. The same conditions (load, temperature and lubrication) were used in all tests.

The tests were performed during up to 30 hours for the carbonitrided samples and up to 65 hours for the shot-peened parts. At the end of the tests, the carbonitrided samples showed some pitting on both the gear and the pinion, but the gear presents also spalls with a surface of some square millimetres. On the shot-peened samples, the duration of the tests was sufficient to produce pitting everywhere on both the pinion and the gear surface, and some punctual spalls on the pinion surface.

We have observed that the damage appears after 20-30 hours of testing for both surface treatments, but with different forms: pitting and spalling for the carbonitrided samples and only pitting for the shot-peened samples. In the carbonitrided samples the pitting appears in a restricted area, usually with the form of a thin line near the lower limit of the working surface, whereas in the shot-peened ones it affects an important surface, sometimes all the extension of the teeth length. The shot-peened samples present also spalling but much later and in a much lower extension.

5 CHARACTERISATION OF MECHANICAL PROPERTIES IN THE SURFACE HARDENED LAYERS

5.1 *Four points monotonic bending*

Mechanical tests of four points monotonic bending were performed on laboratory samples of 27CrMo4 steel, submitted to the same surface treatments than the gears, to obtain the behaviour laws of the hardened layers. The evolution of the stresses in the surface of the samples during the tests was obtained by X-ray diffraction analysis. The applied macroscopic strains were measured by strain gauges.

The hardened layers were tested in compression because the brittle behaviour under tensile conditions always originates the rupture of the samples. Tensile tests were also performed with the bulk material. The results are presented in figure 2 for the bulk material, the carbonitrided layer and the shot-peened layer.

There is a good agreement between the tensile and the four points bending tests, which validates the used methodology. In table 2 are presented the hardness, the yield strength σ_y and the kinematic hardening modulus h of the different layers. Two different values of the yield strength are presented: the conventional one at 0.2% of strain, and another called *true* determined in conjunction with the modulus h assuming that the materials obey a linear kinematic hardening law. The last values will be used in the finite element calculations. As the materials do not obey exactly a linear kinematic law, we have several different choices in the definition of the coupled σ_y and h values, mainly for the bulk material. This will be taken into account later, in its application to the finite element model. We can see that the behaviour of both the surface layers are quite similar between them, but very different from the bulk material. This leads to a considerable increase in the h modulus of the surface layers, which is in agreement with the measured hardness.

Table 2. Characteristics of different layers of the 27CrMo4 steel, from four points bending and tensile tests.

Material	σ_y 0.2% [MPa]	σ_y true [MPa]	h [GPa]	HV$_{0.3}$
Bulk	1150	700	98	550
		1100	48	
Carbonitrided	1600	700	370	900
		1100	270	
Carbonitrided and shot-peened	1600	700	370	980
		1100	270	

5.2 Spherical indentation tests

The mechanical properties of the materials can be obtained from a comparison between the experimental indentations profiles (characterised by indentation radius and depth) and profiles obtained using the finite element model. The mechanical properties are then progressively adjusted until calculated profile is sufficiently close to the experimental profile. For a multilayer material, the bulk and each layer characteristics are successively determined.

In order to avoid singularities and to have an easily modelled contact, we have chosen a spherical indentation. For the indentations of the surface treated materials we suppose that the sample consist of only two different materials: one for the bulk material and another one for the hardened surface layer. This hypothesis is reliable in our experimental tests, taking into account the hardness profiles in surface treated parts. A perfect adherence is assumed to exist between the layers.

Indentation tests have been performed on samples of the bulk material, carbonitrided samples and carbonitrided + shot-peened samples, for maximal pressures between 3000 and 7000 MPa using a hard metallic ball with a diameter of 20 mm. This maximal pressures have been calculated using Hertz theory, taking into account the normal loads applied on the ball. From each test the experimental residual indentation radius and depth have been determined, from the indentation profile measured on a Perthometer S6P perfilometer.

The axi-symmetric model of the 2D finite element code is a rectangle of 4×6 mm. The depth of the treated layer was assumed to be 0.3 mm. The mesh was defined with about 2000 triangular elements with a high density in the contact zone (figure 3).

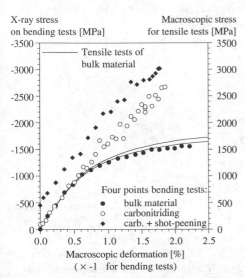

Figure 2. Stress-strain laws for different layers of the 27CrMo4 steel, from bending and tensile tests.

Table 3. Values of the linear kinematics hardening modulus h for different layers of the 27CrMo4 steel, from indentation modelling.

Material	Maximal pressure [MPa]	h [GPa]
Bulk	3400	340
	4200	195
	5600	133
	6500	99
	7200	78
Hardened layer	3400	2000
	4200	1000
	5600	314
	6500	272
	7200	195

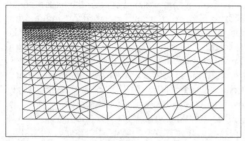

Figure 3. Finite element mesh used in the indentation modelling.

· For both the bulk material and the hardened layers we assumed the same values of the elastic characteristics: 210 GPa for the Young modulus E, and 0.3 for the Poisson ratio v. The yield strength σ_y was assumed to be about 700 MPa for the bulk material and 900 MPa for the hardened layer. The values for the linear kinematics hardening modulus h can then be obtained from the indentations modelling. The results of this procedure are presented in table 3. We can see that in both materials different values of the modulus h have been obtained, which means that the material does not really follow a linear kinematic hardening law.

From this results we can evaluate the tensile curve of the materials, considering that the maximal equivalent stress (calculated from Von Mises yield criterion) applied in the contact volume is equivalent to the stress applied in a section during a simple tensile test. From each contact test the determined values of σ_y and h in conjunction with the maximal equivalent stress allows to determine one point of the

tensile curve. The results of this procedure are presented in figures 4 and 5, for the bulk and the hardened materials respectively. For the bulk material we observe a good agreement between the results of this procedure and the results of the tensile tests if we do not exceed 1700 MPa. After this value the method does not give precise values, maybe because the stress-strain law does not follow a linear hardening law in this zone. For the hardened materials, we observe always a good agreement between the modelled stress-strain curve and the experimental one determined from four points bending tests.

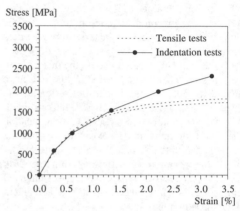

Figure 4. Comparison between the modelled stress-strain curve obtained from the indentations modelling and the results of tensile tests, for the bulk material.

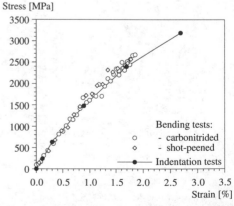

Figure 5. Comparison between the modelled stress-strain curve obtained from the indentations modelling and the results of four points bending tests, for the hardened layers.

6 RESIDUAL STRESS STATE EVOLUTION DURING CONTACT FATIGUE TESTS

The behaviour of the residual stresses and the diffraction peak broadness of the martensitic phase was studied as a function of the contact fatigue tests duration. The residual stress determination was performed with a Set-X apparatus, equipped with a position sensitive detector. We have used the {211} diffraction planes of the α-Fe phase, and Cr-K$_\alpha$ X-ray radiation with a vanadium filter in the diffracted beam. The centred gravity centre method was used for the peak position determination. The irradiated surface was about 4 mm^2. A more detailed description was presented in a recent publication (Batista 1994).

At the starting state the residual stress profiles for the carbonitrided gears are quite similar for the two longitudinal and transversal directions of the teeth working surface. The residual stresses are constant with a value of about -300 MPa up to a depth of 300 μm, followed by a slow decrease of the compressive stresses over a depth superior to 700 μm. After 30 hours of contact fatigue test we can observe that the compressive residual stresses have only increased in a layer very close to the surface, and do not show any evolution at depths superior to 10 μm. One example of this behaviour is shown in figure 6. The gradual increase of the compressive stresses at the surface, induced by the rolling-sliding contact, is more important in the longitudinal direction.

For the shot-peened gears we observe that shot-peening developed compressive stresses which reach a maximum value of -1200 MPa at a depth of about 30-50 μm, followed by a constant value similar to those of the carbonitrided samples, but corresponding to higher compressive stresses. We verify that the peak in compression observed below the surface, characteristic of the shot-peening treatment, presents some relaxation after 15 minutes of testing and disappears after 5 hours, showing no further evolution with the continuation of the tests. A maximum level of compressive stress have been developed at the surface, induced by the rolling-sliding contact. An example of this behaviour is presented in figure 7. The residual stresses at the surface show an initial decrease followed by an increase of the compressive stresses, which shows the competition between the relaxation induced by the cyclic loading and the increase induced by the contact.

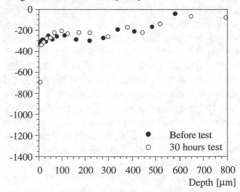

Longitudinal residual stress [MPa]

Figure 6 - Residual stress profiles of the carbonitrided gear, before and after 30 hours of contact fatigue testing.

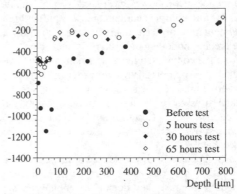

Transversal residual stress [MPa]

Figure 7 - Residual stress profiles of the shot-peened gear, versus contact fatigue test time.

7 DISCUSSION

The characterisation of the mechanical properties of the hardened layers has been performed with monotonic four points bending tests, and spherical indentation tests coupled with a numerical elasto-plastic analysis.

The monotonic four points bending tests have been performed with the treated surface in compression, because the tests under tensile conditions always originate the rupture of the hardened layers after a certain amount of strain. The comparison between the results of bending tests under tensile and under compressive conditions, up to the rupture, is in good agreement. This validates the results obtained under compressive conditions.

We have also compared the results of bending tests with those of tensile tests for the bulk material. The good agreement between these two kinds of tests shows that bending tests are an interesting and useful method to obtain some mechanical characteristics of hardened layers of surface treated materials. In the present study we have only considered one hardened layer, but we can apply this technique to get the mechanical properties as a function of the depth. We only have to remove progressively the superficial material with a non-hardening method, and apply this technique to the material at different depths. Due to the low penetration of X-rays in the material and to the use of strain gauges for the measurement of strains, the results are characteristic of the superficial layer.

The results of the indentation tests are in good agreement with the results of the other tests for all the studied materials, except at high loads for the bulk material. This can be due to the fact that the bulk material does not follow a linear hardening law in this zone. A more complex hardening law is presently in implementation on the numerical model, to allow a more precise modelling of this type of situations.

From the contact fatigue tests with real gears, we observe a progressive degradation of the microstructure, which leads to gradual changes in the residual stresses of the tested samples, mainly the shot-peened ones. Comparing the final state of residual stresses for the two surface treatments we can see that:

1. The residual stresses induced at the surface by the contact are the same for the two surface treatments, because they depend essentially on the load characteristics that are the same on all tests.

2. In the first 100 μm depth the stabilised compressive residual stresses of the shot-peened samples are stronger than those of the carbonitrided samples, which stabilise at a constant value of about -300 MPa.

8 CONCLUSIONS

The steel 27CrMo4 was submitted to two different surface treatments:

1. Carbonitriding and quenching in oil.

2. Carbonitriding and quenching in oil followed by a shot-peening treatment.

Monotonic four points bending tests, and spherical indentation tests coupled with a numerical elasto-plastic analysis, were used to identify the mechanical behaviour of the bulk material and the hardened layers. X-ray diffraction analysis was used to obtain the evolution of the stresses in the surface of the samples during bending tests. No significant differences were observed for the behaviour of the two types of surface treatments. A good agreement was observed between the results of the different procedures.

The two materials present a different behaviour with respect to the residual stress relaxation, under contact fatigue tests. This may explain some differences observed in the damage evolution, notably in the behaviour of spalling in the materials with different surface treatments.

A numerical simulation is presently being developed to model a real gear of an automotive gearbox, in order to predict the residual stresses evolution of the gear under service and his working life. It includes the actual loading cycle of the gears and the elasto-plastic characteristics of the different layers of material obtained experimentally.

9 REFERENCES

Batista, A.C., A.M. Dias, J.C. Le Flour & J.L. Lebrun 1994. Residual Stresses Evolution During Pitting Tests of Automotive Gears. *4th International Conference on Residual Stresses ICRS4*. Baltimore: 8-10 June.

Inglebert, G. & J. Frelat 1989. Quick analysis of inelastic structures using a simplified method, *Nuclear Engineering and Design* 116: 281-291.

Inglebert, G. & al 1991. Fatigue dans les roulements, *Mécanique Matériaux Electricité - Revue du GAMI* 441:11-13.

Johnson, K.L. 1985. *Contact mechanics*. Cambridge: Cambridge University Press.

Virmoux, P., G. Inglebert & R. Gras 1993. Characterisation of elastic-plastic behaviour for contact purposes on surface hardened materials. *Proc. Conference Leeds-Lyon*. Lyon: 7-10 September.

10 ACKNOWLEDGEMENTS

The authors would like to thank the JNICT Programa Ciência for his financial support.

Recent Advances in Experimental Mechanics, Silva Gomes et al. (eds) © 1994 Balkema, Rotterdam, ISBN 90 5410 395 7

Mechanical properties and residual stresses of WC-Co thermal-sprayed coatings

J. Pina, A. Dias & M. J. Marques
Faculdade de Ciências e Tecnologia, Universidade de Coimbra, Portugal

A. Gonçalves
Centro Tecnológico da Cerâmica e do Vidro, Coimbra, Portugal

J. L. Lebrun
Laboratoire de Microstructure et Mécanique des Matériaux, LM3, UA CNRS1219, ENSAM, Paris, France

ABSTRACT: This work deals with the influence of the thermal-spraying processes, APS (atmospheric plasma spraying) and HVOF (high-velocity oxygen fuel), and also the nature of the sprayed powders, on the structure and properties of the cermets WC-Co coatings. It compares the differences of microstructure, microhardness and residual stress states in both types of coatings.

It was observed that the APS process leads to the decomposition of WC into W_2C and to the amorphysme of the cobalt, whereas the HVOF induces less decomposition and oxidation. HVOF coatings have more homogeneous microstructures and higher microhardness and stiffness values than APS coatings.

The results of coatings' stiffness shown very low values, compared with those of the sintered materials.

The evaluation of the residual stresses was carried out by X-ray diffraction on samples with different coating and substrate thicknesses. The results found with HVOF and APS are compared. A model describing the nature of the residual stresses is proposed.

1. INTRODUCTION

Thermal-spraying technology is an important method to enhance the performances of materials over a wide range of functions: wear protection, corrosion, thermal insulation, etc.

Atmospheric Plasma Spraying (APS) has been developed during last 50 years. Its energy source is an electric arc established between a tungsten cathode and a water cooled copper anode which also confines the heating gas(es) as they pass around the arc. The plasma-forming gases are superheated by the arc column and are expanded through the nozzle. The powder is injected into the plasma jet where the particles are melted (5000 - 15000 °C) and accelerated. The plasma-jet velocity is of the order of 300 m/s at the exit of the spray gun [1].

High Velocity Oxygen Fuel (HVOF) was recently developed and now competes with plasma spraying for many coating applications. It involves the combustion of oxygen and a fuel gas at high pressure to produce a high-velocity exhaust jet. The spray powder is injected axially into the jet where it is melted and accelerated. The maximum flame temperature can reach 3000°C and the exhaust-jet velocity is about 2000m/s [2].

The high-temperature plasma flame is capable of completely melting even high-melting-point ceramic powders. However, the HVOF flame accelerates the powder particles to considerably higher velocities than APS and is, therefore, better suited to produce dense and well-bonded coatings, using metal and low-melting alloy powders, as well as cermets.

Thermally-sprayed coatings are composed of several layers, each one formed from a stream of molten particles propelled to a substrate, where they impact, spread and solidify rapidly, forming lamellae. Thermal-mismatch strains arise during the process at two different times: during deposition and in subsequent cooling to room temperature. In the studied coatings, these two stages are sources of different stress states [3, 4], which we call *quenching stress* and *secondary-cooling stress*.

The *quenching stress* was well studied by Kuroda and Clyne [3]. It occurs during the contraction of each molten particle, while it cools to the temperature of the underlying material and is always tensile. Its magnitude is so high that mechanisms of relaxation are required, such as plastic yielding, creep, interfacial sliding, microcracking. In the case of brittle materials, microcracks are generally initiated, especially if it involves high temperature.

The *secondary-cooling stress* arises from the thermal mismatch strains between substrate and deposit as a consequence of different thermal-expansion coefficients. It is especially confined to the interface region, but can affect all the thickness of the coating if its structure is compact.

Since HVOF deposits produce very dense coatings, it will be possible to separate the influence of the two origins of residual stresses, comparing the results for adherent and detached coatings.

In this work the microstructure and some mechanical properties were studied in order to understand the calculated residual stress values.

The X-ray diffraction technique was used for residual stress evaluation.

2. EXPERIMENTAL

2.1. Coatings elaboration

The coatings were produced by two types of thermal-spraying technologies: Atmospheric Plasma Spraying (APS) and High Velocity Oxygen Fuel (HVOF). Three types of commercial powders, with a cobalt content of 12%, were sprayed onto the surface of grit-blasted structural ferritic steel substrates, 50×25×1.5 mm. Table 1 lists same properties of the powders and the thicknesses of the coatings. The substrates were not cooled on the back side, but the temperature was reduced in the coated side using compressed air. The depositions were performed continuously, i.e., without delay between successive layers.

Table 1. Powder specifications and thicknesses of the coatings

Spraying process	APS	HVOF	
Powder reference	A	B	C
Process	agglom. sintered	fused	agglom. sintered
Grain size [μm]	45-6	45-15	45-11
Particle shape	mainly spheric	irregular blocky	mainly spheric
Coating thickness [μm]	290	200	210

In order to study the influence of both the coating and substrate thicknesses on the residual stress level, a second lot of samples was produced by HVOF, using the powder B and the initial spraying conditions. The substrates were made of the same material, with a surface of 100×9 mm and thicknesses of 10, 20 and 40 mm. The thicknesses of the deposited coatings were: 200, 400 and 600 μm.

2.2. Experimental procedures

Microhardness Vickers tests were performed on the polished cross sections of the samples. A testing load of 50 g was applied, for 15 s. Ten measurements were used to calculate the average value.

The microstructure of the initial powders and of the sprayed coatings was investigated by optical metallography and by scanning electron microscopy, using equipment provided for EDS analysis.

Phase structures of the powders and of the coatings were analysed by X-ray diffraction, using Cu-Kα radiation and a germanium detector.

The apparent porosity of the sprayed coatings was determined by quantitative image analysis of the cross sections.

Samples of the coatings included in the first lot, with a surface of 50×10 mm, were removed from the substrates in order to study their stiffness. The coatings were bent as an end-loaded cantilever beam. In these experiments, the samples are fixed at one extremity and a load is applied to the other extremity. The tip deflection is measured by a LVDT device. According to elasticity theory, the stiffness of the coatings can be studied by equation (1):

$$E = \frac{FL^3}{3fI} \qquad (1)$$

where F is the applied load, L is the length of the beam, f is the deflection at the loaded end, E is the elasticity modulus of the material, and I is the moment of inertia of the cross section of the cantilever.

Residual stress evaluation was accomplished by X-ray diffraction, studying the peak shift of the {112} WC diffracting planes, at $2\theta = 145.5°$, with Fe-Kα radiation. The first lot of the samples was analysed in a difractometer equipped with a proportional detector, and the second lot in one provided with a germanium detector. Due to the macroscopic elastic isotropy of the material, the $\sin^2\psi$ method could be used. For calculation, we considered the values of the elasticity constants, $E = 700$ GPa and $\nu = 0.20$, given elsewhere for isotropic WC polycristals [5]. Measurements were carried out in the interface of the samples included in the second lot, after dissolving the substrate in a bath composed of 1/4 HNO_3, 1/4 HCl and 1/2 H_2O.

In detached coatings, the value of the curvature radius was used to calculate stress relieving. The technique is based on the formulation of the elasticity theory giving the stresses - σ - on a beam subjected to a pure bending moment:

$$\sigma = \frac{E}{r} y \qquad (2)$$

where: E is the elasticity modulus of the material, r is the radius of curvature, and y is the distance to the neutral fibre.

3. RESULTS

3.1. Microstructural study

Figures 1 and 2 show micrographs of cross sections of the coatings. They exhibit a structure of superposed and oriented layers, related to the forming process. A microcrack network is observed in figure 1 of an APS coating. In APS coatings, the higher-porosity content is marked. This was confirmed by the results of image-analysis: about 10% of the porosity content for APS, and 5% for HVOF. The HVOF coatings are denser, this phenomenon being explained by the greater kinetic energy of the particles. EDS analysis shown that the grey regions of the micrographs are associated to the presence of cobalt. The same technique allowed the identification of tungsten in the white ones. In the HVOF coatings the carbides have a fine distribution. No influence in the microstructure was detected due to the different powders.

Figure 1. Cross section of APS coating

Figure 2. Cross section of HVOF coating

The main phases detected by X-ray diffraction are presented in table 2. The results show that the phase composition depends on the spraying process and is not affected by the sprayed-powder type.

The APS coatings exhibit a content of the WC phase similar to the W_2C one, and with lesser intensity than in the patterns of the powders. This reduction of the proportion of WC phase is explained by the decarbonisation and oxidation of the sprayed particles during flight, when the following reactions probably take place [6]:

$$2WC \rightarrow W_2C + C \qquad (3)$$
$$W_2C \rightarrow W_2(C,O) \qquad (4)$$

In the HVOF coatings, the decomposition doesn't occur, perhaps because of the spraying conditions [7]. The content of the WC phase is dominant in these coatings.

Table 2. Main phases of the powders and coatings

Powder	Phases in the powder	Phases in the coating
A	WC, Co	WC, W_2C, W, WC_{1-x}
B	WC, Co	WC, $W_6C_{2,54}$
C	WC, Co, $W_6C_{2,54}$	WC, $W_6C_{2,54}$

The cobalt present in the powders (12%) was detected in the coatings by XRD as an amorphous material. In fact, the diffractogrames exhibit a broad and high background in the region of $2\theta = 44°$, which corresponds to the planes {111}, the dense planes of the Co. This was more evident in APS.

Phase analysis was performed both in the surface and in the interface of the coatings included in the second lot, sprayed by HVOF and removed from the substrates. The diffracted intensity is smaller in the interface than in the surface. This is correlated with a poor crystallisation in the interface where the molten particles strike the substrate material and solidify in a shorter time. The decrease in temperature for these particles is greater than for those impinging in a previous deposited layer.

3.2. Mechanical properties

Table 3 presents the results of micro-hardness tests and porosity, averaged over the depths of the coatings. Comparing the spraying processes, the densification arising from HVOF is evident. No significant changes of the hardness values were noted in the depth of the coatings.

Table 3. Mechanical properties of the coatings

Coating	A - APS	B - HVOF	C - HVOF
$[HV_{50g}]$	800 ± 150	1000 ± 200	1500 ± 150
Porosity	10	1	1
Elasticity stiffness [GPa]	45 ± 5	120 ± 15	180 ± 20

The elasticity-stiffness values are included in table 3. HVOF coatings, denser than the APS, show higher values. Compared with the Young's modulus for WC isotropic polycrystal (700 GPa [5]) and for WC-Co sintered material (530 GPa [8]), these values are very low. This effect is related to the microstructure of the coatings, where the porosity and the microcracks play an important role, especially in APS coatings.

3.3. Residual stresses

The residual stresses were first evaluated in the surface, and the results compared for APS and HVOF coatings. In a second stage, the residual stresses of HVOF samples, with different thicknesses of coatings and of substrates, were studied in three ways:
1. in the surface of the coatings adherent to substrates,
2. in the surface of the coatings detached from the substrates, and
3. in the interface of the coatings detached from the substrates.
In some specimens the relieved stress could also be evaluated by the measurement of the radius of curvature, after separation from the substrates.

Residual stress and FWHM experimental data

Table 4 lists the values of residual stresses and Full Width at Half Maximum (FWHM) of the X-ray interference lines, calculated in the surface, from 5 specimens of each coating. They show that the APS coatings are unstressed, and that the stress values in the HVOF coatings are independent of the sprayed powders.

Table 4. Residual stresses and diffraction peak width

SAMPLE	σ [MPa]	FWHM [°]
A - APS	20±50	1.42±0.07
B - HVOF	160±95	1.74±0.06
C - HVOF	160±90	0.67+0.04

Figure 3 shows the residual stress values in the surface of HVOF coatings with different thicknesses of deposits (200, 400, 600 and 800 μm) and of substrates (10, 20 and 40 mm). These values are of the same order as the ones presented before, calculated on the first lot of samples. The residual stress state in the surface of the coatings is a plane-equiaxial one (within the error bars) over the depth penetrated by the X-rays (≈ 2 μm). The stress level seems to be slightly affected by the thickness of the coating, but is independent of the substrate thickness.

The values of residual stress in the surface of the coatings detached from the substrates are presented in figure 4. Compared with the ones determined when the coatings were adherent, they show a reduction of about 80 MPa.

The residual stress values in the interface of the coatings, after removal from the substrate, presented in figure 5, show considerable relaxation of the residual stresses. It must be pointed out that, after detaching, all the coatings were concave on the coating side. Even before detaching chemically, the deposit 800 μm thick and the one 600 μm sprayed on the thicker substrate (40 mm), were slightly debonded in the extremities.

The FWHM results presented in table 4 allow the comparison of APS and HVOF sprayed coatings. A higher cold-work of the particles is pointed out in the case of the HVOF process.

Figure 6 compares the experimental FWHM values for the HVOF-sprayed coatings. Significant differences are noticed between the results determined on the surface, for adherent and detached coatings, and on the interface of the detached coatings. The values seem to be independent of the thickness of the coatings.

Two factors are generally considered to have a great influence in the FWHM values: the crystallite size and the elastic distortion of the crystallites. Keeping that in mind, the results of figure 6 mean that the difference between curves 1 and 2 corresponds to the recovery of the elastic distortion of the material after detachment from the substrate. The differences between curves 2 and 3 can be explained by a smaller crystallite size in the interface, evidencing a more cold-worked material.

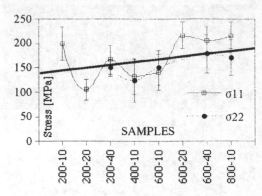

Figure 3. Residual stresses in the surface of the coatings adherent to the substrates

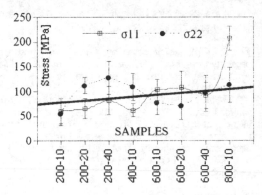

Figure 4. Residual stresses in the surface of the coatings detached from the substrates

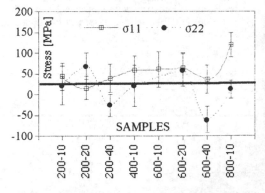

Figure 5. Residual stresses in the interface of the coatings detached from the substrates

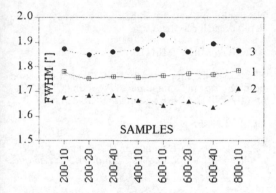

Figure 6. FWHM values from the samples:
1 - surface of the coatings adherent to the substrates
2 - surface of the detached coatings
3 - interface of the detached coatings

Discussion

The analysis of the experimental values can be based on the two mechanisms suggested in § 1.

The study of the results of stress measurements in coatings adherent and detached, taking into account the superposition principle in an elastic regime, allows the separation of the influences of the two sources of residual stress. In order to easily explain the results, figure 7 shows a schematic representation of the residual stress generation.

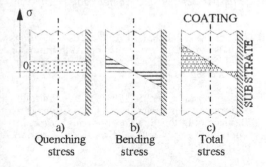

a) Quenching stress b) Bending stress c) Total stress

Figure 7. Schematic explanation of residual stress generation

The stress state in the coatings after debonding can be related to the *quenching stress* (figure 7 - a)). These stresses will be roughly constant in depth, as observed in measurements performed on ceramic APS coatings, where only the quenching stress remains in the finished deposit [3, 9]. The magnitude of the quenching stress is equal to the values of the residual stress measured in the surface of the detached coatings (figure 4).

The *secondary-cooling stress* is related to the thermal mismatch strains between substrate and deposit during the cooling to room temperature. It may have an important role on the bending of the coating. If the stiffness of the coating is high enough, its effects are denoted in the surface and produce a stress gradient (fig 7.b)), that is responsible for the curvature, or the fracture, of the coatings, after substrates' removal.

The low values of residual stress on the surface of the APS coatings (table 4) are explained by the higher porosity content (table 3) and, above all, by microcracking (figure 1). According to cracks and porosity induced compliance, the final stress state on the coating can be considered as a result only of the *quenching stress*.

The HVOF coatings showed different values when they were adherent or detached. They also exhibited a curvature, or were broken, after removing the substrate. These events mean that the constriction of the coating, induced by the substrate, affects the macroscopic stress state. In the kind of coatings studied, with high stiffness, this influence can be related to secondary-cooling stress.

The *secondary-cooling stress* can be estimated from the difference between the stress results evaluated by X-ray diffraction in the surface of the coatings, when they are adherent and when detached. This difference is of the order of 80 MPa. To confirm this assumption, this value has to be compared with the calculated one from the differential thermal contraction.

The stresses arising from differential thermal contraction can be estimated by equation (5):

$$\sigma_C(T_0) = (\alpha_C - \alpha_S).(T_S - T_0).E_C(T_0) \quad (5)$$

where: T_0 - room temperature (30°C),
α_C - coefficient of linear thermal expansion of WC ($\alpha_{WC} = 6.10^{-6}°K^{-1}$ [6]),
α_S - coefficient of linear thermal expansion of Fe ($\alpha_{Fe} = 11.10^{-6}°K^{-1}$),
T_S - average temperature in the interface of the substrate ($\approx 200°C$),
and E_C - elasticity stiffness calculated for the coating (table 6 - E = 120 GPa).

The values proposed for the coefficients of linear thermal expansion are handbook values. For the temperature of the substrate, in former studies we concluded that its value does not exceed the interval of 150 - 200°C, at 1 mm depth, even in the most severe conditions of spraying ceramics [10]. The calculation shows a compression stress value of - 100 MPa in the interface region of the coating. The equilibrium of the stresses throughout the volume of the coating implies that a symmetric value arrives in the surface. This value can be considered of the same order of magnitude as the 80 MPa, referred to above.

The explained results were also corroborated by the stress relieving, calculated after measuring the radius of curvature on some detached coatings. For this purpose, equation (2) was used, assuming a value of 120 GPa (table 3) for the elasticity modulus. The results are presented in table 5, and compared with the ones calculated by the other two methods. They can be considered in good agreement, showing that the secondary-cooling stress is related to the mismatch in thermal expansivities between coating and substrate.

Table 5. Comparison of the stress values [MPa]

Experimental Method	SAMPLE		Stress
	COAT	SUB	[MPa]
Stress recovery calculated from the bending of the detached coatings	200	10	60
	400	10	90
	200	1.5	100
Stress calculated from differential thermal contraction			100
Difference of the X-ray results on adherent and detached coatings			≈ 80

The secondary-cooling stress generation explains the debonding of the thicker coatings, observed in the end of spraying, after extinction of the thermal source. In fact, the stress values in the interface will be increased by the accumulation of heat in the substrate, when the thickness of the coatings is large enough or the substrate is very thin.

4. CONCLUSIONS

- WC-Co HVOF sprayed coatings are denser than the APS deposited, showing more homogeneous microstructures and higher microhardness and stiffness values. APS leads to the decomposition of WC into W_2C and to the amorphysme of the cobalt, whereas HVOF induces less decomposition and oxidation.

- The results of coatings' stiffness shown very low values, compared with the sintered materials ones.

- The residual stress state in the WC-Co very-dense coatings, produced by HVOF, can be separated into two principal sources: *quenching stress* and *secondary-cooling* stress.

- The quenching stress occurs during the spraying and is due to the contraction of each molten particle. It is always tensile in the depth of the coating. Its magnitude is equal to the stress values measured by X-ray diffraction in the surface of the coatings detached from the substrate.

- The secondary-cooling stress arises after spraying and is related to the mismatch in thermal expansivities between coating and substrate. It is the main responsible for the stress gradient in the core of the deposit, that explains the curvature, or the debonding of the coatings.

- The WC-Co coatings produced by APS exhibit low stress values, according to the cracks and porosity induced compliance. The final stress state can be considered as a result only of the *quenching stress*.

No significant stress gradients will be present in the depth of such coatings.

ACKNOWLEDGEMENTS

This work was supported by JNICT within the scope of the European Communities "Programa CIENCIA" and by JNICT-CNRS within the PICS nº 117.

REFERENCES

1. Fauchais, P., Desmaison, J., Machet, J. 1987. "Les Dépôts Céramiques". *L'Industrie Céramique*, 812: 48 - 56.

2. Kreye, H. 1991. "High Velocity Flame Spraying - Process and Coating Characteristics". *Proc. 2nd Plasma-Technik Symposium*, Lucerne, Switzerland. 1: 39 - 47.

3. Kuroda, S., Clyne, T.W. 1991. "The Quenching Stress in Thermally Sprayed Coatings". *Thin Solid Films*. 200: 49 - 66.

4. Kingswell, R., Scott, K.T., Sorensen, B. 1991. "Measurement of Residual Stress in Plasma Sprayed Ceramic Coatings". *Proc. 2nd Plasma-Technik Symposium*, Lucerne, Switzerland. 3: 377 - 388.

5. Krawitz, A.D., Crapenhoft, M.L. 1988. "Residual Stress Distribution in Cermets". *Materials Science and Engineering*. A105/106: 275-281.

6. Tronche A., Fauchais, P 1987. "Hard Coatings (Cr_2O_3, WC-CO) Properties on Aluminium or Steel Substrates". *Materials Science and Engineering*. 92: 133 - 134.

7. Wang, Y., Kettunen, P. 1992. "The Optimization of Spraying Parameters for WC-Co Coatings by Plasma and Detonation-Gun Spraying". *Proc. Int. Thermal Spray Conf. & Exposition*, Orlando, Florida, USA. 575 - 580.

8. Waterman, N.A., Ashby, M.F. (eds.) 1991. Elsevier Materials Selector, vol. 2. London: Elsevier applied Science.

9. Pina, J., Costa, V., Dias, A.M., Zaouali, M., Lebrun, J.L. 1992. "Residual Stresses on Plasma Spraying Coatings. Influence of HIP and LASER Treatments". *Proc. Int. Conf. on Residual Stresses 3 (ICRS3)*. ed. H. Fujiwara et al., Elsevier Applied Science. 1: 686 - 691.

10. Houhou, M., Zaouali, M., Lebrun, J.L., Pina, J., Dias, A.M. 1992. "Genèses des Contraintes Résiduelles dans le Procédé Plasma. Experimentation et Modélisation", *Proc. Coloquio Tensões Residuais Portugal - França*, Luso, Portugal. 139 - 148.

Recent Advances in Experimental Mechanics, Silva Gomes et al. (eds) © 1994 Balkema, Rotterdam, ISBN 90 5410 395 7

Industrial optimisation of stress gradient determination by X-Ray diffraction and incremental hole-drilling on Beryllium and Aluminium Alloy

C. Le Calvez & J. L. Lebrun
LM3 URA CNRS1219, ENSAM, Paris, France

C. Cluzeau & P. Harcouet
CEA, Bruyere Le Chatel, France

ABSTRACT : Incremental hole drilling and X-Ray diffraction can be used to evaluate stress gradients. X-Ray analysis is based on a wave length variation if observed dimensions correspond to mean diffraction depth (case of Beryllium), or based on successive chemical polishing in other case (case of Aluminum Alloy). We optimised these methods for Aluminum (6061T6) and Sintered Beryllium, and we established their limits and precision.

INTRODUCTION:

The knowledge of stress gradient at the surface of mechanical parts is be very important. The measurement of this gradient is not a simple problem. We consider two methods : Incremental Hole Drilling [1] and X-Ray diffraction [2]. Hole drilling is a semi destructive method, as X-Ray diffraction if electro-chemical polishings are needed between each measurement. X-Ray diffraction can also be a non destructive method if different wave lengths are used to analyse different mean diffraction depths [3]. In this case, the observable gradient depth depends on the penetration of X-Ray beam in materials.(about 15 μm for Aluminum, some millimeters for Beryllium [4]). If the gradient depth is higher than the mean X-Ray penetration, successive electro-chemical polishings are needed, and the observable depth is some 100 μm. Incremental hole drilling gives informations on few 100 μm from the surface of the sample.

The two methods have been used on Sintered Beryllium and Aluminum Alloy (6061T6), two materials which present different physical characteristics. Hole drilling is not often used on Beryllium, X-Ray diffraction has been used on both metals.

Experimental conditions have been validated , the precision and the limits of analysis have been evaluated through measurement on samples with controlled stress state.

1. HOLE DRILLING METHOD

1.1 Introduction :

The incremental hole drilling is the first technique used for measuring residual stresses in aluminum Alloy. The application of this method involves incremental drilling of a small shallow hole (depth = 0.30 diameter) in the specimen. The removal of stressed material causes localised stress and strain relaxation around the hole location. The strain relaxations are measured after successive small increment of hole depth, by using strain gauges (type 062RE-120). The stresses are then calculated with CETIM's "metro" data program [5].

1.2 Experimental procedure :

In this study, we analyse the effects of parameters such as :

- influence of the hole drilling method on the stress level developed by machining (electro discharge machining (EDM) or traditional drilling (3000tr/min))
- influence of the increment (10 μm or 100 μm) on residual stress calculation in the very beginning of drilling (0<x<200μm) when a well known stress level is applied (100 MPa).

- influence of the increment (10 µm or 100 µm) on residual stress calculation at the end of drilling (200<x<600 µm) when a well known stress level is applied (100 MPa).

Four aluminum plate specimens (150x35 mm), 8 mm thick, without initial stress (with a chemical polishing 100 µm) have been studied. A four points bending equipment has been used to induce controlled stress state in the samples

1.3 Results

1.3.1 influence of the hole drilling method (EDM or traditional drilling) :

On a free sample, a 1.9 mm radius hole has been drilled in five equal 100µm depth increments, using EDM and traditional drilling. Figure 1 shows the comparison between the results of the two drilling methods.

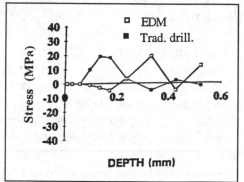

Figure n°1 : Influence of the hole drilling method.

The two methods induce few stress (at most 25 MPa) on Aluminum material. This measured strain is unchanged with depth up to 600µm. Deeper than 600µm strain gauges are not sensitive enough.

We take into account this result during stress determination when bending is applied.

1.3.2 Influence of increment (10 µm, 100 µm) under bending

Every specimen has been stressed under a well known level (100 MPa) with the four points bending equipment.

Figure n°2 shows the resulting curves obtained with a 10 µm drilling increment and Figure n°3 those obtained with a 100 µm drilling increment (uncertainty

on each measured point is ±20 MPa). On this two figures the second curve represents the theoretical bending stress evolution with depth.

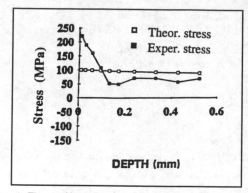

Figure n°2 : Influence of 0,1 mm increment

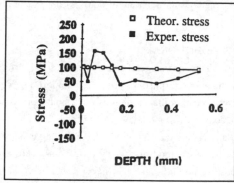

Figure n°3 : Influence of 0,01 mm increment.

Drilling with small increments (10 µm) leads to a better accuracy with theoretical gradient between 0 and 50 µm. Between 150 µm and 600 µm, the 100 µm drilling increments gives a better accuracy with theoretical stressing.

1.4 Conclusion of results obtained with Incremental Hole Drilling :

We used two drilling methods to estimate uniform stresses in aluminum sample. We established that the two drilling methods induced few stress in aluminum (25 MPa) and that this value is constant with depth.

Under a 100 MPa stress state, accuracy with theoretical bending is good between 150 µm and 600 µm. Influence of the drilling increment (10 µm, 100 µm) is not negligible : 100 µm drilling increment gives less accuracy between 0 and 150 µm, and better accuracy between 150 and 600 µm.

Measurement on Beryllium have not been yet realised but experiments are planed. Our goal is to define accuracy and sensitivity of incremental hole drilling on this material which presents a very important Young modulus (310 GPa) and very low Poisson coefficient (0.07) compare to Aluminum (72 GPa and 0.3).

2. X-RAY DIFFRACTION ANALYSIS :

X-Ray diffraction stress analysis method is based on lattice variation under stress field in the sample [2]. It is well applied on aluminum material but it is something quite new on Beryllium, because of its X-Ray low absorption [4] (the penetration depths for an absorption of 63% are shown in table n°3).

2.1 X-Ray diffraction on Beryllium :

The important penetration of X-Ray must be taken into account during measurement and during the analysis of the results.

2.1.1 Interpretation of measurement :

During peak acquisition the information comes from different depths. 2θ measures take into account 2θ at each depth level headed by diffracted intensity.

$$2\theta_{mes} = \frac{\int_0^e I(z).2\theta(z).dz}{\int_0^e I(z).dz} \quad (1)$$

$$\sigma_{mes} = \frac{\int_0^e \sigma(z).I(z).dz}{\int_0^e I(z).dz} \quad (2)$$

Formula n° 1 shows the expression of the measured angle where '2θ' is the diffraction Bragg angle, 'e' is the sample thickness, 'I(z)' diffraction intensity at 'z' millimeter deep from the surface. If during measurement, the variation of the mean penetration depth (Zr : formula n° 3) is small (smaller than 15 %), and the stress gradient is not too high, the results of the stress analysis could be seen as a mean stress (σ_{mes}) which expression is given in formula n°2 versus $\sigma(z)$, the stress at every depth.

Usually, in X-Ray stress analysis, measurements are done just at the surface of the

sample, so that σ_{33} is supposed to be zero. In Beryllium, measurements are done much deeper from the surface, so this hypothesis can't be verified. For triaxial analysis, Bragg angle of stress free material must be known to estimate σ_{33}. For uniaxial analysis, if σ_{33} is not supposed to be zero the result of measurement in a Φ direction is ($\sigma_\Phi - \sigma_{33}$).

2.1.2 Problem of Z-missetting :

During the measurement if the surface of the sample is placed at the centre of the goniometer a Z-missetting occurs (Figure n°4).

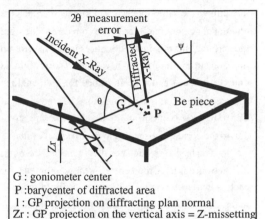

G : goniometer center
P :barycenter of diffracted area
l : GP projection on diffracting plan normal
Zr : GP projection on the vertical axis = Z-missetting

Figure n°4 : Z-missetting due two low X-Ray absorption of Beryllium. The case of ψ setting

The effect of a well known Z-missetting on the measured 2θ can easily be calculated, formula n°4 gives its expression in the case of ψ goniometer setting (R is the goniometer radius). Formula n°3 shows the developed form of the mean diffraction depth Zr (µ/ρ value is given in table n°3).

The expression of the measurement error on 2θ in the case of Beryllium and a ψ goniometer setting is given in formula n°5.

$$Zr = \frac{\int_0^e z.I(z).dz}{\int_0^e I(z).dz} = \frac{1}{2\alpha\mu}\left(1 - \frac{2\alpha\mu e.e^{-2\alpha\mu e}}{1 - e^{-2\alpha\mu e}}\right) \quad (3)$$

$$For \quad \psi \quad setting$$

$$\alpha = \frac{1}{\sin(\theta).\cos(\psi)} \quad (3 \text{ bis})$$

$$\delta 2\theta = \frac{2.Zr.\cos(\theta)}{R.\cos(\psi)}.\frac{180}{\pi} \quad (4)$$

$$\delta 2\theta = \frac{\sin(2\theta).180}{2\mu R\pi}\left(1 - \frac{2\alpha(\psi,\theta)\mu e.e^{-2\alpha(\psi,\theta)\mu e}}{1 - e^{-2\alpha(\psi,\theta)\mu e}}\right) \quad (5)$$

2.1.3 Correction of Z-missetting:

In case of a ψ setting, if sample thickness is very important compare to $2\alpha\mu$, $\delta 2\theta$ becomes independent on the incidence angle ψ. So a constant error is made on the measured 2θ value. If the value of 2θ powder used for calculation is shifted from the error value, the stress analysis would be standard.

For a small thickness a correction must exist. A simple idea consists in a mean correction of Z-missetting (the same for each ψ) : for measurement the goniometer center is placed at the corrected value from the surface. But this method induces negative error on the measured 2θ for small ψ angle (correction is underestimated) and a positive one for high angle ψ (correction is then overestimated). The true slope of $2\theta=f(\sin^2(\psi))$ is modified, and a compressive stress due to the correction is added to the true stress value.

The best correction consists in placing the mean diffraction point at the centre of the goniometer for each ψ angle. A special equipment has been made to realise that automatically. The result of measurement is directly usable for stress analysis.

2.1.4 Analysis of beryllium powder :

In order to validate our approach, a 10 mm thick Beryllium powder (density 0.90) has been analysed with the three missetting correction methods. The error induced on stress estimation can be calculated for each method and compared with measured value. Table n° 1 shows a good adequation between the calculated and measured results.

For an X-Ray triaxial stress analysis it is necessary to know precisely the Bragg angle of the stress free material. Powder measurement has been made and experimental values are shown in table n° 3.

Table n° 1 : Stress evaluated in a Beryllium powder versus the correction of X-Ray transparency adopted. Predictive calculation and X-Ray measurement

	Zr = 0	mean Zr	Zr = f(ψ)
Calculation	0	-195	0
measurement	-20±20*	-190±40*	-5±25*

(* Uncertainty on stress determinate by X-Ray diffraction is independent of stress level (§2.1.5)).

2.1.5 Stress analysis on Bending sample :

Bending sample, without initial stress (thermally relaxed), has been analysed; comparison is made between the measured stress and the deformation imposed which is controlled by strain gauge. Iron and Vanadium anodes have been used for triaxial analysis.

$$\sigma_{Fe.initial} = \begin{pmatrix} -35\pm 25 & -10\pm 20 & -15\pm 5 \\ & -10\pm 25 & 5\pm 5 \\ & & 15\pm 5 \end{pmatrix} (MPa)$$

$$\sigma_{Fe.bending} = \begin{pmatrix} -135\pm 25 & 10\pm 20 & 0\pm 5 \\ & 35\pm 25 & -5\pm 5 \\ & & 30\pm 5 \end{pmatrix} (MPa)$$

$$\sigma_{V.initial} = \begin{pmatrix} -25\pm 30 & 5\pm 20 & 10\pm 5 \\ & -55\pm 30 & -10\pm 10 \\ & & -35\pm 5 \end{pmatrix} (MPa)$$

$$\sigma_{V.bending} = \begin{pmatrix} -110\pm 30 & 10\pm 20 & 0\pm 5 \\ & -30\pm 30 & -5\pm 10 \\ & & -30\pm 5 \end{pmatrix} (MPa)$$

The bending stress gradient is well known, so we can calculate the theoretical stress value which would be measured using formula n°2. Results are presented in the table n°2.

Table n° 2 : Analysis of bending gradient. The difference between stress free and bended stress determinated with Fe anode and V anode is compared with theoretical values.

	Vanadium	Iron
Calculated difference	-100 MPa	-80 MPa
Measured difference	-85±30 MPa	-100±25 MPa

Taking into account the measurement uncertainty the difference between calculated and measured values is quite good. Measurement could not be more precise because of the high Young modulus of Beryllium which increase the uncertainty. Indeed Beryllium Young modulus (310 GPa) is 32% higher than steel one, so for the same precision in strain determination, stress uncertainty is 32% higher. With Beryllium, uncertainty is about 30 MPa, which correspond to 22 MPa in the case of steel. It is a common error value in X-Ray stress analysis of steel material.

In X-Ray analysis stress uncertainty is due to material characteristics and diffraction conditions and it is not dependant on the stress level. So for low stress uncertainty is of the same order than the measured stress (20±20 MPa), and it represents a low percent for a high stress level (135±25 MPa).

2.1.6 Reproducibility and accuracy :

The results (§2.1.5) show that σ_{33} obtained with Iron anode on bended sample is not very important (less than 30 MPa). If we suppose σ_{33} is zero we can make unidirectional analysis and obtain σ_Φ (§2.1.1) (Φ is chosen as the longitudinal direction of the sample).

To know the reproducibility and the accuracy of X-Ray analysis on Beryllium, we made numerous uniaxial analysis corresponding to different bending stressing. For each loading we calculate the theoretical value taking into account Beryllium X-Ray transparency, using formula n°2. On figure 5 we compare these values to those determined with an Iron anode.

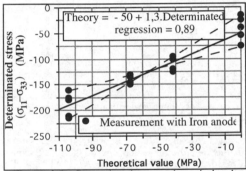

Figure n° 5 : Comparison between theoretical and measured values of stress for numerous stress levels in the case of Fe anode.

The global dispersion for each loading is less than 75 MPa, so the experimental reproducibility is better than ±40 MPa.

Each determined value is 50 MPa more in compression. This gap can easily be understood looking tensorial analysis on the same sample (§2.1.5) and the meaning of uniaxial analysis when σ_{33} is not exactly zero (§2.1.1). Initially σ_{11} is quite in compression (-35 MPa), and σ_{33} isn't exactly zero as supposed, it is +15 MPa. Taking into account these two gaps, the displacing of measured value is expected to be -50 MPa ($=\sigma_{11}-\sigma_{33}$)$_{initial}$.

The slope of the linear fitting is 1,3 which represents 30% gap with an ideal adequation. Considering the experimental reproducibility and the little extend of measured stress (0 to 105 MPa) this result is satisfactory. Reproducibility alone (±40MPa) could induce 50% mistake on the slope of the linear fitting.

2.1.7 Conclusion of results obtained on Beryllium :

For Beryllium we validate an experimental procedure to determine stresses by X-Ray diffraction. The result of the analysis is a mean value of the stress level headed by X-Ray intensity at each depth. Experimental reproducibility is better than ±40 MPa and the accuracy of the stress determination is good.

Measurements with the two anodes (Iron and Vanadium) give an information on the importance of the stress gradient. A more precise analysis could be imagined : using more wave lengths for X-Ray diffraction, it could be possible to give a polynomial approximation of the stress gradient.

2.2 X-Ray diffraction on Aluminum :

With Aluminum X-Ray diffraction gives information from only the surface (mean penetration depth is about 15 µm). So to analyse the stress gradient we use local successive chemical polishings.

The anode used is Chromium, diffraction plane is 222 for a Bragg angle of 156°.

Measurement has been done on a bended sample the deformation of which is controlled by strain gauges. The applied bending correspond to a 100 MPa stress at the surface of the sample. The sample thickness is 8 mm, so the gradient is weak. We analyse the sample on 520 µm depth.

Electro-chemical polishing has been made on a small surface (Ø 5 mm) compared to the sample surface and the attacked depth is also small compare to the sample thickness (6.5%), so the stress relaxation due to electro-chemical preparation is negligible. Results are presented on figure n° 6.

Table n°3 : Selectioned diffraction conditions on Beryllium.

Wavelength	Diffraction plan (h k l)	2θ (°)	μ/ρ (cm² g⁻¹)	P *	infinite depth (mm)
K_α V	1 0 2	140.44	4.166	613	10
K_α Fe	2 0 0	155.97	1.916	1383	25

*P : Penetration depth for attenuation of 63%(μm)

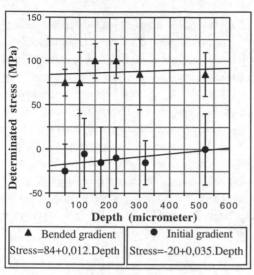

Figure n° 6 : X-Ray analysis of bended stress gradient on 6061T6 Aluminum sample.

The determined uncertainty is quite important, but the fitting of measurement is very good. The step between bending and initial fitting at surface is 105 MPa, and it is 90 MPa at 500 μm. Theoretical steps are respectively 100 MPa and 87.5 MPa.

3. CONCLUSION

In the precedent parts we have established the limits of hole drilling stress gradient analysis on Aluminum, and X-Ray diffraction stress gradient analysis on Aluminum and Beryllium. Complementary experiments with incremental hole drilling on Beryllium will soon be done, to complete present conclusions.

For Beryllium, the observed depth by X-Ray diffraction is equal to the observation scale of stress gradient. X-Ray stress analysis in Beryllium takes always into account the stress gradient on some millimeters. On Beryllium, incremental hole drilling would give more local information.

For Aluminum, hole drilling is efficient between 150 and 600 micrometers. X-Ray analysis gives a surface information corresponding to a mean value of stress on 15 micrometers depth. With local electro-chemical polishing (dimensioned in order not to disturb stress distribution), more accurate results could be found between 0 and 150 micrometers.

The two methods seem to be complementary to evaluate stress gradient with accuracy. During the analysis of stress gradients the observation depth must be compared to the accuracy limits of each method in order to choice one or the other approach or there combination.

REFERENCES :

1.   M. T. Flaman, *Experimental mechanics* **22**, 26-30 (1982).

2.   G. Maeder, J. L. Lebrun, in *"Methodes usuelles de caracterisation des surfaces"* , Eyrolles, Ed.§Eds. (SOCIETE FRANCAISE DE METALLURGIE, PARIS, 1988), vol. 1, pp. 251 269.

3.   J. M. Sprauel, M. Barral, S. Torbaty, in *"Advances in X-Ray Analysis"* , P. K., D. B. Leyden, Ed.§Eds. (PLENUM PUBLISHING CORPORATION, 1988), vol. 26, pp. 217 224.

4.   J. L. Lebrun, C. Bonnet, D. Brousse, C. Girard, *"Determination of Residual Stresses in Beryllium by X-Ray Diffraction"*, H. Fujiwara, T.Abe, K. Tanakas, Eds., ICRS - III (ELSEVIER APPLIED SCIENCE, TOKUSHIMA - JAPAN, 1991), vol. 2, pp. 638 643.

5.   J. Lu, A. Niku-Lari, J. F. Flavenot, *"Matériaux et techniques "*, 709-717 (1985).

Recent Advances in Experimental Mechanics, Silva Gomes et al. (eds) © 1994 Balkema, Rotterdam, ISBN 90 5410 395 7

New methods of X-ray tensometry

S.A.Ivanov & V.I.Monin
Saint Petersburg Technical University, Russia

J.R.Teodosio
COPPE-EE, Federal University of Rio de Janeiro, Brazil

ABSTRACT: An analysis of the new possibilities of x-ray tensometry using portable x-ray apparatus is presented. The characteristics of the apparatus and the methods of stress measurement are discribed. The results of residual stress measurements in railway wheels in the as-fabricated and post-service conditions are presented. The problems of x-ray tensometry related to stress measurements in surface layers with both concentration and stress gradients and with surface relief are analysed.

1 INTRODUCTION

The use of portable x-ray apparatus provides the possibility of applying x-ray tensometry to the nondestructive control of stresses in structures and components. Moreover, it is possible to carry out the in-service control of important technologies, such as:
1. The creation of thin layers with predetermined physico-mechanical properties by various deposition techniques or ion implantation to increase strength, wear and corrosion resistance.
2. The creation on the surface of materials of a stable microstructure with predetermined residual stresses to increase durability during cyclical loading, to decrease the tendency to fissure forming during neutron irradiation of materials etc.
3. The creation of new welding technologies capable of producing special stress states that can prevent the nucleation and propagation of cracks.

In all these cases, new problems of steess measurement can arise. These are connected with necessity of taking into account the surface concentration gradient of impurities or the presence surface relief after machining. These problems are discussed below.

2 CHARACTERISTICS OF PORTABLE X-RAY APPARATUS AND METODOLOGY OF STRESS MEASUREMENT

Developed by the authores, the portable x-ray apparatus consists of the following parts:
1. A high voltage source coupled to an air cooled doubleanode x-ray tube, collimator and film cassette mounting.
2. A power and control unit.

The weights of these parts are 2.5 kg and 1.5 kg, respectively. The size of the first is (37x7x5) cm^3 and the second one is (20x12x8) cm^3. The improved "ψ - goniometer" geometry is applied here for stress measurement. According this geometry the traditional expression for strain is applied:

$$\varepsilon(\phi,\psi) = \frac{1+\nu}{E}\,\sigma\phi\,.\,Sin^2\psi - \frac{\nu}{E}(\sigma_1+\sigma_2) \quad (2.1)$$

where E, ν=elastic constants; ϕ, ψ = azimuthal and polar angles; $\varepsilon(\phi,\psi)$ = deformation in arbitrary direction. This deformation is related to the diffraction angle θ as:

$$\varepsilon(\phi,\psi) = \frac{d\phi,\psi - d_0}{d_0} = ctg\,\theta_0(\theta_{\phi},\psi - \theta_2) \quad (2.2)$$

where d = interplanar distance. The index (o) refers to the unstressed material, but if we measure the dif

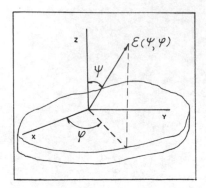

Figure 2.1
Coordinate system of measured sample
and definition of ϕ - and ψ - angles

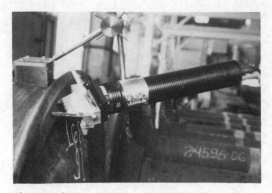

Figure 3.1
Procedure of stress measurements with portable x-ray apparatus

fraction angle Θ at different azimuthal angles ψ we can calculate the
stress tensor component in the ϕ direction. For a definition of all the
stress tensor components it is necessary to apply the variations of ψ
and ϕ angles.
The procedure of stress measurement
can be carry out under "in-laborato-
ry" and "in-field" conditions. Exposure time in the case of stress measurements in steels takes approximately 3 minutes.

3 RESIDUAL STRESS MEASUREMENT IN RA-
ILWAY WHEELS

The portable x-ray apparatus was successfully applied to stress measurements in railway wheels. The problem is that after service with heavy braking processes residual tensile stresses can arise in railway wheels. These undesirable stresses can
be the cause of wheel failure. Therefore, the periodical control of
the wheels stress state can predict
the onset of dangerous accumulatio-
ns od such tensile residual stresses.
The stress state in railway wheels
was investigated after manufacturing and after service with braking.
Figure 3.1 shows the process of the
stress measurements under "in-field"
conditions.
The tangential components of the stresses were measured in the flange section of the wheels. It was discovered that, after manufacturing the flange of railway wheels is subjected
to compressive residual stresses. The

average values of stresses measured
in 16 wheels lay within the range ze-
ro up to $\sigma_{max.} = -80$ MPa; the measurement erroy was $\Delta\sigma = \pm 20$ MPa.
After service during 24 days, with
heavy braking and in volving service
temperatures for the wheels to up to
205°C the stress state was modified.
The temperature range during the bra-
king process was 114°C - 205°C. Depending on the temperature, the new va-
lues of residual stresses lay typically in the range zero to $\sigma_{max.} = -50$
MPa.

4 EFFECT OF UNHOMOGENEITY IN THE SUR-
FACE STRAIN DISTRIBUTION

4.1 Influence of surface concentrati-
on and stress gradients

As mentioned, the above technologi -
es can create surface impurity concentration gradients and residual stresses. When an impurity concentration
gradient exists together with a biaxial residual stress state it is necessary to separate the influence of
residual stresses on the lattice parameters. In the case of a biaxial
stress state when $\sigma_1 = \sigma_2$ the problem
is solved by measurement of the diffraction angles Θ_ψ at fixed values
of the angle ϕ. The formula for calculation of the cubic lattice parameter will have these following form:

$$\Theta_0 = \frac{\Theta_{\psi_1} - \Theta_{\psi_2}\, B}{1 + B}; \quad B = \frac{2\nu - (1+\nu)\sin^2\psi_1}{2\nu - (1+\nu)\sin^2\psi_2} \qquad (4.1)$$

In the case of an existing surface stress gradient the x-ray stress measurement gives the averange experimental information. Mathematically this can be writthen as:

$$<\sigma_{ik}> = \int_v \sigma_{ik}(z)\mu e^{-\mu z}\,dz \qquad (4.2)$$

where $<\sigma_{ik}>$ = the average value of the stress component; and μ = the effective absorbtion coefficient. The definition of the stress tensor components includes in this case the following procedures:

1. The measured every the deformation $\varepsilon(\phi, \psi)$ (figure 2.1) for approximately 20 values of ϕ and 10 values of ψ for every ϕ.

2. The measurement of deformations $\varepsilon(\phi, \psi)$ and average component stresses are discribed as a finite Fourier series limited by a double angle term (Ivanov, Kolotov 1987):

$$\frac{E}{1+\nu}\varepsilon(\phi,\psi) = A_0(\psi) + A_1(\psi)\cos\phi + A_2(\psi)\cos 2\phi +$$
$$+ B_1(\psi)\sin\phi + B_2(\psi)\sin 2\phi \qquad (4.3)$$

where coefficients $A_0(\psi)$, $A_1(\psi)$ are expressed by formulas:

$$A_0(\psi) = \frac{1}{2}[\bar{\sigma}_{11}(\psi) + \bar{\sigma}_{22}(\psi)](\sin^2\psi - 2\nu) +$$
$$+ \bar{\sigma}_{33}(\psi)(\frac{1}{1+\nu} - \sin^2(\psi); \quad A_1(\psi) = \bar{\sigma}_{13}(\psi)$$
$$\sin^2\psi; \quad A_2(\psi) = \frac{1}{2}[\bar{\sigma}_{11}\psi - \sigma_{22}(\psi)] - \sin^2\psi;$$
$$B_1(\psi) = \sigma_{23}(\psi).\sin^2\psi; \quad B_2(\psi) = \bar{\sigma}_{12}(\psi)$$
$$\sin^2\psi \qquad (4.4)$$

The expression (4.3) has only 5 independent linear functions of angle ϕ and therefore the definition of six tensor components of stress is required for the equilibrium and deformation compatibility equations. The excess equation system (4.2) is solved by the least-squares technique.

3. The real value of $\sigma_{ik}(z)$ is limited by the strength limit of material. Moreover the real stress gradients g_σ will also be limited. The estimation gives the value of $g_\sigma \leq 200$ MPa/mkm for most materials. In order to use this a priori information, equation (4.2) is solved by the regularization method. This method assumes the minimization of the linear combination of the squared sum of the difference between left and right parts of the equation (4.2), and square of the norm square of the desired function $\sigma(z)^2$:

$$R + \alpha \| S(z) \|^2 \to min. \qquad (4.5)$$

The parameter α defines the a priori information part. A norm $\sigma(z)^2$ is used for the functional:

$$\| S(z) \|^2 = \sum_{i=1}^{M} \int_0^\infty (f\sigma(z)|^2 + \frac{1}{\mu^2}| d\frac{S(z)}{dz}|^2)$$
$$e^{-\mu_i z}\,dz \qquad (4.6)$$

where M is the order of digitisation of equation (4.2).

4. For a definition of the corridor of errors and effective x-ray penetration, the worse noise method is developed. This signifies that the function $f(z)$ in the capacity of the worse noise is represented as an isosceles triangle. If the slope of triangle side is equal to the stress gradient $\sigma_{ik}(z)$ then at any point z the height of this triangle at any point z_0 found from the conditions

$$\forall\mu : \int_\infty^\infty f(z)e^{-\mu z}\mu\,dz \leq \Delta \qquad (4.7)$$

will be equal to the error of the restoring of $\sigma_{ik}(z)$ at any point z_0. In expression (4.7) the value Δ is equal to the apparatus error of the stress definition. Thus, the error of the restoring of $\sigma_{ik}(z)$ depends on the minimal and maximal values of the effective absorbtion coefficients, from the error Δ of stress measurement $\bar{\sigma}(M)$ and from the gradient of restored dependence of $\sigma_{ik}(z)$. In this case the concept of x-ray penetration depth differes from traditional sense. It implies limiting value of z when the uncertainty in the measured stresses become so high that the accuracy of calculated results becomes extremely small.

4.2 Influence of a surface relief

Mechanical surface treatment causes microrelief. The problem of stress measurements on rough surfaces is related to a principal stress axis rotation and a stress variation in the cavities and protuberances (figure 4.1). The relief of this figure is related to the sinusoidal relief with the following parameters: A=amplitude of the relief, T=period of the relief; $\vec{n}$=normal to the any point of sinusoidal relief. The applied stress has direction along x axis.

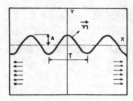

Figure 4.1
Parameters of the sinusoigal relief
for stress calculation.

Stress state depends on the degree of
relief (Husu, Vitenberg, Palmov 1975).
This situation exerts an effect on
the results of the measurements. Ac
tually, for surface roughness fact-
or of 5 mkm and 0,4 mkm, the stress
measurements across the relief direction show a difference of 30% for
applied stress and 33% for residual
stress (Motoyama, Enani, Hozizawa
1963).
The measurements along the relief
direction do not indicate stress di
fferences. The experimental results
for different roughness factors indicate that the influence of the re
lief parameters is visible when the
protuberance height is comparable
with the effective x-ray penetration (Ruhs, Sturm, Stuwe 1984). This
result is explained by the fact that,
in case of deep penetration, the st
ress information is connected only
with the undistorted material layers. In this case the influence of
the relief is negligable. In the pa
per (Ivanov, Tchistiakov, Zavodskaia
laboratoria, 1987) the numerical solution of stresses in the case of si
nusoidal relief is made. Tne calculation results for the stress compo
nent σ_{xx} are show in the figure 4.2.

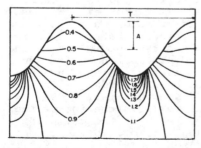

Figure 4.2
Distribution of the stress ratio σ_{xx}/σ_O for sinusoidal relief

It is possible to use stress concen-
tration coefficient $K = \sigma_{max.}/\sigma_O$. The
calculated dependence of this coeffi
cient versus the relief factor $\alpha = 2nA/T$ is show in figure 4.3.

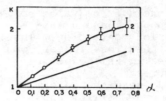

Figure 4.3
Stress concentration coefficient ver
sus relief factor

1 - refered data (Husu, Vitenberg,
    Palmov 1975)
2 - author's data

Thus stress measured on the rough sur
faces are always less than the appli
ed stresses. The correction of measu
red values can be carried out by usi
ng the curve series (figure 4.4) for
different α and M that is equal:

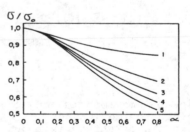

Figure 4.4
Stress ratio σ/σ_O versus relief factor for different values of normali-
zed absorbtion coefficient M; 1-M=4;
2-M=2; 3-M=1; 4-M=0,75; 5-M=0,5.

where A = amplitude of a relief, μ_O
= linear absorbtion coefficient. The
data of 4.3 and 4.4 permite to define the maximal stress value acting
in the cavities of relief surface.
The value of $\sigma_{max.}$ in this case is
equal to $\sigma_{max.} = K\sigma_O$. When we use
small penetrating x-rays with M< 0.5
the formula (Ivanov, Kolotov, Vasili
ew 1985) $\sigma_O = \sigma(1+\alpha^2)3/2$ can be appl
ied for correction of measured stre-
sses.

REFERENCES

Husu, A.R.; Vitenberg, Y.R.; Palmov,
  V.A. 1975. Sherokhovatost poverkh-
  nostei. M., Fizmatgiz, 344-348.

Ivanov, S.A.; Kolotov, A.Z. 1987.
  Zavodskaia laboratoria. 53:12
  37-41.

Ivanov. S.A.; Tchistiakov, A.M. 1987.
  Zavodskaia laboratoria. 53-12
  41-44.

Ivanov, S.A.; Vasiliew, D.M.; Kolotov, A.Z. 1985. Fizika y Tekhnolo
  gia Uprotcchnenia materalov. L.,
  FTI, 56-61.

Motoyama, M.; Enani, T.; Hzizawa, H.
  1963. J.S. Mater. Sci. 12: 882-888.

Ruhs, W.; Sturm, F.; Stuwe, H. 1984.
  Z. Metall. 75: 384-389.

Recent Advances in Experimental Mechanics, Silva Gomes et al. (eds) © 1994 Balkema, Rotterdam, ISBN 90 5410 395 7

Metrological characterization of X-ray stress methods by means of cybernetics

Dominik Senczyk
Technical University of Poznan, Poland

ABSTRACT: In all experiments two fundamental elements may be distinguished: subject of the experiment and the experimental method. Despite the importance of experimental method in every experiment, in general the analysis of its possibilities and their usefullness in real experimental conditions is underestimated. This comment refers also to X–ray stress methods. Thus it seems reasonable to comment on at least such properties which play an important role in the selection or estimation of the method in the projection of the experiment. The properties are characterized first of all by the following parameters [1]–[9]: accuracy and precision of the method, its sensitivity, detectability of physical quantities, reproductibility and repeatability of results of measurements of determined quantities, efficiency and velocity of X–ray stress methods. In this paper the above mentioned parameters of X–ray methods in policrystalline materials are described and analyzed. For the description of different X–ray stress methods a basic cybernetic concept was used.

INTRODUCTION AND AIM OF THE PAPER

In all experiments two fundamental elements may be distinguished: subject of the experiment and the experimental method. This approach implies an understanding of the method as a complete set principles at work together with the technical means and also the purpose of this set is acceptance, amplification, recording and possible preliminary transformation of signals generated by the material investigated (Senczyk 1980). Despite the importance of experimental method in every experiment, in general the analysis of its possibilities and their usefullness in real experimental conditions is underestimated. This comment refers also to X–ray stress methods. Thus it seems reasonable to comment on at least such properties which play an important role in the selection or estimation of the method in the projection of the experiment or even in looking for the source of failure in inaccurately projected or realized experiments. The properties are charac- terized first of all by the following parameters (Senczyk 1988a): accuracy and precision of the method, its sensitivity, detectability of physical quantities, reproductibility and repeatability of

results of measurements of determined quantiteis, efficiency and velocity of X–ray stress methods. In this paper the above mentioned parameters of X– ray methods in policrystalline materials are described and analyzed. Different X–ray stress methods are considered: diffractometric method of uniaxial stress measurement, the Sachs–Weerts method of measurement of the sum of principal stresses, the methods of measurement of stress in a given direction ($\sin^2\psi$ method, Glocker–Hess– Schaaber method, Gisen–Glocker–Osswald method, Δ method) and the methods of measure- ment of principal stresses ($\sin^2\psi$ method, Δ method).

For the description of above–mentioned X–ray stress methods a basic cybernetic concept was used (Senczyk 1990a). An investigation method can be regarded as an object to which two kinds of signals are delivered: signals from an object investigated x_i and signals of a disturbance z_j. In this way we obtain a cybernetic model of investigation method (Senczyk 1992). This method transfers signals y, consisting of informations about the investigated object. Cybernetics use such terms as: signal (as an energetic carrier of information) as well as input and output. It way be stated that in an input of an experimental method signals x_i and z_j occur, and an output a signal y is created. This means that the experimental method transforms input signals x_i, generated by the investigated object, and disturbance signals z_j into output signals y. Thus the experimental method is a function $y = f(x_i, z_i)$.

PRECISION OF X–RAY STRESS METHODS

Considering the investigation method as a generator of random errors one can characterize its precision by a standard deviation. In order to determine the practicaly obtainable precision of stress measurements in a set of steels the stresses in a value of 100 MPa and 200 MPa were measured by a $\sin^2\psi$ method. Standard deviation for 11 measurements of given stress values were evaluated. These experiments show that the standard deviation values of measured stresses are between 5 MPa and 18 MPa.

ACCURACY OF X–RAY STRESS METHODS

Accuracy of a method characterizes an agreement of the average measurement results with a value considered as a comparative. In the case of stress measurements the mean arithmetical sum of a series of measurements is taken as a comparative value. The upper boundary of an absolute error Δ_x can be assumed to be triple the value od standard deviation s_x: $\Delta_x = 3s_x$. The values of error Δ_x for steels investigated are between 15 MPa and 53 MPa which indicates the possibilities of this methods (Senczyk 1991).

SENSITIVITY
OF X–RAY STRESS METHODS

Using a cybernetic terms such as signal, input and output, the sensitivity of a method is defined as the ratio of the change of the signal at the output of this metod to the change of the signal at the input of this method:

$$C = \frac{dS_{OUT}}{dS_{INP}} \tag{1}$$

where S is the value of the signal and indicators refers it relatively to the output (OUT) and input (INP). In the case of X–ray stress methods the input signal is the measured amount of stress with appropriate value and a sign. The definite change of a certain diffraction angle is determined by the signal at the output. This leads to the following expression for sensitivity of the X–ray stress methods:

$$C = \left| \frac{d(\Delta\Theta)}{d\sigma} \right|. \tag{2}$$

Using this formula the sensitivity of different X–ray stress methods were de termination:

a) for diffractometric method of uniaxial stress measurement:

$$C_\sigma = - s_1 tg\Theta_0, \tag{3}$$

b) for Sachs–Weerts method of principal stresses sum measurement:

$$C_{\Sigma\sigma} = - s_1 tg\Theta_0 - \frac{1}{4}s_2 tg\Theta_0 \sin\eta, \tag{4}$$

c) for $\sin^2\psi$ method of stress measurement in a given direction:

$$C_\sigma = \frac{1}{2}s_2 tg\Theta_0 \sin\psi, \tag{5}$$

d) for Glocker–Hess–Schaaber method of stress measurement in a given direction:

$$C_\sigma = \frac{1}{2}s_2 tg\Theta_0 \sin^2\psi \sin 2\Theta_0, \tag{6}$$

e) for Gisen–Glocker–Osswald method of stress measurement in a given direction:

$$C_\sigma = \frac{1}{2}s_2 tg\Theta_0 \sin^2\eta, \tag{7}$$

f) for Δ method of stress measurement in a given direction:

$$C_\sigma = - \frac{1}{2c}s_2 \sin^2\eta, \tag{8}$$

g) for $\sin^2\psi$ method of principal stress measurements for biaxial state of stresses:

$$C_{\sigma 1} = \left| - tg\Theta_0 \left(\frac{1}{2}s_2 \cos\varphi^2 \sin^2\psi + s_1 \right) \right| \tag{9}$$

$$C_{\sigma 2} = \left| - tg\Theta_0 \left(\frac{1}{2}s_2 \sin\varphi^2 \sin^2\psi + s_1 \right) \right|, \tag{10}$$

h) for Δ method of principal stress measurements for biaxial state of stresses:

$$C_{\sigma_1} = \left| \frac{1}{c}\left(\frac{1}{2}s_2\cos^2\varphi\sin^2\psi + s_1\right)\right| \qquad (11)$$

$$C_{\sigma_2} = \left| \frac{1}{c}\left(\frac{1}{2}s_2\sin^2\varphi\sin^2\psi + s_1\right)\right| \qquad (12)$$

where s_1 and s_2 – elastic constants of materials investigated (Senczyk 1977), Θ_0 – angular position of a definite diffraction line for a specimen without stresses, ϕ – angle between a projection an a plane XOY of a coordinate system of stress measurement direction and direction of stress activity, $\eta = 90 - \Theta_0$, ψ – the slope of X–ray axis according to the flat specimens surface, σ_1 and σ_2 – principal stresses, s_ϕ – stress measured in direction determined by angle ϕ, c – constant for Δ method.

The above Equations enable to estimate the sensitivity of chosen method in real experimental conditions. For example for Glocker–Hess–Schaaber method a maximal sensitivity is obtained for $\psi = 45^\circ$. In the previous experiments it was found that in this case measurement error is minimal. Both these facts account for the usefulness of this ψ value in measurements. A choice of this value, till now, was only accidental.

STRESS DETECTABILITY
BY X–RAY METHODS

Detectability of stresses with X–ray methods is defined as the smallest value of stress, which can be measured with a certain probability by given method in certain conditions. Using a cybernetic approach, the detectability is defined as the smallest value of input signal which brings about a measurable change of signal on output. As a measure of the stresses detectability by X–ray methods a certain multiplicity of standard deviation can be assumed:

$$W = z \cdot \sigma(\sigma_1), \qquad (13)$$

where z – standarized variable. In this way the detectability of uniaxial stress is given by the Equation:

$$W^2 = 1.960\,\sigma_1\left(\frac{2}{\sin2\Theta_0} - \frac{ctg\Theta_0}{\sigma_1 s_1}\right)^2 \sigma^2(\Theta_0) \qquad (14)$$

$$+ \frac{\sigma^2(s_1)}{s_1^2} + \left(\frac{tg\Theta_0}{\sigma_1 s_1}\right)^2 \sigma(\Theta_1)^{\checkmark},$$

where Θ is an angular position of the diffraction line of a stressed specimen at normal fall of X–ray. Calculations for a carbon steel containing 0.73 wt–pct C, for which $s_1 = -1.8457.10$ MPa (Senczyk 1975), leads to following conclusions:

a) stress detectability depends really on the value of the standard deviation of diffraction angle: at constant stress its value increases as the deviation increases,

b) the values of sectioned–accuracy characterization depends on the stress and grows a little by its increase.

These conclusions are supported by experiments performed on carbon steel containing 0.65 wt–pct C.

REPRODUCTIBILITY
OF X–RAY STRESS DETERMINATION

Another characteristic of each measuring method is reproductibility of results. This may by characterized as the ratio of concordance of results of following measurements of the same value performed by one observer in the same laboratory. The results must be obtained under the same conditions, using the same tools and measuring methods. The measure of concordance is the methods precision and therefore it is unnecessery to do separate investigations. Taking into account the above mentioned results it may be stated, that the X–ray stress methods are easily reproducible.

REPEATABILITY
OF X–RAY STRESS DETERMINATION

As the repeatability of measurements the precision of a method is understood. The precision is a measure of agreement of results obtained by independent investigators reserching in different laboratories or in the same laboratory in different periods of time. The necessery conditions for this is that each investigator obtain individual results by measurements of the same physical quantity using different methods and tools.

The experimental investigations of repeatability of measuring stress equals 100 MPa and 200 MPa for 0.25 wt–pct C and 0.45 wt–pct C carbon steels in annealed and tempered states shows that the standard deviation of stresses measured in these steels lies between 5.5 MPa and 8.0 MPa (Senczyk 1988b). The values of the upper boundary of an absolute error for investigated materials are between 16.5 MPa and 24.0 MPa. This means that the repeatability of stress measurement results at stable experimental conditions is good.

EFFICIENCY AND VELOCITY OF X–RAY STRESS DETERMINATION

The next indications of measuring methods especially in industrial conditions are their efficiency and velocity.

A measure of the method efficiency is the number of results obtained by measuring a value during a limited period of time. A measure the velocity of the method is time needed for conducting one measurement this quantity.

The analyses performed show that the X–ray stress method realized by conventional X–ray diffractometers is slow and has a low efficiency rate, however by using special diffractometers it becomes an effective and quick method.

SUMMARY

The above presented characterization of X–ray stress methods in policrystalline materials with a cybernetical approach shows that the precision and accuracy of this method is compatible with the values obtained by other experimental stress methods. The given relations enables us to evaluate the sensitivity of the presented methods for definite experimental conditions. These relations lead to a scientific (not intuition) selection of the right method. Analysis of sensitivity and detectability od stresses leads to many conclusions which indicates the possibility of increasing or keeping the accuracy of measuring stresses by X–ray methods.

This research was supported in part by Polish State Committee for Scientific Research (KBN) under Grant No. 7–7031–92–03.

REFERENCES

Senczyk, D. 1975. Einige Probleme der Messung elastischer Konstanten polykristalliner Werkstoffe mittels der rentgenographischen Spannungs-messung. *Proc. 2nd Konferenz "Struktur– und Gefügeanalyse:* Vorträge RG–7. Freiberg: Bergakademie.

Senczyk, D. 1977. Measuring and anisotropy of elastic constants of polycrystalline materials in X–ray strain gauging. *Proc. "Kovove materialy s vysokymi mechanicko–fizykalnymi vlastnostami":* v. 2 177–186. Bratislava:

Senczyk, D. 1980. *The analysis of error sources and experimental determination of precision in measurements of macrostresses by X–ray methods for chosen polycrystalline materials.* Poznan: Technical University of Poznan.

Senczyk, D. 1988a. *Macrostress measurements in polycrystalline materials by X–ray methods.* Poznan: Technical University of Poznan.

Senczyk, D. 1988b. Macrostress investigations in grey cast iron frames. *Proc. 13th Conference on Applied Crystallography:* v.1 260–264. Cieszyn: Silesian University in Katowice and Institute of Ferrous Metallurgy in Gliwice.

Senczyk, D. 1990a. A general method of investigation of influence of different factors on accuracy of macrostresses measurement by X–ray methods. *Proc. 14th Conference on Applied Crystal-lography:* v. 2 430–432. Cieszyn: Silesian University in Katowice and Institute of Ferrous Metallurgy in Gliwice.

Senczyk, D. 1990b. Computerization of meas-urement of components of a stress tensor. *Proc. 14th Conference on Applied Crystallography:* v. 2 441–444. Cieszyn: Silesian University in Katowice and Institute of Ferrous Metallurgy in Gliwice.

Senczyk, D. 1991. Computer programs for X–ray macrostress measurements. *Proc. European Conference "Physics for industry – industry for physics":* 179–180. Cracow: European Physical Society and Polish Physical Society.

Senczyk D. 1992. Fundamentals of metrological characterization of X–ray stress measurement methods in polycrystalline materials. *Proc. 15th Conference on Applied Crystallography:* 30. Cieszyn: Silesian University in Katowice and Institute of Ferrous Metallurgy in Gliwice.

Recent Advances in Experimental Mechanics, Silva Gomes et al. (eds) © 1994 Balkema, Rotterdam, ISBN 90 5410 395 7

Residual stress measurement in a thick-section steel weld

N.W.Bonner & D.J.Smith
University of Bristol, UK

R.H.Leggatt
TWI, Cambridge, UK

ABSTRACT: Significant levels of residual stress are developed in the production of thick-section welds. To assess adequately the integrity of such components, accurate information concerning the magnitude and direction of the residual stresses must be obtained. Residual stresses can be determined by variety of methods, but few offer the ability to determine the complete through thickness distribution in thick-section steel welds. The deep hole measurement method can provide through thickness distributions, though conversion of measured displacements to residual stresses can be complex. In this paper improvements to the deep hole method are described. The technique is employed to measure residual strains in a butt-welded steel plate of 105mm wall thickness. An analysis is presented for converting measured strains into an original residual stress field. The results are compared with those expected in a weld of the geometry examined.

1 INTRODUCTION

The presence of significant levels of residual stress can play an important role in assessments of the structural integrity of welded engineering components. In particular, it is crucial that proper attention is paid to estimating the magnitude and direction of these stresses. Since there is a paucity of reliable information about residual stress distributions in steel welds of thickness greater than around 50mm, conservative estimates are used in the structural integrity assessments.

Reviews of the various techniques for measuring residual stresses have been given by Parlane (1977), Keller et at (1989) and De Angelis et al (1990). Classification of the methods can be conducted on the basis of the degree to which any specimen is damaged. X-ray or neutron diffraction, ultrasonics and electromagnetism are examples of non-destructive methods; centre-hole drilling, ring-coring and deep hole drilling exemplify semi-destructive methods; and, block removal, splitting and layering (BRSL) and Sach's boring illustrate destructive methods.

However, measurement of the complete through thickness distribution of residual stresses in thick-section steel welds can be achieved by few of the above techniques. Neutron diffraction (Holden et al 1988) can be used to depths as great as 25mm, but only if access allows examination of opposite faces. The methods of BRSL and Sach's boring (Ueda et al 1975) can be used to sufficient depths, but completely destroy the specimen, and in the case of former, are only suitable where no significant thickness changes or curvature are present.

One method, initially devised by Beaney (1978) and Zhandov et al (1978), which measures through thickness residual stresses semi-destructively is the deep hole measurement method (Fig.1): a gun drill is used to make a small straight reference hole through the thickness of the specimen; accurate measurements of the hole diameter are made at numerous positions; co-axial machining is conducted on an annulus surrounding the reference hole until a column of material is freed from the remaining specimen (any changes in axial hole length are detected during this process); comprehensive strain data surrounding the hole is provided on remeasurement of diameter, allowing calculation of the residual

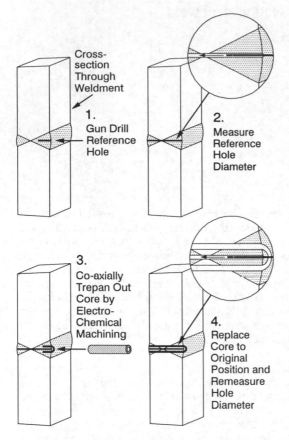

Figure 1. Basic process of deep hole residual stress measurement.

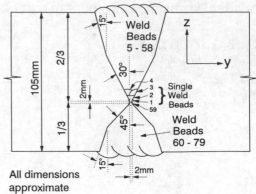

All dimensions approximate

Figure 2. Cross-section of welded plate wall.

stress distribution.

Following on from earlier developments made by Procter et al (1987) and Mitchell (1988), significant improvements to the deep hole experimental process have been achieved and are described here. These improvements include using non-contacting air probes to measure reference hole diameters easily and accurately, measuring such diameters at a greater axial density, and measuring the axial position of this data rather than relying on the system to make the required displacements accurately.

In this paper a single double-V butt-welded plate specimen with a thickness of approximately 105mm is used to demonstrate these further developments. An analysis technique is described for predicting the residual stresses from the measurements of strain.

2 EXPERIMENTS

2.1 Specimen

Measurements were conducted on a nominally one metre square BS 1501-224-490B ferritic steel plate of approximately 105mm thickness, bisected by a single asymmetric double-V butt-weld (Fig.2). The weld consisted of seventy nine weld beads in twenty three layers. Manual metal arc welding was employed for the first and second passes, and submerged arc welding for the remainder.

2.2 Experimental Procedure

The experimental equipment comprised five major components: a substantial frame to support large test specimens; a common reference frame connected to the front of the frame on which the remaining equipment was placed in turn; a gun drilling head to facilitate hole drilling; a measuring frame used to provide diametral measurements; and, an electro-chemical machining (ECM) head to allow co-axial trepanning around the hole.

Firstly the reference frame was positioned at an appropriate height and made horizontal. The gun drill head was placed in position and a laser aligned to its drilling axis. The ECM apparatus then replaced the gun drill and was aligned to the laser, thereby providing co-axial translation of the two machining heads. The test specimen was then located on the frame and electrically insulated from it. The position of the specimen was then adjusted until the co-axis passed through the point where residual stress measurement was required.

Reference bushes were attached to the front and rear of the specimen before gun drilling commenced. These bushes served several functions: tending to minimise any 'bell-mouthing' in the specimen; providing a reference for measurement of hole diameter; and, ensuring that the reference hole was relocated precisely to its original position following ECM.

The gun drill was then installed. Three features characterise this specialised drill:

1. High pressure oil is supplied to its tip through a hole passing down its centre.

2. A straight channel runs down its outside to ease the discharge of both swarf and oil.

3. The point at its tip is off axis causing its rotation to produce self-centralising forces.

The effect of these features was to provide a very straight hole with a highly uniform specific diameter (3.175mm). On completion of drilling, the gun drill head was removed to allow subsequent diameteral measurement.

The measurement frame was then placed upon the reference frame. The former contained locating holes on a pitched circle which facilitated reference hole diameter measurement at different angles using a Mercer air probe. This probe comprised a thin hollow tube, capped at one end and with two small diametrically opposed holes in the curved surface. Pressurised air supplied to the uncapped end of the probe jets from the small holes. If the probe is placed in a circular hole then it is self-centred by the air jets and the air pressure inside the probe rises to a value which is dependent on the hole diameter. Conversion of the changes in air pressure to voltages was achieved using a transducer. Voltage readings were calibrated against diameter using several gauge rings.

Output voltages from both the pressure tranducer and a linear voltage displacement transducer (LVDT) detecting axial position were logged using a data recording system, which was also used to move the probe via a stepper motor.

Following hole measurement, the front reference bush was removed and the ECM head replaced the measurement frame. The ECM electrode comprised a copper tube, with approximately 15mm and 25mm internal and external diameters respectively, that was electrically insulated using nylon sheaths. Sodium nitrate solution (with a specific density of about 1.15) was used as the electrolyte. This flowed along the tube to the end of the electrode before it was recycled via a reservoir. The electrode co-axially removed an annulus of material completely through the specimen thickness, resulting in a deep hole 'core'.

Changes in the axial dimension of the reference hole were monitored during the trepanning process using LVDTs. Both this change in the axial length and the local temperature of the test specimen were recorded as a function of the axial position of the ECM electrode during trepanning.

When trepanning was complete, the ECM head was removed, the 'core' and front bush were restored to their former positions, and the measurement frame reinstalled. Remeasurement of the reference hole in the specimen and both bushes was then conducted at the same positions as before.

3. RESULTS

Deep hole measurement was carried out at a position located 20mm from the weld centre line and 60mm from the perpendicular plate centre line. Measurements of reference hole diameter were conducted in the plate and both bushes at three angles θ_1, θ_2 and θ_3 where $\theta_2 = \theta_1 + 45°$ and $\theta_3 = \theta_2 + 45°$. Repeatability was ensured by taking three measurements at each of these angles after both of the machining operations. An axial distance of around 0.2mm separated individual diametral measurements. It was not possible to take diametral measurements at specific axial positions, however, but since the axial locations were accurately measured, comparison between diameters could be effected by averaging over blocklengths (the interval between specific axial locations).

Diametral strain data at each angle was obtained from the measurements of diameter by determining the differences in mean value over 5.2mm blocklengths. The changes in mean diameter within each blocklength, before and after trepanning, were determined based on an average of the three measurements at each angle. Changes in the axial dimension of the deep hole 'core' were measured, for every millimetre of travel of the ECM electrode. The average axial strain corresponding to each blocklength was then determined from the total extension detected whilst the ECM electrode passed that blocklength, divided by its length. The thermal data was used to predict the axial extension that would be expected from the temperature changes alone. A previous run on a specimen believed to contain very low levels of residual stress was also used as a means of quantifying the effect on axial extension due to the ECM process alone. A deduction from the measured changes in axial dimension was then made to allow for such effects.

An analysis technique is presented in the next section for converting the diametral and axial strains into the through thickness residual stress distribution.

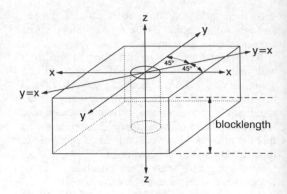

Figure 3. Orientation of axes used in theoretical analysis.

4. THEORETICAL ANALYSIS

The following analysis is based on a number of assumptions.
1. The specimen has isotropic material properties.
2. Each blocklength contains a uniform stress field.
3. A state of plane stress is present normal to the reference hole axis.
4. Shear stresses other than normal to the hole axis are ignored.
5. The core is stress free after trepanning.
6. The strain measured at one blocklength is unaffected by the residual stress of another.

These assumptions allow the distortion of a plate containing a circular through hole, and subjected to a uniform in-plane stress at infinity, to be described by an exact solution (Williams 1973):

$$u_{rr}(r,\theta) = \frac{\sigma a}{E} \left\{ \begin{bmatrix} \frac{a(1+v)}{2r} + \frac{r(1-v)}{2a} \end{bmatrix} + \begin{bmatrix} \frac{r(1+v)}{2a}\left(1-\frac{a^4}{r^4}\right) + \frac{2a}{r} \end{bmatrix} \cos 2\theta \right\}$$

(1)

where $u_{rr}(r,\theta)$ is the radial displacement at point with polar coordinates (r,θ), E is Young's Modulus, v is Poisson's ratio, and a is the hole radius. On letting r tend to a in Eq.(1) and then dividing by a, the radial strain at the hole edge is produced:

$$\varepsilon_{rr}(a,\theta) = \frac{\sigma}{E}(1+2\cos 2\theta)$$

(2)

which equals the diametral strain at the hole edge since $u_{rr}(r,\theta)$ has a period of 180°. Therefore, the strains in the x, y, $y = x$ and z directions (Fig.3), for relaxation of stress σ_{xx}, are given by:

$$\underline{\varepsilon} = \begin{bmatrix} \varepsilon_{xx} \\ \varepsilon_{yy} \\ \varepsilon_{zz} \\ \varepsilon_{y=x} \end{bmatrix} = -\frac{\sigma_{xx}}{E} \begin{bmatrix} 3 \\ -1 \\ -v \\ 1 \end{bmatrix}$$

(3)

For the general case of three-dimensional plane stress, the strains resulting from the independent stresses σ_{yy}, σ_{zz} and τ_{xy} must also be considered:

$$\underline{\varepsilon} = -\frac{\sigma_{yy}}{E}\begin{bmatrix} -1 & 3 & -v & 1 \end{bmatrix}^T$$

(4)

$$\underline{\varepsilon} = -\frac{\sigma_{zz}}{E}\begin{bmatrix} -v & -v & 1 & -v \end{bmatrix}^T$$

(5)

$$\underline{\varepsilon} = -\frac{\tau_{xy}}{E}\begin{bmatrix} 0 & 0 & 0 & 4 \end{bmatrix}^T$$

(6)

where in Eq.(5) σ_{zz} produces uniaxial tensile strain in the z direction over a gauge length equal to the blocklength, combined with Poisson contraction normal to this direction, and in Eq.(6) the shear stress τ_{xy} is composed of a tensile stress of magnitude τ_{xy} in the $y = x$ direction and a compressive stress of equal size in the $y = -x$ direction. Combining equations (3) to (6) provides an overall relation between the initial residual stress $\underline{\sigma}$ and the relaxed strain $\underline{\varepsilon}$:

$$\underline{\varepsilon} = \frac{-1}{E}[C] \cdot \underline{\sigma} \qquad (7)$$

where $\underline{\sigma} = \begin{bmatrix} \sigma_{xx} & \sigma_{yy} & \sigma_{zz} & \tau_{xy} \end{bmatrix}^T$ and $[C]$ is a compliance matrix of influence coefficients that is given by:

$$[C] = \begin{bmatrix} 3 & -1 & -v & 0 \\ -1 & 3 & -v & 0 \\ -v & -v & 1 & 0 \\ 1 & 1 & -v & 4 \end{bmatrix} \qquad (8)$$

Hence finally, rearranging Eq.(7) produces:

$$\underline{\sigma} = -E[D] \cdot \underline{\varepsilon} \qquad (9)$$

where $[D] = [C]^{-1}$, termed an elasticity matrix, is the inverse of the matrix $[D]$, and is given by:

$$[D] = \begin{bmatrix} Q & P & vR & 0 \\ P & Q & vR & 0 \\ vR & vR & 2R & 0 \\ -S & -S & 0 & 2S \end{bmatrix} \qquad (10)$$

where:

$$\left. \begin{aligned} P &= \left(1 + v^2\right) \big/ 8\left(1 - v^2\right) \\ Q &= \left(3 - v^2\right) \big/ 8\left(1 - v^2\right) \\ R &= 1/2\left(1 - v^2\right) \\ S &= 1/8 \end{aligned} \right\} \qquad (11)$$

Equations (9) to (11) directly enable conversion of the measured blocklength strains to a pre-existing residual stress field as required.

To obtain the residual stresses from Eqs.(9) to (11), the averaged blocklength strain data obtained by experiment were assembled in vectors of four components, equivalent to $\underline{\varepsilon}$ of Eq.(9). This was achieved by considering the diametral data at each of the three angles, together with the axial data. The residual stresses at each blocklength were then provided by Eq.(9), on calculation of the elasticity matrix. The results in the directions longitudinal and tranverse to the weld are presented in Fig.4 using $E = 200\text{GPa}$ and $v = 0.3$.

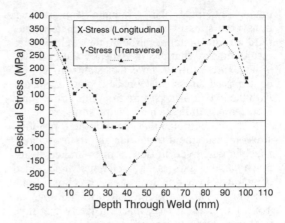

Figure 4. Through thickness residual stress distribution in a butt-welded steel plate.

5. DISCUSSION

Limitations in the strain data which could be obtained experimentally were the principal motivation for the assumptions made in order to develop the analysis above. It was not possible to measure tangential strain data in the plane of the hole, but the assumptions make this information redundant to the diametral strains.

On considering the geometry of the welded plate specimen examined, the results for the transverse through thickness residual stress would be expected to follow the well known tensile-compressive-tensile distribution for unrestrained multi-pass welds (Leggatt 1986). This distribution arises since the contraction of each cooling weld bead as it is laid down results in residual tension in the bead, and has the effect of producing compressive stress on the already cool beads of previous passes.

The deep hole results shown in Fig.4 appear to confirm this prediction. Transverse residual stresses of about 300MPa were present near to one surface of the plate passing into compression at a depth of about 15mm, and reaching a maximum compressive value above 200MPa at a depth of about 35mm. The transverse stress then returns into tension at a depth of about 60mm, reaching a local maximum above 350MPa at about 90mm, before reducing in magnitude towards the rear face.

Since the region of weld examined first was the 'one third' part of the weld preparation, one might anticipate the minimum compressive transverse stresses to occur approximately one third of the distance through the weld. Our results display that this is evidently the case. It is also interesting to note that states of almost equi-biaxial stress exist on both plate surfaces, with the extent of departure from this stress state increasing with depth beneath the surfaces.

It should be noted that the deep hole residual stress measurement technique and its associated analysis employed in this work is limited in as much as the axis of the reference hole must coincide with a principal stress direction. The specimen geometry with a hole drilled normal to a welded plate surface, makes it likely that this is at least close to being the case. However, in more complicated arrangements, for example involving T-welds, such a situation could not be assumed to exist.

6. CONCLUSIONS

(a) A technique for converting deep hole strain data into the pre-existing residual stress distribution has been presented. Analysis has been conducted on the basis that measurement of both diametral and axial strains has been possible.

(b) The deep hole method has been used to measure the through thickness distribution of residual stress at one location off the weld centre line of an approximately 105mm butt-welded steel plate. Results are presented based on the analysis described.

(c) The results for the transverse residual stress distribution follow the classic tensile-compressive-tensile distribution expected in welds of this type. Maximum compressive transverse stress was found at approximately one third depth, accounted for by the asymmetric double-V weld preparation.

REFERENCES

Beaney, E.M. 1978. *Measurement of Sub-Surface Stress*. CEGB Report No. RD/B/N4325.

De Angelis, V. & C. Sampietri 1990. *Report on the State of the Art Regarding the Problem of Residual Stresses in Welds of LMFBR Components*. CISE Tecnologie Innovative, Report No. 5495.

Holden, T.M., J.H. Root, R.A. Holt & G. Roy 1988. *Proc. 7th Int. Conf. on Offshore Mech. & Arctic Eng.*: 127-131.

Keller, H.P., H. Kerkhoff, R. Giffeler & J. Meinhardt 1989. *Residual Stresses and Their Influence on the Integrity of Pressure Vessels*. Institute for Material Testing, TÜV Rheinland, Report No. SB 203/89.

Leggatt, R.H. 1986. Residual Stresses in Science and Technology, Vol.1. *Proc. Int. Conf. on Residual Stress:* 997-1004. Oberursel: DGM.

Mitchell, D.H. 1988. *R6 Validation Exercise: Through Thickness Residual Stress Measurements on an Experimental Test Vessel Rig*. CEGB Report No. RD/B/6088/R88 (Unrestricted).

Parlane, A.J.A. 1977. Residual Stresses in Welded Construction and Their Effects. *Proc. TWI Int. Conf., London:* 63-78..

Procter, E. & E.M. Beaney 1987. Advances in Surface Treatments: Technology-Application-Effects, Vol.4. *International Guidebook on Residual Stresses:* 165-198. Oxford: Pergamon.

Ueda, Y., K. Fukuda & S. Endo 1975. *Trans. Jap. Weld. Res. Inst.* Vol.4 No.2: 13-17.

Williams, J.G. 1973. *Stress Analysis of Polymers*. London: Longman.

Zhdanov, I.M. & A.K. Gonchar 1978. *Automatic Welding* Vol.31 No.9: 22-24.

Recent Advances in Experimental Mechanics, Silva Gomes et al. (eds) © 1994 Balkema, Rotterdam, ISBN 90 5410 395 7

Research into residual stresses in ceramic products

V. Dolhof
ŠKODA, Research, Pilsen, Czech Republic

ABSTRACT: The contribution summarizes the experience in verification and applications of semi-destructive hole-drilling strain-gauge method on the example of eight ceramic products and two testing samples. The results of evaluation of residual stresses incl. their behaviour along the depth are also mentioned. Regarding the influences of production process, the influence of speed of the products passage on the level of residual stress and their mutual interaction of glaze and porcelain during their firing in the circular furnace was investigated. The experimental research proved that with respect to the variability of sense, values, and behaviours of residual stresses along the thickness, it is necessary to know their level in products after their firing so that the composition, or respectively, thickness of glaze and body can be optimized in case of higher tension stresses.

1 INTRODUCTION

In our Department of Elasticity and Strength of the ŠKODA, Research, Pilsen, various methods of residual stress measurements in metals were developed, but as for ceramic materials there was no experience in the beginning of this research work. There were pieces of information from the professional literature available about residual stress in rocks and polycrystallic diamond (Schwartz 1990).

The ceramics is classified as a polycrystallic anorganic substance with heterogeneous structure and numberous cavities air voids, microcracks, and other concentrators. There are especially so-called micro-stresses i.e. residual stresses of the 2nd order which are due to the volume changes of crystal components in the mixture and which are either accompanying modification transformations of some phases at certain temperatures or are arising during cooling after their burning due to different thermal coefficients of expansion of various components. In addition to it, the residual stresses of the 1st order which were subject of the research and which bring about the rise, or respectively, the development of magistral cracks during cooling of the pieces after their burning, and/or during the storage of finished products in the store of manufacturer, or respectively, with the customer.

2 METHODS OF RESIDUAL STRESS MEASUREMENT

The decision on use of a suitable semi-destructive drilling method was made after the verification tests on glazed and unglazed ceramic samples. The following processes were used such as cutting through by means of a carborundum disk, drilling method by means of a carbide-tip drill and then also by means of a diamond drill with continuously controlled revolutions with regard to the used tool, out of which the spot-drilling by means of the diamond drill was selected as the best method. The drill was clamped into the drilling bushing guided by a drilling fixture and driven by a drilling machine. The procedure of residual stress evaluation along the drill depth was elaborated, which enabled to determine the residual stress values in the proper glaze layer and adjacent area of porcelain.

The strain gauge rosettes type 3/120 RY21 from Hottinger Baldwin Messtechnik, GmbH, were used, which were bonded by the Schnell-klebstoff Z 70 fast-acting single-component adhesive. The strain gauges and outlets were protected against the influences of cold water by insulating Abdeck Kitt AK 22 mastic. The guide sleeve attached to ferromagnetic materials by spot welds, was mounted on the surface of ceramics by clamps, bonded by two-component X 60 fast acting adhesive. Strains were measured by a static strain gauge bridge of MK Manuall Kompenzator.

The evaluation was carried out in conformity with the Brochures of Measurements Group Messtechnik, GmbH. It complied with ASTM E837-85. The measured increments of relative strains were inserted into the formulas (1), (2) and by their digitization the value and directions of stresses along the measured depth were determined.

The relationships for evaluation of main stresses and angle of their orientation:

$$\sigma_{12} = \frac{\varepsilon_a + \varepsilon_c}{4A} \pm \frac{\sqrt{2}}{4B} \sqrt{(\varepsilon_a - \varepsilon_b)^2 + (\varepsilon_b - \varepsilon_c)^2} \quad (1)$$

$$\varphi = \frac{1}{2} \, arctg \, \frac{2\varepsilon_b - \varepsilon_a - \varepsilon_c}{\varepsilon_a - \varepsilon_c} \quad (2)$$

The constants A, B can be calculated out of the following relationships:

$$A = -\frac{1 + \mu}{2E} \cdot \bar{a} \; ; \qquad B = -\frac{1}{2E} \cdot \bar{b} \quad (3)$$

wherein $\bar{a}$, $\bar{b}$ are also constants derived from Schajer's study of proportion of the stress as the depth of drilled hole (1981) for ratio of the medium diameter of grids of strain gauge rosettes to the drilled hole diameter D/Do = 3.4 .

3 MATERIAL PROPERTIES OF CERAMICS

The main objective it was to determine the Young's modulus E and Poisson's ratio μ of the porcelain and glaze for calculation of residual stresses during the strain gauge measurements on products. These material constants were determined on 8 tension samples (cross-section 15 x 20 mm), 4 compression samples and 6 bent samples (cross-section 15 x 15 mm). The samples for each test were of non-glazed porcelain , glazed porcelain and pure glaze.

Summarizing the results it followed that either determined material constants showed approx. the same values during the tension and compression both for porcelain and glaze. The average value of Young's modulus E determined out of 12 values is 71,500 MPa and average value of Poisson's ratio μ=0.18. The scattering of both constants did not exceed + 8 %. For detailed results see (Dolhof 1992). The maximum values of strain gauge measurements of the surface stress reach up to + 35 MPa in the area of strain gauges and no crack occurred under the strain gauges. Therefore the tension strength can be assumed being higher. The values of tensile, compressive, and ultimate bending strengths were calculated out of the force, which is necessary for de-

struction of the sample and out of the sample cross-section.

4 RESIDUAL STRESSES IN CERAMIC PRODUCTS

The experimental programme for determination of residual stresses was performed on 8 pieces of porcelain vessels (PV) and on two circular plates of dia. 185 mm and thickness 12 mm. For better overview all products are summarized in succession as was their order of measurement - see Table 1 and 2 - incl. results of residual stresses.

With the first porcelain vessel PV 0, the method of drilling using the diamond drill tube with outer diameter of 4 mm manufactured by DIAZ Turnov Company and delivered by NAREX Praha Company was verified and residual stresses were evaluated after drilling of a 5 mm depth.

During the next stage the residual stress in three products provided with white glaze burnt at various speed of passage through the furnace and therefore at various burning and cooling speeds were investigated. The PV 1 product was burnt at the passage speed of 24 minutes, the PV 3 at 26 minutes and PV 6 at 28 minutes. The results of evaluated residual stresses are set forth in Table 1. In all three cases residual compressive stresses between -2 and -10 MPa were determined in white glaze (the drill depth of 1 mm had already hit about 0.2 to 0.3 mm into porcelain). In the proper porcelain of the PV 1 sample small values of compressive stress, and in the PV 3 and PV 6 small values of tensile stress were determined. This findings cannot be overestimated with regard to small level of residual stresses, which is practically comparable with their measurement accuracy. Therefore, the influence of the burning speed had not been investigated any longer.

During the next stage the programme of residual stress determination was focused on ceramic products provided with pink glaze First the stress in the PV 9 and PV 11 samples with dark pink glaze, and in the PV 10 sample with bright pink colour was determined because during the verification of the production of these products some problems occurred during their manufacture. Based upon the experience with evaluation of the residual stresses along the depth of tested samples, the consecutive drilling of the hole (in annularrings) at steps of 0.25 mm each up to the depth of 1 mm and then at steps of 0.5 mm each up to the final depth of 5 mm was selected. The results of evaluated main residual stresses are summarized in Table 2.

In the PV 9 a very high level of residual

Table 1. Residual stresses in porcelain vessels.

Depth	White product PV 0			White product PV 1			White product PV 3			White product PV6		
z /mm/	σ_1 /MPa/	σ_2 /MPa/	φ /o/	σ_1 /MPa/	σ_2 /MPa/	φ /o/	σ_1 /MPa/	σ_2 /MPa/	φ /o/	σ_1 /MPa/	σ_2 /MPa/	φ /o/
1	-	-	-	-6.4	-8.3	0	-5.1	-8.2	35.8	-1.9	-10.0	40.9
2	-	-	-	-0.1	-1.4	22.5	1.1	-1.1	13.3	1.9	-1.9	38.0
3	-	-	-	-0.3	-2.1	22.5	-0.5	-2.0	31.7	3.1	0	35.8
4	-	-	-	-1.0	-3.2	30.5	0.7	-1.3	22.5	4.1	1.3	35.8
5	6.2	-3.8	43.3	-0.5	-3.1	31.7	1.9	0.6	7.0	3.8	1.6	33.4

Table 2. Residual stresses in porcelain vessels.

Depth /mm/	Dark-pink product PV 9			Bright-pink product PV 10			Dark-pink product PV 11			White product PV 8		
z /mm/	σ_1 /MPà/	σ_2 /MPa/	φ /o/	σ_1 /MPa/	σ_2 /MPa/	φ /o/	σ_1 /MPa/	σ_2 /MPa/	φ /o/	σ_1 /MPa/	σ_2 /MPa/	φ /o/
0.25	15.5	29.7	-22.5	-28.0	-36.1	-35.8	0.2	7.3	22.5	-14.9	-7.8	-22.5
0.5	13.7	26.6	-35.8	-9.7	-17.7	28.2	3.6	6.5	-22.5	3.5	4.9	-22.5
0.75	13.0	30.7	-38.5	-1.6	-4.6	31.7	6.3	8.2	-22.5	3.1	7.3	-9.2
1.0	13.8	33.3	-42.1	1.2	0.3	0	14.8	16.1	-22.5	3.8	9.4	-15.5
1.5	11.6	29.9	-43.1	5.9	4.5	-31.7	11.4	16.0	-33.4	4.3	10.8	-10.9
2.0	12.7	27.8	-44.1	6.8	5.7	-31.7	11.7	15.6	-27.2	3.7	11.8	-8.2
2.5	18.1	40.9	41.7	8.5	7.7	-31.7	13.1	16.8	-26.6	4.8	12.1	-12.0
3.0	22.8	43.1	41.3	10.1	9.4	0	13.4	17.7	-28.8	5.4	13.5	-10.7
3.5	23.2	44.9	39.8	11.4	10.6	-35.8	14.0	17.9	-28.8	5.6	13.1	-10.7
4.0	25.0	46.2	38.6	12.0	10.9	-28.2	14.5	19.3	-30.1	6.2	13.1	-10.0
4.5	25.9	47.7	38.0	12.7	11.4	-31.7	16.2	20.6	-30.8	5.1	12.4	-10.7
5.0	26.0	48.2	37.7	12.7	11.4	-31.7	14.3	21.3	-38.9	3.9	12.4	-11.7

Notes for Tables 1 and 2: 1/ $\measuredangle\varphi$ for $(\sigma_{1,2})_{max}$ is related to the orientation of ε_a

2/ Sense of orientation of the angle

tension stresses up to + 30 MPa was determined, yet in a proper dark-pink glaze and higher level of tension stress was determined in the proper porcelain. The determined residual tension stresses explained the production difficulties unambiguously. Similar results were determined also in the PV 11 product.

In the PV 10 product on the surface of the glaze compressive stress of about -30 MPa was determined, in the direction to the depth of the glaze the residual stress decreased and on the passage between the glaze and porcelain it passed over to tension stress of about +5 MPa. The resulting residual stress in the depth of 4 to 5 mm was tension stress with values of main stress + 13 and +11 MPa.

For verification of the glaze influence on the porcelain, two circular plates were made with a diameter of 185 mm and thickness 12 mm. This thickness corresponds to the thickness of the glaze applied on products. The plate had sufficient rigidity and therefore no deflection occurred due to the glaze after its burning. In both glazes the compressive stress occurred with high rate where the highest values of about -35 MPa were determined on the surface of the glaze. In the plate provided with bright-pink glaze there was no residual stress up to the depth of 3 mm, in the plate provided with white glaze compressive residual stress of about -4 MPa was determined.

Since both testing plates were made of the same porcelain material (the composition of the body) it can be said that white glaze is causing more favourable conditions with respect to the residual stress and so would be probably also in more complicated products.

5 SUMMARY OF THE TESTED PRODUCTS

The results of all performed experiments showed the following:

1. The most favourable values and stress curves, i.e. compressive stresses of about - 15 MPa in the glaze and very small stress of about + 3 MPa in the porcelain were determined in PV 0, 1, 3, and 6 provided with white glaze. The influence of the passage speed of burning at 24, 26, and 28 min. on the level of residual stresses was not proven.

Quite different values of stresses were found in PV 8 provided with white glaze, which was delivered with typical crack, which had already two magistral cracks in the support part of the vessel in the time of determination of the stresses. Although the stress level was decreased due to the cracks, the tensile stress in the incriminated spot were still reaching the value of + 12 MPa.

2. The acceptable stresses i.e. compressive stress of about - 30 MPa in the glaze and relatively higher values of tension stress of + 11 to + 13 MPa in the porcelain were determined in PV 10 provided with bright-pink glaze.

3. Unacceptably high stresses i.e. high tension stress up to + 30 MPa, or respectively + 16 MPa, in the glaze and higher tension stresses in the porcelain were determined in PV 9 and similar tension stresses in PV 11 provided with dark-pink glaze.

4. The results of experiments with circular plates are in conformity with the results of the products.

6 CONCLUSIONS

The conclusions can be made based upon the results of experiments that especially the type of glaze, its chemical composition specifying the thermal coefficient of dilation and its thickness are decisive for rise of residual stress of the 1st order. After the application on product surface the glaze is melted during the heating and is firmly bound to the porcelain surface. In case of different thermal expansion coefficient of the glaze α_g and porcelain α_p the internal stresses occur in the product during the cooling down after burning which remain there as residual stresses. If the coefficient α_g is lower α_p, compressive stresses arise on the poduct´s surface (in this case probably white, or respectively bright-pink glaze), in opposite case the tensile stress is arising on the product´s surface.

In addition to the kind of the glaze the level of residual stress may be influenced by the geometrical shape of the product (difference between ceramic vessel and circular plate) and probably also some precisely unidentified influence of the production process (e.g. composition of the porcelain body, phase transformation of its components during the burning, or respectively cooling a.o.), the determination of which would require detailed investigation.

The experimental research of stress in ceramic porcelain products showed that with regard to the variability of the sense, values and courses along the thickness, it is necessary to know the value of residual stresses in the product after its burning so that in case of high tension stress the composition, or respectively the thickness of the glaze and body, can be optimized.

The verified semi-destructive method of measurement of residual stress in ceramic materials are recommended for use in the following applications:

1. Evaluation of possible use and suitability of use of newly developed glazes on simple samples before their application on products.

2. Determination of the residual stresses on real products provided with new glaze before starting the series production.

3. Quality inspection by random verification of residual stress level on selected products taken out of the series production.

During the systematic utilization of this experimental method the residual stresses can be optimized in products, which can significantly contribute to the solution of production problems and to lead to a significant economic benefit subject to a systematic approach.

REFERENCES

Dolhof, V. et al. 1992. Research into residual stresses in ceramic products. Research report. Pilsen: ŠKODA, Research.
Schwartz, I.F. 1990. Residual stress determination in hardmetal and polycrystalline diamond using the blind-hole drilling technique. Powder Metall. Int. 22(5):5-9.
Schajer, G.S. 1981. Application of finite element calculations to residual stress measurements. J. Eng. Technol. 103:157-161

Recent Advances in Experimental Mechanics, Silva Gomes et al. (eds) © 1994 Balkema, Rotterdam, ISBN 90 5410 395 7

Residual stresses by hole-drilling in curved components

Wu-Xue Zhu & D.J.Smith
University of Bristol, UK

ABSTRACT: A theoretical analysis is developed to apply the incremental hole-drilling method to components with curved surfaces. To interpret the relaxed strains, a procedure is presented to calculate stiffness matrixes using three-dimensional finite element analyses. The analysis developed is general and in this paper the analysis is used to obtain near surface residual stress distributions in 8mm diameter hot forged steel round bars. The results are shown to compare well with residual stress measurements obtained using X-ray and neutron diffraction techniques.

1 INTRODUCTION

The hole-drilling method was developed as a semi-destructive method to measure near surface residual stresses in components and consists of drilling a hole to a depth of a few millimetres, with the diameter about the same size as the depth. The residual stresses are calculated from the measured surface strain change caused by relaxation of the residual stresses. The conventional method of hole-drilling assumes that the stress field does not vary with depth. However non-uniform residual stresses can be determined by using an incremental hole-drilling method together with finite element (FE) analyses to interpret the incrementally relaxed strains. This technique includes the conventional one-step method that assumes that the residual stresses are uniform with depth, Mathar (1934) and incremental hole-drilling method to determine near surface distribution of residual stresses, Schajer (1988a). There are a number of methods for interpreting the incrementally relaxed strains (Nickola, 1986, Schajer, 1981, and Bijak-Zochowski, 1978), but these recent developments have been confined to flat specimen surfaces.

In this paper, an analysis is developed for interpreting relaxed strain data from the incremental hole drilling method for components with curved surfaces. Three-dimensional FE calculations are used for the interpretation of the strain data. The analysis is applied to 8mm diameter forged steel bars and for this geometry, extensive FE calculations were carried out. Experimental results from measurements on the forged steel bars are given and compared with residual stress results obtained from X-ray and neutron diffraction measurements.

2. ANALYSIS

In order to determine the stress distribution from the hole drilling method it is assumed (following Niku-Lari, 1987) that:
i) The material is elastic. This means the stress released during drilling is less than the elastic limit and the plastic deformation caused by drilling is negligible.
ii) The stress components normal to the surface are very small compared to the other stress components.
iii) The stress in each layer of material relaxed by drilling is uniform. A mean value is therefore used.
iv) The material is isotropic and only two elastic constants are required, Young's modulus, E, and Poisson's ratio, υ. If the material behaves anisotropically, more than two elastic constants are required.

In addition to these assumptions we assume

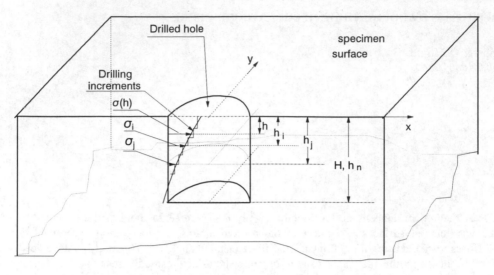

Fig. 1 Hole drilling and its parameters

that plane stress conditions exist in the area where a hole is drilled and that stresses only vary with depth (in the z-direction normal to the surface), so that the residual stresses can be expressed by using three independent parameters

$$\sigma_x = \sigma_x(h), \; \sigma_y = \sigma_y(h) \; and \; \tau_{xy} = \tau_{xy}(h) \quad (1)$$

where h is the normal distance from the surface. When a hole is being drilled to depth H, the strain changes on the surface at three locations, say a, b, c, are measured

$$\varepsilon_a = \varepsilon_a(H), \; \varepsilon_b = \varepsilon_b(H), \; \varepsilon_c = \varepsilon_c(H) \quad (2)$$

Using matrix notation, we obtain

$$\vec{\sigma} = \vec{\sigma}(h) = \begin{Bmatrix} \sigma_x(h) \\ \sigma_y(h) \\ \tau_{xy}(h) \end{Bmatrix} \quad (3)$$

and

$$\vec{\varepsilon} = \vec{\varepsilon}(H) = \begin{Bmatrix} \varepsilon_a(H) \\ \varepsilon_b(H) \\ \varepsilon_c(H) \end{Bmatrix}. \quad (4)$$

Assuming that only stresses on the hole surface in the range $(h, h + dh)$, where $h \le H$, are removed then the strain change at the surface should be

$$d\vec{\varepsilon} = \hat{A}(h, H)\vec{\sigma}(h)dh \quad (5)$$

where $\hat{A}(h, H)$ is a 3×3 matrix that is a function of material, hole diameter and geometry of the strain gauges relative to the hole. Integrating Eq (5), we obtain the integral equation

$$\vec{\varepsilon}(H) = \int_0^H \hat{A}(h, H)\vec{\sigma}(h)dh \quad (6)$$

which is similar to the integral method described by Bijak-Zochowski (1978). If $\hat{A}(h, H)$ is known and $\vec{\varepsilon}(H)$ is measured as a function of H, then the unknown distributions of residual stress $\vec{\sigma}(h)$ can be theoretically determined by solving the integral equation.

With the average strain method, the actual stress distribution is to be replaced by a step-wise-constant distribution as shown in Fig 1. For n drilling steps and h_n the corresponding depth of the hole, the distribution is assumed to be

$$\bar{\sigma}(h)\big|_{h_{i-1}\le h\le h_i} = \bar{\sigma}_i = \begin{Bmatrix} \sigma_{x,i} \\ \sigma_{y,i} \\ \tau_{xy,i} \end{Bmatrix} \qquad (7)$$

$$i = 1,\cdots n \text{ and } \begin{aligned} h_0 &= 0 \\ h_n &= H \end{aligned}$$

The integral equation (6) then becomes

$$\bar{\varepsilon}(h_j) = \sum_{i=1}^{j} \int_{h_{i-1}}^{h_i} \hat{A}(h,h_j)\bar{\sigma}(h)dh$$

$$= \sum_{i=1}^{j} \tilde{A}_{ji}\bar{\sigma}_i \qquad (8)$$

$$i \le j \le n, \quad j = 1,\cdots,n$$

where

$$\tilde{A}_{ji} = \int_{h_{i-1}}^{h_i} \hat{A}(h,h_j)dh. \qquad (9)$$

$\tilde{A}_{ji}$ is a 3×3 matrix and can be found by FE analysis. If $\tilde{A}_{ji}$ is known and $\bar{\varepsilon}(h_j)$ is measured, then the unknown residual stresses $\bar{\sigma}_i$ can be determined by solving the linear algebra equations (9). However $\tilde{A}_{ji}$ for all i and j is yet to be determined.

3 FE CALCULATIONS

Niku-Lari et al (1987) and Schajer (1988b) calculated coefficients only for flat plates using 2D FE analyses. Since a round bar has a cylindrical surface, the geometry is no longer axisymmetric after hole drilling and three-dimensional FE analyses are necessary.

From Eq (8), we obtain the strain change caused by only removing $\bar{\sigma}_i$ from ith layer

$$\bar{\varepsilon}_{ji} = \tilde{A}_{ji}\bar{\sigma}_i. \qquad (10)$$

Using a matrix format, Eq (10) becomes

$$\begin{Bmatrix} \varepsilon_a^{ji} \\ \varepsilon_b^{ji} \\ \varepsilon_c^{ji} \end{Bmatrix} = \begin{bmatrix} a_{11}^{ji} & a_{12}^{ji} & a_{13}^{ji} \\ a_{21}^{ji} & a_{22}^{ji} & a_{23}^{ji} \\ a_{31}^{ji} & a_{32}^{ji} & a_{33}^{ji} \end{bmatrix} \times \begin{Bmatrix} \sigma_x^i \\ \sigma_y^i \\ \tau_{xy}^i \end{Bmatrix}. \qquad (11)$$

The procedure to obtain the coefficient matrix $\tilde{A}_{ji}$ using FE analyses includes:
i) Generate FE mesh with a hole depth being h_j.
ii) Let $\sigma_x^i = 1$, $\sigma_y^i = 0$ and $\tau_{xy}^i = 0$, then calculate the surface traction of the hole in ith layer corresponding to this stress state.
iii) Convert the surface traction into nodal forces acting on the FE model.
iv) From FE calculations find the strain values in three locations at which the strain gauges are located.
v) Using Eq (12), we obtain the first column of the coefficient matrix $\tilde{A}_{ji}$

$$\begin{bmatrix} a_{11}^{ji} \\ a_{21}^{ji} \\ a_{31}^{ji} \end{bmatrix} = \begin{Bmatrix} \varepsilon_a^{ji} \\ \varepsilon_b^{ji} \\ \varepsilon_c^{ji} \end{Bmatrix}_{\sigma_x^i=1}. \qquad (12)$$

vi) Let $\sigma_y^i = 1$, $\sigma_x^i = 0$ and $\tau_{xy}^i = 0$, and using the same procedure as above obtain

$$\begin{bmatrix} a_{12}^{ji} \\ a_{22}^{ji} \\ a_{32}^{ji} \end{bmatrix} = \begin{Bmatrix} \varepsilon_a^{ji} \\ \varepsilon_b^{ji} \\ \varepsilon_c^{ji} \end{Bmatrix}_{\sigma_y^i=1}. \qquad (13)$$

vii) Let $\tau_{xy}^i = 1$, $\sigma_x^i = \sigma_y^i = 0$. In this case, the loading condition is asymmetric and the constraint boundary condition should be different from the above cases. This yields the third column of the matrix $\tilde{A}_{ji}$

$$\begin{bmatrix} a_{13}^{ji} \\ a_{23}^{ji} \\ a_{33}^{ji} \end{bmatrix} = \begin{Bmatrix} \varepsilon_a^{ji} \\ \varepsilon_b^{ji} \\ \varepsilon_c^{ji} \end{Bmatrix}_{\tau_{xy}^i=1}. \qquad (14)$$

viii) In turn, let $j = 1,\cdots,n$ and $i = 1,\cdots,j$, then all the coefficient matrixes $\tilde{A}_{ji}$ are determined.

The FE element type chosen was a twenty-node isoparametric element. A refined mesh surrounding the hole was used. Assuming the number of drilling steps are ten, then the number of loading cases is

$$N = 3 \times \sum_{j=1}^{10} j = 165. \qquad (15)$$

Extensive FE analyses were carried out and FORTRAN programs were written for procedures (iii) and (iv) to convert surface traction into nodal forces and nodal displacement results into strains. These results were then used to obtain the coefficient matrixes $\tilde{A}_{ji}$.

4 EXPERIMENTS AND RESULTS

The specimen was a hot forged shot blasted round En15 steel bar with a diameter of 8 mm. Young's modulus and Poisson's ratio was 208 GPa and 0.28 respectively. The model and type of strain gauge rosette used was the TEA-06-062RK-120 manufactured by the Measurements Group Inc. designed for the hole-drilling method. The strain gauge was carefully wrapped and bonded around the cylindrical part in the centre of the bar.

The hole was drilled incrementally using a high-speed air turbine. The total number of steps was ten and the increment in each step was about 0.08mm. The hole diameter D_0 was 1.926mm.

All results of strain measurements were smoothed using curve fitting. Results for the axial and tangential residual stresses are shown in Figs 4 and 5 respectively. Also shown in these figures are residual stresses obtained using the X-ray and neutron diffraction techniques (Zhu et al, 1994) in which the X-ray measurements were obtained on the surface and neutron diffraction measurements from the interior of six hot forged round bar specimens.

5 DISCUSSION

An analysis has been presented that interprets measured strains from hole drilling at surface into relaxed residual stresses. With the use of finite element analyses, this analysis has allowed the hole drilling method to measure residual stresses in components with curved surfaces for any elastic materials. The results are fairly good agreement with those obtained by X-ray and neutron diffraction techniques.

From the FE analysis, the determination of the matrix $\tilde{A}_{ji}$ decreases with hole depth and even drops to less than 1% of $\tilde{A}_{11}$ when the hole reaches a depth of about 1 mm. Therefore for

the same error in strain measurement for any depth, the error for the calculated stress could be about one hundred times greater than in the first layer. This implies that there is a limitation to the depth to which the hole can be drilled. This depth for the round bar specimen is roughly the same as described by Schajer (1988b) for plane surfaces This is about 0.25 times the mean radius of the strain rosettes D and practically this limit is reduced to $0.15D - 0.2D$. For $D = 5.13mm$, the limit is about $0.8 - 1.0mm$.

On the other hand, by interpreting Eqs (8), it may be concluded that the calculated stresses $\bar{\sigma}_i$ accumulate all the errors before this step. This can be simply shown by taking $n = 2$ in Eqs (8)

$$\left. \begin{array}{l} \bar{\varepsilon}(h_1) = \tilde{A}_{11}\bar{\sigma}_1 \\ \bar{\varepsilon}(h_2) = \tilde{A}_{21}\bar{\sigma}_1 + \tilde{A}_{22}\bar{\sigma}_2 \end{array} \right\} \tag{16}$$

The solution for the above equations is

$$\left. \begin{array}{l} \bar{\sigma}_1 = \tilde{A}_{11}^{-1}\bar{\varepsilon}(h_1) \\ \bar{\sigma}_2 = \tilde{A}_{22}^{-1}\bar{\varepsilon}(h_2) - \tilde{A}_{22}^{-1}\tilde{A}_{21}\tilde{A}_{11}^{-1}\bar{\varepsilon}(h_1) \end{array} \right\} \tag{17}$$

where $\tilde{A}_{11}^{-1}$ and $\tilde{A}_{22}^{-1}$ are the inverse matrixes of $\tilde{A}_{11}$ and $\tilde{A}_{22}$, respectively. Assuming each matrix has an error from the FE analysis and measurement, we obtain the error for the stresses

$$\left. \begin{array}{l} \delta\bar{\sigma}_1 = \delta\tilde{A}_{11}^{-1}\bar{\varepsilon}(h_1) + \tilde{A}_{11}^{-1}\delta\bar{\varepsilon}(h_1) \\ \delta\bar{\sigma}_2 = \delta\tilde{A}_{22}^{-1}\bar{\varepsilon}(h_2) + \tilde{A}_{22}^{-1}\delta\bar{\varepsilon}(h_2) \\ \quad - \delta\tilde{A}_{22}^{-1}\tilde{A}_{21}\tilde{A}_{11}^{-1}\bar{\varepsilon}(h_1) - \tilde{A}_{22}^{-1}\delta\tilde{A}_{21}\tilde{A}_{11}^{-1}\bar{\varepsilon}(h_1) \\ \quad - \tilde{A}_{22}^{-1}\tilde{A}_{21}\delta\tilde{A}_{11}^{-1}\bar{\varepsilon}(h_1) - \tilde{A}_{22}^{-1}\tilde{A}_{21}\tilde{A}_{11}^{-1}\delta\bar{\varepsilon}(h_1) \end{array} \right\} \tag{18}$$

where δ expresses the error of each matrix. Obviously, the error $\delta\bar{\sigma}_2$ is much bigger than $\delta\bar{\sigma}_1$. Since the error accumulates, more increments do not imply improved results for the interpreted residual stresses. Therefore, the total number of increments and the depth of the hole need to be optimized.

From the results, shown in Fig 4 and 5, we see that all residual stresses are compressive in the surface layer and become tensile in the interior of the bar. This general trend was also observed from the measurements using X-ray and neutron diffraction techniques.

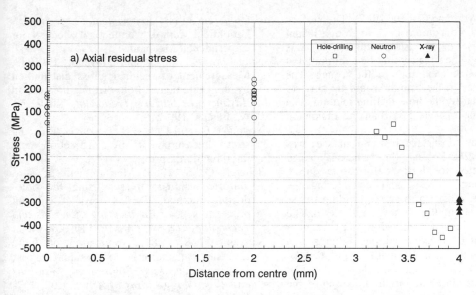

Fig. 2 Axial residual stresses in hot forged round bars measured
by X-ray, neutron diffraction [12] and hole-drilling methods

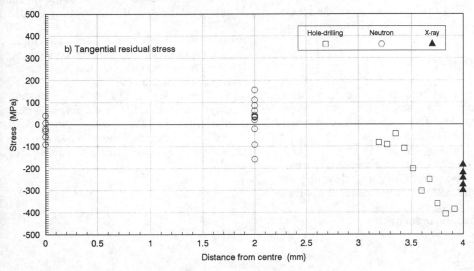

Fig. 3 Tangential residual stresses in hot forged round bars measured
by X-ray, neutron diffraction [12] and hole-drilling methods

Comparing results obtained using the X-ray diffraction technique with those from the hole-drilling technique, it is observed that the residual stresses on the surface measured by X-ray diffraction are higher in absolute value than those observed from the hole-drilling technique. This difference was also observed by Niku-Lari et al (1987). This may be due to two aspects:
i) The residual stress measured by X-ray diffraction is an average value over a certain

volume (Noyan and Cohen, 1987). The surface roughness R_t (maximum peak to trough height of the profile in the assessment length) was about 54 μm. The stress would be a minimum at the peak and a maximum in the trough. The residual stress in the trough would be close to that measured by the hole drilling method. The average value may be less than the maximum value.

ii) The value of surface residual stresses was about 300 MPa. Because the surface layer had been subjected to plastic deformation during manufacturing process, and considering the Bauschinger effect, a reverse yielding stress occurred at about 300 MPa (Shatil, 1990). The material in the surface region may be at a critical point for reverse yielding. When a hole is drilled with a small depth, some part of material will yield. Therefore the deformation including this plastic deformation caused by the hole will be greater than that for elastic deformation alone. Consequently, elastic FE analyses will obtain larger residual stresses than would be obtained using elastic-plastic analyses. Nickola (1986) has observed positive errors up to 30% (and greater) when the initial residual stress is equal to the yield strength.

6. CONCLUSIONS

The incremental hole drilling technique has been developed for components with curved surfaces, and the stiffness coefficients obtained using finite element analyses.

Measurements of near surface residual stresses have been made on a hot forged steel round bar. The results are found to be in general agreement with residual stress measurements obtained using the X-ray and neutron diffraction techniques.

REFERENCES

Bijak-Zochowski, M. 1978. A semi-destructive method of measuring residual stresses, *VDI-Berichte*, 313: pp. 469-476

Mathar, J. 1934. Determination of initial stresses by measuring deformation around drilled hole, Trans ASME, 56: 249-254

Nickola, W. E. 1986. Practical subsurface residual stress evaluation by the hole-drilling method., *Proceedings of 1986 SEM Spring Conference on Experimental Mechanics*, Brookfield Centre, Connecticut: Society for Experimental Mechanics.

Niku-Lari, A., Lu, J. and Flavenot, F. 1987. Measurement of residual stress distribution by the incremental hole-drilling method, in *Advances in Surface Treatment*, Niku-Lari, A., Eds., 4: 199-219.

Noyan, I. C. and Cohen, J. B. 1987. Residual stress measurement by diffraction and interpretation, Springer-Verlag.

Schajer, G. S. 1988. Measurement of non-uniform residual stresses using the hole-drilling method: Part I-stress calculation procedures, *J. of Engineering Materials and Technology, Trans of ASME*, 110: 338-343.

Schajer, G. S. 1981. Application of finite element calculations to residual stress measurements, *J. of Engineering Materials and Technology, Trans of ASME*, 103: 157-163.

Schajer, G. S. 1988. Measurement of non-uniform residual stresses using the hole-drilling method: Part II-Practical application of integral method, *J. of Engineering Materials and Technology, Trans of ASME*, 110: 344-349.

Shatil, G. 1990. High strain multiaxial fatigue and life prediction of service component, *Ph.D. Thesis*, Dept. of Mech. Engng., University of Bristol.

Zhu, W.X. and Smith, D.J. 1994. Estimating residual stresses in hot forged steel components, in *Proceedings of Second Biennial European Joint Conference on Engineering Systems Design & Analysis*, London.

Recent Advances in Experimental Mechanics, Silva Gomes et al. (eds) © 1994 Balkema, Rotterdam, ISBN 90 5410 395 7

Numerical and experimental determination of residual stress in CFRP laminates

Marcelo M. S. Moura & António Torres Marques
DEMEGI, Porto, Portugal

José J. L. Morais
UTAD, Quinta dos Prados, Vila Real, Portugal

ABSTRACT: An analytical procedure based upon the classical laminate theory (CLT) was used in order to work out the residual stresses for carbon fiber reinforced epoxy laminates. It was also used a tridimensional 20 nodes finite elements (FEM 3D) from a commercial code ABAQUS; an algorithm was employed which considers that all adjacent layers with the same orientation are modelized as a unique layer.

To compare these techniques an experimental work was performed: (1) use of non-symmetric laminates and measuring curvature after curing, (2) using symmetric laminates and by machining layer by layer which causes curvatures and (3) a first-ply failure technique was also applied.

A final comparison between numerical FEM3D, CLT and experimental results was performed.

1 INTRODUCTION

Carbon fiber reinforced epoxy laminates, after cooling from the maximum curing temperature, present residual stresses. This is due to the anisotropy of the coefficient of thermal expansion associated with the different orientations of each layer. The residual stresses may cause rupture (transverse matrix cracking and delamination) or reduce significantly the laminate strength.

Hence, prediction and measurement of residual stresses are important in relation to production, design and perfomance of composite components.

In this paper the residual stresses were predicted by two different ways: classical laminate theory (CLT) and numerically, using 3D finite elements with 20 nodes (FEM3D). To test the applicability of these methods, three types of tests [1, 2, 3] were performed: curvature measurement of non-symmetrical laminates, first-ply failure and layer removal method.

2 EXPERIMENTAL PROCEDURE

2.1 Material and specimens

The material system chosen was AMPREG 75 which is a prepreg supplied by SP Systems. The material was cured using the vaccum-bag moulding technique (maximum vaccum -0.5 bar), according the cure cycle recommended by the supplier: (i) heating at a rate of 3°C/min up to 125°C, (ii) 15 minutes at 125°C and (iii) cooling at the same rate until ambient temperature.

Two types of lay-up were made: $[O_4/9O_4]_t$ and $[9O_2/O_2]_s$. The nominal specimen dimensions were 200 x 20 x 1 mm. Table 1 lists the mechanical and thermal properties of the unidirectional layer.

Table 1. Mechanical and thermal properties

E_1 (GPa)	$E_2 = E_3$ (GPa)	$G_{12} = G_{13}$ (GPa)	G_{23} (GPa)
144.1	9.66	5.25	3.17

$v_{12} = v_{13}$	v_{23}	α_1 (K^{-1})	α_2 (K^{-1})
0.31	0.52	- 0.45x10^{-6}	27.5x10^{-6}

2.2 Experimental Tests

One of the experimental procedures referred to in litterature [1, 2] in order to evaluate residual stresses is the determination of the first-ply failure (FPF). For this purpose, tensile tests were carried out in an INSTRON 4208 machine, at a cross-head speed of 1mm/min. An INSTRON extensometer was also used, having a gauge length of 50mm. The specimens tested had end tabs made of polyester resin reinforced with chopped strand mat.

Another experimental technique used to assess the level of residual stresses was the measurement of the $[0_n/90_n]_t$ laminate curvature [2, 3]. Here, the curvature of two laminates was measured: $[0_4/90_4]_t$ and $[90_2/0_2]_t$. The latter was obtained from a laminate $[90_2/0_2]_s$, removing layers by using a plane abrading machine. The removing process was controlled by sucessive measurements made with a travelling microscope.

Using a laminate $[0_4/90_4]_t$ an estimation of the stress-free-temperature(SFT) was made, by gradual heating in an oven until the specimens appeared to be flat. From the observations made, it can be assumed that the SFT coincides with the maximum temperature for the curing cycle, i.e., 125°C.

3 ANALYTICAL AND NUMERICAL EVALUATION OF RESIDUAL STRESSES

In the following, the procedures used to determine the residual stresses (CLT and FEM3D methods) due to cooling down from SFT (125°C) to ambient temperature (20°C), are briefly described. It was assumed, for simplification, that thermoelastic properties are independent of temperature.

On figure 1, the global and local systems used are presented, having the following notations:

Fig. 1 - Global and local systems

x, y, z → global system

1, 2 → local system

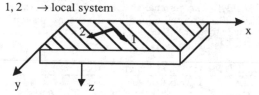

3.1 Classical laminate theory

The constitutive equation of a single layer under the assumptions of the classical laminate plate theory, including thermal effects is [4]:

$$\sigma_i = \overline{Q_{ij}}\left(\varepsilon_j^0 + z k_j - \varepsilon_j^T\right) \ , \ i, j = 1, 2, 6 \tag{1}$$

where

$\overline{Q}_{ij} = $ reduced stiffnesses at the final temperature of interest

$\varepsilon_j^0 = $ central plane strain measured from the stress-free state

$k_j = $ curvatures measured from the stress-free state

$\varepsilon_j^T = $ thermal strains

The laminate constitutive equation given by the CLT is, for the case of thermal loading only [4]:

$$\left(N_i^T, M_i^T\right) = (A_{ij}, B_{ij})\, \varepsilon_j^0 + (B_{ij}, D_{ij})\, k_j \tag{2}$$

In this equation we have:

$$\left(A_{ij}, B_{ij}, D_{ij}\right) = \int_{-h/2}^{h/2} \overline{Q_{ij}}\left(1, z, z^2\right) dz \tag{3}$$

and

$$\left(N_i^T, M_i^T\right) = \int_{-h/2}^{h/2} \overline{Q}_{ij}\, \varepsilon_j^T\, (1, z)\, dz \tag{4}$$

From the above equations, an expression for the residual stress σ_2^T in a 90-degree layer of a $[90_2/0_2]_s$ laminate can be obtained:

$$\sigma_2^T = \frac{\Delta T\,(\alpha_2 - \alpha_1)\, E_2\, E_1}{E_2 + E_1} \tag{5}$$

Using the maximum rupture stress criterion, the average stress applied to the laminate when FPF occurs is given by:

$$\sigma_X^F = \frac{1}{h}\, \frac{A_{xx}\, A_{yy} - A_{xy}^2}{Q_{22}\, A_{yy} - Q_{12}\, Q_{xy}}\left(X_T - \sigma_2^T\right) \tag{6}$$

being X_T the transverse strength of a lamina.

The equations 1 to 4 are used to determine the $[0_n/90_n]_t$ laminate curvatures:

$$\frac{1}{R} = \frac{24}{h}\, \Delta T\, K_a\, E\, (\alpha_1 - \alpha_2) \tag{7}$$

where

$$E = E_2 / E_1$$

and

$$K_a = \frac{1 - \upsilon_{12}^2\, E}{1 + 14\,E + \left(1 - 16\upsilon_{12}^2\right)E^2} \tag{8}$$

With these equations one can't get good results when large displacements relatively to thickness are involved, since they are obtained considering a linear relation between strains and displacements. However, the displacements observed were much bigger than the plate thickness ($12 \le w/h \le 46$), which implies a non-linear geometric analysis [5]. A correction to the CLT [6] that accounts for the non-linear geometric effects, is basically the replacement in equation (7) the factor K_a by:

$$K_c = \frac{1 + \left(1 - 2\upsilon_{12} - \upsilon_{12}^2\right)E + \left(1 - 2\upsilon_{12}\right)\upsilon_{12}^2 E^2}{1 + 15E\left(15 - 16\upsilon_{12}^2\right)E^2 + \left(1 - 16\upsilon_{12}^2\right)E^3} \quad (9)$$

As experimentally one measures the deflections and not curvature radius, it is necessary to relate the two quantities, using the following equation:

$$\frac{1}{R} = \frac{8 \ w}{a^2 + 4 \ w^2} \quad (10)$$

where w is the deflection at the central point and a is the length of the chord of the arc considered.

3.2 Numerical Analysis

In the numerical analysis a tridimensional 20 nodes finite element were used with the ABAQUS software. Each element "joins" adjacent layers equally oriented, which allows a reduction of the number of elements to be used. The determination of the residual stresses is made from the thermal strains and using the well known relation:

$$[\sigma] = [D] \ [\varepsilon] \quad (11)$$

where [D] is the tridimensional elasticity matrix. The thermal strains $[\varepsilon^T]$ are determined using the product of the coefficient of thermal expansion vector, $[\alpha]$, and the temperature difference (ΔT) between SFT and the ambient temperature:

$$[\varepsilon] = [\alpha] \ \Delta T \quad (12)$$

It was the used Newton's method in order to solve the problem when non-linear geometric effects is considered.

4 RESULTS AND DISCUSSION

The geometric non-linearity related to the problem of displacements in non-symmetric laminates is emphasized in figure 2, where a view of the deformed strip is presented.

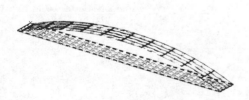

Fig. 2 - Deformed laminate, $[O_4/9O_4]_t$

In table 2 the deflections determined by the different methods are presented. The experimental result is the average of three measurements, being the amplitude of the results 0.5mm for the case of $[O_4/9O_4]_t$ laminate and 1.5mm for $[O_2/9O_2]_t$ laminate. This major amplitude is a consequence of the difficulty to remove the different layers.

Table 2 - Deflections of a non-symmetric laminates, w (mm)

Laminate	Experimental	Numerical	CLT
$[O_4/9O_4]_t$	12.9	12.8	12.4
$[O_2/9O_2]_t$	22.5	23.4	24.7

We take the FPF as the failure of the weakest ply, implying a multiple transverse cracking, detected by a change in the load-displacement curve-Figure 3. The strain jump in the load-displacement diagram is

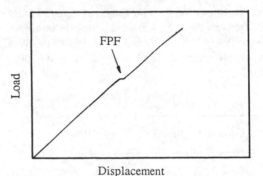

Fig. 3 - Typical load - displacement curve

probably atributed to that damage in the 90-degree plies [7,8]. A photomicrograph, showing a typical transverse crack is presented in figure 4.

Fig. 4 - Photomicrograph showing a transverse crack (X500)

In figure 5 it can be seen the σ_2^T stress distribution in the 90-degree layers, obtained by the CLT and by FEM3D, where the bord effects are clearly visible. It can also be seen that the two solutions agree very well in the central portion of the specimens (in 50% of its width). The bord effects explains a better

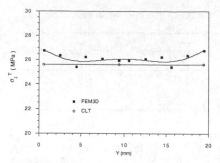

Fig. 5 - σ_2^T stress distribution in the 90 degree plies

agreement between the prediction of σ_X^F by FEM3D method and the experimental results, in relation with the prediction of CLT-Table 3. In the prediction of σ_X^F using equation 6, X_T was considered as 50.1 MPa.

Table 3 - Values of σ_X^F (MPa)

Experimental	Numerical	CLT
165.2 *	173	198

* Average for 4 values

The stress distribution obtained by FEM3D - Figure 5 - sugests that tranverse matrix cracking begins in the free bord, allowing the detection of FPF by using a microscope.

5 CONCLUSIONS

1. The layer removal method, in despite of its difficulty of execution, gave reliable results.

2. The FEM3D and CLT, corrected to account for non-linear geometric effects, gave results for the curvatures of $[O_2/9O_2]_t$ and $[O_4/9O_4]_t$ laminates which agree very well with the experiment. For its simplicity, the CLT is the more performant method.

3. The FEM3D predicted more accurately than CLT the rupture stress for the first layer, σ_X^F. The CLT

method underestimates in about 20% the value of σ_X^F, as it doesn't cater for bord effects. The FEM3D gave results differing 5% from the experimental ones.

REFERENCES

1. Kim, R. Y., Hahn, H. T., *Effects of Curing Stresses on the First-ply Failure in Composite Laminates*, J. Comp. Mat., Vol. 13, pp. 2-16, 1979.
2. Jeronimidis, G., Parkyn, A. T., *Residual Stresses in Carbon Fibre-Thermoplastic Matrix Laminates*, J. Comp. Mat., Vol 22, pp. 401-415, 1988.
3. Manson, J.A. E., Seferis, J. C., *Process Simulated Laminate (PSL): A Methodology to Internal Stress Characterization in Advanced Composite Materials*, J. Comp. Mat., Vol. 26, Nº. 3, pp. 405-431, 1992.
4. Tsai, S. W., Hahn, H. T., *Introduction to Composite Materials*, Technomic Publishing Company, Westport, Connecticut, 1980.
5. Hyer, M. W., *The Room Temperature Shapes of Four-Layer Unsymmetric Cross-Ply Laminates*, J. Comp. Mat., Vol. 16, pp. 318-340, 1982.
6. White, S. R., Hahn, H. T., *Process Modeling of Composites: Residual Stress Development. Part I. Model Formulation*, J. Comp. Mat.,Vol. 26, Nº 16, pp. 2402-2421, 1992.
7. Talreja, R., *Internal Variable Damage Mechanics of Composite Materials*, Yelding, Damage and Failure of Anisotropic Solids, EGF5 (Edited by J. P. Boehler), Mechanical Engineering Publications, London 1990, pp. 509-533.
8. Ohira, H., *Difference in the Failure Processes of CFRP and GFRP Cross-Ply Laminates*, Composites-Design, Manufacture and Application, ICCM VIII, Tsai and Springer eds., SAMPE, July 1991.

Recent Advances in Experimental Mechanics, Silva Gomes et al. (eds) © 1994 Balkema, Rotterdam, ISBN 90 5410 395 7

Residual stress and strain in a resin/metal laminated beam caused by cooling

Shigeo Matsumoto
Toyama Polytechnic College, Japan

Suguru Sugimori
Kanazawa Technical College, Japan

Yasushi Miyano & Takeshi Kunio
Kanazawa Institute of Technology, Ishikawa, Japan

ABSTRACT: The residual stress and strain in a resin/metal laminated beam caused by cooling was obtained by experimental and numerical hybrid analysis using photoviscoelastic birefringence. It was clear that the residual stress and strain distributions in this laminated beam indicate characteristic viscoelastic behaviors. That is, the residual shearing stress and strain distributions at the interface between resin and metal layers change remarkably with the cooling conditions, gradual or rapid cooling. Furthermore, in the case of rapid cooling, the shape of the residual shearing stress distribution is very different from that of the residual shearing strain distribution, and Saint-Venant's principle can not always be applied to the residual shearing strain distribution.

1 INTRODUCTION

Plastics bonded with other materials are being widely used for precision machine parts, electronic components and others. In the forming process, they are sometimes rapidly heated up and cooled down during the molding process. In this process, large amounts of residual stress and strain are generated at the interface between the plastics and the other materials as well as in these materials, because their mechanical properties show viscoelastic response. These residual stresses and strains result in degrading the strength of the plastics, especially the bonding strength of the interface between plastics and other materials. Thus, for the reliability of these polymer components, the development of an estimating method of the residual stress and strain in a viscoelastic body is strongly desired.

It is well known that the photoviscoelastic technique can be used as an experimental technique for viscoelastic stress and strain analysis. The practicality of this technique has been demonstrated by the authors. This photoviscoelastic technique can be also used for solving the thermoviscoelastic problem. However, this technique is very complicated, because the transient birefringence and temperature should be continuously measured over the period of the loading process in order to determine the residual stress and strain.

On the other hand, FEM and other methods for numerical viscoelastic analysis are ordinarily easier and simpler than the experimental techniques. However, the unreliability of the numerical solution still essentially exists, because many conditions such as material properties, configuration and boundary conditions and others should be numerically modeled as precisely as possible.

An approach to the straightforward and reliable experimental and numerical hybrid analysis of the stress and strain in a viscoelastic body was proposed by the authors. In this hybrid method, the experimental photoviscoelastic birefringence and the numerical birefringence are compared for the purpose of confirming the reliability of numerical results of stress and strain in a viscoelastic body.

In this paper, the residual stress and strain concentrations at the interface in the resin/metal laminated beam caused by gradual or rapid cooling are analyzed by this hybrid method, which are industrial interest.

2 PROCEDURE OF HYBRID ANALYSIS

The practical procedure of the experimental and numerical hybrid analysis of the stress and strain in a viscoelastic body is shown in Fig.1.

The right side of Fig.1 is the experimental procedure for measurement of birefringence. The loading and heating are applied to the surface and body of the specimen with a two-dimensional configuration made of photoviscoelastic material. The fringe order distribution $N_{exp}(x,y,t_a)$ in the specimen is measured at an arbitrary time t_a, where fringe order can be easily measured in the same manner as the ordinary two-dimensional photoelastic technique.

On the other hand, the left side of Fig.1 is the

numerical procedure for calculation of stress, strain and birefringence. The mechanical and optical properties, that is the relaxation modulus $E_r(t,T)$, the relaxation Poisson's ratio $\nu_r(t,T)$, the relaxation birefringence strain coefficient $C_r(t,T)$, the coefficient of thermal expansion $\alpha(T)$, and the thermal diffusivity $a(T)$, are characterized and expressed as the functions of t and/or T for the numerical analysis. The same configuration and the same boundary conditions as the experiment are given, and time and space are discretely divided for the numerical analysis. The numerical calculations by FEM and others are operated in the following procedure. First, transient temperature distribution $T(x,y,t)$ is calculated based on the law of thermal conduction. Second, transient stress distribution $\sigma_{ij}(x,y,t)$ and transient strain distribution $\varepsilon_{ij}(x,y,t)$ are calculated based on the linear-viscoelastic theory. Third, the fringe order distribution $N_{num}(x,y,t_a)$ at the time of t_a, where experimental birefringence N_{exp} has been measured, is calculated based on the linear-photoviscoelastic theory.

If the calculated $N_{num}(x,y,t_a)$ agrees well with the measured $N_{exp}(x,y,t_a)$, the validities of transient stress distribution $\sigma_{ij}(x,y,t)$ and strain distribution $\varepsilon_{ij}(x,y,t)$ obtained by the numerical analysis are confirmed for any time in this analysis. However, if N_{num} and N_{exp} do not coincide with each other, " Modelling" in Fig.1 should be modified. Thus, the numerical analysis is repeated until N_{num} coincides with N_{exp}.

3 EPOXY/STEEL LAMINATED BEAM BY COOLING

A resin/metal laminated beam shown in Fig.2, which is composed of an epoxy resin layer and a steel layer was used. Among these materials, the epoxy resin shows remarkable photoviscoelastic behavior, while the steel layer shows elastic behavior. The laminated beam, which is initially free from stress and strain, was held at high holding temperature T_h for a long time, then it was cooled down gradually or rapidly to the cooling temperature T_c. The residual stress and strain generated in the beam were analyzed by the proposed hybrid method.

3.1 *Experimental Analysis*

The laminated beam was held at $T_h(=180°C)$ for a long time in a heating oven. A higher temperature than the glass transition temperature T_g of the epoxy resin was selected as T_h. Two kinds of cooling conditions were employed; one was gradual cooling, in which the beam was cooled down gradually at the rate of 0.5°C/min to room temperature, the other was rapid cooling, in which the upper surface of the laminated beam was quickly immersed in running water having a cooling temperature $T_c(=10°C)$ while insulating the other surfaces and keeping therein until the temperature of the entire beam became T_c.

After gradual and rapid coolings, the residual fringe patterns were measured with an ordinary photoelastic apparatus. Figures 3 (a), (b) show photographs of the isochromatic fringe pattern in a dark field after gradual and rapid coolings.

3.2 *Numerical Analysis*

The mechanical and optical coefficients $E_r(t,T)$, $\nu_r(t,T)$, $C_r(t,T)$, $\alpha(T)$ and $a(T)$ of epoxy resin used have been characterized and given by functions of t and/or T in the previous paper.

The Cartesian coordinate axes x, y and z are taken in the direction of the length, height and thickness respectively as shown in Fig.2. It is assumed that the transient temperature distribution in the beam during rapid cooling is one dimensional in the y direction. It is also assumed that the beam is rapidly cooled on the top surface at the moment t=0 and two-dimensional plane stress state is caused in the beam. The method of space descritization is shown in Fig.4 as a schematic diagram. We divide the entire rapid cooling period 1200 sec into 130 sub-intervals and the entire gradual cooling period 20400 sec into 85 sub-intervals in limit such that the maximum temperature change in a time interval at any location was less than 5 °C.

The transient temperature distribution $T(y,t)$ was calculated by the equation of unsteady state heat conduction. The two-dimensional transient thermal stress and strain distributions are numerically calculated based on linear-viscoelastic theory by the FEM program. The fringe order distribution was calculated based on the linear-photoviscoelastic theory.

Figures 3 (c), (d) show the numerical results of the residual fringe order distributions obtained under the conditions mentioned above. The numerical distributions agree well with the experimental ones shown in Figs. 3 (a), (b).

3.3 *Residual shearing stress and strain distributions*

The transient shearing stress and strain distributions of resin layer at the interface indicated as hatching area in Fig.4 is focused on in this section. Figures 5 and 6 show these transient distributions calculated in the beam of l=40mm during gradual and rapid coolings. It is clear that transient shearing stress and strain concentrate at the end of the interface in both cooling conditions. The transient shearing strain generates at the early stage, however the transient shearing stress generates scarcely at the early stage and generates suddenly at the final stage in both cooling conditions.

Figures 7 and 8 show the residual shearing stress and strain distributions in the beams of l= 40, 80, 120 mm after gradual and rapid coolings. The residual shearing stress for gradual cooling is smaller than that for rapid cooling. The shape of residual shearing stress distribution does not change with length of beam and cooling condition.

From Fig.8, the residual shearing strain for gradual cooling is also smaller than that for rapid cooling. The shape of residual shearing strain distribution changes with length of beam and cooling condition. Especially, the residual shearing strain for rapid cooling does not decrease at a position far from the end of interface. Therefore, Saint-Venant's principle can not be applied to the residual shearing strain distribution for rapid cooling.

4 CONCLUSIONS

An approach to the straightforward and reliable experimental and numerical hybrid analysis of stress and strain in a viscoelastic body was proposed. The stress and strain in the laminated beams composed of epoxy resin and steel, which were cooled down gradually or rapidly, were analyzed by this hybrid analysis. As results, the residual shearing stress and strain at the interface of the beam are very characteristic of a thermoviscoelastic body.

REFERENCES

Sugimori, S., Miyano, Y. and Kunio, T. 1984. Photo-viscoelastic analysis of thermal stress in a quenched epoxy beam, Experimental mechanics, **24**: 150-156.
Shigeo, M., Sugimori, S., Miyano, Y. and Kunio, T. 1993. Hybrid stress analysis of viscoelastic body using photoviscoelastic birefringence, Proc. ATEM '93: 55-59.

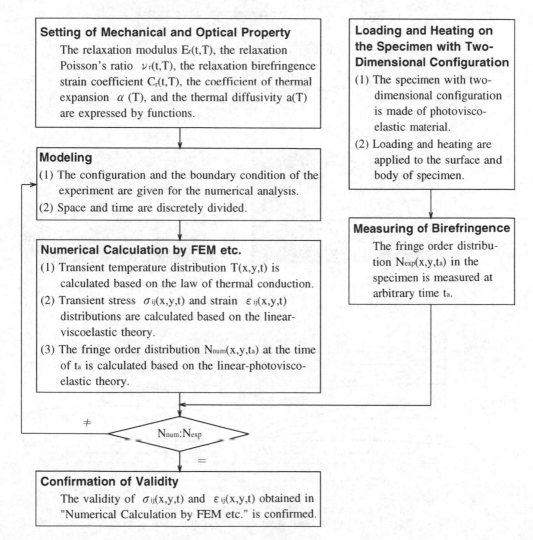

Fig.1 Procedure of experimental and numerical hybrid analysys.

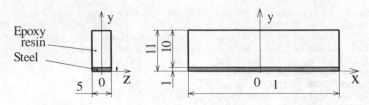

Epoxy resin

Steel

(a) Configuration of laminated beam

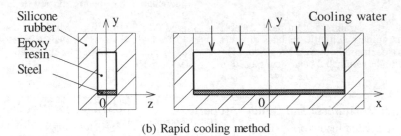

Silicone rubber

Epoxy resin

Steel

Cooling water

(b) Rapid cooling method

Fig.2 Configuration of laminated beam and rapid cooling method.

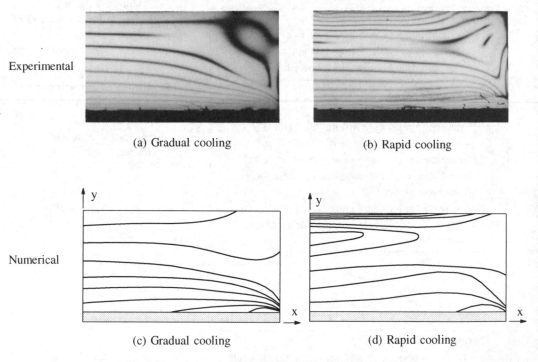

Experimental

(a) Gradual cooling

(b) Rapid cooling

Numerical

(c) Gradual cooling

(d) Rapid cooling

Fig.3 Experimental and numerical fringe order distributions after gradual and rapid coolings. (l=40 mm)

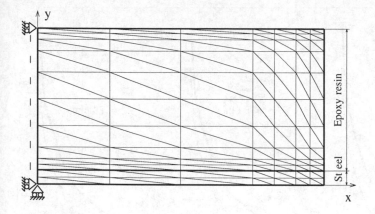

Fig.4 Schematic drawing of space discritization for FEM model.

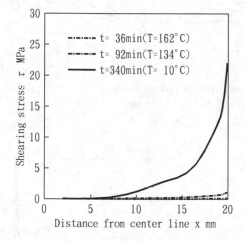

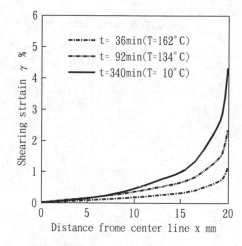

Fig.5 Transient shearing stress and strain distribution of resin layer at interface during gradual cooling. (l=40 mm)

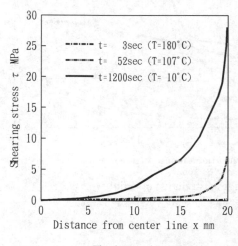

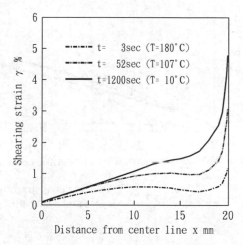

Fig.6 Transient shearing stress and strain distribution of resin layer at interface during rapid cooling. (l=40 mm)

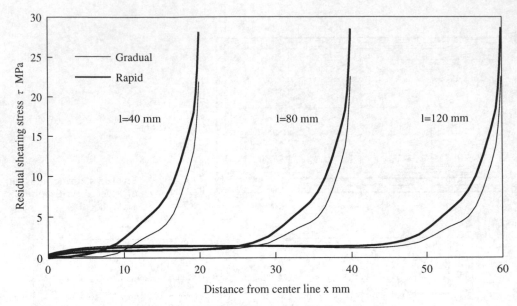

Fig.7 Residual shearing stress distribution of resin
layer at interface after gradual and rapid coolings.

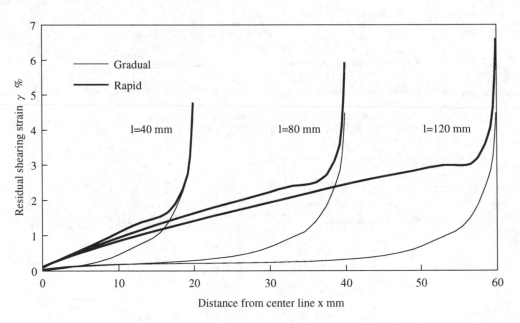

Fig.8 Residual shearing strain distribution of resin
layer at interface after gradual and rapid coolings.

Recent Advances in Experimental Mechanics, Silva Gomes et al. (eds) © 1994 Balkema, Rotterdam, ISBN 90 5410 395 7

Residual stress in thermoplastic resin plates by extrusion molding

Suguru Sugimori
Department of Mechanical Engineering, Kanazawa Technical College, Japan

Minoru Shimbo, Yasushi Miyano & Takeshi Kunio
Materials System Research Laboratory, Kanazawa Institute of Technology, Japan

ABSTRACT : The residual stress in thermoplastic resin plate under extrusion molding is studied in this paper. First, a high density polyethylene (HDPE) is selected as the material for specimen and the viscoelastic behavior and thermal expansion of resin which have a direct effect upon residual stress are investigated carefully. Second, the residual stresses in HDPE plates as molded, annealed plate and rapidly cooled plate are measured accurately by use of the layer-removal method. Finally, using measured thermo-viscoelastic datum, the residual stress distributions in the plates are numerically calculated based on the linear-viscoelastic theory. The growth mechanism of residual stress in thermoplastic resin by extrusion molding is clarified by the comparison of experimental results with theoretical ones.

1 INTRODUCTION

The residual stress in plastics generated during the molding process has direct effects on the strength and deformation of the molded parts. The study of the growth mechanism and a prevention of residual stresses has recently become a serious subject through the development of new materials and the advancement of processing techniques. It is very difficult to treat the residual stress of plastics because there are many factors that cause residual stresses. However, the factors which introduce residual stresses in thermoplastic resin can be classified widely into two groups; one is due to fluid of polymer during molding, the other is due to the non-uniformity of the temperature distribution during the cooling process. The former residual stresses are generated by the heterogeneous and anisotropy of the modulus and thermal expansion due to molecular orientation in the fluid. On the other hand, the latter residual stresses are generated by the crystallization and the thermo-viscoelastic behavior with non-uniformity temperature distribution during cooling.

Authors have already performed the theoretical and experimental analysis of the residual stress induced in thermosetting resin and amorphous thermoplastic resin caused by thermo-viscoelastic behavior. Heretofore, the theoretical work on thermoviscoelasticity has been developed in accordance with thermo-rheologically simple material behavior. The layer-removed method has been developed to obtained experimentally the residual stresses in both the thermosetting and the amorphous thermoplastic resins. The relation between the residual stress and thermo-viscoelastic behavior has been verified.

The purpose of this paper is to study the growth mechanism for residual stresses in thermoplastic resin generated under the process of extrusion molding by using above mentioned technique. First, a high density polyethylene (HDPE) are selected as the material for specimen and the viscoelastic behavior and thermal expansion of resins which have a direct effect upon residual stress are investigated carefully. Second, the residual stresses in HDPE plates as molded, annealed plate and rapidly cooled plate are measured accurately by use of the layer-removal method. Finally, using measured thermo-viscoelastic datum, the residual stress distributions in the plates are numerical calculated based on the linear-viscoelastic theory. The growth mechanism of residual stress in thermoplastic resin by extrusion molding is clarified by the comparison of experimental results with theoretical ones.

2 VISCOELASTIC BEHAVIOR

The HDPE plate used in this study has a thickness of 4 mm. It was molded by extrusion molding under the conditions of die temperature of 220°C and mean temperature of 90°C of cooling roll.

The storage modulus and thermal expansion of HDPE as thermo-viscoelastic behavior which have direct effects upon residual stress were investigated carefully. Preparation of the specimens for measuring the storage modulus and the thermal expansion is illustrated in Fig.1. The specimens for storage modulus had a thickness of 1.3 mm and were cut from the direction of extrusion (x direction) and its vertical direction(y direction), and divided into three parts in the direction of the thickness. The storage moduli were measured by using a viscoelastic analyzer. Likewise, specimens of thickness 0.8 mm for thermal expansion were prepared in the same way. The expansion and contraction of a gage length of specimen during heating and cooling was measured to obtain the thermal expansion.

Figure 2 shows the master curves of the storage modulus $E'(t',T_0)$ on x direction for the upper, middle and lower surfaces. $E'(t',T_0)$ was obtained by shifting the storage modulus $E'(t,T)$, measured at various constant temperatures T, parallel to the logarithmic scale of time. The time-temperature shift factors $a_{T0}(T)$ for HDPE are shown in Fig. 3. It is evident from these figures that the storage modulus for every specimen in each thickness direction agree fairy well with each other, and also that the same time-temperature equivalence holds. It was also confirmed that the storage modulus of specimens on y direction and any portion were almost equal to those shown in Figs. 2 and 3. This indicates that the modulus of the HDPE used here changes with time and temperature, but is homogeneous and isotropic over the whole of plate.

Figure 4 shows the thermal expansion ε_t with temperature for each layer in the x and y directions. These values were obtained after annealing specimens for sufficiently long time to release residual strain generated during the molding process. In the x direction, the thermal expansions ε_t of upper and lower layers agree well with those other, but ε_t of the inside layer differs apparently with those of outside layers. The ε_t's in the y direction show a similar trend to those in the x direction, but the magnitude of inside is larger than those of outside layers. By comparison of ε_t in x direction with that in y direction, ε_t's of inside layer in x and y both directions agree fairy well, but those of outside layers, in both directions, differ remarkably with each other. It is obvious from the above results that the thermal expansion changes considerably through the thickness of the plate by the effect upon molecular orientation during extrusion molding, and that thermal expansion shows anisotropy remarkably on the vicinity of the upper and lower surfaces.

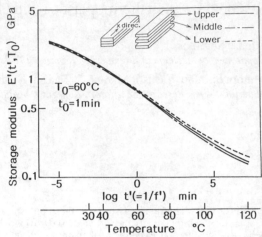

Fig. 2 Master curves of storage moduli of HDPE.

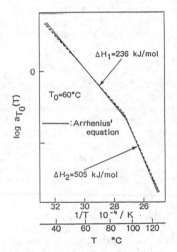

Fig. 3 Time-temperature shift factors of HDPE.

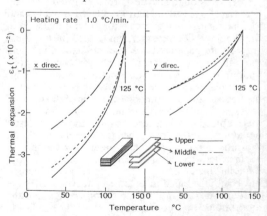

Fig. 4 Thermal expansion with temperature for each layer in the x and y directions.

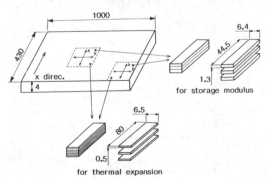

Fig. 1 Preperation of the specimens for measuring the storage modulus and the thermal expansion.

3 EXPERIMENTAL ANALYSIS

Three kinds of specimens, plate as molded, annealed plate and rapidly cooled plate, were prepared in order to identify the growth mechanism of residual stresses. The annealed and rapidly cooled specimen were made by the following method. The configuration of specimen, which was cut from the plate as molded, is shown in Fig. 5. The annealed specimen was annealed in order to release the residual stresses and strains introduced during the molding process. The rapidly cooled specimen was cooled by exposing both surfaces to the cooling water under the various temperature conditions, after keeping the specimen at the holding temperature T_h for a sufficiently long time. The water flow was in the direction of extrusion (x-direction) as shown in Fig. 5.

The residual stress distributions in the plate was measured by using the layer-removal method. Two strips of width 5 mm and length 100 mm as shown in Fig. 5 were cut in x and y directions from the three kinds of specimens described above. The lower surfaces of the strips were removed successively with emery paper at room temperature. The changes of curvature of the strips were measured using a microscope. The residual stresses of the plate were obtained by substituting measured curvatures into the equations of the layer-removal method.

4 THEORETICAL ANALYSIS

4.1 *Fundamental Equation*

Consider the viscoelastic plate shown in Fig. 5 which is initially at a uniform temperature T_h and free of surface tractions and body forces. Assume that Poisson's ratio ν of material is constant and the temperature T begins to change at the time t=0, the relationship between the stress component σ_{ij} is in the position x at the time t and the resultant strain component $\varepsilon_{\sigma ij}$ is expressed by the hereditary integration of the following equations :

$$\sigma_{ij}(x,t) = 2 \int_0^t G_r(t'-\tau',T_0) \frac{d[\varepsilon_{\sigma ij}(x,\tau)]}{d\tau} d\tau \quad (1)$$

$$\varepsilon_{\sigma ij} = \varepsilon_{ij} + \delta_{ij} \frac{\nu}{1-2\nu} e - \delta_{ij} \frac{1+\nu}{1-2\nu} \int_{T_h}^T \alpha(T) dT \quad (2)$$

$$e = \varepsilon_{xx} + \varepsilon_{yy} + \varepsilon_{zz} \quad (3)$$

$$t' = \int_0^T \frac{du}{a_{T0}[T(x,u)]} \quad (4)$$

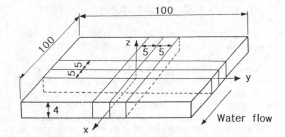

Fig. 5 Configuration and cooling method.

Table 1 Physical properties of HDPE.

Item			
Young' modulus	E	GPa	2.3
Specific heat	c	kJ/kg°C	2.30
Density	ρ	kg/m³	0.96×10^3
Thermal conductivity	λ	W/m°C	0.494
Thermal diffusivity	a	m²/s	2.19×10^{-7}

where, t' and τ' are the reduced time at the reference temperature T_0, which correspond to the physical time and $G_r(t',T_0)$ is the relaxation shear modulus at T_0. The relationship between t and t' is represented by eq.(4) according to the time-temperature conversion law. In eq.(4), $a_{T0}(T)$ is a time-temperature shift factor at T_0. In eq.(2), $\alpha(T)$ is the coefficient of linear thermal expansion, and δ_{ij} is the Kronecker's delta. The integro-differential equation (1) forms the system of equations to be solved.

4.2 *Numerical Calculation*

The numerical calculation of residual stress was carried out by using the method of successive substitution for eq. (1). The physical properties of the HDPE required for numerical calculations are tabulated in Table 1. The Poisson's ratio of HDPE measured at room temperature is $\nu=0.3$. The thickness of plate is d=4 mm as shown in Fig.5.

The temperature conditions for numerical analysis was selected as $T_h=120°C$ and T_c equals room temperature for the annealing, and $T_h=120°C$, 80°C and $T_c=10°C$ for the rapid cooling. The temperature distributions T(z,t) in the plate under various temperature conditions were calculated by the equation of unsteady-state heat conduction. The relaxation shear modulus $G_r(t',T_0)$ at a reference temperature T_0 was obtained by mathematical conversion of the storage modulus shown in Fig.2. The time-temperature shift factor $a_{T0}(T)$ in Fig.3 was represented by two Arrhenius' equations with different activation energies. The coefficient of linear thermal expansion $\alpha(T)$ was approximated by polynomial equation based on measured values in Fig. 4.

5 RESULTS AND DISCUSSION

Figure 6 shows the experimental and theoretical values of residual stress distributions in the plate as molded and annealed on the direction of extrusion (x direction) and its vertical direction (y direction). The experimental value is the average of a couple of measurements, and the repeatability of experiments is confirmed. The experimental values agree nearly well with the theoretical ones and the residual stress distribution in the x direction of both specimens is tensile in the vicinity of upper and lower surfaces, and is compressive in the inner part. On the other hand, in the y direction, compressive residual stress develops in the vicinity of upper and lower surfaces, while in the inner part tensile residual stress is produced. The difference in the residual stress distribution in the x and y directions corresponds directly to that of thermal expansion as shown in Fig. 4. Furthermore, since the residual stress distribution in the plate as molded is similar to that in the annealed plate in both the x and y directions, it is suggested that the residual stress induced during the molding process is due to the difference in thermal expansion in the direction of thickness.

Figure 7 shows the experimental values of residual stress distributions in the rapidly cooled plate for cooling conditions of holding temperature $T_h=80°C$ and $120°C$ and cooling water temperature $T_c = 10°C$. The residual stress in the annealed plate is also shown in this figure. The residual stress in the rapidly cooled plate under the condition of $T_h = °C$ C and $Tc = 10°C$ increases only slightly compared with that in annealed plate. While, the residual stress in the plate under $T_h=120°C$ and $Tc = 10°C$ where viscoelastic behavior is predominant increases considerably in both the x and y directions.

Figure 8 shows the experimental and theoretical values of the increments of residual stress due to rapid cooling. The experimental values agree nearly well with the theoretical ones and In the vicinity of upper and lower surfaces, compressive residual stress develops, while in the inner part tensile residual stress is produced in both directions. The residual stresses are very small when T_h is low, while they increase suddenly when T_h becomes high. It is evident that these residual stresses induced by rapidly cooling is due to the viscoelastic behavior of the material during the cooling process.

From the above results, it can be found that the residual stresses generated in thermoplastic resin during extrusion molding are mainly due to heterogeneous and anisotropy of thermal expansion by molecular orientation in molding process.

6 CONCLUSIONS

In the present paper, the growth mechanism of residual stresses generated in crystalline thermoplastic resins, in particular for HDPE, during

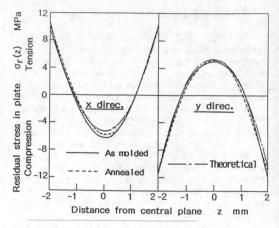

Fig. 6 Residual stress distributions in plate as molded and annealed plates.

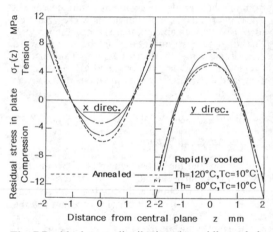

Fig. 7 Residual stress distributions in rapidly cooled plate and annealed plates.

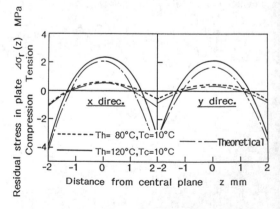

Fig. 8 Increments of the residual stresses in rapidly cooled plate due to thermoviscoelastic behavior.

extrusion molding was investigated experimentally and theoretically. As a result, it can be clarified that the residual stresses generated in HDPE during extrusion molding are mainly due to heterogeneous and anisotropy of thermal expansion by molecular orientation in molding process, and is not due to thermo-viscoelastic behavior with non-uniformity temperature distribution during cooling.

REFERENCES

Birger, L.A.1962, Zavodskya Laboratoriya, vol.28, 599.

Davidenkov, N.M., E.M.Shevandin.1939, Zhurnal Tekhnicheskoi Fizik, 21.

Maneschy, C.E., Y. Miyano, M. Shimbo, T.C. Woo,1986 j. Exp. Mech., vol.26 , 306.

Miyano, Y., M.Shimbo, T. Kunio, 1982, Exp. Mech., vol.22, 310.

Miyano, Y., M.Shimbo, T. Kunio, 1984, Exp. Mech., vol.24, 75.

Miyano, Y., M.Shimbo,1986, Ibid,195.

Miyano,Y.,M.Shimbo,S.Sugimori,T.Kunio,1988, Pro. SEM Conf., Exp. Mech., 619.

Morland, L.W., E.H. Lee, 1960, Tras. Soc. Rheol., vol.4, 233.

Miyano, Y., S.Sugimori, M. Shimbo,T. Kunio, 1990,Pro. 9th Int. Conf., vol.3, 277.

Shimbo, M., Y.Miyano, T. Kunio, 1985, Pro. SEM Spring Conf., Exp. Mech., 459.

Shimbo, M., Y.Miyano, C.E.Maneschy, T.C.Woo, T.Kunio,1986,Pro. APCS-86,131.

Shimbo, M., S.Sugimori, Y.Miyano, T.Kunio, 1990, Pro. KSME/JSME Conf., 37.

Schwarzl, F., A.J.Staverman,1952, J. Apl. Phys., vol.23, 838.

Recent Advances in Experimental Mechanics, Silva Gomes et al. (eds) © 1994 Balkema, Rotterdam, ISBN 90 5410 395 7

Effect of residual stresses upon the structural integrity of advanced airframe alloys

S.A. Meguid & M. Ferahi
Engineering Mechanics and Design Laboratory, Department of Mechanical Engineering, University of Toronto, Ont., Canada

ABSTRACT: The present study examines the beneficial effects of shot-peening residual stresses upon fatigue crack initiation and propagation behavior of recently developed Aluminum-Lithium (Al-Li) as well as conventional 7075 aluminum alloys. The results reveal that the total fatigue life corresponding to ultimate fracture of peened Al-Li 2090 and Al-Zn 7075 specimens is higher than those of unpeened ones. The fatigue life improvement is however more evident at low stress levels. The higher fatigue fracture performance of peened specimens is a direct result of slower crack propagation of small cracks caused by the reduction of the stress intensity factor. At applied stresses approaching the yield stress of the material, the crack growth rate in both peened and unpeened specimen are comparable. The results also reveal that the crack morphology is influenced by the presence of the residual stress field.

1. INTRODUCTION

Shot-peening is employed as one of the most effective surface treatments for improving the fatigue properties of engineering components. The method is used to enhance fatigue strength, reduce fatigue crack growth rate, and retard stress corrosion cracking. Earlier studies showed that the fatigue performance of high strength 7000 Al alloys can be significantly improved by shot-peening (Hammond and Meguid 1990, Oshida and Daly 1990, and Mutoh et al. 1987). This improvement in fatigue life is due mainly to the high surface compressive residual stresses introduced by the peening treatment.

It must be recognized, however, that surface compressive residual stresses are not permanent. They fade-out (relax) with cyclic loading (James 1982). The time dependent relaxation of these residual stresses strongly influence the fatigue fracture behavior of the treated components. The rate at which the residual stress field relaxes is governed by: (i) the applied load and its history, (ii) the mechanical properties and (iii) the microstructure of the treated components.

The effect of compressive residual stresses upon the fatigue fracture performance of high strength Al-Li 2090-T81 have not been investigated. In view of their significant reduction in density, increase in elas-

tic modulus, and compatibility with existing metal forming techniques, Al-Li alloys are currently being considered as replacement for conventional 7000 aluminum series in various applications in the aircraft industry. In this study, attention is devoted to the characterization of the induced shot-peening residual stresses, evaluate their effects on fatigue initiation and propagation of cracks emanating from high stress concentration features, and the interaction of these effects with the cyclic load level.

2. EXPERIMENTAL AND THEORETICAL INVESTIGATIONS

Rolled Al-Li 2090-T8 and Al-Zn 7075-T651 alloys, aged to their peak strength, were received in a 12.7 mm thick plate form. The chemical composition is given in Table 1. The Al-Li alloy exhibits a highly anisotropic microstructure, with an average grain size of $50\mu m$, $500\mu m$ and 3 mm in the short transverse (ST), the long transverse (LT) and the longitudinal (L) orientations, respectively. The pancake microstructure of Al-Zn 7075 had an average grain size of $20\mu m$, $80\mu m$, $150\mu m$ in the ST, LT and L directions, respectively. The L orientation being the rolling direction. The measured longitudinal mechanical properties for both materials are given in Table 2.

TABLE 1 - Chemical Composition of Al-Li 2090 and Al-Zn 7075 alloys (wt%)

Alloy	Cu	Li	Zr	Fe	Si	Mg	Mn	Ti	Zn	Cr
Al-Li	2.4-3.0	1.9-2.6	0.08-0.15	0.12	0.1	0.25	0.05	0.15	0.1	0.05
Al-Zn	1.2-2.0	-	-	0.5	0.4	2.1-2.9	0.3	-	5.1-6.1	0.18-0.28

TABLE 2 - Mechanical Properties of Al-Li 2090 and Al-Zn 7075 alloys

Alloy	Yield Stress (MPa)	Tensile Stress (MPa)	Elongation (%)
Al-Li	552	586	8.2
Al-Zn	530	598	13

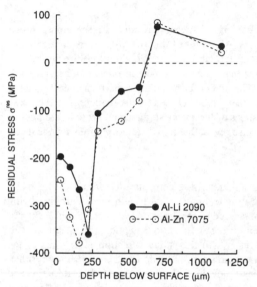

Figure 1. Shot-peening residual stress distributions in Al-Li and Al-Zn alloys

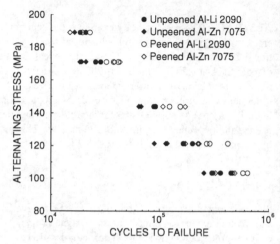

Figure 2. S-N curves for Peened and unpeened specimens in Al-Li and Al-Zn alloys: R = 0.1

A digitally controlled electro-hydraulic servo-controlled fatigue test equipment was used to conduct uniaxial fatigue tests of specimens containing a semi-circular notch through the thickness with a stress concentration factor of 3.26. The uniaxial loading was applied parallel to the rolling orientation. The tests were carried out under load control with a sinusoidal waveform at a frequency of 20 Hz. All tests were conducted at room temperature at a stress ratio R = 0.1.

The notch surface in both alloys was shot-peened to an Almen intensity of 12-14 A, with S330 shots and a coverage of 100%. The peening parameters were selected in compliance with the Military Specifications 13165C for Al-Zn 7075 alloy.

The resulting residual stress field was measured using the incremental hole-drilling technique (ASTM 1985). The method was modified to accommodate the non-uniformity of shot-peening residual stresses. The modified approach, which is based on the integral method of (Schajer 1988), adopts the finite element to evaluate the calibration constants associated with a non-uniform stress distribution, and takes into account the unequal contributions to the measured surface strain from stresses at different depths. The calibration constants were evaluated using the Ansys finite element package for a mesh geometry that complies with two commercial strain gage rosettes. The coefficients obtained for increments of 0.005 in. were found to be comparable to the results obtained in (Schajer 1988) with an error no greater than 2%. To account for the high gradient in the shot-peening residual stress profile close to the surface, calibration coefficients for increments of 0.0025 in. were also evaluated. These coefficients can be found in (Lindenschmidt et al. 1993).

Crack initiation and growth was monitored by examining cellulose acetate replicas of lightly polished

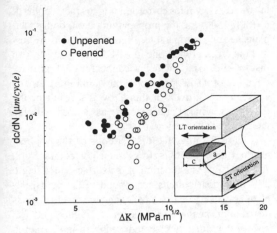

Figure 3. Comparison of crack propagation rates in peened and unpeened Al-Li alloy

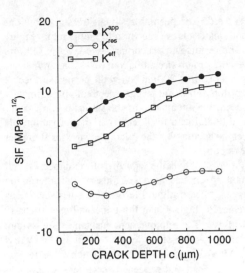

Figure 5. Applied, residual and effective stress intensity factor distribution

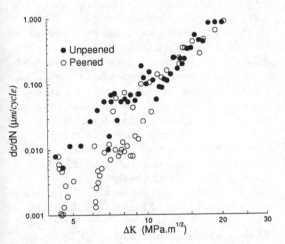

Figure 4. Comparison of crack propagation rates in peened and unpeened Al-Zn alloy

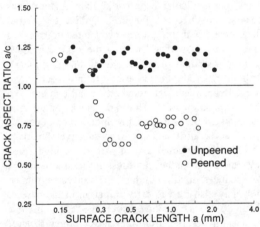

Figure 6. Crack aspect ratio variation in peened and unpeened Al-Li alloy

surfaces in a light optical microscope. Both the surface of the notch as well as the side surfaces ahead of the notch root were replicated. This allowed the monitoring of the length of the corner crack (a) in the direction of the thickness (ST orientation), and the crack depth (c) in the long transverse orientation (LT). The replicas were taken at a load equal to 80% of the maximum load. A number of tests were interrupted and the specimens were broken and observed in a scanning electron microscope (SEM) to study the crack morphology.

In view of the three-dimensional nature of the encountered corner cracks, three-dimensional stress in-

tensity factor solutions under general loading were developed. A new method for obtaining discretized weight functions of cracked geometries (Ferahi and Meguid 1993), along with the slice synthesis technique were used to obtain the applied and residual stress intensity factors.

3. ANALYSIS OF RESULTS

Despite the anisotropic nature of the Al-Li 2090 alloy, the measured strains revealed an equal biaxial residual stress field. The peening residual stress distributions in both materials are shown to be comparable (Figure 1). The results show a compressive layer of 600μm, with a maximum stress of about 370 MPa. and occurring at a depth of 200μm below the peened surface. It is worth pointing out that at a depth of approximately 1000μm, the attenuation of the strains becomes considerable and as a result errors in the measurements affect the accuracy of the calculated stresses at greater hole depths.

Figure 2 shows the total life of the unpeened and peened specimens as a function of the alternating stress σ_a. The fatigue life of the treated specimens is generally higher than the unpeened specimens in both alloys. The improvement due to peening is more evident for lower stress levels. At an applied stress (189 MPa) inducing localized plasticity, the fatigue lives of peened and unpeened specimens are comparable.

In both alloys, and for most load levels, the fatigue process began with a corner crack at the root of the notch. These small cracks were initiated and grown perpendicular to the loading axis. As the cracks became longer (c $>$ 1000 μm), significant crack deflection and branching were encountered in the Al-Li alloy. In the case of Al-Zn 7075, the cracks remained relatively straight during the entire fatigue process. In both materials, the initiation life in unpeened and peened specimens was relatively small compared to the propagation life. The early crack initiation in these materials, which are precipitation hardened alloys, can be attributed in part to the development of precipitate-free zones and in part to the high stress level arising from the high stress concentration factor. The reduced ductility in the short transverse orientation may also lead to early initiation in the case of Al-Li 2090. The improvement in the fatigue life due to peening resulted from the slower propagation of short and relatively long cracks (150-2000 μm) which grew through the compressive layer of the residual stress field.

The crack propagation rates dc/dN of the surface crack depth in the peened and unpeened specimens are compared as a function of the applied stress intensity factor range ΔK^{app} in Figures 3 and 4. These results were obtained under a constant stress amplitude of 121.5 MPa. The Figures show that for the same ΔK, the short crack growth rates in the peened specimen are reduced by one to three orders of magnitude in the Al-Li alloy (Figure 3), and up to six orders of magnitude in the Al-Zn alloy (Figure 4). In both alloys, the crack growth rates in the peened specimen are reduced for crack depths beyond the compressive

layer. At ΔK values corresponding to crack depths of c=1300-1500 μm, the propagation rate in both specimens became comparable. Figures 3 and 4 also show that the crack growth rate in Al-Li 2090 is lower than that in Al-Zn 7075, which accounts for the higher fatigue life of Al-Li specimens shown in Figure 2. The slower crack growth in this alloy is a result of the roughness induced crack closure associated with intense crack deflection.

Figure 5 shows the distributions of the applied, residual and effective stress intensity factors K^{app}, K^{res} and K^{eff} as function of the surface crack depth c. These distributions correspond to the actual applied stress ($\sigma_a = 121.5$ MPa) and the residual stress distribution in the Al-Li specimen (Figure 1). The results show that K^{app} is reduced even in the region of tensile residual stress (beyond $c = 600\mu m$), which accounts for the crack growth retardation in that region. It is also found that the maximum retardation in the peened specimens (Figure 3) corresponds roughly to the maximum absolute value of K^{res}.

Another reason for the slower crack advance in the region beyond the compressive layer is the crack shape difference in the peened and unpeened specimens. Figure 6 shows the crack aspect ratios (a/c) in the Al-Li specimens as a function of the crack length a. Cracks in the unpeened specimen initiated with an aspect ratio $a/c > 1$ and increased with increasing crack length a. This is due to the fact that as the crack size increases, it becomes more sensitive to microstructure effects (Venkateswara et al. 1988), and since the short transverse (ST) orientation had lower crack growth resistance, the crack length a grows faster than the crack depth c advancing in the LT orientation which exhibits higher resistance to crack propagation. In the case of the shot-peened specimens, the aspect ratio remained relatively smaller resulting in a lower crack driving force for the same value of crack depth c. The lower crack aspect ratio in the peened specimen is a consequence of the high compressive residual stress at the notch base where the crack length a is growing. Scanning electron micrographs of the fractured surface of partially fatigued specimens revealed a quarter elliptical morphology of the crack, with a smooth crack front in the unpeened specimens and a relatively tortuous front in the peened ones. The fracture surface in Al-Zn 7075 was relatively smooth and featureless compared to that of Al-Li 2090 where delamination in the rolling direction and fretting debris were clearly apparent.

Crack propagation rates in peened and unpeened specimens were investigated for three different load levels. The results presented in Figures 3,4 and 5 were obtained at an alternating stress of σ_a=121.5 MPa with a stress ratio of R=0.1. Similar results were observed for σ_a= 144 MPa (R=0.1). However, the crack depth at which the rate of growth in peened specimens approached that of the unpeened samples was reduced to c=600-700 μm compared to $c = 1300 - 1500\mu$m for σ_a=121.5 MPa. This may be due to the residual stress relaxation resulting from the increase in the alternating load. The fatigue data of Figure 2 showed that the total fatigue life for these stresses is of the order of 10^5 cycles. This fatigue regime is between the extremes of low cycle fatigue (LCF) where residual stresses may relax rapidly, and high cycle fatigue (HCF) where the relaxation may not be significant. At a test stress of σ_a=189 MPa (R=0.1), for which the total life corresponds to LCF regime, the propagation rates in both specimens were comparable. Finite element calculation of the stress distribution for this load level resulted in a localized plastic zone ahead of the notch root, which caused the residual stress field to redistribute in the early stage of the fatigue process, resulting in little or no account of peening.

4. CONCLUSIONS

Fatigue life of notched specimens made of Al-Li 2090-T81 and Al-Zn 7075 can be improved by shot-peening. The improvement in life is due to the slower short crack propagation which is reduced within and beyond the compressive layer of the residual stress field. This behavior is well predicted using the superposition of the applied and residual stress intensity factors. The presence of the peening compressive residual stresses in the vicinity of the exposed surface layers reduces the crack aspect ratio a/c, resulting in a smaller crack surface for the same crack depth c. The relaxation of residual stresses at load levels high enough to introduce local plasticity resulted in comparable fatigue crack propagation rates in unpeened and peened specimens. Further studies relating to the optimization of the peening parameters for Al-Li alloys, as well as the relaxation of residual stresses are currently being conducted.

REFERENCES

ASTM Standards 1985. Standard test method for determining residual stresses by the hole drilling strain gage method. ASTM E 837.

Ferahi, M. & S.A. Meguid 1993. On the effect of residual stresses upon the fatigue fracture behavior of advanced alloys. Internal Report: utme 187-93.

Hammond, D.W. & S.A. Meguid 1990. Crack Propagation in the presence of shot-peening residual stresses. Engng Fract. Mech. 37: 373-387.

James, M.R. 1982. The relaxation of residual stresses during fatigue. In Residual Stress and Stress Relaxation: 297-314.

Lindenschmidt, K.E., M. Ferahi, S.A. Meguid & P.C. Xu 1993. On the Effect of Plasma Spray Coatings on the Residual Stress State of Peened Compressor Blades. Proc. 2nd Int. Conf. Surface Engineering.

Luo, W., G.H. Fair, B. Noble & R.B. Waterhouse 1987. The effect of residual stresses induced by shot-peening on fatigue crack propagation in two high strength Aluminum alloys. Fatigue Fract. Engng Mater. Structs. 10: 261-272.

Oshida, Y. & J. Daly 1990. Fatigue damage evaluation of shot- peened high strength Aluminum alloy. Proc. 1st Int. Conf. Surface Engineering, 404-416.

Schajer, G.S. 1988. Measurement of non-uniform residual stresses using the hole drilling method. J. Engng Mater. Tech. 1: 338-343.

Venkateswara Rao, K.T., W. Yu, & R.O. Ritchie 1988. Fatigue Crack Propagation in Aluminum-Lithium Alloy 2090. Metallurgical Transactions 19: 549-569.

ACKNOWLEDGMENTS

The financial support of the Manufacturing Research Corporation of Ontario and the Natural Sciences and Engineering Research Council of Canada is gratefully acknowledged. Thanks are also due to Ms. S. Verzasconi of ALCOA (Pitsburgh) for provision of the material, and to Mr. B. Billings and the staff of the Metal Improvement Company of Canada for performing the peening treatment.

Recent Advances in Experimental Mechanics, Silva Gomes et al. (eds) © 1994 Balkema, Rotterdam, ISBN 90 5410 395 7

Stress corrosion cracking in a free residual stress field

A. Amirat
Mechanical Engineering Institute, University of Annaba, Algeria

W. J. Plumbridge
Mechanical Department, Faculty of Engineering, Queen's Building, University Walk, University of Bristol, UK

ABSTRACT : Stress corrosion crack growth rates have been investigated in single edge notched specimens prepared from a heat treated bar of a high strength low alloy steel. Residual stresses in the parent bar after quenching and tempering have been found to be compressive at the surface and tensile within the core. Manufacture of thin test pieces relieved most of these residual stresses. Apart from the general observations that K_{ISCC}, decreases as the yield strength of the material increases, a significant difference in K_{ISCC} is observed for the 500°C temper condition according to the specimen location in the heat treated bar. This effect is associated with the hardness difference existing between the inner and outer specimens where the hardness values in the inner ones were slightly higher.

1 INTRODUCTION

Failure of engineering components operating in corrosive environments due to cracking remains a safety and economic problem, despite the effort which has been devoted in recent times to understanding the phenomena involved. The need for a means of estimating service life of many structural steels so that failure is avoided during operation exists for many industries (Parkins 1976).

Stress corrosion cracking is a failure process which attracts much attention. It requires an acting static tensile stress and the presence of a corroding environment. Removal of either will prevent the initiation of cracking or stops cracks that are already propagating. The acting stress can be a residual stress which is just as effective as an externally applied load (Nelson 1982). The corroding media are highly specifique to each alloy (Scully 1966). The threshold for stress corrosion cracking K_{ISCC} (stress intensity factor below which stress corrosion cracking does not occur) varies widely in structural steels with respect to strength levels or microstructure (McIntyre and Priest 1972). The general observation for steels shows that the value of K_{ISCC} diminishes as the yield strength of the material is increased but little has been demonstrated about the factors which control this threshold.

In the present work, stress corrosion cracking is investigated on a mining chainsteel used for haulage chains within the mining industry. Residual stresses and hardness are measured after quenching and tempering. The values of K_{ISCC} are determined from the crack growth rate da/dt, curves.

2 EXPERIMENTAL DETAILS

A high strength low alloy steel (DIN 17115) Werkstoff 1.6753 or W 1.6753 of the specification given in Table 1 was used.

Table 1. Specification for DIN 17115

Element	Specification	
Carbon	0,20 to	0,26
Silicon	0.15 to	0.35
Sulphur	0.020 max	
Phosphorus	0.020 max	
Manganese	1.40 to	1.70
Nickel	0.90 to	1.10
Chromium	0.20 to	1.40
Molybdenum	0.40 to	0.55
Aluminium	0.02 to	0.05

Single edge notched uniaxal specimens prepared from a 31 mm diametre heat treated bar were used, Figure 1.

The heat treatment similated that used in the manufacturing process; after one hour quench from 890°C in water the bar was tempered for three hours at temperatures of 200°C and 500°C which were regarded as lower and upper bounds of the

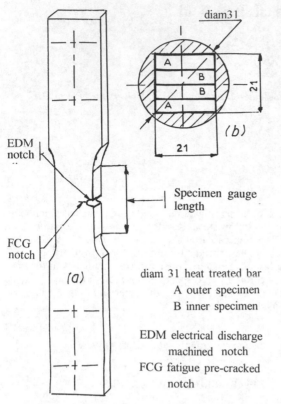

EDM notch

FCG notch

(a)

Specimen gauge length

diam 31 heat treated bar
A outer specimen
B inner specimen

EDM electrical discharge machined notch
FCG fatigue pre-cracked notch

(b)

Figure 1. Specimen geometry (a) and specimen location in the parent bar (b).

3 RESULTS

3.1 Residual stresses

Residual stresses measured in the parent bar after quenching and tempering have been found to be compressive at the surface and tensile within the core. However manufacture of thin test pieces relaxed most of these residual stresses, Figure 2.

3.2 Hardness values

Hardness values of three batches of the material

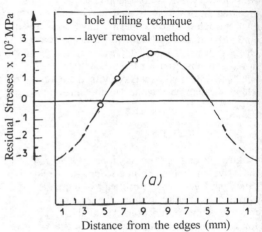

(a)

Distance from the edges (mm)

tempering temperatures. Hardness tests were carried out after the machining process on each specimen.

Measurements of the residual stresses were mainly made through the different geometry stages that a specimen undergoes during its manufacturing process by using the air/abrasif hole drilling technique (Beany & Procter 1974).

Basically two specimens were simultaneously tested on double lever uniaxial load machines. Open perspex boxes were constructed with wall height sufficient to allow the wetting of the specimen gauge length by a solution of 3.5% NaCl in distilled water. The solution was supplied from a 15 liter capacity circulating loop at a rate of 1.5 to 2 liters per minute. All tests were conducted at ambiant temperature, and at a pH of 6.8 to 7.2. Stress corrosion crack velocities were measured using direct current potential drop method and crack opening displacement clip gauge technique enhanced by optical observations. With both methods, crack growth rates da/dt, were obtained by the slope of a straight line connecting adjacent pair of data points on the crack length versus time to failure curves.

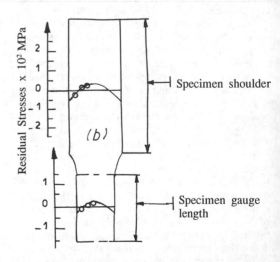

Specimen shoulder

(b)

Specimen gauge length

Figure 2. Residual stress distribution after quenching and tempering at 200°C (Amirat 1987). (a) in the parent bar, (b) through specimen shoulder and gauge length.

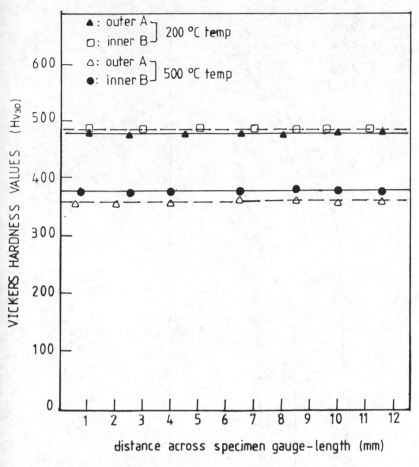

Figure 3. Hardness values through the specimen gauge length for the 200°C and 500°C temper conditions

are presented in Figure 3. Basically, when increasing the tempering temperature from 200°C to 500°C the steel softens of 120 Hv_{30}. Consistent values appear across the specimen gauge length, which indicate a constant material. The hardness values in the inner specimens are slightly higher than those in the outer specimens. The highest difference is observed in the 500°C temper condition, about 8 to 15 Hv_{30}.

3.3 Stress corrosion cracking tests

Figure 3 displays plots of stress corrosion crack growth rate versus stress intensity factor K, for the outer and inner specimens in the two heat treatment conditions. In the 200°C temper condition, the stress corrosion crack growth rate increased continuously over the entire range of the applied stress intensity factor K. The values of k_{ISCC}, calculated from the

applied load, initial crack length and specimen geometry were found to be 38 to 39 $MNm^{-3/2}$ in both specimens (outer and inner). Increases in tempering temperature up to 500°C raised K_{ISCC} to either 63 $MNm^{-3/2}$ or 88 $MNm^{-3/2}$ according to the specimen location in the parent bar. For instance in the outer specimens, K_{ISCC} was very high and little subcritical crack growth occured. In the inner specimens K_{ISCC} was 28% lower. Above 90 $MNm^{-3/2}$ the growth curves joined together and grew undistinguishably until failure occured.

4 DISCUSSION

4.1 Residual stress analysis

Residual stresses in crack growth specimens are almost zero through the gauge length. Figure 2

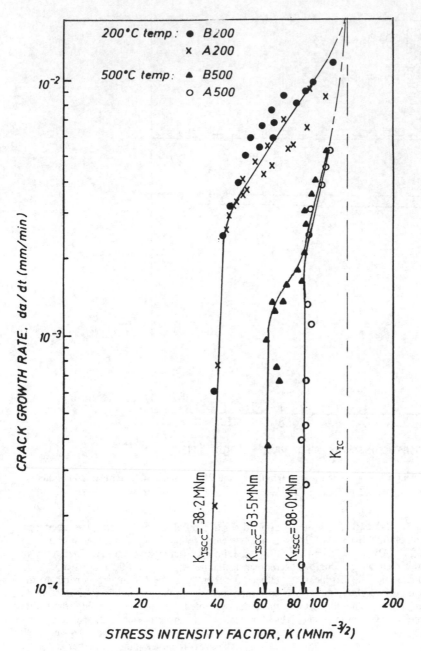

Figure 4. Effect of tempering temperature on K_{Iscc}

shows that residual stresses produced by quenching and tempering in the square bar are lost as the specimen blank is manufactured. However, small compressive stresses continued to act at the surface and tensile within the core. Therefore, the investigation of stress corrosion cracking is conducted in a free-residual stress specimen.

4.2 Effect of mechanical properties on K_{ISCC}

The observation of the effect of yield strength on K_{ISCC} is again demonstrated by the present results. The higher the yield strength the lower K_{ISCC} and vice-versa. In the 200°C temper condition, K_{ISCC} is about 38.2 MNm$^{-3/2}$ in both specimens (outer and

Table 2. Effect of side grinding on hardness

TT °C	specimen	original hardness Hv_{30}	one side Hv_{30}	two sides Hv_{30}
500	A500	363	363	369
500	B500	368	369	375
200	A200	482	485	492
200	B200	484	488	496

inner). In the 500°C, K_{ISCC} is 88 MNm$^{-3/2}$ for the outer and 63 MNm$^{-3/2}$ for the inner. The K_{ISCC} drop, in the 500°C temper condition, from 88 MNm$^{-3/2}$ to 63 MNm$^{-3/2}$ might be due to the hardness differences of 5 to 15 Hv_{30} existing between the outer and inner specimens; Figure 3. This has apparently been due to the machining process since in the outer specimens there had been one saw cut and on the inner two. This has been demonstrated on four selected specimens, Table 2. On each specimen, the hardness was first measured on its shoulders. Then on one end a layer of 1 mm was ground down from one side, and on the other end of the specimens, a layer of 0.5 mm was ground down from each side.

The hardness measurements show, effectively that on the two side ground shoulder, the hardness is about 5 to 7 Hv_{30} higher than that on the one side.

This approach of associating small variations in hardness with large differences in K_{ISCC} can be supported by previous data. Table 3 gives K_{ISCC} data for several high strength steels in sodium chloride solution. Shahinian and Judy (1976) reported a large discrepancy between K_{ISCC} values (a drop of K_{ISCC} from 93 to 75 MNm$^{-3/2}$) for two 4340 steels samples, heat treated in the same conditions, with a hardness difference of only 2 HRC (5 to 15 Hv_{30}). It may be deduced from the data reported by Clark and Landes (1975) that the slight difference of tempering temperature conditions (560°C and 580°C) should have induced a small difference in the hardness of the two tested samples and which yielded a significant drop of K_{ISCC} from 95 MNm$^{-3/2}$ to 40 MNm$^{-3/2}$. A similar phenomenon was observed by Gilleland (1986), in En26 steel; the 14 Hv_{30} hardness difference may also have caused the K_{ISCC} drop from 90 to 72 MNm$^{-3/2}$.

5 CONCLUSIONS

The residual stress pattern in W1.6753 chainsteel was found to be compressive at the surface balanced by tensile stresses in the core. Residual stresses in

Table 3. K_{ISCC} in high strength steel in relation with the hardness.

Authors	Steel	σ_y MPa	TT °C	Hv_{30}	K_{ISCC}
McIntyre & Priest 1972	835M30	1600	200	--	13.2
	817M30	1300	600	--	52.8
S.Carter 1971	4340	1650	260	--	17.6
		1490	496	--	20
Clark & Landes 1975	4340	1240	560	--	40
		1100	580	--	95
Shahinian 1976	4340	890	--	--	75
		890	--	--	93
Gilleland 1986	En 26	--	530	396	72
		--	550	382	90
		--	591	351	142
Present work	W1.6753	1240	A200	480	38
			B200	488	38
		1100	A500	360	88
			B500	375	63

tension test specimens are relieved during the machining of the specimen gauge length.

Susceptibility to stress corrosion cracking decreases as the tempering temperature is increased. The significant difference of K_{ISCC} in the 500°C temper condition is attributed to the stress corrosion cracking sensitivity to material hardness.

REFERENCES

Amirat,A. 1987. MSc. Thesis. *The effect of a corrosive environment on crack growth rate in chainsteel*. Bristol University, Bristol, UK.

Beany, E.M & E.Procter 1974. A critical evaluation of the centre hole technique for the measurement of residual stress. *Strain* 10:7-14.

Carter, C.S. 1971. Stress corrosion crack branching in high strength steels. *J. Eng. Fracture Mechanics* 3:1-13.

Clark,W.G & J.D.Landes 1976. An evaluation of rising load K_{ISCC} testing. *ASTM STP* 610:108-127.

Gilleland,A. 1986 Private communication.

McIntyre,P. & A.H. Priest 1972. Accelerated test technique for the determination of K_{ISCC} in steels. *British Steel Co.* Report MG/31/72.

Nelson,D.V. 1982. Effects of residual stress on fatigue crack propagation. *ASTM STP* 776:172-181

Parkins,R.N. 1976. *Corrosion*. London. Newnes-Butterworth.

Scully,J.C. 1966. *The fundamentals of corrosion* London:Pergamon Press ltd.

Shahinian,P. & R.W.Judy Jr. 1976. Stress corrosion crack growth in surface-cracked panels of high strength steels. *ASTM STP* 610:128-142.

Recent Advances in Experimental Mechanics, Silva Gomes et al. (eds) © 1994 Balkema, Rotterdam, ISBN 90 5410 395 7

Consequences of residual stress on crack propagation in MDPE pipes

K.Chaoui
Mechanical Engineering Institute, University of Annaba, Algeria

A.Moet
Macromolecular Science & Engineering Department, Case Western Reserve University, Cleveland, Ohio, USA

A.Chudnovsky
Department of Civil Engineering, Mechanics & Metallurgy, University of Illinois at Chicago, Ill., USA

ABSTRACT: Residual stress distribution in medium density polyethylene pipe is evaluated using the layer removal method in longitudinal and circumferential directions. It is found that the strain gauge technique combined with layer removals shows realistic values. A tensile residual stress dominates about 25% of the inner section of the pipe wall and then, gradually becomes compressive towards the outermost layer. Crack propagation behavior in single edge notched (SEN) specimens is examined under creep conditions. After calculation of the elastic energy release rate taking into account residual stress distribution, it is concluded that differences in crack propagation rates are also influenced by morphological variances. Finally, it is shown experimentally that non-equilibrium residual stress state is basically a consequence of the long-term residual strain relaxation.

1 INTRODUCTION

Residual stresses in plastic pipes are a consequence of the thermomechanical history imparted by extrusion. Different cooling rates on inner and outer surfaces of the pipe wall create a gradient of temperature which results in residual stresses and morphological variances (Hodgkinson & Williams 1983 and Struik 1978).

Microstructural gradients coupled with residual stresses strongly alter material properties directionnally and particularly its resistance to crack growth. Therefore, predictions of material behavior on the basis of data obtained from studies on isotropic compression molded plaques may not represent the actual service conditions. In spite of this serious practical concern, little attention is directed to elucidate this effect.

This paper presents results of experiments to study the effects of pipe extrusion condition on the material resistance to crack propagation. In addition, long-term strain relaxation is monitored to explain non-equilibrium residual stress distribution.

2 EXPERIMENTAL PROCEDURE

The materials used in this investigation are PE2306-IIC and PE2306-IA which are medium density polyethylenes supplied by Plexco Inc. (Tennessee) and Philadelphia Electric Co. (Pensylvania) respectively. The pipes are 11.1 mm thick and 113.7 mm in outside diameter. Longitudinal specimens are cut using a double circular saw assembly as shown in Figure 1 and they are used for residual stress measurements and crack propagation tests. Turning and milling of different specimens (strips and rings) are performed at low speed while cooling the latter with a stream of pressurized service air. Four 3 mm thick rings are prepared to examine strain relaxation after slicing at one point.

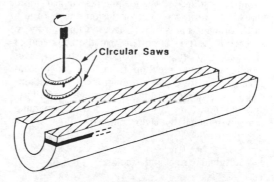

Figure 1. Double saw assembly for longitudinal specimen preparation at low speed.

3. RESULTS AND DISCUSSION

3.1 Residual stress analysis

The modified layer removal consists in considering the rectuangular strips as plaques and using the equation developed by (Treuting & Read 1950). The ring slitting technique uses wide pipe rings which are sliced to measure the resulting ring closure. For the strain gauge method, turned rings are employed and the corresponding strains are measured directly (Chaoui, Moet & Chudnovsky 1988). The maximum values obtained in each method are summarized in Table 1.

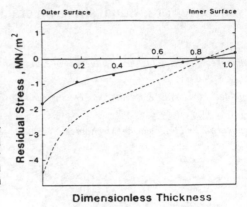

Figure 2. Circumferential residual stress distribution from strain gage measurements after 24h (The broken curve shows the deduced longitudinal stress distribution).

Table 1. Comparison of maximum values of residual stress along longitudinal and circumferential directions.

σ MN/m^2	Modified layer removal method	Ring Slitting method	Strain gauge method
σ_c^{in}	2.6	1.8	0.2
σ_c^{out}	-3.9	-8.0	-1.8
σ_l^{in}	-6.6	-4.5	-0.5
σ_l^{out}	-9.7	-15.0	4.5

Graphical measurements of dimension changes may cause some lack of confidence in the first two methods. The results are in quantitative agreement with those of literature (Broutman, Bhatnagar & Choi 1986 and Hodgkinson & Williams 1983); however, strain gauge results are much lower than those given by layer removal and ring slitting methods. The distribution is depicted in Figure 2.

Here, residual stress magnitudes appear more realistic according to the conclusion drawn from a crack propagation study (Chaoui, Chudnovsky & Moet 1987). It is important that the achievement of a practical stress distribution by strain gauges should not draw attention from the need to examine several unresolved limitations which include localised heating in adhesion, gauge sensitivity, material non-linerarity and, of course, structural changes induced by layer removals.

3.2 Residual stress and fracture

Two identical single edge notched (SEN) specimens of PE2306-IIC, one with an outside notch (inbound

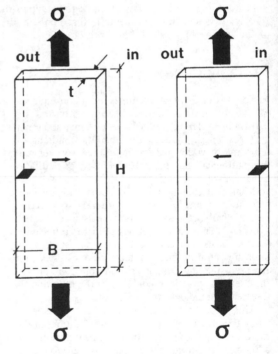

Figure 3 : Sketch of inbound (left) and outbound (right) crack propagation specimens.

crack) and the other with an inside notch (outbound crack), are subjected to a constant tensile load of 6.75 MN/m^2 to ensure crack propagation. Figure 3 represents these specimens and the horizontal arrows show crack propagation directions.

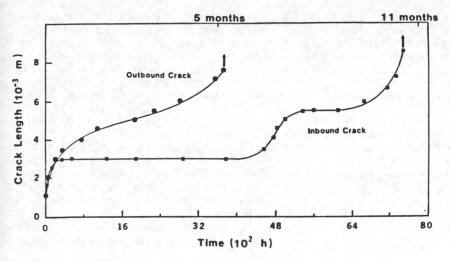

Figure 4. Overall crack propagation. Vertical arrows indicate ultimate failures at 3705 h (outbound crack) and 7709 h (inbound crack).

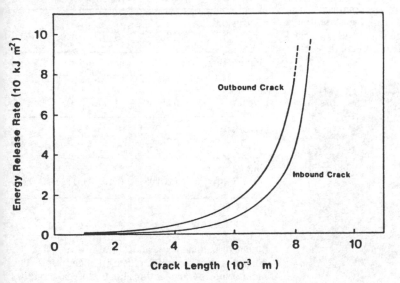

Figure 5. Computed energy release rates for inbound and outbound cracks using elastic considerations.

Crack tip is observed at a 10X magnification with transmitted light. The resulting growth behavior is depicted in Figure 4. It is observed that inbound crack followed a typical creep curve with its three stages and ultimate failure occured monotonically at 3705 h. The outbound crack showed a completely different curve involving two crack arrests at 3 and 5.5 mm. In the final stage, the crack accelerated towards failure in 7709 h.

The markedly different inbound and outbound crack propagation behavior in the same material suggests that the material's resistance is strongly influenced by its processing history.

The non-uniformity of stress distribution along

the crack trajectory is accounted for by calculating the stress intensity factor using Green's fonction derived for a unit dipole force applied at the crack faces in opposite directions at a point x in SEN:

$$K_I = \frac{2}{(\pi a)^{1/2}} \int_0^a P(x) \ F \ (x/a, \ a/B) \ dx \qquad (1)$$

$P(x)$ is the traction distribution due to the applied load and residual stress in a cracked specimen along a line coincident with the crack path.

If one assumes linear elastic behavior and using a modulus of elasticity of 55 GPa, the elastic energy release rate may be computed for both cases as:

$$G_I = \frac{K_I^2}{E} \qquad (2)$$

As shows in Figure 5, the energy release rate associated with the outbound crack is always higher than that associated with the inbound crack. Obviously, this is confirmed by the dominant tensile residual stress towards the inner pipe skin (bore). The critical energy release rate of the outbound crack is only 77 kJ/m². At the same crack length (8 mm),

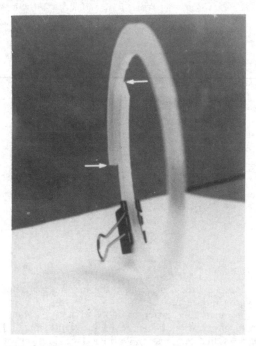

Figure 6. Unpigmented polyethylene pipe ring undergoing closure under the effects of circumferential residual stress.

the inbound energy release rate is 0.58 the outbound value. In fact, the inbound critical crack length is 8.5 mm which leads to 107 kJ/m² as a critical rate. Apparently, the energy release rate is not the unique controlling parameter since fracture surface analysis at similar energy levels exhibits notably different features (Chaoui, Chudnovsky & Moet 1987).

3.3 Long-term residual strain relaxation

Figure 6 shows a 3 mm thick pipe ring experiencing free strain relaxation over a period of 5 months. The overlap distance is included between the two horizontal arrows.

The specimens used in this case are identified in Table 2 together with the instantaneous strains.

Table 2. Identification of ring specimens and respective initial strains.

PE2306	IA	IIC(1)	IIC(2)	IIC(3)
Extrusion year	1967	1983	1985	1986
Pigmention	yes	no	no	no
Service (years)	16	0	0	0
Overlap (mm)	7	11	8.5	10
initial strain (mm/mm)	0.0195	0.0307	0.0237	0.0279

It is noticed that unused and newer pipes have higher instantaneous residual strain recovery in the hoop direction as compared to the aged pipe. The monitoring is extended for about two years and the obtainted strains are depicted in Figure 7 as a function of log (t) where t is in minutes. It is evident that strain difference between samples is widening with increasing time, suggesting especially for the aged pipe, that its microstructure does not contrain anymore the original amount of residual stress. For longer times, the gap between unused pipes becomes noteworthy which may mean that each ring is evolving with its specific form of relieving strains though all rings have gone through very similar steps of processing.

These curves are close to those of long-term creep f polyehylene (Findley 1967). In Findley's creep

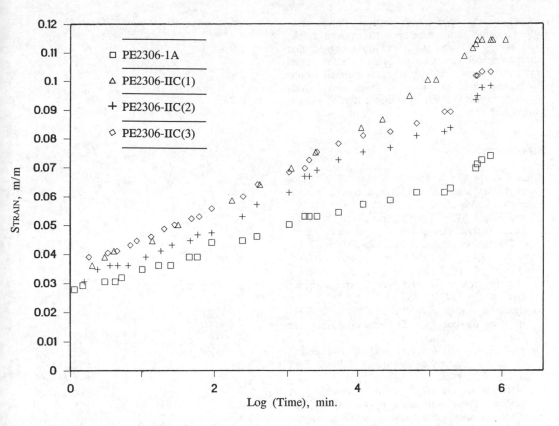

Figure 7. Long-term strain relaxation as a function of time.

Table 3. Summary of calculated and published values of ε^+ and n.

Material	ε^+	n	Duration (years)
PE2306-1A	2.24	0.10	1.8
PE2306-IIC(1)	2.48	0.12	1.8
PE2306-IIC(2)	2.65	0.12	1.8
PE2306-IIC(3)	2.80	0.11	1.8
PE1 (Findley)	0.236	0.154	26
PE2 (Findley)	0.710	0.154	26
PE3 (Crissman)	2.18	0.487	0.2

PE1: Static load of 1.5 MPa.
PE2: Static load of 2.75 MPa.
PE3: Static load of 10.0 MPa.

tests, the polyethylene specimens were subjected to static loads for 26 years and were characterized by a simple power law of the form:

$$\varepsilon = \varepsilon^\circ + \varepsilon^+ \, t^n \tag{3}$$

where ε is the measured strain, t is the time in hours and ε^0, ε^+ and n are constants. Adopting the same formalism for the previous residual strain relaxation, the obtained values of the constants are compared in Table 3. It seems that the values of n for experimental times over a year are comparable; however, from Crissman's study n is drastically different from the expected value even within the first 1900 h. Most probably, this is the direct consequence of the high load level used (Crissman 1986).

4 CONCLUSIONS

Residual stresses in plastic pipes are compressive at the outer layers in hoop and longitudinal directions and the strain gauge technique gives lower values which appear more pratical. Polyethylene pipes exhibit more resistance to inbound cracks and energy release rate calculations suggest that morphological variances participate in the actual material's resistance. Recovery of residual strains in plastic pipes extends over many years which explains in part the non-equilibrium state of residual stresses.

REFERENCES

Chaoui,K., A.Chudnovsky & A.Moet 1987. Effect of residual stress on crack propagation in MDPE pipes. *J. Materials Sci.* 22:3873-3879.

Chaoui,K., A.Moet & A.Chudnovsky 1988. Strain gauge analysis of residual stress in plastic pipes. *J. Testing & Evaluation* 16:286-290.

Crissman, J.M. 1986. *GRI Final Report* 86/0070.

Findley,W.N. 1987. 26-year creep and recovery of polyvinylchloride and polyethylene. *Polymer Eng. & Sci.* 27:582-587.

Hodgkinson,J.M & J.M.Williams 1983. Residual stress in plastic pipes. *Deformation, Yield and Fracture of Polymers* 35:1-7.

Struik,L.C.E. 1978. Orientation effects and cooling stresses in amorphous polymers. *Polymer Eng. & Sci.* 18:799-811.

Treuting,R.G. & W.T.Read, Jr. 1950. A mechanical determination of biaxial residual stress in sheet materials. *J Applied Physics* 22:130-134.

Recent Advances in Experimental Mechanics, Silva Gomes et al. (eds) © 1994 Balkema, Rotterdam, ISBN 90 5410 395 7

Inference of plastic deformation in welding residual stress field from infrared thermography

I.Oda, T.Doi, H.Sakamoto & M.Yamamoto
Kumamoto University, Japan

ABSTRACT: A nondestructive and noncontact technique for evaluation of the plastic and damaged region by using infrared thermography is proposed. Butt welded plates with a through-thickness crack are loaded in tension. The temperature rise of specimen surface near a crack is measured by the infrared thermography. The effect of the welding residual stress on the shape and dimensions of the heated region are examined. The heated regions are compared with the plastic zones obtained by the elasto-plastic finite element analysis.

1 INTRODUCTION

Considerable papers(Kihara and Masubuchi 1959) which experimentally examined the effect of residual stress on the brittle fracture strength of weldment have been published. There have been, however, few works concerned with stress and strain near a crack in the welding residual stress field(Oda and Sakamoto 1992, Oda,Sakamoto and Yamamoto 1992).

Plastic deformation near a crack takes an important role relating to the failure or the fracture of the structure. In the present paper, a nondestructive and non-contact technique for evaluation of the plastic and damaged region near a crack by using infrared thermography is proposed. Most of energy expended in deforming a metal plastically is converted into heat. The temperature rise of specimen surface near a crack is measured by the infrared thermography. It is assumed that the region heated by the work of plastic deformation corresponds to the plastic zone near a crack. The effect of the welding residual stress on the shape and dimensions of the heated region are examined in the present study. The heated regions are compared with the plastic zones obtained by the elasto-plastic finite element analysis. In addition, the usefulness and the limitations of the application of infrared thermography to the evaluation of the plastic and damaged region are discussed.

2 EXPERIMENT

The material supplied is stainless steel, SUS 304 in JIS. Figure 1 shows the shape and dimensions of a test specimen(Type C) used in the experiment. Three types of sheet specimen were used. They are Type N, Type T and Type C. Each type of specimen has a through-thickness center crack of the same length. Type N is not welded. A V-groove and 2 V-grooves without root gap were prepared by cutting on both surfaces, in parallel to the major axis(y axis), in Type T and Type C, respectively. The arc welding beads were laid on the grooves under the conventional welding condition. And then, the pre-crack was prepared after machining down the thickness of parallel part of specimen to 4 mm. Both tips of pre-crack were electrospark-machined with a wire of 0.03 mm diameter. The pre-crack perpendicular to the welding bead was desired to be in the tensile and compressive residual stress field for Type T and Type C, respectively. A uniform tensile load was applied to each specimen in parallel to the y axis. The temperature rise of specimen surface was measured by the infrared thermography. The crack opening displacement was also measured by clip gauge.

3 ANALYSIS

The deformation near the tip of a crack in

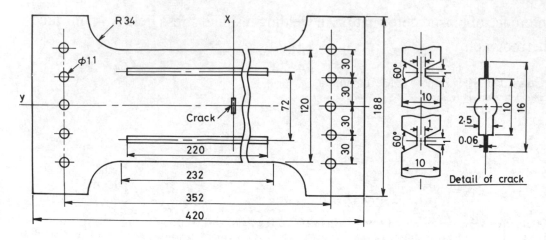

Fig.1. Specimen used in experiment(Type C).

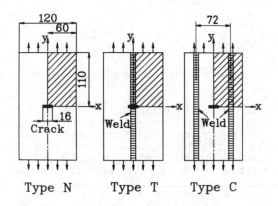

Fig.2. Models for FEM analysis.

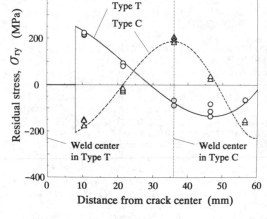

Fig.3. Distributions of longitudinal residual stress along the x axis.

a plane stress field, which is uniaxial tension in the y direction, was analysed. An elasto-plastic element analysis with an incremental theory of plasticity was used. Three types of specimen model shown in Fig.2 were used in the analysis, which correspond to those in the experiment. The material was assumed to harden according to a power law relation between stress and strain suggested by Swift.

For Type T and Type C, the residual stresses redistributed after pre-cracking were considered in the analysis. The distributions of longitudinal residual stress, σ_{ry} , and transverse residual stress, σ_{rx} , were measured by the stress relaxation method with biaxial strain gauges for similar welded and pre-cracked specimens.

Figure 3 shows distributions of σ_{ry} along the x axis for Type T and Type C. In Type T, a high tensile stress exists in the region near the pre-crack tip and a compressive stress is produced in the region away from the pre-crack. A relatively high compressive stress in the vicinity of the pre-crack tip and a tensile stress of high magnitude near the weld center line(x=36 mm) exist in Type C.

4 RESULTS AND DISCUSSION

Figure 4 shows the temperature rise in specimen surface of Type N. A moving speed of crosshead in the present tensile test was kept constant, that is, 30 mm per minute. The distribution of temperature changes as the time proceeds since loading is started, that is, as the applied load increases. The heated region near the pre-crack tip spreads at an angle of about 45 deg. to the x axis. The temperature distribution in the heated region can be also confirmed.

The temperature rise in Type T is shown in Fig.5. The size of heated region is much larger and the absolute value of temperature is much higher for Type T when compared with those for Type N at the same time after the start of loading. At a relatively low applied stress level in Type T, the spread angle of the heated region has a tendency to become wider than that , about 45 deg., for Type N. These phenomena in Type T occur because the plastic deformation progresses under the influence of high tensile residual stress in the vicinity of pre-crack.

Figure 6 shows the temperature rise in Type C during loading. At a low applied stress level, the size of heated region is smaller and the absolute value of temperature at pre-crack tip is lower for Type C because of compressive residual stress near pre-crack, when compared with those for Type N. At a high applied stress level, however, the heated region in Type C spreads to the zone away from pre-crack more easily than that in Type N. The spread angle of the heated region in Type C has a tendency to become wider in the zone away from pre-crack than in the vicinity of pre-crack. These phenomena in Type C are caused by high tensile residual stress near the weld center line which is 28 mm distant from pre-crack tip.

Figure 7, Figure 8 and Figure 9 show the works of plastic deformation during loading analysed for Type N, Type T and Type C, respectively. The behavior of plastic deformation in each specimen type agrees qualitatively with that of temperature rise at relatively low applied stress level. This implies that the effect of residual stress on plastic deformation can be inferred from the temperature rise obtained by infrared thermography. If the considerable time elapses since the start of loading, however, the temperature change due to thermal conduction should be also considered in drawing an inference of plastic deformation from the temperature rise.

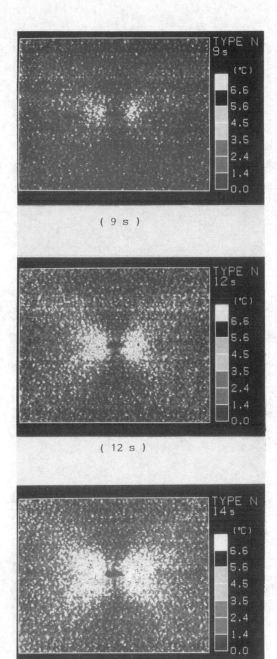

(9 s)

(12 s)

(14 s)

Fig.4. Temperature rise during loading in Type N.

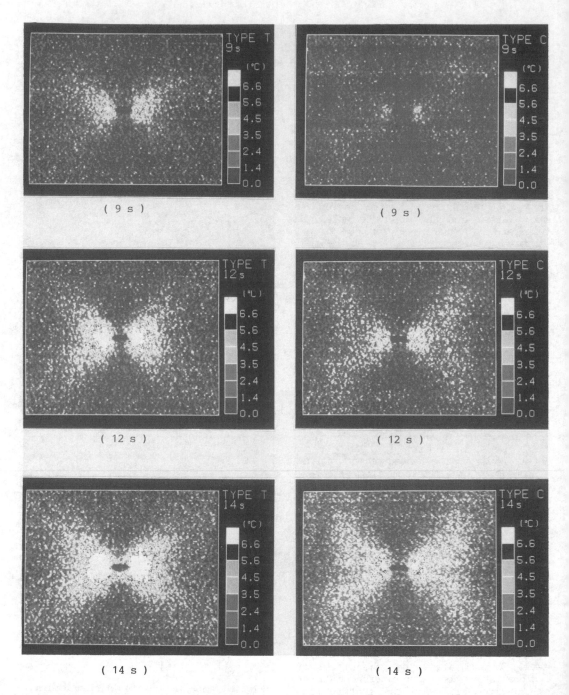

(9 s)　　　　　　　　　　　　　　　　　(9 s)

(12 s)　　　　　　　　　　　　　　　　(12 s)

(14 s)　　　　　　　　　　　　　　　　(14 s)

Fig.5. Temperature rise during loading
in Type T.

Fig.6. Temperature rise during loading
in Type C.

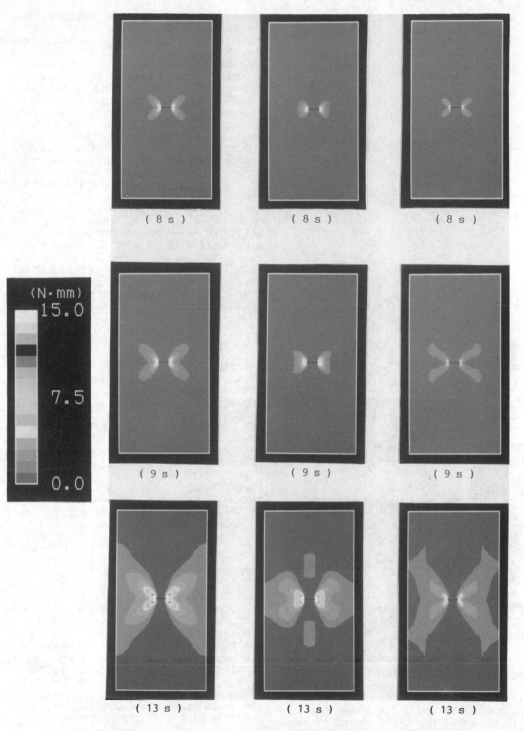

(8 s)   (8 s)   (8 s)

(9 s)   (9 s)   (9 s)

(13 s)   (13 s)   (13 s)

Fig.7. Work of plastic   Fig.8. Work of plastic   Fig.9. Work of plastic
deformation during   deformation during   deformation during
loading in Type N.   loading in Type T.   loading in Type C.

82l

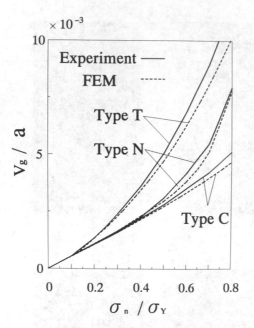

Fig.10. Relationships between crack opening displacement and applied net stress.

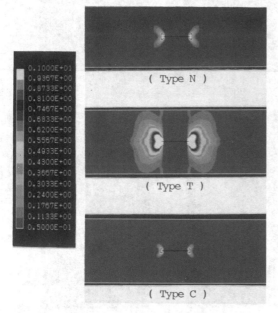

Fig.11. Work of plastic deformation in the case of $\sigma_n/\sigma_Y = 0.8$.

Figure 10 shows the relationship between crack opening displacement V_g at 4 mm inside from the tip and applied net stress σ_n. Although there is a little difference between the analytical results and the experimental ones because of some assumptions in the analysis, both results show the same tendency. The crack opening displacement becomes larger in the order, Type C, Type N, Type T within the range of stress ratio shown in Fig.10.

Distributions of work of plastic deformation for three specimen types, when the stress ratio, σ_n/σ_Y, is 0.8, are compared in Fig.11. The plastic zone size and the absolute value of work of plastic deformation at crack tip become greater in the order, Type C, Type N, Type T. Characteristics of plastic deformation shown here agree well with those of temperature rise at low applied stress level shown in Fig.4 to Fig.6. These phenomena of plastic deformation and temperature rise can explain the effect of residual stress on crack opening displacement shown in Fig.10.

5 CONCLUSIONS

The following conclusions may be drawn.

(1) The effect of residual stress on plastic deformation can be qualitatively inferred from temperature rise obtained by infrared thermography.

(2) The effect of residual stress on the crack opening displacement can be explained by the temperature rise or the plastic deformation near a crack tip.

(3) If the considerable time elapses since the start of loading, the temperature change due to thermal conduction should be considered in drawing an inference of plastic deformation from the temperature rise.

REFERENCES

Oda, I. and Sakamoto,H. 1992. *Residual stress-III*. ed. by Fujiwara et al. 482-487. Elsvier applied science: London and New York.

Oda,I.,Sakamoto,H. and Yamamoto.M. 1992. Effect of mechanical stress-relieving on deformation near a crack in weldment. *Proc. 7th ICEM* : 51-56. SEM,INC.:USA

Kihara,H. and Masubuchi,K. 1959. Effect of residual stress on brittle fracture. *Welding journal* 38: 159s-168s.

6 Measurement of stresses and strains in hostile environments

Recent Advances in Experimental Mechanics, Silva Gomes et al. (eds) © 1994 Balkema, Rotterdam, ISBN 90 5410 395 7

Cure effect on composite material at elevated temperature

Eltahry I. Elghandour & Faysal A. Kolkailah
Aeronautical Engineering Department, Cal Poly State University, San Luis Obispo, Calif., USA

Said A. Khalil & S. R. Naga
Department of Mechanical Design, College of Engineering & Technology Mataria, University of Helwan, Cairo, Egypt

ABSTRACT:The objective of this study is to characterize the material properties of a composite material. The material characteristics are determined at several elevated temperatures. A number of specimens with three different stacking sequences laminates, cured at four different cycles were constructed from an carbon fiber/epoxy resin (IM7) flat plate. A heated test chamber was constructed to heat the test specimens. The elongation and temperature data obtained from the strain gages and thermocouples were used to determine the modules of elasticity and poisson's ratio as functions of temperature.

The results of this study present the complete characterization of the composite material (IM7) such that analytical models can be generated. These models could be used to study the effect of temperature on an aerospace structure's integrity and stability which could be seriously affected by the extreme high surface temperatures and temperature gradients caused by the High speed flight of aerospace vehicles.

INTRODUCTION

Within the past decade, composite materials have been suggested for the primary structure of aircraft and spacecraft. One of the results of high Mach number flight is high skin temperature. High skin temperatures and temperature gradients can seriously effect the structural integrity of these high speed aircraft. Thermal effects can results in variations in structure elastic characteristics cousing a degradation of the aircraft aeroelastic stability. And also, the properties of the thermoset resin composites depends on the process. The cure of thermoset resin composites requires the application of the manufacture's prescribed process for heat and pressure. Heat is used to facilitate and control the chemical reaction between fiber and resin. Pressure is applied to squeeze out excess resin, to consolidate plies, and to minimize void content in laminatc.

To safely design an aircraft for high Mach number flight, the mechanical characteristics must be known at various temperatures. These mechanical properties can be obtained through uniaxial tensile testing.

Modern day composites are formulated in a variety of ways from many different materials. This study focuses on fiber-reinforced composites, namely, continuous carbon-fiber/epoxy -resin composites. This kind of composite require both heat and constant pressure for the curing process which makes the epoxy resin undergo exothermic chemical reaction. The thermal response of these composites to the applied heat during curing is an important aspect for the design of the curing cycle.

The thermal characteristics of the composite during curing could lead to an unevenly cured product which is structurally unsafe. On the other side, a thorough understanding of these thermal characteristics can lead to the design of an optimal curing cycle which maximizes product quality and minimizes cost.

Tang J. (1988) studied the estimation of thermal properties((conductive and volumetric heat capacity) of a carbon/epoxy composite during curing. Eltahry I., Elghandour El.I. (1993) studied the effect of cure of different laminate composite materials on the mechanical properties at room temperature. They concluded that the mechanical properties is constant with different thickness in cross-ply and is more significant in unidirectional laminate. Also, with different cycles, the young's modules increases with increasing temperatures during the cure cycles for different laminates.

Therefore, the main objective of this paper is to evaluate the effect of cure cycle, with change of temperature and constant pressure on composite material characteristics. The evaluated temperatures range from room temperature to 350 °F during tensile tests for different stacking sequences of laminated composite materials.

FABRICATION OF SPECIMENS

A composite air press machine and heating chamber were employed in the experiment tests of this study. The procedures of cooking specimens by using a composite air press machine are described in Elghandour El. I. (1993). The resin plates were cured according to the cure cycles shown in figure (1). This figure shows the four different curing cycles used in this study. The total curing time is about eight hours, after, which the machine opens automatically. Three types of laminates, Unidirectional 8[0°], Cross-ply [0°/2(90°)]$_S$ and Quasi-isotropic [0°/+-45°/90°]$_S$, were fabricated in a composite air press according to the manufacture's recommended cure cycles. A total of twenty specimens were fabricated. First, for the unidirectional laminate, two specimens for each cycle were used to take the average result. Four specimens from 8[0°], [0°/2(90°)]$_S$ and [0°/+-45°/90°]$_S$ were employed in studying the effect of cure cycles with different temperatures and constant pressure at evaluated temperatures: room temperature, 150 °F, 225 °F, 300 °F and 350 °F during tensile tests. Using ASTM 3039-76, as a guide, as in ASTM (1989) and Carlsson L. A. (1987), the

specimens were cut from the panels using a diamond-impregnated saw. All specimens were 8 in. long and 0.5 in wide as shown in figure (2). The side edges of the specimens were grounded with a sand paper as well as the defects on the edges. Tabs were used to increase the thickness at the ends which allowed for a more secure hold by the testing machine's grip. The tabs were made from aluminum and cut to 1.5 in. long and 0.5 in wide. The tabs were bonded to the specimens with an adhesive (structural adhesive 3M Scotch-Weld) and a composite air press was used to cure the bond between specimens and tabs under pressure and room temperature as shown in figure (3). For the experimental test, two strain gauges were used for each experimental test. One strain gauge measured longitudinal strain and the other measured the transverse strain. The strain gages were bonded to the specimens using ARAIDAITE. This slow curing adhesive has a maximum temperature of 400 °F. All specimens were tested on a closed-loop electrohydraulic Instron machine testing with Instron friction type jaw grips. The loading rate during testing was 0.2 in/min. A heat test chamber was built to heat the test specimens, as shown in figure (4). A thermocouple was employed to measure the temperature of the test specimens. Temperature was controlled by varying the voltage to the heat coils. During the experimental test, the specimen was brought to a specific high temperature between room temperature and 350 °F. Figure (5.a) and (5.b) give the block diagram and the control diagram for the heating chamber.

RESULTS AND CONCLUSION

Tensile tests were performed to determine uniaxial tensile properties: Young's modulus in the longitudinal direction, Young's modulus in the transverse direction and Poisson's ratio for composite material laminate plate for different stacking sequence 8[0°], [0°/2(90°)]$_S$ and [0°/+45°/-45°/90°]$_S$ with different temperature and different cure cycles. For

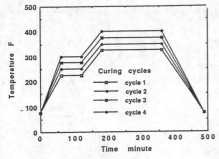

Figure 1 Curing Cycles

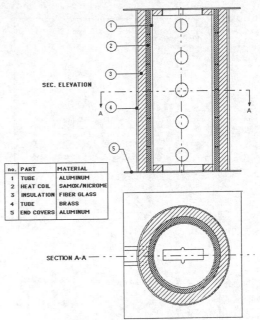

SEC. ELEVATION

no.	PART	MATERIAL
1	TUBE	ALUMINUM
2	HEAT COIL	SAMOX/NICROME
3	INSULATION	FIBER GLASS
4	TUBE	BRASS
5	END COVERS	ALUMINUM

SECTION A-A

FIG. 4 HEATING CHAMBER ASSEMBLY DIAGRAM

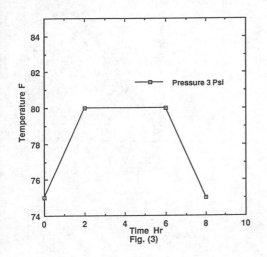

Figure 2 Specimen and Its Dimensions

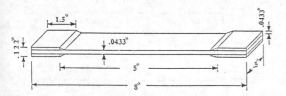

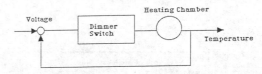

Block diagram of the heating chamber set up

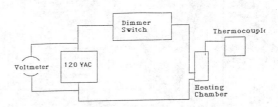

Fig. (5.a) Controls diagram showing how temperature is regulated

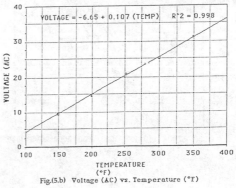

VOLTAGE = -6.65 + 0.107 (TEMP)   R^2 = 0.998

Fig.(5.b) Voltage (AC) vs. Temperature (°F)

827

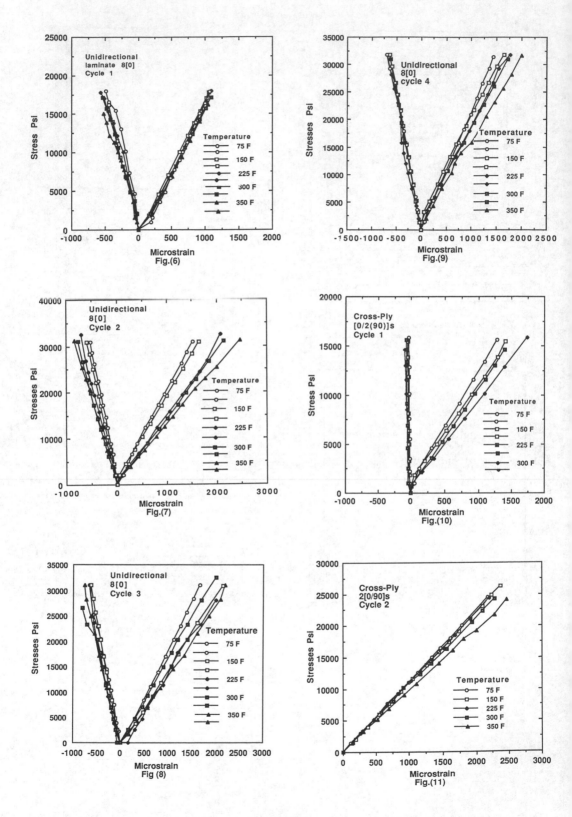

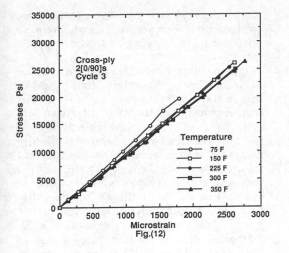

Fig.(12)

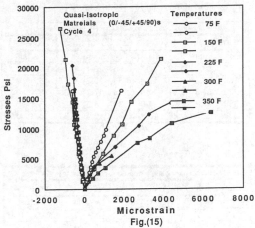

Fig.(15)

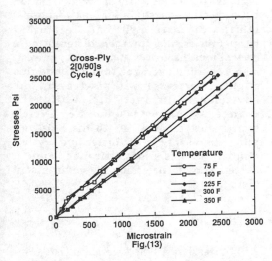

Fig.(13)

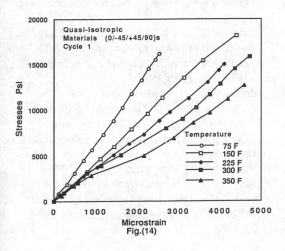

Fig.(14)

Table 1. For Unidirectional laminated

Tem (F)	Longitudinal Young's Modulus and Poisson's Ratio							
	Cycle (1)		Cycle (2)		Cycle (3)		Cycle (4)	
	El(ksi)	Ult	El(ksi)	Ult	El(ksi)	Ult	El(ksi)	Ult
75	18.2	0.32	18.7	0.36	20.1	0.39	21.7	0.45
150	16.5	0.32	17.4	0.36	18.2	0.37	18.8	0.39
225	14.8	0.31	15	0.35	17.7	0.36	18.1	0.39
300	14.2	0.29	14.8	0.34	16.3	0.35	17.4	0.37
350	12.6	0.29	13.9	0.33	15.4	0.33	16.8	0.36

Table 2. For Cross-Ply laminated

Tem (F)	Longitudinal Young's Modulus and Poisson's Ratio							
	Cycle (1)		Cycle (2)		Cycle (3)		Cycle (4)	
	El(ksi)	Ult	El(ksi)	Ult	El(ksi)	Ult	El(ksi)	Ult
75	10.3	0.4	10.4	0.41	10.9	0.42	11.2	0.45
150	9.84	0.37	9.83	0.38	9.93	0.38	11.1	0.39
225	9.71	0.26	9.79	0.29	9.8	0.3	11	0.35
300	9.1	0.2	9.15	24	9.45	0.26	10.1	0.3
350	8.7	0.19	8.74	22	9.41	0.24	9.63	0.28

Table 3. For Quasi-isotropic laminated

Tem (F)	Longitudinal Young's Modulus and Poisson's Ratio							
	Cycle (1)		Cycle (2)		Cycle (3)		Cycle (4)	
	El(ksi)	Ult	El(ksi)	Ult	El(ksi)	Ult	El(ksi)	Ult
75	6.33	0.21	*	*	*	*	8.19	0.23
150	5.63	0.18	*	*	*	*	6.39	0.21
225	4.55	0.17	*	*	*	*	5.62	0.18
300	4.16	0.16	*	*	*	*	4.59	0.17
350	3.59	0.15	*	*	*	*	4.1	0.15

the laminates $[0^o/2(90^o)]_S$ and $(0^o/+45^o/-45^o/90^o)_S$ it was assumed that Young's modulus in the longitudinal is equal to Young's modulus in the transverse direction, Carlsson L. A. (1987).

The tensile tests were performed utilizing wedge-section friction grips. Each specimen was first aligned in the grips and tightened in place. The specimen was then loaded monotonically to 700 Ibs. Different load were applied and strain was recorded using a model P-3500 strain indicator and model SB-10 switch and balance unit.

Figures (6) through (9) show the stress vs strain for unidirectional laminate specimen at different temperatures and for each different cure cycle. These curves were drawn from the data of two specimens. From these curves, average Young's modulus,(E_L)ave. and average poisson's ratio ($1t$)ave. were calculated. From the curves and table 1, one can see that the average Young's modulus and Poisson's ratio decrease with increasing the environment temperature and increase with the increasing of the cycle's temperature.

Figures (10) through (13) show the relation between stress vs. strain for cross-ply laminate specimen at different temperatures and for each different cure cycle. From the curves and table 2, one can see that average Young's modulus slightly decreases with increasing the environment temperature while the change with increasing the cycle's temperature is insignificant. As to the Poisson's ratio, there is a slight increase with the cycle's temperature and it decreases with increasing the environment temperature .

Figures (14) through (15) show the relation between stress vs. strain for Quasi-isotropic laminate specimen at different temperatures and for each different cure cycle. From the curves and table 3 , one can also see that average Young's modulus slightly decreases with increasing the environment temperature while the change with increasing the Cycle's temperature is small. As to the Poisson's ratio, the change with increasing Cycle's temperature is insignificant and it decreases with increasing the environment temperature.

REFERENCES

ASTM standards, 1989 " Standard test method for tensile properties of fiber-resin composite "

Carlsson L. A.and R. B. Pipes" Experimental characterization of advanced composite materials" Prewtice-Hall Inc., Englewood Cliff,New Jersey,1987.

Elghandour, El. I.Said A.Khalil,S R. Naga and Faysal A. Kolkailah " Effect of cure of different laminate composite materials on the mechanical behaviour" the proceeding of the 3rd Japan International SAMPE Symposium , Japan, 1993.

Tang J. and G.S. Springer " Effect of cure and moisture on the properties of fiberite 976 resin" J. of composite materials , Vol. 22. 1988

Recent Advances in Experimental Mechanics, Silva Gomes et al. (eds) © 1994 Balkema, Rotterdam, ISBN 90 5410 395 7

Investigations of pellet-clad mechanical interaction (PCMI) of two dimensional photoelastic models

H.Y.Yang & J.F.Jullien
Institut National des Sciences Appliquées (INSA), Lyon, France

N.Waeckel
Service des Etudes de Projets Thermiques et Nucléaires, EDF, Lyon, France

ABSTRACT : Nuclear fuel rods are composed of a thin cylindrical clad containing fuel pellets. A potential fuel rods mechanism failure in water cooled reactors is pellet-clad interaction (PCI) due to the swelling and the thermal expansion of the fuel pellets. As radial cracks appear in the pellets, a very high stress in the clad would be expected. This may result in clad failure. We present here experimental and numerical analyses of a 2-D photoelastic model of the mechanical pelletclad interaction phenomenon. The stress distribution according to various parameters in investigated. Both experimental and numerical results show that the stress distribution is closely related to the pellet state, continuous or fragmental, and that the clad circumferential stresses in the region of cracks are very high. This reveals macroscopically the stress states of pellet-clad during PCI. It shows that the numerical and the experimental results are in good agreement, therefore it may be concluded that the simulation at the interface between pellet and clad in the numerical models are appropriate.

1 INTRODUCTION

A nuclear fuel rod consists of a thin cylindrical clad containing fuel pellets. One of the main purposes of fuel design is to avoid any cladding failure during the life time of the nuclear plant. This failure may occur when large power transients involve thermomechanical pellet-clad interaction (PCI) ans stress corrosion threshold overstepping.

It is a well-known fact that the parabolic radial remperature distribution in fuel rod causes non-uniforme thermal expansion of the pellet, which leads to develop radial cracks in the pellet. During the first set-up of power, the dilation at the centre of pellet is much greater than the outer region. As the power increases the temperature difference between inner and outer radius of the pellet rises, causing circumferential stresses greater than rupture limit of fuel material. This, in turn, leads the pellet to lie in fragments which move towards the inner skin of the clad and results in high local strain in the region of the crack and consequently may cause cladding failure.

A number of studies has been done in the problem of PCI using numerical (finite element) techniques. A review of the numerical analyses is given by YU & al, however it seems that only a few experimental studies have been carried out. It is presented in this paper that an experimental and numerical investigations of the mechanical aspect of PCI using both photoelasticity and finite element techniques.

The objective of the present study is to investigate the variation of the mechanical stress distribution in the clad during the pellet-clad interaction according to the various parameters : continuous of fragmental pellet, the grap between pellet and clad, and the friction at the interface.

2. MODELS

2.1 Experimental models

A two-dimensional photoelastic model of pellet-clad cross section has been analyzed by using the approximation of plane strain. The photoelastic model consists of a disk representning thepellet and a ring representing the clad. The disk is inserted into the ring.

The experiments are conducted by using both continuous and fragmental disks inserted in a ring. The fragmental disk is made of six equal-sized sectors in ordet to simulate the effect of pellet cracks on the clad. In addition, the influence of the pellet-clad gap on the stress distribution in the clad is studied by using a continuous disk and two values of gap. The effect of friction between pellet and clad is studied by employing a fragmental disk under two contact conditions : one is natural, the other is realised with dry ruggedness.The dimensions of the two types of models are listed on Table 1. Epoxide (PSM-5) and plycabonate (PSM-1) have been selected for the disk and the ring respec-

Table 1 Dimensions of the models

	Int. radius(mm)	Ext. radius(mm)	Gap(mm)
Ring 1(PSM-1)	89.8	81	
Disk 2(PSM-5)	81	25	0(G1)
Disk 3(PSM-5)	80.84	25	0.16(G2)
Disk 4(PSM-5)	80.84	25	0.16(G2)

Table 2 Characteristics of the materials

Characteristics		PSM-5	PSM-1
Modulus of YOUNG(Hbar)		310	240
Coefficient of POISSON		0.36	0.38
Fringe value	White light	0.454	0.383
(kg/mm)	Mercury light	0.502	0.400

tively in order to restitute the fuel pellet-clad moduli ratio. The characteristics fo PSM-5 and PSM-1 are shown in Table 2.

Since the real load in a fuel rod in very complex, we have voluntarily limited our study to the mechanical effects caused by temperature variation. The loading is obtained by applying a radial pressure in the centre of the disk. The value of the pressure is adjusted to achieve an easy measurement. The tests are carried out on a TIEDMANN photoelasticimeter.

2.2 Numerical models

In order to simulate the experimentas described above axisymetric and one sixth two-dimensional numerical modes have been applied for both continuous and fragmental models. The boundary conditions of the two dimensional numerical model is shown in Fig. 1. Here L1 and L2 crack lines are assumed to be free. The joining element of MOHR-COULOMB is used to approximate the compartment of the interface between disk and ring. The numerical analysis is carried out in code INCA of CASTEM 2000 system.

3. RESULTS

3.1 Experimental results

The photoelasticity permits to determine the difference of principal stresses by the measurement of

isoclinic lines and isochromatic lines. The former is the loci of points of the same directions of principal stresses, the later is the loci of the same difference of two principal stresses. The relation can be expressed as :

$$\sigma_1 - \sigma_2 = 2, f/e$$

where :

σ_1, σ_2 principal stresses
n   fringe value
f   fringe value
e   thickness of model

It is necessary to separate the principal stresses in order to determine the state of stresses. The theoretical method based on the equations of equilibrium has been used. The equations of equilibrium are integrated progressively in the radial direction starting from a point where the stresses are known or from a load-free edge ($\sigma r = 0$).

The experimental results are divided into two parts, i.e, continuous disk and fragmental dis with ring.

3.1.1 Continuous disk with ring

The experiments are limited to the elastic domain. Three values of pressures have been used, i.e. 37.5, 47.5, 56.5 bars. Geometry and loading being symetric, the measurements are conducted along 7 radii of a sector of 75 degrees. The disk is divided into 19 points along the radial direction and the ring is divided into 4 points.

The separation of the principal stresses of the disk is realized with a point of known stress obtained form the measurment of photostress separator gage. For ring the separation begins at the load-free edge.

The average stresses distribution for the gap1 (G1) model are shown in Fig. 2. It is generally believed that the stress increase with the pressure, but that is

MEAN RADIAL AND CIRCUMFERENTIAL
STRESSES ON THE PELLET AND THE CLAD
(CONTINUOUS PELLET (G1)

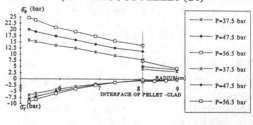

Figure 1

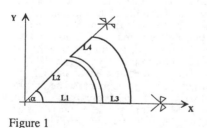

Figure 2

not true at the interface. The experiments show that the radial and circumferential stresses in the disk increase with the pressure, however the radial stress in the disk decreases. The average stress distributions for gap2 (G2) model are shown in Fig. 3. The results are similar to gap1 model.

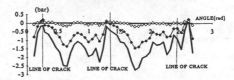

Figure 4

MEAN RADIAL AND CURCUMFERENTIAL STRESSES ON THE PELLET AND THE CLAD (CONTINUOUS PELLET (G2)

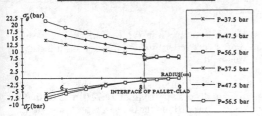

Figure 3

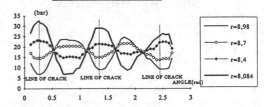

Figure 5

The average stress values at the interface are shown in Table 3 and Table 4. As compared with the measured stresses under two different gap, it can be easily found that the radial and circumferential stresses in the ring and the radial stress in the disk are getting higher as the gap increases. On the other hand, the circumferential stress in the disk seems no to be affected.

Tableau 3 Mean stresses on the interface with gap1

Pression(bar)	Radial stress σr(bar)		Circumferencial stress σθ(bar)	
	Disk	Ring	Disk	Ring
37.5	-0.4883	-0.4208	7.9640	4.0885
47.5	-0.2774	-0.5597	11.3473	5.2311
56.5	-0.0984	-0.7628	13.6876	7.4111

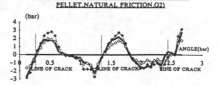

Figure 6

Tableau 4 Mean stresses on the interface with gap2

Pression(bar)	Radial stress σr(bar)		Circumferencial stress σθ(bar)	
	Disk	Ring	Disk	Ring
37.5	-1.0228	-0.8349	7.9916	7.0096
47.5	-1.0054	-0.8532	10.6237	7.0965
56.5	-0.7829	-0.8781	13.9285	7.7137

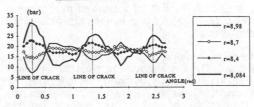

Figure 7

In summary, the photoelastic technic permits to show how the stresses are transfered from the disk to the ring. The squeeze between the disk and the ring is likely to reduce the edge pression of the disk on the ring. Because the gap the squeeze is eased, and hence the stress becomes higher with the increase of the gap.

3.1.2 Fragmental disk with ring

The pressure applied at the centre of the fragmental disk is 5.5 bars and the measurements are focused

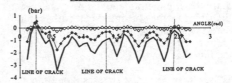

Figure 8

SHEAR STRESS ON THE CLAD ALONG THE
CIRCUMFERENTIAL DIRECTION/FRAGMENTAL
PELLET.DRY FRICTION.G2)

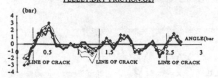

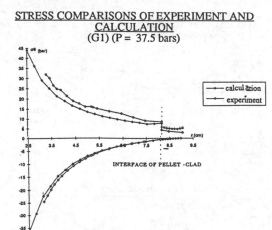

Figure 9

the two cracks, where the radial and shear stresses
reach their maximum values.

It can be seen in Fig. 5 that the stress concentration
factor is over 1.8 in the region of the cracks comparing with the skin mean circumferential stresses. If
the friction factor between the disk and the ring is
increased, the circumferential stresses tend to be
uniform, their maximum and average value remaining unchanged.

Finally, the experimental findings can be summarized as follows :

- the distribution of the circumferential stresses in
the ring is very repetitive and periodical in the cas of
fragmental disk ;

- the circumferential stresses in the ring for the
fragmental disk is approximately 21 times of that for
the continuous disk under the same pressure value ;

- the influence of friction has not appeared because
of the very little displacements.

3.2 Comparisons of numerical and experimental results

3.2.1 Continuous disk model

The comparison of stress distribution for the gap1
(G1) model are shown in fig. 10 and Fig. 11. It can
be seen that the radial stress distributions for the experiment and the calculation are in good agreement.
However, there are some differences in circumferential stress. In the disk, the difference is uniforme
except at the interface, it appears that the mesured
circumferential stress is slightly higher than the calculated one. But in the ring, it shows that the calcu-

STRESS COMPARISONS OF EXPERIMENT AND
CALCULATION
(G1) (P = 37.5 bars)

Figure 10

only in the ring. The separation of the stresses begins at the load-free edge. The ring is divided into 4
points along the radial direction.

The radial stress, circumferential stress and shear
stress distribution for the interface with natural
contact are shown in Fig. 4, Fig. 5 and Fig. 6 and
those with dry ruggedness are presented in Fig. 7,
Fig. 8 and Fig. 9.

It can be seen that the radial stresses along the circumferential direction are non-uniformly distributed.
The singularities fo the stress distribution become
more evident with the dry ruggedness of the interface. The maximum radial stress is located at the
mid area of sector. It has been found from the experiments that the radial stress can be very small ad
even possibly in state of tension in the region of
cracks. This is also found by the numerical model
calculations.

The circumferential stress distribution looks like a
sinusoid. The stress reaches its maximum value in
the region of cracks on the inner side to the ring, i.e.
the contacting surface. The two points of singularity
correspond to changing signs of moment between

STRESS COMPARISONS OF EXPERIMENT AND
CALCULATION
(G2) (P = 37.5 bars)

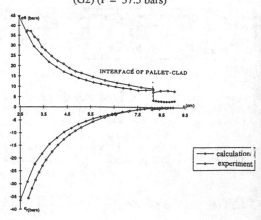

Figure 11

RADIAL STRESSES $\sigma_r \approx$ ANGLE θ IN THE CLAD (p = 5.5 bars)

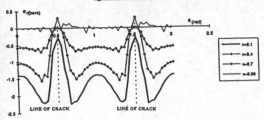

Figure 12

CIRCUMFERENTIAL STRESSES $\sigma_\theta \approx$ ANGLE θ IN THE CLAD (p = 5.5 bars)

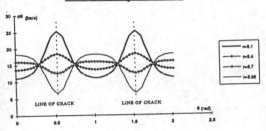

Figure 13

lated stress is greater than the measured for gap1 model, it becames smaller for gap2 model. In addition, the difference increases with the gap of the interface.

3.2.2 Fragmental dik model

The radial and circumferential stress distributions of numerical results for fragmental disk model are shown in Fig. 12 and Fig. 13.

Several phenomena, by comparing with experimental results, can be seen as follows :

1. The main features of the experiments being repetitive and periodical have been successfully reproduced by the numerical models.

2. The minimum of radial stresses are in the same order of magnitude.

3. The calculated mean and maximum circumferential stresses are smaller than the experimental results, which have also been found in continuous disk model with gap2. This may becaused by the difference of the pressure applied and mesured. That means the measured pressure is less than the applied one. The same orders of mean and maximum cir-

cumferential stresses as the experiment are obtained by calculations with P = 6.6 bars.

The effect of frictional coefficient on the maximum circumferential stresses has been numerically analysed. The result shows little increase of the stresses when the coefficient becomes higher. This may be explained by the fact that the circumferential displacement between the fragmental disk and the ring is small.

4 CONCLUSIONS

It is evident that the stress distribution is closely related to the pellet state, continuous or fragmental.

For uniformly distributed cracks (6 craks) without dry ruggedness, the circumferential stresses in the ring are 21 times of the stresses for the continuous one. In addition, the disk cracks results in high modulation of the circumferential and radial stresses on both sides of the ring. It is important to notice that the localised circumferential stress concentration factor due to the disk cracks can be over 1.8 comparing to the mean value obtained with a crackless disk.

The numerical and the experimently results are in good agreement. It may be concluded that the simulations of MOHR-COULOMB at the interface of disk and ring in the numerical models are appropriate.

It seems that the friction influence on the local stress concentration at the interface is not obvious.

The photoelastic measurements open up new horizons to the experimental method on the field of interfacing.

ACKNOWLEDGMENT

The work described in this paper was carried out in the Concrete and Structures Laboratory of INSA, Lyon, France, and was financed by EDF/SEPTEN under the research contract n° ND 2258 MS. The assistance to the laboratory staff and the financial support are gratefully acknowledged.

REFERENCES

Modélisation thermomécanique de la pastille combustible - SEPTEN réf. MS/90.013 (SC/009) PPL/JCC/ND/MCP

A. YU, S.P. WALKER & R.T. FENNER - "Pellet-clad bonding during PCMI" - Nuclear Engineering and Design 121 (1990) 53-58

Recent Advances in Experimental Mechanics, Silva Gomes et al. (eds) © 1994 Balkema, Rotterdam, ISBN 90 5410 395 7

Methods for the investigation of composite materials thermal deformation by optical dilatometers

V.A. Borysenko & L.I. Gracheva
Institute for Problems of Strength of the Academy of Sciences of Ukraine, Ukraine

N.D. Pankratova
Institute of Mechanics of the Academy of Sciences of Ukraine, Ukraine

ABSTRACT: To study composite materials including destructing ones that operate under conditions of high temperatures, the procedures and unique set-ups are described which make it possible to determine thermal strain of specimens along one or three mutually orthogonal axes simultaneously by a non-contact optoelectronic method over the temperature range from 300 to 3000 K in different gas media, as well as under changes of heating rates. Experimental data for linear and volumetric thermal strain are presented, a method is given for calculation of structural elements made of composite materials with the account taken of the results obtained.

When studying coefficients of thermal linear expansion (CTLE), α, for destructing composite materials operating under conditions of high and superhigh temperatures not only their thermal expansion but also compression (shrinkage) associated with phase and structural transformations in polymer matrices are of essential importance. Careful analysis of all these changes is difficult since coking materials are multicomponent systems wherein a large number of interconnected processes occur simultaneously. Therefore, curves for coefficients of thermal expansion α for similar class of materials are of complex (alternating and polyextremal) nature. To measure thermal strains in such class of materials, a special equipment is necessary which takes into account the characteristic behaviour of polymer and carbonized materials in heating.

The present paper describes the procedures for determining CTLE for coking and carbonized nonmetallic materials developed at the Institute for Problems of Strength of the Academy of Sciences of Ukraine. The setups created provide for the investigation of thermal strains of nonmetallic composite destructing materials over the range of 20...3000 C. There exists the possibility of studying the effect of heating rates, chemical composition of the environment and other factors simulating real service conditions for structural components on the deformation process.

The experimental setups i.e. dilatometers are comprised of a high temperature furnace, system for specimen heating, as well as its temperature control and measurement, a system for specimen thermal strain measurements, a system to exert and measure gas pressure inside the furnace or to blow down it with an inert gas, as well as a system for cooling the furnace components.

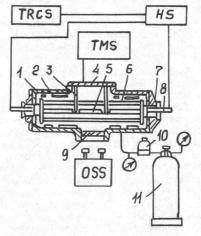

Fig. 1. Block diagram of DTM-1. TCCS - temperature condition control system; TMS - temperature measuring system; HS - heating system; OTS - optical tracing system.

Figure 1 shows a block diagram, of a dilatomer DTM-1 which permits studying the strain properties of specimens made of heat-protective materials under conditions of uniform and unilateral heating under different pressures of the medium and heating rates above 100 K/min. Argon, nitrogen, air are used as a working medium (Tretyachenko 1978).

The above block diagram of the dilatometer illustrates its operation on uniform heating of the specimen. The data on the strain of heat-protective materials in the case of unilateral heating typical of real working conditions for heat-protective coatings

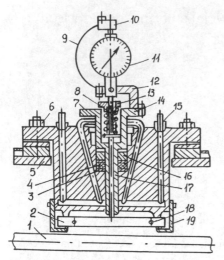

Fig.2. Specimen arrangement in the DTM-1 dilatometer furnace with one-sided heating.

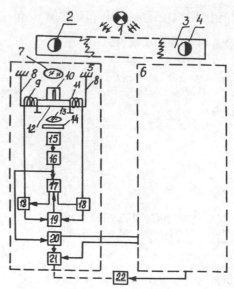

Fig.3. Block diagram of optical follow-up system OSS-50-1300.

are obtained with the use of a carborundum rod placed under the specimen.

Figure 2 schematically shows the specimen mounting in the test chamber on unilateral heating. The cover-grip of the specimen is bolted to the chamber wall 5.

To create the thermal gradient along the thickness of the specimen 2 placed in sliding supports 19 directly over the carborundum heater I the specimen is cooled by flowing water on the reverse of heating side (input unions 15). Horizontal strains are measured by bench marks 18 made in the specimen at the required distance from the heating surface. In this case the specimen is gripped by rod 17 when its round platform rotates clockwise. The sealing system which consists of backup rings 3 and rubber-fabric cups 4 protects the test chamber from loss of sealing when testing under pressure. Measuring rod 8 located inside loading rod 17 serves for transmission of vertical displacements of the specimen when deflecting from the specimen heating surface to dial gauge 11 and simultaneously to the round cramp with resistance strain gauge 9 which is fixed on dial gauge 11 with arm 12 and clamp 10 (nuts 13 and 16 are locking ones). Spring 7 serves as a compensator of gas pressure and prevents rod 8 from pushing out during pressure testing.

On heating, specimen strains are measured automatically by means of specially developed optical follow-up system OSS-50-1300 (Marasin 1983 a,b). Its operation principle is shown in Fig.3.

The measuring optical follow-up system includes light source 1, two identical measuring channels 5 and 6, adder 22. Each measuring channel has lens 7, differential solenoid with windings 9, 10 and core 12 with slit diaphragm 10 attached to it which is suspended by springs 8, collecting lens 13, photoelectric converter 15, photocurrent limiting

amplifier 16, control signal driver 17, current switches 18, search generator 19 and displacement signal shaper 20. The narrow-band light filter 14 is mounted ahead of the input of the photoelectric converter. Specimen 3 has two marks 2 and 4, each is a direct line separating two adjacent areas 20 and 21 of the test specimen which have a good optical contrast range (for example, they may be painted black and white, respectively). The specimen edge against a contrast background may be used instead of the mark. Each measuring channel of the system provides tracking one of the marks. The mark image is projected on the slit diaphragm plane by the objective. The collecting lens directs the light flux toward the input of the photoelectric converter whose output signal arrives at the input of the limiting amplifier which forms voltage of the same level of positive or negative polarity. If the diaphragm is on the white area of the specimen image, then the control signal is applied to that solenoid winding which provides such a displacement of the core with the slit diaphragm fastened on it when the diaphragm will move to the black area of the specimen image passing the mark image. At the moment the diaphragm passes through the mark image, the control signal is applied to the other winding of the differential solenoid giving rise to the diaphragm displacement toward the opposite side, i.e. to the white area of the specimen image. Thus, the continuous scanning occurs across the mark image by the diaphragm.

The d.c. voltages corresponding to mark displacements are applied from the inputs of both measuring channels to the adder where they are algebraically added together. The recording device (e.g. electronic potentiometer) makes continuous record of the

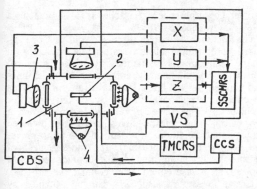

Fig.4. Block diagram of a dilatometer for composite materials, type DCM-3. 1 - testing chamber; 2 - specimen; 3 - optical unit; 4 - illuminant; 5 - SSCMRS - system for measuring and recording changes in the specimen size; VS - vacuum pumping system, TMCRS - temperature measuring, control and recording system; CCS - chamber cooling system; CBS - chamber blowdown system.

specimen strain on coordinates $\Delta\ell$ (displacement) - T (temperature).

A high-temperature dilatometer DCM-3 is developed at the Institute for Problems of Strength of the Academy of Sciences of Ukraine. It is designed to study thermal strain processes in anisotropic composite materials over the temperature range from 293 to 2000 K in air or neutral medium and in vacuum as well. An experimental prototype of DCM-3 consists of a high temperature furnace, a non-contact system for measuring the specimen volumetric thermal strains, a system for heating and recording systems for measuring and recording, the specimen temperature, a system for vacuum pumping and a system for cooling the furnace parts and assemblies. A block diagram of the DCM-3 is shown in Fig.4 (Marasin 1991).

The furnace comprises a body with side covers, current input leads, a graphite heater and thermal graphite shields. The furnace is equipped with six illuminators arranged in pairs in three mutually orthogonal directions which provide free access for measuring the specimen thermal strains by non-contact optical system. The current lead system design makes it possible to seal the furnace and to compensate the heater thermal expansion.

The specimen is heated by radiation with the help of a graphite heater with the power applied from a secondary winding of a power transformer.

The primary winding of the power transformer includes a voltage regulator controlled by a high-accuracy temperature regulator whose input is connected to a thermocouple, type IPP. The heating system has control circuits of lamp signalling when the heater turns on and of audible signalling (ringer) when water pressure in the furnace cooling system drops.

The specimen temperature measuring system over the range of temperatures from 293 to 2000 K contains a thermocouple and an automated potentiometer which records the current value of the specimen temperature.

Provision is made to rise the test temperature up to 3000 K. In this case the specimen temperature is measured by a pyrometer and the measuring circuit

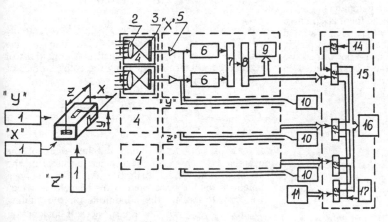

Fig.5. Block diagram of non-contact electron-optical system. 1 - illuminant; 2 - objective, PCCD - photosensitive charge-coupled device, 4 - optical unit; 5 - videosignal amplifier; 6 - videopulse processing unit; 7 - subtractor; 8 - code converter; 9 - digital panel; 10 - oscilloscope; 11 - external code; 12 - register; 13 - timer; 14 - digital panel; 15 - digital printing control unit; 16 - digital printing device, 17 - register control unit.

includes a pyrometric lamp supply unit, a digital device and a digital printing device.

The furnace blowdown by an inert gas is accomplished from a cylinder through a reductor. Air and gas are pump down from the furnace by the vacuum pump.

The dilatometer comprises a system NMCS-3 which is designed for non-contact measurement of the linear change in the specimen size along three mutually orthogonal axes. A block diagram of the system, NMCS-3 is shown in Fig.5. The NMCS-3 is an electron-optical measuring system and structurally consists of three identical measuring channels, each having two illuminants, optical and electronic units. The illuminants and the optical units are arranged in pairs, on the opposite sides of the testing chamber with illuminators. The electronic units are mounted in a unified rack where a control unit with a timer, an digital printing device, an oscilloscope and supply units are mounted as well.

The linear elongation is determined by a value of displacement of working faces image of the specimen placed into the test chamber and by their subsequent algebraic addition. The objectives project images of the specimen faces in the scale of 1:1 on two photosensitive charge-coupled devices (PCCD) each having 500 cells arranged in a row with a pitch of 24 μm. To provide for the required image contrast the LEDs used as illuminants are directed toward the faces.

The control unit generates the required constant and pulse voltages to supply PCCD, as well as clock signals for other devices. Outputs pulses from both PCCDs arrive to the processing unit (PU) where the value of the linear elongation of the specimen is extracted and is sent to the panel, the digital printing device and external devices (e.g. a computer).

The results of studies into linear thermal strain of destructing polymer composites are demonstrated with an example of a carbon- and metal-filled plastic material (Fig.6). Dilatometering is performed on DTM-1 system with heating rates being varied in neutral gas atmosphere (Gracheva 1981, Gracheva 1982, Tretyachenko 1983).

The analysis of cubic thermal strain of phenol-based carbon-filled plastic material is presented in Fig.6. Dilatograms obtained by three axes of anisotropy simultaneously are shown by solid lines and curves of the linear thermal strain for the same specimen material are shown by dotted lines for comparison. The curves of relative thermal strain of the carbon-filled plastic investigated are obtained under equal conditions: the neutral gas atmosphere, the heating rate (50 K/min); the measurement of thermal strains is automated, non-contact; errors of the strain measurement are comparable (2-3%). Thermal stresses in structures can be obtained by calculation making use of the experimental data obtained.

When obtaining thermal strain coefficients the effect of heating rate, gas medium composition, anisotropy, etc. on thermal stresses in composite laminate sheaths was evaluated from accurate solutions to three-dimensional elasticity theory problem for a non-uniform anisotropic body in a quasistatic setting. The method of calculation is based on the combination of equations of elasticity theory, thus of heat conductivity and the method of numerical analysis. The body's material obeys the Hook generalized law with the Dugamel-Neuman hypotheses taken into account (Pankratova 1992).

The temperature field for the i layer of the cylinder is defined from the equation of heat conductivity which in the cylindrical coordinate system is expressed by

$$K_r^i \frac{d}{dr}\left(r\frac{dT^i}{dr}\right) + rK_z^i \frac{d^2T^i}{dz^2} + \frac{K_\theta^i}{r}\frac{d^2T^i}{d\theta^2} = 0,$$

where $K_r^i = K_r^i(r)$, $K_z^i = K_z^i(r)$, $K_\theta^i = K_\theta^i(r)$ are the coefficients of heat conductivity acting in the directions r, z, θ. It is assumed that conditions of rigid conjugation of layers along the entire surface of the contact are fulfilled

$$T^i = T^{i+1}, K_r^i \frac{dT^i}{dr} = K_r^{i+1} \frac{dT^{i+1}}{dr}.$$

A case is considered when the butt-ends of the cylinder $z = 0$, $z = 1$ do not displace in this plane and are free from normal load. By taking the resolving functions for the basic ones with the help of which we can formulate the conditions on the limiting surfaces $r = r_0$, $r = r_N$ and the surfaces of conjugation of the layers r_i. and fulfilling transformations in the initial equations of elasticity and the equations of heat conductivity, the resolving system of equations after division of variables is expressed in the form (Gracheva 1982, Grigorenko 1992).

$$\frac{d\bar{\sigma}_{kn}^i}{dr} = C_{kn}^i \bar{\sigma}_{kn}^i + \bar{f}_{kn}^i,$$

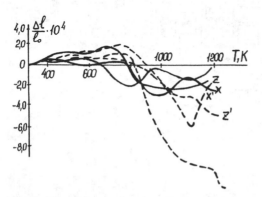

Fig.6. Dilatograms of carbon-filled plastic material which are obtained by three coordinates simultaneously.

840

$$\overline{\sigma}^i_{kn} = \left\{ \sigma^i_{r,kn} \cdot \tau^i_{rz,kn} \cdot \tau^i_{rx,kn} \cdot u^i_{r,kn} \cdot u^i_{z,kn} \cdot u^i_{x,kn} \cdot T^i_{kn} \cdot T'^i_{kn} \right\};$$

$$C^i_{kn} = \left\| c^i_{mq,k}(r) \right\|; \quad f^i_{kn} = \left\{ f^i_{1,kn} \cdot f^i_{2,kn} \cdot \ldots \cdot f^i_{8,kn} \right\},$$

$$m, q = 1, 2, \ldots, 8.$$

Here, σ_r is the radial stress, $\tau_{rz} \tau_{rz}$, $\tau_{r\theta}$ are the tangential stresses, u_r, u_z, u_θ are radial, axial and circumferential displacements, respectively, T is the temperature, T' is the flow of temperature. The matrix elements C^i_{kn} depend on the mechanical characteristics of the layer materials. Integration of the equations is made by means of the steady numerical method allowing to receive the solution with high degree of precision. Selection of the basic values by which contact conditions for layer conjugation are formulated makes it possible to automatically and continuously obtain solutions for the prescribed number of layers.

CONCLUSION

The procedures are developed and the set-ups are created to investigate linear and cubic (by three mutually orthogonal coordinates simultaneously) thermal strains of heat-protective coking and carbonized materials in the temperature range from 20 to 3000 K. Processes of measuring thermal strains are automated by the use of non-contact electron-optical devices and converters which exclude mechanical action on the specimen being investigated. On the basis of the experimental data obtained, the calculations of stress-strain state of specimens made of laminar composite materials in a 3D formulation are performed.

REFERENCES

Gracheva, L.I. 1981. Dependence of thermal trains of carbon- and carbonmetalplastics upon the specimen thickness. Strength of Materials 8: 1018-1022.

Gracheva, L.I. 1982. Peculiarities of stress-strain curves of carbon plastics at high temperatures. Strength of Materials 3: 390-393.

Grigorenko, J.M., A.T.Vasilenko & N.D.Pankratova 1992. The problems of elasticity theory of non-homogenious bodies. Kiev: Naukova Dumka (in Russian).

Marasin, B.V., A.G.Malyj, N.A.Fot, L.I.Gracheva & V.V.Sinchuk 1983. Photographic follow-up system for measuring object strains. Inventor's certificate 998858 (USSR).

Marasin, B.V., A.G.Malyj, N.A.Fot, L.I.Gracheva & V.V.Sinchuk 1983. Photographic follow-up measuring system OSS-50-1300 for measuring thermal strains in specimens made of destructing composites. Strength of Materials 9: 1341-1345.

Marasin, B.V., V.V.Ruban, L.I.Gracheva et al. 1991. Setup for dilatometric tests at high temperatures. Inventor's certificate 1656428 (USSR).

Pankratova, N.D. 1992. The calculation of anisotropic structural elements in space formulation. Proc.XXXI Symposium on Modelling in Mechanics: 327-334.

Tretyachenko, G.I. & L.I.Gracheva 1978. Device for studies of deformation properties of thermal-resistant materials under conditions of different media and pressures upon the uniform and one-sided heating. Strength of Materials 7: 857-860.

Tretyachenko, G.N. & L.I.Gracheva 1983. Thermal strain in non-metallic destructing materials. Kiev: Naukova Dumka (in Russian).

Recent Advances in Experimental Mechanics, Silva Gomes et al. (eds) © 1994 Balkema, Rotterdam, ISBN 90 5410 395 7

Mechanical structure subjected to cyclic thermomechanical loading
Interaction between ratcheting and creep

O. Philip & M. Cousin
INSA, Lyon, France

ABSTRACT : Several phenomena can occur in a metallic structure subjected to cyclic thermomechanical loading in the plastic domain. In addition to fatigue and damage phenomena which can happen over a long periods, we can observe during shorter periods the following phenomena like :
- a regular increase of the strain after each cycle : ratcheting,
- strain increase at constant load : creep,
- stresses decrease at constant strain : relaxation,
- dependence of the material response on the loading rate : viscosity
A good simulation of the cyclic behaviour of the type of structures requires to take into account of all these phenomena which often coexist. Among these, creep at low temperature should not be neglected.

1 INTRODUCTION

The increasing demands of nuclear industrial needs and the necessity to take into account of phenomena induced by thermal loads had led us to imagine a specific experimental device allowing to test a steel structure under cyclic thermomechanical loading.

Such a structure can undergo strain increase with each cycle, refereed to as "progressive strain" or a "ratcheting".

An other phenomenon can occur : the creep depending on time and temperature.

The development of modern technics requires the examination of the consequences of such phenomena with respect to safety.

The first part of this work is devoted to the presentation of experimental analysis of a metallic structure subjected to a thermomechanical cyclic loading. The results show the necessity to take into account the creep at room temperature of the 316 L stainless steel.

In the second part, we study the creep at room temperature in order to carry out some calculus.

2 THE "BITUBE" TEST

2.1 Geometry of the structure

The "bitube" structure is composed of two concentric tubes rigidly connected at their ends in order to have equal elongation at all times (Taleb 1991). The tubes are made of 316L stainless steel, ant they are manufactured such as residual stresses are avoided. The geometry of the tubes is shown on the figure 1.

The two tubes have the same length and section.

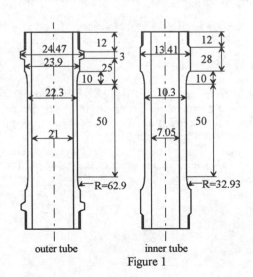

outer tube   inner tube

Figure 1

2.2 Loadings

The loading intensity is such that the tested material is stressed in the plastic domain. The applied loading has two components :
- a constant mechanical one
- and a cyclic thermal one

The primary mechanical load is an axisymetric tensile one. The thermal load is obtained by periodically heating the external tube while maintaining the internal tube at room temperature (see figure 2).

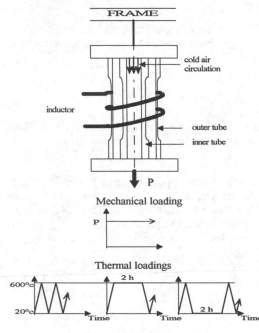

figure 2 : the loadings

2.3 Measurements

In order to see the evolution of the structure, four types of measurements are carried out : temperature, strain, displacement and force. The measure points are indicated in the figure 3.

Note that the stress state in the tubes is obtained by simply measuring the strain on the upper part (the thicker cross-section) as long as the latter remains elastic.

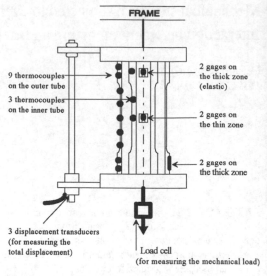

Figure 3 : measurements

2.4 Tests

The external tubes is subjected to a thermal cyclic loading inducing a variation of stress ΔQ in the structure (ΔQ is the variation of the secondary stress elastically calculated). The inner tube is at room temperature. A constant primary P load is applied on both tube.

The bitube structure allows the record of the strain variation created by the application of the loads (P, ΔQ) and to study the phenomenon of ratcheting. To give prominence to different forms of creep, we have done several tests changing the time parameter and the intensity of the loads.

In another paper in the S.M.I.R.T.12 (Philip,1993), we reported tests with P = .50 MPa and ΔQ = 675 MPa (σ_{max} = 600°C) and we see the necessity of taking into account the creep at low temperature for the 316L stainless steel.

In this paper we report the results of our tests with P = 0 and the thermal load applied   is :

TEST	CYCLE	LOADS
DPF4		$\theta_{max.}=600°c$ $\Delta Q=675MPa$
DPF5	2 h	$\theta_{max.}=600°c$ $\Delta Q=675MPa$

In the following, the strain is defined as the axial strain in the central part of the inner tube. The combination of the two types of load create different types of stress state in tension and in compression in the tubes (see figure 4).

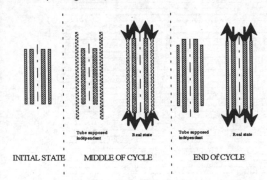

INITIAL STATE   MIDDLE OF CYCLE   END Of CYCLE

Tube supposed independant Real state Tube supposed independant Real state

PHYSICAL BEHAVIOUR OF THE STRUCTURE
Figure 4

With P = 0, in the middle of the cycle the outer tube yells in compression (his mechanical characteristics are lower because of the high temperature). Consequently, the evolution of the structure is controlled in compression by the outer tube. The length of the structure decreases at every cycle.

If we look at the results for the first cycles, we see the strain decrease during the holding time and that the increment of strain after each cycle is more important in the test with holding time .(see graph n°1)

In conclusion, creep of external tube in compression at high temperature is the most important.

We carried out other series of tests with P = 120 MPa and ΔQ = 350 MPa.

TEST	CYCLE	LOADS
DPF6		$\theta_{max.}=350°c$ $\Delta Q=350MPa$
DPF7	2 h	$\theta_{max.}=350°c$ $\Delta Q=350MPa$
DPF8	2 h	$\theta_{max.}=350°c$ $\Delta Q=350MPa$

The results are shown in the graph n°2 :
We see that :
- the stabilised strain is greater for the test with holding time at high temperature than for the other tests. Consequently there is creep at room temperature in traction for the inner tube.

In the same way, in the test with holding time at low temperature, we have creep at room temperature in traction for the outer tube.

3 CALCULATION

The numerical approach is based on the hypothesis of the material behaviour simulated by a mathematical modelisation.

Consequently, if we want a good representation of the behaviour, w must have a complex model.

We can realise calculus using either a classical multilinear model with creep law at all temperatures or a more sophisticated model which takes into account real behaviour of materials at all temperatures.

Chaboche's model can be considered as a good example of a model which allows a qualitative representation of phenomena sufficiently near to reality.

We saw in the paper of the S.M.I.R.T.12 (Philip 1993) that Chaboche's model have great possibilities for time dependant phenomena if these are properly quantified.

Actually, the creep a high temperature for the 316L is quantified. But at room temperature, there are no data because in the codes the creep at room temperature is neglected.

In 1985, P.S. White studied the creep for the 316L material but the characterisation has been done for steel sheet which have different mechanical characteristics in comparison to our material.

In order to quantify the creep at room temperature in tension and in compression, we made tests on the tubes off the bitube's structure.

We applied to the tube $\sigma_1=0.8\sigma_e$, $\sigma_2=0.9\sigma_e$, $\sigma_3=\sigma_e$ with holding time at each loads.(see graph n° 3 and 4)

We see that the result are similar for the inner and the outer tubes.

We found that there is creep since $\sigma_1=0.8\sigma_e$, and the strain by creep increase with the increase of the stress.

We will make other tests with different loads :
- σ_e in order to compare the results to the test with holding times at σ_1, σ_2 and σ_3 , we will compare on the accumulated strain, in order to see if $\varepsilon(\sigma_1,\sigma_2,\sigma_3)=\varepsilon(\sigma_e)$ or not.
- tests in compression to compare creep in tension and compression.

Then we may identify the parameters of some creep law as, for example as the Lemaitre Nadaï Hoff law.

4 CONCLUSION

In conclusion we see that if we want to predicate the evolution of a metallic structure subjected to

Graph n°1 : Tests DPF4-5 : curve strain = f(Time) first cycles

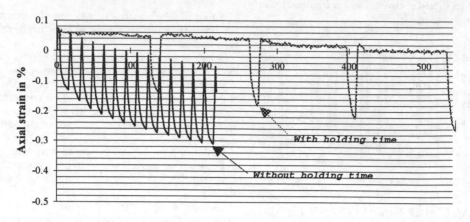

With holding time

Without holding time

Time in mn

Graph n°2 : curve strain=f(number of cycles)

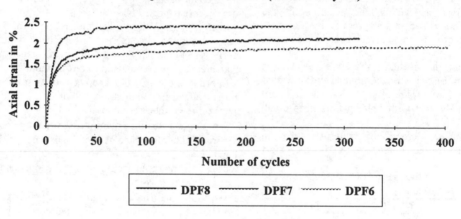

———— DPF8 –––––– DPF7 ········· DPF6

Graph n°3 : 316L stainless steel (20°c)
Outer tube

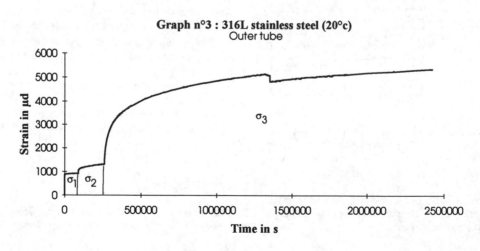

Time in s

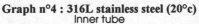

Graph n°4 : 316L stainless steel (20°c)
Inner tube

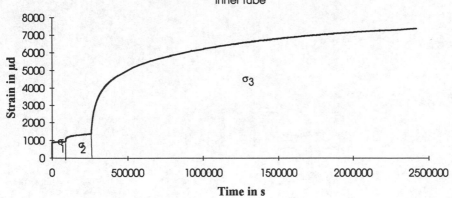

complex loads, we must know all the parameters which control the material's behaviour.

For the 316l stainless steel, the creep at room temperature can not be neglected.

REFERENCES

Chaboche, J.L., Nouailhas, D. 1989. An unified constitutive model for cyclic viscoplasticity and its applications to various stainless steel. J. of Engineering Materials and Technology. Vol. 111.

Philip, O., Cousin, M., Taleb, L.,1993. Interaction between ratcheting and creep. S.M.I.R.T.12, paper L08/5. Stuttgart.

Taleb, L.,1991. Structure métallique sous chargement therrmomécanique. Effet des surcharges mécaniques de courte durée. Thèse INSA de Lyon, Génie Civile, Structure, 273 p.

White, P.S., 1985. The effect of cold creep on finite element calculations of Lyon test. CEC. inelastic benchmark study step 2 phases 2 and 3, Whestone, Leicester, ERC(W) 12.0634, 38p.

Recent Advances in Experimental Mechanics, Silva Gomes et al. (eds) © 1994 Balkema, Rotterdam, ISBN 90 5410 395 7

Expertizing and monitoring of r/c structures in corrosive media

A.Catarig, L.Kopenetz & P.Alexa
Technical University, Cluj-Napoca, Romania

ABSTRACT: Monitoring the corrosion affectiing the R/C structures using special technique is presented together its specific field of application. Two procedures – ring method and electric resistance method – applicable to R/C and prestressed elements and several practical results obtained from practical investigations conducted by authors are also presented.

1 INTRODUCTION

Monitoring of R/C structures in corrosive media, aiming at diminishing the damaging risk is an acute problem that has to be analyzed and interpreted in a more general frame, connected with the global quality of structures.

Regarding the "quality of structure" as the measure of satisfying the users' exigency, expressed in technical terms through performance criteria and associated levels, it is necessary to take into account the following aspects:

1. Due to the intense aleatory feature of the action of corrosive medium and of the phisical – mechanical and deformability properties of R/C structures, the existence of the damaging risk is an objective reality. Complete removing of this risk is in practice impossible while, its reduction beyond a certain limit is economically unreasonable.

2. The extent of the damage i.e. nature, frequency and localization of the physical degradations versus either the available strength capacity level, or the vulnerability degree.

Viewing the above grounds and taking into account the large volume of structures affected by corrosive media, a systematic and continuous activity aiming at reducing the failure risk has to be carried out (Hauptmanns 1991, Misteth 1968).

Such an activity involves the following aspects:

* Examination and monitoring of R/C structures aiming at estimating the failure risk.

* Designing the repairing works.

* Performing the repairing /strengthening works.

The present paper deals with the first of the above aspects.

2 EXPERTIZING R/C STRUCTURES IN CORROSIVE MEDIA

When estimating the bearing capacity of a R/C structure affected by corrosion, the usual structural analysis methods are no longer valid. That is why in very many cases the assesment of the safety level of the structure has either not been done or it has been done via investigator's experience (Catarig 1992). Usually the past behaviour is extrapolated over the future. Such "methods" could very ofen be very dangerous (Bob 1989).

The big difficulty of the assesment task comes from the difficulty of a veridical model of the corroded structure to be used in the structural analysis. The models of a corroded structure are either very sophisticated, becoming inoperative for the current cases, or too approximate leading to results remote from the real behaviour.

The expertizing begins by ascertaining the corrosive actions.

The corrosive actions may come from:

1. Chemical aggressions (the action of carbon dioxide that brings forth a gradual carbonation in the concrete; action of H_2SO_4, HCl, NH_4 and H_2NO_3 that generate disintegration and exfoliation of the concrete and rusting with exfoliation of the reinforcement) (Bob 1990).

2. Microbiological aggressions (certain heterotropic bacillus-bacteria, microscopical fungus, sulphuricant and nitricant bacillus generate an acide action at the concrete surface).

3. Water and marine media aggressions.

4. Actions of the dispersive electric current due to electric transport equipment supplied with c.c.

The next step of an expertize consists of rendering evident the concrete corrosion. It can be achieved through:

- assessing the alterings of mechanical strength,
- assessing the chemical and physical changes,
- evaluating the Young's modulus,
- evaluating the geometrical changes.

With the data obtained from the above assesments, a structural analysis using oriented FEM (Kopenetz 1983) can be carried out.

3 STRUCTURAL ANALYSIS

The analysis aims at studying the structure for both, the current moment and over a period of time applying the technique of numerical simulation of the concrete damage due to corrosion.

The computer program SACORC-01 (Structural Analysis of Corroded R/C Structures) developed by the authors, with the flow-chart shown in figure 1 is a very useful product for such an analysis.

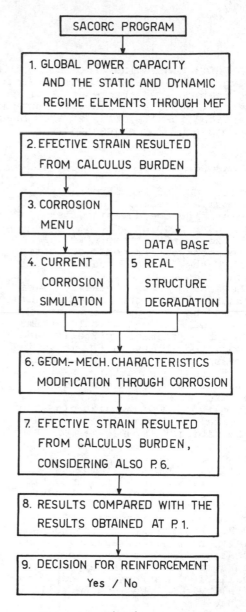

Fig.1

Figure 2 shows an example for corrosion simulationof a R/C beam investigated in three different situations:

1. Concrete partially affected by corrosion having the Young's modulus and geometry unmodified.

2. Concrete globally affected by corrosion having unmodified geometry and 75% of the initial Young's modulus.

3. Concrete partially affected by corrosion with the beam geometry modified, in two

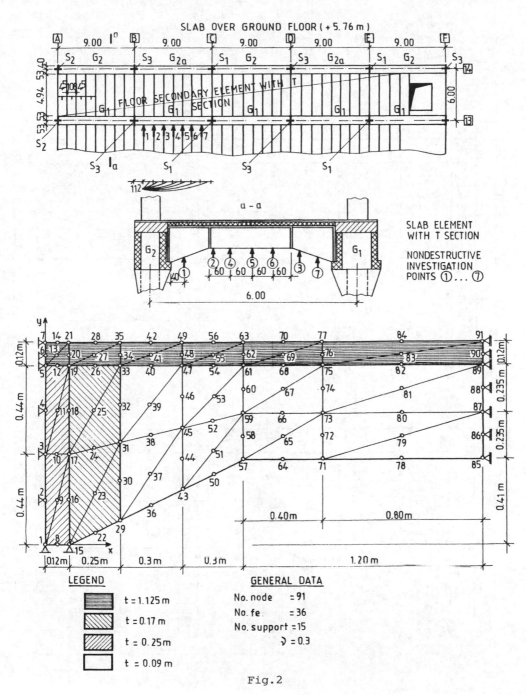

Fig.2

851

different cases (Fig.3).

An important conclusion of these analysis is that a change of maximum 20% in the bending stifness EI does not affect the structural safety (Catarig 1991).

4 MONITORING

According to the objectives, field of application, duration, specificity, technical means and solutions, the monitoring of R/C structures in corrosive media may take two forms: current monitoring and special monitoring.

The current monitoring is usually carried out for the entire structure. It is done visually or by using simple technical means.

The special monitoring is carried out for elements or their parts of increased rôle in

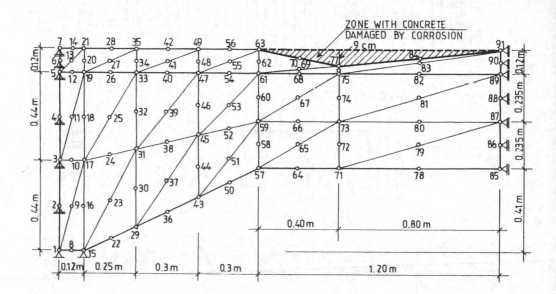

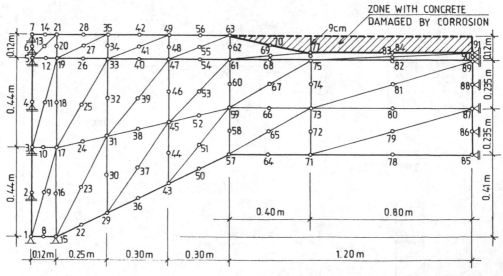

Fig.3

the structural behaviour. It requires complex technical means.

4.1 Monitoring the critical reinforcement R/C structures

The critical reinforcement is that reinforcement whose corrosion may lead to a critical state of the structure in what regards its safety (Denison 1991, Diem 1982, * * * 1989). Such cases can be met at the chemical factories, in the metalurgical industry, nuclear power plants, dams, large water tanks, long span bridges, offshore structures, etc.

The paper presents a monitoring procedure applicable to corrosion phenomena characterized by a decrease of concrete density and its pH and, also, to corrosion phenomena accompanied by concrete expansion.

In the first case due to the reduction in concrete density and of its pH, corrosive micropiles are generated leading to the precipitation of a solid phase, - hydrated iron oxide -the rust- in the enlarged concrete volume.

In the second case the concrete expansion leads to internal pressures due to the hydrated salt that crystallizes in the concrete pores.

The new investigation procedure called "the ring method" applicable to the last cases is based on this very change in the concrete volume.

The critical reinforcement has to be provided with rings equiped with tensometric gauges that are, in their turn, protected against acid action. Any change in the volume is instantaneously signaled (Fig.4).

4.2 Monitoring the corrosion of the cables of prestressed R/C elements

Monitoring of prestressing cables of the prestressed R/C elements is a special monitoring and it is needed in order to assure the general safety of the structure. The prestressing cables have an increased sensibility to corrosion because their very making favours the standstill of humidity. Monitoring and controlling the progress of degradation due to corrosion of the prestressing cables - that changes the physical - mechanical properties and, therefore, the stress and strain states of the structure - is one of the most delicate problems of structural analysis.

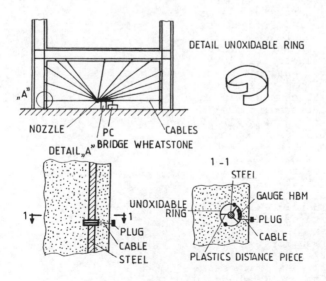

Fig.4

Currently, the monitoring of prestressing cables of R/C elements is done using Rontgen or radiation technique, or by potential method.

The ultrasonic procedure is only applicable to pretensioned elements with only one cable in the tube and having a diameter over 10 - 16 mm.

Yelding good results, the authors applied a technique "electric resistance method". The method is based on the fact that corrosion leads to a decrease of the reinforcement cross section. Therefore, in the absence of intercrystalline corrosion, the diminishing of the cross section leads to an increase of the electric resistance.

In the case of intercrystalline corrosion, the cross section area is not significantly changed, but the specific resistance increases.

Monitoring implies the comparisson of the current measured values with the initial values. The recorded differences signal the outset of corrosion.

REFERENCES

Bob, C. 1989. Testing of quality, safety and durability of constructions. Facla Publishing House, Timisoara (in Romanian).

Bob, C. 1990. Some aspects concerning corrosion of reinforcement. In Proceedings of the Conference THE PROTECTION OF CONCRETE, Dundee.

Catarig, A. & Kopenetz, L. 1992. Investigation, analysis and restauration of reinforced concrete structures in aggresive surroundings. In Proceedings of the FIP'92 Symposium 3, p.59-64, Budapest.

Catarig, A. & Kopenetz, L. 1991. Numerical simulation of concrete corrosion. Journal of Construction Quality 4/5, p.44 - 46, Bucharest (in Romanian).

Denison, C.A. & Harold, R. 1991. Concrete structures: materials, maintenance and repair. John Wiley & Sons, New York.

Diem, P. 1982. Zerstrorungsfreie prufmethoden fur das bauwesen. Bauverlag GmbH, Wiesbaden, Berlin.

Hauptmanns, U. & Webwr, W. 1991. Engineering risks. Springer - Verlag, Berlin, Heidelberg, New York.

Kopenetz, L. 1983. Research report 1. Technical University, Cluj-Napoca (in Romanian).

Misteth, E. 1968. Some safety problems. In Proceedings of the 8 th Congress IABSE, New York.

* * * 1989. Summary report on safety objectives in nuclear power plants. EUR 12273EN, Brussels.

Recent Advances in Experimental Mechanics, Silva Gomes et al. (eds) © 1994 Balkema, Rotterdam, ISBN 90 5410 395 7

Experimental study and numerical simulation of large ore mills

Nicolae Constantin
'Politehnica' University of Bucharest, Romania

ABSTRACT: The paper presents some aspects concerning the research ment to allow the production of a new generation of ore mills in Romania, required by economic reasons. A large ore mill was first designed using the conventional calculus conducted until then for smaller ones. Structural problems appeared at this type of equipment elsewhere imposed experimental tests made by strain gauge measurements. They revealed high stresses in several areas of the ore mill structure where, in fact, failures have occured later on. A modification of the mill design became necessary, which was possible using numerical simulation.

1 INTRODUCTION

In modern milling operations, economics of scale often dictate extremely large autogenous or semi-autogenous ore mills. The usual approach has been simply to scale up conventional mill designs to meet the required production volume. But the complex structure of the first designed large mill was severely stressed in the milling process by the inside load of about 200 tons, his own weight of more than 250 tons and the gearless drive wrap arround electric motor, weighting about 150 tons. As in the meantime bad accidents happened in several plants the world over, careful experimental measurements made in situ became necessary. The most adequate method seemed to be strain gauging, which revealed high stresses in some areas of the mill structure, where failures occured later on. In this situation, the mill had to be modified. This time, numerical simulation by FEM was used to provide a safe design.

2 EXPERIMENTAL

As strain gauge measurements on ore mills were not made until then in Romania, some exploratory experiments were performed on a smaller mill(6 m shell diameter), which was running with no structural problems since more than 15 years. The aim was to fit up the measurement technique, in order to meet the more severe requirements imposed by the same operation carried out on the large mill. Besides, a checking up of the strain-stress state was intended, for a life prediction.

The measuring points were placed along a rib and in its proximity (fig. 1). In this way, the strain-stress state in the entire structure could be measured by turning round the mill. The connections between strain gauges and the switch box were made by external cables for the measurements on the stationary mill and by an internal one for those made in real running conditions. The latter connected successively the most stressed points revealed by the static measurements and the rotating contactor placed at the discharge end of the small mill(fig. 1) and at the feed end of the large mill(fig. 2).

Special care was paid to the protection of the strain gauges against the water contained by the ore, which flawed down from the mill body. For the same reason, an ana-

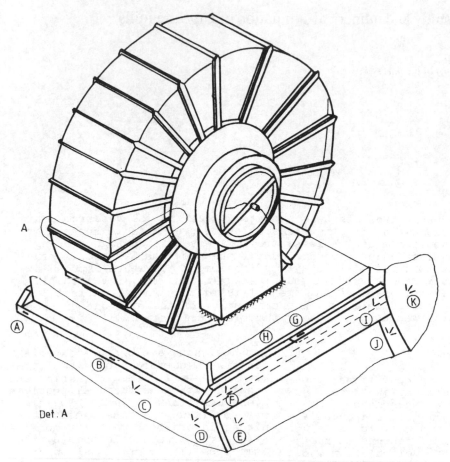

Fig. 1:Location of the measuring points on the small mill

logic amplifier was used, which permitted an easier observation of an ill signal caused by a mass connection.

The internal cable used for measurements on the running mill was particularly protected against the ore boulders and the balls by introducing it in a rubber flexible tube, placed between the inside armour and the shell.

The strain gauge type was chosen function of the stress state supposed to be in every point:

- Single grid for single-axial stress state.

- Two-element 90-deg for plane stress state with known principal axes.

- Three-element 45-deg rozettes for plane stress state with unknown principal axes.

On the stationary mill, the strains were read when the rib on which sides the measuring points were grouped was in three positions: horizontal, top and bottom. The first was the reference position, where zero strains were assumed and the others, positions where the extreme stresses were reached. The measurements were made on empty and full load mill, the highest principal stresses rezulting in points A, B, G and K(fig. 1), in the generatrix direction and on the bottom of the full load mill, excepting point K. These values were under 50 MPa, thus explaining the good behaviour of the mill structure

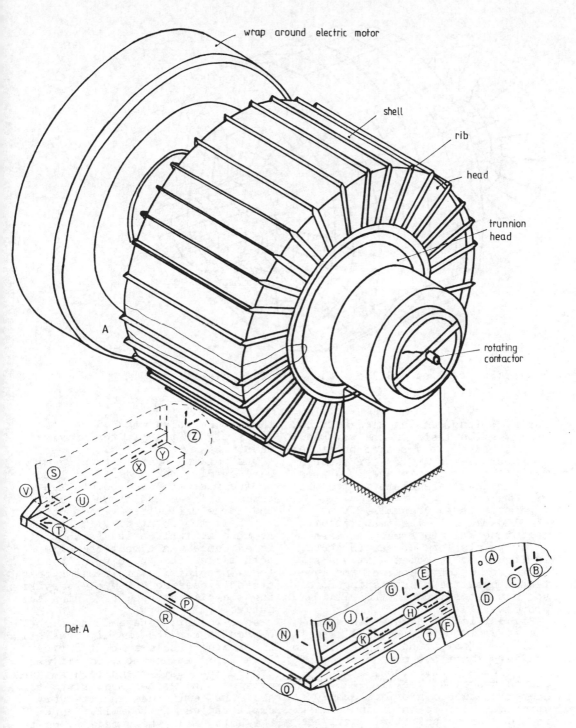

Fig. 2:Location of the measuring points on the large mill

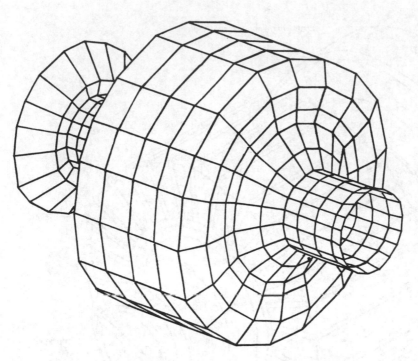

Fig. 3:FE model of the large mill modified design

during a long running time service.

Making the tests in the way mentioned above, it was also possible to separate the effects of the own weight and of the inside load upon the strain-stress state in the mill structure.

For the large ore mill(8.5 m shell diameter), the measurement technique was the same. The measuring points were chosen as shown in figure 2. In point A it was a bolt, for which a control was intended upon the prestressing force and his alteration in the milling process.

Big difficulties were faced in the attempt to put the full load mill in different positions, due to the very important moment produced by the out-of-axis load, which caused dangerous oscillations. The operation needed a bridge crane, which was used to maintain the mill in the desired position, until the hand-moving of the ore and balls assured the equilibrium.

The highest values of the principal stresses were reached in points F, I, 0, P, U and Y. In the last, the stress value was 2.5 up to 3 times higher than in the previous ones.

The readings on the running mills were carried out under full load, in the most stressed points mentioned before. Additionally, for the large mill, they were made on the bolt too(point A, fig. 2). The average multiplication factor put in evidence was almost the same for both mills, the rough value being 1.8. In point Y, this meant a peak stress exceeding 200 MPa - very high for a cyclic stress, reversing in nature.

As a result, severe damages appeared in service, mainly consisting in fatigue fractures of the ribs. Some bolts failed too, but it was proved that this phenomenon was not owing to the rather moderate stresses. The real cause was the very small radius of the fillet under the bolt head, which made it quite sensible to fatigue. It was also confirmed the remark made by Thomala(1978) that low stressed bolts are more sensible to fatigue than highly stressed ones.

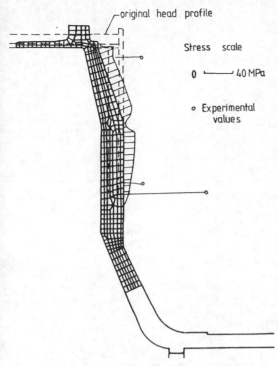

original head profile

Stress scale

0 ⊢———┘ 40 MPa

○ Experimental values

Fig. 4:Stress distribution in the ribs of the FE model, compared with experimental values obtained on the original large mill

3 NUMERICAL

The failures in service of the large ore mill imposed substantial design modifications. As the trunnion heads behaved very well(in accordance with the low stresses measured on them), the modified parts were the heads and the shell. The heads with welded ribs were replaced by cast heads with ribs, while the shell was no more stiffened with ribs. The shape of the heads became conical towards the joint with the shell.

The new design of the mill was studied using FEM, and the global model(fig. 3) included thin shell elements and beam elements for the ribs. The swing bearings and the loading with weight and inertia forces corresponding to the structure, ore and balls and the gearless drive wrap around electric motor were properly simulated. From this 3-D model, a 2-D model was derived,

for a more detailed study of the stress distribution in the shell and the rib in any position of this section, at both ends of the mill.

The numerical rezulted stresses on a rib in the bottom position, at the discharge end, is shown in figure 4, versus the experimental values obtained on the original mill in points U, X and Y.

4 CONCLUSIONS

The experimental measurements, made in heavy conditions(humidity, vibrations etc.), offered the possibility of forecasting severe failures of the large ore mills. The stress distribution found in this way was the background for the mill redesigning, carried out later by numerical simulation.

In figure 4, one can easily observe the significant reduction obtained for the stress level in the ribs. This favourable situation is due to the modified heads geometry and the increased stiffness of the shell, compared to that of the ribs, which had as result an average ratio between the stress level in the ribs and shell of about 1.5, against 3 in the experimentally studied mill.

Consequently, fully conical cast heads were adopted for another large ore mill design, in spite of the better behaviour of flat heads in the grinding process. This entirely new design was numerically and experimentally studied by Constantin and al.(1981,1989).

Both large ore mills proved their reliability in thousands of hours of safe running.

REFERENCES

Constantin, N., M. Blumenfeld, M. Munteanu, I. Constantinescu 1981. Study of the stress distribution in a large ore mill(in Romanian). Studii şi Cercetări de Mecanică Aplicată, Vol. 40, No. 3:409-424.
Constantin, N., M. Găvan, D. L. Constantin 1989. Calculus of the stresses in the structure of a large ore mill(in Romanian). Proc. of the 5th National Symp. on Exp. Mech., Vol. IV:159-166.

Stefănescu, D.M., N. Constantin, R.
Enache 1989. Strain gauge measurements on a large autogenous
ore mill(in Romanian). Proc. of
the 5th National Symp. on Exp.
Mech., Vol. I:239-246.
Thomala, W. 1978. Beitrag zur
Dauerhaltbarkeit von Schrauben-
verbindungen. Diss. TH Darmstadt.

Recent Advances in Experimental Mechanics, Silva Gomes et al. (eds) © 1994 Balkema, Rotterdam, ISBN 90 5410 395 7

Estimation of sulfide stress cracking resistance of low-alloy steels based on fracture mechanics approach

V.I.Astafjev
Samara State University, Russia

S.V.Artamoshkin & T.V.Tetjueva
Research Institute VNIITNeft, Samara, Russia

ABSTRACT: The sulfide stress corrosion cracking resistance of some low-alloy tubular steels for casing was evaluated on the basis of linear fracture mechanics approach. A modification of the NACE K1ssc test method was proposed. Accuracy and simplicity of the proposed method was discussed. Critical flaw sizes which can be used for tube inspection were calculated from obtained values of the threshold stress intensity factor K1ssc.

1 INTRODUCTION

The presence of hydrogen sulfide in the well fluids imposes severe restrictions on the use of high strength tubular goods because of sulfide stress corrosion cracking (SSCC). The appropriate selection of materials for well completions is the important factor in the economic success of oil and gas production. The choice is governed by mechanical properties and is highly dependent on corrosion cracking behaviour of metals for the range of environments expected in service. Such knowledge is gained partly from service experience, but mainly from appropriate laboratory testing at the design stage.

Laboratory evaluation of materials performance for sour service applications has been largely based on using standard test method such as that recommended by NACE (1990). One of the NACE test methods based on the use of fracture mechanics approach has been utilized to identify the threshold stress intensity factor K1ssc below which the material can be used safely and confidently. On the basis of the K1ssc data the operational limits of tubular goods for known defect sizes can be calculated or the critical flaw sizes for the maximum cross section stress can be estimated for a specific grade and size of casing or tubing.

Some recent studies show that the applicabilities of the K1ssc test method are limited due to the facts that the test solution is too severe to represent all possibilities of service exposure (Kermani 1991) and the test procedure is too complicated to determine the K1ssc data in field conditions by simple and correct manner (Astafjev 1990,1993a).

The purpose of the present paper is to outline a simple modification of the K1ssc test method and to evaluate the SSCC susceptibility for some low-alloy tubular steels. Both normalized and quenched and tempered steels for casing with different microstructure and strength level were investigated. A simple modification of K1ssc test method was proposed and the threshold stress intensity factors K1ssc for some casing were determined. The upper bound for critical flaw size was estimated.

2 MATERIALS AND EXPERIMENTAL PROCEDURE

2.1 Materials

Six as-received tubular steels for casing were used in this investigation (Russian steel grades St.45 and 30G2, API low-alloy steel grades SM90-SSU and C75-I). Steel grade 30G2 was prepared by different heat treatment on three strength levels (30G2-(1), 30G2(2) and 30G2(3), respectively). The main mechanical properties and chemical composition of steels investigated are summarized in Tables 1 and 2.

2.2 Heat treatment and microstructure

In the process of pipes manufacture the metal was subjected to different types of heat treatment and different kinds of

Table 1. The mechanical properties of casing steels

No	Steel Grade	Yield Strength (MPa)	Tensile Strength (MPa)	Elongation (%)
1	St.45	476	704	23.2
2	30G2(1)	474	726	20.4
3	30G2(2)	685	882	22.4
4	30G2(3)	736	864	18.0
5	SM90-SSU	693	796	20.0
6	C75-I	513	805	21.2

Table 2. The chemical composition of casing steels

No	C	S	P (wt%)	Si	Mn	Mo	Cr
1	0.45	0.014	0.013	0.24	0.77	<0.1	<0.1
2	0.34	0.024	0.019	0.24	1.50	<0.1	<0.1
3	0.31	0.017	0.019	0.23	1.40	<0.1	<0.1
4	0.33	0.010	0.013	0.24	1.35	<0.1	<0.1
5	0.29	0.005	0.011	0.28	0.53	0.32	0.95
6	0.23	0.010	0.022	0.32	0.96	0.11	0.14

microstructure were obtained. All steel grades can be divided into two groups. The first group (steels 1, 2 and 6 after normalization) was characterized by yield strength σ_{ys} < 520 MPa and ferrite-perlite microstructure. The second group (steels 3, 4 and 5 after quenching and tempering) was characterized by yield strength σ_{ys} > 690 MPa and mixed ferrite-tempered martensite (steels 3 and 4) or tempered martensite microstructure with small recrystallized ferrite grains (steel 5).

2.3 Environment

Corrosion tests were carried out in standard NACE solution (NACE 1990) - the aqueous solution contained 0.5 wt% acetic acid and 5 wt% sodium chloride. To reduce dissolved oxygen the solution was firstly deairated by purging with nitrogen. Hydrogen sulfide was periodically bubbled into solution throughout the test. The hydrogen ions concentration monitored at regular intervals was approximately equal to pH = 3.1~0.2. The tests were carried out at room temperature 24~2°C.

2.4 NACE Standard test procedure

The sulfide stress cracking resistance of each steel grade was characterized by the threshold stress intensity factor K1ssc. To determine the K1ssc value double cantilever beam (DCB) specimens are usually employed , which are the most appropriate and convenient ones to use. The test procedure for K1ssc determination with DCB specimens described firstly by Heady (1977) has been recently proposed in new version of NACE Standard TM 01-77-90.

In brief, the specimens are wedge loaded and exposed in NACE solution during 360-720 hr. After removal from the test environment the equilibrium value of wedge load P and the equilibrium crack length a are determined. These values are used in K1ssc calculation as follows (Heady 1977, NACE 1990):

$$K1ssc = \frac{2\sqrt{3}Pa(1+ch/a)}{Bh^{3/2}} (B/Bn)^{1/\sqrt{3}} \qquad (1)$$

where h, B and Bn are the specimen halfheight, thickness and web thickness, c = 0.687.

2.5 Modified calculation of K1ssc value

As it has been shown by our own experimental practice the direct measurement of the equilibrium wedge load P is not quite accurate especially in field conditions and can lead to certain errors. To avoid it another procedure for K1ssc calculation based on the second Heady's relationship (1977)

$$K1ssc = \frac{2\sqrt{3}Ad(a+ch)h^{3/2}}{(a+ch)^3 - (ch)^3} (B/Bn)^{1/\sqrt{3}} \qquad (2)$$

was used (A is the dimensional constant and d is the specimen arms displacement upon wedge loading, measured over the load line).

As it was pointed out by Heady (1977), the direct use of this equation led to the broad band of scatter for K1ssc values. Probably, it is due to the necessity to place the wedge exactly on the load line and to measure the specimen arms displacement only over this line. But this equation can be rewritten in the following manner (Astafjev 1990,1993a):

$$\frac{K1ssc}{K10} = \frac{a+ch}{a0+ch} \cdot \frac{(a0+ch)^3 - (ch)^3}{(a+ch)^3 - (ch)^3} \qquad (3)$$

where a0 is the initial crack length and K10 is the initial stress intensity factor, calculated for initial values of wedge load P0 and crack length a0 in accordance with equation (1).

In this case it is sufficient to determine only the values of initial and equilibrium crack length and the initial wedge load. Moreover, in this case it is not necessary to place the wedge exactly upon the load line. Following from the beam theory, the wedge load P0 for the wedge placed on some distance a0 from the crack tip and the specimen opening load P1 on the distance a1 are connected by simple manner:

$$2a0P0 = (3a1-a0)P1 \qquad (4)$$

Hence, for K1ssc determination the next practice can be recommended (Astafjev 1993b):
1) specimens are loaded by some opening load P1 and specimen arms displacements are fixed by wedges placed on some distance a0 from the crack tip;
2) all specimens are placed into the NACE solution and exposed there during 360 hr;
3) after removal from the test environment specimens are torn, cut open and crack extension Da=a-a0 is measured directly on the face of specimens;
4) K1ssc value is calculated in accordance with equations (3) and (4).

In our investigation all test specimens were cut from the pipe wall thickness in the longitudinal direction. The specimen configuration was standard and was shown everywhere [1-5]. The specimen dimensions (length, height and thickness) were 100x 25x5.5 mm with ratio Bn/B=0.636 and distance from the initial notch tip to holes for loading a1=30 mm. The opening load P1, the distance from wedge to initial notch tip a0 and the initial value K10 for every steel grade under investigation are summarized in Table 3.

3 RESULTS AND DISCUSSION

Results of K1ssc tests according to the above mentioned procedure for all steel grades investigated are given in Table 3 (three identical specimens for each steel grade were tested and the resulting crack extension Da=a-a0 was averaged). The K1ssc values presented in Table 3 coincide with those received by the original NACE test method. In fact, this modification of K1ssc test method needs only simple loading devices without any high precision measurement of specimen arms displacement. So, it can be utilized as the method of casing and tubing assessment to environment cracking both in laboratory and in field conditions.

Table 3. Results of sulfide fracture toughness tests

No	a0 (mm)	P (kN)	Da (mm)	K10 (MPa√m)	K1ssc (MPa√m)	r (mm)
1	31.5	2.5	1.0	58.3	54.1	8.4
2	31.5	3.3	5.0	75.9	57.1	9.1
3	29.0	3.3	19.0	77.0	34.5	2.0
4	28.0	4.2	50.0	96.9	19.2	0.6
5	29.5	3.8	24.0	87.5	33.6	2.0
6	30.5	3.0	3.0	70.0	59.5	8.1

The K1ssc values measured in sour service environment can be used in prediction of critical flaw size for casing or tubing failure as follows (Cherepanov 1979):

$$K1 = Y\sigma\sqrt{\pi b} < K1ssc \qquad (5)$$

where Y is the crack shape factor, $\sigma = (0.8-0.9)\sigma ys$ is the maximum cross-section stress and b is the depth of the surface crack. For more severe conditions as $\sigma = (\sigma ys+\sigma ts)/2$ the critical flaw depth can be varied from $0.2(K1ssc/\sigma)^2$ for an infinitely long surface crack to $0.33(K1ssc/\sigma)^2$ for an elliptical flaw. Hence, the ratio

$$r = (K1ssc/\sigma)^2 \qquad (6)$$

presented in Table 3 allows to estimate the upper bound of critical flaw depth for casing investigated.

The results obtained show also that steels with different strength level, microstructure and heat treatment have different K1ssc values, i.e. different SSCC resistance. It is observed that K1ssc decreases with increasing yield strength, as usually expected. The K1ssc values obtained in this investigation correspond to those published in many papers concerning SSCC resistance of tubular goods (Tuttle 1981).

4 CONCLUSIONS

The following conclusions can be made on the basis of the present investigation.
1. A simple modification of NACE test method for K1ssc evaluation in field conditions was proposed and the threshold stress intensity factors K1ssc for some casing steels were determined. The upper bound for the critical flaw size for casing investigated was estimated.
2. Both normalized and quenched and tempered steels with different strength levels were investigated. The results obtained show that steels after normalization with ferrite-perlite microstructure and

yield strength σ_{ys} < 520 MPa are more resistant to SSCC process. Among the steels after quenching and tempering with tempered martensite microstructure and yield strength level σ_{ys} >690 MPa only the steel grade 30G2(3) with yield strength σ_{ys} >730 MPa can be regarded as SSCC susceptible steel.

ACKNOWLEDGEMENTS

Authors are grateful to the Russian Foundation of the Fundamental Researches for financial support (Project 93-013-17652).

REFERENCES

Astafjev, V.I., V.K.Emelin & T.V.Tetujeva 1990. Investigation of sulfide stress corrosion cracking in low alloy steels. Proc. ECF8: 478-485. Warley:EMAS.

Astafjev,V.I.,S.V.Artamoshkin & T.V.Tetujeva 1993a. Influence of microstructure and nonmetallic inclusions on sulfide stress corrosion cracking in low-alloy steels. Int.J.Pres.Ves.Piping 55:243-250.

Astafjev,V.I. et al 1993b. Fracture Toughness Evaluation of Tubular Steels under SSCC Conditions Samara:Samara State University (in Russian).

Cherepanov,G.P. 1979. Mechanics of Brittle Fracture NY:McGraw Hill.

Heady,R.B.1977. Evaluation of sulfide corrosion resistance in low-alloy steels. Corrosion 33:98-107.

Kermani, M.B. et al 1991. Experimental limits of sour service for tubular steels in Corrosion-91:pap no 21. Houston:NACE.

NACE Standard TM 01-77-90 1990. Laboratory testing of metals for resistance to sulfide stress cracking in H2S environments Houston:NACE.

Tuttle,R.N.& R.D.Kane (eds) 1981. H2S Corrosion in oil and gas production-A compilation of classic papers Houston:NACE.

7 Quality control and testing of materials and components

Recent Advances in Experimental Mechanics, Silva Gomes et al. (eds) © 1994 Balkema, Rotterdam, ISBN 90 5410 395 7

Report of the activities on smart materials/structures research in Japan

Koichi Egawa, Ippei Susuki & Shinichi Koshide
National Aerospace Laboratory, Tokyo, Japan

Yasufumi Furuya
Tohoku University, Sendai, Japan

Tadaharu Adachi
Tokyo University of Technology, Japan

ABSTRUCT: The activities on smart materials/structures research in Japan, which an innovative, new concept of materials and structures, are stated here. There are research
activities conducted by Japanese government and national laboraories, and also those of
research groups. Today there are seven research groups distributed all over Japan. The
object of this report is to introduce these research organizations including governmental
ones and their activities. At the end a suggestion of the authors concerning the development of smart materials/structures research in Japan is expressed.

1. WHAT IS IN A NAME ?

An astonishing, innovative concept, putting
the functions, which is almost same as these
of living creatures having, into industrial,
manufacturing materials and structures, was
invented in U.S.A. In Fig.1, which is made
by professor C.A.Rogers of Virginia Polytechnic Institute, it is shown that a prevailing concept of smart materials/struc-
tures which plans to put the functions done
by skeleton, muscles and nerve system in
human body into a composite laminate.

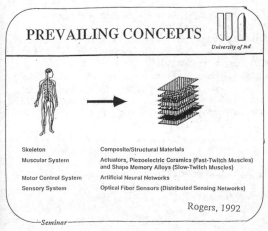

Fig. 1 Prevaling concept

2. GOVERNMENT & 7 RESEARCH GROUPS

2.1 Government and National research laboratories

1. Science and Technology Agency (STA)
   1) National Research Institute for Metal
   2)   ″   ″   ″   for Non-
      organic Materials
   3) National Aerospace Laboratory
2. Ministry of International Trade and
   Industry
   1) National Institute of Materials and
      Chemical
3. Ministry of Education
   1) Institute of Space and Astronautical
      Science

2.2 Research groups

1. Intelligent Material Forum
   Society of Non-Traditional Technology
2. Group of IAS(International Conference
   on Adaptive Strucutres)
3. Working Group on Smart Materials/Stru-
   ctures
   Society for Non-destructive Inspection
4. Intelligent Materials Forum in Tohoku
5. Study Group on Smart Structures/Matri-
   als
   Society of Mechanical Engineers
6. Study Group on Smart Space Structure
   Society of Aeronautics & Astrinautics
7. Study Group on Smart Composite
   Society of Materials Science

3. THEIR ACTIVITIES

3.1 Government and National Research Laboratories

1. Science and Technology Agency (STA)
As the answer for the inquiry of minister of Science and Technology Agency, a report "Promotion of research developement on the creation of new materials which have the capability of intelligent response against environmental change" was submitted by deliberation committee on aeronautics and electronics in July, 1987. Researches on smart materials in Japan, it is possible to say, were started by the publication of this report.

The Japanese definition of smart(intelligent) materials stated in this report is shown in Fig.2. This was made by the esti-

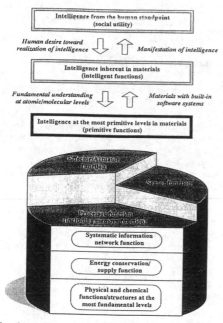

Fig.2 A Japanese definition of intelligent material

mation of the trend of material development, that is, materials were developed from structural ones to functional ones. Regarding this tendency as the compass indicating the future development, the report told us that we will be able to create new materials which have intelligent functions in themselves; they judge their degradation, intercept crack growth, repair themselves by themselves and predict their lives. The report also said that the newly created materials will have the possibility to

functionate more than that of living creatures do. By this report the direction of research is set, that is, to develope new materials. The research of new material creation was appointed as a main project of STA and national research institute for metal and that for non-organic materials began their researches.

1) National Research Institute for Metal
       "   "   " for Non-organic Materials
From 1992 a research project "new material creation by utilization of atomic and molecule harmonic action" was started. This project will be continued up to 1997.
Other two projects are planned now; "new material creation by nano-space technology" and "survey of smart materials/structures".
In Fig.3 an example of research objet, interception of crack growth in high temperature atmosphere is shown.

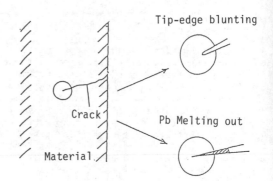

Fig.3 Ideas of interception of crack growth

2) National Aerospace Laboratory
Aiming advanced composite material to be intellectual one, the research of embeddding optical fibers in composite laminate was begun recently.
As another activity three of the authors in the laboratory organized the research group of No.3 and 5 stated in section 2.2, and are playing important roles even in No. 1 group.

2. Ministry of International Trade and Industry (MITI)
By affiliated organization research surveies are conducting especially for aerospace and new materials of next generation.

1) National Institute of Materials and Chemical
They are promoting new project for making material smart and ecological, that is,

they are trying to apply this new concept for treating waste materials efficiently.

3. Ministry of Education
1) Institute of Space and Astronautical Science

The research was started in the earliest time in Japan, and in 1985 they have held the first symposium on space structures. Last year the symposium was held for two days and 40 papers were presented. On these days space structure, especially for unfolding structures(several models) were developed and examined. Variable shape space structures were also studied and as result new concept of these structures "Space Cocoon" was presented by professor K.Miura.

3.2 Research groups

1. Intelligent Material Forum
(Society of Non-Traditional Technology) Under the influence of Science and Technology Agency they are making effort to promote research activities stated in the report for the Inquiry. In 1990 they have founded Intelligent Material Forum, and are making effort to promote research of new material creation even in Japanese industry. The president now is professor emeritus of Kyoto University, Dr. T.Takagi (president of ion engineering laboratory).

As a part of their activity, conference and symposium, as shown in Table 1, were held regularly, and in March in this year the 3rd Intelligent Materials Forum will be held in Tokyo and in June the International Conference held at Williamsberg in U.S.A. The journal "Intelligent Materials" published quarterly.

Table 1. Symposium & Int. Conferences held by SNTT

---

1. 1990, Tsukuba, Japan
   Int. Workshop on Intelligent Materials
2. 1991, Tokyo, Japan
   First Intelligent Materials Forum
3. 1992, Ohiso, Japan
   First Int. Conference on Intelligent Materials
4. 1993, Tokyo, Japan
   Second Intelligent Materials Forum
5. 1994, Tokyo, Japan
   Third Intelligent Materials Forum
6. 1994, Williamsberg, U.S.A.
   Second Int. Conference on Intelligent Materials

For searching the tendency of the member's research activities, papers presented at first and second Intelligent Materials Forum are classified and shown in Fig.4. Judging from this result, researches are directed to new material creation, new creation of functional materials, but there were, take for instance, no space structures, no composite materials related with machine use or design.

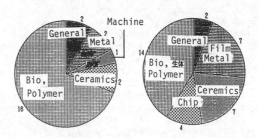

No. of Presentatiion 23   No. of Presentatioin 34

(First) 1991, March   (Second) 1993, March

Fig.4 Classification of papers presented at 1st & 2nd Intelligent Materials Forum

2. Group of IAS
Professors and researchers of Institute of Space and Astronautical Science (ISAS) form the core of this group, and its activity extends from space structure to vibration control of airplane, building etc.

In 1990 they have held First Japan-US Conference on Adaptive Structures and continue this activity as shown in Table 2. The representative of Japan was professor M. Natori of ISAS at third conference in 1993. (U.S. representative was Ben K. Wada of JPL).

Table 2. Int. Conference on Adaptive Structures

---

1. 1990, Hawaii, U.S.A.
   First Japan-US Conference on Adaptive Structures
2. 1991, Nagoya, Japan
   Scond Japan-US Conference on Adaptive Structures
3. 1992, San Diego, U.S.A.
   Third International Conference on Adaptive Strucutures
4. 1993, Kern, Germany
   Fourth International Conference on Adaptive Structures
5. 1994, Sendai, Japan
   Fifth International Conference on Adaptive Structures

---

3. Working Group on Smart Materials/Structures
   (Society for Non-destructive Inspection)
One of the author of this report, Dr. Egawa
visited CIMSS (Center of Intelligent Material
Systems and Structures) and Fiber & Electro-
Optic Research Center at VPI (Virginia Polytechnic Institute and State University) in
May 1991 and then founded this working group
in July. Popularization of the concept of
smart materials/strucutres and the application of this concept to materials and structures of industrial use are two main objects
of the group. The meeting was held every
month during first two years and now four
times a year. More than 100 lectures and
reference studies have been done for these
three years, and in 1993 symposium and one-
day seminar were held by inviting Dr. Jim
Sirkis of Maryland University as lecturer.
Many topics, optical fiber sensor, shape
memory alloy, piezo-electric ceramic, electro-
rheological fluid and their application, composite, concrete parts, vibration of building
etc were presented in the symposium. Dr.
Egawa is the leader of the group, and its
members are now more than 50.

In 1994 another study group, whose object
is to make maintenance inspection of infra-
structure intellectual, will be started in
JSNDI.

4. Intelligent Material Forum in Tohoku
The study group, whose leader is professor
J. Tani in Tohoku University's Fluid Science
Laboratory, is in full activity, especially
in northern part of Japan (Tohoku region).
As shown in Table 3 many international conferences on intelligent materials/structures
were held by this forum at Tohoku University.
Many members of this forum are the faculty
of Tohoku University, and Dr. Y. Furuya, one
of the author, made good achievements in the
application of shape memory alloy.

Table 3. Symposiums & International Conferences held by Intelligent Material Forum in
Tohoku

---

1. 1991, Sendai, Japan
   International Workshop on Intelligent
   Material System and Structures
2. 1992, Sendai, Japan
   Third Int. Workshop on I.M.S.S.
3. 1993, Sendai, Japan
   Fouth Int. Workshop on I.M.S.S.
4. 1994, Akita, Japan
   Symposium on Functional Materials and
   Intelligent Materials/Structures

---

5. Study Group on Smart Structures/Mate-
   rials
   (Society of Mechanical Engineers)
In April 1993 this group was started, and
the meeting was held two times up to now.
The group belongs to Material and Mechanics
Division (about 3000 members), and so the
author can expect that the group will be
strong in intellectualization of advanced
composites, interaction of sensor/actuator
in composite, FEM and other analysis of
these problems. Dr.K. Egawa is the leader
and Dr.I. Susuki, one of the author, is
sub-leader of this group, and the members
are now about 30.

6. Study Group on Smart Space Structure
   (Society of Aeronautics & Astrinautics)
In 1993 this was founded in Kansai (south-
ern part) branch of this society (JSAA).
Professor Y.Sugiyama of Ohsaka Prefecture
University is th representative of this
group, and the meeting was held two times
up to now.

7. Study Group on Smart Composite
   (Society of Materials Science)
Last year this group was founded and the
meeting was held two times up to now.
Professor T.Fukuda of Ohsaka University is
the leader of this group and the main members are in Kansai area.

4. SUMMARY OF THE ACTIVITIES IN JAPAN

Except the study on space structures, there
are no so many research achievements, especially for concrete, materialized ones.
The researches which aim the realization
of the concept totally are growing little
by little.

The study in Japan ranges from material
creation to intellectualization of composite
material, development and application of
optical fiber sensor, shape memory alloy,
electro-rheological fluid, development of
crack detection method of ships and concret
component, earthquake damage mitigation
and vibration control of building, and
maintenance inspection of very large constructions.

Thus we think we are asked to make effort
to spread the concept of smart (intellligen
materials/structures even now as professor
C.A. Rogers of VPI pointed out in 1990,
and foster the occuring of new research
works on this concept.

5. A SUGGESTION OF JAPANESE STYLE RESEARCH DEVELOPMENT

Looking in a application list of the research on this concept made in U.S.A., we can easily find that many of them belongs to aerospace and military ones (1). These are also important even in Japan, but the Japanese industrial construction is quite different from U.S. one's, that is, ours are consisted mainly by articles of public use like automobile, home-appliance, electronic parts, optical and precise parts and machines.

So we should be prudent to import and spread the trend of research and its application directly from U.S.A., but we should think about Japanese style research development based on this industrial formation.

As an example, the concept of Health Monitoring Airplane, it is better for us to apply this concept to automobiles and cruiser boats. But different from the application on airplane, one should be more cautious on cost performance of those production. From this point of view, it is closed up to us as one of the most hopeful topic of this concept to make large constructions intelligent (or smart), like smart building, dam, highway system, tunnel etc.

REFERENCES

1. Sirkis, J. 1993. "Intelligent" Structures : Mechanics & Fablication Seminar Note, Japanese Society for Non-destructive Inspection.
2. Report for the Inquiry No.13, Deliberation Committee on Aeronautics and Electronis, Science and Technology Agency.
3. Gandhi, M.V. & Thompson, B.S. 1992, Smart Materials and Structures, 46-47 London.
4. Watanabe, T. 1991. Summary of first Intelligent Materials Forum, Promuthus, 82:54-55
5. Natori, M. 1991. Dynamics & Intelligence, Newsletter of Dynamics & Measurement Division, JSME. 7:5-7

Recent Advances in Experimental Mechanics, Silva Gomes et al. (eds) © 1994 Balkema, Rotterdam, ISBN 90 5410 395 7

Shrink fit stress analysis between a circular shaft and a shrunk ring

Toshiyuki Sawa & Ken Shimotakahara
Department of Mechanical Engineering, Yamanashi University, Kofu, Japan

Takahide Hamajima
Nippon Denso Co., Ltd, Japan

ABSTRACT:On shrinkage fittings, it is important to know the shrink fit stress distribution at the interface. This paper deals with the shrink fit stress and the deformation of solid and hollow shafts which are shrink fitted by shrunk rings. When the shrink fitted joints are subjected to external force, the strain near the interface and the strength of the joints are measured by these experiments. Moreover, the shrink fit stress at the interface and the deformations of the joints are analyzed by using the axisymmetrical theory of elasticity. When a shrunk ring is fitted to a solid shaft, the shrink fit stress distribution in the axial direction is estimated by these approaches. And the strength of bonded shrink fitted joints is also estimated from the experiment. In the case of a hollow cylinder, the shrunk fit stress distribution and the deformation are clarified from both experimental and analytical approaches.

1 INTRODUCTION

Shrink fitted joints have been used in mechanical structures. In designing shrink fitted joints,it is necessary to estimate the shrinking allowance and shrink fit stress distribution in the joints. In addition, it is important to estimate the shrink fit stress distribution and shrink fitted joint strength when the joints are subjected to external loads. However, up to now, little researches have been done on the stress distribution of shrink fitted joints. Moreover, the stress distributions of shrink fitted joints subjected to external loads have yet been clarified. In this paper,the following shrink fitted joints are investigated. (1)Shrink fitted joint strength is measured experimentally when the joints composed of solid shafts and rings are subjected to push off loads. (2) Strength of shrink fitted joints in which adhesives are bonded between the shrink fit surfaces, is measured experimentally. (3)The characteristics of shrink fitted joints, where a hollow shaft is fitted in one or many rings, is examined experimentally. Analytical approaches are developed in order to estimate the shrink fit stress distribution of the aforementioned shrink fitted joints. The strength of shrink fitted joints are discussed.

2 EXPERIMENTAL PROCEDURE

Figure 1 shows a shrink fitted joint subjected to push off force. After a shaft is fitted in a ring, a push off force is applied to the joint. Figure 2 shows a shrink fitted joint, where a hollow shaft is fitted in a reinforced ring, subjected to internal pressure. Figure 2(a) is the case where a ring is fitted in a shaft and Fig.2(b) shows the case where some rings are fitted.

2.1 Push off tests of shrink fitted joints and bonded shrink fitted joints

2.1.1 shrink fitted joints

Figure 3 shows the dimensions of specimens used in experiments. Figure 3(a) shows the dimensions of rings of which the inner diameter is $2a_2-\delta$,where δ denotes the shrinking allowance. Two types of rings of which the outer diameter is 70.0 and 80.0 mm are prepared. The rings are manufactured from steel of which Young's modulus E_2 is 206 GPa and Poisson's ratio ν_2 is 0.3. Figure 3(b) shows the dimensions of shafts. The shafts are manufactured from steel. The shrinking allowance δ is varied as 0.01 - 0.05 mm. The maximum

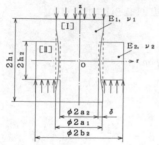

Fig.1 A shrink fitted joint subjected to push off force

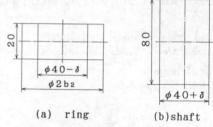

(a) ring   (b) shaft

Fig.3 Dimensions of specimens used in experiments

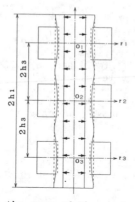

(a) the case of one ring

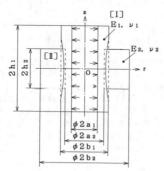

(b) the case of three rings

Fig.2 Shrink fitted joints subjected to internal pressure

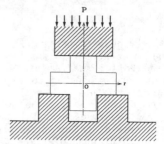

Fig.4 A sketch of experimental setup of push off tests

2.1.2 bonded shrink fitted joints

The dimensions of the specimens are same as those in Fig.3. However, three types of rings of which the outer diameter $2b_2$ 60.0, 70.0 and 80.0 mm are prepared. Adhesives (14486, Loctite Co., Ltd.) are bonded previously at the shrink fitted surfaces of the rings and the shafts. The rings are heated by 250 C in a furnace and the shafts are fitted in the rings. Push off forces of the bonded fitted joints is measured.

2.2 Photoelastic experiments

Photoelastic experiments (stress freezing method) are performed in order to measure the stress distribution at the fitted interface of shrink fitted joints when a push off load is applied to the joints. Figure 5 (a) shows the dimensions of photoelastic specimen. Figure 5 (b) shows an experimental setup in order to apply a push off force W.

2.3 Strain measurement near the contact surface of shrink fitted joints subjected to internal pressure

2.3.1 The case of one ring

Strain measurements are performed near the

surface roughness R_{max} of the shafts and rings was measured between 5.0 -15.0 μm. The rings are heated by about 250 C in a furnace and the shafts are fitted in them. After cooling the shrink fitted joints, the joints are set on the apparatus shown in Fig.4. Then, push off force is applied to the shrink fitted joints using a material testing machine. An applied load and displacement are measured.

874

contact surface of shrink fitted joints subjected to internal pressure. Figure 6 shows the dimensions of rings and hollow shafts. In order to examine an effect of reinforcement of ring, strain measurement on the case where a shaft is fitted in three rings is carried out. The rings and the shafts are manufactured from steel of which Young's modulus and Poisson's ratio are 206 GPa and 0.3, respectively. The shrinking allowance δ is 0.1 mm. The rings are heated by about 700 C in a furnace and the shafts are then fitted in the rings. After the shrink fitted joints are cooled to room temperature, strain gauges are attached to the positions 12, 20, 30, 40 and 60 mm in the Z direction from the origin O of the shafts. In the rings, strain gauges are attached to the positions 69.5, 72.0, 79.0 and 85.0 mm in the radial direction from the origin O.

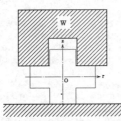

(a) dimensions of specimen

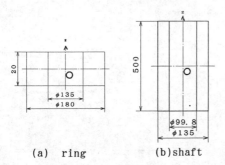

(b) experimental setup

Fig.5 Photoelastic experiment

Internal pressure is applied to the shrink fitted joint as shown in Fig.7 and strains are measured with the strain gauges.

2.3.2 The case of three rings

In the same way, strain are measured with strain gauges in the case where a shaft is fitted in three rings. The interval $2h_3$ between two rings (Fig.2(b)) is 100 mm.

3. ANALYTICAL APPROACH

3.1 Shrink fitted joints subjected to push off force

Shrink fit stress distribution at the interface when the joint is shrink fitted is analyzed as a two-body contact problem using axisymmetrical theory of elasticity. It can be predicted that the boundary condition at the interface of shrink fitted joints in practice is slipless or frictionless. The shrink fit stress distribution of joints is analyzed under the both boundary conditions. The boundary conditions are expressed by Eq.(1). In the analysis Michelle's stress functions are used.

Slipless

$$
\begin{aligned}
z = \pm h_1 &: \quad \sigma_z^{\mathrm{I}} = \tau_{zr}^{\mathrm{I}} = 0 \\
z = \pm h_2 &: \quad \sigma_z^{\mathrm{II}} = \tau_{zr}^{\mathrm{II}} = 0 \\
r = b_2 &: \quad \sigma_r^{\mathrm{II}} = 0 \\
&\quad \tau_{rz}^{\mathrm{II}} = 0
\end{aligned}
$$

$$
-h_2 \leq z \leq h_2
$$

$$
\begin{aligned}
(\sigma_r^{\mathrm{I}})_{r=a1} &= (\sigma_r^{\mathrm{II}})_{r=a2} \\
(\tau_{rz}^{\mathrm{I}})_{r=a1} &= (\tau_{rz}^{\mathrm{II}})_{r=a2} \\
(u_r^{\mathrm{II}})_{r=a2} - (u_r^{\mathrm{I}})_{r=a1} &= a_1 - a_2 \\
(w_z^{\mathrm{I}})_{r=a1} &= (w_z^{\mathrm{II}})_{r=a2}
\end{aligned}
$$

Frictionless

$$
\begin{aligned}
z = \pm h_1 &: \quad \sigma_z^{\mathrm{I}} = \tau_{zr}^{\mathrm{I}} = 0 \\
r = a_1 &: \quad \tau_{rz}^{\mathrm{I}} = 0 \\
z = \pm h_2 &: \quad \sigma_z^{\mathrm{II}} = \tau_{zr}^{\mathrm{II}} = 0
\end{aligned}
$$

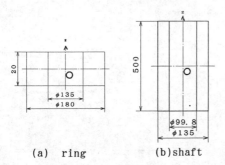

Fig.6 Dimensions of specimens used in internal pressure tests

Fig.7 Experimental apparatus in internal pressure tests

$$r = a_2 \quad : \quad \tau_{rz}^{\,!}=0$$
$$r = b_2 \quad : \quad \sigma_r^{\,!}=0$$
$$\tau_{rz}^{\,!}=0$$
$$-h_2 \leqq z \leqq h_2$$
$$(\sigma_r^{\,!})_{r=a1} = (\sigma_r^{\,!})_{r=a2}$$
$$(u_r^{\,!})_{r=a2} - (u_r^{\,!})_{r=a1} = a_1 - a_2 \qquad (1)$$

In addition, the analysis is done when a push off force is applied.

3.2 Bonded shrink fitted joint

An adhesive bond is replaced with a hollow cylinder, where Young's modulus and Poisson's ration denote E_3 ν_3, respectively. In the same way mentioned in 3.1, the shrink fit stress distribution at the interface is analyzed as a three-body contact problem using axisymmetrical theory of elasticity.

3.3 Shrink fitted joints subjected to internal pressure

The shrink fit stress distribution of joints subjected to internal pressure is analyzed as a two-body contact problem using axisymmetrical theory of elasticity. The shrink fit stress distribution is analyzed under the conditions of slipless and frictionless at the interface. Michelle's stress functions are used in the analysis.

4 RESULTS AND DISCUSSIONS

4.1 Push off force measurement of joints

Figure 8 shows an example of measurement of push off force. The ordinate is an applied push off force and the abscissa is displacement. The value F_r is determined as the push off force. Figure 9(a) shows the experimental results of push off force in shrink fitted joints. The abscissa is the normalized shrinking allowance $\delta/2a_1$. The mark o indicates when the outer diameter of ring is 70 mm and the mark o is the case of 80 mm. From the results, it is noticed that the push off force increases with an increase of the normalized shrinking allowance and that the push off force in the case where $2b_2$ is 80 mm is larger than that in the case where $2b_2$ is 70 mm.

Figure 9(b) shows the experimental results of push off force in the case where adhesives are bonded at the interfaces. From the results, it is seen that the push off force increases with an increase of the outer diameter $2b_2$ of rings and that the value of push off force is independent of the shrinking allowance $\delta/2a_1$. From the comparisons between the results of Fig.9(a) and Fig.9(b), it is shown that the push off forces in bonded shrink fitted joints are larger than those in shrink fitted joints. This is due to the adhesive bonding force. Figure 10(a) shows the analytical result of shear stress distribution τ_{rz} at the interface between the shaft and the ring, where push off force of 100 kN is applied. It can be supposed that slip may be initiated from the end ($Z=h_2$) of the interface. Figure 10(b) shows the comparison between the numerical and the photoelastic results concerning the principal stress difference ($\sigma_1 - \sigma_3$). A good agreement is seen.

4.2 Results of shrink fitted joints subjected to internal pressure

Figure 11 shows the comparisons between the experimental and the analytical results concerning strains of joints when internal pressure of 20 MPa is applied to the joints. The solid lines are the analytical results when the boundary condition at the interface is assumed to be frictionless and the dotted lines the case of slipless. From the results, it is seen that they are in fairly good agreement and that a difference between the slipless and the frictionless is small concerning the strains. Figure 12 shows the analytical results of Fig.2 concerning the effects of Young's modulus E_2/E_1 and the outer diameter of ring b_2/b_1 on the shrink fit stress distributions when internal pressure P=0. From the results, it is seen that singular stress causes at the end $Z_1=h_2$ of the interfaces and that the shrink fit stress

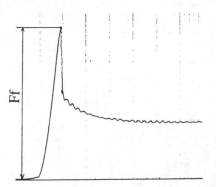

Fig.8 An example of relationship between push off force and displacement

σ_r decreases as the value E_2/E_1 decreases. Thus, it is necessary to increase the stiffness of rings in order to increase the shrink fit stress. From Figure 12(b), it is seen that the shrink fit stress σ_r increase as the outer diameter $2b_2$ increases.

Figure 13 shows the analytical results of the shrink fit stress distribution when internal pressure P is applied. From the results, it is seen that the shrink fit stress σ_r increases with an increase of the internal pressure P. Figure 14 shows the comparisons between the analytical and the experimental results concerning the strains of shrink fitted joints in Fig.6(c). A good agreement is seen between

the analytical and the experimental results.

5 CONCLUSIONS

This paper dealt with the shrink fit stress distribution and the strength for push off force in shrink fitted joints subjected to push off force and internal pressure. The following results are obtained.

(1) Push off tests are performed for shrink fitted joints and bonded shrink fitted joints. It is shown that the push off force increase with an increase of the shrinking allowance in shrink fitted

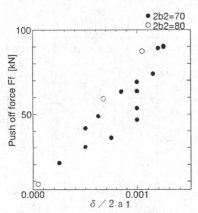

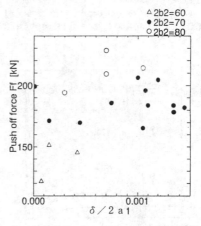

(a) push off force of shrink fitted joints

(b) push off force of bonded shrink fitted joints

Fig.9 Experimental results of push off force

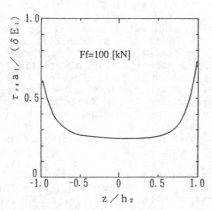

a1=20.015[mm], b2=35[mm], h1=40[mm], δ =0.015[mm], E1=206[GPa]

Fig.10(a) Analytical result of shear stress distribution at the interface (Push off force of 100 kN is applied)

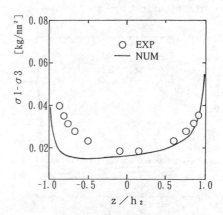

a=20[mm], b=35[mm], h=40[mm]

Fig.10(b) Comparisons between the numerical and the experimental results (r=a,-h2<z<h2)

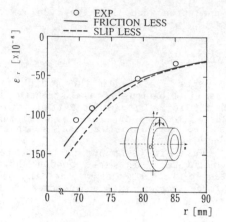

Fig.11 Comparisons between the numerical and the experimental results concerning Strains

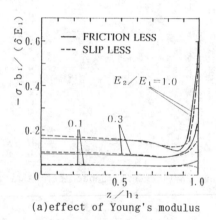

(a)effect of Young's modulus

Fig.12 Numerical results of shrink fit stress distributions(case of P=0)

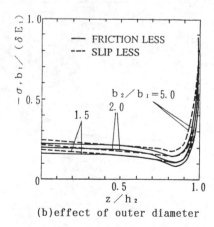

(b)effect of outer diameter

Fig.12 Numerical results of shrink fit stress distributions(case of P=0)

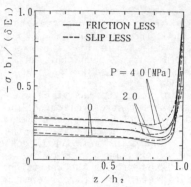

Fig.13 Numerical results of shrink fit stress distribution when internal pressure is applied

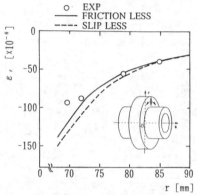

Fig.14 Comparisons of strains where a shaft is fitted in three rings

joints. In addition, it is seen that the push off force of bonded shrink fitted joints is larger than that in shrink fitted joints.

(2)Shrink fit stress distribution of joints in which a hollow shaft is fitted in one or three rings is analyzed using axisymmetrical theory of elasticity. In the numerical calculations, the effects of the stiffness and the outer diameter of rings on the shrink fit stress distribution are clarified. It is seen that shrink fit stress increases as the stiffness and the outer diameter of rings increase.

(3) Strains on the ring and the shaft are measured with strain gauges when internal pressure is applied to the joints. Good agreements are seen between the analytical and the experimental results.

References

Oda,J. and Shibahara,M.,Contact stress between solid and hollow cylinders, Transaction of JSME,38-306(1972),241-242

Recent Advances in Experimental Mechanics, Silva Gomes et al. (eds) © 1994 Balkema, Rotterdam, ISBN 90 5410 395 7

An inverse analysis among finite strips

Toshiyuki Sawa & Katsuyuki Nakano
Department of Mechanical Engineering, Yamanashi University, Kofu, Japan

ABSTRACT:This paper deals with an analysis of contact stress among finite strips and a determination of material properties of the finite strips as an inverse problem. Strains measured with strain gauges and displacements measured with electric micrometer are used as the boundary condition, where their measurement values are expanded with Fourier series. In the analysis, Airy's stress functions are used. The contact stress is analyzed as a contact problem using a two-dimensional theory of elasticity. In this study, the effect of measured point numbers and of locations of measured strains on contact stress distribution are examined. Photoelastic experiments were performed. Fringe patters obtained by photoelastic experiments are compared with those obtained by the inverse analysis. A good agreement is obtained.

1 INTRODUCTION

In designing mechanical structures,it is necessary to know contact stress distribution between machine elements. However,it is difficult to measure the contact stress experimentally. Recently, some investigations have been carried out on an inverse analysis, in which the contact stress is determined by using the displacements and the strains measured experimentally. But, in the most of them, finite element method and boundary element method were used.

In this paper,the contact stress distributions among finite strips and the material constants of the strips are determined by an inverse analysis using a two-dimensional elasticity. The contact stress among finite strips is obtained by using the following two measured values,that is,(1)strains measured with strain gauges,(2)displacement measured with electric micrometer. Then, a method for estimating Young's modulus is demonstrated by using the measured strains. Moreover, a direct analysis is done and the obtained contact stress distribution is compared with that obtained from an inverse analysis. The effects of the measured locations and numbers on accuracy of the contact stress distribution are examined. In addition,photoelastic experiment is carried out and then the obtained

results from the inverse analysis are compared with the experimental results.

2 PROCEDURE OF INVERSE ANALYSIS

Figure 1(a) shows the case where finite strip [II] is compressed symmetrically by two finite strips [I] and Fig.1(b) shows the case where finite strip [II] on rigid foundation is compressed by finite strip [I]. The length of finite strip [I] is designated as $2l_1$, the height of it as $2h_1$, Young's modulus as E_1 and Poisson's ratio as ν_1. Those of finite strip [II] are designated as $2l_2$,$2h_2$,E_2, ν_2,respectively. The origins of the coordinate are designated as o_1 and o_2,respectively. In this paper, the contact stress distribution in the both cases is determined as an inverse analysis by using measured strains and displacements,where a stress distribution F(x) is unknown. Figure 2 shows the dimensions of specimens used in the experiments. Figure 2(a) shows finite strip [I] manufactured from steel and (b) is finite strip [II] manufactured from epoxide resin.

2.1 Inverse analysis using measured strains

Figure 3 shows a sketch of experimental

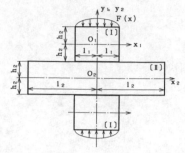

(a) A model where a strip is compressed by two strips

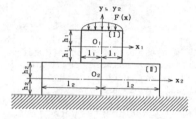

(b) A model where a strip on rigid foundation is compressed by a strip

Fig.1 Finite strip compressed by finite strips

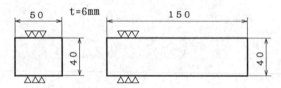

Fig.2 Dimensions of specimens used in experiments

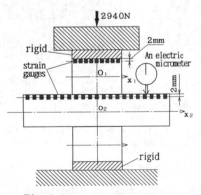

Fig.3 Experimental setup

setup. Strain gauges are attached to the position (y_1=18 mm) 2 mm from the upper end and the position (y_2=18 mm) 2 mm from the contact surface. A compression of 2940 N is applied to the finite strips [I] by rigid strips. The measured values of strain gauges are developed by Fourier series. The boundary conditions are express by Eq.(1) in an inverse analysis of finite strips [I] and [II], where the displacements in the x and the y directions are denoted as u and v, respectively.

A finite strip[I]
$x=\pm l_1$: $\sigma_x = \tau_{xy} = 0$
$y=h_1$: $\tau_{xy} = 0$
$y=-h_1$: $\tau_{xy} = 0$
: $(\sigma_y)_{y=-h1} = (\sigma_y)_{y=h2}$ $(-l_1 \leqq x \leqq l_1)$
: $\left(\dfrac{\partial v}{\partial x}\right)_{y=-h1} = \left(\dfrac{\partial v}{\partial x}\right)_{y=h2}$ $(-l_1 \leqq x \leqq l_1)$
$y=h$: $\left(\dfrac{\partial v}{\partial y}\right)_{y=h} = b_0 + \sum_{s=1}^{\infty} b_s \cos \dfrac{s\pi}{l_1} x$

A finite strip[II]　　　　　　　　　　　　(1)
$x=\pm l_2$: $\sigma_x = \tau_{xy} = 0$
$y=\pm h_2$: $\tau_{xy} = 0$
$y=h$: $\left(\dfrac{\partial v}{\partial y}\right)_{y=h} = a_0 + \sum_{s=1}^{\infty} a_s \cos \dfrac{s\pi}{l_2} x$

The following condition is added in the analysis of Fig.1(b)
$y=-h_2$: $\left(\dfrac{\partial v}{\partial x}\right)_{y=-h2} = 0$

Airy's stress function X is used in order to analyze finite strips [I] and [II] under the boundary conditions Eq.(1). Each stress and displacement is expressed by Eq.(2) (plane stress)

$\sigma_x = \dfrac{\partial^2 \chi}{\partial y^2}$, $\sigma_y = \dfrac{\partial^2 \chi}{\partial x^2}$

$\tau_{xy} = -\dfrac{\partial^2 \chi}{\partial x \partial y}$

$u = -\dfrac{1+\nu}{E}\dfrac{\partial \chi}{\partial x} + \dfrac{1}{E}\dfrac{\partial \phi}{\partial y}$　　　(2)

$v = -\dfrac{1+\nu}{E}\dfrac{\partial \chi}{\partial y} + \dfrac{1}{E}\dfrac{\partial \phi}{\partial x}$

where

$\nabla^2\nabla^2\chi = 0$ $\left(\nabla^2 = \dfrac{\partial^2}{\partial x^2} + \dfrac{\partial^2}{\partial y^2}\right)$

$\dfrac{\partial^2 \phi}{\partial x \partial y} = \nabla^2\chi$, $\nabla^2\phi = 0$

where, E is Young's modulus and ν is Poisson's ratio.

Airy's stress function X^I for the analysis of finite strip [I] is selected as Eq.(3) from solutions of separation of variables, in consideration of the boundary conditions Eq.(1) and X^{II} for finite

880

$$\chi_1\{\overline{A}_0^{\rm I}, \overline{A}_n^{\rm I}, \overline{B}_s^{\rm I}, \overline{A}_n^{\rm I}, \overline{B}_s^{\rm I}, \alpha_n^{\rm I}(h_1), \alpha_n^{\rm I'}(h_1), \lambda_s^{\rm I}(l_1), \overline{\Delta}_n^{\rm I}(l_1), \overline{\Delta}_n^{\rm I}(l_1), \overline{\Omega}_s^{\rm I}(h_1), \overline{\Omega}_s^{\rm I}(h_1)\}$$

$$= \frac{\overline{A}_0^{\rm I}}{2}x^2$$

$$+ \sum_{n=1}^{\infty} \frac{\overline{A}_n^{\rm I}}{\alpha_n^{\rm I\,2}\overline{\Delta}_n^{\rm I}}[\{sinh(\alpha_n^{\rm I}l_1) + \alpha_n^{\rm I}l_1 cosh(\alpha_n^{\rm I}l_1)\}cosh(\alpha_n^{\rm I}x) - sinh(\alpha_n^{\rm I}l_1)\alpha_n^{\rm I}x sinh(\alpha_n^{\rm I}x)]cos(\alpha_n^{\rm I}y)$$

$$+ \sum_{s=1}^{\infty} \frac{\overline{B}_s^{\rm I}}{\lambda_s^{\rm I\,2}\overline{\Omega}_s^{\rm I}}[\{sinh(\lambda_s^{\rm I}h_1) + \lambda_s^{\rm I}h_1 cosh(\lambda_s^{\rm I}h_1)\}cosh(\lambda_s^{\rm I}y) - sinh(\lambda_s^{\rm I}h_1)\lambda_s^{\rm I}y sinh(\lambda_s^{\rm I}y)]cos(\lambda_s^{\rm I}x)$$

$$+ \sum_{n=1}^{\infty} \frac{\overline{A}_n^{\rm I}}{\alpha_n^{\rm I'\,2}\overline{\Delta}_n^{\rm I}}[\{sinh(\alpha_n^{\rm I'}l_1) + \alpha_n^{\rm I'}l_1 cosh(\alpha_n^{\rm I'}l_1)\}cosh(\alpha_n^{\rm I'}x) - sinh(\alpha_n^{\rm I'}l_1)\alpha_n^{\rm I'}x sinh(\alpha_n^{\rm I'}x)]sin(\alpha_n^{\rm I'}y)$$

$$+ \sum_{s=1}^{\infty} \frac{\overline{B}_s^{\rm I}}{\lambda_s^{\rm I\,2}\overline{\Omega}_s^{\rm I}}[\{cosh(\lambda_s^{\rm I}h_1) + \lambda_s^{\rm I}h_1 sinh(\lambda_s^{\rm I}h_1)\}sinh(\lambda_s^{\rm I}y) - cosh(\lambda_s^{\rm I}h_1)\lambda_s^{\rm I}y cosh(\lambda_s^{\rm I}y)]cos(\lambda_s^{\rm I}x) \qquad (3)$$

where

$$\alpha_n^{\rm I}(h_1) = \frac{n\pi}{h_1}, \quad \alpha_n^{\rm I'}(h_1) = \frac{2n-1}{2h_1}\pi, \quad \lambda_s^{\rm I}(l_1) = \frac{s\pi}{l_1}$$

$$\overline{\Delta}_n^{\rm I}(l_1) = cosh(\alpha_n^{\rm I}l_1)sinh(\alpha_n^{\rm I}l_1) + \alpha_n^{\rm I}l_1, \qquad \overline{\Omega}_s^{\rm I}(h_1) = cosh(\lambda_s^{\rm I}h_1)sinh(\lambda_s^{\rm I}h_1) + \lambda_s^{\rm I}h_1$$

$$\overline{\Delta}_n^{\rm I}(l_1) = cosh(\alpha_n^{\rm I'}l_1)sinh(\alpha_n^{\rm I'}l_1) + \alpha_n^{\rm I'}l_1, \qquad \overline{\Omega}_s^{\rm I}(h_1) = cosh(\lambda_s^{\rm I}h_1)sinh(\lambda_s^{\rm I}h_1) - \lambda_s^{\rm I}h_1$$

$$(n, s = 1, 2, 3, \cdots\cdots)$$

$$\chi_2 = \chi_1\{\overline{A}_0^{\rm II}, \overline{A}_n^{\rm II}, \overline{B}_s^{\rm II}, \alpha_n^{\rm II}(h_2), \lambda_s^{\rm I}(l_2), \overline{\Delta}_n^{\rm I}(l_2), \overline{\Omega}_s^{\rm I}(h_2)\} \qquad (4)$$

$$\chi_2' = \chi_1\{\overline{A}_0^{\rm II}, \overline{A}_n^{\rm II}, \overline{B}_s^{\rm II}, \overline{A}_n^{\rm II}, \overline{B}_s^{\rm II}, \alpha_n^{\rm I}(h_2), \alpha_n^{\rm I'}(h_2), \lambda_s^{\rm I}(l_2), \overline{\Delta}_n^{\rm I}(l_2), \overline{\Delta}_n^{\rm I}(l_2), \overline{\Omega}_s^{\rm I}(h_2), \overline{\Omega}_s^{\rm I}(h_2)\} \qquad (5)$$

strip [II] as Eqs.(4) and (5),where $\overline{A}_0^{\rm I}, \overline{A}_n^{\rm I}, \overline{B}_s^{\rm I}, \ldots, \overline{B}_s^{\rm II}$ (n,s=1,2,3,...) are unknown coefficients determined from the boundary conditions. The boundary conditions (1) are given by strains. The strains are obtained by differentiate the displacements expressed in Eq.(2). Suffixes I and II on right shoulder denote finite strips I and II,respectively.

2.2 Inverse analysis using displacement

An inverse analysis is done when the rigidity of finite strip [I] is comparatively larger than that of finite strip [II],that is finite strip [I] assumed to be rigid. The displacement v at the surface $y_2 = h_2$ of finite strip [II] is measured with an electric micrometer as shown in Fig3. It is assumed that the deformation at the contact surface($y_2 = h_2$, $|x| \le l_1$) is flat because the deformation equals the shape of the end surface ($y_1 = -h_1$) of finite strip [I]. Developing the displacement v into Fourier series, the boundary condition is expressed as Eq.(6)

For Fig.1(a) and (b)

$$x = \pm l_2 : \sigma_x = \tau_{xy} = 0$$
$$y = \pm h_2 : \tau_{xy} = 0$$
$$y = h_2 : (v)_{y=h_2} = c_0 + \sum_{s=1}^{\infty} c_s cos\frac{s\pi}{l_2}x \qquad (6)$$

The following condition is added in the analysis of Fig.1(b)

$$y = -h_2 : \left(\frac{\partial v}{\partial x}\right)_{y=-h_2} = 0$$

The method for analysis is the same as mentioned in 2.1.

2.3 Inverse analysis of Young's modulus

In order to determine Young's modulus E_2 of finite strip [II], strain distributions ε_y in the y_2 direction in finite strip II are used. The value of E_2 is obtained from Eq.(7),where the load $P = \int_{-l_2}^{l_2}F(X)dx$ is known.

$$E_2 = P \Big/ \int_{-l2}^{l2} \varepsilon_y(x)dx \qquad (7)$$

3 PHOTOELASTIC EXPERIMENTS

Photoelastic experiments are performed in the both cases shown in Fig.1 using the specimens shown in Fig.2. Figure 4 shows an experimental setup. A compression of 2940 N is applied to the apparatus. Photograph of isochromatic fringes are taken.

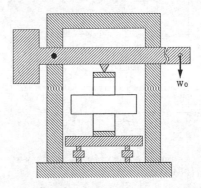

Fig.4 A sketch of photoelasitc experiment

4 DIRECT ANALYSIS OF CONTACT STRESS DISTRIBUTION

When the stress distribution F(x) at $y_1=h_1$ is known or the boundary condition at $y_1=h_1$ is given as dv/dx=0, the contact stress distribution is analyzed directly. The method for analysis of each finite strip is the same as shown in 2.1.

5 RESULTS AND DISCUSSIONS

For verification of inverse analysis, a direct analysis is carried out in the case of Fig.1. The numerical calculations are done under the condition dv/dx=0 at $y_1=h_1$ and $\int_{\ell}^{\ell} F(x)dx=2940$ N. Young's modulus E_1, E_2 and Poisson's ratio ν_1 and ν_2 are obtained as 205.8 kN/mm^2, 3.43kN/mm^2,0.33 and 0.376 from the experiments. A number of terms in series is set at 50 in the numerical calculations. Using the numerical results of the strains and the displacements obtained from the direct analysis, inverse analyses are performed. It is seen that the difference between the numerical results from the direct analysis and from the inverse analysis is below 1%. From the results, it is confirmed that inverse analysis is valid.

5.1 Results of inverse analysis using measured strains

Figure 5 shows the numerical results of the contact stress σ_y obtained from the inverse analysis when the number of meas-

ured strain is 150 and 30. In addition, the numerical results obtained from the direct analysis is also indicated. From the result, it is seen that singular stress causes near the edge ($x_2=+l_1,y_2=h_2$) of finite strip [I]. Figure 6 shows the effects of the location of measured strain on the contact stress distribution. In the inverse analysis, the values of strain obtained from the direct analysis is used,where Young's modulus E_2 is put to equal E_1. The ordinate indicates the contact stress σ_y and the abscissa is the distance x_2. From the results, it is seen that accuracy of the contact stress decreases with an decrease of the location y_2 of measured strain. When the value y_2 of measured strain becomes under 12 mm,the contact stress distribution becomes to diverge.

5.2 Results of inverse analysis using the displacements

Figure 7 shows the effects of numbers of measured displacements on the contact stress distribution. It is seen that the stress distribution obtained from inverse analysis approaches that obtained from the direct analysis with an increase of measured displacements.

5.3 Results for determining Young's modulus

Using the measured strains, Young's modulus E_2 is determined as $E_2=3.50$kN/mm^2.

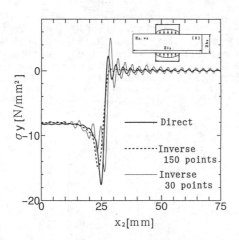

(a)The case where a strip is compressed by two strips(Fig.1(a))

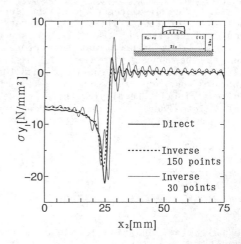

(b)The case where a strip on rigid foundation is compressed by a strip(fig.1(b))

Fig.5 Effects of measured point numbers of strain on contact stress distributions

On the other hand, the measured values of E_2 is 3.43kN/mm². From the result, it is shown that the result from inverse analysis is in fairly good agreement with the measured result.

6 COMPARISION BETWEEN THE RESULTS OBTAINED FROM INVERSE ANALYSIS AND EXPERIMENTAL RESULTS BY PHOTOELASTICITY

Figure 8 shows an example of isochromatic fringes in the case of Fig.1(a). Figure 9 shows the comparisons between the analytical and the experimental results concern-

ing the principal stress difference at the position of y_2=19.5 mm. Figure 9(a) is the case of Fig.1(a) and Fig.9(b) is the case of Fig.1(b). In addition, the results from the direct analysis are also indicated by the solid lines. From the results, good agreements are seen between the analytical and the experimental results. Figure 10 shows the principal stress difference obtained from the inverse analysis. From the comparison between Fig.8 and Fig.10, a good agreement is observed and it is said that the inverse analysis is valid.

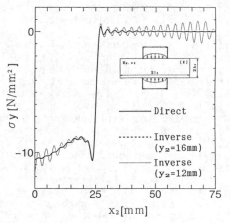

(a)The case where a strip is compressed by two strips(Fig.1(a))

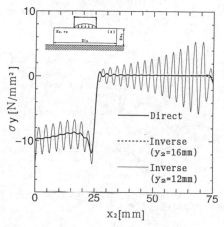

(b)The case where a strip on rigid foundation is compressed by a strip(Fig.1(b))

Fig.6 Effects of locations of measured strains on contact stress distributions

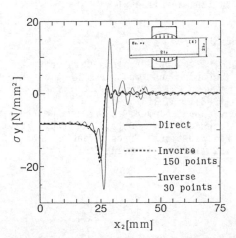

(a)The case where a strip is compressed by two strips(Fig.1(a))

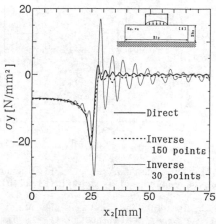

(b)The case where a strip on rigid foundation is compressed by a strip(Fig.1(b))

Fig.7 Effect of measured point numbers of displacement on contact stress distributions

Fig.8 An example of isochromatic photo (experiment)

Fig.10 Isochromatic fringes obtained from inverse analysis using displacement(numerical result)

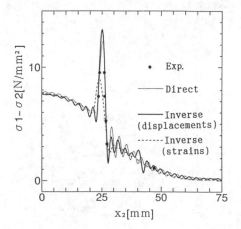

(a)The case where a strip is compressed by two strips(Fig.1(a))

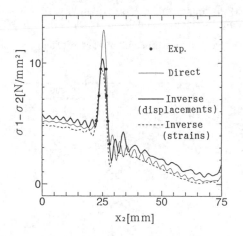

(b)The case where a strip on rigid foundation is compressed by a strip(Fig.1(b))

Fig.9 Comparisons between the numerical and the experimental results(y_2=19.5mm)

7 CONCLUSIONS

This paper dealt with the contact stress distribution among finite strips as an inverse problem. The following results are obtained.

(1)A method for analyzing the contact stress distribution and Young's modulus using strains and displacements is demonstrated using a two-dimensional theory of elasticity, in the cases where a finite strip is compressed by two finite strips and where a finite strip on a rigid foundation is compressed by a finite strip.

(2)The numerical results as an inverse analysis are compared with the results obtained from direct analysis. The numerical results are in fairly good agreement with the direct analytical results. In addition, the effects of numbers of measured strains and displacements on accuracy of the inverse solutions are examined. As a result, it is shown that the proposed method is useful.

(3)Photoelastic experiments are carried out. The numerical results are consistent with the experimental results.

REFERENCES

Kubo,S.,Inverse Problems Related to the Mechanics and Fracture of Solids and Structures, JSME Int.J.,Ser.I,31-2 (1988), 157-166.

Oda,J. and Shinada,T., On Inverse Analysis Technique to Obtain Contact Stress Distributions, Transaction of JSME ,53-492 (1987),1614-1621.

Recent Advances in Experimental Mechanics, Silva Gomes et al. (eds) © 1994 Balkema, Rotterdam, ISBN 90 5410 395 7

On the approximation of the fatigue stress diagrams

I.Goia & E.Carai
University Transilvania, Brasov, Romania

ABSTRACT: A new approximation of the fatigue stress diagram is presented along with some other known approximated diagrams. A method of classification of these diagrams, based on the correlation with experimental results is proposed. Following this method it is possible to assert which diagram approximates better one sample set of measurements.

1. FATIGUE STRESS DIAGRAMS

The following parameters define fatigue loading:

$\sigma_m = (\sigma_{max} + \sigma_{min}) / 2$ mean stress

$\sigma_v = (\sigma_{max} - \sigma_{min}) / 2$ alternating stress amplitude

$R = \sigma_{min} / \sigma_{max}$ stress ratio

Suppose fatigue tests are performed for various σ_m, σ_v on similar specimens. The result can be represented in a system of co-ordinates σ_m, σ_v (Haigh). The points that limit the fatigue strength determine a curve which is termed the fatigue diagram (Fig. 1).

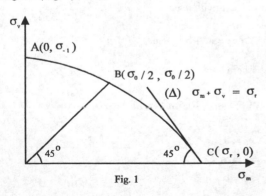

Fig. 1

One property of the fatigue diagram comes from

$$\sigma_m + \sigma_v \le \sigma_r$$

It means that the diagrams lies in the positive halfplane defined by (Δ).

2. CLASSICAL APPROXIMATION SCHEMES

The points A, B, and C represent results of tests for the reversed symmetrical stress cycle, repeated stress cycle and ultimate strength in simple tension. The real fatigue diagram is usually approximated according to the following classical schemes:

2.1 Geber

$$\sigma_v = \sigma_{-1}(1-(\sigma_m/\sigma_r)^2)$$

2.2 Goodman

$$\sigma_v = \sigma_{-1}(1-\sigma_m/\sigma_r)$$

2.3 Soderberg

$$\sigma_v = \sigma_{-1}(1-\sigma_m/\sigma_c)$$

2.4 Serensen

$$\sigma_v = \sigma_{-1}(1-\psi\sigma_m/\sigma_{-1}) \quad R \ge 1$$
$$\sigma_v = \sigma_c(1-\sigma_m/\sigma_c)/\delta \quad R < 1$$

where

$$\psi = (2\sigma_{-1}-\sigma_0)/\sigma_0$$
$$\delta = (2\sigma_c - \sigma_0)/\sigma_0$$

The parameters which defines the approximated diagrams are:

σ_{-1} reversed symmetrical stress

σ_{-0} repeated stress

σ_r static breaking stress

σ_c yield stress (used instead of static breaking stress for materials with a tension diagram having a zone of general yielding).

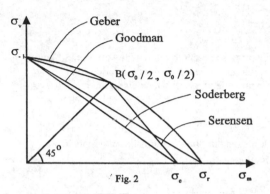

Fig. 2

3. NON CLASSICAL APPROXIMATED FATIGUE DIAGRAMS

3.1 Elliptical approximation (G. Buzdugan)

$$(\sigma_v / \sigma_{-1})^2 + (\sigma_m / \sigma_c)^2 = 1$$

3.2 Parabolic approximation (A. Petre)

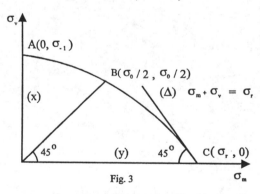

Fig. 3

$$f(x,y) = x^2 + a_1 xy + a_2 y^2 + a_3 x + a_4 y + a_5 = 0$$

The coefficients of $f(x,y)$ result from the following conditions:

(3.2.1) $a_2 - a_1^2 / 4 = 0$ - $f(x,y)$ is a parabola

(3.2.2) $\sigma_{-1}^2 + a_3 \sigma_{-1} + a_5 = 0$ - $f(\sigma_{-1}, 0) = 0$

(3.2.3) $(\sigma_0 / 2)^2 (1 + a_1 + a_2) + (\sigma_0 / 2)(a_3 + a_4) + a_5 = 0$ - $f(\sigma_0 / 2, \sigma_0 / 2) = 0$

(3.2.4) $\sigma_r^2 a_2 + \sigma_r a_4 + a_5 = 0$ - $f(0, \sigma_r) = 0$

(3.2.5) $(a_1 - 2a_2)\sigma_r + a_3 - a_4 = 0$

           - (Δ) is tangent in $C(0, \sigma_r)$

We have:

$a_1 = (S_1 \pm S_3) / S_2$

$a_2 = (S_1 \pm S_3)^2 / (2S_2)^2$

$a_3 = [(S_1 \pm S_3)(S_1 \pm S_3 - 4S_2)\sigma_r^2 + 4\sigma_{-1}^2 S_2^2] / [4(\sigma_r - \sigma_{-1})S_2^2]$

$a_4 = -[-(S_1 \pm S_3)^2 \sigma_r^2 + 2(S_1 \pm S_3)(S_1 \pm S_3 - 2S_2)\sigma_r \sigma_{-1} + 4\sigma_{-1}^2 S_2^2] / [4(\sigma_r - \sigma_{-1})S_2^2]$

$a_5 = -\sigma_r \sigma_{-1} [4\sigma_{-1} S_2^2 + (S_1 \pm S_3)(S_1 \pm S_3 - 4S_2)\sigma_r] / [4(\sigma_r - \sigma_{-1})S_2^2]$

where

$S_1 = 4\sigma_0 \sigma_r (\sigma_{-1} + \sigma_r) - 2\sigma_0^2 (\sigma_r - \sigma_{-1}) - 8\sigma_r^2 \sigma_{-1}$

$S_2 = \sigma_0^2 (\sigma_r - \sigma_{-1}) + 4\sigma_{-1} \sigma_r (\sigma_0 - \sigma_r)$

$\Delta = 64\sigma_r^4 \sigma_{-1}^2 + 16\sigma_0^2 \sigma_r^2 [(\sigma_r + \sigma_{-1})^2 + 2\sigma_{-1}(\sigma_r - \sigma_{-1})] - 64\sigma_0 \sigma_r^3 \sigma_{-1}(\sigma_r + \sigma_{-1}) - 64\sigma_r \sigma_{-1}^3 (\sigma_0 - \sigma_r)^2 - 16\sigma_0^2 (\sigma_r^2 - \sigma_{-1}^2)[\sigma_0 \sigma_r + \sigma_{-1}(\sigma_0 - \sigma_r)]$

$S_3 = \sqrt{\Delta}$

3.3 Piecewise cubic polynomial (proposed by the authors)

The fatigue diagram could be approximated by two cubic polynomials

$$Y_i(x) = a_i + b_i u_i + c_i u_i^2 + d_i u_i^3$$

where

$$u_i = (x - x_i) / (x_{i+1} - x_i) \quad i = 0,1$$

$x_0 = 0$

$x_1 = \sigma_0 / 2$

$x_2 = \sigma_c$

The conditions for the two polynomials are:

$Y_i(x_i) = y_i$

$Y_i(x_{i+1}) = y_{i+1} = (a_i + b_i + c_i + d_i)/(x_{i+1} - x_i)$

$Y_i'(x_i) = D_i = b_i/(x_{i+1} - x_i)$

$Y_i'(x_{i+1}) = D_{i+1} = (b_i + 2c_i + 3d_i)/(x_{i+1} - x_i)$

where

$y_0 = \sigma_{-1}$

$y_1 = \sigma_0/2$

$y_2 = 0$

The values for the first derivatives are imposed as follows:

$D_0 = 0$

$D_1 = -\sigma_{-1}/\sigma_c$

$D_2 = -1$

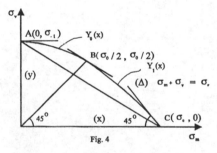

Fig. 4

Following the calculations the values of the coefficients are:

$a_i = y_i$

$b_i = D_i(x_{i+1} - x_i)$

$c_i = 3(y_{i+1} - y_i) - (2D_i + D_{i+1})(x_{i+1} - x_i)$

$d_i = -2(y_{i+1} - y_i) + (D_i + D_{i+1})(x_{i+1} - x_i)$

$a_0 = \sigma_{-1}$

$b_0 = 0$

$c_0 = 3(\sigma_0/2 - \sigma_{-1}) + \sigma_{-1}\sigma_0/(2\sigma_c)$

$d_0 = -2(\sigma_0/2 - \sigma_{-1}) - \sigma_{-1}\sigma_0/(2\sigma_c)$

$a_1 = \sigma_0/2$

$b_1 = -(\sigma_c - \sigma_0/2)\sigma_{-1}\sigma_c$

$c_1 = -3\sigma_0/2 + (2\sigma_{-1}/\sigma_c + 1)(\sigma_c - \sigma_0/2)$

$d_1 = \sigma_0 - (\sigma_{-1}/\sigma_c + 1)(\sigma_c - \sigma_0/2)$

4. CLASSIFICATION

Suppose we have a sample set of experimental results

(x_i, y_i^o) $i = 1, n$

For each approximated scheme it is possible to build

(x_i, y_i^*) $i = 1, n$

where y_i^* is the corresponding y for x_i according to the selected scheme.

It is also possible to consider the correlation coefficient

$$\rho = \frac{\sum(y_i^o - \bar{y}^o)(y_i^* - \bar{y}^*)}{\sqrt{\sum(y_i^o - \bar{y}^o)^2 \sum(y_i^* - \bar{y}^*)^2}}$$

as a measure of the approximation: as ρ is closed to 1 the scheme better approximates the experimental results.

REFERENCES

Bartels R.H., Beatly J.C., Barsky B.A. 1987. An Introduction to Splines for Use in Computer Graphics and Geometric Modelling. California: Morgan Kaufmann.

Buzdugan G. 1986. Rezistenta Materialelor. Bucuresti: Editura Academiei.

Cioclov D. 1975. Rezistenta si Fiabilitate la Solicitari Variabile. Timisoara: Facla.

Goia I. 1982. Rezistenta Materialelor. Brasov: Universitatea Transilvania.

Gnedenko B.V. 1988. The Theory of Probability. Moscow: MIR.

Petre A., Draghici I. 1990. Tehnici de Calcul al Coeficientului de Siguranta la Solicitari Variabile. Bucuresti: Studii si Cercetari de Mecanica Aplicata, Tom 49, nr. 4, pag. 365-374.

Recent Advances in Experimental Mechanics, Silva Gomes et al. (eds) © 1994 Balkema, Rotterdam, ISBN 90 5410 395 7

Adherence, friction and contact geometry of a rigid cylinder rolling on the flat and smooth surface of an elastic body (natural rubber) and influence of uniaxial prestrains

Michel Barquins
Centre National de la Recherche Scientifique, Paris, France

ABSTRACT: Static equilibrium conditions, the contact geometry and the rolling friction force of a hard and polished cylinder in contact with the clean surface of an elastic body are studied using the classic thermodynamics, the concepts of fracture mechanics, such as the stress intensity factor or the strain energy release rate and the plane elasticity theory. It is shown that the rolling resistance, the contact geometry and the pressure are linked to the three interfacial quantities: Dupré's energy of adhesion and the dissipative forces resisting the rupture and the formation of the contact. Experiments carried out with PMMA (polymethyl-methacrylate) cylinders rolling on smooth NR (natural rubber) surfaces confirm all the theoretical predictions, in particular, one can observe that the experimental value of the contact length, which strongly increases with the rolling speed, remains perfectly comprised between the theoretical ones deduced from the measured rolling resistance and the extreme values of the dissipative force resisting the formation of the contact. Moreover, it is shown that, due to the intervention of molecular attraction forces, a cylinder can roll under an inclined NR surface without falling down, and it is displayed that the rolling speed is the same when the cylinder rolls upon the same inclined surface. Also, this study clearly proves that, if the rubber surface is stretched, the hypothesis concerning the preservation of the global surface energy, for uniaxial extensions in the range 1-2, is valid. So, in spite of strong stresses created by rolling, it is demonstrated that imposed extensions do not provoke the appearance of rubber material from the bulk to the surface.

1 INTRODUCTION

Previous works (Maugis and Barquins 1980, 1983, Barquins 1988, Felder and Barquins 1989) have shown that an approach derived from the Griffith's theory allows one to describe the evolution of the contact area between a hard punch and an elastic solid for different geometries such as plane and axisymmetric peeling experiments or contacts of plan, spherical or cylindrical hard punches. The edge of the contact area is assumed to be a crack tip which propagates in the interface, moving backwards or forwards as the applied load is increased or decreased. This approach, which takes into account the strain energy release rate G and its derivative $\partial G/\partial A$ with respect the area of contact A, enables the determination of the elastic adherence force, to study the kinetics of the crack propagation speed V and predict the evolution of the system whatever the geometry of contact and loading conditions. It has been shown that, in a wide range of crack propagation speed V (if V is not too small, *i.e.* $G \gg w$), the strain energy release rate varies as a power function of the crack speed as follows:

$$G-w = kwV^n \qquad (1)$$

w being the Dupré's energy of adhesion and k a parameter depending only on the temperature.

Although more complicated geometries like rolling (Kendall 1975, Barquins *et al.* 1978, Roberts 1979, Fuller and Roberts 1981, Roberts 1989) and sliding (Barquins and Courtel 1975, Roberts and Thomas 1975, Savkoor and Briggs 1977, Barquins 1987) friction conditions have already been widely studied, it appears that their theoretical analyses are incomplete. This paper aims to show that the thermodynamical approach, which generalizes Griffith's theory together with the plane elasticity theory (Johnson 1985) allow us to solve problems of adherence and rolling of a hard cylinder in adhesive contact with an elastic solid, such as natural rubber.

2 THERMODYNAMICAL APPROACH

Let us consider a long and rigid cylinder with weight W, radius R and length L in contact with the flat and smooth surface of an elastic body, characterized by Young's modulus E and Poisson ratio ν, and let $E^*=E/(1-\nu^2)$. All the forces, like normal applied loads or rolling resistances, are expressed per unit axial length of cylinder.

When a cylinder rolls upon a rubber surface (Kendall 1975, Barquins 1988), the edges of the contact area, perpendicular to the motion, can be seen as two crack tips propagating in the same direction, with the same mean speed but their behaviours are different: there is a closing crack in the front region and an opening crack at the rear where the major part of the energy is dissipated by a peeling mechanism, so that a large extension is observed at the trailing edge. This is the reason for considering an asymmetrical contact (Figure 1).

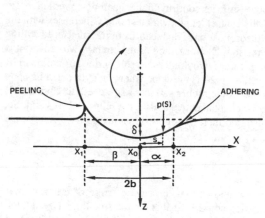

Figure 1. Contact geometry of a rigid cylinder rolling on an elastic solid.

At a given time t, the relative position of the two solids in contact is determined by the elastic displacement δ and abcissas X_0, X_1 and X_2 of the cylinder axis, the trailing and leading edges of the contact area, respectively (Figure 1). According the second principle of thermodynamics, evolution of the system is only possible if:

$$P\dot{\delta}+F\dot{X}_0+df/dt \geq 0$$

where P and F are respectively the normal force and rolling resistance per unit axial length, $\dot{\delta}=d\delta/dt$, $\dot{X}_0=dX_0/dt$ and f is the component of the free energy that depends on the stored elastic energy U_E together with the Dupré energy of adhesion w which are a function of the contact size (Figure 1), so that:

$$f = U_E(\delta,\alpha,\beta)-w(X_2-X_1)$$

with $\alpha=X_2-X_0$ and $\beta=X_0-X_1$. Taking into account the two strain energy release rates G_α and G_β at the leading and trailing edges respectively:

$$G_\alpha = \partial U_E/\partial\alpha \quad \text{and} \quad G_\beta = \partial U_E/\partial\beta$$

Equation (1) can be written:

$$(P-\partial U_E/\partial\delta)\dot{\delta}+(F+G_\alpha-G_\beta)\dot{X}_0+$$
$$(G_\alpha-w)\dot{X}_1+(w-G_\beta)\dot{X}_2 \geq 0 \qquad (2)$$

This condition means that a sum of terms, each one being equal to the product of variables of force type by evolution velocities of the system cannot be smaller than zero. Equation (2) is satisfied using the normal principle of dissipation and assuming that every force is equal to the partial derivative of a function Ω, called dissipative function, with respect to the associated evolution speed:

$$\Omega - \omega(\dot{X}_1)+\omega(-\dot{X}_2)$$

where ω is a derivable function, which can be possibly discontinuous at zero and which verifies the condition:

$$X\omega'(\dot{X}) \leq 0 \qquad (3)$$

So, the behaviour law of the system is defined by the four relations:

$$P = \partial U_E/\partial\delta \qquad (4)$$

$$F = G_\beta-G_\alpha \qquad (5)$$

$$G_\alpha = w+\omega'(-\dot{X}_2) \qquad (6)$$

$$G_\beta = w+\omega'(\dot{X}_1) \qquad (7)$$

so that the rolling resistance is:

$$F = \omega'(\dot{X}_1)-\omega'(-\dot{X}_2)$$

For the simple case of a static contact under a normal applied load, obviously one has $F=0$, hence $G_\beta=G_\alpha$ and $\alpha=\beta=b$. Regarding rolling conditions at constant speed V, as noted earlier, the leading and trailing edges of the contact area move at the same velocity $\dot{X}_1=\dot{X}_2=V\geq0$. Taking into account Equations from (4) to (7), the rolling resistance is equal to the sum of two positive terms which are the dissipative components $[\omega'(\dot{X}_1)]$ and $[-\omega'(-\dot{X}_2)]$ of molecular attraction forces acting during the detachment of the elastic solid at the trailing edge of the contact area and during the re-adhering in the front part, respectively. As the energy release rate is always positive (or equal to zero) and as $\omega'(-\dot{X}_2)\leq0$ from Equation (3), one deduces, according to experimental results obtained by Kendall (1975) for plane peeling conditions:

$$0 \leq G_\alpha \leq w \quad \text{and} \quad -w \leq \omega'(-\dot{X}_2) \leq 0$$

So, it can be predicted that the front half-length α

is smaller than the rear one β, as confirmed by rolling ball experiments (Barquins *et al.* 1978, Roberts 1979, 1989).

3 PLANE ELASTICITY THEORY APPLICATION AND CONTACT GEOMETRY

Let us consider the asymmetrical contact, shown in Figure 1, between a hard cylinder and an incompressible elastic half-space ($\nu=0.5$), such as a rubber-like material. This particular two-dimensional problem in which displacements are specified over the interval $-\beta \leq X \leq \alpha$ can be easily written from general equations (Johnson 1985) under the following form:

$$\int_{-\beta}^{\alpha} [p(S)/(X-S)]dS = \pi E^* X/2R \qquad (8)$$

where $p(S)$ is the normal pressure and R the cylinder radius. Inverting Equation (8), the pressure at the distance X from the cylinder axis, is:

$$p(X) = [1/\pi(X+\beta)^{1/2}(\alpha-X)^{1/2}]\{-P+(E^*/2R)$$
$$\int_{-\beta}^{\alpha} [(S+\beta)^{1/2}(S-\alpha)^{1/2}/(X-S)]dX\} \qquad (9)$$

In order to simplify calculations, one symmetrizes the contact area, with the length $2b=\alpha+\beta$, by choosing the axis origin at the centre of the loaded region; distances are thus changed from X to $r=X-(\alpha-\beta)/2$ and from S to $s=S-(\alpha-\beta)/2$. Thus, Equation (9) becomes:

$$p(r) = [1/\pi(b^2-r^2)^{1/2}]\{-P+(E^*/2R)$$
$$\int_{-\beta}^{\alpha} [(b^2-s^2)^{1/2}(s+(\alpha-\beta)/2)/(r-s)]ds\} \qquad (10)$$

as proposed by Johnson (1985) with some misprints. The integration is easier if displacements r and s are expressed in terms of b:

$$p(r) = [1/(b^2-r^2)^{1/2}]\{-P/\pi+(E^*/2R)[r^2+$$
$$r(\alpha-\beta)/2)-b^2/2]\}$$

The stress intensity factor K_α at the leading edge of the contact area, calculated as follows:

$$K_\alpha = \lim_{r \to +b} \{p(r)[2\pi(b-r)]^{1/2}\}$$

is equal to:

$$K_\alpha = -P/(\pi b)^{1/2}+E^*(3\alpha-\beta)(\pi b)^{1/2}/8R \qquad (11)$$

At the trailing edge, the corresponding stress intensity factor K_β takes the similar form:

$$K_\beta = -P/(\pi b)^{1/2}+E^*(3\beta-\alpha)(\pi b)^{1/2}/8R \qquad (12)$$

For a plane strain state, the strain energy release rate is linked to the stress intensity factor by:

$$G_i = K_i^2/2E^* \qquad (13)$$

in which the factor (1/2) is introduced in order to take into account the cylinder rigidity.

So, the relation linking the normal applied load P per unit axial length, the half-length b of the contact area and the strain energy release rates G_α and G_β in front and rear parts of the contact zone can be easily deduced from Equations (11), (12) et (13):

$$P = \pi E^* b^2/4R-(\pi E^* b/2)^{1/2}[(G_\alpha)^{1/2}+(F+G_\alpha)^{1/2}] \qquad (14)$$

where F is the rolling resistance per unit axial length of the cylinder (Equation (5)).

4 STATIC CONTACT

For an equilibrium static contact, without rolling *i.e.* $F=0$ and $G_\alpha=w$, Equation (14) takes the simple form:

$$P=\pi E^* b^2/4R-(2\pi E^* bw)^{1/2} \qquad (15)$$

as already proposed by Barquins (1988). This equilibrium relationship is represented on Figure 2, in which forces and lengths are normalized as:

$$|P|=P/(\pi E^* w^2 R/16)^{1/3} \qquad (16)$$

$$|b|=b/(2wR^2/\pi E^*)^{1/3} \qquad (17)$$

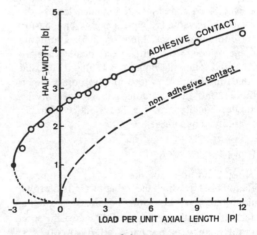

Figure 2. Half-width of the equilibrium contact area between a rigid cylinder and the smooth surface of an elastic solid as a function of the normal applied load per unit axial length, in reduced coordinates. Experimental data (symbols) were obtained with a PMMA cylinder in contact on a NR surface. The curve deduced from the classical theory (non adhesive contact) is given for comparison.

Figure 2 clearly shows that molecular attraction forces, which act through the Dupré energy of adhesion w, increase the length of the contact area for the same normal applied load (the non-adhesive contact case, $w=0$, is given for comparison). Due to this intervention of surface energy effects, an equilibrium contact area exists under zero and negative applied loads, the ultimate value that corresponds to the limit of stability of the contact, $(\partial G/\partial b)_P=0$ with $G=w$, is $|P_c|=-3$, and the associated smallest stable contact area is $|b_c|=1$.

The experimental device used in order to verify the theoretical curve (heavy line in Figure 2) is fundamentally the same as that the one used by Barquins *et al.* (1978) to study the rolling friction of a glass ball in contact with a rubber plate. The ball is replaced by a PMMA cylinder, 6.3 mm long, with a radius of curvature $R=2.5$ mm. This cylinder is applied against the smooth surface of a soft rubber-like material plate with the help of a balance arm. The area of contact and its immediate vicinity are observed in reflexion through the specimen with a low power microscope equipped with a video camera. The elastic specimen is a transparent plate, 2 mm thick, of an unfilled natural rubber cured with 2% dicumyl peroxide ($E=0.89$ MPa).

In a first set of experiments, the PMMA cylinder is applied during 10 min against the NR plate under several normal loads varied corresponding to forces per unit axial length from -6.228 to +31.143 N.m^{-1}. For every load and associated contact length, the corresponding value of Dupré's energy of adhesion w is computed from Equation (15). From the average value $w=75$ mJ.m^{-2}, experimental data are normalized following Equations (16) and (17) and plotted on Figure 2 where all the points (open symbols) fall in the immediate vicinity of the theoretical equilibrium curve (heavy line) confirming the validity of hypotheses.

5 ROLLING CONTACT

Rolling conditions are defined by the general Equation (14) in which the value of G_α is narrowly limited ($0 \leq G_\alpha \leq w$). Equation (14) clearly shows that an increase in rolling resistance F, at increasing speed without change in normal load P, involves an increase in contact area length. From Equations (12) and (13) the size of the contact area is given by:

$$\beta=3b/2-(2R/\pi E^*b)[P+(2\pi E^*bG_\alpha)^{1/2}] \tag{18}$$

$$\alpha=2b-\beta \tag{19}$$

so that at increasing rolling speed, a large extension of the rear part of the contact area can be easily predicted.

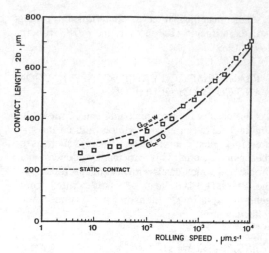

Figure 3. Total contact length versus rolling speed: experimental data (symbols) and theoretical curves for the two extreme values of the strain energy release rate at the leading edge of the contact area.

Lengths of contact area between the PMMA cylinder and NR plate, at various speeds, are shown on Figure 3 (symbols). Every point represents the average value of 10 measurements at constant speed along a rolling distance 15 mm long, *i.e.* for about one cylinder rotation. As already observed with ball rolling experiments (Barquins *et al.* 1978, Roberts 1979, 1989), the contact length strongly increases at growing speed, with a marked extension of the rear part of the contact area, according to Equation (18). Figure 3 also shows the two theoretical curves that can be deduced from Equation (15) for the extreme values of the strain energy release rate at the leading edge of the contact area, $G_\alpha=0$ and $G_\alpha=w$ (with $w=75$ mJ.m^{-2}), taking into account measured values of the rolling resistance F *versus* corresponding rolling speeds.

The very good position of experimental data (symbols) between the two theoretical curves mainly shows that the increase in the rolling contact length together with the increase in the rolling resistance at increasing speed are two concomitant consequences of the viscous energy dissipated during the separation of the rubber at the trailing edge of the contact area.

6 ROLLING ENERGY

As the rolling friction can be seen as a particular form of a $\pi/2$ peeling test (Kendall 1975, Roberts and Thomas 1975), and because, whatever the peeling angle the relation between the peeling force and the crack propagation speed does not depend on the length of the not peeled trip or on the area of contact in a rolling experiment, we have study the rolling speed of a PMMA cylinder in contact on a natural rubber strip for various incline angles θ in the range 0°-180° with respect to the horizontal.

If the weight P of the cylinder per unit axial length is smaller than the absolute value of the adherence force P_c, firstly, the cylinder rolls under the inclined rubber trip and, secondly , the rolling speed takes the same value as if the cylinder is in contact upon the same inclined rubber surface. In Figure 4 experimental data obtained with a PMMA cylinder, are calculated from the strain energy release rate G calculated following the simple form (Kendall 1975, Roberts and Thomas 1975):

$$G = P\sin\theta \qquad (20)$$

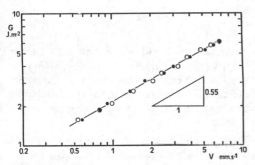

Figure 4. G strain energy release rate *vs* V rolling speed of a PMMA cylinder along an inclined soft natural rubber flat surface (open symbols θ≤90°; filled symbols θ≥90°).

Figure 4 clearly proves that G is computable by Equation (20) for every slope (0°≤θ≤180°) of the cylinder trajectory, so that the energy dissipated by rolling arises solely from the work of tensile stresses at the trailing edge of the contact area, independent of the weight per unit length of the cylinder, provided that this weight is smaller than the absolute value of the elastic adherence force P_c. Moreover, Figure 4 shows that the strain energy release rate G varies as the n power fonction of the crack propagation speed with $n=0.55$ for the soft unfilled natural rubber tested:

$$G = \kappa V^{0.55} \qquad (21)$$

with κ=96.5 SI units for environmental conditions (temperature and humidity ratio). Thus, using Equations (20 and 21), it is possible to compute and draw the variation in the rolling speed V as a function of the incline angle θ.

7 INFLUENCE OF PRESTRAINS ON ROLLING

The main goal of the last part of this study is to demonstrate that the increase in the rolling speed of a rigid and smooth cylinder on the pre-stretched surface of a natural rubber sheet at increasing strain is solely due to the corresponding decrease in the Dupré's energy of adhesion, as already shown in the case of static contacts of cylinders (Barquins and Felder 1991), in spite of strong surface stresses provoked by rolling.

Experiments were carried out, at constant temperature (22°C) and humidity ratio (61%), using three rigid cylinders, made of PMMA, with the same length l=20 mm and with the same radius of curvature R=30 mm, complete and with two axial circular groovings in order to obtain three loads per unit length: P_1=33.25 N.m^{-1} (complete cylinder), P_2=18.47 N.m^{-1} and P_3=6.318 N.m^{-1} (grooved cylinders). Before each experiment, pre-strain sheets were carefully placed on a 8 mm thick sheet of unstrained rubber in order to create a continuous elastic half-space with regards lengths of the contact areas between rubber surfaces and cylinders.

In a first set of experiments, Dupré's energy of adhesion w_0 between an unstrained natural rubber sheet and cylinders was determined from Equation (15). From measured equilibrium values: $2b_1$=2.75 mm, $2b_2$=2.35 mm and $2b_3$=1.90 mm, corresponding to P_1, P_2 and P_3, respectively, and from $2b'_3$=1.20 mm for the load P'_3=-P_3, because this is smaller than the equilibrium critical value (Barquins 1988), the average value w_0=67 mJ.m^{-2} was calculated.

In a second set of experiments, the rolling speeds of the three cylinders were measured for different inclinations of the rigid plate sustaining pre-streched natural rubber sheets. Experimental data are shown on Figure 5, in which the strain energy release rate G, calculated from Equation (20), is expressed as a function of the rolling speed V. G varies as the n=0.55 power function of the rolling speed V whatever the imposed pre-strain, in the range ε=0-100% or λ=1-2 for which the very small variations of Young's modulus are not significant (Treloar 1958), hence, the value n=0.55 wholly characterizes the viscoelastic behaviour of the rubber sample tested, as already seen (Figure 4).

A parallel set of experiments was carried out in order to establish the variation law of Dupré's energy of adhesion w as a function of the imposed pre-extension λ. Figure 6 shows the variation of the area S of an initial square, with area S_0, carefully drawn on the unstrained rubber surface *versus* the extension λ in the range 1-2. For comparison, the variation

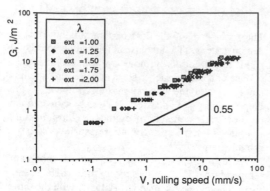

Figure 5. Strain energy release rate as a function of the rolling speed, for five pre-extensions λ in the range 1-2.

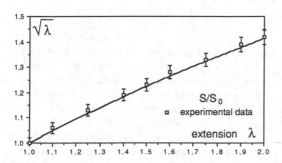

Figure 6. Area, normalized to its initial value, of a square initially drawn on the rubber surface, as a function of extensions imposed to rubber sheets. For comparison, the square root of the extension is computed in order to show the incompressive behaviour of the rubber sample tested.

of the square root of λ illustrating the incompressive behaviour of the rubber is drawn, and all the experimental points fall on the computed curve, so that the variation of the area of a square versus the uniaxial extension is given by the simple relation:

$$S/S_0 = \lambda^{1/2}$$

Assuming that the global surface energy U_S is conserved if the rubber is stretched, that is to say, $U_S = w_0.S_0 = w.S$, the value of Dupré's energy associated to a given extension λ can be easily deduced from the relation:

$$w = w_0.(S_0/S) = w_0.\lambda^{-1/2} \qquad (22)$$

the corresponding values of w, with $w_0 = 67$ mJ.m^{-2}, are given in the table drawn in Figure 7.

Using the equation of the kinetics of adherence (Equation (1)), Figure 7 shows that all the

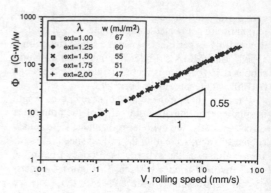

Figure 7. Dupré's energy of adhesion values associated to imposed extensions and master curve showing the dissipative function $\Phi = (G-w)/w$ of the natural rubber, as a function of the crack propagation speed in opening mode at the interface with a rigid solid.

experimental points fall on the same master curve illustrating the dissipative function $\Phi = (G-w)/w$ versus V, which expresses the viscoelastic properties of the natural rubber sample tested as a function of the crack propagation speed V at the interface with a rigid solid.

These results clearly prove that the hypothesis concerning the preservation of the global surface energy, for uniaxial extensions in the range 1-2, is valid. So, in spite of strong stresses created by rolling, it is demonstrated that imposed extensions do not provoke the appearance of rubber material from the bulk to the surface.

8 CONCLUSION

Static equilibrium conditions, the contact geometry and the rolling friction energy of a hard and polished cylinder in contact with the clean, smooth and flat surface of an elastic body are studied using the classic thermodynamics, the concepts of fracture mechanics, such as the stress intensity factor or the strain energy release rate and the plane elasticity theory. It is shown that the rolling friction energy, the contact geometry and the pressure are linked to the three interfacial quantities: Dupré's energy of adhesion and the dissipative forces resisting the rupture and the formation of the contact. Moreover, if the rubber surface is subjected to an uniaxial tensile strain, it is proved that the increase in rolling speed at increasing pre-extension is closely linked to the corresponding decrease in Dupré's energy of adhesion.

Experiments carried out with PMMA (polymethylmethacrylate) cylinders rolling on NR (natural rubber) surfaces confirm all the theoretical predictions.

REFERENCES

Barquins M. 1987. Adhérence, frottement et usure des élastomères. *Kautschuk Gummi + Kunstsoffe.* 40: 419-438.

Barquins, M. 1988. Force d'adhérence et cinétique de roulement d'un cylindre rigide en contact adhésif avec la surface plane et lisse d'un matériau élastique souple (caoutchouc naturel). *C. R. Acad. Sci. Paris, série II.* 306: 509-512.

Barquins M. & R. Courtel 1975. Rubber friction and the rheology of viscoelastic contact. *Wear.* 32: 133-150.

Barquins M., D. Maugis, J. Blouet & R. Courtel 1978. Contact area of a ball rolling on an adhesive viscoelastic material. *Wear.* 51: 375-384.

Barquins M. & E. Felder 1991. Influence des précontraintes sur l'adhésion du caoutchouc naturel. *C. R. Acad. Sci. Paris, série II.* 313: 303-306.

Felder E. & M. Barquins 1989. Adhérence, frottement et géométrie de contact d'un cylindre rigide roulant sur la surface plane et lisse d'un massif élastique. *C. R. Acad. Sci. Paris, série II.* 309: 1101-1104.

Fuller K. N. G. & A. D. Roberts 1981. Rubber rolling on rough surfaces. *J. Phys. D: Appl. Phys.* 14: 221-239.

Johnson K. L. 1985. *Contact mechanics.* Cambridge: Cambridge University Press.

Kendall K. 1975. Rolling friction and adhesion between smooth solids. *Wear.* 33: 351-358.

Maugis D. & M. Barquins 1978. Fracture mechanics and adherence of viscoelastic bodies. *J. Phys. D: Appl. Phys.* 11: 1989-2023.

Maugis D. & M. Barquins 1983. Adhesive contact of sectionally smooth-ended punches on elastic half-space: theory and experiment. *J. Phys. D: Appl. Phys.* 16: 1843-1874.

Roberts A. D. 1979. Looking at rubber adhesion. *Rubber Chem. Technol.* 52: 23-42.

Roberts A. D. 1989. Rubber adhesion at high rolling speeds. *J. nat. Rubb. Res.* 4: 239-260.

Roberts A. D. & A. G. Thomas 1975. The adhesion and friction of smooth rubber surfaces. *Wear.* 33: 45-64.

Savkoor A. R. & G. A. D. Briggs 1977. The effect of tangential force on the contact of elastic solids in adhesion. *Proc. R. Soc. A.* 356: 103-114.

Treloar L. R. G. 1958. *The physics of rubber elasticity.* Oxford: Clarendon Press.

Recent Advances in Experimental Mechanics, Silva Gomes et al. (eds) © 1994 Balkema, Rotterdam, ISBN 90 5410 395 7

Experimental stress analysis of normally intersecting pipes subjected to moments

Nuno Ferreira Rilo & José Maria O. Sousa Cirne
Departamento de Engenharia Mecânica da Faculdade de Ciências e Tecnologia da Universidade de Coimbra, Portugal

Joaquim F. Silva Gomes
Laboratório de Óptica e Mecânica Experimental, Secção de Mecânica Aplicada, DEMEGI da Faculdade de Engenharia da Universidade do Porto, Portugal

ABSTRACT: Stresses in the neighbourhood of T cylinder-cylinder intersections concern to piping designers, particularly in chemical industries and power plants.

The authors investigated Stress Intensification Factors (SIF) in unreinforced fabricated Tees. Three different geometries were considered (d/D = 1, 0.52, 0.27), using cantilever like models subjected to a complete set of six individual moment loading.

A loading framework was construted and care was taken to ensure that the loads applied were such that the moments transmitted through the junction were pure moments with minimum extraneous axial or shear forces.

Results with data from strain gauge measurements were compared with FEM analysis using a DKT shell element of MODULEF program and also with values predicted from ASME III, ANSI 31.3 and BS 806 codes.

INTRODUCTION

Intersecting cylindrical shells are commonly used in structural components of chemical industries and power plants. Since the early investigations of Schoessow and Koistra (1945) and of Bijlaard (1954, 1955a, b), that based WRC 107 (Whichman *et al.*, 1965), there have been many studies on this subject. We refer works of Steele *et al.* (1980, 1981, 1983, 1986a, b, Khathlan *et al.* 1986) whose Shelltech Reports supported WRC 297 (Mershon *et al.*, 1984). In this "cook book" stresses in both the nozzle and vessel can be determined and the range of vessel diameter-to-thickness ratio covered is increased over that of WRC 107. The analytical method used was derived and developed on the basis of thin shell theory. These two bulletins provide the basis for current ASME pressure vessel codes (ASME, 1980, 1983).

During the 1970's several authors carried out in Oak Ridge National Laboratory relevant experimental and numerical studies (Corum *et al.*, 1972, Gwaltney *et al.*, 1975) whose models are commonly used as ORNL reference models.

In the 1980's quite a good number of studies concerning stresses in cylinder-cylinder intersections were developed. We mention works of Khan *et al.* (1984a, b, 1987) in the Univ. of Oklahoma and those of Moffat *et al.* (1986, 1989) in Univ. of Liverpool.

This is only a summary of the work to date on stresses in the vicinity of cylindrical shell intersections. A more detailed presentation of previous work can be found in Rilo (1992) where more than seventy publications are discussed.

From a study of all these previous references, it is seen that not much effort has been expended into the problem of maximum stresses in the nozzle-cylinder intersection for d/D>0.5. British (BS 806, 1982) and American (ASME, 1980) piping codes include design procedures for d/D up to 1.0. There is however room for improvement as well as debate in this codes. It is also clear that a great deal of work needs to be done before the design methods can be considered reliable.

Low-temperature plants often use unreinforced fabricated tees (U.F.T) of thin-walled pipes. This paper presents results of Stress Intensification Factors (SIF) using cantilever like models subjected to a complete set of six individual moment loading (Fig 1).

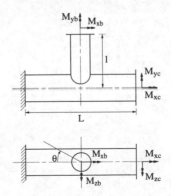

Figure 1. Cantilever model and moment loads.

They were obtained by data from strain gauge measurements and from the finite element method on a series of three fabricated branch junctions of d/D=1.0, 0.52 and 0.27 (Table 1).

Table 1. Dimensional parameters for experimental models.

Model	d/D	d [mm]	D [mm]	t [mm]	T [mm]	l [mm]	L [mm]
T1	1.0	216.3	216.3	2.8	2.8	433	866
T2	0.52	112.1	216.3	2.8	2.8	433	866
T3	0.27	58.7	216.3	1.6	2.8	300	866

EXPERIMENTAL DETAILS

One end of the main shell was clamped in 25 mm thick plate, and the six moment loading on the end of the branch or the attached shell were applied by means of pairs of loading cells. The stresses in the intersection were determined for the boundary condition of one end clamped and one end free. Micro Measurements electrical resistance three-element rosette gauges type CEA-06-125-UR-120 and two--element rectangular rosettes type CEA-06-125-UT-120 were used for the determination of strains. The bulk of gauges were spaced at 22,5 deg. intervals around the junction (Fig. 2) in T1 and T2 models and 45 deg. in T3 model.

Pure moments were applied in the manner illustrated schematically in Fig 3. All loads were applied at equal increments and strain gauge were switched and balanced in MM SB10 units and readings were obtained in a MM P3500 strain indicator.

The readings were subsequently processed using a least-square fit for all gauges for each of the six moment categories. Strains from 1KN×m moments were calculated and conventional experimental stress analysis procedures were followed; strains from each rosette were combined to give the principal strains. The principal stresses were then obtained from the generalized Hooke's Law and the Mises effective stresses were then obtained.

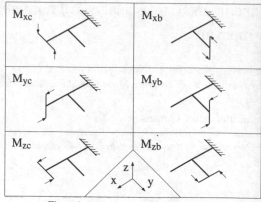

Figure 3. Schematic of loading techniques.

FINITE ELEMENT ANALYSIS DETAILS

The finite element analysis has been performed using Modulef program developed by Modulef club at the INRIA, Paris. In the analysis, the DKT shell element TRIA CQFA was used (Fig. 4 and 5). This element has six degrees of freedom at each node.

Due to symmetry of the loading and geometry, only one-half or one-quarter of the cylinder-cylinder intersection geometry is analyzed.

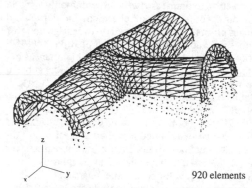

920 elements

Figure 4. Finite element mesh for one-half modelT1.

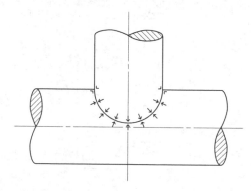

Figure 2. Strain gage locations.

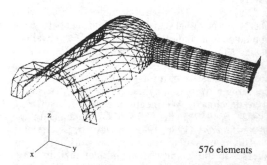

576 elements

Figure 5. Finite element mesh for one-quarter modelT2.

898

RESULTS AND DISCUSSION

Stress Intensity Factors (SIF) for all moment loadings are based on the nominal von Mises stress intensity Mr_c/I arising from a bending moment $M=1KN\times m$ acting on an equivalent plain pipe. It should be born in mind that i factors, used in several codes, are fatigue stress intensities and their empirical relation with Stress Intensity Factors is $SIF=2i$.

The experimental and finite element SIFs around crotch line were plotted and compared in graphs like those of figures 6 to 9 for the three models and six load moment categories.

Considering the fact that finite element results do not include stress intensification effects due to the presence of the welding, the agreement between stress ratios obtained by the use of finite elements and experimental techniques is remarkable for model T1, as can be seen from these graphs. In two other models increase in the relative wall branch thickness an the effect of weld are proeminents and agreement is not so good.

Tables 2, 3 and 4 summarize the maximum stress intensities and compares these values with those given in design codes (ASME, 1980; BS 806, 1982; ANSI/ASME 1980a, b).

Table 2. SIF in model T1.

Moment	Exp.		F.E.		ASME	ANSI	BS
	Cil.	Bran.	Cil.	Bran.			
M_{xc}	16.58	28.10	18.48	17.83	18.28	1.00	1.00
M_{yc}	2.02	4.53	2.99	3.40	18.28	20.56	20.22
M_{zc}	4.30	2.96	5.77	8.01	18.28	15.92	15.67
M_{xb}	8.11	8.84	10.62	9.73	33.84	20.56	20.22
M_{yb}	17.52	31.28	18.38	18.22	33.84	1.00	1.00
M_{zb}	13.31	12.37	8.27	8.04	33.84	15.92	15.67

Table 3. SIF in model T2.

Moment	Exp.		F.E.		ASME
	Cil.	Bran.	Cil.	Bran.	
M_{xc}	1.21	0.58	0.75	0.64	9.51
M_{yc}	0.30	0.41	0.19	0.14	9.51
M_{zc}	0.69	0.62	0.89	0.19	9.51
M_{xb}	16.35	16.95	16.16	17.78	19.05
M_{yb}	1.76	2.27	1.35	2.32	19.05
M_{zb}	2.48	2.87	5.69	5.47	19.05

Table 4. SIF in model T3.

Moment	Exp.		F.E.		ASME
	Cil.	Bran.	Cil.	Bran.	
M_{xb}	7.46	4.29	8.03	11.93	9.91
M_{yb}	1.26	1.37	1.01	1.00	9.91
M_{zb}	2.62	2.31	5.04	5.76	9.91

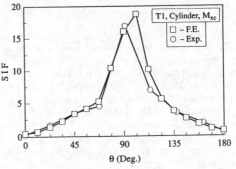

Figure 6. Model T1, SIF in cylinder part under M_{xc}.

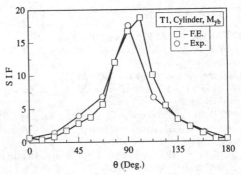

Figure 7. Model T1, SIF in cylinder part under M_{yb}.

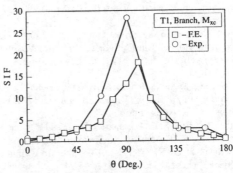

Figure 8. Model T1, SIF in branch part under M_{xc}.

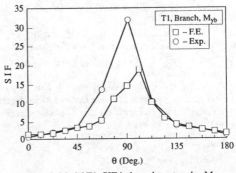

Figure 9. Model T1, SIF in branch part under M_{yb}.

CONCLUSIONS

Experimental and finite element results show that decreasing of d/D ratios makes stress intensity less and less significant for moments apllied in the run pipe limb and may be neglected for d/D<0.5.

Both ANSI/ASME (1980a, b) and BS 806 (1982) codes do not consider twisting moments which is unconservative. The ASME Boiler and Pressure Vessel Code, Section III (ASME, 1980) on the other hand gives unconservative estimating.

All three codes are clearly conservatives when considered bending moments.

REFERENCES

ANSI/ASME B 31.1, 1980. *Power Piping*, ASME.

ANSI/ASME B 31.3, 1980. *Chemical Plant and Petroleum Refining Piping*, ASME.

ASME *Boiler and Pressure Vessel Code, Section III, Division 1 – Nuclear Power Plant Components*, ASME, 1980.

ASME *Boiler and Pressure Vessel Code, Section VIII, Division 2 – Alternative Rules for Pressure Vessels*, ASME, 1983.

Bijlaard P. P., 1954. Stresses from Radial Loads in Cylindrical Pressure Vessels, *Welding Journal*, 33 (12), Research Supplem., 615-623.

Bijlaard P. P., 1955a. Stresses from Local Loading in Cylindrical Pressure Vessels, *Trans. ASME 77*, 805-816.

Bijlaard P. P., 1955b. Stresses from Radial Loads and External Moments in Cylindrical Pressure Vessels, *Welding Journal*, 34 (12), *Research Supplem.*, 208-217.

BS 806, 1982. *Specification for Ferrous Pipes and Piping Installations for and in Connection with Land Boilers*, British Standars Institution.

Corum, J. M., Bolt, S. E., Greenstreet, W. L. and Gawltney, R. C., 1972. Theoretical and Experimental Stress Analysis of ORNL Thin-Shell Cylinder to Cylinder Model 1, *ORNL*-4553.

Gwaltney, R. C., Bolt, S. E., Corum, J. M. and Bryson, J. W., 1975a. Theoretical and Experimental Stress Analysis of ORNL Thin-Shell Cylinder to Cylinder Model 3, *ORNL*-5020.

Gwaltney, R. C., Bolt, S. E. Corum, J. M. and Bryson, J. W., 1975b. Theoretical and Experimental Stress Analysis of ORNL Thin-Shell Cylinder to Cylinder Model 2, *ORNL*-5021.

Khan A. S. and Hsiao C., 1984a. Research Note; Strain field in straight cylindrical shells due to applied forces on an attached shell. Part I: no hole in the intersection region, *J. Strain Analysis*, vol. 19, nº 4.

Khan A. S., Chen J. -C., Hsiao Ch. and Woods G., 1984b. A Comparative Study of the Stress Field Around a Reinforced and an Unreinforced Normal Intersection of Two Cylindrical Shells, *Int. J. Pres. Ves. & Piping*, 15, 79-92.

Khan A. S., 1987. A Study of Stress Intensification Factors of Integrally Reinforced Shell to Shell Intersections, *Int. J. Pres. Ves. & Piping*, 29, 83-94.

Khathlan A. A. and Steele C. R., 1986. Analysis of Large Openings in Cylindrical Vessels, Including Nozzle Flexibility, *ASME-PVP Meeting*.

Mershon J. L., Mokhtarian K., Ranjan G. V. and Rodabaugh E. C., 1984. Local Stresses in Cylindrical Shells Due to External Loadings on Nozzle-Supplement to WRC Bulletin nº 107, *Welding Research Council Bulletin* nº 297.

Moffat D. G., Mistry J., 1986. Interaction of External Moment Loads with Internal Pressure on an Equal Diameter Branch Pipe Intersection, *J. Pres. Vess. Technology*, vol. 108, 443-452.

Moffat D. G., Mistry J. and Mwenifumbo J. A M., 1989. Moment Loads on Branch-pipe Junctions: The Effect of Run-pipe Fixing on Elastic Stress Levels and its Implications on Plastic Limit Loads, *Int. J. Pres. Ves. & Piping*, 38, 249-259.

Rilo, N. F., 1992. *Local Stresses in Vessels an Piping*, PhD these, 287-322, Coimbra.

Steele C. R., 1980. Evaluation of Reinforced Openings in Large Pressure Vessels, *Shelltech Report* 80-2.

Steele C. R. and Steele M. L., 1981. Reinforced Openings in Large Pressure Vessels: Effect of Nozzle Wall Thickness, *Shelltech Report* 81-5 (Revised with Appendum, August 1982).

Steele C. R. and Steele M. L., 1983. Stress Analysis of Nozzles in Cylindrical Vessels with External Loads, *J. Pres. Ves. Technology*, Vol. 105.

Steele C. R., 1986a. Comparison of FAST 2 Calculations with Experimental Results for Thick and Thin Shells, *ASME-PVP Meeting*.

Steele C. R., Steele M. L. and Khathlan A., 1986b. An Efficient Computational Approach for a Large Opening in a Cylindrical Vessel, *J. Pres. Ves. Technology*, Vol. 108, 436-442.

Schoessow G. J. and Kooistra L. F., 1945. Stresses in Cylindrical Shell Due to Nozzle and Pipe Connections, *Transitions ASME*, 67, A-107.

Whichman K. R., Hopper A. G. and Mershon J. L., 1965. Local Stresses in Spherical and Cylindrical Shells due to External Loadings, *Welding Research Council Bulletin* nº 107.

Recent Advances in Experimental Mechanics, Silva Gomes et al. (eds) © 1994 Balkema, Rotterdam, ISBN 90 5410 395 7

Construction of bending moment diagram of ferromagnetic beam-plate subjected to magnetic field using strain gages

Sin Min Yap
Valparaiso University, Ind., USA

Carl R.Vilmann, Milton O.Peach & Nels S.Christopherson
Michigan Technological University, Houghton, Mich., USA

ABSTRACT: Magnetic fields which are invisible and without substance deform ferromagnetic structures. The manner by which these magnetic fields are translated into mechanical loading is explored in this paper. In order to have a better understanding of the basic relationship of the elastic response of a ferromagnetic beam-plate under the influence of an applied magnetic field, the bending moment diagram was constructed using strain gages. This bending moment diagram will shed more light on the possible equivalent mechanical loading which acts on the beam-plate caused by an applied magnetic field. A valid magnctoclastic bending theory is developed using classical elasticity theory, once this loading condition is understood.

1 INTRODUCTION

Most of the structural mechanics problems which have been dealt with successfully by engineers have only mechanical and gravitational forces acting on them. With a magnetic force which is invisible and without substance, the engineer will no longer have the luxury of just drawing a simple free body diagram to represent the force system on the deformable bodies and using existing strength of materials or elasticity theory to calculate the stress and strains induced by such a force system. There is no way to visualize how the magnetic forces are distributed on a solid body. But such forces would cause deformation on solid bodies as if they were being produced by mechanical or gravitational forces (See Figure 1).

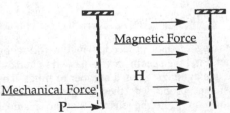

Figure 1: Comparison Between Mechanical Forces and Magnetic Forces

This investigation experimentally explores the relationship between magnetic fields and ferromagnetic thin plates. A thin cantilevered plate which bends cylindrically is used. Assuming classical beam theory is valid whether the load is of mechanical or magnetic origin, the familiar relationship between strain (ε) and bending moment (M) is employed:

$$M = \frac{\varepsilon EI}{c(1 - v^2)} \qquad (1)$$

where E = Young's Modulus, I = Second Area Moment of Inertia, c = distance from the neutral axis to the surface of the beam-plate, and v = Poisson's Ratio.

The bending moment diagram was constructed from the strain gages data and the experimental bending moment diagram compared to the one predicted theoretically. For the most part, the calculated bending moment values agrees extremely well with those measured. As a result of this research, a much better insight into the actual force distribution on a magnetic structure is attained. Most of the magnetic forces are centered at the edges and corners, which can be attributed to the concentration of the magnetic field near the *edges* and *corners* of a finite specimen.

2 EXPERIMENTAL APPROACH

2.1 Specimen Setup

The loading of the ferromagnetic (low carbon steel) thin plate is accomplished through the use of a Direct Current (DC) magnetic field. This eliminates from consideration such factors as time dependence, fatigue of specimens, and interaction with necessary instrumentation. Also the specimens are initially demagnetized before testing to insure that no residual magnetism was present. To achieve this, a coil approximately 0.15 m in diameter and 0.20 m long was employed. This was powered by AC current from a VARIAC transformer, and by passing the specimens in and out slowly they were demagnetized.

To eliminate the influence of gravity, the specimens will be mounted as cantilevered thin plates and oriented so that the action of gravity is normal to the plane of bending. The resulting configuration is one in which the specimen is assumed to act as a beam in cylindrical bending. To obtain measure of deformation, strain gages were mounted along the length of the beam to measure the strain present when the beam was deformed by the magnetic field. (See Figure 2). The specimen was mounted in a fixture constructed of aluminum with non-magnetic brass hardware. By causing the angle between the undeformed specimen and the flux lines to be five degrees, magnetoelastic bending and not magnetoelastic buckling occurred.

2.2 Magnetic field measurement

A static magnetic field ranging between 10^{-3} Tesla to 0.15 Tesla with less than one percent flux variation with time was used as the source of magnetic flux.

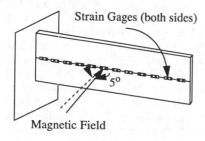

Strain Gages (both sides)

5°

Magnetic Field

Figure 2: Specimen Orientation and
Strain Gage Locations

The measurements of magnetic field magnitude were taken using an F.W. Bell Model No. 640 Gaussmeter. The output of the Gaussmeter was monitored on a digital voltmeter due to limited resolution on the Gaussmeter meter face. The probe used for the readings was a Hall Effect element at the end of a 1.25 m flexible cable and was calibrated using an F.W. Bell reference magnet. The active element in the tip of the probe is 0.10 cm in diameter. This element is located in a holder 0.05 cm thick by 0.33 cm wide. The calibration was checked with an internal calibration on the Gaussmeter which has a maximum error of 0.3 percent.

2.3 Strain measurements

Various specimens were caused to deform in the magnetic field. The sizes range from 7.62 cm to 17.78 cm in length, all of 2.54 cm width, and two thicknesses, 0.04 cm and 0.07 cm. After observing the behavior of various size specimens, one was strain gaged to obtain final data on bending strains. This specimen when mounted as a cantilever beam was 17.78 cm long, 0.07m thick, and 2.54 cm wide. All specimen s were of low carbon steel (SAE 1020). A sketch of this specimen is shown in Figure 2, showing the location of strain gages. The strain gages used were Micro-Measurements Model No. EA-06-062AP-120. They are constructed of Constantan and were all selected from the same lot to maintain nearly identical characteristics. The strain gages (twenty eight of them) were connected to a Northern Technical Service (NTS) switch and balance unit. The output of this was in turn connected to NTS digital strain indicator. Constantan gages were chosen because they will not interact with the magnetic field. The NTS equipment was also selected for this reason because it powers the external portion of the stain gage bridge with a direct current.

2.4 Data Taking Scheme

The taking of data proceeded in the following manner. First the specimen was passed in and out of the demagnetizing coil a number of times. The Gaussmeter probe was then placed at various positions on it, to be ensure that no residual magnetism remained. The specimen was then mounted in the aluminum fixture and secured. At this point in time, another set of Gaussmeter readings was taken to insure that no flux was present. The strain gages were then all balanced to give a relative zero reading. The motor generator was then started which energized the electromagnet. The magnetic field created then caused the specimen to move to a static but deformed configuration. Now strain readings were taken, and immediately after this, the magnetic field readings at various locations on the beam were also taken. This sequence would conclude the taking of one data set because the specimen becomes magnetized. To obtain another set of readings, the whole sequence would be repeated.

3 EXPERIMENTAL RESULTS

The values of strain present every 1.25 cm along each side of the beam were combined (averaged) to give the values of both bending and axial strain along the length of the beam. The axial strains at all cross sections were so small as to be negligible. Using simple wide beam theory (Equation 1), the value of bending strain is easily converted to bending moment along the length of the beam. These values are tabulated in Table 1. The flexural strain values from the strain gage measurements were then used to construct the

Table 1: Strain data and Bending Moments.

X cm	Strain Top	Strain Bot.	Bend. Strain	Axial Strain	Moment N-m
0.6	182	-185	183	-1.5	0.086
1.9	172	-170	174	1.0	0.08
3.2	155	-164	160	-4.5	0.075
4.5	144	-151	148	-3.5	0.069
5.7	124	-130	127	-3.0	0.059
7	108	-123	116	-7.5	0.054
8.3	92	-101	97	-4.5	0.045
9.5	76	-84	80	-4.0	0.037
11	59	-72	66	-6.5	0.031
12	46	-61	54	-7.5	0.025
13	29	-40	35	-5.5	0.016
15	17	-27	22	-5.0	0.01
16	7	-14	11	-3.5	0.005
17	1	-6	4	-2.5	0.002

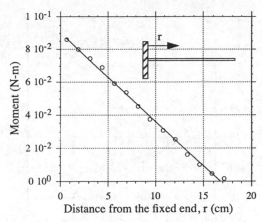

Figure 3: Bending moment diagram constructed from strain gage measurements

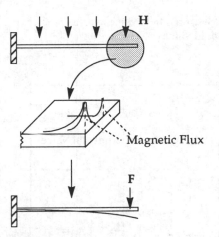

Figure 4: Closeup of Magnetically Loaded Plate

bending moment diagram (see Figure 3), caused by an applied magnetic field.

The bending moment diagram (after a linear fit through the data), looks remarkably similar to bending moment diagram of a end loaded beam. Consequently, a better insight into the actual force distribution on a magnetic structure is attained. Most of the magnetic forces are believe to be concentrated at the edges and corners of the plate, which can be attributed to the concentration of the magnetic field near the *edges* and *corners* (see Figure 4).

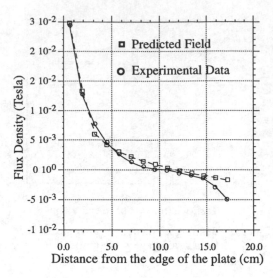

Figure 5: Measured field values versus predicted field values

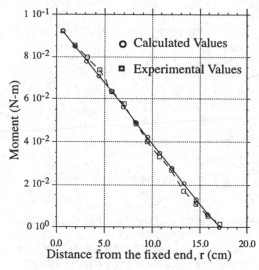

Figure 6: Comparison of bending moment diagram obtained using strain gages and bending moment diagram predicted theoretically

4 FORCE CALCULATIONS

Using **magnetic thread theory** (Yap,Vilmann, and Peach, 1992), a method of calculating the equivalent mechanical forces acting on a ferromagnetic structure caused by an applied magnetic field was derived. The fundamental magnetic element is postulated as a magnetic thread rather than a magnetic dipole. The force equation based on this model is:

$$F = \int_S \frac{\left[H(H \cdot n) + \frac{H^2 n}{2} \right]}{4\pi} dS \qquad (2)$$

where H is the applied field.

Thus, in a magnetic deformation problem, if the resulting magnetic field distribution, H, can be predicted, given the applied field, then the equivalent mechanical loading can be easily deduced from Equation 2. Unfortunately, H is difficult to determine. There is a narrow strip of surface adjacent to the edges (edge effect) and the corners (corner effect) of a beam-plate having a rectangular cross-section, where the magnetic field becomes very large (see Figure 4).

5 DETERMINATION OF MAGNETIC FIELD DISTRIBUTION

In order to evaluate the character of magnetic field H near a plate's edge, an approach similar to that used for the determination of stresses near a crack tip was followed. The results are the magnetic field equations at the edges and corners of a plate surface. For field along the midline of a plate, the equation which govern the magnetic field distribution is:

$$H_n = H_o \left[\frac{(41.4 - 41.17\cos\theta)}{r^{1/3}} + (-7.5 + 7.45\cos\theta)r^{\frac{1}{3}} \right] \quad (3)$$

where H_o is the applied field and θ is the angle between the unit normal component of the field with respect to the plate.

For field along the corner of a plate the equations which govern the magnetic field distribution are:

$$H_n = H_o \left[\frac{125.9 - 125\cos\theta}{r^{0.36}} + (-18.8 + 18.7\cos\theta)\, r^{0.28} \right]_{7.5^o}^{*}$$

$$\vdots$$

$$H_n = H_o \left[\frac{80.2 - 80\cos\theta}{r^{0.353}} + (-15.4 + 15.3\cos\theta)\, r^{0.292} \right]_{45^o}$$

$$\vdots$$

$$H_n = H_o \left[\frac{120.4 - 120\cos\theta}{r^{0.36}} + (-25.9 + 25.7\cos\theta)\, r^{0.28} \right]_{82.5^o}$$

$$(4)$$

Three - Dimensional Field Equations
at the corner area at different radial directions

***Only the first, middle, and last of the
field equations are shown here**

Equations (3) and (4) provide a means of predicting the net magnetic field intensity at the edges and corners of a plate. The measured field and the predicted field values are in excellent agreement as shown in Figure 5

6 BENDING MOMENT CALCULATION

Using Equation 2 and the equations for magnetic field distribution (Equation 3 and 4), the equivalent mechanical forces acting on a plate caused by an applied magnetic field is predicted. The bending moment diagram of the magnetically loaded plate was then constructed based on a resultant force acting on the centroid location of the distributed forces (concentrated load near the free end). This bending moment diagram was then compared to the one calculated based on the measured bending strains. The results are shown here in Figure 6

For the most part, the calculated bending moment diagram values agree extremely well with those measured. The average difference is approximately four percent. This can easily be attributed to the much simplified force system (concentrated load near the free end).

7 CONCLUSION

What has been demonstrated here is the use of some relatively simple measurement techniques to help explain the basic mechanics of magnetoelastic bending of thin plates. This investigation also formulated a few new and important ideas:

(a) a new model (Magnetic Thread Theory) to calculate the equivalent mechanical forces caused by an applied magnetic field.

(b) a new method of predicting the magnetic field distribution at the corners and edges of a ferromagnetic plate.

This paper shows a useful and simple method for calculating forces and deformation of a rectangular plate built-in along one edge exposed to a magnetic field.

REFERENCES

Brown, W.F., (1966). "Magnetoelastic Interactions," Springer-Verlag, New York Inc.

Christopherson, Nels Sy, (1984). "An Experimental Study of the Magnetoelastic Bending of Thin Plates Made of Soft Ferromagnetic Material," Ph.D. Dissertation, Michigan Technological University, Houghton, Michigan, University Microfilms International Ann Arbor, MI, 1984.

Yap, S.M., Vilmann, C.R., and Peach, M.O., (1992). "How Good is the Postulate ρ = Div M = 0?," Society of Engineering Science 29th Annual Technical Meeting, University of California San Diego, La Jolla, California, September 14-16, 305.

Yap, Sin Min, (1993). "An Investigation Into Non-Classical Magnetoelastic Bending and Buckling of Ferromagnetic Thin Plates," Ph.D. Dissertation, Michigan Technological University, Houghton, Michigan, University Microfilms International Ann Arbor, MI.

Recent Advances in Experimental Mechanics, Silva Gomes et al. (eds) © 1994 Balkema, Rotterdam, ISBN 90 5410 395 7

On the second order bending of the thin circular plates

M.Gh.Munteanu & Gh.N.Radu
Transilvania University, Brasov, Romania

ABSTRACT: The paper deals with the second order bending of thin circular plates subjected to different membrane stress fields due for instance to an uniform rotational speed, to a non-uniform distribution of the temperature on the radius, etc. If the plate is thin, membrane stresses strongly influence the bending behaviour of the plate. The problem is solved by means of an original method for which the authors wrote several computer programs. The theoretical results were verified by experimental measurements, a very good agreement being noticed.

1 INTRODUCTION

Circular plates are widespread in machine building: grinding disks, circular saws, clutches disks, turbine disks and others. As it is known, the plane plates loading can be divided in two parts: a loading given by forces contained in the plate plane, so called membrane loading, and bending loading given by forces acting perpendicularly to the plate plane. As a result of Kirchhoff's hypothesis, the stresses produced by the in-plane loading, the membrane stresses, are constant on the thickness of the plate and those resulted by bending have a linear distribution on the thickness. The material is linear, that is it the Hooke's law is valid. In general it may assume that the two loadings are independent and they can be studied separately. But if the plate is thin, the membrane stresses have a strong influence on the bending of the plate and also on the natural frequencies or buckling stress field level. In this paper it will be considered only the axy-symmetrical membrane stress fields, produced for instance by an uniform rotational speed, by a non-uniform temperature distribution on the radius, etc. The bending of the circular plate is not necesseraly axi-symmetric.

The Kármán's equations describe the behaviour of the thin plates subjected to membrane stresses as well as to bending loads and their reciprocal influence. In this paper it will be considered that membrane stress field is not influenced by the bending. Hence, if in addition the membrane stress field is constant during the bending loading, the dependence between the bending loading and the displacements measured perpendicularly on the plate plane is linear. This assumption is in very good agreement with the experimental results if the buckling level of the membrane stress field is not exceeded. Otherwise the Kármán's equations have to be used. This paper deals only with the pre-buckling behaviour.

2 THEORETICAL SOLUTION

The equation which describes the bending of the thin plates in presence of a membrane stress field is:

$$D \nabla^4 w - G(w) = p , \qquad (1)$$

where in polar coordinate, r, θ, the G operator is:

$$G(w) = h(\sigma_r \chi_r + \sigma_\theta \chi_\theta + \tau_{r\theta} \chi_{r\theta} - R\frac{\partial w}{\partial r} - \Theta\frac{\partial w}{r\partial\theta}) , \quad (2)$$

the curvatures having the expressions:

$$\chi_r = \frac{\partial^2 w}{\partial r^2}; \ \chi_\theta = \frac{1}{r}\frac{\partial w}{\partial r} + \frac{\partial^2 w}{r^2\partial\theta^2}; \ \chi_{r\theta} = 2\frac{\partial}{\partial r}(\frac{1}{r}\frac{\partial w}{\partial\theta}).$$

In the above relations p is the force acting perpendicularly on the plate plane, w is the displacement perpendicular to the plate plane, D=Eh³/12(1-v²), E is the Young's modulus, v -

the Poisson ratio, h - the plate thickness, σ_r, σ_θ, $\tau_{r\theta}$ - are the membrane stresses and R, Θ are the specific in-plane inertial forces.

The equation (1) may be solved by means of numerical methods like finite difference method (FDM) or finite element method (FEM). The authors used both methods. The bending loading is decomposed in Fourier series, thus the solution becomes (if a plane of symmetry exists):

$$w(r,\theta)=w_0(r)+\sum_{q=1}^{\infty} w_q(r)\,\cos q\theta \; .$$

In practice the Fourier series was limited to few terms, this method being rapidly convergent.

It may define an eigenvalue problem, the buckling problem. The equation (1) becomes for n different membrane stress fields and for no bending loads:

$$D\,\nabla^4 w - \sum_{i=1}^{n}\lambda_i\,G_i(w)=0 \; . \qquad (3)$$

In this equation all coefficients λ_i are known,

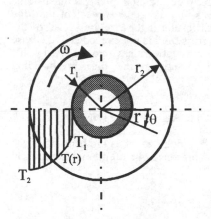

Fig. 1. Clamped annular disk

except one for which the buckling value is desired. By solving this eigenvalue problem it will be found eigenvalues and eigenvectors which for circular plates have nodal circles and nodal diameters. It was noticed that for annular disks clamped on their inner boundary, Fig. 1, the problems:

$$D\,\nabla^4 w - \lambda_i\,G_i(w) = 0 \; , \; i=1...n \; , \qquad (4)$$

have the eigenvectors very similar for each of the n membrane stress field, especially for big

values of r_1/r_2 and for no nodal circles case. Thus it is easy to obtain:

$$\sum_{i=1}^{n} \frac{\lambda_i}{\Lambda_i} = 1 \; , \qquad (5)$$

where Λ_i is the eigenvalue of the problem (4), that is the buckling level for the i-th membrane stress field, λ_i is the actual level for the same i-th membrane stress field. When the condition (5) is fulfilled, the buckling under several superposed membrane stress fields occur. This conclusion was verified in some previous papers (Munteanu, Gogu, Radu 1987, Munteanu, Gogu 1991, Munteanu 1992, Munteanu 1993) using several original computer programs (FDM and FEM), as well as experimental measurements. A very good agreement was observed.

For instance, in figure 2 is represented the dependence between the uniform rotational speed ω and the buckling level of a linear variation of

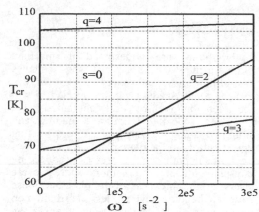

Fig . 2. Critical temperature versus ω^2

the temperature on the radius, T(r). T is constant on the thickness. The disk has the dimensions: r_1=29 mm, r_2=125 mm, h=1.5 mm. The equation (5) becomes:

$$\frac{T}{T_{cr}} - \frac{\omega^2}{\Omega^2} = 1 \; ,$$

where it may consider that the stresses produced by the rotational speed have negative eigenvalue, $-\Omega^2$. The diagram in figure 2 was determinated by exact computations, using original computer programs.

The first order behaviour of the plate is described by:

$$D\,\nabla^4 w = p \; . \qquad (6)$$

908

The solution of this equation may be decomposed after the eigenvectors, u, of the problem (4), which are very similar for different membrane stress fields and thus almost independent to the form of G operator . For the circular plates the solution of the equation (6) may be written:

$$w(r,\theta) = \sum_{s=0}^{\infty} \sum_{q=0}^{\infty} c_{sq} u_{sq} , \qquad (7)$$

s being the number of the nodal circles, and q the number of the nodal diameters. The coeficients c_{sq} are simply to determinate because of their orthogonality. Also it is easy to demonstrate that the solution of the equation (1) is:

$$w = \sum_{s=0}^{\infty} \sum_{q=0}^{\infty} \frac{c_{sq}}{1 - \sum_{i=1}^{n} \frac{\lambda_i}{\Lambda_{sq_i}}} u_{sq} , \qquad (8)$$

if n different membrane stress fields act at a time. In practice, regarding nodal circles, the term for s=0 suffices and for nodal diameters a maximum of 5 is more than satisfactory. For circular disks clamped on their inner boundary, Fig. 1, the maximum of the w displacement is reached on the outer boundary and thus it may be written:

$$w_{max} = \sum_{q=0}^{q_{max}} \frac{c_{0q}}{1 - \sum_{i=1}^{n} \frac{\lambda_i}{\Lambda_{0q_i}}} , \qquad (9)$$

because $u_{0q} (r_2 , 0) = 1$. The eigenvectors are normalised, that is $u_{max}=1$ for each eigenvector. In the second order theory, when the buckling occurs, i.e. equation (5) is satisfied, w_{max} tend to infinite.

3 EXPERIMENTAL MEASUREMENTS

The validity of the equations (8) and (9) for perfectly plane disks was verified in several papers (Munteanu, Gogu, Radu 1987, Munteanu, Gogu 1991, Munteanu 1992, Munteanu 1993). In this paper an other aspect will be studied. The disks cannot be exactly plane, small imperfections always do exist. Hence the membrane stress field produce w displacements even if the level of the field is under the buckling value. This dependence is not linear.

Let's consider $w_0(r,\theta)$ a small imperfection, in absense of any load.. The equation (1) becomes:

$$D \nabla^4 w - G(w) = p + G(w_0) . \qquad (10)$$

The initial displacement w_0 may be decomposed:

$$w_0(r,\theta) = \sum_{s=0}^{\infty} \sum_{q=0}^{\infty} a_{sq} u_{sq} ,$$

u_{sq} being the eigenvectors of the problem (4). In the case of p=0, the solution of the equation (10) is:

$$w_0(r,\theta) = \sum_{s=0}^{\infty} \sum_{q=0}^{\infty} \frac{a_{sq}}{1 - \frac{\lambda}{\Lambda_{sq}}} u_{sq} , \qquad (11)$$

where Λ_{sq} are the eigenvalues and λ is the level of membrane stress field. If several membrane stress fields exists at a time the relation (11) becomes similar to (8).

The authors studied thin disks in rotational motion and heated on their outer contour. Some difficulties had the determination of the temperature distribution on the radius. Because the temperature distribution depends on too many factors, experimental measurements were prefered. Many disks of different dimensions were used. It was found that a parabola fit very well to experimental data:

$$T(r) = T_1 + (T_2 - T_1)[\frac{r - r_1}{r_2 - r_1}(1-c) + (\frac{r - r_1}{r_2 - r_1})^2 c] ,$$

where T_1 and T_2 are the temperature on the two contours of the disk and c is an coefficient. For c=0 the linear distribution is obtained. From experiments it result that c=-0.65....+0.6. Different values for c were found for the same disk with different angular speed.

Further on , the results for a disk, Fig.1, with r_1=29 mm, r_2=125 mm, h=1.5 mm will be presented. For n=1400 rot/min it resulted c=-0.65. Table 1 contains some values of buckling temperatures ($\Delta T = T_2 - T_1$). The figure 3 contains an example of " imperfection", w_0 , measured on the r_2 radius. This imperfection has no "nodal circles", that is w_0 increases monotonically from 0 to w_{max} on each radius. The function $w_{0\,max}$ was decomposed in Fourier series:

$$w_{0\,max}(\theta) = \sum_{i=0}^{\infty}(a_i \cos(i\theta) + b_i \sin(i\theta)) . \qquad (12)$$

The disk was heated on the outer boundary and the value of ΔT was determinated by measuring. It was found that ΔT=62 K. At the same time the

Table 1. Buckling temperatures

ΔT_{cr} [K]	q = 0	q = 1	q = 2	q = 3	q = 4	q = 5
s = 0	-81.8	-202.4	+71.1	+80.3	+120.9	+173.3
s = 1	-463.7	-605.4	-704.6	-984.1	-1210	-1473

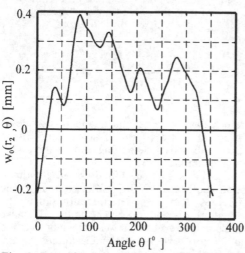

Fig. 3. Imperfection of the outer boundary

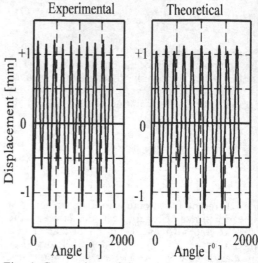

Fig. 4. Comparison of the results

disk had a rotational speed of n=1400 rot/min. On the left of the Fig. 4 the results of the measurement of the displacement $w(r_2)$ were represented. The imperfection increased due to the membrane stress field.

For the theoretical estimation, in the relation (12) the coefficients a_i and b_i were divided by the factor: $1-\Delta T/\Delta T_{cr\,i}$, according to the relation (11). The theoretical result is represented on the right of the Fig. 4. It is obvious that the two results, experimental and theoretical, are in very good agreement.

4. CONCLUSIONS

The study presented in this paper allows a fast analysis of the second order bending of circular plates, as well as an analyse of the influence of different membrane stress fields acting at a time.

The well known comercial finite element programs are not suitable for this purpose being expensive and very time consuming. The determination of the eigenvalues and eigenvectors is not an disadvantage, because plane axi-symmetrical problems are mono-dimensional and thus the eigenvalue problem is easy to solve . In addition relations like (8), (9) and (11) give very quick information, by simple modifying of the level λ of the membrane stress fields or of the eigenvalues Λ for the different membrane stress fields acting at a time.

REFERENCES

Munteanu, M.Gh., Gr. Gogu & Gh. Radu 1987. Study of displacements and natural frequencies of thin circular plates in the presence of membrane stresses. *Mécanique Appliquée - Revue Roumaine des Sciences Technique*, Bucuresti.No.2: p. 159-175.

Munteanu, M.Gh. 1992. On stresss and displacement study of the thin circular plates. Second order calculations. *XXI Convegno Dell'Associazione Italiana per l'Analisi delle Sollecitazioni*, Genova, p. 389-396.

Munteanu, M.Gh. 1993. On the bending of thin plates. Second order calculations. *ELFIN2 Congress*, Sibiu, Romania.

Munteanu, M.Gh. & Gr. Gogu 1991. Elastic stability and vibration of thin circular rotors processed by hooping. *Eighth World Congress of the Theory of Machines and Mechanisms*, Praga: p. 209-212.

Recent Advances in Experimental Mechanics, Silva Gomes et al. (eds) © 1994 Balkema, Rotterdam, ISBN 90 5410 395 7

Earth mechanical modelling taking into account its mass, participating in the vibration

Gheorghe Petre Zafiu & Constanţa Herea-Buzatu
Institutul de Construcţii, Bucureşti, Romania

ABSTRACT: Earth layers dynamic modelling taking into account its mass was analyzed during compacting by vibration. Five dynamic models taking into account the soil nature and the development in the compacting process were supplied.

1 VIBRATIONS'ACTION IN THE EARTH LAYER SUBJECTED TO COMPACTING

During the compacting by vibration process, the whole earth mass lying in the vibrations'influence zone takes part at the oscillating movement with a decreasing intensity to this zone's bounds. Some particles, detached during the vibration process, carry out solitary, aleatory oscillations, related to the vibration acceleration and frequency, to their mass and to the physical structure of the neighborhood skeleton.

It can be therefore assessed, that a part of the earth mass lying in the equipment neighborhood participates in the vibration process, building up a whole dynamic system: road-roller-earth.

In the compacting by vibration case, the equipment acts upon the earth layer, producing its remanent deformation as result of two simultaneous actions:

1. by repeated vibrations, the material granules are being set going by inertic forces proportional to their masses, fiction between them being diminished, what permits a denser seating, with smaller emptiness;

2. by the pressure due to the roll of the steel drums, as result of the static mass of the equipment occurs the layer's cylindering.

Following the composition of the external forces, in their application zone, occurs the deformation of the material's layer subjected to the compacting.

The deformation has two components: the remanent deformation and the elastic deformation (figure 1).

The remanent deformation can appear in two forms: with the diminishing of the volume, by particles resetting (reducing of the pores) and with the modification of the shape, the volume remaining constant. As a rule, the remanent deformation comprises both simple forms of deformation in different proportions, depending on the nature and the size of the external stresses. By relatively small external forces prevails the deformation with volume reduction, while by big forces occurs earth sinking and repression under the working device of the compacting equipment. It was experimentally established that the highest compacting degree is secured in the layer subjected it compacting, at a certain depth, depending on the working parameters and the nature of the earth. As this depth is usually situated over the mass center of the layer, and taking into account that the acquired density is proportional to the deformation, it follows that the earth, layer deformation is larger near the surface.

As a result, it can be considered that the vertical displacement of the earth layer's center of gravity under the external forces's action will be:

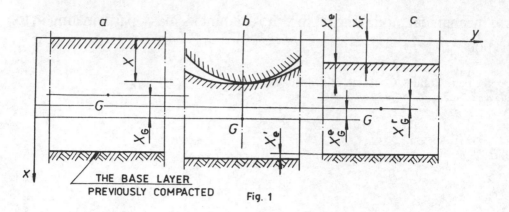

THE BASE LAYER
PREVIOUSLY COMPACTED

Fig. 1

$X_G = X_G^e + X_G^r \leq 1/2(X_e + K_r); \quad X_G^r \leq 1/2X_r \quad (1)$
where X_G - represents the displacement of the center of gravity during the action of the external stress; X_e - represents the elastic deformation; X_r - represents the remanent deformation; X_G^r - represents the remanent displacement of the center of gravity (after the disappearance of the stress) and X_G^e - represents the elastic displacement of the center of gravity.

2 DYNAMIC MODELLING OF THE EARTH, TAKING INTO ACCOUNT ITS MASS

To find a model of the capable to take into account all the considerations concerning the deformability of the layer subjected to compacting, as well as its structural modifications at the same time with the considerations of its mass, there were carried out studies of the rheological behavior's mechanical models worked out till now. Given the Kelvin model, by which the earth mass has a displacement of the center of gravity equal with the layer's deformation, it was tried to conceive a model by which the earth mass participating in the vibration wouldn't be considered as an isolated mass, heated besides the model, and its center's of gravity displacement would be smaller than the deformation.

It was aimed also to model the connections of the mass with the layer to which it belongs, connections that have a direct implication in the dynamic behavior of the model. By the models making, there were kept in mind the two distinct layers:

1. The base layer, previously compacted, with a quasielastic behaviour, enjoying stable dynamic characteristics;
2. The layer subjected to compacting, with variable characteristics tending to a quasielastic behaviour, proportionally with the increasing passing number.

To model the elastic deformation of the earth, taking into consideration the modification of the center's of mass position, we propose the device from figure 2. By pushing the button, the spring is deformed and the liquid pushed by the piston 1 is redistributed setting in movement the pistons 2. After the removal of the external stress, the system comes back vessel's inside there is no friction, neither between the liquid and the vessel, neither between the pistons and the vessel, it is granted to the model the property not to transmit forces to the vessel's support base.

In this way, a new modelling element is introduced, named "THE FREE DEFORMABLE MASS". Certainly of course, by the taken assumptions, this model is a theoretical, idealized model. To simplify the models, we propose to use only one symbol for the deformable mass, indifferent of the relation between the displacement of the center of gravity and the deformation X, this relations remaining to be specified for each model, by a constant ratio "a" (figure 3).

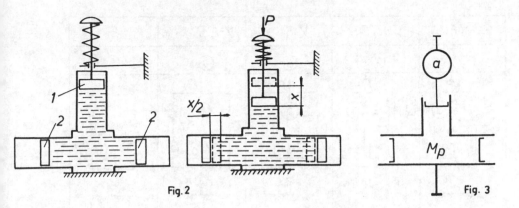

Fig. 2   Fig. 3

$$a=(X''_e+X_r)/(X_a-X'_e) \qquad (2)$$

where, X''_e - represents the elastic deformation of the layer subjected to compacting; X_r - the remanent deformation of the layer subjected to compacting; X_a - the highest displacement of the center of gravity under the action of the external stress; X'_e - the elastic deformation of the previously compacted layer under the action of the external force.

The total elastic deformation produced by the external stress comprises the elastic deformations of the two distinct layers:

$$X_a=X'_e+X''_e \qquad (3)$$

The deformation x''_e and x_r will be introduced in the compound models, by combinations of simple mechanical models, parallel connected with the deformable mass, and the deformation x'_e will be taken into consideration by the elastic connection of these combinations. In this way the model will have a behaviour close to the real behaviour of the earth.

For the modelling of the main dynamic characteristics specific for the different earth types and stages in the compacting process, there were conceived five model types: (figure 4).

1. model elasto-plasticity with dry friction, E/P(Fu)
2. model viscous-elasto-plasticity with dry friction V/E/P(Fu)
3. model viscous-elasticity V/E
4. model viscous-elasto-plasticity V/E/P
5. model elasticity E

3 DYNAMIC STUDY OF THE COMPOUND MODELS, TAKING INTO ACCOUNT THE MASS OF THE EARTH PARTICIPATING IN THE VIBRATION

The proposed models represent new models, which take into consideration the initial properties of the earth, reproducing its behaviour depending on the compacting process stage. The dynamical study of all the presented models was carried out considering that on the models acts a variable force, with a periodical character, produced by a vibrating road roller, $P=P(t)$.

For instance, for the model nr. 1 E/P (Fu), the elasto-plastic properties with dry friction of the earth layer subjected to the compacting process are schematically reproduced by the helicoid arc with the elastic characteristic K_2, parallel connected with the columbian damper, which develops a resistant force R.

The deformable mass of the earth is parallel bounded with the columbian damper, rigidly connected. The previously compacted earth layer is reproduced by the helicoid arc with the elastic characteristic K_1.

According to the correspondence relation between the deformation of the earth layer subjected to the compacting and the displacement of the center of gravity, defined by the ratio "a" results:

$$X_a=(x_2-x_1)/a+x_1 \qquad (4)$$

Using the notation from the figure 5, we can write the differential equations of the mathematical

913

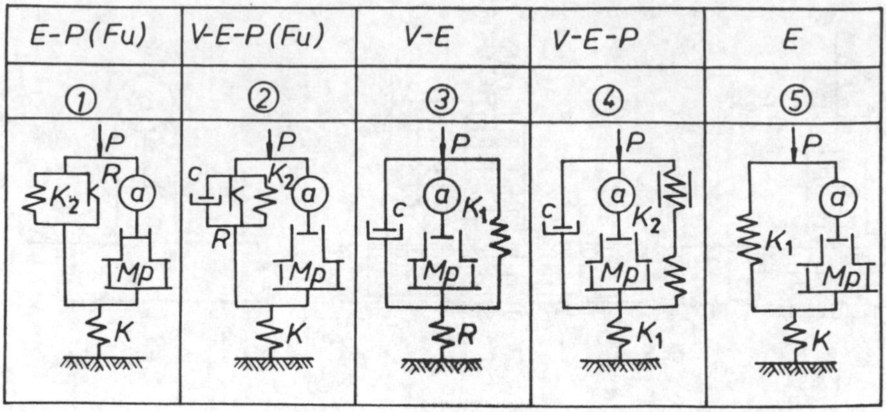

E-P(Fu)	V-E-P(Fu)	V-E	V-E-P	E
①	②	③	④	⑤

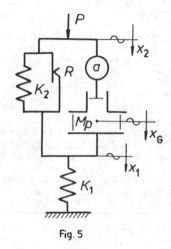

Fig.4

Fig.5

modelling of the dynamic system:

$$M_p\ddot{X}_G+R\,sign(\dot{x}_2-\dot{x}_1)+K_2(x_2-x_1)=P$$
$$R\,sign(\dot{x}_2-\dot{x}_1)+K_2(x_2-x_1)-K_1x_1=0 \qquad (5)$$
$$M_p\ddot{X}_G+K_1X_1=P$$

The equations can be written also:

$$M_p\ddot{x}_2/a-M_p\ddot{x}_1/a+M_p\ddot{x}_1+R\,sign(\dot{x}_2-\dot{x}_1)+K_2(x_2-x_1)=P \qquad (6)$$
$$R\,sign(\dot{x}_2-\dot{x}_1)+K_2x_2-(K_2+K_1)x_1=0$$

Taking out x_1 from the second equation and introducing it in the first one, we obtain the modelling equation of the dynamic system as a quadratic differential equation with the variable x_2.

$$M_p(K_1+aK_2)\ddot{x}_2/a(K_1+K_2)+RK_1sign(\dot{x}_2-\dot{x}_1)/(K_1+K_2)+K_1K_2x_2/(K_1+K_2)=P \qquad (7)$$

We note:
$M_p(K_1+aK_2)/a(K_1+K_2)=m$ (equivalent mass)
$RK_1/(K_1+K_2)=F_t$ (equivalent dry friction force)
$K_1K_2/(K_1+K_2)=K$ (equivalent elastic constant)

The differential equation becomes:

$$m\ddot{x}_2+F_t sign(\dot{x}_2-\dot{x}_1)+Kx_,=P \qquad (9)$$

This equation is that of a columbian damper (with dry friction), with a constant friction force, equivalent to the modeler dynamic system, the relation (8) being the equivalency relation between the two system.

The friction force F_t is constant, irrespective of the deformation or speed, and has always an direction opposite to the relative speed between the particles, this fact being represented in the equation by $sign(\dot{x}_2-\dot{x}_1)$.

The discontinuity of the damp down force which emerges at each half-period, due to the change of the speed direction, requires the solving of the equation step by step.

4 CONCLUSIONS

The introduction of the new idealized modelling element, named "FREE DEFORMABLE MASS" and of the ratio "a" that takes into account the relation between the displacement of the center of the mass and the total deformation x leaded to the creation of new models

914

for earth, able to take into consideration its mass, participating in the vibration.

According to the group it which belongs the earth and to the stage of the compacting process, five model types, with specific dynamic c h a r a c t e r i s t i c s , w e r e differentiated. From the analysis of the dynamic behaviour of the achieved models, based on modeler mathematical equations system, results that some of them have movement equations identical with known damper types (columbian damper or dynamic system without damping).Between the studied models, the viscous-elasto-plastic model represents better than all the real behaviour of the earth, but is more difficult to explain from the mathematical point of view, as a result of the remanent deformations which lead to a system dependent on the perturbation force's phase.

The new models that where obtained give a closer to reality representation of the behaviour during the compacting by vibration process.

REFERENCES

Bărdescu, Ioan. Studiul vibraţiilor la plăcile vibratoare pentru compactarea pământurilor. Teză de doctorat, Bucureşti 1971-I.C.B.
Cyril, M.H. & Crede, C.E. Socuri şi vibraţii, Bucureşti E.T. 1968
Vaicum, Al. Studiul reologic al corpurilor solide, Bucureşti Ed. Academiei R.S.R. 1978
Zafiu, Gheorghe, Petre. Studiul optimizării parametrilor tehnologici ai rulourilor compactoare vibratoare în concordanţă cu caracteristicile mecanice şi fizice ale pământurilor. Teză de doctorat, Bucureşti 1984-I.C.B.

Recent Advances in Experimental Mechanics, Silva Gomes et al. (eds) © 1994 Balkema, Rotterdam, ISBN 90 5410 395 7

Numerical-experimental investigation of a pre-twisted bar under buckling load

E. Guglielmino, G. La Rosa & S. M. Oliveri
Institute of Machines, Faculty of Engineering, University of Catania, Italy

X. X. Zhai
Department of Mechanical Engineering, Qinghai University, People's Republic of China

ABSTRACT: This paper reports a numerical-experimental approach to pre-twisted bars under buckling load. Based on theoretical results obtained by other authors, a numerical model was developed, in order to analyze different boundary conditions. Several twisted specimens, varying in geometry of the rectangular cross section and in pitch, were tested in order to compare the results for different combinations of constraints. An image analysis system was also used to measure the position and the direction of the buckle. The results defined the applicability range of the numerical model, showing it to be restricted to the lower values of bar thickness and pre-twisted angle.

1 INTRODUCTION

The pre-twisted bar has been the object of numerous theoretical and experimental studies which have led to the definition of the relationship between the stress-strain state of the bar and the applied external load. These studies have also defined the field in which the theory of pre-twisted solids is applicable (Göhner 1940, Giovannozzi 1942) in terms of the geometry of the bar. The pre-twisted bar is defined geometrically by means of two characteristic parameters: the ratio between the two semi-axes of the orthogonal cross-section, ß=b/a, and the pitch, p=2Ðk (1/k=dø/dz, i.e. the angular variation of the cross-section along the geometric axis of the bar, Figure 1). The verification of the theoretical results has, therefore, confirmed the possibility of analysing and resolving cases in which pre-twisted bars are employed in modern engineering projects, e.g. propeller, compressor or turbine blades.

The present research proposes the numerical-experimental verification of the behaviour of pre-twisted bars under buckling load. The authors proceeded from the theoretical results obtained in the buckling of rectangular cross-section pre-twisted bars simply supported at the ends. The problem has been resolved in a complete theoretical manner based on a direct variational method using the Fourier series and a rapidly convergent algorithm has been developed (Serra 1993). Other equations, derived from the principle of virtual work and computed using the Rayleigh-Ritz method, again referring to rectangular cross-section, have provided results for bars with fixed-free ends and with fixed-spherically hinged ends (Tsuiji 1981).

Tests were performed on several series of specimens with different values of a/k and ß in order to verify the range of the theoretical results. The experimental analysis thus allowed the definition of the

Table 1

Series	k (mm)	a/k	ϕ (rad)	ϕ (deg)
0	0	0.000	0	0
1	900	0.017	0.55	31.8
2	450	0.033	1.11	63.7
3	300	0.050	1.67	95.5
4	200	0.075	2.50	143.2
5	150	0.100	3.33	190.9
6	120	0.125	4.17	238.7
7	100	0.150	5.00	286.5
8	85.7	0.175	5.83	334.2

Table 2

	L (N)	t=3 (mm)	t=4 (mm)	t=5 (mm)	t=6 (mm)	t=8 (mm)
(a)	P_{Et}	551	1306	2551	4409	10452
	P_{En}	551	1307	2552	4410	10454
(b)	P_{Et}	1102	2613	5103	8819	20904
	P_{En}	1134	2687	5248	9069	21498
(c)	P_{En}	551	1307	2253	4411	10455
	P_{Ee}	550	1350	2620	5850	19350

Table 3

τ	t (mm)	ϕ (Degrees)			
		0	63.7	95.5	143.2
	3	1	3.93	4.12	5.03
	4	1	3.59	3.38	3.47
P_{crn}/P_{Et}	5	1	3.45	3.05	2.76
	6	1	3.35	2.87	2.37
	8	1	°2.35	2.68	1.98
	3	1	3.64	3.55	4.35
	4	1.1	3.43	3.49	3.81
P_{cre}/P_{Et}	5	1.1	3.06	3.01	2.80
	6	1.3	3.20	2.90	3.42
	8	1.8	2.06	2.91	3.45

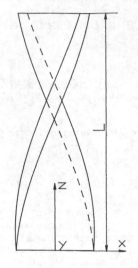

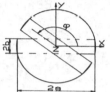

Figure 1

influence that the geometric parameters have on the deformation of the bar on varying the applied axial load. Using a finite element program, a numerical analysis was performed in order to compare the theoretical results and to determine the range of a/k and ß

for which the numerical analysis, referring to plate elements, presented values closest to those obtained experimentally. The values of some important buckling parameters (direction and position relative to the mid-point), were also measured using an image acquisition and analysis system to evaluate the lateral displacement of the bar under buckling.

2 NUMERICAL-EXPERIMENTAL APPROACH

In order to compare the data obtained by authors who have resolved the problem of the buckling load for two different boundary conditions theoretically, and to extrapolate the results for other constraints, numerical FEM models were prepared for the following three boundary conditions: (a) simply supported ends, (b) fixed-spherically hinged ends and (c) both ends spherically hinged (Figure 2). The configurations represent the two different theoretical (a) and (b), and experimental (c) conditions.

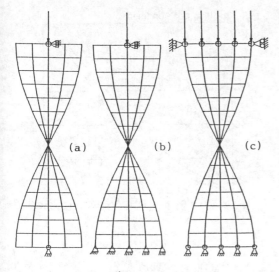

Figure 2

Figure 4

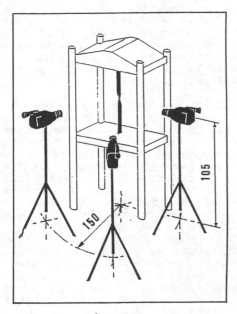

Figure 3

The bars, of identical length L (500 mm) and width 2a (30 mm), were of five different thicknesses 2b (3, 4, 5, 6 and 8 mm). In the models, all the bars were placed in orthogonal X, Y, Z co-ordinate system where the Z axis coincides with the bar length L and the X axis with the width 2a at the bottom of the bars. The first model

was realised with about 2000 thin shell elements using the SDRC I-DEAS structural FEM program working on a SUN workstation. Very similar results were obtained with a more simple FEM program (SuperSAP) working on PC and using fewer plate elements: 4x80=320 (for pre-twisted angles less than 180 degrees) or 5x90=450 (for pre-twisted angles greater than 180 degrees), by dividing the width 2a in 4 or 5 parts and the length L in 80 or 90 parts. A uniform 100 N compressive force along Z was applied to every upper node. Using plate elements, the numerical boundary conditions differ from theoretical values for the specimens of greatest thickness but are quite representative of the experimental conditions.

The FEM models were constructed with constant 2a, five different thicknesses 2b (with values of ß ranging from 3/30 to 8/30) and nine different values of a/k (ranging from 0 to 0.175). Correspondingly, forty-five series of steel test specimens were produced. Five bars were realized for each series, characterised by the same nine different a/k values and five ß values of the numerical models, so that mean values could be obtained from the results of five test measurements. Table 1 shows the parameters of twisting and the corresponding pre-twisted angles

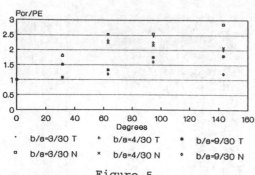

Figure 5

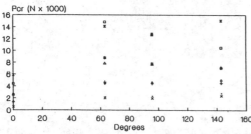

Figure 6

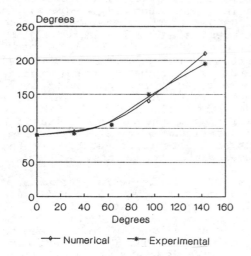

Figure 7

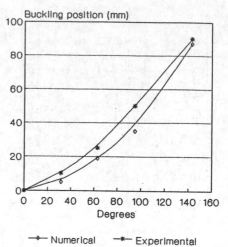

Figure 8

load was applied to the bars under the (c) boundary condition. The specimens were positioned between two fixed plates allowing displacement along the Z axis only. Translation, in the X-Y plane, of the edges of the specimen was thus completely constrained.

In order to evaluate the position and direction of the buckling, three videocameras were positioned at 0, 45 and 90 degrees, following the scheme given in Figure 3. The telecameras were placed so that they filmed a suitably marked part of the test specimen. It was thus possible to define the components of the deflection by the same equations used for the 0-45-90 strain gauge rosettes. An Imaging Technology VP 1320 Color Frame Grabber board, 512 x 512 pixels with 24 bits of depth, equipped with Optimas software for morphological and densitometric image analysis was used. The images recorded during the test were compared with the initial (zero load) images in order to evaluate the lateral displacement as a function of load, ß and a/k. Figure 4 shows the superimposed images of the specimen (thickness 6 mm and ø=95 degrees) at the start and the end of the test.

between the two extreme cross-sections for the test series.

Using a Metro Com Engineering universal static testing machine, with load cells of 1 t and 5 t full scale, an linearly increasing axial

3 ANALYSIS OF RESULTS

The first calibration of the numerical-experimental system was performed with the pre-twisted angle equal to zero. In this case, the configurations (a) and (c) can both be considered as single end supported (at least for lower thicknesses). The good agreement between the numerical (P_{En}) and theoretical (Euler, P_{Et}) buckling loads is shown in Table 2, which also gives the experimental results (P_{Ee}) for configuration (c). The theoretical Euler critical loads for configuration (c) are, in fact, the same as those of configuration (a). The experimental results confirmed the theoretical values for the lower thicknesses. It should be noted that for higher thicknesses (t=6 and 8 mm), the behaviour of the specimens tends to the boundary condition (b), due to the constraint on the rotation of the edges: the greater the thickness, the more difficult it is for the bar to rotate along the axis of the smaller cross-section moment of inertia.

The second stage was to compare the theoretical and numerical results for the two configurations (a) and (b). For (a), the theoretical values were obtained calculating the first 5 terms of the Fourier series (Serra 1993) for pre-twisted angles from 0 to 90 degrees. For higher values, more terms are required and, also, the numerical program does not always give reliable results. The differences between the results increase with thickness due to the changing constraint of the edges. In fact, the plate FEM model is not provide an accurate reproduction of the simply supported boundary condition of the theoretical model. Therefore, for this configuration, the correspondence cannot be considered acceptable.

Instead, the comparison between the theoretical-computed (Tsuiji 1981) and numerical FEM results for configuration (b), shown in Figure 5, where τ (P_{cr}/P_E) is the ratio between the critical load for the pre-twisted bar and the theoretical Eulerian (P_{Et}) load, there is a

good correspondence for pre-twisted angles up to 90 degrees. The results were compared for the parameter $D=(a/b)^2$ equal 10, 50 and 100, corresponding to ß values of approximately 9/30, 4/30 and 3/30, respectively.

The latter results allow the model to be applied to the (c) boundary condition, not simply supported at the central node, but with all the end nodes constrained. Figure 6 shows the comparison between the numerical and experimental critical loads for configuration (c). In this case, the results again show good agreement up to about 90 degrees, especially for the lower thicknesses. Table 3 shows the values of τ for the same conditions, including the thickness 8 mm (for clarity, omitted from Fig. 6). Figures 7 and 8 show, respectively, the direction and position (referring to the mid-point) of the buckling section. The values are the same for all the thicknesses examined, except for minor differences, but with a high degree of dispersion, especially for the higher pre-twisted angles. The values were obtained by measuring the deformation parameters on all the specimens at the end of the test. For some specimens (series ß=6/30 and pre-twisted angle up to 143 degrees), the image analysis technique was also used and the preliminary results show acceptable agreement.

4 CONCLUSIONS

Proceeding from the data found by other authors who have resolved, in a theoretical or theoretical-numerical manner, the problem of the buckling load in pre-twisted bars of rectangular cross-section with different boundary conditions (simply supported or with fixed-spherically hinged ends), a numerical and experimental analysis was undertaken.

A numerical model using plate elements was developed and verified using theoretical models. The results showed unacceptable differences for the simply supported ends boundary condition, since it was impossible to reproduce the same constraint with plate elements. However, good agreement was found for the buckling load in the range of a/k from 0 to 90 degrees and for the lower values of ß (from 3/30 to 5/30) for the fixed-spherically hinged ends boundary condition. The good agreement between the results confirmed the theoretical values which have been previously obtained in this range of a/k and ß. Further, it allows the FEM model to be used in the analysis of the behaviour of buckle loaded pre-twisted bars under different constraints. The numerical model was then used to analyse other boundary conditions, some more complicated than those usually chosen. In particular, the configuration with simply supported and constrained translation at the ends was investigated both numerically and experimentally. The results obtained showed an acceptable correspondence of the FEM model to the experimental results in terms of buckling load. A better agreement was found correlating other parameters of the buckle; in particular, the direction and position of the buckling. These parameters were measured both at the end of the test on the buckled bars and also, for some specimens, during the test using a system of three videocameras at 0, 45 and 90 degrees to the principal axis of the specimen. The recorded images were studied using an image analysis system combining the deflections with the same equation employed in strain-gauge rosette techniques. The results showed a good agreement for low values of the pre-twisted angle a/k (up to 90 degrees) and of the thickness 2b (ß=b/a up to 5/30) which are those most common in engineering practice.

The authors intend to develop the research, studying other cross-section shapes and boundary conditions in order to optimize the image analysis technique and to calibrate a numerical-experimental model which is also suitable for more general parameters.

REFERENCES

Fischer, W. 1970. Buckling of the knife-edge supported and twisted bar: *Ingenieur-Arch* 39: 28-36.

Frisch-Fay, R. 1973. Buckling of pre-twisted bars: *Int. J. Mech. Sci.* 15: 171-181.

Giovannozzi, R. 1942. Trazione, torsione e flessione pura di solidi svergolati a sezione costante: *Pontificia Academia Scientiarum*.

Göhner, O. 1940. Spannungsverteilung in einem an den Enden belasteten Schraubenstab beliebiger Steigung: *Ingenieur-Archiv*.

Leipholz, H. 1960. Buckling of twisted bars subjected to the pressure of conservative continuous and uniformly distributed loading: *Ingenieur-Arch* 29: 262-279.

Rosen, A. 1991. Structural and dynamic behaviour of pre-twisted bars and beams: *Appl. Mech. Rev.* 44, 12.

Serra, M. 1993. Flexural buckling of pre-twisted columns: *Journal of Engineering Mechanics* 119, 6.

Tsuiji, T. Yamashita, T. 1981. The buckling of pre-twisted bars: *31st Japan National Congress for Applied Mechanics*: 131-137.

Ziegler, H. 1948. Buckling of a twisted bar: *Schweizerische Bauzeitung* 66: 463-465.

Recent Advances in Experimental Mechanics, Silva Gomes et al. (eds) © 1994 Balkema, Rotterdam, ISBN 90 5410 395 7

Full scale six degrees of freedom motion measurements of two colliding 80 m long inland waterway tankers

L.J.Wevers, J.v.Vugt & A.W.Vredeveldt
Dutch Organisation for Applied Scientific Research TNO, TNO Centre for Mechanical Engineering, Delft, Netherlands

ABSTRACT: The TNO Centre for Mechanical Engineering in Delft, the Netherlands, carried out full scale ship collision experiments with 80 m long inland waterway tankers. This paper mainly describes the computation methodology to obtain pure ship motions, Pitch, Roll, Heave, Surge, Sway and Yaw, from acceleration measurements during ship collisions. Some information about the tests in general is given. Some overall results are shown.

1 INTRODUCTION

Inland waterway navigation in Europe is becoming increasingly important for safety- and environmental reasons. Several bulk chemicals are not allowed in inland vessels. The Dutch government has carried out a research study concerning the general safety and environment risk of inland waterway traffic [1]. One subject of this study refers to the risk involved when calamities with ships, such as collisions and/or groundings, occur. Here the crashworthiness, energy absorbing resistance, of the ship structure is of interest [2].

Several calculations methods are available to determine the crashworthiness of ship structures. However no full scale experimental data was available which could be used to verify these methods.

Experiments and computer simulations were performed in this joint Dutch Japanese project (see: ACKNOWLEDGEMENTS). Four collision tests were carried out with two 80 m long inland waterway tankers of 1500 Tonnes each.
At each collision the bow of the striking ship hit the side of the struck ship at an angle of 90 degrees (see Figure 3). Single- and double side structures were tested. Ship's speed was 5 km/h for the first test, and 15 km/h for test 3, 4, 5.

The experiments were focused on measuring the collision forces between striking and struck ship, motions of both ships, penetration depths in the struck ship and strain in the hull.

The tests were carried out in an inland waterway near the city of Moerdijk in the Netherlands.

This paper describes the motion measurements and especially the calculation algorithm and method to obtain the pure six motions from acceleration measurements: Roll, Pitch, Heave, Surge, Sway and Yaw. Some examples of results are given.

For more general information on the subject see [3]..[7] of References.

2 MEASUREMENT SYSTEM

The penetration of the bow of the striking ship into the side of the struck ship was, for reasons of safety and security, measured by means of several different devices. Two wire gauges, measuring displacement and velocity as a function of time, were fitted at the trunk deck of the struck ship. Ultra sonic devices were use as backup system for displacement. High speed cameras were used at the bow of the striking ship and in the tanks. The colliding forces were measured with 19 special made load cells fitted between the cut off bow of the striking ship and the main hull. The vertical load between bow and main hull was measured with two load cells. Strain gauges were fitted at several location in the struck ship.

In figure 3 the position of the acceleration sensors for the motion measurements of both ships are given. In both ships eight sensors were mounted.

The locations of the four collisions are also shown in figure 3 (see end of paper).

3 CALCULATION METHODOLOGY

3.1 Coordinate system for motion calculations

For the computation of the motion, the ship is considered as a rigid body, with six degrees of freedom. It is usual to define the six degrees of freedom in a fixed coordinate system to the ship (body coordinate). The six degrees of freedom in body coordinates are for the translation: surge, sway and heave. For the rotation: roll θ, pitch ϕ and yaw Ω. In order to describe the position of the ship a second, earth fixed coordinate system is required.
The body coordinate system can be derived from the fixed coordinate system by executing of four coordinate transformations. One translation and three rotations.
The translational coordinate system used is given in figure 1.

3.2 The relation between vectors in fixed and local coordinates

If Uo is the translational velocity, fixed coordinate, and U $\{Ux, Uy, Yz\}$ is the translational velocity in body coordinates then the relation is described as:

$$U = \overline{R} \ (\theta, \phi, \Omega) \ Uo \tag{1}$$

and if Po is the rotational velocity in fixed coordinates and P $\{p, q, r\}$ is the rotational velocity in body coordinates then this can be described as:

$$P = \overline{A} \ (\theta, \phi, \Omega) \ Po \tag{2}$$

The matrices R and A are known.

3.3 The measured acceleration of one point in the ship

The position vector in fixed coordinates for a point P with body coordinates is:

$$Xop = Xo + \overline{R}^{-1} Xp \tag{3}$$

Xp $\{xp, yp, zp\}$ is the position in the ship in body coordinates.

The acceleration vector of point P, in fixed coordinates, is found by differentiating Xop twice to the time. The following formula is found:

$$\frac{d^2 \ Xop}{dt^2} = \overline{R}^{-1} \frac{dU}{dt} + \frac{d\overline{R}^{-1}}{dt} U + \tag{4}$$

$$\frac{d^2 \ \overline{R}^{-1}}{dt^2} Xp + G$$

Xop is the kinematic displacement vector in fixed coordinate system. Normally measured the acceleration transducer the gravity. This is expressed by the vector G.
In body coordinates the acceleration of point P is:

$$\frac{d\vec{Up}}{dt} = \overline{R} \frac{d\vec{Xop}}{dt} \tag{5}$$

Up $\{Upx, Upy, Upz\}$ is the velocity vector in body coordinates.

The practical acceleration of a transducer in one point and three directions is described as:

$$\begin{bmatrix} \dot{U}px \\ \dot{U}py \\ \dot{U}pz \end{bmatrix} = \begin{bmatrix} \dot{U}x \\ \dot{U}y \\ \dot{U}z \end{bmatrix} +$$

$$\begin{bmatrix} 0 & -r & q \\ r & 0 & -p \\ -q & p & 0 \end{bmatrix} \begin{bmatrix} Ux \\ Uy \\ Uz \end{bmatrix} + \begin{bmatrix} 0 & -\dot{r} & \dot{q} \\ \dot{r} & 0 & -\dot{p} \\ -\dot{q} & \dot{p} & 0 \end{bmatrix} \begin{bmatrix} xp \\ yp \\ zp \end{bmatrix} +$$

$$\begin{bmatrix} -p^2-q^2 & pq & rp \\ pq & -p^2-r^2 & rq \\ rp & rq & -p^2-q^2 \end{bmatrix} \begin{bmatrix} xp \\ yp \\ zp \end{bmatrix} + \begin{bmatrix} -\sin\phi \\ \sin\theta & \cos\phi \\ \cos\theta & \cos\phi \end{bmatrix} g - \begin{bmatrix} 0 \\ 0 \\ g.off \end{bmatrix}$$

$$\tag{6}$$

For the vertical acceleration transducer in most cases an intrinsic mechanical correction for gravity is performed. g.off has a value in case a mechanical correction is used.

3.4 The transducer positions

The acceleration transducer is normally sensitive in one direction. So one equation is used from (6) for every transducer. If we use several transducers then we have a set of equations. If the vectors U|P = $\{Ux, Uy, Uz, p, q, r\}$ and PP = $\{-(q^2+r^2), -(p^2+r^2), -(p^2+q^2), rq, rp, pq\}$ are chosen then we can recognize some repetition in the matrix notation for more than one transducer. In compact notation the matrix presentation for the different acceleration transducers is:

$$Up = (\overline{E},\overline{Xp}) \, \frac{dU|P}{dt} + (\overline{Xp},\overline{Xp}) \, PP + \qquad (7)$$
$$\overline{E} \, PxU + \overline{E} \, \overline{RG} - G.off$$

The matrix E is a sub matrix of the matrix (E,Xp) and consist of ones in the sensitivity direction. The sub matrix Xp in (E,Xp) gives the position of the transducer in the not sensitive directions. The matrix (Xp,XP) looks like (E,Xp) except that the one in the E matrix is replaced by de sensitivity position of the transducer.

3.5   The solution fore more than six transducers

For six transducers the non linear differential equation (7) is square. For more than six transducer a combination selection matrix SM is introduced. First equation (7) must be multiplied with SM. If the inverse of {SM (E,Xp} exists a solution is possible. If we those:

$$\overline{T} = [\overline{SM} \, (\overline{E},\overline{Xp})]^{-1} \, \overline{SM} \qquad (8)$$

equation (7) becomes:

$$\frac{dU|P}{dt} = \overline{T} \, \frac{dUp}{dt} - \overline{T} \, (\overline{Xp},\overline{Xp}) \, PP - \qquad (9)$$
$$\overline{T} \, \overline{E} \, (PxU + \overline{RG}) + \overline{T} \, G.off$$

U|P is found by integration of dU|P. For every time step in the calculation PP, PxU and RG must be calculated. The matrix T, matrix product T (Xp,Xp) and T G.off are not changed during the calculation. Because the differential equation (9) is not linear, an explicit integration method is chosen. A calculation flow chart is given in figure 2.

4 RESULTS

In figure 4 the test-4 motion results versus time of the struck ship are shown. Here the double side structure of starboard tank-3 was tested. Maximum roll motion of approximately 1 degree was found at the end of the collision at 0.819 s after pre-trigering. Due to the shape of the bow of the striking ship, the roll was found to be small for all tests. Pitch and and Heave were very small. Yaw was relatively large depending on the location of collision. Figure 5 shows the motion results of the striking ship of the same test-4.

Validation of computer simulation was one of the objectives of the full scale collision tests with inland waterway tankers. Figure 6 shows results both of measurement and simulation of forces and penetrations. Figure 7 shows deformations in the side structure and the deck of the struck ship.

5 CONCLUSIONS

General conclusions of the full scale collision tests with two inland waterway tankers are given. Conclusion 5.7....5.11 refer to the results of the motion measurements.

5.1   The collision tests carried out have been succesful. The objectives of the project as listed below were achieved:
- measurement of total collision forces
- determination of collision durations
- measurement of motions of both ships during collision
- measurement of penetration depths and penetration durations
- measurement of strain in the sides of the ship due to collision

5.2   The estimates made in the development phase of the project were found to be within the limits set

5.3   The full scale collision experiments yield valuable information with regard to ship's collision damage resistance. This refers especially to the behaviour of the ship's side/deck structure when subjected to a collision with a ship's bow.

5.4   The acquired test data can be used for validation of calculation methods in this field.

5.5   With a validation calculation method it is possible to predict penetration depths in a single or double sided ship in relation to the speed of the vessels and their mass. Also striking forces and collisions can be predicted.

5.6   In all test cases initial cracking of the shell occurred at the weld.

5.7   Using the system with very sensitive accelerometers, calibrated on a large horizontal shock testing device, was found to be accurate. The computations in the computer easily divided the complex mixture of motion signals of the sensors into pure Pith, Roll, Heave, Surge, Sway and Yaw of the ship.

5.8   Because of the short duration of the

collisions, approximate 500 ms, there was no problem with drift of the signals after double integration of complete 4096 point time blocks. After 1 s some drifting started.

5.9 The motion algorithm in the computer calculations of the time histories has proved to be sufficiently accurate to obtain motions up to 2 degrees rotation and 1 m translation.

5.10 An advantage of the motion system with very sensitive accelerometers is the simplicity of the measurement system in combination with good use of sophisticated computers.

5.11 In a motion system as described there are no moving parts, this is an advantage.

5.12 The main motion of the struck ship during collision were found to be sway and yaw. Remarkable is that almost no roll motion of the struck ship was measured. Reason for this is the shape of the bow of the striking ship resulting in a horizontal and a vertical component of the collision force. Apparently the resulted force acted through the axis of roll motion.

5.13 Approximately 1/3 of the kinetic energy of the striking ship is dissipated by fracture and deformation of the ship structure.

ACKNOWLEDGEMENTS

The authors wish to acknowledge many individuals of this joint Dutch Japanese project in which TNO Centre for Mechanical Engineering did the experiments. The Netherlands Foundation for the Coordination of Maritime Research (CMO: Mr. Carlebur) coordinated the project. The main sponsor in the Netherlands was the Ministry of Transport and Public Works (RW: Mr. Stipdonk). The McNeal Schwendler Company in the Netherlands (MSC: Mr. Lenselink) did the computer simulations. For Japan the Japanese Association for the Structural Improvement of the Shipbuilding industry (ASIS: Mr. Kameyama and Mr. Inoue) was the sponsor represented by Mitsubishi Heavy Industries (MHI: Dr. Kuroiwa and Mr. Kawamoto). A special word of thanks is directed to all persons mentioned and many others for their valuable discussions during the tests.

REFERENCES

[1] Safe waterway transport. Netherlands Ministry of Transport and Public Works. VVW-N-89040, March 1989, Rotterdam, The Netherlands.

[2] Vredeveldt A.W. Wevers L.J. Full scale ship collision test results. TNO Centre for Mechanical Engineering, April 1992, Delft, The Netherlands.

[3] Vugt J. van, Collision Motion Monitoring, see [2] Appendix-A

[4] Vredeveldt A.W. and Journee J.M.J. Roll motion of ships due to sudden water ingress, calculations and experiments, Second Kummerman International Conference on Ro-Ro Safety and Vulnerability - The way ahead, London April 1991

[5] Prevention of oil spills from tankers, November 22, 1991, Mitsubishi Industries

[6] Peterson M.J. Dynamics of ship collisions, Ocean Engineering, Vol.9 No.4 1982

[7] Lemmen P.P.M. Brockhof H.S.T. Collision resistance of a 3000 Tonnes chemical tanker, TNO rep. 5087007-90-1, June 1990

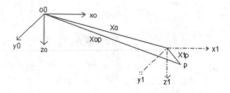

Figure 1. Translational coordinate system

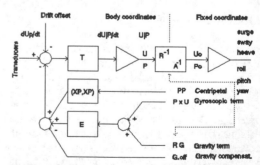

Figure 2. Flow chart of motion calculations

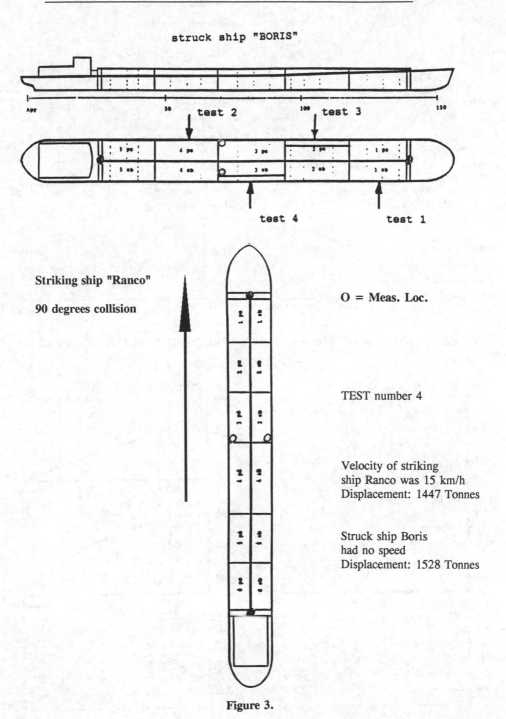

Figure 3.

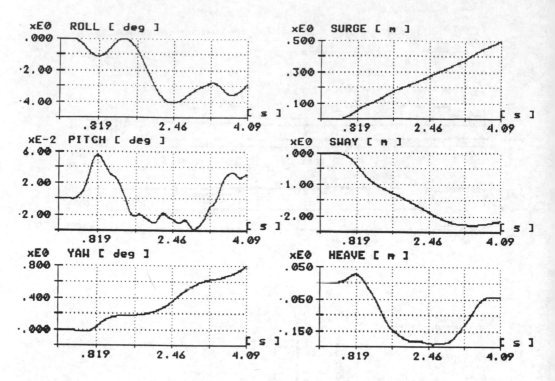

FIGURE 4. MOTIONS OF STRUCK SHIP, DOUBLE SIDE STRUCTURE SB TANK-3

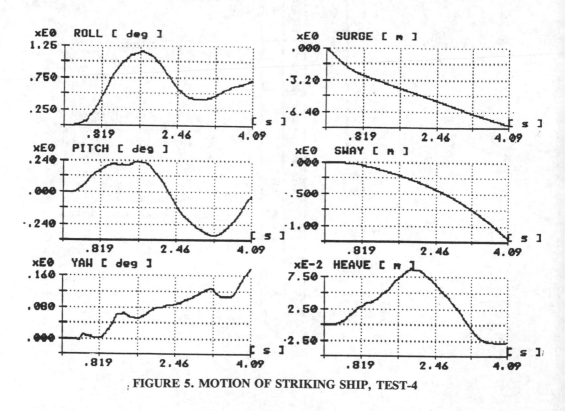

FIGURE 5. MOTION OF STRIKING SHIP, TEST-4

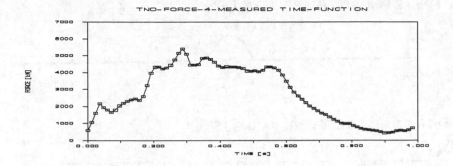

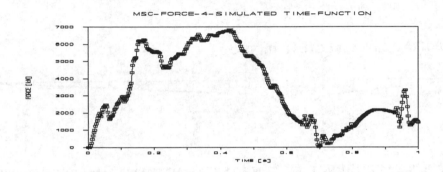

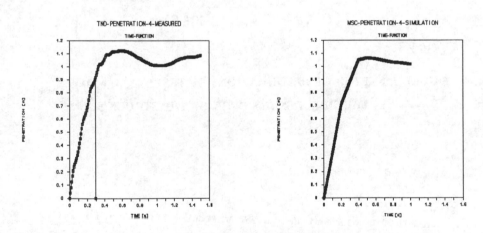

FIGURE 6. FORCES AND PENETRATIONS OF TEST-4 MEASURED AND SIMULATED

TEST-1: SINGLE SIDE STRUCTURE, TANK-1 SB

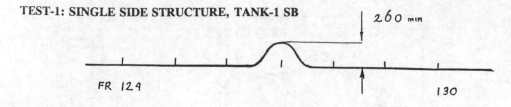

260 mm

FR 124

130

TEST-2: SINGLE SIDE STRUCTURE, TANK-4 PS

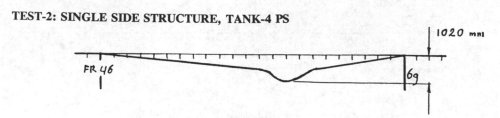

1020 mm

FR 46

69

TEST-3: DOUBLE SIDE STRUCTURE, TANK-2 PS

770 mm

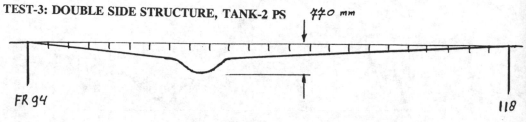

FR 94

118

TEST-4: DOUBLE SIDE STRUCTURE, TANK-3 SB - - - WITH SPECIAL SIDE STIFFENING

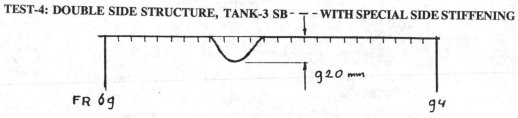

920 mm

FR 69

94

FIGURE 7. SKETCH OF DEFORMATIONS OF TEST 1, 2, 3, AND 4
IN THE SIDE AND THE DECK OF THE STRUCK SHIP

Recent Advances in Experimental Mechanics, Silva Gomes et al. (eds) © 1994 Balkema, Rotterdam, ISBN 90 5410 395 7

Critical analysis of a steel tower experimental test

Ana M.S. Bastos & J. Sampaio
Civil Engineering Department, Faculty of Engineering, University of Porto, Portugal

F.M.F. de Oliveira
Mechanical Engineering Department, Faculty of Engineering, University of Porto, Portugal

A.G. Magalhães
Instituto Superior de Engenharia do Porto, Portugal

ABSTRACT: The present work describes an experimental test carried on a latticed steel transmission tower under the actions of the supported high voltage electric lines, wind and self weight actions. The layout of the loading and instrumentation used with data acquisition system is presented. The experimental response of the speciment is illustrated and compared with the analytical design. Some considerations are made about the efficiency of the shear bolted connections, that at least, conditioned the ultimate capacity of the structure.

1 - INTRODUCTION

The steel transmission tower, usually denominated "Poste de 2tf" was designed to support high voltage electric lines which introduce, in working conditions, a 20 kN load (nominal load) on the top of the structure, in both horizontal and vertical directions.

The portuguese specification recommends that the design, adequacy of connections and efficiency of constructive disposals be verified by experimental analysis.

For an optimized tower design an experimental test was provided to validate the analysis method and evaluate the load distribution through the structure and local load capacity for the specified load case.

The structure is a spatial lattice steel tower (Fig. 1) with 30 m high, produced from equal leg angles with shear bolted connections.

The steel strength characteristics are $F_y = 300$ MPa and $F_u = 395$ MPa, for the yield and tensile strength.

Bolts class 5.6 were used in connections.

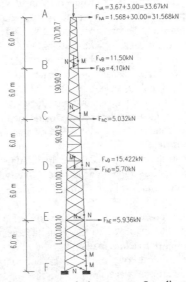

Figure 1 - Steel transmission tower. Loading scheme and instrumented members (N, M).

2 - EXPERIMENTAL PROGRAM

The NP341 portuguese specification [1] recommends that loads should be applied by loading and unloading cycles of increments of 25%, 50%, 75%, and 100% of the maximum specified load.

The loads applied to the steel tower were due to high voltage steel lines, wind and self weight of the structure loads, and agreed with loads specified for design.

The loading scheme is shown in Fig. 1:
- loads due to high voltage lines - two 30 kN concentrated loads at the top of the tower in both horizontal and vertical directions introduced by stressed strands corresponding to 20 kN nominal load times load factor 1.5.

Figure 2 - Photo of the experimental test. Horizontal loads at the top and wind loads.

- wind loads - pointed loads equivalent to distributed horizontal loading, acting normal to the axis of the tower at end sections A-E of different pannels. The loads were provided by dead weights, aluminium plates lying in baskets suspended by cables embracing the structure, so that loads could be resisted by the main structural system, Fig. 2.

- self weight load - three concentrated loads introduced by strands along the structure axis, on sections A, B and D, stressed up to load levels equal to the weight of the pannels between them.

Deflection measurements were made before and after load conditions as well as at all intermediate load increments.

As the structure deflected under load, adjustments were made in the strand device in order to ensure the correct load and directions specified in the loading scheme.

Two previous load tests were carried out up to the unfactored nominal load 20 kN, simultaneously with the wind and self weight loads.

On a third test, two loads in the horizontal and vertical directions were applied on the top of the structure, up to the factored nominal load, 30 kN.

After that, the horizontal load was increased until the failure of the structure.

The wind and self weight were kept constant during the test.

3 - EXPERIMENTAL SETUP

The test specimen was a full-scale steel tower in the horizontal position for easier application of the loads and deformation measurements.

The structure was rigidly supported at the base to a vertical concrete wall. It was also supported on platforms on the limit sections of the different pannels, by spherical lubricated metalic rolls to minimize friction in the supports and allow free horizontal displacements.

Instrumentation setup was used to evaluate the stress distribution and deformations of the critical sections of the structure, for the designed loads:

- mechanical dial gauges and scales for measuring the deflection of the pannels end sections;

- electric resistance strain gauges, with dummy (temperature compensation) gages, for measurement of member loads and moments introduced by eccentricity of loading;

- load cells for measuring axial force in strands (Fig. 3).

- data acquisition system consisting in six multi-channel data collecting stations, remotely operated by two host computers.

Figure 3 - View of the experimental setup.

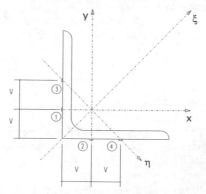

Figure 4 - Axial force and bending moment strain gage configuration.

The strain gage configuration in member angles is shown in Fig. 4. Locations 3 and 4 were used for evaluation of axial load on members N (Fig. 1); locations 1, 3 e 4 for bending moments on sections M (Fig. 1). Location 2 was not used.

The axial force and bending moment in respect to the x - y axis may be computed by:

$$N = (\sigma_3 + \sigma_4)*A/2$$

$$Mx = 1/4v[(I_1\text{-}I_2)\sigma_4 - (I_1\text{-}3I_2)\sigma_3 + 4I_2\sigma_1]$$

$$My = 1/4v[-(I_1\text{+}I_2)\sigma_4 + (I_1\text{-}3I_2)\sigma_3 + 4I_2\sigma_1]$$

4. TEST RESULTS

In the first test, only one loading and unloading cycle up to 20 kN was applied. Deflections of the top (A) of the tower.are plotted in Fig. 5 .showing a residual of 60.3% of the maximum deflection. This value was due to slides and friction on the joints that did not allow the replacement of initial geometric conditions.

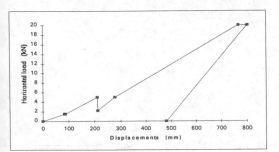

Figure 5 - Deflections at the top of the tower on the first loading test.

In the second test, the residual deflections up to the applied load cycle of 20 kN, were considerably lower than those observed in the first test, increasing for higher loads at the top of the tower. The permanent displacement at the top was $\delta_r = 18.3\%$ of the maximum displacement observed, Fig. 6.

After these tests the structure has been disassembled for rectification of the pannels. Larger diameter bolts were used to minimize slide movements at the connections.

In the third test, displacement and strain measurements were made for loading cycles up to 30 kN, simultaneously with wind and self weight loads. Horizontal load at the top was then increased until

Figure 6 - Deflections at the top of the tower for the second loading test.

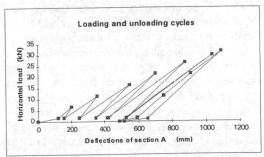

Figure 7 - Photo of shear failure of the bolt connections.

colapse of the structure, by shear failure of the bolt connections at the base support, Fig. 7.

Failure ocurred under 51.85 kN, 1.73 times the nominal factored load (30 kN), without any visible buckling of the angle members.

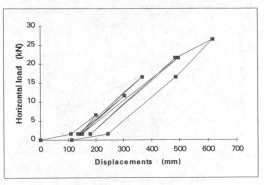

Figure 8 - Deflections at the top of the tower on the third loading test.

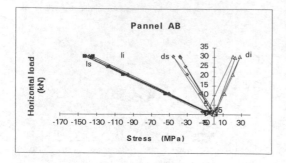

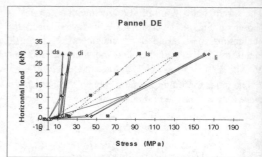

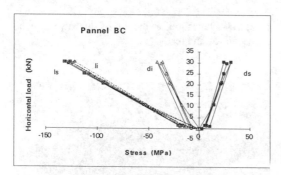

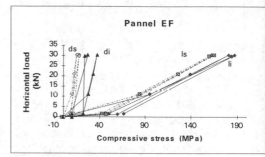

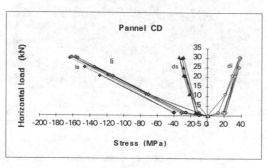

Figure 9 - Stresses in the instrumented members for third loading test.

Deflection of the pannel limit sections are plotted in Fig. 8. Residual displacements observed after each loading cycle were due to sliding at bolt connections. In fact, at the last cycle, where the reached load value was the same of the former cycle, 30 kN, displacement recovery was almost complete.

Stresses at instrumented members on pannels AB to EF are plotted in Fig. 9 (where **l** stands for longitudinal, and **d** for diagonal members; **s** stands for upper and **i** for lower location)

Up to the nominal factured load stresses measured in instrumented members were under steel yield stress limit, showing an elastic behaviour of the structure.

The values of bending moments were very small, leading to eccentricities less then 0.5 cm, so those

Table 1 - Comparison of analytical and experimental load members.

PANNEL	MEMBER	ANGLE	AXIAL LOAD (kN)		Nanal./Nexp.
		(Type)	Anal.	Exp.	
EF	Long.	L100.100.10	366,1	331,9	1,10
	Diag.	L40.40.4	3,9	7,9	0,87
DE	Long.	L100.100.10	322,4	248,0	1,30
	Diag.	L40.40.4	7,0	5,7	1,23
CD	Long.	L90.90.9	281,0	250,7	1,12
	Diag.	L40.40.4	13,2	10,9	1,21
BC	Long.	L90.90.9	227,3	202,6	1,12
	Diag.	L40.40.4	11,4	11,3	1,01
AB	Long.	L70.70.7	153,5	134,1	1,14
	Diag.	L40.40.4	12,6	11,4	1,11

may be considered as secondary effects on the structure.

Experimental results showed quite a good agreement with analytical design values [2], with average differences of 11.5%, Table 1.

5- COMMENTS ON BOLTED CONNECTIONS BEHAVIOUR

The portuguese code "Regulamento de Estruturas de Aço para Edifícios" [3] states, for bolted connections, that the hole diameter can not exceed 2 mm of nominal body size of the bolt, or 3 mm when this is larger than 24 mm. There will always be a certain amount of play on the holes, which compensates for small misalignements in the hole location.

At common bolted joints, stresses induced by fastening setup will be different from bolt to bolt, consequently different behaviour will be expected.

Considering two plates connected by two common differently tightened bolts and equal opposed forces N acting on them, if a load increment ΔN overpasses frictional resistance induced between one of the bolts and the washer, it will slide over the washer and the load increment will be supported only by the other bolt. At this moment, shear stresses in the bolts may be computed by:

N_0 - load equal to the frictional resistance of first bolt

ΔN - increment load

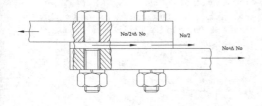

Reducing the load by No, there will be the following resultants on shear sections:

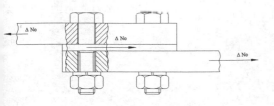

When unloading is performed, a residual shear force $\Delta N/2$ will remain in the bolts. Compression and tension restrain stresses will apear in upper (Bs) and lower (Bi) plates, between the bolts (length a).

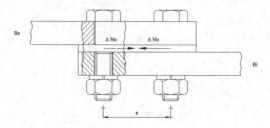

The displacements will be the same on future loadings and have very few influence on node displacements of the main structure. Those slide displacements about the same value of elastic deformations, may therefore spare some of the bolts in the joints.

For higher load values up to the friction resistance of all the bolts in the connection, these will be contact with the holes. Ovalization of holes and permanent shear deformations of the bolts may occur on the connections. This will lead to irreversible deformations in the members and residual node displacements on unloading.

This was observed at the tower loading tests, where high residual displacements were detected under relativey low loads applied to the struture.

Even without dinamometric fastening possibilities, it is always possible to fasten the nuts up to stresses that can guarantee no sliding at the joints, using apropriate wrenches. Indeed, using this process, yield stress of the bolt can be reached.

The work induced by moment by fastening operation, M_b, up to the yield limit tension of the bolts, may be computed by the sum of the work due to bolt extension and the work induced by friction moment over the nut:

$$M_b \cdot d\theta = \frac{e}{2\pi} \cdot d\theta \cdot N_y + \frac{\mu N_y}{2\pi} \cdot (\rho_1 + \rho_2) d\theta$$

e - bolt pitch,
μ - friction factor,
ρ_1 - average radius of contact surface circle,
ρ_2 - radius of nominal bolt,
$d\theta$ - rotation.

Considering a 20mm diameter bolt, Fe 360 ($Rm = 235$ MPa), with $\mu = 0,3$, $\rho_1 = 15mm$, $\rho_2 = 8mm$, $e = 2mm$, the value of the moment M_b will be $M_b = 104.5$ kN.mm, which may be easily reached by the operator.

Comparing bolt designs, one by friction and one by shear:

$$F = n\frac{\pi d^2}{4}\sigma_y \cdot \mu \quad \text{by friction,}$$

$$F = n\frac{\pi d^2}{4}\tau_s \quad \text{by shear.}$$

The two methods will be equivalent when:

$$\sigma_y \cdot \mu = \tau_s$$

For Fe 360, the former expression will lead to:

$\sigma_y \mu = 0.3 \cdot 235 = 70.5$ MPa
$\tau_s = 0.7 \cdot 135 = 94.5$MPa, very close to each other.

In fact, the stresses induced on bolts by tightening setup provides friction resistance, which may balance shear stresses on the bolts, under working loads. Consequently, a decrease of the shear stresses in the bolts will be expected and a delay of the slide displacements on connections.

6- CONCLUSIONS

The present work concerns an experimental test carried out on a steel transmission tower under the nominal loads introduced by electric lines, the wind and self weight loads.
The following conclusions may be summarized:
Up to nominal load of 30 kN at the top of the tower,
-The strains measured on instrumented members were above the yield limit values of the material. Consequently there were no residual deformations after all the loading and unloading cycles
On the contrary, the residual values of the deflections at the top of the tower were percently high, creating permanent disalignements of the pannels.
The maximum values of stresses observed were above 200 MPa, (85% of yield stress limit), showing an elastic behaviour of the struture;

Failure ocurred by shear of the bolts in the base support connection, under the load of 51.85 kN, leading to an overload capacity factor of 1.73. No visible buckling of members was noticed during the loading test;
When, in a shear bolt design, an accurate tightening of bolts is provided, the friction resistance mobilized may be equal to the bolt shear stress, under service conditions. Consequently, for working loads, there will be no sliding or residual node displacements and a higher security level may surely be expected.

ACKNOWLEDGEMENT

The study was requested by EDP Electricidade de Portugal, S.A., and the prototype was construted by A. Silva e Silva Lda.
The authors wish to acknowledge the facilities granted by the both companies, who assisted during the experimental program.

REFERENCES

[1] Norma Portuguesa NP-341. "LINHAS ELÉCTRICAS. Postes de aço reticulados, dimensionamento, fabricação e ensaios. Portaria n°21816 de 20/1/1966.
[2] Freitas J. M., Guimarães M., "Postes de suporte de linhas aéreas. Projecto optimizado de um conjunto de dez elementos e suas fundações", para E.D.P. Electricidade de Portugal S.A. Out 1992.
[3] R.E.A.E., Regulamento de Estruturas de Aço para Edificios, (Decreto-Lei n° 211/86, 31 Julho 1986). Imprensa Nacional Casa da Moeda E P Lisboa 1986.

Recent Advances in Experimental Mechanics, Silva Gomes et al. (eds) © 1994 Balkema, Rotterdam, ISBN 90 5410 395 7

Cable forces measurements on the externally prestressed and cable-stayed structures

Miodrag Pavisic

Institute Kirilo Savic, Belgrade, Yugoslavia

ABSTRACT: The two simple but different technics of cable forces measurement have been analyzed. Through the real example of the long span, externally prestressed structure, both of them have been applied and compared. The results are in favor of slope, rather than frequency measurement technics.

1. INTRODUCTION

Cables have become today very widely used and prospective structural members. Their presence is particularly evident on the large long span structures. In the same time, there is a rising interest in the media of experts, designers and contractors for the exact evaluation of cable forces on the basis of 'in situ' measurements.

There are today two methodologically different methods of cable forces prove and measurement. Both of them determine the forces indirectly, by the measurement of some other relevant parameters, e.g. slopes and frequencies.

In the paper, these two methods have been discussed and their efficiency, limitations and accuracy analyzed through the real example of measurement.

2. MEASURING TECHNICS

2.1 Slope measurement method

This method is founded on the simple equilibrium condition between externally applied point force and cable force (Figure 1).

$$2 S \sin \alpha - P = 0 \qquad (1)$$

$$S = P / 2 \sin \alpha \qquad (2)$$

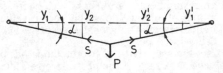

Figure 1. The slope measurement method

Since the applied point force intensity is known, one need to measure only the cable slope change (α) under given force.

The boundary condition require the cable ends immobility.

The cable force intensity is usually very high (compared with applied point force) and therefore the measured slope is low enough. It means that, the cable force could be calculate with the next approximation:

$$\sin \alpha = \alpha \quad , \quad \sin \alpha = \operatorname{tg} \alpha \qquad (3)$$

Actually, it means that the cable elongation under point force action is negligible.

The cable force change under action of externally applied point force is also negligible being that the intensity of the point force is many times lower than the existing cable force.(In the given example its intensity was only 0.06% of the measured cable force.

The angle α, however, is dependent on the value of P. Thus, cable force will not be a linear function of P. From the pure geometrical approach we have:

$$L + \Delta L = \sqrt{L^2 + \delta_c^2} = L \sqrt{1 + (\delta_c/L)^2} \qquad (4)$$

Since, $\delta_c / L \ll 1$ (in the given example $c/L = 0.0002$, we have: $\delta_c /L \approx \alpha$

In accordance with expression (2) only one parameter need to be measured. Since this value can be measured with high precision, consequently the cable force obtained by this way, is of the great exactness.

In practice, the best results can be obtained using two pairs of deflectometers on the both side of the point force (Fig. 1). By this way, a cable inflection, as a consequence of stiffness, can be avoided.

Also, whenever is possible the rule of symmetry should be used and by that way overall measured accuracy increased.

The testing results, obtained in the laboratory using the real cable tensioned to the force level of 150 kN., showed max. error of 4.0% . We should note, however, that the error of measurement is greater as the cable force is smaller. Obviously, this is a consequence of the initial assumption, that the cable has to be under high rate of tension.

the basis of cement, increases the overall cable rigidity and by that way increases the unavoidable mistake because of the neglected second part of expression (3). Also, we have to take in account that the grout present the dead weight distributed along the cable. Therefore, in essence, the grout has the rule of dumper. Nevertheless, for the low sag cables (as the stay cables are) it was discovered that in plane and sway mode frequencies are the same and the cable behaves like a taut string.

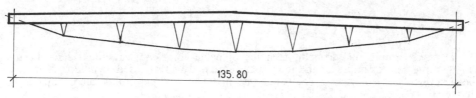

135.80

Figure 2. The main girder on the structure Hangar-2 at the Belgrade airport

2.2 Method of frequency measurement

This method is based on the expression relates the stress in wire with frequency of free oscillation.

$$f^2 = \sigma g/4l^2 + \pi^2 EIg / 4l^2 A\gamma \qquad (5)$$

for the first mode of oscillation.

$$S = 4l^2 \gamma f^2 A/g \qquad (6)$$

where is: l-length of cable; γ-specific weight; f-frequency; A-area; g-gravity acceleration; σ-stress; E-module elasticity, I-moment of inertia, S-cable force.

3. EXAMPLE

As an example, we will present the cable forces measurement carried out on the structure of Hangar 2 at the Belgrade aeroport.

The main girders, with the span of 135.80m., are built up as a concrete girders with external prestressing cables(Fig.2). The nine parallel cables per girder, are leaded by the help of the seven steel "chars". Each of the cables is formed of eleven strands with diameter of 15.2mm. filed with cement grout and protected with polyethylene tube ∅ = 142 mm.

The free length of cables between two neighboring "chars" is 17.22 m. The applied point force intensity was 2.60kN.

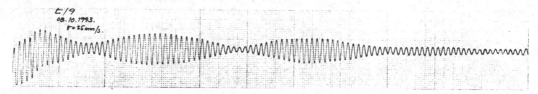

Figure 3. The response spectrum for one cable

The second part of expression (5) is too small value and therefore neglected. The method is widely used in the case of tensioned wires, but some problems arise relate to cables. Namely, cable is usually composite, consisting of strands, grout and polyethylene protective tube. Very often the grout is on the basis of cement emulsion and, as a consequence, its weight could be significant comparing with other components.(In the given example, the weighing proportion between strands and grout was 70/-30%). In addition, the grout, particularly on

The two above mentioned measuring procedure has been applied. The results are shown comparatively in Table 1. The response spectrum for one of the cables is shown on the Figure 3. It is evident that the values obtained by frequency measurements are slightly higher(with exception of one cable). The mean value of differences is 208 kN. what presents 4.5% of the cable force per girder.

Also, the sensitivity of two methods has been analyzed in regard of the possible errors during measurement. The result is clear from the Fig-

Table 1. Cable forces

Cable	Slope Method	Frequency Method
1.	4555	4766
2.	4607	4766
3.	4496	4735
4.	4705	4977
5.	4760	5116
6.	4598	4945
7.	4792	4913
8.	4444	4642
9.	4616	4581

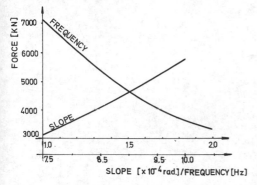

Figure 4. Cable force as a function of measured slopes and frequences

ure 4. which presents the cable force variability in function of the measured slopes(curve a) and frequencies(curve b).

We can see from the curves that the slope measurement error of 1×10^{-6} rad causes corresponding error in cable force intensity of 18.00 kN. Since, the cable slope measurement in the given example was accomplished by the instrument of very high precision ($p_{min} = 0.178 \times 10\text{-}6$ rad) the corresponding measurement error in the cable force intensity was only 1.60kN.(0.03%).

On the other side, the values of cable forces obtained by the frequencies measurement have the greater degree of sensitivity. The measurement error of 0.1 Hz. causes corresponding error of 97.00kN.

4. CONCLUSIONS

The two main methods of cable forces measurement has been analyzed and applied on the real example.

The slope measurement method is simpler to apply in practice and the obtained results are more accurate. Also, the slope measurement method is less sensitive to the errors of measurement.

The presence of grout is essential cause of inaccuracy concerning the frequency measurement method. Nevertheless, the method is acceptable particularly when the high rates of cable force is expected.

5. REFERENCES

Budynas, R.G. 1977. Advanced Strength and Applied Stress Analysis, McGraw-Hill Inc., New York.

Stiemer, S.F., Taylor, P.,Vincent, D.H.C. 1988. Full Scale Dynamic Testing of the Annacies Bridge, IABSE Periodica, February, 1988

Testing Reports 1985-1993. Inspection and Testing of the Main Girders on the Structure of Hangar-2 at the Belgrade Airport, Doc. Institute "Kirilo Savic", Belgrade.

Recent Advances in Experimental Mechanics, Silva Gomes et al. (eds) © 1994 Balkema, Rotterdam, ISBN 90 5410 395 7

The 'optimality degree' of powered supports in coal mine stopes

E. R. Medves, M. Ionita & A. Petrescu
s.c. IPROMIN s.a., Bucharest, Romania

D. Fota
Wilnsdorf, Germany

ABSTRACT: In order to determine the level where a powered support agrees to the technical-mining and geomechanical conditions, specific to a coal stope, we have resorted to a synthetical parameter called "degree of optimality". The paper presents the experimental methodologies (the modelling by optically active materials, optical interferometry, the resistive tensometry, etc.) as well a computer assisted analysis, by the means of which a number of parameters are determined: the state stress in the support subassemblies and in surrounding rocks, in conditions of rock/support interaction, as well as the variation of the kinematic parameters of the support. By correlating all these results the support "degree of optimality" is determined. Using this investigation method we could choose and/or design the kind of supporting equipment which best corresponds to the specific conditions in the stope.

1 INTRODUCTION

In most cases, the mining of coal seams is realised by means of long wall stopes and complex mechanization flux operations. One of the basic component of a mechanized complex is the powered support, consisting of several support sections, where the main supporting element is represented by the hydraulic prop. The beam, shield and floor plate of the powered support intermediate the action of this on the surrounding rocks the concordance between the constructive-functional parameters of the existing equipment and the "seam-surrounding rocks" system interaction mode is the priority and actual problem for understanding and mastering the laws cocerning the mining stress manifestation.

The geometry of powered support section, their general kinematics, web width, as correlated to the depth of slices cutted by the shearer, generate new geomechanical states, which determine new failure laws, orientation and layout of the mined area with different levels of mining pressure manifestation, during the thechnological operation are being carried out. This paper describes a short presentation of a complex methodology for the analysis of the correlation between the powered support and the geomechanical conditions in coal mines, a methodology based on the analysis of the rock/support interaction. The suitability of powered supports technico-functional parameters under the geomechanical conditions which are specific to a certain stope is materialized by the "optimality degree" of the support.

According our opinion, the "optimality degree" is a synthetic parameter which includes the results of a analysis concerning the state of stress in the rock and in the support subassemblies.

The working research which has been carried out for this problem has led us to stucturing this analysis, as follows:

a. the determination of the pressure acting on the respective support based on mining pressure in the stope, which is specific to the lythology of the overlaying rocks and the functional type of support;

b. the analysis of the stress field changes in the main support subassemblies and in the surrounding rocks, depending on the changes of support geometry at various working heights;

c. the analysis of the stress state changes in the lignite seam in front of the stope, depending on the main technological operations in the stope and the support geometry at various working heights;

d. the analysis of the interaction mode between the support lower and the rocks in the stope floor;

e. the kinematic analysis of the support and the stress it is subjected to, as compared to the existing bearing capacities.

Finally, the results of this analysis

are intercorrelated and respectively correlated with the hydrological conditions, which can change the mechanical parameters of rocks as well as with some of technological and constructive parameters of the machines in the stope, which can change the mining stress pattern in the stope.

2. METHODOLOGY OF THE ROCK/SUPPORT INTERACTION ANALYSIS

The analysis of the mining pressure pattern acting on lignite stopes, has led us to considering two notions:
- mining pressure acting on the stope;
- mining pressure acting on the powered support.

In this context, the mining pressure on the stope is the pressure generated by the rocks in the roof, as an effect of the geomechanic mechanism specific to coal deposits, while the mining pressure on the support is part of the mining pressure on the stope, which means it represents the effective pressure acting on the safeguard-buckler part of the support.

In order to determine the stope pressure pattern as well as its distribution mode on the main subassemblies of the powered support, three main research methods are used:
- theoretical determinations based on the mathematical modelling of the geomechanical phenomena;
- laboratory measurements, using tests on physical models;
- field measurements, with direct or indirect determination of pressure and strains.

Each of the above mentioned methods has advantages and disadvantages. For this reason we recommend to resort to determinations using all these methods.

According to the results of the researching work which has been carried out to determine the pressure pattern which is specific to a certain stope, of great importance is a strict analysis of the geomechanical phenomena, corelated to the working height of the supporting equipment, to the constructive and kinematic parameters of the support, as well as to the forces generated in the hydraulic props of the supporting sections. Fig.1 presents the pressure diagrams for the subassemblies of supports which are hypothetically used in the same stope (similar geomechanical conditions, identical face height, a strain force of the supporting props equal to 800 kN) only support geometry being different.

The pressure analysis should also be carried out in correlation with the variations in the working height of the supporting equipment in a certain stope. Fig.2 presents variation of pressure distribution and of strain forces with working height, in similar geomechanical conditions.

In conclusion we could say that the determination of the stope pressure is not only a problem depending on the action of the rocks, but also a more complex one involving the rocks/support interaction, which makes necessary its separate study.

The change of a powered support working height brings about the change of the lining part angle, an aspect which slows down or speeds up the roof control, changes the resultant bearing capacity, which corresponds to the value of the resultant force of the mining pressure which be borne by the support, and, especially, modifies its point of aplication.

These facts in the case of some machines, produce moments of force tending to capsize, respectivelly, stress concentration on the front part of the floor plate. The above mentioned change also leads to a variation of the state of stress in the support subassemblies and in the surrounding rocks.

In order to analyse the variation of the state of stress in the subassemblies of a given support, as well as that of variation of the stress field generated in the rocks around the support, brought about by the change of the geometry of the support section when the working height is modified, our researching work has led us to very good results using photoelasticity and interferometric methods.

These studies were carried out on plane models, made of optically active materials or on spatial ones, which reproduce to scale very accurately the support section, respectively on photoelastic gel models, which reproduce the rock. By recording isocromatics, isoclimes fields and interferometrical holograms and by procesing them, the values of the working height of machines, the state of stress and the evolution of stresses in the support subassemblies are determined, as well as the load ratio of the supporting props and the evolution of the stress in the roof, floor and face rocks.

Fig.3 presents several isocromatics fields as well as the variation diagrams, in the case of those supports which are commonly used in the lignite mines in Romania. The development of these analysis allowed us to characterize the rock/support interaction mode, but these analysis do not have a restrictive character and should be also correlated with the kinematic analysis of the support.

The choice of the support type as well

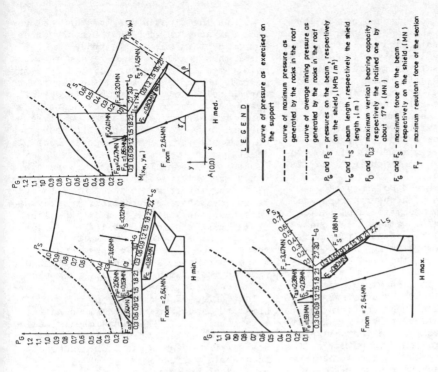

L E G E N D

L E G E N D

— curve of pressure as exercised on the support

–––– curve of maximum pressure as generated by the rocks in the roof

–·–·– curve of average mining pressure as generated by the rocks in the roof

P_G and P_S – pressures on the beam, respectively on the shield, (MPa/m^2)

L_G and L_S – beam length, respectively the shield length, (m)

F_0 and $F_{0,3}$ – maximum vertical bearing capacity, respectively the inclined one by about 17°, (MN)

F_G and F_S – maximum force on the beam, respectively on the shield, (MN)

F_T – maximum resultant force of the section

Fig. 2   Distribution of pressures and forces on the support CMA-4L

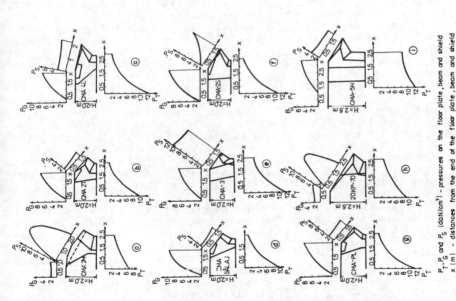

P_T, P_G and P_S (daN/cm^2) – pressures on the floor plate, beam and shield

x (m) – distances from the end of the floor plate, beam and shield

Fig. 1   Distribution of mining pressure on the subassemblies of powered supports in lignite mines in similar working conditions

943

as of the mining coal technology in the stope strictly depends on the stability of the working front, thus on the possibilities to maintain that stability. The stability of the working face depends on the mechanical characteristics of the coal, the presence of steril bands with in the coal seam, the thickness, position and strengths of these bands, the height of the front, the mining depth, as well as the distribution of the abutment pressure on the coal seam in front of the stope, the degree of coal fissuring, the fissure layout, a.s.o..

All these factors can be analysed in correlation, by in situ observations and measurements, laboratory tests on photoelastic models and/or mathematical ones, which reproduce - by constitutive equations - rocks and respectively the geomechanical mechanism of the area being analysed.

Mainly by means of laboratory tests, the evolution of the state of stress appears in the working front depending on the lithological structure and its geometry as well as effect on the abutment pressure, from a rheological viewpoint, of the rate of advance in the front and the type of support are also under observation.

The results of these investigations allowed us the determination of the type of interaction between the rocks in the front and the supporting equipment. One of the main factors on which depends a good and profitable technological activity in a longwall stope equiped with powered supports and mechanized cutting and conveing machines is represented by the manner in which the rocks in the floor receive the stresses generated by the support floor plate, respectively the way in which the floor rocks and the support floor plates interact.

Quantitatively, the above mentioned interraction is determined by means of the lifting power capacity of the rocks in the floor and by the specific pressure, due to the floor plate effect on the rocks in the floor. For the calculus of the lifting power capacity of rocks the literature specifies several theories and calculus methods created by several authors. For this kind of calculus one should resort

to more precise methods, by means of the mathematical modelling phenomena and the utilization of numerical calculus with finite elements methods.

In order to determine the real specific pressure on the floor, a complex analysis has to be carried out, which will better approximate it, both from a qualitative and a quantitative viewpoint. The method which has been used consists in determi-

ning, in the first place, of the real contact surface, as well as the study of evolution of this surface in time.

For this reason, we had resort to laboratory tests on physical models, by similitude, where the rocks in the floor were reproduced by equivalent materials, while the support was reproduced by means of a force transducer, which makes possible the measurement of the specific pressure on some portions of the support floor plate.

Once these aspects are known, one could study the interraction between the floor plate and the rocks on the floor. In order to improve this interraction, in many situations, the modification of the floor plate shape should be carried out, depending on the mechanical parameters of the rocks in the floor. Both the cross and plane shape can be studied and optimized, interactively with the rock, using photoelastical models.

The degree of optimality of support, besides the specific geomechanic conditions, also consists of the change of kinematic parameters of the support, depending on the working heights which are necessary in a certain stope.

An ideal support should not change dramatically the geometry of its part which is in contact with the surrounding massif, when the working heights varies. In order to achieve this desideratum, powered supports should respect the following conditions:

- the shield inclination should correspond to the angle of rock disposition in the roof and should vary as little as possible around the values of this angle;
- the distance between the working front and the end of the beam should be as little as possible and practicaly constant;
- the moment of the resulting forces, which act on the support, should be equal to those developped by the hydraulic props of the support system;
- the inclination of the hydraulic props should correspond to the direction of the vectors of the resulting forces, which act on the support, being approximately constant for any working height;
- the position of the center of gravity of the respective section should vary with in limits which should not modify its equilibrium state.

In order to characterize a powered support from a kinematic viewpoint, an algorithm was elaborated, on the basis of wich the "SUMINA" computer programme was conceived.

Fig.2 shows the results of analyses carried out in the case of the CMA-4L support.

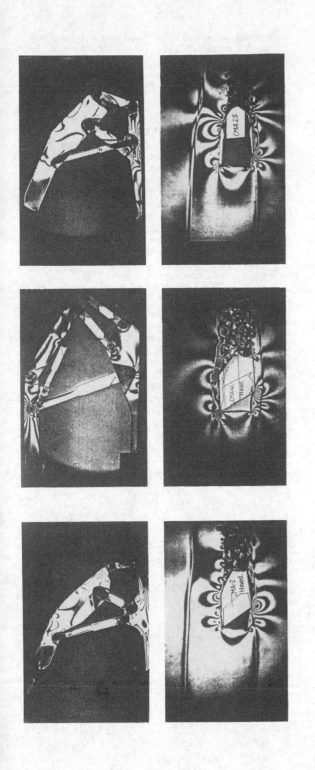

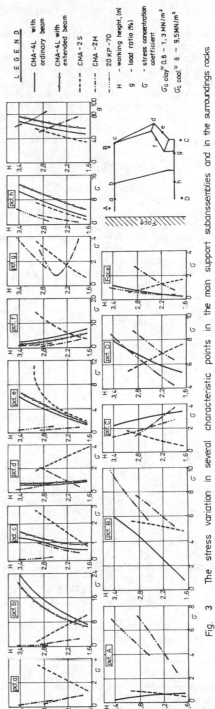

Fig. 3   The stress variation in several characteristic points in the main support subassemblies and in the surroundings rocks

3. "OPTIMALITY DEGREE" DETERMINATION

The analysis of the proposed methodology are complex enough and the results offer sufficient elements for a very good characterization of the rock/support interraction in coal mine stopes.

By synthetizing the results which have already been achieved and by a correlative analysis, the "optimality degree" may be determined, acording to the following equation:

$$O_g = O_s p_s + O_i p_i + O_t p_t + O_f p_f + O_c p_c \qquad (1)$$

The significance of parameters in (1) is the following:

p_s, p_i, p_t, p_f, p_c - weighting coefficients used to establish the influences (depend on the technological aspects and the type of support);

O_s - optimum indicator for the state of stress of support subassemblies:

$$O_s = \sum_{i=1}^{n} o_i^2 \, w_i \qquad (2)$$

where w_i = weighting coefficients, inversely proportional to the degree of importance of the respective point; o_i = indicator which characterizes the variation of stress in point i (for point a - h, see fig.3), for the working field under consideration; n = number of points where the analysis is carried out;

O_i - indicator of optimum for the state of stress in surrounding rocks:

$$O_i = \sum_{k=1}^{n} o_k^2 \, w_k \qquad (3)$$

where w_k and o_k have the same significance as in (2), (for points A -D, see fig.3)

O_t - indicator of optimum for the state of stress of rocks in the floor:

$$\theta_t = \left(-\frac{C_p}{P_s} - A \right)^2 W \qquad (4)$$

where C_p = bearing capacity of rocks in the floor; P_s = specific pressure exercised by the support; $0.7 \leqslant A \leqslant 0.8$ = coefficient depending on the shape on the floor plate;

O_p - optimum indicator for the state of stress of the rocks in the face:

$$O_f = \sum_{l=1}^{n} o_l^2 \, w_l \qquad (5)$$

where o_l and w_l have the same significance as in (2);

O_c - optimum indicator of support kinematics:

$$O_c = (\Delta x_M)^2 w_1 + (\Delta \theta_2)^2 w_2 + f(y_p - y_i) w_3 + g(\gamma - 17^\circ) w_4 +$$
$$+ h(M_{F_T} - M_{F_{0.3}}) w_5 \qquad (6)$$

where Δ = deviations quantums; f, g, and h = function of ..., (see fig.3); M = the moment of force ..., and w_i have the same significance as in (2).

REFERENCES

Fota, D., Medves,E., Ionita, M. & Petrescu, A. 1988. The influence of the geometry and kinematics of stope powered supports on the distribution of stress in the main constructive subassemblies and in the surrounding rock massif. In Mine, Petrol, Gaze, no. 7.

Ionita, M.,& Medves, E. 1988. The utilization of jells as photoelastic materials in the rock-support interraction study in coal mines. At The 4-th National Simposion of Tensometry, Brasov, 24-27 september.

Medves, E. 1990. The classification of roof type, characteristic to the coal deposits of Romania. In Mine, Petrol, Gaze, no. 3.

Medves, E. & Nicolescu, C. 1989. Some consideration concerning the interraction between the rocks in the stope floor and the sole of the powered supports in lignite mines. In Mine, Petrol, Gaze, no.5.

Medves, E. 1991. Powered supports utilized in the lignite mines in the Oltenian basin and the rock support interraction mode. Doctor's Degree paper, the Technical University Petrosani, mai.

Recent Advances in Experimental Mechanics, Silva Gomes et al. (eds) © 1994 Balkema, Rotterdam, ISBN 90 5410 395 7

Full-scale testing of a GRP minesweeper hull to underwater explosions

R. A. Bourgois & B. Reymen
Royal Military Academy, Brussels, Belgium

ABSTRACT: The paper deals with the final batch of tests to develop a GRP minesweeper hull for a new class of minesweepers for the Belgian and Dutch Navies. A real section of the ship, instrumented with a large number of transducers, was tested by means of large charges on a test site in the North Sea, simulating the real working conditions. The presentation deals particularly with strain and pressure measuring equipment and the measuring problems related to the very large dynamic deformations and the very short events.

1. INTRODUCTION

Eight years ago the Belgian and the Dutch governments decided to build a new class of GRP minesweeper, based on the experience gained with the Tripartite-minehunters (Belgium - France - Netherlands).

A preliminary phase was started in 1987 to establish the SDS (Structural Design Specification). One of the important points was the composition and the required thickness of the hull. From the beginning and related to the Tripartite-program, sandwich hulls were rejected and only a solution of a relatively thin GRP skinplate with vertical GRP stiffeners was retained. The thickness of the GRP skin and the dimensions of the stiffeners are important, not only for the dynamic resistance but also from an economic point of view.

The designers proposed, on a practical base, different alternative solutions. It was suggested to study them theoretically and to verify their behaviour experimentally during three test series. The first two series, realized in 1987 and 1988, are related to underwater explosions on relatively small specimens of the hull (3 x 3 m) (Fig. 1) and are described in a previous paper (Bourgois, Reymen 1991). These tests led to a practical solution for the GRP hull. The third and final batch of shock tests was realized on a real section of a prototype-ship in 1992-93 and will be described in this paper.

The aim of these tests was first to verify the general concept of the design and secondly to obtain information about the behaviour under shocks of construction details.

2. GENERAL DESCRIPTION

The third batch of tests was subdivided in three phases characterized by increasing severity of the underwater explosions.

The tests were conducted by the Naval Construction Service (SCHEBO) of the Dutch Royal Navy and four laboratories from Belgium and the Netherlands were involved in this shock test program.

Two Dutch laboratories were in charge of the measurement of accelerations (TNO-CMC) and displacements (R.N-MEOB). High speed photography was realized by a specialized service of the Dutch Army.

The laboratory of experimental stress analysis from the Royal Military Academy of Brussels (Belgium) was designed to realize first the underwater blast pressure measurements in order to calculate the real shock factor of the explosion and secondly the stress measurements on the hull for the verification of the calculations of the design.

The prototype section was equipped with dummy loads simulating the real loads of the minesweeper (diesel engine, sweeping winch, ...). Figure 2 gives a general view of the test section on the ship-elevator. The electrical autonomy of the section is assured by a generator located on the deck.

Figure 1: Testplate used during the first two series

Figure 2: General view of the test section

Figure 3: Strain-gages on the stiffeners from the bottom

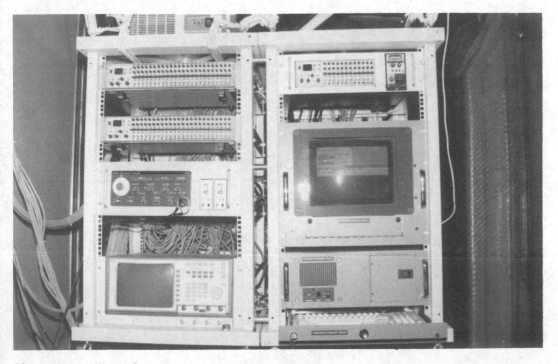

Figure 4: Measuring equipment

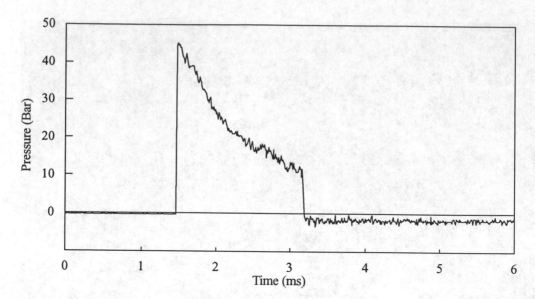

Figure 5: Typical pressure registration

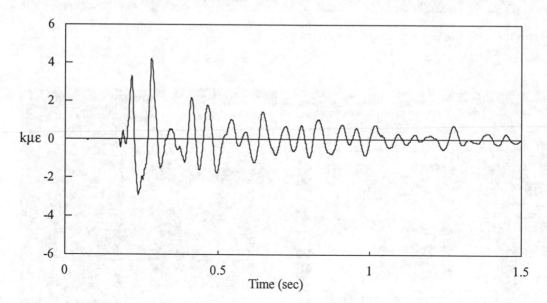

Figure 6: Typical strain registration on a stiffener of the bottom

The prototype section, with a length of 14 m, was constructed by the shipyard Van Der Giesen-De Noord at Alblasserdam (NL).

3. STRAIN MEASUREMENTS

3.1 *Strain-gages*

Stress problems are suspected in the new concept of the hull, especially in the bottom, on the top deck and in the webframe below the sweeping winch. Eleven measurement zones where selected, with a total number of 60 strain-gages. Several gages on the bottom are situated on both sides of the skinplate, requiring very good electrical insulation (marine environment).

Very large strains are expected (1 to 10 k$\mu\epsilon$). For this reason post-yield strain-gages of TML are chosen. Regarding the presence of glassfibers, a gagelength of 10 mm was taken. All the strain-gages are mounted with cyanoacrylate adhesive.

The protection of the gages is realized with an HBM insulation type ABM75-kit for the inside gages and a glasfiber roofing with polyester for the outside gages. Despite the wet environment no serious insulation problems were encountered during the measurement campaign.

Because of the large inertial forces, all measuring cables were firmly attached to the hull. Figure 3 represents some strain gages on the stiffeners from the bottom.

3.2 *Measuring equipment*

Simultaneous short dynamic measurements (max. duration 1 sec) are requested for the 60 strain gages during each shot. Temperature sensivity and drift are not critical, a classic 1/4 bridge configuration is used with a shielded 3-wire connection. The frequencies involved in the vibrations are not very high (max. 1000 Hz).

The measurements are done with a signal conditioner system Micro Movements, type M1000, with amplifier moduls M1060. The bandwidth of these amplifiers is 2,5 kHz. Because of the low thermal conductivity of GRP, a 3V bridge exitation is used.

The registration of the signals is done by an industrial computer and data-acquisition boards Bakker, type BE490, with a 12 bit resolution and a measuring frequency of 10 kHz per channel.

A generator is installed on the top deck of the test section and the measuring system is electrically protected by an U.P.S. Meta System, type GPC501.

The equipment is installed on board and suspended elastically to avoid excessive vibrations (fig. 4).

Firing and triggering of the measurement equipment is remote controlled.

4. PRESSURE MEASUREMENTS

To determine the shock factor of the explosions, measurements of the free-field pressures in the water are done on three locations outside the hull at depths of 4 m to avoid reflections on the skin-plate. The pressure transducers, PCB type 138 A05, are suspended in the water by means of a weight.

The registration of the pressure signals is also done on the industrial computer with a complementary data-acquisition board Bakker with a sample rate of 100 kHz.

5. RESULTS AND CONCLUSION

The results being confidential for both Navies it is not possible to give extensive measurement results.

Figures 5 and 6 represent typical registrations of pressure and strain.

Nine explosion tests were done with increasing shock factors. The test section resisted the explosions.

ACKNOWLEDGEMENT

The autors acknowledge their sincere gratitude to the Belgian and the Dutch Navies for the authorization to publish this article.

REFERENCE

Bourgois, R.A. & Reymen, B 1991. Resistance of a GRP minesweeper hull to underwater explosions. *20° Convegno Nazionale AIAS, Palermo 25-28 September 1991, Italy, pp. 367-376*

Recent Advances in Experimental Mechanics, Silva Gomes et al. (eds) © 1994 Balkema, Rotterdam, ISBN 90 5410 395 7

Experimental investigations of active acoustic detection method for LMFBR steam generator

H. Kumagai & T. Sakuma
Central Research Institute of Electric Power Industry, Tokyo, Japan

F. Yoshida
Department of Mechanical Engineering, Hiroshima University, Japan

ABSTRACT: In the steam generators (SG) of LMFBR, it is necessary to detect the leakage of water from tubes of heat exchanger as soon as leakage is occurred. The active acoustic detection method has drawn general interests owing to its short response time and reduction of the influence of background noise. In this paper, the application of the active acoustic detection method for SG is proposed and the sound attenuation by bubbles is investigated experimentally. Furthermore, using SG sector model, sound field characteristics and sound attenuation characteristics due to injection of bubbles are studied. As a result, it is clarified that the sound attenuation depends upon bubble size as well as void fraction, that the distance attenuation of sound in the SG model containing heat transfer tubes is 6dB at each two fold increase of distance, and that emitted sound attenuates immediately upon injection of bubbles.

1 INTRODUCTION

Fast breeder reactor employs sodium-heated steam generator. It is necessary to detect precisely and immediately the leakage of water from tubes of heat exchanger in order to prevent the propagation of sodium-water reaction.

For detection of leakage, a hydrogen meter and a pressure meter have been used. These detectors take too long a time for detection. Thus, the acoustic method is now drawing attentions.

According to this method, the detection is done very quickly and very small amount of leakage can be detected. There are two methods in this acoustic method. One is the passive method (Greene 1981, Kuroha 1981, Kong 1986, Tanabe 1990, Rowley 1990) which detects leakage receiving sounds generated at the time of sodium-water reactions. This method is not yet employed for practical use because of interference by background noise.

Another is the active method (Girard 1990, Kumagai 1992). As well known, hydrogen bubbles are generated at the time of sodium-water reaction. According to the active method, an emitter and a receiver sensor are attached to the steam generator, creating sound field, and the attenuation of sounds due to passing of bubbles through the sound field is detected and measured. The active method is least affected by background noise and has the potentiality of achieving high sensitivity.

The study of the active acoustic method is still in the primitive stage. In this paper, the attenuation characteristics of sound attenuated by bubbles are investigated. Furthermore, by using the SG sector model simulating the actual SG heat transfer tube arrangements, the profile of sounds, the attenuation of sounds according to the numbers of layers of tubes, and the attenuation of sound according to the bubble injection volumes is discussed.

2 SOUND ATTENUATION BY BUBBLES

2.1 *Experimental procedure*

Fig.1 shows the experimental apparatus of sound attenuation due to bubbles. Nitrogen gas bubbles are generated in the water bath using the bubble generator. The bubble sizes and void fractions are controlled by means of adjustments of water and gas volumes. In order to prevent the micro size bubbles from floating in the water bath and also to alter the thickness of layers, a thin vinyl film is placed in the sound passage space.

Experimental conditions are divided into 3 categories according to difference of average bubble sizes, that is, $\bar{a}$=3.8mm, 0.34mm, and 0.09mm on an average. Thickness of bubble layer t is arranged to reach 3.8-50mm for $\bar{a}$=3.8mm, 5-60mm for $\bar{a}$=0.34mm, and 5-15mm for $\bar{a}$=0.09mm. The bubble size is measured by means of both the ascending rate measurement method and the digital image processing method by the bubble photographic picture. The void fraction is measured by collecting ascending bubbles into pipe.

Emitter sensor is attached to the outside of the water bath. Receiver sensor is placed at the inside of water bath in case of measurements of sound attenuation through the sound field, and in all other cases the receiver sensor is attached to the outside of the water bath. For emitter and receiver sensors, piezoelectric element sensors with 20mm diameter and 30kHz, 50kHz, and 70kHz resonance frequencies are used.

Emitted sounds used are burst sound, continuity sound and pulse sound. Emitted frequencies are 20-140kHz for burst sound and continuity sound. Received sound is power amplified, and filtered, and then its spectrum is analyzed and root mean square voltage is obtained.

2.2 Sound field and bubble conditions

Sound in the water bath is attenuated during its propagation from emitter side to receiver side. The distance attenuation P_0 of sounds in the water bath is shown in Fig.2. Distance x is the position of receiver sensor from the emitter sensor. The solid straight line is obtained by the calculation on the assumption that the emitted sound propagates through the water as spherical wave. It attenuates rectilineally by 6dB at each 2 fold increase of distance from the sound source.

Each sound attenuates almost proportionally to the distance rectilineally in the same pattern as the solid straight line.

Fig.3 shows the relationship between bubble volume rate Q, bubble size a and void fraction γ.

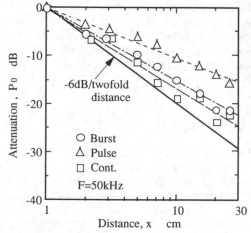

Fig.2 Distance attenuation in water bath.

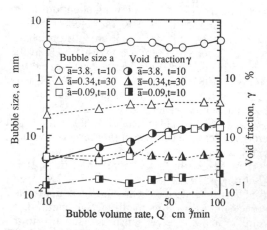

Fig.3 Relationship between bubble volume rate, bubble size and void fraction.

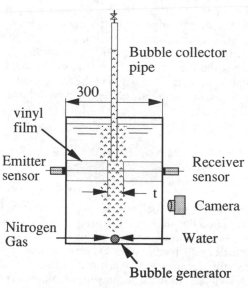

Fig.1 Experimental apparatus.

In case of the average bubble size $\bar{a}$=3.8mm, bubble size a is almost constant, but void fraction γ increases, both in relation to bubble volume rate Q. In case of the average bubble size $\bar{a}$=0.34mm and $\bar{a}$=0.09mm, the void fraction γ shows almost no change but the bubble size a makes slight increase, both in relation to bubble volume rate Q. Constancy of void fraction γ is due to dispersion of bubbles and an increase of bubble size a is caused by the mixture of large size bubbles as the bubble volume rate Q increases.

2.3 Experimental results

2.3.1 Influence of emitted sound

Fig.4 shows the relationship between the bubble volume rate Q and the sound attenuation P, using as parameters the emitted sound, and an average bubble size $\bar{a}$. The sound attenuation P is defined as following equation (1).

$$P = 20\log\frac{E_b}{E_O} \qquad (1)$$

where E_0 is received sound without bubble, and E_b is received sound with bubble. The attenuation P of average bubble size $\bar{a}$=3.8mm is small. The attenuation P makes slight increase with the bubble volume rate Q. An increase of the attenuation P is owing to an increase of void fraction γ, inasmuch as the bubble size a is almost constant. In case of $\bar{a}$=0.34mm, and $\bar{a}$=0.09mm, the attenuation P becomes larger. The attenuation P de-

creases with the bubble volume rate Q. This reason is considered to be owing to an increase of a bubble size a, inasmuch as the void fraction γ is almost constant. There is almost no change of the attenuation P due to difference of emitted sounds.

2.3.2 Relationship between the bubble size and the void fraction

Fig.5 shows the attenuation P in relation to the bubble size a ($\bar{\gamma}$=1.9$\times$10^{-1}, const.,$\bigcirc$), and void fraction γ ($\bar{a}$=3.8mm, const., $\triangle$). The attenuation P decreases with increases of bubble size a. The decrease of sound attenuation with increase of bubble size is also confirmed. In case the bubble size becomes larger, the attenuation P depends upon void fraction. In case of $\bar{a}$=3.8mm, the attenuation P increases with increases of void fraction. The sound attenuation changes according to the void fraction and also to the bubble size.

2.3.3 Influence of emitted sound frequency

Fig.6 shows the relationship between the emitted sound frequency F and the sound attenuation P. In case of bubble size of $\bar{a}$=3.8mm, the sound attenuation P is least altered by the emitted frequency F. In case of bubble size of $\bar{a}$=0.38mm, P becomes the minimum at around F of 20-30kHz and in case of bubble size of $\bar{a}$=0.15mm, P becomes the minimum at around F of 60-100kHz. The approximate value of resonance frequency F_0 of bubble is obtained by the following equation (2)(Urick 1975).

$$Fo = \frac{10}{\pi a}\sqrt{\frac{3kP_s}{\rho}} \qquad (2)$$

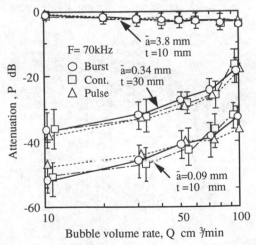

Fig.4 Relationship between bubble volume rate and sound attenuation.

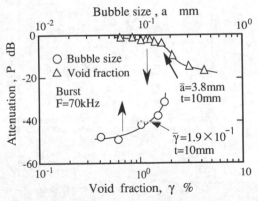

Fig.5 Relationship between bubble size, void fraction and sound attenuation.

where, κ is specific heat ratio of gas, P_s is static water pressure, ρ is density of water, and a is bubble size (mm). In equation (2), resonance frequency F_0 at bubble size $\bar{a}=3.8$mm is 1.7kHz, at $a=0.1$-0.5mm($\bar{a}=0.38$mm) is 14-65kHz, at $a=0.06$-0.25mm($\bar{a}=0.15$mm) is 27-110kHz respectively. The attenuation of sound reaches maximum when bubbles resonate with the emitted frequency.

2.3.4 *Influence of thickness of bubble layer*

Fig.7 shows the relationship between thickness of bubble layer t and sound attenuation P. In case of

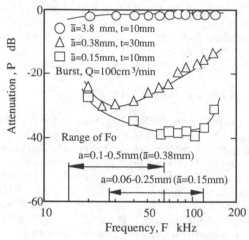

Fig.6 Relationship between emitted frequency and sound attenuation.

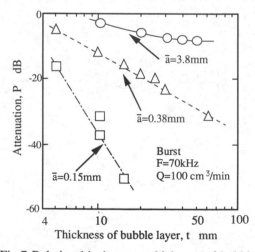

Fig.7 Relationship between thickness of bubble layer and sound attenuation.

bubble of $\bar{a}=3.8$mm, P increases with t, but an increment of P gradually slows down. In case of bubble of $\bar{a}=0.38$mm, P increases with increases of t. In case of bubble of $\bar{a}=0.15$mm, the increase of P with increase of t is much greater. From Fig.7, it is observed that the smaller the bubble size $\bar{a}$ is, the more the sound attenuation P increases in relation to the bubble layer t. The same results are obtained with the pulse sound and continuity sound. The increase of sound attenuation with the increase of bubble layer is considered due to the increase of the void fraction.

3 SOUND ATTENUATION BY SG SECTOR MODEL

3.1 *Experimental procedure*

Fig.8 is the water bath model containing the heat transfer tubes of SG. The side walls and the bottom are made by SUS plate with 10mm thick. The front and rear plates are 25mm thick, made of transparent acrylic resin, enabling to see the inside. The heat transfer tubes are made of SUS, with the size of 31.8mm diameter and of 3mm thick, arrayed squarely, 13 pieces horizontally and 13 pieces vertically, each with 45mm pitch. The number of heat transfer tubes N can be altered. Bubbles are generated from the bubble generator. Bubble injection location is the bottom of the model.

Emitter sensor is attached to the outside of the side the wall. Receiver sensor is placed at the inside of the model in case of measurements of sound attenuation through the sound field, and in

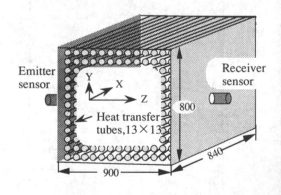

Fig.8 Water bath model containing the heat transfer tubes of SG.

all other cases receiver sensor is attached to the outside of the side wall. The XYZ coordinate axes of receiver sensor location, when the sound attenuation through the sound field is measured inside the model, is shown in Fig.8. X axis is horizontal along, Y axis is vertical to, and Z axis is horizontally rectangular to, the heat transfer tubes respectively. The axes origin is located in the water 5cm from the inside of the emitter sensor wall.

Emitted sounds are pulse sound, burst sound and continuity sound. The resonance frequency of both emitter and receiver sensors is 70kHz. In the case of pulse sound, the maximum impressed voltage to the emitter sensor is 250V, and the cycle is 125Hz. In the case of burst sound, the maximum impressed voltage to the emitter sensor is 150V.

Received sound is amplified, and the sound level is read by the RMS voltage meter and recorded by the pen recorder. Emitted and received sounds are analyzed by the spectrum analyzer and their waveforms, spectrum and sound travelling time are measured.

3.2 Sound field

Fig.9 shows the variation of received pulse sound at $z=800$ in comparison with the same at $z=0$. The sound can be regarded as emitted at $z=0$ because of location of $z=0$. The sound at $z=800$ can not be clearly recognized at an early stage, and its amplitudes increase after around 0.65msec. Contrary to waveforms at $z=0$, the variation of waveform at $z=800$ remains long period of time due to multiple reflections caused by the heat transfer tubes.

The variation of burst sound at $z=800$ in comparison with $z=0$ is almost same pattern as pulse sound.

Fig.10 shows the distance attenuation P_t along X axis distance x measured at $z=0$ and $z=800$. The receiver sensor is traversed along X axis in water. Emitted sounds are pulse sound, burst sound and continuity sound. The attenuation P_t of sound at $z=0$ increases within the distances around +10cm and -10cm along X axis. This is caused by the directivity of emitted sound. At beyond 10cm, the attenuation is not correlated with distance.

The attenuation P_t of sound at $z=800$ along X axis is about 25dB. Owing to dispersion of sounds, the attenuation seems to have been averaged. In case of continuity sound, there are locations where the attenuation becomes especially great that seems to be due to interaction of sounds. At such a location, the sound attenuation due to bubbles is small and the detection sensitivity will be lower.

Fig.11 shows the sound attenuation with and without heat transfer tubes. The receiver sensor is traversed along Z axis in water. With heat transfer tubes, sound attenuates proportionately to the distance. The distance attenuation is almost 6dB at each two fold increase of distance. Without heat transfer tubes, sound attenuates proportionately to the distance until around $z=50$cm, and then, the attenuation ratio becomes weak.

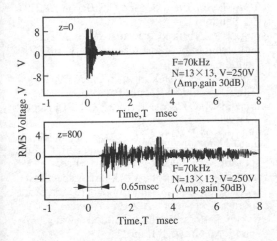

Fig.9 Variation of received pulse sound.

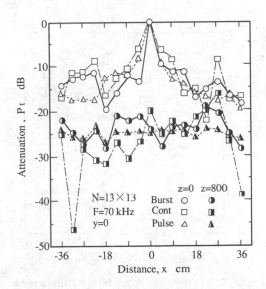

Fig.10 Distance attenuation along X axis.

3.3 Sound attenuation due to bubbles

Fig.12 shows the relationship between bubble volume rate Q and sound attenuation P due to bubbles. The range of bubble size $\bar{a}$ at less than bubble volume rate $Q=400\text{cm}^3/\text{min}$ is 0.5-1.5mm. The bubble size $\bar{a}$ becomes larger with increase of Q. The range of bubble size $\bar{a}$ at more than bubble volume rate $Q=400\text{cm}^3/\text{min}$ is 2.0-4.1mm. Emitted sound attenuates immediately upon injection of bubbles. With the bubble volume rate Q of less than $400\text{cm}^3/\text{min}$, the attenuation P decreases with increase of Q. With the bubble volume rate Q of more than $400\text{cm}^3/\text{min}$, the attenuation P increases with increase of Q. Bubble size is small when bubble volume rate Q is small and attenu-

ation P is becomes greater. In case bubble size become big, P increases owing to increase of void fraction γ. The sound attenuation due to bubble in the SG model shows the same results as shown in Fig.5.

4 CONCLUSION

In order to study the applicability of active acoustic method for detection of water leakage in the SG of fast breeder reactor, the sound attenuation characteristics due to bubbles are investigated under various bubble conditions and emitted sound conditions. Furthermore, using SG sector model, sound field characteristics and sound attenuation characteristics due to injection of bubbles are studied.

As a result, it is clarified that the sound attenuation due to bubbles varies dependent upon size of bubbles, emitted frequency and void fraction. The distance attenuation of sound in the SG model containing heat transfer tubes is around 6dB at each two fold increase of distance. Emitted sound attenuates immediately upon injection of bubbles, and sound attenuation depends upon void fraction as well as bubble size.

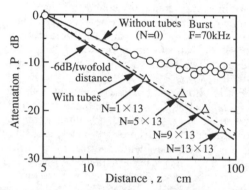

Fig.11 Distance sound attenuation with and without heat transfer tubes.

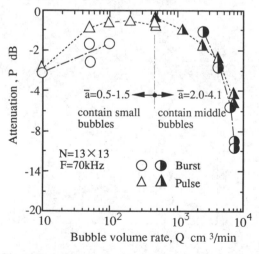

Fig.12 Relationship between bubble volume rate and sound attenuatione due to bubbles.

REFERENCES

Girad, J.P., P. Garnaud & R.Demarais 1990. IAEA specialists meeting on acoustic/ultra sonic detection of in-sodium water leaks, Aix-En-Provence.

Greene, D.A, P.M. Magee, A.W. Thieleand & T.N.Claytor 1981. Second U.S./Japan LMFBR SG Seminor

Kong,N. A.Lebris & M.Brunet 1986. Science and technology of fast reactor safety, BNES, London.

Kuroha, M., M.Nishikimi & Y.Daigo 1981. IEICE, Technical Report, EA81-52.

Kumagai, H & K.Yoshida 1992. CRIEPI Report, T91041.

Rowley,R. & J.Airey 1990. IAEA specialists meeting on acoustic/ultrasonic detection of in-sodium water leaks, Aix-En-Provence.

Tanabe,H. & M.Kuroha 1990. IAEA specialists meeting on acoustic/ultrasonic detection of in-sodium water leaks, Aix-En-Provence.

Urick, R.J. 1975. Principles of under water sound. McGraw.Hill, Inc.

Recent Advances in Experimental Mechanics, Silva Gomes et al. (eds) © 1994 Balkema, Rotterdam, ISBN 90 5410 395 7

Applied research and development in the design of railway vehicles

A. L. Salgado Prata
SOREFAME, Amadora, Portugal

ABSTRACT: This work concerns the design, construction and erection a new fatigue tests equipment sets as well as the first tests done. The equipment sets is intended for applied research in the design of railway vehicles, has capacity to bear large dimensions test specimens, can apply uni, bi or tridirectional loads, in a large range of frequencies, pratically with any regimen of variable loads. The design adopted for the basis of work and structure has proved to achieve a satisfactory operationality, a good rigidity, a good capacity to weakening vibrations.

The first tests executed proved not to be neccessary full penetration in the welds of the box beam bicelular type, forming one of the super structure critical components of some railway vehicles.

1 INTRODUCTION

In a general way, one can say that the I&D work, implying fatigue tests, which have been done in Sorefame has had as main goals: weight reduction, easier construction solutions, confirmation or validation of design, satisfaction of some technical specifications.

Within the scope of the work that is described here it was intended, on the one hand to equip that company with a more powerful and versatile "tool" than the old fatigue tests equipment and, on the other, to make in the present time, with the aims already mentioned above, some I&D implying fatigue tests, the result of which could be applicable to some orders already asked for and others.

2 THE TARGETS OF I&D MADE WITH THE NEW FATIGE TESTS EQUIPMENT SETS

The super-structure of the railway vehicles contains sub-assembling we call end-under-frame, to which the bogies are linked. The principal resistent elements of the end-underframe are frequently of the box beam type, being the joinings of each one of the flanges to the webs made with full penetration welds.

There has been a doubt for a long time whether the full penetration is really necessary or whether the partial penetration or even T fillet welds without groove and without penetration would be enough. However, the analysis methods, even the most sophisticated, are unable to answer to this question. So, by confidence and safety reasons, we go on using full penetration.

Now the production in these conditions requires more time and costs, becoming less economical. So it was decided to choose, as the first subject of the I&D, implying tests with the new fatigue tests equipment set, to make clear above mentioned question.

The so called UQE's (quadruple electric units) which the company has produced a considerably quantity, have been chosen as typical vehicles.

3 LARGE DIMENSION SPECIMENS AND TESTS CONDITIONS

The two large dimension specimens reproduce in the dimensions, materials and fabrication conditions (except for the welding) the main resistance element, that is, the bolter beam of the end-underframe of the UQE's. The relative position of this element in the end-underframe is illustrated in the croquis (Fig. 1), used in the

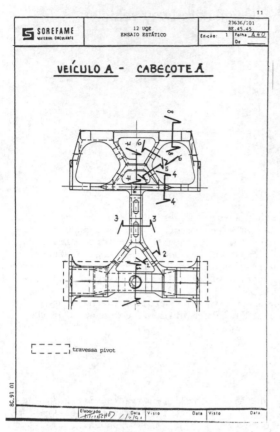

SOREFAME MATERIAL CIRCULANTE | 12 UQE ENSAIO ESTÁTICO | 23636/:101 8E.45.45 | Edição: 1 Folha A40 Da

VEÍCULO A - CABEÇOTE A

⌐ ¬
⌐ ¬ travessa pivot

FIG. 1 - VEHICLE A - END-UNDERFRAME

static test of the vehicle A, in a series of 12 UQE's. In what concerns the welds, the two specimens are different from each other, they are different from what we have used in the vehicles until now, and in principle, they are statically and dynamically less resistent than those in the vehicles. In fact, and as it has already been told, we have used full penetration in the welds joining the flanges to the webs, forming the box beam.

Now the specimen Nº. 1 has incomplete penetration in these welds because though it has had bevels of 45° and root face 1 mm, gaps were not allowed in these. The specimen Nº. 2 is nominally without penetration because the T fillet welds were simply made without grooves and without gaps.

One of the specimens is illustrated in Figs. 2 and 3.

Concerning the design of the bolter beam there are some differences, which are necessary so that the linking of test specimens to the basis of the test equipment set

may be made as well as the linking of these (specimens) to the actuators and also so that the fabrication of the central king -pin may be simplified.

One has admitted that the differences have no influence in the fatigue behaviour of the test specimens.

Concerning the specification of the tests conditions, the following are to be registered:

The loads have been determined for the situation of the vehicle being complete as regarding the sitting places. It is obvious that in the rush hours the vehicle also carries standing passengers, but, on the other hand, on other occasions the sitting places are not fulfilled, by which under the point of view of accummulated fatigue, the hypothesis formulated seems to be reasonable. It is, besides, the hypothesis that other constructors of railway material do, for the purpose of fatigue tests.

So it was determined that during the test each specimen will be subjected to three variable loads, being two of them of compression in the vertical direction,from up to downwards and of an average value of about 12,4 T (each), and another, alternate, acting in a horizontal plan, in an obliquous direction regarding the simetry plan (vertical) of the bolter beam and of a maximum value of about ± 5 T.

As a variation of the compression vertical loads, we have adopted our usual criterium of ± 25% around the medium value. It should be remarked that from the acceleration spectra which we will refer to later on it seems it may be concluded that these extreme values of the loads are already little frequent.

In order that the effective frequency and effective accelerations at work can be observed, measurement and register of acceleration in some zones of UQE vehicles were made, when circulating in the lines of Sintra and Carregado. In Fig. 4, one extract of these registers is shown. Such registrations could make us conclude namely, that the peaks of the vertical accelerations can present frequencies of about 2Hz while the horizontal ones are about 200 times smaller.

Such observation made us conclude the horizontal load with litle importance under the point of view of fatigue.

So we have decided to make the tests by using only the two variable, vertical, compression loads, taking into account that this simplification would have an insignificant influence in the results.

Concerning the number of cycles to be used in the tests, we thought it would be enough to arrive at 2×10^6 without fracture, as we

FIG. 2 - FATIGUE TEST EQUIPMENT SET AND SPECIMEN

FIG. 3 - A TEST EXECUTION

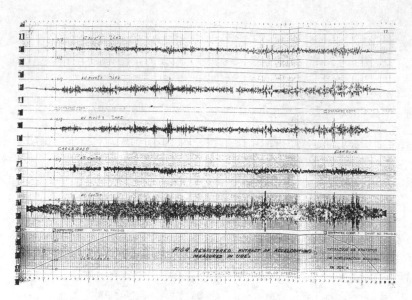

Figure 4

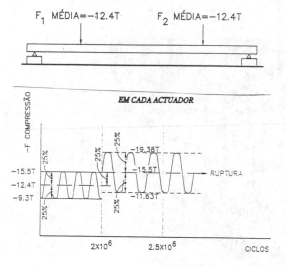

F_1 MÉDIA=−12.4T F_2 MÉDIA=−12.4T

FIG. 5 – SCHEMATIC DIAGRAM OF TESTS CONDITIONS

DIAGRAMA ESQUEMÁTICO DAS CONDIÇÕES DOS ENSAIOS

have admitted that the applied materials, have already reached the step of their infinite life at the end of these cycles number. Being so, such criterium would assure endless life for the vehicles, by which reason it has been adopted by some fabricators of railway material.

However, by greater safety reasons, we use to follow the criterium of, once rea-
ched the $2x10^6$ cycles, of making the tests longer $0,5x10^6$ cycles but by increasing now the medium load 25% relatively to the last one, by maintaining a variation of ± 25% around the average load. In this work we continued using the same criterium.

In what regards the frequencies to be used in the tests, they do not have, as it is already known, any significant influence in the results.

In short, the tests conditions are syn thetized in the schematic diagram in Fig. 5.

4 TESTS AND RESULTS OBTAINED

For the effects of an eventual relationship in an hypothetical fracture of the test specimens with pre-existing faults, one has done a 100% control dye penetrant and ultrasonics in the main welded joints of the test specimens, which correspond to the joining of the flanges with the webs. The observed faults, which are not relevant, were signaled in the specimens themselves.

Preliminary tests have been done for a better knowledge of the fatigue test equipment set and on the specimen Nº. 1, working particularly on: mastering the control software of the actuators, variation of conditions of the actuators work, variation in the vibration frequency.

About these tests we have only presented

962

Teste n.11	Carga média -124 KN	Amplitude 25%	N. Ciclos 2000	Frequência 20Hz

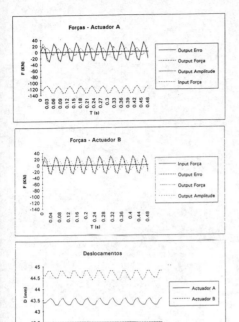

Forças - Actuador A
(Output Erro, Output Força, Output Amplitude, Input Força)

Forças - Actuador B
(Input Força, Output Erro, Output Força, Output Amplitude)

Deslocamentos
(Actuador A, Actuador B)

20 Hz

Figure 6

HORA	N.º CICLOS	TORRE	FACTOR	MÍNIMO	MÁXIMO	DIA
16-30	≈1.190.000	A	FORÇA	-155,5 KN	-96,50 KN	93/08/03
			DESLOC.	43,13 mm	43,88 mm	
		B	FORÇA	-156,2 KN	-95,40 KN	
			DESLOC.	44,32 mm	45,11 mm	
18-20	≈1.290.000	A	FORÇA	-155,3 KN	-96,70 KN	93/08/03
			DESLOC.	43,13 mm	43,88 mm	
		B	FORÇA	-156,1 KN	-95,40 KN	
			DESLOC.	44,33 mm	45,13 mm	
20-30	≈1.410.000	A	FORÇA	-155,0 KN	-96,80 KN	93/08/03
			DESLOC.	43,07 mm	43,83 mm	
		B	FORÇA	-156,0 KN	-96,03 KN	
			DESLOC.	44,29 mm	45,07 mm	
10-50	≈1.495.000	A	FORÇA	-155,7 KN	-96,4 KN	93/08/04
			DESLOC.	42,99 mm	43,76 mm	
		B	FORÇA	-156,4 KN	-95,70 KN	
			DESLOC.	44,25 mm	45,04 mm	
12-30	≈1.586.000	A	FORÇA	-155,5 KN	-96,60 KN	93/08/04
			DESLOC.	43,09 mm	43,85 mm	
		B	FORÇA	-156,2 KN	-95,60 KN	
			DESLOC.	44,29 mm	45,08 mm	
16-30	≈1.710.000	A	FORÇA	-155,3 KN	-96,81 KN	93/08/04
			DESLOC.	43,09 mm	43,85 mm	
		B	FORÇA	-156,1 KN	-95,50 KN	
			DESLOC.	44,32 mm	45,12 mm	
20-00	≈1.760.000	A	FORÇA	-155,6 KN	-96,64 KN	93/08/04
			DESLOC.	42,99 mm	43,76 mm	
		B	FORÇA	-156,4 KN	-95,62 KN	
			DESLOC.	44,24 mm	45,03 mm	

Figure 7

the graphics, corresponding to 20 Hz frequency (Fig. 6).

In the tests themselves, the test conditions synthetized in Fig. 5 were followed and, concerning the gain, integral and derivative, the values determined in the preliminary tests. In the programme safety loads and safety displacements were imposed. In the Nº. 1 specimen test 5,16 and 22 Hz frequencies were used. The test specimen Nº. 2 was tested with 22 Hz.

During the tests, the maximum and minimum loads were read every two hours, as well as the maximum and minimum displacement in each actuator. Fig. 7 presents an extract of the respective registers.

The tests of the two specimens have been concluded without any mark of fracture start being notice, either directly or indirectly (for example an increase in the displacements).

5 CONCLUSIONS

The fatigue test equipment set was erected and checked, having been proved that it presents the necessary resistence and rigidity, an efficient weakening of the vibrations and noise absortion (this one relatively to the oilmatic source) that works well with more than one applied load, with several frequencies and large dimensions specimens. It has been confirmed that it can also execute several kinds of cycles and load regimens, allowing loads readings and register loads and displacements, either manually or by computer, to vary the conditions imposed during the test work, etc.

Thus, it can be told that the first target, that is, to dispose of a powerful and versatile "tool" was fully reached.

On the other hand, the results obtained through the test is very interesting.

In fact, both the specimen with incomplete penetration and the specimen without penetration have presented a good resistence to fatigue.

According to the results above exposed it seems one may conclude that full penetration is not necessary and even welds without penetration can be used, at least in the resistent main element, that is in the bolter beam of the UQE's end-underframes.

The fabrication, in these conditions, will need less time and costs, becoming more economical. In fact, it will not be necessary to make grooves, the relative positioning of the flanges and webs will be easier and quicker, as it will not be necessary to keep constant gaps, the welding execution will be easier and it will not impose so many limitations either to the maximum intensity one can use or to use of automatization, being less the quantity of faults and consequent repairs.

Considering the responsability involved it is considered wise to complement the tests done with one (or more) confirmation tests so that the confidence may be enlargened.

Recent Advances in Experimental Mechanics, Silva Gomes et al. (eds) © 1994 Balkema, Rotterdam, ISBN 90 5410 395 7

Multiaxial fatigue testing machine under variable amplitude loading of bending and torsion

Thierry Palin-Luc & Serge Lasserre
LAMEF (LAboratoire Matériaux Endommagement Fiabilité), ENSAM-CER de Bordeaux, Talence, France

ABSTRACT : Great improvements have been made in methods of forecasting multiaxial fatigue in components with a long operationnal life, and it is now necessary to study cumulative damage under variable amplitude loadings. With this aim, L.A.M.E.F. has developed an original new machine for testing bending and torsion in combination. The design of the machine allows the specimen to remain fixed, while subject to a range of very different servo-controlled forces.

1 INTRODUCTION

Over the last twenty years, there has been a tremendous development in the study of models to forecast fatigue behaviour in metals. Much progress has been made with multiaxial sinusoidal loadings of constant amplitude or in blocks of different amplitudes. With a first prototype which has been built in the 1980's, three Ph.D.thesis have been defended and about 2000 specimens have been tested (Bennebach 1993 ; Dubar, 1992 ; Froustey, 1987).

Nowadays, there is a need for much more reliable materials and a more economic use of resources, and industry is looking more closely at studies of loadings which reflect more closely conditions found in real components in operational conditions ; the automobile industry is a case in point. Research and development departments want to be able to forecast fatigue damage under loadings of variable amlpitude. At the request of Renault S.A., L.A.M.E.F. turned its attention to the design and développment of a new fatigue machine for testing bending and torsion in combination, where the specimen was subject to stresses similar to those found in operational conditions at the critical point of an internal combustion engine crankshaft.

Funding for the prototype presented in this article was provided by the Région Aquitaine and Renault S.A., with further assistance from D.R.E.T., Ponticelli and C.E.S.T.A.

2 OPERATIONAL EQUIPMENT

2.1 *Mechanical principles*

This machine is similar in principle to that patented by L.A.M.E.F. in 1977 (No. 77-37146; Lasserre, Lizarazu, Séguret, 1977). Generation of the bending moment applied to the specimen is based on the transposition into mechanical terms of the rotating fields principle (Figure 1). The specimen does not rotate, two hydraulic jacks impose two moments of plane bending in two perpendicular planes. The resultant bending load (over four points of support) may be one of three types :

- plane bending (symmetrical or not) : where the two plane bending moments are absolutely identical (same amplitude, frequency, mean value) and evolve in phase ;

- rotative bending : where the two plane bending moments have a mean value null and a phase difference of 90° ;

- dissymmetrical rotative bending : where the two plane bending moments have the same non-null mean value and a phase difference of 90°.

The symmetrical or alternating torsion moment is imposed on the specimen by a third hydraulic jack.

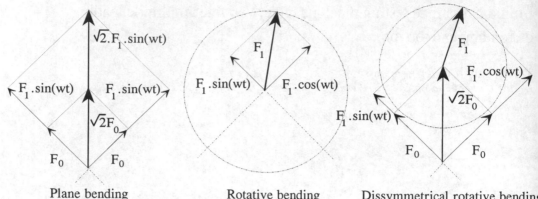

Plane bending   Rotative bending   Dissymmetrical rotative bending

Figure n°1 : Composition of forces in bending

The significance of this arrangement lies in the fact that bending and torsion loadings can be totally separate one from the other. Moreover, the scattering of experimental results experienced when several different machines are used is no longer a problem.

2.2 *Mechanical design*

The originality of the principle described above led us to adopt the mechanical design outlined in figure 2, a design similar to that used in the first prototype (Froustey, Lasserre, Séguret, 1986).

end in supple sheets of steel (3) and (3'). The bending forces FA(t) and FB(t) are applied to the tubular components via steel sheets (5) and (5'). This method of contact via a supple barrier produces the equivalent of ball and socket joints with no play in relation to bending, and firmly embedded elements in relation to torsion. The torsion moment is applied on component (4), which is guided as it rotates within the frame (0) by two prestressed oblique contact bearings.

Transmission of the forces FA(t), FB(t) and F(t) between the jacks and the mechanical components is ensured by flexible rods.

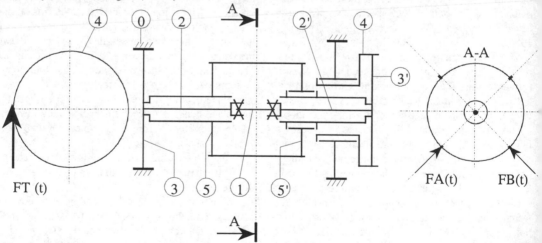

Figure n°2 : Scheme of the mechanical part

The equipment is based on a central beam consisting of the specimen (1) attached firmly to two tubular components (2) and (2') including the moments sensors. This beam is embedded at each

Thus the beam has four supports ; the specimen and the components for measuring the moments are in an area where moments are constant (no sharp force).

The specimen is fixed to the two tubular components by means of two hemispheres screwed together. The ends of the specimen are first placed inside two split tubular parts, which spread the strain of the clamp and adapt the diameter and the shape of the specimen to that of the jaw.

3 CONTROL EQUIPMENT

3.1 *Principle of servo-control*

The servo-control of the present prototype is based on the principle developed by Mr. Lizarazu for the first prototype. Each jack is piloted by means of two nested loops (Figure 3). The first loop, analogical in relation to the position of the jack rod, is contained within a numerical loop controlled by micro-computer. For each jack, the control software considers that the system to be controlled is all the mechanical and hydraulic equipment and the analogical loop. By regulating this loop, it is possible to obtain a frequency behaviour close to a second order behaviour with no overflow. Using this system, the jacks are never in an open loop ; specimens can be put in place and remove in safety.

better than one per cent of the full scale of the sensors (better than 1,65 N.m.). In the sinusoidal endurance mode, amplitude, mean value and the phase differences in the command signals are servo-controlled. The phase difference between the loadings on each jack can always be ajust at the beginning of a test between 0° and 90°. In pseudorandom mode, the form of the signal is servo-controlled.

With this software it is also possible to monitor in real time any movement of the ends of the specimen. Thus crack initiation can be detected, and the test stopped.

3.2 *Detection of cracks*

The method selected is that of monitoring the rigidity of the specimen. As fatigue tests with a large number of cycles are carried out with imposed stresses, thus with an imposed load, by monitoring movement at the ends of the specimen, any variation in rigidity can be detected. In order to do this, our prototype is fitted with three displacement sensors with Foucault current (no contact) which measure any such movement. As there is no contact between sensors and specimen, there is no risk of

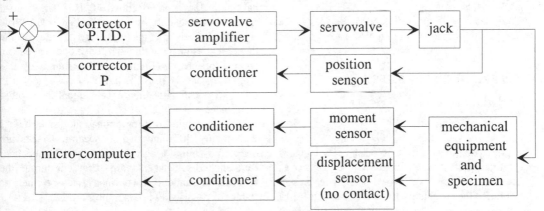

Figure n°3 : Principle of servo-control for one jack

The software for piloting the jacks was developed by Alliance Automation and is extremely user-friendly. The user selects one of the three pilot modes described in paragraph 4. Those used most often are the sinusoidal endurance mode and the pseudorandom mode. Those operations that are servo-controlled are usually the bending and torsion moments and occasionaly moving the jack rods. The moments are servo-controlled with a precision

disrupting experimental results.

The piloting software compares this movement with that recorded at the beginning of the test during the first stable cycles, when the specimen is considered to contain no cracks. A test is ordered to be stopped when at least one of the displacements measured has moved from its corresponding reference point by more than a certain percentage, preset by the user. Using this method, cracks 0,5 mm in depth can be be detected in a resistant section of 80 mm^2.

3.3 *Security System*

We have given this prototype a triple security system.

The first level is governed by the piloting software, which in real time compares measurements it is making with the acceptable force and movement limits for the machine. It can stop a test at any time if any one of these measurements exceeds the programmed limits. The piloting computer is also equipped with a "watchdog" device which checks that the software loops are functioning properly.

Over and above this system is another independent, entirely analogical device which fulfils a similar purpose. In order to protect the machinery, an order to stop from this device produces an emergency stop, which results in the loss of the test in progress.

Lastly, the transmission rods linking the jacks to the equipment act as mechanical fuses. They represent a final passive form of security, protecting the mechanical part of the equipment from accidental overload.

Equipped with these safety devices, the prototype can operate without human supervision.

4 DIFFERENT TYPES OF TEST

4.1 *Modes of functioning*

The principal purpose of this machine is to carry out fatigue tests with a large number of cycles (more than 10^5 cycles), so we shall here present only briefly the machine's limited capacity to carry out monotone tests ; we shall look more closely at its capacity for endurance tests under sinusoidal and pseudorandom loadings.

1.Monotone tests

The maximal forces produced by this prototypes are rather weak for monotone tests, but it is nevertheless possible to carry out bending and torsion tests of this type. The imposed moment is continuously servo-controlled, developing along a straight line or series of straight lines with a positive or negative slope. In this case, the specimen must be fitted with deformation gauges

linked to an existing scale of measurements in the laboratory, in order to measure the physical dimensions characteristic of this type of test.

2.Sinusoidal endurance

This is the type of experiment which best suits with this machine. Frequency is between 10 Hz and 60 Hz (selected by the user). Bending and torsion moments evolve as a sinusoidal function of time. They may be in phase or out of phase, at the same frequencies or at different frequencies. The degree to which the imposed moments are out of phase is servo-controlled, preselected by the user at between 0° and 90°.

Successive blocks of loadings are possible. Blocks are defined by duration, amplitude, mean value, frequency and degree to which each moment is out of phase. Transitions between successive blocks occur without overloading the specimen and without discontinuing the force.

3.Pseudorandom endurance

The specimen is subject to bending and torsion moments, which may or may not be simultaneous and which evolve throughout the test as a non-sinusoidal function of time. When this function is broken down into Fourier series, six harmonics are produced, at the most. The basic frequency should be between 10 Hz and 50 Hz, so that the highest harmonic does not go above 60 Hz. The control signal is constantly servo-controlled. The true signal applied to the specimen follows the theoretical signal within an envelope called an "tunnel of error" (Figure 4). The range of this tolerance zone is fixed by the user ; we generally select 5% of the theoretical signal maximum amplitude.

It is possible to programme sucessive blocks with totally different loading spectra. Cracks are detected by monitoring the rigidity of the specimen. Reference displacements are recorded during the first pseudorandom tests when the specimen is stable, as described in paragraph 3.2.

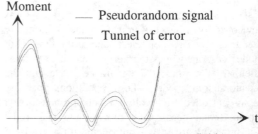

Figure n°4 : Tunnel of error definition

Endurance tests enable us to study the behaviour of metal materials under multiaxial fatigue and with a large number of cycles and loads similar to those found in operationnal conditions. Depending on the combination of variables, i.e. shape of specimen and composition of imposed moments, the tensor of the stresses at the critical point on the specimen can include up to four non-null terms. This does not take into account any residual stresses that may be introduced by treating the specimen before testing.

4.2 Geometry of the specimens

The anchoring system on the prototype enables us to test cylindrical specimens with a clamp zone of between 16 and 20 mm diameter. The useful test zone on the specimen should be less than 30 mm in length, as shown in figure 5. The specimen may be of any shape between plane P1 and P2 ; a notch, for instance, enables us to study the influence of geometrical feature on the behaviour of the material under fatigue.

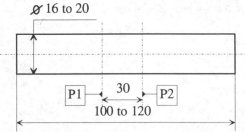

Figure n°5 : Dimensions of specimens

Figure 6 shows a usual specimen. With bending, the median torus generates a theoretical stress concentration factor of 1,07 (Peterson, 1974) enabling us to pinpoint the area of crack initiation. This machine also accepts specimens which are parallelepiped in shape, with dimensions within the limits defined for figure 5. In order to fix such specimens into the machine, special parts are required, which can be produced on demand.

4.3 Technical specifications

Moment maxima :
Plane bending :   212 N.m
Rotative bending :   150 N.m
Torsion :   150 N.m
Specifications of the control signal :

Maximum frequency :   60 Hz
Maximum number of harmonics :   5
Precision of servo-control :
Servo-control of moments :   ±1,6 N.m
Servo-control of phase :   ±1°
Crack detection :   through variations in rigidity

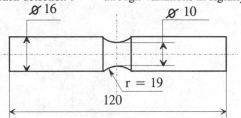

Figure n° 6 : Example of a usual specimen.

4.4 Examples of loads

1. Sinusoïdal endurance
As it is possible to programme successive blocks of quite different loads, a possible load combination could be that shown in diagram form figure 7. With this type of load, the cumulated damage can be studied when the size and type of loadings vary according to time.

2. Pseudorandom endurance
In order to study the behaviour of a material under loadings similar to those sustained by a real

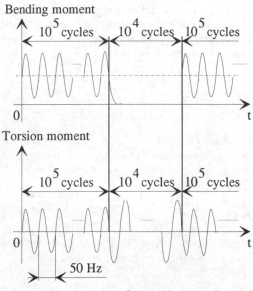

Figure n°7 : Example of a combined bending and torsion load in blocks.

component under working conditions, a pseudorandom load like that shown in diagram form in figure 8 can be applied to the specimen. This example represents the effect of the bending moment applied to a working crankshaft.

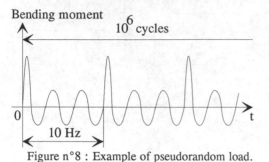

Figure n°8 : Example of pseudorandom load.

5 CONCLUSION

This multiaxial fatigue testing machine enables us to study cumulated damage under loads which resemble those to which components are subject in true working conditions. Thus we hope to be in a better position to test the validity of existing methods for calculating the lifetime of materials under fatigue with a large number of cycles, and possibly to suggest new ones.

This prototype is presently an experimental tool which has many potential uses and should open the way to improved techniques for research and development departments to set the ideal dimensions for components.

REFERENCES

Bennebach M. 1993.
Fatigue multiaxiale d'une fonte G.S. Influence de l'entaille et d'un traitement de surface.
Thèse E.N.S.A.M.

Dubar L. 1992.
Fatigue multiaxiale des aciers. Passage de l'endurance limitée à l'endurance illimitée. Prise en compte des accidents géométriques.
Thèse E.N.S.A.M.

Froustey C. 1987.
Fatigue multiaxiale en endurance de l'acier 30NCD16.
Thèse E.N.S.A.M.

Froustey C., Lasserre S. 1986.
Machine d'essai de fatigue multiaxiale. Flexion rotative et torsion.
6ème Colloque international, Cercle d'étude des métaux. Saint-Etienne, France.

Froustey C., Lasserre S., Séguret J. 1987.
Fatigue sous sollicitations combinées de flexion et torsion sur un acier 30 NCD 16.
Conférence IITT.

Lasserre S., Lizarazu F., Séguret J. 1977.
Machine d'essais de fatigue des matériaux sous sollicitations combinées et éprouvette fixe.
Brevet d'invention n° 77-37146, E.N.S.A.M. Bordeaux, France.

Peterson R.E. 1974.
Stress concentration factors.
A Whiley-interscience publication.

Recent Advances in Experimental Mechanics, Silva Gomes et al. (eds) © 1994 Balkema, Rotterdam, ISBN 90 5410 395 7

Proposal for a standardized experimental set-up to study multimaterial joints under dynamic loading – Application: Behavior of the windscreen during a car crash

D. Soulas, F. Collombet, J. L. Lataillade & A. Diboine
LAMEF (LAboratoire Matériaux Endommagement Fiabilité), ENSAM-CER de Bordeaux, Talence, France

ABSTRACT: The aim of this work is the development of an experimental set-up to investigate the local behaviour of adhesive joints under dynamic loadings. In laboratory, the dynamic response of the adhesive is objective if the test re-creates the load conditions met by the joint within the real structure. The final aim is to define a standardized test for the dynamic characterization of adhesive bonded joints by means of parametric studies.

The experimental device is designed to study the mechanical behaviour of a steel/glass assembly close to the one of a bonded windscreen. The first results show that the apparatus is able to reproduce the backward and forward motion which usually characterizes the motion of the windscreen during a car crash. Nevertheless, it is unable to re-create accurately the steel-frame deceleration recorded during such a collision.

1 INTRODUCTION

Sticking is now a rival technique to fasten materials contrary to the more traditional techniques such as brazing, welding, riveting and bolting. The multitude of advantages that adhesives can offer (the ability to join dissimilar materials, an improved stress distribution in the joint, a very good dynamic-fatigue resistance of the bonded component, lightening of bonded structures, an improvement in corrosion resistance,...) interest several sectors of activity [Kinloch 1987]. Furthermore, industries like the car industry have adopted this new technology because the bonding operation can often be automated and the appearance of the fastened structure is improved.

During their use, adhesive bonded joints are submitted to severe conditions. So, a detailed study is needed to understand their mechanical behaviour and their durability.

For many years, the cars' windscreens have been bonded on the steel-frame to improve the stiffness of the bodywork. During a car crash, the windscreen is submitted to dynamic loading: the kinetic energy due to the deceleration is released for a short time. Passenger car safety may depend on the behaviour of the adhesive joint. Moreover, the windscreen surface increase and thus, the weight increase incites the manufacturer to integrate the shock resistance of windscreen to the design of cars. For that reason, an original experimental set-up is developed in the laboratory to study the impact behaviour of a steel/glass assembly. The new test procedure must reproduce the shock conditions induced by a frontal car crash.

2 PRESENTATION OF THE PROBLEM

During the dynamic loading of an adhesive bond, the adhesive local response depends on the load transmission within different structural components. Consequently, laboratory testing machines must not only load dynamically the adhesive joint, but also re-create fundamental phenomena met by the adhesive bonded joint within the real structure.

There are several experimental devices to test shock resistance of adhesive bonded joints (Chappy test-machine, Hopkinson bar principle, inertia wheel, drop-weight set-up,...) [Ziane 1987]. These testing methods characterize fracture behaviour of adhesive joints under high strain rates loading by changing parameters of adhesion and of adherence [Keisler 1994]. In this case, study is focused on the mechanical behaviour of the "adhesive" material without taking the real loading conditions into account.

The aim of this work is to achieve a new experimental device which should be able to characterize the effects of dynamic progressive load transmission on the local behaviour of the adhesive.

The study of a bonded windscreen during a car crash illustrates the necessity to imagine such a set-up. The conventional crash tests are too expensive to realize a parametric study of the windscreen joint behaviour. Moreover, crash test instrumentation is not able to measure the evolution of the adhesive versus time. So, it is a need to design a new apparatus to investigate in laboratory and thus, on a reduced structure, the local behaviour of the adhesive between the windscreen and the steel-frame under dynamic

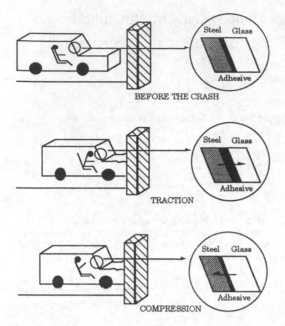

Fig.1 : Global behaviour of a car during a frontal crash

loading. In order to obtain a valid dynamic response of the adhesive, the experimental set-up has to reproduce the conditions of load transmission representative of a complete vehicle frontal collision.

Analysis of the driver motion during a low speed frontal collision allows us to apprehend mechanical behaviour of the windscreen joint (Fig.1).

- Before the crash, the velocity of the car is constant. The driver is in a right position, the joint is not loaded.
- With the shock, the front brutally decelerates and bends. The cockpit continues to move forward: the driver is projected against the wheel and the joint is in tensile.
- During the crash, the deceleration is transmitted to the entire vehicle. The driver is projected against his seat, the joint is in compression.

Figure 2 shows the deceleration recorded on the bodywork near the windscreen during a conventional crash test (impact velocity = 15 m/s).

The steel-frame is submitted to a succession of deceleration peaks. Amplitude of these peaks increases with the time; the most important decelerations occur when the car velocity is about 10 m/s. The maximum deceleration recorded during such a crash is about 600 ms^{-2}.

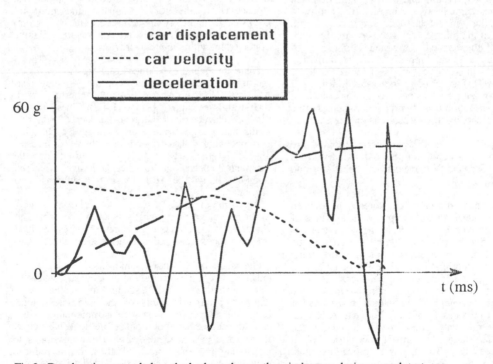

Fig.2 : Deceleration recorded on the bodywork near the windscreen during a crash test

A numerical process has defined a test configuration able to re-create fundamental phenomena met by the joint during a crash [Bonini 1993]. According to this numerical study, the sharp deceleration of the windscreen and the traction-compression behaviour induced in the adhesive joint can be experimentally reproduced by the impact of a reduced structure against a rigid body (Fig.3).

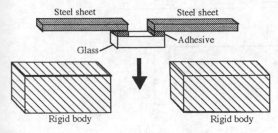

Fig.3 : Principle of the test configuration

3 EXPERIMENTAL TECHNIQUE

3.1 Test specimen and materials

The specimen is a multimaterial bonded structure close to a windscreen joint. It is made of two "U" steel sheets, a piece of glass and a polyurethane adhesive. Its configuration is presented in figure 4.

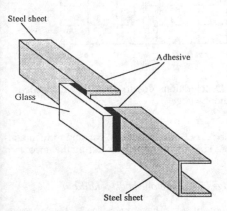

Sheet: length = 85 mm
         height = 25 mm
         width = 20 mm
         thickness = 0,8 mm

Glass: length = 50 mm
         width = 25 mm
         thickness = 5 mm

Adhesive: length = 25 mm
           width = 10 mm
           thickness = 4 mm

Fig.4 : Schematic of the specimen

The adhesive is a polyurethane mastic with a low elastic modulus: E = 5,5 MPa at 23°C, 10^{-2} Hz

The steel sheets are automobile sheets. The "U" reproduces the rigidity of a steel-frame. In fact, the cockpit must keep its shape in case of a low speed frontal collision. Only the front of the car has to bend to absorb energy.

The piece of glass is a homogeneous glass. A "primary" is applied to glass areas in contact with the adhesive. The "primary" improves adhesion mechanics and protects adhesive joint against sunbeams.

The sample is not exactly representative of a bonded windscreen section. Actually, car windscreens are laminated and slightly curved. Furthermore, the geometry of the adhesive joint is not the one of a real windscreen joint. The joint measurements are those defined by the manufacturer for its tests.

3.2 Experimental set-up (Fig.5)

The experimental set-up is designed to re-create the brutal deceleration and the backward and forward motion of the windscreen noted during a real crash test. In order to reproduce these phenomena, the specimen is moved before it hits a fixed obstacle.

A compressed air gun projects a trolley on a guide-rail with the sample on it. During the motion of the trolley, specimen is kept in position by wedges. The wedges must have a low fracture energy in order not to disturb the contact conditions during the shock. The specimen in speed before impact, runs onto a fixed body whereas the trolley crosses under the body.

The impact velocity varies according to the pressure in the gun in the range of 5 to 15 ms^{-1}. The rigidity of the still body and the impact velocity govern the shock intensity.

3.3 Metrology

The device is instrumented to follow the kinematics of different specimen components (Fig.6).

A piezoelectric accelerometer located one of the specimen steel sheets measures the sheet deceleration due to the shock. A second piezoelectric accelerometer is put on the piece of glass to measure the deceleration of this element. A double integration of these two signals gives the sheet displacement and the glass displacement during the impact.

A fibre-optic sensor completes the metrology to measure the velocity of the specimen just prior to impact.

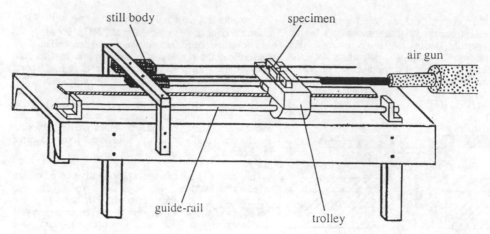

Fig.5 : Experimental set-up

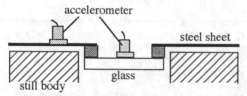

Fig.6 : Position of the accelerometers on specimen

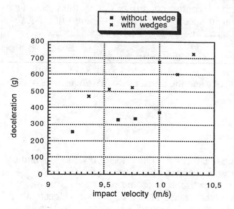

Fig.7 : Variation of the maximum deceleration versus impact velocity for a shock with wedges and for a shock without wedge

4 VALIDITY OF THE EXPERIMENTAL SET-UP

The objectives of the experimental device validation are three in number. In the first place, we assess the effect of the wedges on the shock conditions. After that, we calibrate the apparatus to obtain levels of deceleration comparable to those recorded during a car

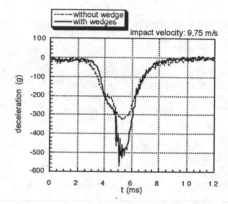

Fig.8 : Decelerations due to a shock with wedges and without wedge

crash (about 600 ms^{-2}). Finally, we study kinematics of various specimen components during the impact.

4.1 Wedges influence on shock conditions

Two kinds of tests have been made to determine influence of the wedges: with wedges and without wedge. In the second case, the wedges are deliberately broken just before the impact (about 5 ms).

Figure 7 shows the variation of specimen maximun deceleration versus impact velocity for a shock with wedges and for a shock without wedge.
Figure 8 superimposes the deceleration due to an impact with wedges and the deceleration due to an impact without wedge. The impact velocity is 9,75 m/s.

The decelerations given here, have been measured with the accelerometer put on the sample steel sheet (see figure 6).

The maximum deceleration due to an impact with wedges is greater than the maximum deceleration due to a shock without wedge. In the two cases, the maximum deceleration (in absolute value) increases with the impact velocity (Fig.7). On the other hand, wedges have no influence on the shock duration and on the general aspect of the deceleration (Fig.8).

Consequently, we can continue to use these wedges to keep specimen in position during trolley motion.

4.2 Experimental set-up calibration

The specimen deceleration depends on the impact velocity and on the ability of the still body to get out of shape during the shock.

The still body is covered with a foam material in order to reproduce the deceleration levels noticed during a frontal collision (about 600 ms^{-2}). In this way, it is possible to calibrate shock intensity by changing the foam thickness.

Figure 9 shows the decelerations obtained with four foam thickness (70 mm, 85 mm, 130 mm, 180 mm) for an impact velocity of 9,5 m/s.

These decelerations have been measured with the accelerometer put on the specimen steel sheet (Fig.6).

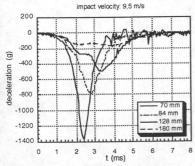

Fig.9 : Variation of the deceleration for four foam thickness

The maximum deceleration decreases when the foam thickness increases. On the other hand, the impact duration increases with the foam thickness.

It is impossible to record a correct signal of deceleration with a foam thickness larger than 180 mm. At the present time, the experimental set-up is unable to re-create the level of deceleration noted during a crash test.

4.3 Kinematics of various specimen components and mechanical behaviour of the adhesive joint

The aim of this paragraph is to verify that the experimental set-up is able to re-create the backward and forward motion of the windscreen observed during a frontal collision. So, we follow the kinematics of each components of the sample during the test (see figure 6). The relative displacement between the steel sheet and the glass gives the mechanical behaviour of the adhesive versus time.

In order to get a better view of the relative displacement, we subject the specimen to a severe shock ($\approx$ 20.000 ms^{-2}).

Figure 10 shows the displacements of each components of the sample versus time. The impact velocity is about 8 m/s.

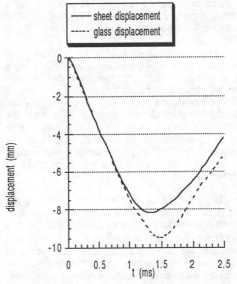

Fig.10 : Displacements of each constituents of the specimen versus time.

Analysis of these displacements reveals the presence of a tensile phase which usually characterizes the behaviour of windscreen during car crashes. About 0,8 ms after the beginning of the impact, the glass displacement is actually greater than the one of the steel sheet; consequently, the adhesive joint is in traction.

The measured relative displacement between the glass and the steel sheet is about 1,35 mm. A rapid calculation (1) gives the maximum strain of the adhesive due to such a shock.

$$\Delta l = \frac{M \gamma e}{2 S E} \qquad (1)$$

with E: Young Modulus of the adhesive,
e: adhesive thickness,
M: mass of the glass with the accelerometer,
S: bonded surface,

γ: maximum deceleration of the glass,

Δl: maximum adhesive strain

The equation (1) gives a maximum deformation of 1,17 mm. The discrepancy between the calculated value and the measured value is about 15%.

The experimental device allows to reproduce accurately the backward and forward motion of the windscreen noted in a real car crash. Moreover, the measured adhesive strain is comparable to the calculated deformation.

5 CONCLUSION

The first tests emphasize the difficulty to re-create experimentally, on a reduced structure, shock conditions comparable to those met by the adhesive bonded joint within a real structure.

The experimental set-up is able to reproduce the mechanical behaviour of the adhesive noticed during a car crash and particularly, the backward and forward motion which usually characterizes the behaviour of the windscreen during such a collision. On the other hand, it is impossible to re-create accurately the deceleration recorded on a windscreen steel-frame during a conventional crash test. In fact, the succession of deceleration peaks, the duration and the level of the deceleration can't be reproduced experimentally.

The relative displacement between the piece of glass and the steel sheet can not be precisely measured with low intensity shocks (maximum deceleration inferior to 10.000 ms^{-2}). To have a better measurement of this displacement for low deceleration levels, we must reduce the width of the adhesive joint. In this way, we will increase the strees level in the adhesive and thus, its strain.

ACKNOWLEDGEMENTS

The authors would like to acknowledge Renault S.A. and the French Ministery of Education and Research for their financial support.

REFERENCES

Bonini J., Collombet F., Diboine A., Lataillade J.L., 1993. "Multimaterial joints under dynamic loading - Behaviour of the windscreen during a car crash - Numerical determination of industrial test", *Proceeding of the 6th Computation Methods and Experimental Measurements*, Sienne. Elsevier Applied Science, vol.2, pp 353_364, ISBN 1-85166-968-x.

Keisler C., Lataillade J.L., 1994. "The effect of roughness characteristics on wettability and on the mechanical properties of adhesive joints loated at medium strain rates". Journal of Adhesion, *Sciences and Technologies,* to be published 1994.

Kinloch A.J., 1987. "Adhesion and adhesives", *Science and Technology*, Chapman and Hall.

Ziane E., Coddet C., Béranger G., 1987. "Caractérisation des joints collés - Revue critique des principaux essais mécaniques", *Matériaux et Techniques - Special collage - Contrôles et essais,* juin 87.

Recent Advances in Experimental Mechanics, Silva Gomes et al. (eds) © 1994 Balkema, Rotterdam, ISBN 90 5410 395 7

A study on the image processing measurement of behaviors near the fatigue crack tip

Yasuo Ochi, Akira Ishii & Hitoshi Yoshida
Department of Mechanical and Control Engineering, University of Electro-Communication, Chofu, Tokyo, Japan

ABSTRACT: For the purpose of the establishment for automatic measurement system of fatigue crack propagation behaviours in real time, a new method by using of computer image processing technique was proposed for measuring a crack opening displacement (COD) and a displacement amount near arround the crack tip in this study. And the radiated projection (RP) method as the image processing algorithm was applied to the detection of arbitrary micto-marks (pits) near arround the crack tip, and the changes of COD and the displacement distribution near crack tip were measured.

1 INTRODUCTION

Some automatic measurement methods by means of image processing technique have been applied to many kinds of engineering and manufacturing area.In the study of strength evaluation of materials, there have been some reviews by Y.Seguchi(1986), Y.Morimoto(1988) and K.Ogura et al(1990).Especially in the fatigue test, as it needs too much time and effort for experimentalist in order to detect small cracks on the test specimens and to measure the crack length during whole fatigue process, then it is desirable to develop the automatic measurement system for such operations. For the measurements of long-through cracks of CT specimens, I.Nishikawa et al(1991) and A.Ueno et al(1992) have tried to measure automatically by the image processing technique using a binary image method basically. However, it is insufficient to use only the traditional method including binary images, mask operators and edge detection for the measurements the small cracks initiation the crack opening displacement and the displacement behaviour near the crack tip, and it needs the higher level recognition method as well that the skilled experimentalist can measure in detail with using the higher level microscope and the replication technique.

A.Ishii et al (1993) proposed the radiated projection(RP) method for detection of small defects as a new image processing algorithm.The RP method used a radiated frame whose size can be changed step by step depending on the size of defects, and whose shape and structure can be carefully designed as considering the shape and the direction of defects. And also A.Ishii et al (1993) have applied the RP method to the recognition of small fatigue crack initiation and crack tip on the specimen surface.

In this study, the fatigue test on aluminum alloy plate specimen was carried out by electro-hydraulic servo system with three dimensional(XYZ)pulse stage on which the optical microscope setting the CCD camera.The images of micro-marks(pits) near the crack tip on the specimen surface were taken, and the personal computer and the image processor were used for the measurement of changes of COD of the displacement amount distribution near crack tip.

2 EXPERIMENTAL EQUIPMENT AND IMAGE-INPUT METHOD

Fig.1 shows the construction of fatigue testing machine and image processing system in this study.

Fatigue test was carried out by micro-computer controlled electro-hydraulic servo machine of 147kN, and the total strain amplitude was controlled in low cycle fatigue region. The material used in this study was aluminum alloy A7075 for the use of aircraft structure. The shape of the specimen was the plate with 12mm in width, 5mm in thickness and 12mm in gauge length, having a small drilled hole (3mm in diameter ×

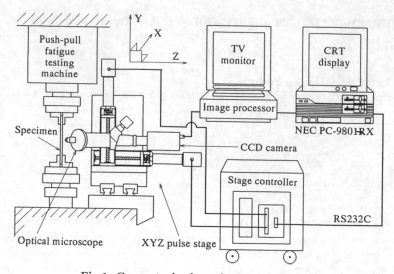

Fig.1. Computerized crack observing system.

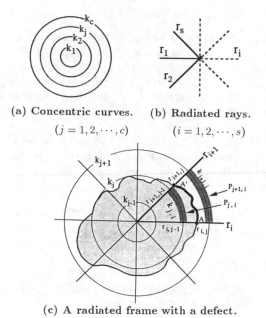

(a) Concentric curves.

$(j = 1, 2, \cdots, c)$

(b) Radiated rays.

$(i = 1, 2, \cdots, s)$

(c) A radiated frame with a defect.

Fig.2. Structure of a radiated frame.

image processor by using CCD camera. The one pixel of the input image ($512 \times 480 pixel$, 256 gray scale) corresponds to about $0.3 \mu m$ when the objective lens of 40 times is used.The minimum feed of XYZ pulse stage having the decelerated device of 1/50 in three axis is $0.015 \mu m/pulse$.

3 IMAGE PROCESSING PROCEDURE

3.1 Radiated projection(RP) method

The RP method is the recognition algorithm of objective for the purpose of automatic detection to small defect having the fixed level by human eyes.

Fig.2 shows the structure of radiated frame using in the RP method. The frame is constructed with concentric curves $k_j (j = 1, 2, 3, \cdots, c)$ and radiated rays; $r_i (i = 1, 2, \cdots, s)$ as shown in Fig.2(a) and (b), respectively.When the objective for detection is shown as a weak net region in Fig.2(c), the existence of the objective is proved by confirming the existence of the boundary of the objecticve in the all divided boundary by radiated rays, and shape and the size of the objective is recognized by the shape and the size of concentric curves.

3.2 Detection method of micro-marks

Fig.3 shows the flow chart of the detection method of micro-marks. The RP method was applied to the recognition of the micro-marks.

3mm in depth) which was starting portion of fatigue crack.The surface of the specimen was buff-finished with aluminum solution.

Automatically controlled three-dimensional XYZ pulse stage on which optical microscope attatched was equiped to the fatigue testing machine as shown in Fig.1. The images of the surface of specimen near the crack tip were taken to frame memory of commercial

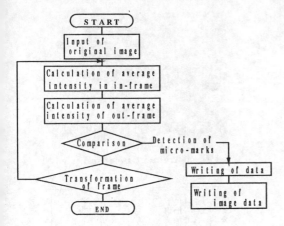

**Fig.3. Flow chart of method
for detecting micro-marks.**

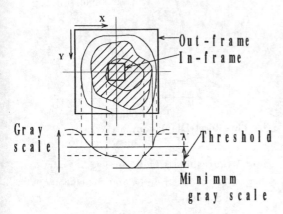

Fig.4. Shape of micro-mark.

Fig.4 shows the square frame used for the recognition of micro-marks. The frame was scanned on the input image from the top of the left side of the image, and the following processes were performed.

(1) The average of intensity in in-frame($= g_{in}$) and the average of intensity on the out-frame($= g_{out}$) were calculated, and the threshold value Th was difined.

(2) if $g_{in} + Th \geq g_{out}$, then it was recognized that a micro-mark was not there, and the processing moves the next position.

(3) if $g_{in} + Th < g_{out}$, the it was recognized that a micro-mark is there, and the coordinate was recorded.

The following three methods were applied to the detection of the center of micro-marks.

(1) Method(I) (The method of constant threshold); As being constant in threshold Th in Fig.4, the center was calculated by taking account of the intensity level.

(2) Method(II) (The method of variable threshold); As the calculating area was also changed when the threshold level was varied, then the area(S) was obtained as constant and the threshold(Th) is determined. After that, the center was calculated by taking account of the intensity level.

(3) Method(III) (The binary method of variable threshold); The threshold was determined by the method of (II), and the center of marks was calculated without taking account of the intensity level.

3.3 Calculation of COD and detection of displacement near crak tip

Fig.5 shows the definition of COD. In the figure, COD was defined as the opening displacement being perpendicular to crack. Then, COD was shown as Eq.(1).

$$COD = (Y_2 - Y_1) \times cos\theta \cdots (1)$$

The points (X_1, Y_1) and (X_2, Y_2) were the arbitrary micro-marks near the crack side. The displacement measurement near crack tip was performed from the input image at the maximum and the minimum load. After several common marks near crack tip of two image were selected, the changes in coordinate of each mark were measured. In this case, it was supposed that the mark in front of crack was fixed.

4 RESULTS AND DISCUSSION

4.1 Detection of micro-marks

Fig.6 shows the original image and the detected result of micro-marks near crack tip area. The image was taken at the condition of total strain range $\Delta\varepsilon_t = 1.7\%$, and the number of cycles of 80, and the crack length $2a = 170 \sim 180\mu m$. In the Fig.6(b), the points enclosed by small square (out-frame) were the recognited marks. These marks were detected with the condition of $Th = 20$, out-frame size$= 10 \times 10 pixel$, in-frame size$= 3 \times 3 pixel$, and the scanning step$= 1 pixel$.

Fig.7 shows the relation between the distance of two point Δr and the threshold level K_T by the method (II) (with taking account the intensity level) and the method (III) (without taking account the intensity level as mentioned above section 3.2).The Δr was defined as the equation (2)~(4).

$$\Delta x_i = x_i - x_0 \cdots (2)$$
$$\Delta y_i = y_i - y_0 \cdots (3)$$
$$\Delta r = \sqrt{\Delta x_i^2 + \Delta y_i^2} \cdots (4)$$

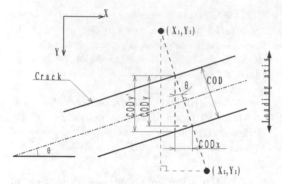

Fig.5. Definition of COD.

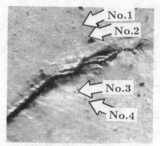

(a) Measurement position.

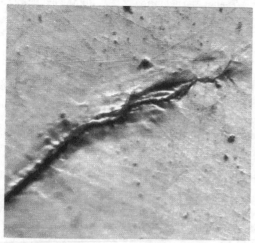

(a) Before detecting micro-marks.

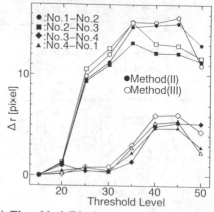

(b) Threshlod-Displacement(Δr) curve
between two points.

Fig.7. Threshold-Displacement(Δr) curve
between 2 points and measurement
positions.

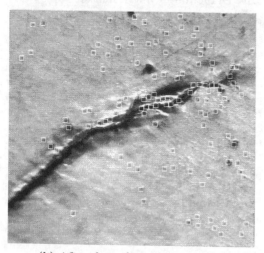

(b) After detecting micro-marks.

Fig.6. The image of crack tip
before and after detecting micro-marks.

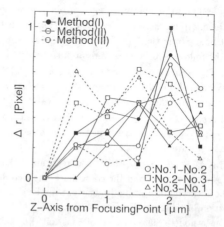

Fig.8. Z-Axis-Displacement(Δr) curve
between two points.

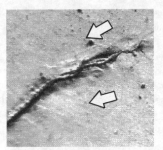

(a) Measurement position of COD.

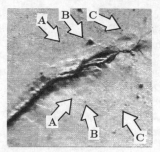

(a) Measurement position of COD.

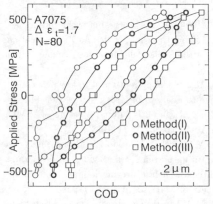

(b) Changes of COD by three methods.

Fig.9. Changes of COD by three methods
and measurement position of COD.

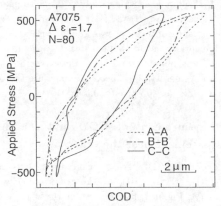

(b) Changes of COD of three different positions.

Fig.10. Changes of COD of three different
positions and measurement position
of COD.

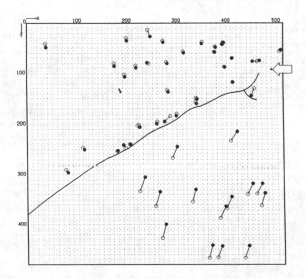

Fig.11. Displacement of crack tip sorrounding.

where, x_0: comparative x axis, (at focus, and at minimum threshold level), x_i: i_{th} x axis, y_0: comparative y axis, (at focus, and at minimum threshold level), y_i: i_{th} y axis

From this figure, it was clear that the Δr increased rapidly at the range over $K_T = 30$, and the difference between two method was not obvious.

Fig.8 shows the relation between the Δr and distance of Z-axis from focusing point by the method of (I),(II) and (III) in section 3.2. The images were taken by traveling at $0.5\mu m$ step in Z-direction from the focusing point. From this figure, it was clear that the method (I) showed the wide variety of Δr on comparison with the method (II) and (III). In this experiment, the discrepancy of focusing was about $1\mu m$, and the Δr value correspond to about $0.5pixel$ when the discrepancy was about $1\mu m$. From these results, this detection system will be expected to measure the Δr in the accuracy of sub-pixel($\approx 0.5pixel$).

4.2 Measurement of COD and the displacement near the crack tip

Fig.9 shows the COD measurement results by three method (I), (II) and (III) at the cycles of 80 with $\Delta\varepsilon_t = 1.7\%$. Fig.9(a) show the two marks which were used for measurement. From the figure, the very good $\sigma - COD$ relation curve can be obtained by the method (II) in comparison with the other method (I) and (III).

Fig.10 shows the COD measurement result at the three positions near the crack by the method (II) in the same cyclic condition as in Fig.9. From this results it was clear that the COD decreased with approaching the crack tip.

Fig.11 shows the measurement result of the displacement distribution near the crack tip at the cyclic condition as the same in Fig.9, and 10. In the figure, the arrow in the front of tip is the fixed point. The open circle shows the maximum load point and the solid circle shows the minimum load point.

5 CONCLUSIONS

For the establishment of automatic measurement system by means of image processing technique to apply the fatigue crack behaviours, the new detection method of micro-marks near the crack was proposed, and the changes of COD and the displacement distribution near the crack tip were measured in this study. The results obtained were summarized as follows,

(1) The micro-marks near the crack could be detected by means of the new image processing algorithm, RP method. An arbitrary marks could be detected by changing the frame size and the threshold level.

(2) As the method of detecting the center variable threshold was effective and the method could recognize the micro-marks in the accuracy of about $0.5pixel$.

(3) The measurement of COD and the displacement distribution near the crack tip could be obtained by the method proposed in this study.

REFERENCES

Seguchi,Y. 1986. Image processing technique for research of strength of materials and its applications. Journal of the Society of Materials Science, Japan. 35:95-105.

Morimoto,Y. 1988. Application of image processing technique to analysis of stress, strain and deformation. Proc. of the 63th Annual Meeting of JSME, Kansai Area. No.884-1, 111-118.

Ogura,K., Y.Miyoshi, S.Kubo & K.Minoshima. 1990. Progress in measurement and Monitoring method of experiments in fracture mechanics. Journal of the Society of Material Science, Japan. 39:575-581.

Nishikawa,I., H.Morikawa, H.fujiwara, Y.Miyoshi & K.Ogura 1991. New crack growth test system and its application to K-controlled fatigue crack growth tests at elevated temperatures. Journal of the society of Material Science, Japan. 40:235-241.

Ueno,A., H.Kishimoto, T.Kondo & M.Uchida. 1992. Automated crack growth tracking system with image processor. Proc. of the 1992 Annual Meeting of JSME/MMD. No.920-72:641-642.

Ishii,A., V.Lachkhia & Y.Ochi. 1993. An image processing algorithm for small defect detection. Journal of Japanese society for Non-destructive Inspection. 42:199-205.

Ishii,A., V.Lachkhia, Y.Ochi & T.Akitomo.1993. Recognition of small surface fatigue crack initiation and crack tip. Journal of the Society of Materials Science, Japan. 42:1231-1237.

Recent Advances in Experimental Mechanics, Silva Gomes et al. (eds) © 1994 Balkema, Rotterdam, ISBN 90 5410 395 7

Computation particularities for the elastic elements of strain gauge transducers

D. M. Stefanescu
Institute of Fluid Mechanics and Flight Dynamics, Bucharest, Romania

Adriana Sandu & M. Al. Sandu
St. Sorohan Polytechnic University of Bucharest, Romania

ABSTRACT: Specific requirements are presented in the order to obtain computation models for the elastic elements utilized in strain gauge transducers. By means of appropriate adaptive networks for finite element generation, the shape of the flexible structures is optimized after several computation stages.

The results obtained by means of the finite element method permit a good estimation of the transducer performances such as: nominal load, sensitivity, fundamental frequency, interaction coefficients for multicomponent transducers. Three practical examples are presented.

1. DESIGN ESTIMATION OF THE TRANSDUCER PERFORMANCES

The most important component of a strain gauge transducer for electric measurement of a mechanical quantity is its elastic element.

The basic metrological characteristics for such transducers are mainly determined by the shape and the size of the flexible elements as well as by the places where strain gauges are bonded on [1], [3] – [5].

The shape design of the elastic element must satisfy some important requirements such as: nominal load, sensitivity, linearity, hysteresis, fundamental frequency, necessary space.

The determination of the stress and strain states for the transducer elastic elements is made as a rule by means of the finite element method (FEM).

It is a numerical (approximate) method, best suited to the static and/or dynamic calculation for every elastic structure [2].

Based on this computation some transducer performances are estimated: rated load, sensitivity, fundamental frequency, interaction coefficients (for multicomponent transducers).

The multicomponent transducers permit the simultaneous measurement of resultant forces F_x, F_y, F_z and moments M_x, M_y, M_z on a stressed structure.

They can measure two to six components, the information being easier to process while the interaction is lower [3].

Ideally, applying a single component (force or torque), a big and unique reading is obtained; zero or negligible readings for other chanels are desired.

Even the real characteristics of the transducers are finally established by calibration, a good design is essential to optimize the shape of sensible elements.

2. SHAPE OPTIMIZATION CRITERIA FOR THE ELASTIC ELEMENTS OF TRANSDUCERS

The measurement of every force (torque) for multicomponent transducer is performed by one of the six full Wheatstone bridges, each having four or eight strain gauges (Fig.1). This bridge has the property to add the effects from the opposite arms and to substract the effects from the adjoining ones.

The nominal (rated) load could be reach only if the elastic element has such a shape to ensure strains with maximum modulus and opposite signs. If

$$\varepsilon_1 = \varepsilon_3 = \varepsilon_{max} \quad \text{and}$$
$$\varepsilon_2 = \varepsilon_4 = -\varepsilon_{max}, \quad \text{then one}$$

could read on the four-active-arm bridge:

$$\varepsilon_r = 2\,(\,\varepsilon_1 + |\varepsilon_2|\,) = 4\,\varepsilon_{max}.$$

Starting from a chosen shape and preliminary dimensions established by analytical computation on simple models, a first discretization for the deformable structure and FEM computation are made.

The results are examined and, if necessary, the initial dimensions are modified and FEM computation is made again. After several statical analyses, the transducer elastic element reaches its optimum shape.

The most important criteria for geometrical optimization of the transducer elastic elements are:
- strain gauges bonded on the places and on the directions having strains closer to the maximum ones,
- half from the strain gauges connected in full Wheatstone bridge to be extended, another half to be shortened, so ε_r to be as great as possible (double push-pull action),
- the fundamental frequency of the transducer to be much greater than the frequency of the parasitic vibration from the environment, to avoid the resonance,
- the coefficient matrix for multicomponent transducers to prove low coupling effects,
- the overall dimensions inserted in the imposed limits.

3. EXAMPLES

The elastic element for a mounting robot sensor (shaft-bore type) is shown in Figure 1.
Depending on the values of the

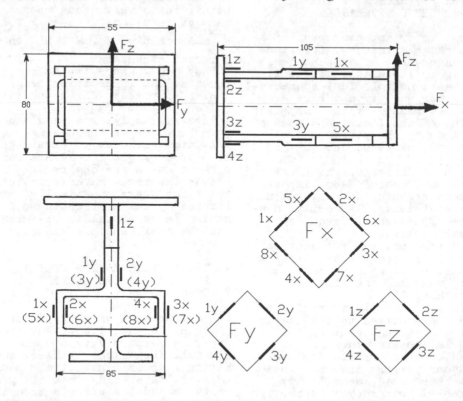

Fig.1. The elastic element for a mounting robot sensor

resistant forces, the transducer commands the gripper's motion, in order to obtain a correct position of the shaft within the bore.

The above-mentioned sensor permits the control of the three components F_x, F_y, F_z of the total resistant force.

Starting from the imposed size limits and the applied nominal loads $F_x = 200$ N, $F_y = F_z = 50$ N, some variants of deformable structure have been analysed.

A computation model containing beams as finite elements has been built. A series of statical analyses has been made modifying the length and the cross-sections of the beams. The best variant is presented in Figure 1.

The following readings under nominal loads have been previsioned:

$$\mathcal{E}(F_x) = 696 \ \mu m/m \ ; \ \mathcal{E}(F_y) = 893 \ \mu m/m;$$

$$\mathcal{E}(F_z) = 1148 \ \mu m/m.$$

Due to the deformation mode of the elastic element, the components F_x, F_y, F_z are theoretically uncoupled.

But, owing to some small misalignments of the strain gauges, little influences between components have been registred by calibration, but within acceptable limits. By FEM computation the fundamental frequency of the structure (195 Hz) has also been determined.

The second example of elastic

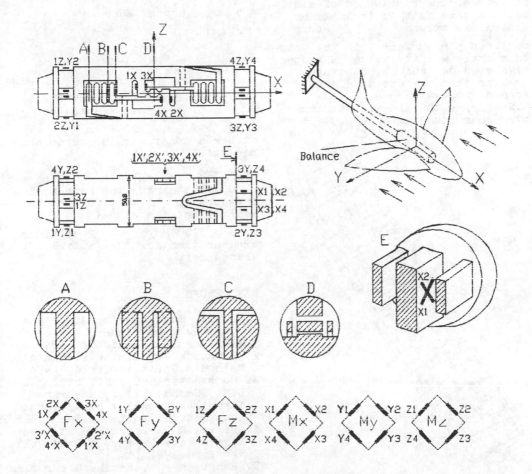

Fig.2. Compact internal strain gauge balance for tunnel

structure is a six-component internal balance for aeronautical models
placed in a wind tunnel (Fig.2).

A FEM computation model containing bricks (tridimensional solid
elements) is presented in Figure 3.

Six load cases have been studied,
corresponding to the forces F_x, F_y,
F_z and the torques M_x, M_y, M_z, obtaining the maximum stresses from
the structure and the optimal places to attach the strain gauges.

The same model permits the determination of the vibration modes
and the natural frequencies for the
elastic structure.

The initial geometry of the
flexible element for a piezoresis-
tive pressure transducer is presented in Figure 4,a. The sensible
body has been discretized by means
of bricks. The conclusion of statical analysis is that the maximum
equivalent stress $\sigma_{eq\ max}$ occurs
on the upper surface, in the points marked * ; its value is respectively 2.84 and 3.58 times
greater than the stresses under
strain gauges 1 and 2.

Modifying the elastic structure
by corner rounding (Fig.4,b), the

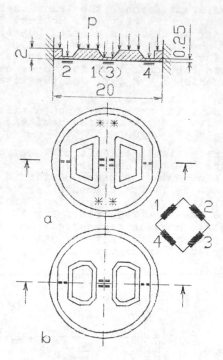

Fig.4. Elastic element for
pressure transducer

stresses under strain gauges have
become:

$\sigma_1 = .94\ \sigma_{eq\ max}$; $\sigma_2 = .53\ \sigma_{eq\ max}$.

So, maintaining the same nominal
pressure p = 1 MPa, for the improved variant the signal

ε_r = 5200 μm/m, is 2.34 times
greater than for the initial
transducer.

REFERENCES

1. Avril, J., Encyclopedie d'analyse des contraintes,
   Micromesures, Malakoff 1984
2. Bathe, K.J., Finite Elemente
   Methoden, Springer-Verlag 1990
3. Constantinescu,I.N., Stefanescu
   D.M., Sandu, M.Al., Masurarea
   marimilor mecanice cu ajutorul
   tensometriei, Ed.tehn., Bucharest '89
4. Perry, C.C., Lissner, H.R., The
   Strain Gage Primer, McGraw Hill,
   New York 1962
5. Rohrbach, C., Handbuch für elek-
   trisches Messen mechanischer
   Grössen, VDI-Verlag 1967

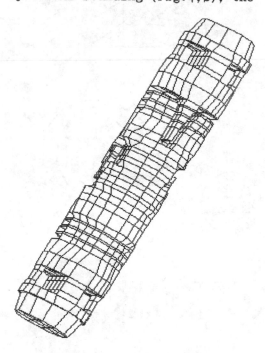

Fig.3. FEM computation model

Recent Advances in Experimental Mechanics, Silva Gomes et al. (eds) © 1994 Balkema, Rotterdam, ISBN 90 5410 395 7

Modern trends in mechanical testing of materials and components

Erhard Bauerfeind
MTS Systems GmbH, Berlin, Germany

ABSTRACT: The forthcoming years and decades will be characterized by extensive search for new technologies in the development of materials and their economical use in design and manufacturing. Innovative, especially man-made, materials and the ecological implications of their application will influence the competitiveness of our research activities and our industries.
State of the art testing and simulation technology will be one of the most important prerequisites to obtain the necessary information for effective implementation of these new materials in new design concepts.
This presentation is intended to provide some insight into general requirements for equipment which can meet present and future needs in the area of mechanical testing of materials, components and structures.

1 DEFINITION OF MECHANICAL TESTING

There is an extremely large range of destructive testing methods. Consequently a large variety of instruments and systems has been developed and is offered in the market. (The term "destructive", however, is sometimes misleading. In a more generic sense, "mechanical testing" might be a better description of the nature of the investigation in this field, since the destruction, i.e. the mechanical separation of the test piece, in many cases, is not the aim of the tests.)
In any case, however, loading (or stressing) is the primary action and the reaction of the object to this is observed and measured with the aim to detect physical relationships among the technically relevant parameters and to draw conclusions and/or to verify whether the criteria for a certain application are being met.

One way to classify the various mechanical testing methods is the sequence in which the test piece is stressed related to time. By this, one can distinguish among: Static, monotonic and dynamic procedures. The latter in particular have gained increased importance during the second half of this century.
The number of applied loads/stresses can further be used to describe the methodology, f.e. uniaxial, biaxial etc.. There is a pronounced tendency towards "multiaxiality" in view of a "closer to real life" way of research and development. (In these cases, the term "multichannel systems" is frequently used.)

Such grouping of methods is, however, not at all sufficient to describe the related test equipment. Additional criteria need to be used - a problem for both the suppliers and the users.
Except for the few cases, where a so called "universal test machine" can be used effectively and efficiently, today's requirements are generally very specific, so that very close interaction between user and supplier has become an essential factor for the success of both. This has lead to specialisation of the individual test equipment manufacturers either as a company in total or within these companies. In the case of my company, MTS, such specialisation in the form of highly focused divisions serving the different industries has been the basis for continued growth over the last three decades.
This broad spectrum of technologies used in mechanical testing presents a problem also for structuring a lecture like this. I must, therefore, limit my presentation and will concentrate on dynamic testing for the most part.

2 MAIN ACTIVITIES DURING THE LAST DECADE

Research and development on new materials is one of the main areas of activities since the beginning of the 80´s, in view of obtaining fundamental knowledge about their behavior in general. Application oriented tests and the verification of concepts (both for production and design), and in particular the simulation of real life conditions are and will be the scope of work, especially related to components and complex structures.
Driven by this, the requirements for testing systems has increased considerably. Just to name the most important: Measuring and control accuracy,

operating frequency, superposition of applied loads/strains, data acquisition and data analysis, ease of use, availability and safety.

Important impulses originate from the aircraft and aerospace industry, the ground vehicles industry and the energy and power generation market. The continual striving for shorter development and production cycles combined with increased awareness of ecological problems, safety and liability issues, requires that in the future, materials and products need to be tested faster more reliably and more intelligently.

By using examples of recently built and successfully delivered systems, I will try to give a comprehensive overview of the present state of mechanical test systems and attempt to envision future trends in this field.

3 EXAMPLES

3.1 Testing of Elastomers and Elastomeric Components

Modern vehicles incorporate a great number of critical parts made in part or in total of elastomers (e.g. engine mounts, parts of the suspension and drive train, tires, rail fixtures, dampers). For their exact characterization, selection and final tuning servohydraulic test systems are being used in many different versions. Their performance has had to be increased to levels unthought of ten years ago in order to reproduce conditions that the parts are subjected to in practice. The operating frequency of those test machines, for instance, was pushed to 1000 Hz, which is more than twice the level that could be handled before.

Fig. 1 shows an elastomer system for axial-torsional testing under simulated environmental conditions. It certainly can give an idea about the amount of hardware needed to meet above mentioned testing needs.

With this system, axial forces of +/- 10 kN can be applied within a frequency range up to 400 Hz. Torques can be excerted up to +/- 225 Nm up to 100 Hz. Within these limits every combination of loading (individual or combined) can be reached.

Complex tests of this nature are almost impossible to execute without the aid of modern digital controls and specific application software. Details of this aspect will be addressed later.

3.2 Thermo-Mechanical Investigations on High Temperature Alloys

An efficient design of the essential parts of for example turbine blades can only be undertaken if the behavior of the selected materials under the relevant conditions are sufficiently known. To simulate these conditions and to analyse the governing parameters requires multiaxial test equipment with an extremely high degree of precision in the area of controls and sensors. Such test systems must allow for accurate phase control of stress and strain channels to study, f.e., the effects of strain hardening/softening, the effects of different, so called, strain paths and the results on the fatigue and fracture behavior of such materials. All this is done to improve the efficiency of turbines by using alloys that can operate at higher temperatures.

Fig. 2 presents a total view of such a system used at the Bundesanstalt für Materialforschung und -prüfung (BAM), Berlin, for basic and applied research.

This servohydraulic multiaxial system has a +/- 250 kN dynamic axial and a 2260 Nm dynamic torsional capacity. Simultaneously hollow cylindrical specimens can be charged with internal pressure of max. 210 bar by means of a servocontrolled pressure intensifier which uses inert gas to pressurize the specimen. An induction heater can bring the temperature of the test piece up to 1100 °C. All these parameters are separately commanded and controlled.

It is obvious that for a system like this all the complex interactions during the course of such tests are almost impossible to manage without adequate assistance from computer hard- and software.

3.3 Large Scale Tests on Specimens and Components

Experience shows that results obtained from small specimens or small scale models cannot be applied in all cases to sizes used in practice. It is, therefore, necessary that large scale tests are performed, either to verify the validity of extrapolation concepts or to perform proof tests during the development phase of components and structures (f.e in the areas of power plants, mining, off-shore constructions, aircraft and aerospace, building engineering).

In these cases, forces in the Meganewton range are required. If the additional desire exists to test dynamically in reasonable time, those systems would require power supplies of a size that could be justified only in exceptional cases.

For those applications, the resonance-concept can be an ideal solution. For servohydraulic test machines, this was developed in the early 60's and was frequently used during the following two decades but neglected during the last ten years.

Quite a few experts claim that this technology receive re-newed attention since it provides ideal opportunities for fatigue testing at high force levels with minimal energy input.

Fig. 3 is a partial view of the largest dynamic "Universal Test System" used at the BAM, Berlin. It was designed for axial forces up to +/- 20 MN and has a hydraulic resonant option which allows dynamic tests up to approx. 30 Hz (depending on the combined resonant behavior of system and specimen).

The principle applied here is that of the so-called "hydraulic lever", invented by Russenberger and further developed by MTS, which reduces the amount of energy input to 1/10 - 1/50 of what would be required under "normal" servohydraulic closed loop conditions.

This system has been successfully used since it was installed about 15 years ago. Major test programs included fracture mechanics studies on wide plate specimen (materials for pressure vessels and tubes) and on large tube connectors for off-shore pipes.

3.4 Materials for Very Low Temperature

Exploration and usage of alternate sources of energy will (have to) gain more and more importance, such as the generation of hydrogen using solar techniques. In that context, elements for storage and transport of liquified gases will be needed and will have to be safely designed, manufactured and maintained. Materials for such applications have to be selected and carefully tested to determine their behavior under the condition of extremely low temperatures (near 0 °K).

A test system, capable of performing monotonic (tension, compression) and dynamic (low-cycle and high-cycle fatigue, fracture mechanics) tests, is shown in Fig 4. It was designed based on the specifications of MPA Stuttgart. Its main test frame is rated for +/- 250 kN axial force, the secondary frame - reaching into the cryo dewar - can bear upto +/- 150 kN. The entire system was designed such that liquid hydrogen can be used as a cooling medium contacting the specimen.

3.5 Fatigue Life Simulation on Vehicle Components

It has already been mentioned how important simulation is for the entire complex research - development - engineering/design - production. Simulation tests are considered more an integral part of this chain, especially because complete and reliable theories for fatigue life prediction do not exist. In particular, there is a need for more research in the field of multiaxial fatigue under random loading conditions. Testing of materials and components under those conditions will play an important role in the future.

Fig. 5 is a typical example . It shows a test rig for testing a passenger car subframe to determine fatigue life. Using commercially available components, like a servohydraulic actuator connected by swivel joint to a reacting bracket, a support frame for the structure to be tested. A T-slotted plate is the main mounting surface and becomes an integral part to react the applied forces. In this case, the researchers from Gesamthochschule Paderborn, were interested in an experimental stress/strain analysis in the context of a comprehensive study of problems associated with the welding of thin steel sheet. The resulting stress/strain fields were sensed by strain gauges.

3.6 Durability Testing of Automobiles

One of the most interesting and most challenging tasks is the simulation of real life loading conditions on complete vehicles. Most of the design verifications nowadays are being performed in the laboratory using so-called "road simulators".

The following photo (Fig. 6) shows a simulator during the final tuning in our check-out area in Berlin. In this case it is a test system with 18 channels. Four channels are connected to each wheel (to introduce vertical, longitudinal, lateral motions and braking moments). Two additional channels can restrain the vehicle in a controlled manner to better reproduce larger accelerations/ decelerations.

In most cases, it is desired to reproduce signals measured in the field at critical points of the vehicle. The principal problem with this is that the exact command signals for the individual actuators are not known. Using iterative techniques, the test engineers first have to establish the response characteristics of the entire test arrangement before the real simulation test can take place. All these processes require powerful computers and sophisticated software to work real time with the test rig, the related controllers and the many sensors in both the simulator and the test piece. MTS has been a forerunner in developing such complex systems since many years. The computer programs needed for such tests are known worldwide under their trade name "RPC" (Remote Parameter Control).

4 CONTROL ELECTRONICS AND COMPUTER AUTOMATION

Over the last 20 to 30 years the development of the mechanical and hydraulical parts of test systems was relatively steady, with a few exceptions. Control electronics and automation (and to some part sensors), however, followed the general trend in industry, which, in this field, can best be described as revolutionary, expressed by "quantum leaps" development.

Fig 7 shows a servohydraulic test system used for a fracture mechanics tests in 1967. The control electronics occupy several large cabinets. The great number of components and the complexity of their arrangement, gives an idea of the expertise required from the staff that operated such a system. There was only a rudimentary step towards automation present at that time.

A control and automation system of today, with incomparably better performance, finds its place on a small desk and requires no more than an average knowledge in PC technology. Figures 8 and 9 show this remarkable difference in size. They speak for themselves.

The aspect of simplicity of use (intuitive learning) and the quality of the individual application programs will gain more and more importance in the future. Apart from the initial investment for the purchase of testing systems, it is the daily investment in personnel costs that determine the expenses for the individual tests to be performed. To utilise the possibilities that information technology offers, and will offer, to continuously improve the software and simplify its use is one of the biggest challenges for test machine manufacturers.

Another good example of a synthesis between greatest possible flexibility and simplicity of use is given in Fig. 10. An industry standard personal computer with a few special boards integrated is the complete control and automation center for an electromechanical test machine. Systems of this kind are typically used for monotonic tests (in tension, compression, bending, shear etc.) - to a large extent in routine quality control. Therefore, it is expected that these systems perform tests according to national and international standards precisely and efficiently with detailed documentation of the test results; however, without requiring specialised staff.

The tools that software development engineers can utilise today allow the creation of test machine software that meet such expectations. Those tools include "Window" - techniques, "Point and Click" interactions, "Pull Down Menus" and many more. Combined with modern data bank managing programs, spreadsheet software and powerful graphics packages, the user interfaces can be created such that intuitive operation of even very demanding test systems is not out of reach any longer.

With such systems, further peripheral equipment necessary in the past (like numerical or analog indicators, recorders, oscilloscopes etc.) does not need to be acquired, since these functions should be included in any viable test system software and can then be displayed on the terminal or documented by means of a printer/plotter at any time, if required.

The task to master complex systems, like the road simulator described above and shown in Fig. 6, is, however, still beyond the capabilities of today's even most powerful PCs.

This is the domain of the workstation computers. But again, it is amazing how little space is required for all the controls in such case, compared to for example 1980. At that time, a whole office was needed to house everything, today an ordinary control console can take the hardware, despite of the improved performance. (Fig 11). The test engineer has full control over such a simulator and over the tests to be performed via a screen terminal, a mouse and/or a keyboard.

Despite of all the positive aspects of fully software controlled test set-ups, there are cases, where a complete step towards computer automation might be too early, too expensive or not appropriate for other reasons. This could hold true for example for simple cyclic tests on specimens or components without the need for sophisticated data acquisition and data analysis. Adequate solutions for these requirements need to be available.

Fig. 12 shows a control unit that contains all the modules required to command and control a servohydraulic actuator (either in a load frame or in a test rig). It includes: Programmable function generator, transducer conditioner, servocontroller, hydraulic controls, limit detectors. All the important functions and data can be displayed on an LCD screen. The functions are supervised by internal software routines, which also lead the operator through the various steps required to set up a test, to inspect its progress and to watch the integrity of the data.

In many cases, such a control unit is the first step into servohydraulic testing technology. It is also often a very cost effective way to modernise existing equipment.

5 CONCLUSIONS

Mechanical testing will be very important in the future, in my opinion, as important as it is today.

Under the general requirement for more productive, more effective and more efficient execution, there a two governing trends:

1. Complex test- and simulation technology will have to support the unavoidable search for materials and products that save natural resources and energy and that meet more stringent ecological and economical criteria.

2. Tighter safety and product liability requirements will force suppliers on all levels to provide quality proofs. The consequence will be that mechanical testing (amongst many other aspects) will have to be an integral part of every service/production.

At the same time, the amount of documentation will increase as well as the national and international connection between companies, institutions, surveillance organisations etc.. The era of information and information processing will influence the field of mechanical testing to a much greater extent than in today's world.

This presentation was intended to provide a brief impression about the present state of the mechanical testing technology. Some of the present important aspects and some of the future trend have, hopefully, been displayed using examples of typical equipment.

Fig. 1   Servohydraulic axial/torsional test system for the characterization of elastomers and elastomeric components (Photo: MTS)

Fig. 2   Axial/torisonal/internal pressure test system for the investiation of high temperature alloys. (Photo: Courtesy BAM, Berlin)

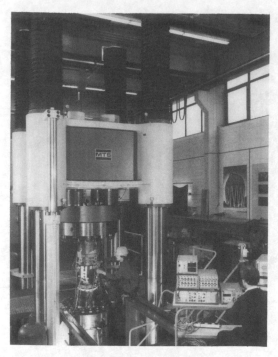

Fig. 3 Servohydraulic universal test system +/- 20 MN with hydraulic resonance option
(Photo: Courtesy BAM, Berlin)

Fig. 4 Test system for static and dynamic material characterization at cryogenic temperatures
(Photo: Courtesy MPA Stuttgart)

Fig. 5   Test rig for fatigue life simulation on vehicle components
(Photo: Courtesy Gesamthochschule Paderborn)

Fig. 6   18-channel road simulator (Photo: Courtesy Mercedes Benz AG)

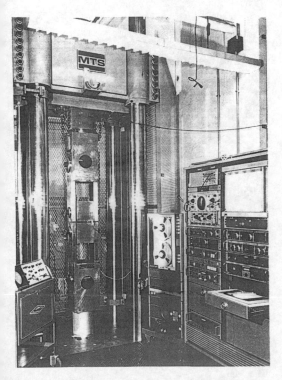

Fig. 7   1967 servohydraulic test system, 2000 kN, for fracture mechanics studies

Fig. 8   MTS TestStar digital control and automation system

Fig. 9 MTS TestStar controller chassis with hardware modules

Fig. 11 Terminal and interface for road simulators with RPC-software (Photo: MTS)

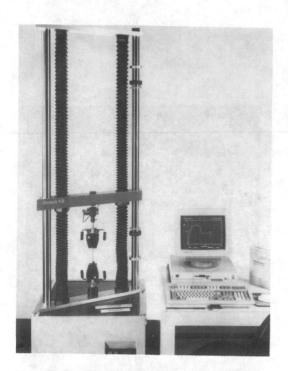

Fig. 10 Electromechanical universal test system, 5 kN, with PC control and automation (Photo: MTS)

Fig. 12 MTS 407 compact controller for servohydraulic actuators or systems

Recent Advances in Experimental Mechanics, Silva Gomes et al. (eds) © 1994 Balkema, Rotterdam, ISBN 90 5410 395 7

Forming limit diagrams for modern Al and Al-Li alloys and MMCs

N. P. Andrianopoulos & S. K. Kourkoulis
National Technical University of Athens, Department of Engineering Science, Section of Mechanics, Zografou, Athens, Greece

ABSTRACT: An attempt is described to predict the limiting strains causing failure in materials subjected to metalforming processes. It is based on the T-criterion of failure according to which a material fails when its capacity to change either its volume or shape is surpasssed because of high external loads. Four materials, i.e. 8090 Al alloy and 2124 Al-Li alloy and their respective Metal Matrix Composites (MMCs) are used to check the validity of this approach. Experimental results show excellent agreement with theoretically constructed Forming Limit Diagrams (FLDs). In addition some comments are made concerning the strain nomenclature in FLDs which enlight intriguing conclusions on the behaviour of materials under biaxial tension.

1 INTRODUCTION

Geometrically there exist only two modes of deformation of a material under external loading. These are volume and/or shape changes. The amount of volume or shape change depends on material properties, specimen geometry and type of loading. However, neither of these two changes can be infinite. Otherwise, certain combinations of material properties and geometry could resist to infinite loads without failure, a situation too desirable to be true. Consequently, both modes of deformation must have an upper limit, which when surpassed failure is caused. In strain energy density terms, the two limits of deformation modes can be described as a maximum value of dilational strain energy density, $T_{v,o}$, for volume changes and a respective value $T_{d,o}$, of distortional strain energy density for shape changes. Both quantities $T_{v,o}$ and $T_{d,o}$ are, by assumption, considered as material properties independent of specimen geometry. Consequently, failure of a material can be described in terms of these two characteristic quantities $T_{v,o}$ and $T_{d,o}$, at least in case of isotropic, homogeneous materials satisfying the Mises yield condition.

This is exactly the basis of the generalized T-criterion of failure (Andrianopoulos and Boulougouris 1990) initially introduced for the description of crack initiation conditions (Theocaris and Andrianopoulos 1982). The main advantage of the T-criterion, beyond its good agreement with experimental evidence, is that it is based on two failure parameters with clear physical content, which are independent from geometry, in contrast to other criteria based on critical stress intensity factors (Andrianopoulos and Boulougouris 1993).

The experimental determination of $T_{v,o}$ and $T_{d,o}$ in case of Mises materials is based on two independent experiments. The first is a torsion one where, by definition $T_{v,o}=0$ and, thus, the total elastic strain energy density at the moment of failure equals to $T_{d,o}$. The second experiment is equal hydrostatic tension where, by definition, $T_{d,o}=0$ and thus, elastic strain energy density at failure equals to $T_{v,o}$. The latter experiment, being practically impossible, can be replaced by a simple tension one, provided that at failure in tension, T_d is smaller than $T_{d,o}$ (Andrianopoulos and Atkins 1992).

This experimental procedure is applied in case of two modern Al and Al-Li alloys and their respective metal matrix composites (MMCs), used in aircraft industry. The critical values of $T_{v,o}$ and $T_{d,o}$ for these four materials are evaluated and they are used for the construction of their Forming Limit Diagrams (FLDs) according to the T-criterion (Andrianopoulos 1993). Finally, comparison is made between the so obtained theoretical predictions and results obtained from a series of in-plane and out-of-plane metalforming experiments for the same materials.

2. EXPERIMENTAL PROCEDURE

2.1 *Tension and torsion experiments*

All specimens after machining were heat treated at 530 $^{\circ}$C for 90 minutes and tested within two hours after oil quenching. Tension specimens of circular cross section with diameter D=0.008m were cut at different directions from the same sheet of material. Mechanical properties measured at specimens of different directions, showed a negligible directional sensitivity and, thus, all the four materials are considered as isotropic. Furthermore, preliminary compression experiments with carefully lubricated contact surfaces showed that first yield stress in compression was, within experimental errors, of the same magnitude as in tension and, so, these materials can be considered as satisfying the Mises yield condition. None of these materials showed localized necking and diffuse necking being insignificant, allowed the assumption of uniaxiality up to failure. Hence, equivalent Mises stress, $\bar{\sigma}$, coincides with good accuracy to uniaxial tensile stress, $\sigma 1$. Similarly, equivalent Mises strain, $\bar{\varepsilon}$, in elastic region equals to:

$$\bar{\varepsilon} = \frac{2(1+v)}{3} \varepsilon_1 \qquad (1)$$

where v is Poisson's ratio of the material ($v \approx 0.275$ for all four materials under study). After yielding is Eq.(1) is accurately approximated by $\bar{\varepsilon} = \varepsilon 1$.

In torsion, two different experimental methods were adopted. The first of them was that of differential specimen diameter (Fields and Backofen 1957), according to which two solid cylindrical specimens of slightly different diameters and identical gauge lengths are used to simulate a thin walled specimen of thickness equal to the difference of solid specimens radii. From a pair of such experiments, values of equivalent stress, $\bar{\sigma}$ and equivalent strain, $\bar{\varepsilon}$, are obtained through the following simple relations:

$$\bar{\sigma} = \frac{3\sqrt{3}\,(M_1-M_2)}{2(r_1^3-r_2^3)}, \quad \bar{\varepsilon} = \frac{(r_1+r_2)}{2\sqrt{3}} \frac{\varphi}{L} \qquad (2)$$

where $(M1,M2)$ are the twisting torque for the thick and thin specimens respectively, $(r1,r2)$ their radii, L the gauge length and φ the twisting angle.

The second method uses one solid cylindrical specimen of radius r (Nadai 1950). Then the required pair $(\bar{\sigma},\bar{\varepsilon})$ is given by:

$$\bar{\sigma} = \frac{\sqrt{3}}{2\pi r^3}\left(M+\varphi\frac{dM}{d\varphi}\right), \quad \bar{\varepsilon} = \frac{r}{\sqrt{3}} \frac{\varphi}{L} \qquad (3)$$

where $dM/d\varphi$ was numerically evaluated from the experimental M-φ curves.

It was concluded that results from the second method were, always, between those of the first one for various pairs (r1,r2). However, the first method requires small differences between r1 and r2 in order to reduce theoretical errors which, in turn, reduce or even change the sign of (M1-M2) arriving, at certain occasions, to unacceptable results. The instability of the first method along with the cost effectiveness of the second in terms of material, drove us to adopt, finally, the second one with specimen dimensions (rxL)=(0.008x0.004)m2.

Experimental pairs ($\bar{\sigma}$-$\bar{\varepsilon}$) from both tension and torsion experiments are pasted together for each material in Fig.1, where characteristic points are, also, indicated. In Table 1, numerical values are given, as well. Each curve after a linear elastic part, continues with a non-linear elastic

Table 1. Mechanical properties of materials

Material → Property ↓	8090 ALLOY	MMC	2124 ALLOY	MMC
Elastic Modulus E (GPa)	92.50	114.0	80.00	109.6
Linearity Limit $\bar{\sigma}_1$ (MPa)	160.0	225.0	180.0	290.0
Elasticity Limit $\bar{\sigma}_y$ (MPa)	295.0	370.0	272.0	392.0
Tension Strength $\bar{\sigma}$ (MPa)	428.0	575.0	435.0	622.0
Torsion Strength $\bar{\sigma}$ (MPa)	554.0	698.0	505.0	760.0
Tension Strain $\bar{\varepsilon}$ (Strains)	0.045	0.040	0.067	0.049
Torsion Strain $\bar{\varepsilon}$ (Strains)	0.172	0.130	0.186	0.145
Dilatation Limit Tv,o (MPa)	0.990	0.890	1.120	1.095
Distortion Limit Td,o (MPa)	10.47	7.890	9.120	9.660
Constant a (MPa)	704.2	886.2	635.7	981.5
Constant b (MPa)	86.90	97.10	74.80	116.1

part and completes with a final elastic-plastic one. For future use, a least squares method was applied for the algebraic description of these curves. It was concluded that an equation of the form:

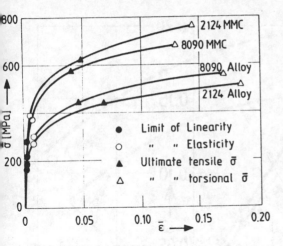

Fig.1 Equivalent stress-strain relations
for the four materials.

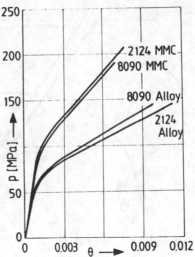

Fig.2 Pressure-dilatation relations for
the four materials.

$$\bar{\sigma} = \begin{cases} E\bar{\epsilon}, & \bar{\sigma} \leq \sigma_y \\ a+b\ln\bar{\epsilon}, & \bar{\sigma} \geq \sigma_y \end{cases} \qquad (4)$$

was the most adequate to describe all four
materials. Values for (a,b) are given in
Table 1.

An iterative procedure based on Flow
Theory of Plasticity was, then, applied in
order to evaluate stresses / strains along
the loading path given by Eq.(4) (Andrianopoulos 1993). By means of this numerical
method, the curves shown in Fig.2 were
plotted. These curves describe the relationship between hydrostatic pressure, p, and
unit volume change, θ, in tension. Obviously, for a Mises material in torsion, p-θ
curve is represented by the single point
(p,θ)=(0,0), apart from a small variation
caused by parasitic axial forces.

Finally, the required values of Td,o and
Tv,o were evaluated numerically through the
relations:

$$Td,o = \int_0^{\bar{\epsilon}_{e,f}} \bar{\sigma}d\bar{\epsilon}_o, \quad Tv,o = \int_0^{\theta_f} pd\theta \qquad (5)$$

where $\bar{\epsilon}_e$ is the elastic part of equivalent
strain, $\bar{\epsilon}_{e,f}$ is its final value at failure
under tension and θf is the unit volume
change at failure in tension. Both $\bar{\epsilon}_{e,f}$ and
θf were computed numerically by means of
the same iterative procedure previously
mentioned (Andrianopoulos 1993). Values of
the critical quantities Tv,o and Td,o are
given in Table 1.

2.2 Theoretical Construction of FLDs

Maximum deformations in metalforming processes are usually described in terms of
pairs of extreme strains (ε1,ε2) attainable
by the material without macroscopic crack
nucleation. Most of such processes are considered as satisfying generalized plane
stress conditions where, however, three
strain components are present. The problem
of strains-nomenclature arising, is faced
in case of sheet forming by disregarding
the through the sheet-thickness strain, a
decision supported by practical reasons,
only. However, experimental evidence for
many materials under various forming processes indicates that limiting in-plane
strains follow a straight line with slope
dε1/dε2 approximately equal either to -1/2
or to -1 (Atkins and Mai 1985). This apparently unstable limiting behaviour of materials gave rise to the development of a
huge number of empirical criteria.

However, according to the T-criterion
there is not any instability of this type.
Materials fail because their capacity either for shape changes or for volume changes is surpassed. In the former case limiting strains belong to the line with slope
equal to -1/2 and in the latter one to the
second with slope equal to -1 (Andrianopoulos 1993).

The requested two limiting lines can be
drawn through the satisfaction of Eqs.(5)
under plane stress conditions. The same
abovementioned numerical procedure is used
and triplets (ε1,ε2,ε3) satisfying Eqs.(5)

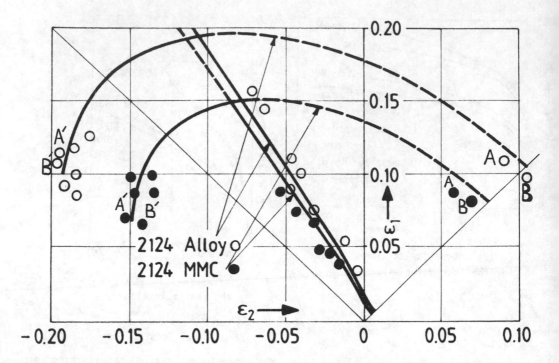

Fig.3: Forming Limit Diagrams for 2124 Al-Li alloy and MMC and experimental results.

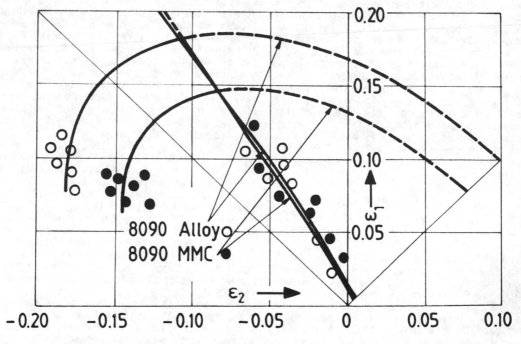

Fig. 4: Forming Limit Diagrams for 8090 Al alloy and MMC and experimental results.

are obtained. Then maximum, $\varepsilon 1$, and mini-mum, $\varepsilon 3$, strains are used to form curves shown in Figs.3 and 4. In these figures $\varepsilon 3$ is renamed to $\varepsilon 2$ for compatibility reasons to existing FLDs. Strain pairs satisfying the first of Eqs.(5) belong to an ellipti-cal arc (part of the Mises ellipse) which for $\varepsilon 1=0$ has a slope $d\varepsilon 1/d\varepsilon 2=-1/2$, coincid-ing with the slope of the first of the two experimental lines. The whole Mises ellipse in the plane $(\varepsilon 1,\varepsilon 2)$ can be obtained by cy-clic permutation of $\varepsilon 1$, $\varepsilon 2$ and $\varepsilon 3$. Limiting strain pairs $(\varepsilon 1,\varepsilon 2)$ satisfying the second of Eqs.(5) belong to a roughly straight li-ne a little steeper than the experimental one i.e. with slope equal to -1.2 up to -1.5 varying with Poisson's ratio.

The relative position of the two limit-ing curves in the plane $(\varepsilon 1,\varepsilon 2)$ depends on the specific values of Td,o and Tv,o. In any case, the combined failure curve is closed either by points satisfying the first of Eqs.(5) or points satisfying the second of the same equations. In this way theoretical predictions according to the T-criterion were obtained in Fig.3 for 2124 Al-Li alloy and the respective MMC and in Fig.4 for 8090 Al alloy and its MMC.

2.3 *Experimental Construction of FLDs*

To see whether or not the above predictions are experimentally supported, two additio-nal series of experiments for each material were performed. The procedure is similar to the one described in the pioneering work of Keeler (Keeler 1965, Keeler and Backofen 1963). According to this method the ducti-lity of metal sheets under stretching is studied by measuring the deformation of a grid, previously marked on the surfaces of the test sheets,to get the limiting strains in the plane of major and minor in-plane strains, along different strain paths. The off-plane third strain is obtained by a-posteriori measuring the thickness of the specimen.

Two types of punch stretching techniques were used in the present study, i.e. in-plane and out-of-plane. In the out-of-plane stretching test, the clamped test pieces are stretched directly over the hemispheri-cal punch. By changing the width of the test pieces and the frictional conditions between the punch and the specimen, diffe-rent strain paths can be obtained. Especi-ally for obtaining the equi-biaxial strain-ing, however, a frictionless interface is required. To achieve this a piece of PTFE film of thickness 0.0001m was placed bet-ween the punch and the specimen.

During the in-plane stretching process, the punch does not contact directly the specimen. A driving plate, with a hole in its centre, is placed between the punch and the specimen causing different forming speeds between the two plates. The diffe-rence in speeds and the friction between the two plates produces the stretching de-formation in the central part of the speci-men. The driving plates used were 0.001m thick ductile steel sheets. Specimens were circular disks with thickness 0.0012m for 2124 Al-Li alloy and 8090 alloy and 0.0009m for their respective MMCs. All experiments were carried out at a stretching speed of 5mm/min.

The experimental points obtained are included in Fig.3 for 2124 Al-Li alloy and MMC and in Fig.4 for 8090 alloy and MMC. Some of them are very close to the respe-ctive theoretically obtained steep line of dilatational strain energy density equal to Tv,o according to the T-criterion. The rest follow with equally satisfactory agreement the elliptical arc of elastic distortional strain energy density equal to Td,o. Conse-quently, the whole range of theoretical predictions of the T-criterion, concerning limiting behaviour of materials subjected to metalforming processes, is clearly sup-ported by experimental evidence.

However, a remark related to the problem of strain nomenclature is worth mentioning here. Really, in Fig.3 and refering to 2124 MMC, points A and B corresponding to in-plane limiting strains lie outside the fai-lure locus for this material. Failure locus in the first quadrant is limited by the line $Tv=Tv,o$ and apparently points like A,B violate the predictions of T-criterion. But, if extreme instead of in-plane strains are considered, then these points are tran-slated to A' and B' respectively. These latter points clearly belong to the elli-ptical arc of constant Td. For such points the algebraically minimum strain is measur-ed along the specimen thickness and not in its plane. This observation resolves the, at least intuitively, inconvenient experi-mental conclusion that, in strain - terms, a material can cope with higher equal bia-xial tensile strains than a uniaxial one. This false conclusion is, simply, the re-sult of considering the intermediate strain in place of the minimum one.

3. CONCLUSIONS

The main idea smouldering under the present approach to materials failure is that this phenomenon obeys to general laws containing easily recognizable and physically inter-pretable parameters which must be part of the identity of materials but not of the geometry of the specimen under consideration. In that sense, empirical metalforming criteria must be, a priori, rejected, not only because their physical content is dark (if existing at all) but for an additional reason, also. The lack of physical meaning of a statement prevents rationally its general application to cases not experimentally tested previously. But if experimental data are necessary to describe the limiting behaviour of a material under a "new" loading or geometry configuration then, where is the necessity to know a priori its behaviour under "other" configurations? This is the reason of asking for general approaches to failure of materials. It seems that such a general approach is possible at least in case of friendly isotropic, homogeneous, Mises materials.

It was indicated that well-known and generally accepted fundamental geometrical reactions of materials to external stimuli (i.e. volume and/or shape changes) can be used to describe their failure behaviour in a quantitatively adequate manner. Although work to this direction is at its initial stages it seems possible that it can be expanded to cover more complicated materials exactly because of the fundamental character of the mechanical parameters (T_v,o and T_d,o) used in the present approach. Any material under any loading and geometry configuration changes volume and/or shape exclusively.

ACKNOWLEDGEMENTS

The work described here is financially supported by European Union under the contract BRITE / EURAM 4500 / 1991. This support is kindly acknowledged.

Metalforming experiments were performed by the second author at the University of Reading (UK). The help and assistance of Dr B.Dodd and the staff of the Dept. of Engineering are kindly acknowledged.

REFERENCES

Andrianopoulos, N.P. 1993. Metalforming limit diagrams according to the T-criterion. J. Mat. Proc. Technology. 39:213-226.

Andrianopoulos, N.P. & A. Atkins 1992. Experimental determination of the two failure parameters Td,o, Tv,o in mild steels according to the T-criterion. Proc. ECF9. Varna Bulgaria, p.624-629.

Andrianopoulos, N.P. & V. Boulougouris 1990. A generalization of the T-criterion of failure in case of isotropically hardening materials. Int. J. Fracture. 44:R3-R6.

Andrianopoulos, N.P. & V. Boulougouris 1993. A numerical elastic-plastic approach of the crack initiation problem. Proc. ICF8. Kiev Ukraine (in press).

Atkins, A.G. & Y.W. Mai 1985. Elastic and plastic fracture. Chichester: Ellis Horwood and Wiley.

Fields Jr, D.S. & W.A. Backofen 1957. Proc. ASTM. 57:1259-1272.

Keeler, S.P. 1965. Determination of forming limits in automotive stampings. SAE paper No 650535. p:1-9.

Keeler, S.P. & W.A. Backofen 1963. Plastic instability and fracture in sheet stretched over rigid punches. Trans. ASM. 56: 25-48.

Nadai, A. 1950. Theory of flow and fracture of solids. New York: McGrow-Hill.

Theocaris, P.S. & N.P. Andrianopoulos 1982. The Mises elastic-plastic boundary as the core region in fracture criteria. Engng. Fract. Mechanics. 16:425-432.

Recent Advances in Experimental Mechanics, Silva Gomes et al. (eds) © 1994 Balkema, Rotterdam, ISBN 90 5410 395 7

Equations of the flow curve for a strain-induced dual phase metal and an analysis of the balancing condition in stress and strain of the coexisting two phases (Results on a SUS 304 foil under tension)

A.Takimoto
Yamaguchi University, Japan

ABSTRACT: Equations for a variable volume model (VVM.) of the strain-induced martensite are proposed to express the stress-strain relation for a tensile specimen of the metastable austenite. The equations are based on stress-strain relations of austenite and martensite and the transformation strain, in addition to the relations between volume fractions of martensites and plastic strain. An analysis of the experimental stress-strain data of a SUS 304 foil under tension using the equations is reported, and a mechanically balanced condition for both phases and the other significant relations are discussed.

1. INTRODUCTION

The metastable austenitic stainless steel is known to show an unusual flow curve which is re-sulted from hardening due to the strain-induced martensite and strain-hardenings of austenite and martensite below some temperature (for example, Seetharaman 1981). The flow curve shows a more complicated form as the volume fraction of martensite increases more at the lower tem-perature deformation. The curve is controlled by stress-strain relations of austenite and marten-site and their volume fractions at a given condi-tion i.e., plastic strain and temperature, in ad-dition to the accompanied transformation strain which depends supposedly on the strain of austen-ite and the amount of the transformed marten-site. The constitutional equations including an expression for the transformation strain have not been reported yet, and here we propose them. The experimental stress-strain data of a SUS 304 foil under tension is suitably regressed by the equations and this result is modified by regress-ing again above the data with reference to the experimental phase stresses obtained by a study on the x-ray residual stress measurement carried out in parallel with this study (Takimoto 1993). We analyse the strain and stress components con-tributed by both phases to the flow curve of a dual-phase, the variations of other useful terms and a

mechanically balanced condition of both phases at a given condition using the equations with the suit-ably regressed coefficients.

2. CONSTITUTIONAL EQUATIONS FOR A VARIABLE VOLUME MODEL OF MARTENSITE

The stress-strain relation of the dual-phase alloy is complicated because the volume fractions and the mechanically balanced condition of austenite (γ) and martensite (α') change as the trans-formation proceeds while plastic strain increases. In addition, the following terms also affect it, i.e., the transformation strain and the transformation characterisity of an intermediate phase of the ϵ' martensite. Here, we firstly assume the flow curve of austenite without the strain-induced transfor-mation at the temperature range from -30 ℃ to -196 ℃ , referring to the experimental changes of a strength coefficient (K_γ), a strain-hardening ex-ponent (n_γ) and a constant (G) in a flow curve equation, $\sigma_\gamma = K_\gamma(\epsilon_\gamma + G)^{n_\gamma}$, from the experimen-tal results of this material at temperatures of 50 ℃ , 20 ℃ , 0 ℃ and the experimental decreases of these constants for a f.c.c. metal of an aluminum from -196 ℃ to 20 ℃ . The stress-strain relation of a 100 % martensite is hardly obtained in exper-iment and we suppose that it deforms plastically a very little with an equation, $\sigma_{\alpha'} = K_{\alpha'}(\epsilon_{\alpha'})^{n_{\alpha'}}$, where $K_{\alpha'}$ and $n_{\alpha'}$ are constant. Then, we construct the constitutional equations for the material as follows. The total plastic strain composes of the strains of austenite (ϵ_γ) and martensite ($\epsilon_{\alpha'}$) and the transformation strain (ϵ_s) as,

$$\epsilon_T = \epsilon_\gamma V_\gamma + \epsilon_{\alpha'} V_{\alpha'} + \epsilon_S \tag{1}$$

where V_γ , $V_{\alpha'}$: the volume fraction of each phase.

And we set hypothetically a functional relation between ϵ_T and ϵ_γ , $\epsilon_\gamma = r \cdot \epsilon_T$, where $r = 1 + X\epsilon_T^Y$, (X , Y : const.) and a functional relation between ϵ_T and $\epsilon_{\alpha'}$, $\epsilon_{\alpha'} = p \cdot \epsilon_T$, where $p = U\epsilon_T^W$, (U , W : const.). The transformation strain ϵ_S is assumed to be proportional to the product of strain of austenite and an incremental volume fraction of martensite ($dV_{\alpha'}$) transformed from austenite and is given by,

$$\epsilon_S = Z \int_0^{\epsilon_T} \epsilon_\gamma \left(\frac{dV_{\alpha'}}{d\epsilon_T} \right) d\epsilon_T \tag{2}$$

where Z : const. (Z is assumed to be 1.2 for an example).

Here, $dV_{\alpha'}/d\epsilon_T$ in the equation is given by differentiating Eq.(4) given later and ϵ_S is expressed by the next equation which is integrated by computer.

$$
\begin{aligned}
\epsilon_S = {} & Z \int_0^{\epsilon_T} \epsilon_\gamma (1 - V_{\alpha'0}) \alpha\beta n \exp(-\alpha\epsilon_T) \\
& \times \exp\{-\beta[1 - \exp(-\alpha\epsilon_T)]^n\} \\
& \times \{1 - \exp(-\alpha\epsilon_T)\}^{n-1} d\epsilon_T \tag{2 a}
\end{aligned}
$$

Similarly, the total plastic stress of the dual phase composes of the stress components of both phases which are affected by the volume fractions of two phases and strain. The ϵ' martensite will also affect the flow curve of the dual phase but it is treated basically as a quasi-two phase material to avoid further complication since its stress-strain relation is hardly predicted, and here, the volume fraction of it is included into V_γ i.e., $V_{\alpha'} + V_\gamma = 1$. And its relative volume fraction to the total plastic strain, $D(V_{\epsilon'}/\epsilon_T)$ [D: const.], is assumed to contribute the stress of the dual phase, then, the total flow stress is given by,

$$\sigma_T = \sigma_{\alpha'} V_{\alpha'} + \sigma_\gamma V_\gamma + D(V_{\epsilon'}/\epsilon_T) \tag{3}$$

in which $\sigma_{\alpha'}$ and σ_γ are given by

$$\sigma_{\alpha'} = K_{\alpha'}(p\epsilon_T)^{n_{\alpha'}} \tag{3 a}$$

$$\sigma_\gamma = K_\gamma(r\epsilon_T + G)^{n_\gamma} - H(dV_{\alpha'}/d\epsilon_T) \tag{3 b}$$

here, $-H(dV_{\alpha'}/d\epsilon_T)$ expresses relaxation in stress of austenite by transformation.

3. ANALYSIS OF EXPERIMENTAL DATA AND DISCUSSION

A rectangular specimen of 45mm length (a gage length is 30mm) × 5mm width was cut out from the rolled sheet of a SUS 304 foil (as bright annealed condition) with 20 μm thickness and the cut edges of it were carefully polished off under the running hot-water of about 65℃ . The specimen was fractured or elongated plastically to the expected strain at temperatures of 20℃ , -30℃ , -50℃ , -70℃ , -196℃ with a special jig under tension. The stress-strain data and the volume fractions of the strain-induced α' and ϵ' martensites are obtained at a given condition [by the x-ray diffraction method, Mo target, K_α line, 35kV and 20mA, the multi-peak analysis (Dickson 1969) in which the peak-intensity was analyzed both by the planimeter and Gauss-Newton methods and both of them gave the very close data. The former results are reported here.]. The tensile test is performed normally at 7 ∼ 9 steps just before the plastic instability for each temperature and each data by a specimen is examined to concide with the continuous flow curve by another specimen which was fractured at the same temperature in the preliminary test. The test is at least duplicated at a condition to confirm the reproducibility of data.

Fig.1 shows the volume fractions of the α' and ϵ' martensites with the total plastic strain and each data point corresponds to a specimen. The data in Fig.1(a) gives the volume fraction of the α' martensite ($V_{\alpha'}$) which includes an initial volume fraction of martensite ($V_{\alpha'0} \simeq 3.2\%$ at $\epsilon_T = 0$) and the solid curve is the one by the modified equation of a shear band intersection mechanism (SBIM., where α, β, n : const.) (Olson 1975),

$$
\begin{aligned}
V_{\alpha'} = {} & (1 - V_{\alpha'0}) \\
& \times [1 - \exp\{-\beta[1 - \exp(-\alpha\epsilon_T)]^n\}] \\
& + V_{\alpha'0} \tag{4}
\end{aligned}
$$

The each curve indicates the experimental data well at the four different temperatures. The data in Fig.1(b) shows the volume fraction of the ϵ' martensite ($V_{\epsilon'}$) which is unstable and the intermediate phase, transforming into the α' martensite as plastic strain increases. The solid curves in this figure are the ones by a function, $V_{\epsilon'} = A\epsilon_T^B \exp(C\epsilon_T)$, (A, B, C : const.). The curves indicate the variations of experimental data suitably.

The stress-strain data at various temperatures are given in Fig.2 in which each symbol corresponds to a data by a specimen. The data at 20 ℃ does not show a strain-induced transformation and the solid curve is given by the n-th power law. The largest strain in the figure is shown by the data at -30℃ where the larger plastic strain of austenite will be required for transformation in addition to some accompanied transformation strain. At -50℃ , the flow stress curve changes its curvature at the strain larger than about 18 % where the value of $V_{\alpha'}$ increases to be larger than about 12 % . The data at -70 ℃ shows the above trend more strongly and the flow curve appears over the previous one. The one at -196℃ shows the most characteristic variation with plastic strain and the stress at about $\epsilon_T \leq 0.12$ is seemed to be controlled by both mechanisms of strengthening due to formation of the α' martensite and strengthening due to the relative volume fraction of the ϵ' martensite ($V_{\epsilon'}/\epsilon_T$) in addition to the one due to austenite. At the larger strain than this, the strength will be predominantly controlled by the α' martensite. These experimental stress-strain data are regressed by Eqs. of VVM. and they are modified by regressing again the above data with reference to the experimental phase stresses by the x-ray study, except the one at -196℃ where the phase-stresses are not obtained experimentally. These results are shown by the solid curves. All curves in Fig.2 describe the experimental data well and the coefficients in the equations are seemed to be prop-

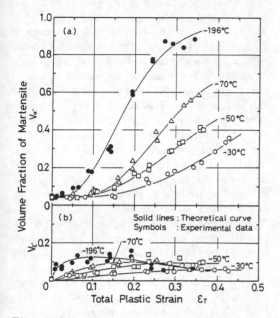

Fig.1 Volume fractions of martensites are expressed as functions of total plastic strain and test temperature.

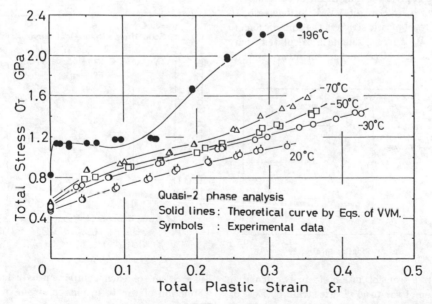

Fig.2 The total flow stress is expressed as a function of the total plastic strain and test temperature.

erly determined. These coefficients are examined to change continuously with the test temperature by expressing them as functions of temperature.

Substituting these coefficients into the various equations given so far, we analyze the strains and stresses of both phases, strength due to the ϵ' martensite $[D(V_{\epsilon'}/\epsilon_T)]$ and relaxation of stress by transformation $[-H(dV_{\alpha'}/\epsilon_T)]$, etc.. We also calculate the contributions of both phases to the total stress and plastic strain. Furthermore, we analyze a mechanically balanced condition of both phases including or excluding the transformation strain at a given condition. Some of them are given below. Fig.3 gives the relation between ϵ_S/ϵ_T and ϵ_T in which the ϵ_S value increases as plastic strain increases and as the test temperature decreases. The symbols on the curves are the data calculated from the experimental phase stresses by the x-ray residual stress measurement. The data at -30 ℃, -50 ℃, -70 ℃ give a good agreement with the theoretical curves. The curve at -196 ℃ is not be compared with the phase stress data since the x-ray residual stresses of both phases are not measured at a strain (i.e., either the stress of austenite or martensite is measured at a strain), but this curve is believed to be correct as well as the other three. The constant Z in Eq.(2) is assumed to be 1.2, here, as an example, but this may be changed with the test condition (e.g., the larger Z value may be more probable at the smaller strain at -30 ℃ test). Fig.4 gives the magnitude of each plastic strain analyzed by Eqs. of VVM. for the test at -30 ℃ . The total plastic strain is the summation of $\epsilon_\gamma \cdot V_\gamma$, $\epsilon_{\alpha'} \cdot V_{\alpha'}$ and ϵ_S. The data calculated from the x-ray residual stress measurement are shown

for each theoretical curve and it is clear that the data show a good agreement with the theoretical curves. The results at other temperatures are also analyzed in the same way and we obtain the similar relations as a matter of course. The magnitude of each stress in the total flow stress is analyzed similarly and expressed in Fig.5 for the test at -196 ℃ . The each circular symbol (○) is the experimental data obtained by a specimen and the square symbol (□) is the one taken out of a continuous load-elongation curve to fracture for another specimen under tension and the solid curve is the regressed one by Eqs. of VVM. as shown in Fig.2. The theoretical curves for $\sigma_{\alpha'} \cdot V_{\alpha'}$, $\sigma_\gamma \cdot V_\gamma$ and $D(V_{\epsilon'}/\epsilon_T)$ are given by the equations presented so far with the regressed coefficients. The stress component contributed by austenite (including the effect of the initial martensite $\simeq 3.2$ %) to σ_T starts at the yielding stress at $\epsilon_T = 0$ and increases a little, and then decreases as plastic strain increases. The effect of stress relaxation of austenite due to transformation is included in the above. The stress contribution by the strain-induced martensite starts from zero and increases rapidly with an increment of ϵ_T as shown by the curve. The volume fraction of the ϵ' martensite increases to about 15% at $\epsilon_T \simeq 0.1$ and it decreases as ϵ_T increases as in Fig.1(b) and the effect of it increases initially and it decreases gradually. All three components in

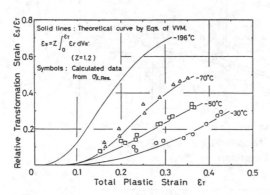

Fig.3 Relative transformation strain is calculated as a function of the total plastic strain by Eqs. of VVM. at each temperature.

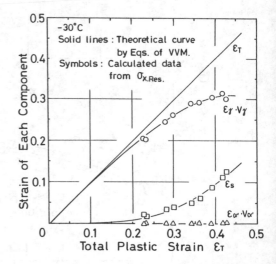

Fig.4 Strain of each component is calculated as a function of the total plastic strain by Eqs. of VVM.

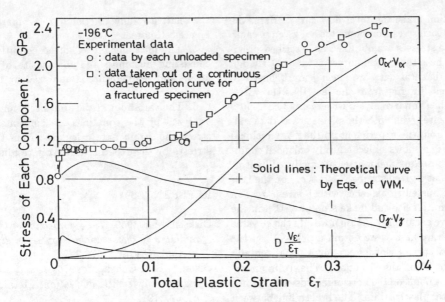

Fig.5 Stress of each component is calculated as a function of the total plastic strain by Eqs. of VVM.

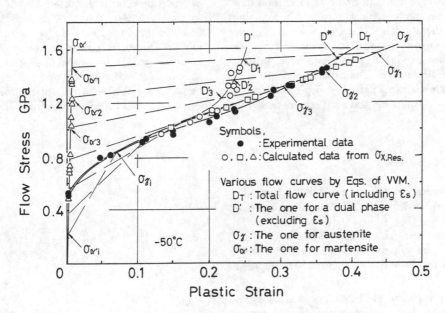

Fig.6 The mechanically balanced condition of the two phases at a given condition is analyzed by Eqs. of VVM. at -50℃ . The calculated phase stress data from the experimental x-ray residual stress are given.

summation make the total flow curve. Here, the data for each stress component is not given because no phase stresses are obtained as described previously. These theoretical curves at -196°C (though no experimental data are given) are believed to be correct, judging from the fact that the variation of the similar analyses with ϵ_T at -30°C , -50°C , -70°C agree well with the phase stress data calculated from the experiment on the x-ray residual stress. Fig.6 shows a mechanically balanced condition for austenite and martensite analyzed by Eqs. of VVM. at a given strain. A stress $\sigma_{\gamma 1}$ on the flow curve of austenite balances with a stress $\sigma_{\alpha' 1}$ on that of martensite and make a point of D'$_1$ on the flow curve excluding ϵ_S. When we add the ϵ_S value to the strain at D'$_1$, we obtain a point D* on the total flow curve (shifting a point D'$_1$ horizontally to a point D* on the σ_T curve). The strain underneath this point on the horizontal axis corresponds to its total plastic strain. The broken line between $\sigma_{\gamma 1}$ and $\sigma_{\alpha' 1}$ shows a mechanically balanced condition. The other balanced conditions at various total plastic strains are shown by the broken lines which have the different gradients. The total flow stress-strain curve is, thus, constructed from the locus of these points. The symbol on each curve shows the experimental data calculated from the phase stresses and they give a good agreement with the theoretical curves. The balancing condition of the two phases varies naturally, depending on the total plastic strain and test temperature as shown in the example.

4. CONCLUSIONS

The following conclusions are derived from the analysis of results on a SUS 304 foil under tension,

1. The stress-strain curves at -30°C , -50°C , -70°C and -196°C are well expressed by the constitutional equations for the variable volume model proposed here.

2. The variation of the transformation strain with the total plastic strain and temperature is estimated by the equations.

3. The stress and strain contributions by the coexisting austenite and martensite to the total plastic stress and strain of the dual-phase are analyzed at a given strain and temperature.

4. The terms expressing relaxation in stress of austenite by transformation and strength due to

the relative volume fraction of the ϵ' martensite to ϵ_T are estimated.

5. The mechanically balanced condition of the coexisting austenite and martensite at a given strain and temperature is analyzed by the equations.

6. The variations of the above relations with ϵ_T analyzed by the equations are confirmed by the stress and strain data calculated from the phase stresses by the experimental x-ray residual stress.

REFERENCES

Dickson, M.J. 1969. *The significance of texture parameters in phase analysis by x-ray diffraction.* J. Appl. Cryst. 2: 176.

Olson, G.B. & Cohen, M. 1975. *Kinetics of strain-induced martensitic nucleation.* Metall. Trans., A., 6A :791.

Seetharaman, V. &d Krishnan, R. 1981. *Influence of the martensitic transformation on the deformation behaviour of an AISI 316 stainless steel at low temperatures.* J. Mater. Sci., 16: 523.

Takimoto, A. 1994. *An Analysis of the Phase Stress by the X-ray Residual Stress Measurement and the Discussion on the Balancing Condition in Stress and Strain of the Coexisting Austenite and Martensite Phases under Tension.* To be presented at the Fourth International Conference on Residual Stresses, June, Baltimore, USA.

Recent Advances in Experimental Mechanics, Silva Gomes et al. (eds) © 1994 Balkema, Rotterdam, ISBN 90 5410 395 7

The electrical control of tribology characteristics

Yoshio Yamamoto
Institute of Science and Technology, Kanto Gakuin University, Yokohama, Japan

Kuniaki Iwai
Tokyo Metropolitan Institute of Technology, Japan

Susumu Takahashi
Faculty of Engineering, Kanto Gakuin University, Yokohama, Japan

ABSTRACT: In this paper we describe an experimental study on the electrical control of tribology characteristics. The relative specimens are electrical insulated from the main body of the wear tester.
The authors take note of generated induced voltage, eliminated the effect of induced voltage on wear surface were recognized indicate remarkably wear-resistance compared with other method.

1 INTRODUCTION

The wear surface between relative specimens does not merely generate heat, it also generates electrical phenomena. The authors focused their attention on this point to devise a method which control induced voltage generated artificially by the friction between different kinds of metals, and found a method to enable adjustments of the friction coefficient, wear ratio and wear energy.

Therefore, experiment with original developed device that the adjust equipment of induced voltage, the wear test equipment, the equipment of impressed magnetic field and management system of automatic calculation for combination of circumferential apparatus were obtained detailed measurement value.

In the measuring test, continuously recording the friction force and the induced voltage(AC,DC), afterward testing to do analysis of line and surface by X-ray micro-analyzer(EPMA), were valued information an atom and molecular level.

The wear surface were conceivable the contact interface of metal and semiconductor with complicated electron phenomena.

The technical method that its impressed filling other voltage from outside and controlled a flow of electricity of the wear interface were contributed to improve the tribology characteristics.

In this paper, we deal with the adjustment method of induced voltage of the sliding dry wear test, put to a test in magnetic field and non magnetic field used combination of material different conductivity were excellent results obtain publish.

2 EXPERIMENTAL METHOD

2.1 Experimental devices

The unique feature of this research is that the double test pieces are electrically insulated from the main body of tester.

Fig.1 shows the experimental equipment, adjustment circuit of induced voltage, impressed equipment of magnetic field and electrical measuring circuit of wear test. As in figure the induced voltage(AC,DC) generated between the double test piece was continuously measured.

To obtain the friction coefficient(μ), the friction velocity was changed using a variable motor and torque was measured using torque meter directly connected to the rotation axis.

The wear test was performed use two electrical circuit that the one is adjust circuit it purpose of eliminate the effect of the induced voltage generated between relative test piece an electric source of outside was impress to the wear surface and others non adjust circuit.

The various testing result were transacted and recorded by the computer through an interface.

Then performed test at impress magnetic field (AC,DC) in the direction of parallel to wear surface.

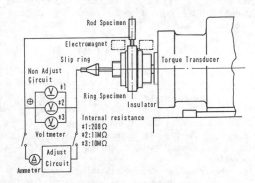

Fig.1 Experimental equipment.

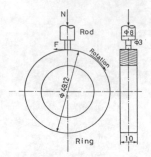

Fig.2 Dimension of the Ring and Rod test piece.

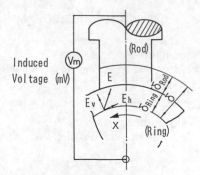

Fig.3 Electrical energy.

Table 1 Experimental condition and Relative material of Ring and Rod.

Environment	Rod Specimen	Ring Specimen	Friction Distance (m)
Atmosphere	AlB3-0	S55C carbon steel	6000
	Zn (99. 99)	S15CK carbon steel	6000
Alternating Field, B=6mT	CuB2-1/2H	S45C carbon steel	6000
Magnetic Field, B=18mT	S15C steel	S45C steel	6000
	SUJ2 steel	Cr_2O_3 coated S45C steel	12000
	SUJ2 steel	Cr_3C_2 coated S45C steel	12000

2.2 Test piece and experimental condition.

The specimen type was used the Ring-Rod type wear test piece. **Fig.2** shows the dimension of the Ring and Rod test piece. **Tab.1** shows contact surface pressure of 0.33MPa and friction distance was set to 6000m.

The experiment were performed for the non adjusted and adjusted induced voltage circuit will changing the friction velocity from 0.17,0.42,0.83 and 1.7m/s.

And **Tab.1** shows the relative material of Ring and Rod, the impress magnetic field and the adjust circuit of induced voltage.

2.3 Energy

As shown in **Fig.2**, the mechanical energy was obtained integrate friction force by friction distance.

$$\mathcal{E}_m = \int FdL = \int \mu NdL = \int \mu PSdL = PS \int \mu dL \quad (J) \quad (1)$$

F:Friction force(N),L:Friction distance(m),
μ:Friction coefficient,N:Normal force(N),
P:Surface pressure(N/m²),S:Rod section area(m²)

The component of Direct current and Alternating current of generated induced voltage on wear surface was detect value measuring at regular 50 m intervals.

The electrical energy was obtained from sum of electrical work, integrate there value by friction distance.

As shown in **Fig.3** the electric intensity of metal surface oxide was obtained by induced voltage and thickness of oxide of Ring-Rod contact interface.

$$E = V_m / \delta \quad (V/m) \quad (2)$$

The electrical energy of tangential direction on rotary Ring can be expressed equation(3).

$$\mathcal{E}_{eh} = \int eE_h dx = e E_h L \quad (J) \quad (3)$$

But, it is difficult to obtain the E_h value by theoretically and experimentally therefore, the E is will be considered to E_h, the electrical energy($\mathcal{E}_e$) was obtained by equation(4).

$$\mathcal{E}_e = e \int EdL = \int e(V/\delta)dL = (e/\delta) \int VdL \quad (J) \quad (4)$$

e:Electric charge(1.6×10^{-19}C), E:Electric field(V/m),
L:Friction distance(m), V_m:Induced voltage(V),
δ:Thickness of oxide (m)(10^{-7}m)

2.4 Adjusting method of induced voltage

In the relative motion of metal, the condition of contact interface is changing every moment by receive various types of adhesion, shearing and oxidation.

So that, its observed studying with coating of semiconductor oxide on wear surface.

Truth point of contact of wear surface were thinkable various combination of metal part and semiconductor part.

Between these micro point were generated potential difference and through electric current flow.

As a consequence, wear surface were generated magnetic field during friction.

The authors were impress from outside voltage that its reverse voltage(phase) to the induced voltage and try to made decrease the friction and wear by control method of electron movement (call as adjustment method of induced voltage).

The wear interface are electric condition that its be capable of impress voltage of positive and negative direction. Moreover, wear interface shaped

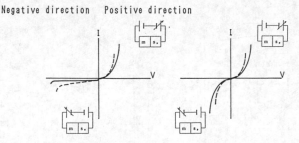

Negative direction   Positive direction

(a) φm>φsₙ, Rectify contact   (c) φm>φsₚ, Ohmic contact

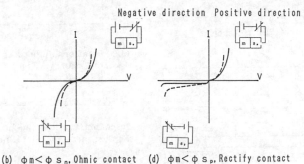

Negative direction   Positive direction

(b) φm<φsₙ, Ohmic contact   (d) φm<φsₚ, Rectify contact

(a), (b)=metal:n-type semiconductor
(c), (d)=metal:p-type semiconductor

Fig.4 Contact interface of metal and semiconductor.

coats from conductor to nonconductor. According to authors observation, oxide semiconductor is partially formed on wear interface. In the case of metal and semiconductor, band structure of semiconductor surface are changes by generated surface level.

Semiconductor to metal contact that it be decided either rectify contact on Ohmic contact depend on band structure. As shows in **Fig.4** the contact interface of metal and n-type semiconductor. When one is the work function of metal is larger than semiconductor, flows electron from semiconductor to metal side. As results of rectifying characteristics appear in near contact interface that its forming peak of potential energy. In the case of contact of negative direction, friction are raise, the surface temperature of wear interface. The carrier made due to friction temperature are increase the amplification by avalanche reaction as point out in broken line in the diagram, therefore, flows electron can be control by applying suitable external voltage. Then the work function of metal is smaller than semiconductor, flows electron from semiconductor to metal side. So Ohmic characteristics appear in near contact interface forming trough of potential energy, electron is flow in any direction. For that reason, electron flows of contact interface can be adjustment by applying suitable external voltage.

The phenomena is essential principle of keep electric energy of friction under control by adjustment of induced voltage. Moreover its contact interface of metal and p-type semiconductor, the carrier is positive hole. The peak of potential energy is non potential barrier and the trough of potential energy is potential barrier.

Consequently, the work function of metal more than p-type semiconductor is Ohmic contact and invert the work function is rectify contact, electron flow be possible control by impress other voltage from the outside. Authors have also clarified the fact that a technical method to control the surface current by applying suitable external voltage.

3 EXPERIMENTAL RESULT

3.1 Induced voltage

The contact potential difference between specimens of relative combination was measured purpose of compared with the generated induced voltage in friction motion. **Fig.5** shows the results of the investigation of contact potential difference.

In the experiment ,anode of galvanometer connect to high temperature side ,its investigated still 50 degree corresponding amount of rise in temperature between specimens surface by friction heat. **Fig.6** shows the record of carbon steel(S15C, Rod and carbon steel(S45C) Ring, copper(CuB) Rod and carbon steel(S45C) Ring from beginning to end of generated induced voltage between wear surface.

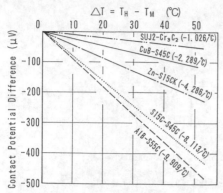

Fig.5 Contact potential difference.

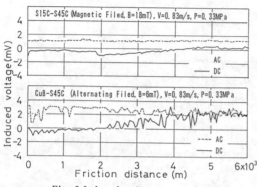

Fig. 6 Induced voltage.

The induced voltage is composed of alternating current composition is larger than direct current composition,the polarity of direct composition were indicated change polarity,straight polarity and reverse polarity.

The condition of contact interface of relative metal in sliding friction were observed changes complexity.

3.2 Mechanical energy and electrical energy

Fig.7 shows the relationship between mechanical energy and electrical energy used non adjust circuit of induced voltage generated during friction. Relate the both was obtained regression analysis according to exponential regression for four points,the result as a follows.

In combination of aluminum(AlB) and carbon steel (S55C),the electrical energy($\mathcal{E}_e$)is as follow equation(5).

$$\mathcal{E}_e = 4.1086 \times 10^{-9} \cdot \exp(-0.9288(\mathcal{E}_m/10^3)) \ (J) \qquad (5)$$

In combination of copper(CuB) and carbon steel (S45C), the electrical energy ($\mathcal{E}_e$) is as follow equation(6).

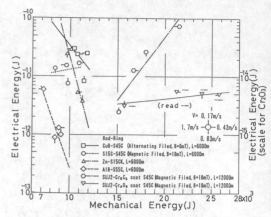

Fig.7 Relationship between mechanical energy and electrical energy.

$$\mathcal{E}_e = 366.02 \times 10^{-11} \cdot \exp(-0.470(\mathcal{E}_m/10^3)) \ (J) \qquad (6)$$

In combination of zinc(Zn) and carbon steel(S15Ck), the electrical energy ($\mathcal{E}_e$)is as follow equation(7).

$$\mathcal{E}_e = 18.73 \times 10^{-6} \cdot \exp(-1.37(\mathcal{E}_m/10^3)) \ (J) \qquad (7)$$

In combination of carbon steel(S15C) and carbon steel(S45C),the electrical energy ($\mathcal{E}_e$) is as follow equation(8).

$$\mathcal{E}_e = 6.19 \times 10^{-12} \cdot \exp(0.076(\mathcal{E}_m/10^3)) \ (J) \qquad (8)$$

In combination of bearing steel(SUJ2) and thermal spray coating(Cr$_3$C$_2$),the electrical energy ($\mathcal{E}_e$) is as follow equation(9).

$$\mathcal{E}_e = 1.9 \times 10^{-15} \cdot \exp(0.508(\mathcal{E}_m/10^3)) \ (J) \qquad (9)$$

In combination of bearing steel(SUJ2) and thermal spray coating(Cr$_2$O$_3$),the electrical energy ($\mathcal{E}_e$) is as follow equation(10).

$$\mathcal{E}_e = 1.81 \times 10^{-15} \cdot \exp(0.0413(\mathcal{E}_m/10^3)) \ (J) \qquad (10)$$

The results of measurements suggests an adequate correlation of positive and negative ,electrical energy is small compared with mechanical energy.

However,eliminate the value of electrical energy of infinitely small,its can be plan to make reduction of friction and decrease of wear.

The authors was eliminate induced voltage by impressed outside source of electricity.

3.3 The adjustment effect of induced voltage

Fig.8 shows the relationship of friction coefficient and wear ratio between specimens of relative combination. The specimens made an experiment of use adjust circuit with regard to the induced voltage generated during friction was observed decrease of

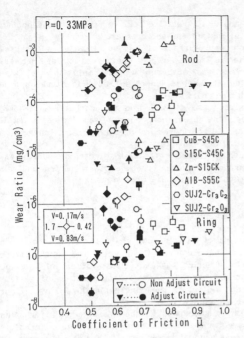

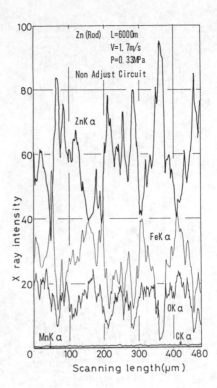

Fig.8 Relationship of friction coefficient and wear ratio.

friction coefficient and wear ratio compared with non adjust circuit.

As shown in figure,white point is indicate experimental value of non adjust circuit and black point is indicate experimental value of adjust circuit.

Then ,except on relative combination of pure Zinc and carbon steel(S15CK),aluminum(AlB) and carbon steel(S55C),other specimens were indicated the result of experimental value in magnetic field .

In magnetic field ,between the were surface was exist among powder of the contact wear out by magnetic motive force was observed a condition comparatively similar solid lubricant.

The eliminate effect of generated induced voltage by impress outside voltage was clear confirmed in magnetic field.

3.4 Surface analysis

The wear surface of sliding dry wear test in non magnetic field and magnetic field was correct observed in consequence of directness surface analysis by EPMA with move specimen in the constant velocity and fix electron probe.

Line analysis and surface analysis of eliminated induced voltage of wear surface was confirmed wide difference compare non adjustment.

For example, **Fig.9,10** shows the result of line analysis of wear surface of zinc(Zn) Rod and carbon steel (S15CK) Ring.

Rod wear surface of non adjust circuit was observed adhesive Fe and Fe_3Zn_7 and streaked thicken oxide, Ring wear surface was observed distribution thicker

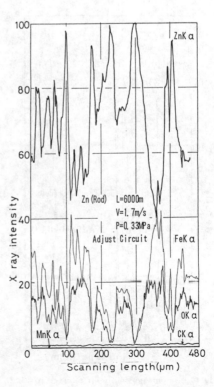

Fig.9 Line analysis in wear surrface of Rod.

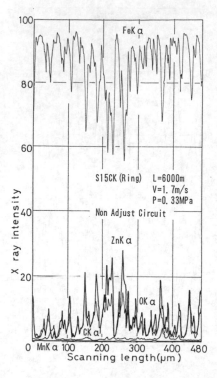

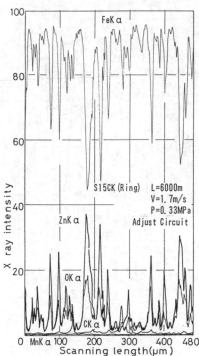

Fig.10 Line analysis in wear surface of Ring.

Zn and O, its relative wear surface showed aspect of metal contact and shiver wear.

Rod wear surface of adjust circuit was observed adhesive ZnO, Fe_3Zn_7 and coat of Fe oxide, Ring wear surface was observed uniform thin coat of zinc oxide, its showed aspect solid lubricant through an oxide and mild wear.

In the case of zinc Rod and carbon steel(S15CK) Ring, the electrical energy was indicated small value 10^{-11} degree of the mechanical energy on sliding dry wear test.

But, the control effect of electrical energy was confirmed influence the results of micro analysis component of relative wear surface.

4 CONCLUSION

The effect on tribology characteristics of electric field and magnetic field generated wear surface during friction was made an experiment of Ring-Rod type dry wear test by relative metals of various conductivity. As the conclusion the following four case are possible.

1. The component of induced voltage has alternating current and direct current composition, alternating current composition is larger than direct current, it have direct effect upon tribology.

2. The control of electric energy during friction is conformed enable to delay and improvement of wear phenomena.

3. The method of impress magnetic field combined with the adjustment method of induced voltage were recognized excellent reduction wear ratio and friction coefficient.

4. The wear surface was considered the contact surface between metal and oxide semiconductor depend on the result of analysis by electron probe micro analysis.

As a result of this experiment, the active control of tribology characteristics was confirmed to be possible realization by the electrical control.

REFERENCES

Y. Yamamoto, K, Iwai & S. Takahashi, Conf. Ad. Technol. Exp. Mech., (1993-7), 371.

Y. Yamamoto and S. Takahashi, Proceeding of The 4th Conference of Asian Pacific Congress on Strength Evaluation: 2, p1459, (1991-10).

Y. Yamamoto, K, Iwai & S. Takahashi, Proceeding of International Symposium on Nondestructive Testing & Stress-Strain Measurement, FEND92(NDT & SSM, FENDT92), (1992-10), 251.

Recent Advances in Experimental Mechanics, Silva Gomes et al. (eds) © 1994 Balkema, Rotterdam, ISBN 90 5410 395 7

Measurement of experimental factors influencing the constitutive equation of a high-strength steel

J.Toribio
ETSI Caminos, Campus de Elviña, La Coruña, Spain

ABSTRACT: This paper deals with the influence on the stress-strain curve of a prestressing steel of such geometric variables as the gauge length and the sample length, mechanical variables as the the gripping conditions and spurious variables as the superficial state of the sample. Small variations in the yield strength and the flow step were noticed when the sample length changed. Gripping conditions and superficial state may also influence the results. No significant influence of the gauge length on the stress-strain curve was detected.

1. INTRODUCTION

The stress-strain curve of a material can be obtained through a tension test according to standards (e.g. ASTM A370, ASTM E8, ASTM E111), whose specifications refer to the gripping conditions, the strain rate and the length of the sample, as well as the limit gauge length depending on the diameter of the specimen.

In a prestressing steel more difficulties arise in determining the stress-strain curve, since it is commercially supplied in the form of small diameter wire (e.g. 7 mm), wound around a coil, with certain small curvature.

Many experimental variables can influence the stress-strain curve in a prestressing steel: geometric variables as the gauge length and the sample length, mechanical variables as the gripping conditions, and spurious variables as the superficial state of the specimen. This paper analyzes their possible influence.

2. EXPERIMENTAL PROCEDURE

The material was a commercial high-strength prestressing steel supplied in wire form (7 mm diameter) around a coil.

The experimental equipment included two universal machines (one of them equipped with rigid clamps and the other with hinges) and four extensometers.

A selection was made of four gauge lengths (B = 12.5, 25, 50 and 300 mm), three sample lengths (L = 0.12, 0.40 and 1.00 m) and two gripping conditions (commercial prestressing anchorage and rigid clamps). Special samples with surface damage were also tested to analyze the influence of the superficial state of the specimen.

To achieve a gauge length of 300 mm it was necessary to design an original experimental device, since the standard extensometers do not reach such a length. The device is depicted in Fig.1, and consists of a mechanism to transmit the relative movement of two points A and B of the sample (separated by a distance equal to the chosen gauge length) to a thin rod able to slide inside a tube. The transmission is attained through two grips with a circular hole. Each grip is connected to the steel wire by an elastic join and a small ring. The small ring localizes the reference measurement points A and B, whose movement is transmitted to the extensometer ends (points A' and B').

The three extensometers were carefully calibrated over the range from 0% to 4% of strain before each test, by using a micrometer with an accuracy of 1 μm. The calibration of the load cell was also checked before each test. All tests were performed at room temperature, under displacement control. Tests were interrupted before necking to avoid damage to the measurement device.

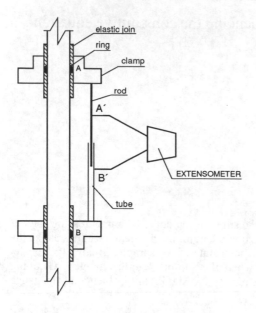

Fig. 1. Experimental device

3. METHOD OF ANALYSIS

To study the influence on the stress-strain curve of geometric variables such as the gauge length, and the sample length, the following parameters were analyzed:

• Young's modulus (E), which was calculated by least squares linear fitting on the basis of the recorded points.

• 0.2 offset yield strength ($s_{0.2}$), defined as the engineering stress which would produce a plastic engineering strain of 0.2%. It was obtained by intersection of the stress–strain curve of the steel and a straight line parallel to the linear elastic segment.

• Ramberg-Osgood parameters (P, n), were used to study the shape of the true stress-strain curve in the plastic regime, according to the following equation:

$$\varepsilon = \varepsilon^e + \varepsilon^p = \frac{\sigma}{E} + \left(\frac{\sigma}{P}\right)^n \qquad (1)$$

where σ and ε are the true stress and the true strain, respectively. These variables are calculated from the engineering stress (s) and strain (e) through the expressions (Hill 1983, Johnson, and Mellor 1983):

$$\sigma = s\,(1+e) \qquad (2)$$

$$\varepsilon = \ln\,(1+e) \qquad (3)$$

Equation (1) represents a straight line in a bilogarithmic scale. The steel under study shows three straight parts in a plot $\ln \sigma$ vs $\ln \varepsilon^p$: nonlinear elastic or pre-plastic (I), flow step or transition zone (II) and strain hardening (III), whose coefficients (P_I, n_I, P_{II}, n_{II}, P_{III}, n_{III}) can be calculated by least squares linear fitting in the bilogarithmic scale. A first observation of the experimental results suggests that steps I and III do not present strong variations with the test parameters. For this reason, only the flow step or transition zone II is analyzed to show the influence of the mechanical variables.

The influence of each variable is studied by observation of the trend and calculation of the correlation coefficient of the least squares fitting. Two simultaneous conditions are required in this work to assure that there is an influence of the variable on the stress-strain curve: monotonic trend and linear correlation coefficient greater than 0.9.

4. GEOMETRIC VARIABLES

4.1. Influence of the gauge length

The four graphs in Fig. 2 represent the variation of the four constitutive parameters under study (E, $s_{0.2}$, P_{II}, n_{II}) as functions of the gauge length B. The intervals corresponding to the standard deviation for each are also plotted.

The corresponding regression lines and correlation coefficients are:

$$E\ (MPa) = 202063 - 17.3\ B\ (mm) \qquad (4)$$
$$r = 0.63$$

$$s_{0.2}\ (MPa) = 1468 + 0.008\ B\ (mm) \qquad (5)$$
$$r = 0.59$$

$$P_{II}\ (MPa) = 1721 - 0.043\ B\ (mm) \qquad (6)$$
$$r = 0.59$$

$$n_{II} = 41.41 + 0.0089\ B\ (mm) \qquad (7)$$
$$r = 0.61$$

The correlation coefficient is always very low, and in no case is there a monotonic trend. Only those tests for the shortest length (L =

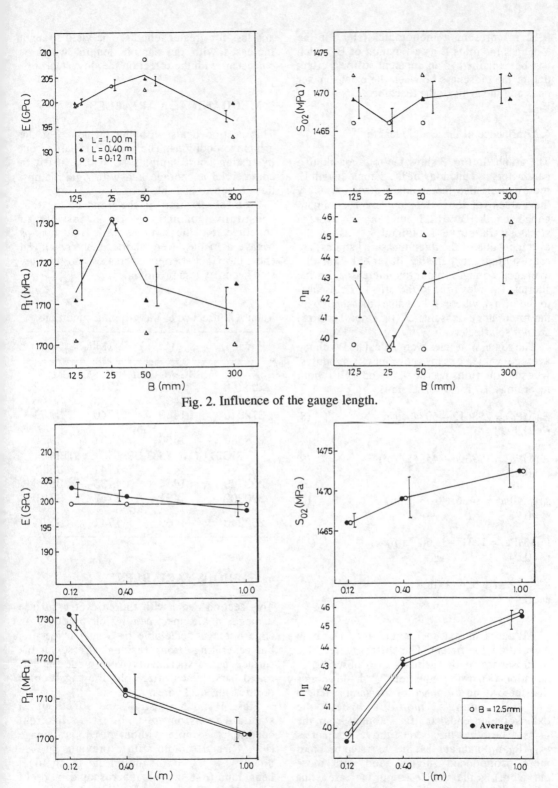

Fig. 2. Influence of the gauge length.

Fig. 3. Influence of the sample length.

0.12 m) present a monotonic trend for the Young´s modulus E as a function of B, which can be explained as an apparent stiffness effect due to the closeness between the extensometer ends and the grips of the machine.

4.2. Influence of the sample length

The graphs in Fig. 3 show the four constitutive parameters as functions of the sample length L, representing also the standard deviation. Only the values for the extensometer of minimum gauge length (B = 12.5 mm) and the average stress-strain curve (obtained by using the average value of the three measured strains) are represented. It can be observed that the average Young´s modulus increases in the shortest samples (due to the effect of closeness to the grips), whereas the values measured with the minimum extensometer (far from the grips) are not so affected.

The regression lines were calculated with the average value, except the Young's modulus, whose equation was calculated with both recordings (L* for B = 12.5 mm):

$$E \text{ (MPa)} = 203412 - 5074 \ L(m) \qquad (8)$$
$$r = 0.9998$$

$$E \text{ (MPa)} = 199694 - 458 \ L^*(m) \qquad (9)$$
$$r = 0.92$$

$$s_{0.2} \text{ (MPa)} = 1466 + 6.53 \ L(m) \qquad (10)$$
$$r = 0.98$$

$$P_{II} \text{ (MPa)} = 1731 - 31.60 \ L(m) \qquad (11)$$
$$r = 0.94$$

$$n_{II} = 39.34 + 6.59 \ L(m) \qquad (12)$$
$$r = 0.95$$

All cases show a monotonic trend and high correlation coefficient. The stiffness effect due to closeness to the grips is also observed in equation (8), while equation (9) demonstrates that there is no influence of the sample length on the Young's modulus when the extensometer ends are far enough from the grips. The other three constitutive parameters ($s_{0.2}$, P_{II}, n_{II}) indicate that the flow step becomes more pronounced (higher, and with more curvature) as the sample length increases, this effect being more significant for short lengths,

and less for greater lengths. The yield strength increases with the sample length, which is consistent with the effect on the flow step.

5. MECHANICAL VARIABLES

This section deals with the influence of the gripping conditions. The analysis is carried out by testing short samples (L = 0.12 m) with commercial anchorage and with rigid clamps. Results appear in Table 1.

From observation of the trend of the four constitutive parameters it is possible to conclude that the samples with rigid clamps behave as if they were shorter than for hinged ends: the yield strength decreases and the flow step becomes less pronounced.

Table 1. Influence of the gripping conditions

E (MPa)	$s_{0.2}$ (MPa)	P_{II} (MPa)	n_{II}
COMMERCIAL GRIPPING SYSTEM			
205750	1468	1719	41.12
207651	1462	1733	38.66
212340	1469	1740	38.71
AVERAGE			
208580	1466	1731	39.50
RIGID CLAMPS GRIPPING SYSTEM			
203070	1461	1741	36.88
202749	1464	1735	38.08
200902	1461	1747	36.19
AVERAGE			
202240	1462	1741	37.05

6. SPURIOUS VARIABLES

This section deals with the case of a surface damage in the specimen which produces an edge notch undetectable by visual inspection but generating strong residual stresses at the surface bar. Experiments were made on hot-rolled bars, which after cold-drawing become prestressing steel wires.

Figs. 4 and 5 illustrate the effect with a standard tension test in two different conditions: sample without previous damage (Fig. 4), and sample with a previous surface notch (Fig.5), both including mechanical behaviour (σ–ε curve), macroscopic mode of fracture and SEM topography after failure.

Fig. 4. Standard tension test of a bar without previous damage: mechanical behaviour (σ–ε curve), macroscopic mode of fracture and microscopic topography (SEM micrograph).

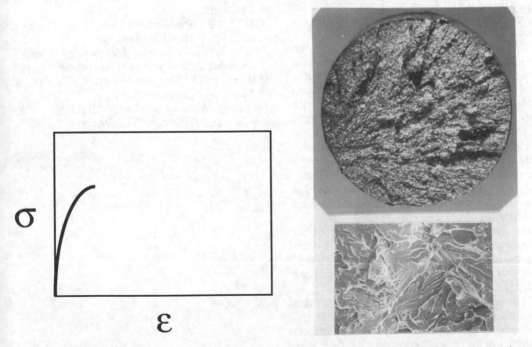

Fig. 5. Standard tension test of a bar with surface damage: mechanical behaviour (σ–ε curve), macroscopic mode of fracture and microscopic topography (SEM micrograph).

Mechanical behaviour is clearly dissimilar in both cases, since the stress-strain curve is completed up to the decrease of engineering stress in the first case, but it is suddenly interrupted in the second case. Necking takes place in the first but not in the second situation, which breaks without previous external signal.

Microscopic fracture is MVC (micro-void coalescence) in the absence of previous damage, but is cleavage-like when a surface notch introduces strong residual stresses in the vicinity of the surface and therefore in the critical region, increasing the triaxiality.

Therefore the previous localized damage in the skin-notched smooth bar produces a local increase of triaxiality, thus changing both the macroscopic and the microscopic modes of failure, and clearly lowering the fracture load (sudden interruption of the stress-strain path). Residual stress measurements in the surface (e.g. X-ray diffraction) could help in preventing this situation.

7. SUMMARY COMMENTS

The gauge length does not influence the stress-strain curve, but the extensometer must be placed far enough from the grips to avoid an apparent higher stiffness, with the result of a greater Young´s modulus.

The sample length has a small influence on the yield strength and a large influence on the flow step. As the sample length increases, the yield strength also increases and the flow step becomes more pronounced (higher and with more curvature). This effect is more significant for short samples.

The gripping conditions are relevant in very short specimens. If the gripping system is stiff (i.e. rigid clamps) the wire behaves as if it were shorter. In this sense the yield strength and the flow step are modified, the first decreasing and the latter becoming less pronounced

The superficial state of the sample does not influence the stress-strain curve, but the test may be suddenly interrupted in the presence of local damage inducing surface residual stresses in the skin of the specimen.

The main conclusion to be drawn from this research is that the stress-strain curve of the prestressing steel under study has an intrinsic character in all its sections except the flow step, which is more pronounced as the length

increases and the stiffness of the gripping system decreases, mainly for short lengths. The slight effect on the yield strength is a direct consequence of the previous consideration. In any case the numerical differences in the constitutive parameters are very small, and the stress-strain curve of the steel can be considered as an intrinsic characteristic of the material.

Acknowledgements

The author would like to thank Professor M. Elices, Head of the Materials Science Department of the Polytechnical University of Madrid, for his encouragement in the development of this work. Acknowledgement is also given to Professor J. Planas and Dr. G. Guinea, Polytechnical University of Madrid, for helpful discussions and assistance in performing the tests.

REFERENCES

ASTM A370. *Standard Methods and Definitions for Mechanical Testing of Steel Products.*

ASTM E8. *Standard Methods of Tension Testing of Metallic Materials.*

ASTM E111. *Standard Methods of Young's Modulus Test at Room Temperature*

Hill, R. 1983. *The Mathematical Theory of Plasticity.* Clarendon Press.

Johnson, W. and Mellor, P.B. 1983. *Engineering Plasticity,* John Wiley and Sons.

Recent Advances in Experimental Mechanics, Silva Gomes et al. (eds) © 1994 Balkema, Rotterdam, ISBN 90 5410 395 7

Loading rate and temperature dependences on the flexural static behavior of unidirectional pitch based CFRP laminates

Yasushi Miyano & Masayuki Nakada
Materials System Research Laboratory, Kanazawa Institute of Technology, Japan

Norimitsu Kitade & Michael K. McMurray
Graduate School, Kanazawa Institute of Technology, Japan

Michihiro Mohri
Central Technical Research Laboratory, Nippon Oil Company, Ltd, Yokohama, Japan

ABSTRACT: The mechanical behavior of polymer resins seriously depends on time and temperature, which is called viscoelastic behavior. Therefore, it is presumed that the mechanical behavior of FRPs also depends on time and temperature because of the viscoelastic behavior of the matrix resin. The flexural static behavior of three kinds of unidirectional pitch based CFRP laminates, which are a combination of two types of fiber and two types of resin, were evaluated at several levels of loading rates and constant temperatures. The flexural static behavior was found to be remarkably dependent on time and temperature. The reciprocation law of time and temperature of the mechanical behavior of the matrix resin held for the flexural static strength of the CFRP laminates. Clearly, the flexural static behavior of the CFRP laminates was dominated by the viscoelastic behavior of the matrix resin.

1 INTRODUCTION

Pitch based carbon fibers have a high tensile modulus, because they are made from coal-tar pitch or petroleum pitch which have good graphitizability. Therefore, FRPs using a pitch based carbon fibers (pitch based CFRP) are being used as structural components of aircraft, spacecraft, etc., where high modulus is required. However, the mechanical behavior of pitch based CFRPs has not yet been fully clarified.

It is well known that the mechanical behavior of polymer resins exhibits time and temperature dependence, called viscoelastic behavior, not only above the glass transition temperature T_g but also below T_g. Thus, it can be presumed that the mechanical behavior of FRP using polymer resins as matrices also significantly depends on time and temperature. It has been confirmed that the viscoelastic behavior of polymer resins as matrices is a major influence on the time and temperature dependence of the mechanical behavior of FRP.

In this paper, the flexural static behavior of three kinds of unidirectional pitch based CFRP laminates, which are a combination of two types of fiber and two types of resin, was evaluated at several levels of loading rates and constant temperatures. The time and temperature dependence of flexural static behavior of these CFRP laminates is discussed from the viewpoint of the viscoelastic behavior of the matrix resin.

2 EXPERIMENTAL PROCEDURE

2.1 *Preparation of specimens*

Unidirectional CFRP laminates using two types of pitch based carbon fiber and two types of thermosetting resin as matrices were prepared. The two types of pitch based carbon fiber were XN40 and XN70 (Nippon Oil, brand name Granoc). The matrix resins were a general purpose epoxy resin 25C and a high glass transition temperature polycyanate resin RS3. The glass transition temperature T_g of 25C and RS3 is 140 and 228°C, respectively. The three kinds of prepreg sheets made from these fibers and resins, which are XN40/25C, XN40/RS3 and XN70/25C (fiber/resin), were hot pressed into 3 mm thick unidirectional CFRP laminates. Both XN40/25C and XN70/25C were molded at 120°C for 1 hour and postcured at 130°C for 2 hours and then at 160°C for 2 hours. XN40/RS3 was molded at 180°C for 2 hours and postcured at 230°C for 2 hours. The fiber volume fractions of all the CFRP laminates were approximately 55%.

2.2 Test procedures

Three point flexural static tests for these unidirectional CFRP laminates XN40/25C, XN40/RS3 and XN70/25C, were conducted by using an Instron type testing machine with a constant temperature chamber. The nominal dimensions of the test specimens are 80x10x3 mm (length, width, thickness). The specimens were placed in a 3-point flexural test set-up as seen in Figure 1, with a span of 60 mm and an upper loading point radius of 5 mm. The testing conditions of temperature and deflection rates are shown in the lower portion of Figure 1. The experimental procedures were carried out according to Japanese Industrial Standard (JIS).

3 RESULTS AND DISCUSSION

3.1 Viscoelastic behavior of the matrix resin

Figure 2 shows the master curves of storage moduli E' versus reduced time t' at a reference temperature of $T_0=25^\circ C$, for the two types of resins, 25C and RS3. These master curves were constructed by shifting the storage moduli E' horizontally and vertically at various constant temperatures T until they overlapped. The horizontal time and temperature shift factor $a_{T_0}(T)$ and the vertical temperature shift factor $b_{T_0}(T)$ are defined as follows.

$$a_{T_o}(T) = \frac{t}{t'} \tag{1}$$

$$b_{T_o}(T) = \frac{E'(t,T)}{E'(t',T_o)} \tag{2}$$

where: t': reduced time (min)

T0: reference temperature (K).

Since E' at various temperatures can be superimposed so that a smooth curve is created, the modified reciprocation law of time and temperature is applicable for the stress-strain relation of these resins.

Figures 3 and 4 show the time-temperature shift factors $a_{T_0}(T)$ and temperature shift factor $b_{T_0}(T)$ obtained experimentally for the master curve of storage moduli E'. The shift factors $a_{T_0}(T)$ are quantitatively in good agreement with Arrhenius' equation by using two different activation energies ΔH:

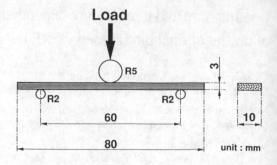

Load

unit : mm

Static Bending Test

Material	Deflection Rate mm/min	Temperature °C
XN40/25C	200	-60, -30, 0, RT, 60, 80, 100, 120, 140, 160, 180
XN70/25C	2	
XN40/RS3	0.02	-50, 0, RT, 50, 100, 150, 200, 230, 250

Fig.1   Test configuration and testing conditions

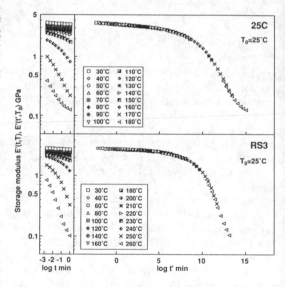

Fig.2   Master curves of the storage moduli for the two types of resin

$$\text{Log } a_{T_o}(T) = \frac{\Delta H}{2.303R}\left[\frac{1}{T} - \frac{1}{T_o}\right] \tag{3}$$

where: ΔH: activation energy (kJ/mol)

R: gas constant 8.314×10^{-3} (kJ/(Kxmol))

T: testing temperature (K).

1020

The temperature at the knee point of the two Arrhenius' equations corresponds closely to the glass transition temperature of each resin.

The horizontal temperature shift factor $b_{T_0}(T)$ can be described by two straight lines from the following equation:

$$\text{Log } b_{T_0}(T) = 1 + K[T_0 - T] \qquad (4)$$

where K is a material constant ($1/°C$). The knee point temperature of the two straight lines also corresponds to the glass transition temperature T_g of the resins.

The master curves of storage modulus E' for two resins were combined to produce Figure 5. Figure 6 is storage moduli E' versus temperature T for two resins at a time t = 1 min, obtained from the left side of Figure 2. The storage moduli E' for all the resins decrease in a similar manner as time increases as shown in Figure 5. In the intermediate time region the resins become more time dependent. Although two resins show the same time dependence, E' for RS3 remains almost constant till a much higher temperature than 25C as shown in Figure 6. Near each of the resin's T_g the E' value decreases significantly for two resins.

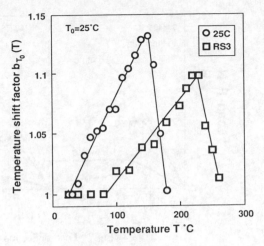

Fig.4   Temperature shift factor $b_{T_0}(T)$ for the two types of resin

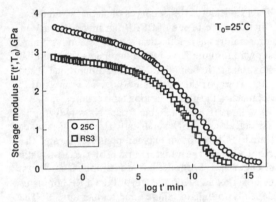

Fig.5   Storage moduli versus reduced time at a reference temperature $T_0 = 25°C$ for the two types of resin

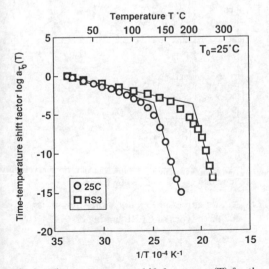

Fig.3   Time-temperature shift factors $a_{T_0}(T)$ for the two types of resin

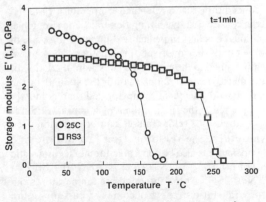

Fig.6   Storage moduli versus temperature at a time t=1min for the two types of resin

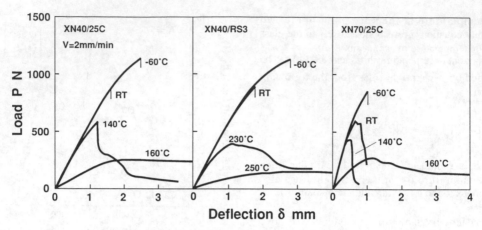

Fig.7 Load-deflection curves on the flexural static
test for the three types of CFRP laminates

3.2 *Flexural static strength*

The load P-deflection δ curves of static flexural tests
for the three types of CFRP laminates at various
temperatures at a deflection rate of V=2 mm/min are
shown in Figure 7. All three laminates show similar
flexural static behavior. The maximum load P and
the slope of the curve decrease as temperature
approaches the glass transition temperature T_g.

The three figures in Figure 8 each show two graphs.
The left side is the flexural static strength σ_{bs} versus
time to failure t at various temperatures T, and the
right side is the master curve of flexural static
strength σ_{bs}. From the left side graphs it is shown
that the flexural static strength of the CFRP laminates
depends on time to failure t and temperature T. The
σ_{bs} master curves of the CFRP laminate on the right
side were constructed in the same way as the storage
modulus E' master curves of the matrix resins. The
master curve of flexural static strength σ_{bs} versus
reduced time t' at a reference temperature $T_0=25^{\circ}C$
was constructed by shifting flexural static strength at
various constant temperatures T along the log scale of
time t until they overlapped each other. The vertical
shift applied in the case of the stress-strain relation of
the matrix resin is not needed for the strength of the
CFRP laminate because the applied load is mostly
transferred to the fiber of the CFRP laminate. Since
σ_{bs} at various temperatures can be superimposed
smoothly, the reciprocation law of time and
temperature is also be applicable for the flexural static
strength $\sigma_{b\ s}$ of XN40/25C, XN40/RS3 and
XN70/25C. At the intermediate temperatures the
flexural static strength becomes remarkably more
dependent on t and T. Each of the master curves can
be divided into three distinct curves.

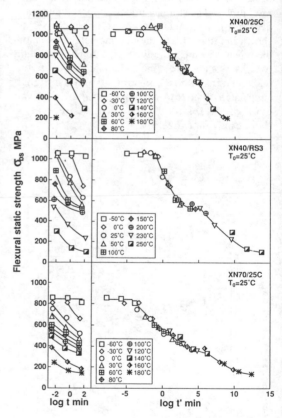

Fig.8 Master curves of flexural static strength for
the three types of CFRP laminates

Figure 9 shows the time-temperature shift factors $a_{T_0}(T)$ obtained experimentally for the master curves of flexural static strength of the CFRP laminates. These shift factors are quantitatively in good agreement with Arrhenius' equation, so the shift factors for each CFRP laminate can be approximated by using two different activation energies ΔH.

Table 1 shows the activation energies ΔH_1 below the glass transition temperatures T_g and ΔH_2 above T_g, T_g and the knee point temperature T_ξ for the three composites and two resins. The activation energies ΔH_1 and ΔH_2 for the time-temperature shift factors $a_{T_0}(T)$ for the matrix resin's storage moduli E' and the flexural static strength σ_{bs} of the CFRP laminates with the same matrix resin are approximately equal. The table also reaffirms that the composites with the same matrix resin have the same knee point

temperature T_ξ of the matrix resin.

A summary of flexural static strength σ_{bs} versus reduced time t' and reduced temperature T' is in Figures 10 and 11. The time dependence of the flexural static strength σ_{bs} of the three CFRP laminates corresponds well with the time dependence of the storage moduli E' for the resins as shown in Figure 5. The temperature dependence of the flexural static strength σ_{bs} of the three CFRP laminates also corresponds well with the temperature dependence of the storage moduli E' as shown in Figure 6.

In the case of the combination of different fibers and the same matrix resin, XN40/25C and XN70/25C, the flexural static strength σ_{bs} coincides with each other in the region of long time and high temperature, and σ_{bs} is different for each in the other two regions as shown in Figures 10 and 11.

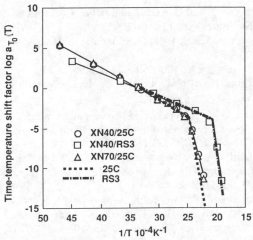

Fig.9 Time-temperature shift factors for the three types of CFRP laminates

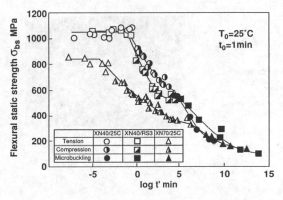

Fig.10 Flexural static strength versus reduced time at a reference temperature $T_0=25°C$ with fracture modes for the three types of CFRP laminates

Table 1 Activation energies, knee point temperatures and glass transition temperatures

Material	ΔH_1 kJ/mol	ΔH_2 kJ/mol	T_ξ °C	T_g °C
25C	75	872	130	140
XN40/25C	75	522	130	–
XN70/25C	75	540	130	–
RS3	56	917	212	228
XN40/RS3	55	1047	208	–

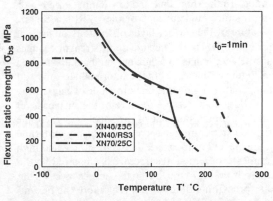

Fig.11 Flexural static strength versus reduced temperature at reference time $t_0=1$min for the three types of CFRP laminates

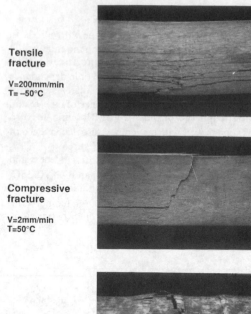

Tensile fracture

V=200mm/min
T= –50°C

Compressive fracture

V=2mm/min
T=50°C

Microbuckling fracture

V=200mm/min
T=250°C

1mm

Fig.12 Fractographs for XN40/RS3

3.3 *Flexural fracture mode*

The CFRP laminates fracture behavior is shown in Figure 12 which exhibits the three kinds of fracture mode. In the short time and low temperature region the fracture mode of each of the CFRPs is tensile fracture of the surface layer on the tension side of specimen where the flexural static strength changes scarcely with time to failure and temperature as seen in Figures 10 and 11. In the intermediate time and temperature region, the fracture mode of each of the CFRPs is compressive fracture of the surface layer on the compression side of the specimen where the flexural static strength decreases suddenly with time to failure and temperature. In the long time and high temperature region, the fracture is triggered by the micro-buckling of the fiber in the surface layer on the compression side of specimen where the flexural static strength decreases distinctly with time to failure and temperature.

4 CONCLUSION

The flexural static strength of three kinds of unidirectional pitch based CFRP laminates, which are a combination of two types of fiber and two types of resin, were measured at several levels of loading rates and temperatures. It was found that the flexural static strength as well as the flexural fracture mode of the CFRP laminates depends remarkably on the loading rate and temperature. The reciprocation law of time and temperature for the matrix resin also held for the corresponding CFRP laminate. The time and temperature dependence of the flexural static strength and fracture modes are unique and mainly controlled by the viscoelastic behavior of the matrix resin.

REFERENCES

Aboudi, J. and G. Cederbaum. 1989. Composite Structures, Vol 12, 243-256.

Gates, T. March 1992. Experimental Mechanics, 68-73.

Miyano, Y., M. Kanemitsu, T. Kunio, and H. Kuhn. Nov 1986. Journal of Composite Materials, Vol 20, 520-538.

Miyano, Y., S. Amagi, and M. Kanemitsu. April 1988. Proceedings of International Symposium on FRP/CM, 7.C.1-7.C.11.

Mohri, M., Y. Miyano, and M. Suzuki. July 1991. International Conference on Composite Materials, 8. 33.B.1-33.B.9.

Nakada, M., M. McMurray, N. Kitade, M. Mohri, and Y. Miyano. 1993. Proceedings of International Conference on Composite Materials 9. Vol 4, 731-738.

Schwarxl, F., A. Staverman. Aug 1952. Journal of Applied Physics, Vol 23, 838-43.

Sullivan, J. 1990. Composite Science and Technology, Vol 39, 207-232.

Williams, M., R. Landel, J. Ferry. July 1955. Journal of American Chemical Society, Vol 77, 3701-06.

Recent Advances in Experimental Mechanics, Silva Gomes et al. (eds) © 1994 Balkema, Rotterdam, ISBN 90 5410 395 7

Some results on the nonlinear viscoelastic behavior of fiber reinforced composites

O. S. Brüller & M. Katouzian
Technical University of Munich, Department 'A' of Mechanics, Germany

S. Vlase
University of Transilvania, Brasov, Romania

ABSTRACT: The nonlinear viscoelastic behavior of both neat and carbon fiber-reinforced Polyether-etherketone (PEEK) and epoxy resin (EP) has been studied under creep loading conditions on $[90_4]_s$ laminates and non reinforced specimens. Series of 10- hour isothermal tensile creep tests were conducted at four temperatures (up to $140°C$ for the epoxy system and up to $120°C$ for the PEEK system) and different stress levels. The Mori-Tanaka method was used in conjuction with the Schapery's nonlinear viscoelastic formulation for the evaluation of the creep response. It has been demonstrated that the theoretical results are in good agreement with the experimental data.

1 INTRODUCTION

The effective viscoelastic behavior of a two phase composite body is dependent upon the elastic/viscoelastic properties of the constituent materials. The nonlinear time dependent behavior of viscoelastic materials has been studied by many investigators (s. for example [1,2]). Most of the works [3-5] which have been done to estimate the bounds on effective elastic/viscoelastic property of fiber reinforced composites, assumed that both phases possess isotropic material behavior. Little work has been presented in the literature for those cases where for instance the reinforcing phase has anisotropic or transversely isotropic properties. An example of this could be a gaphite/epoxy composite where the matrix material is considered to be isotropic but the fibers show anisotropic behavior. In the following, an attempt will be made to estimate the characteristic parameters of a composite material using the Mori-Tanaka method [6] and to obtain experimental data to verify the above theoretical procedure.

2 THEORETICAL BACKGROUND

Let us consider the general case of a cylindrical specimen containing elliptic cylindrical fibers in a polymeric matrix. The fibers are considered to be transversely isotropic, while the matrix is isotropic, having viscoelastic properties. The shape of the elliptic fibers is characterized by the ratio $\alpha = t/w$ as shown in Fig. 1. For the aspect ratio $\alpha = 1$, a composite with cylindrical fibers considered in the current study is obtained.

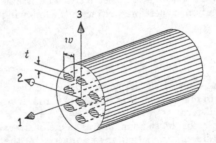

Figure 1. Schematic representation of a composite with monotonically aligned elliptic cylindrical fibers

The solution of the problem is based on the Eshelby's [7] approach for an ellipsoidal inclusion in conjunction with the Mori-Tanaka's [6] mean-field theory for which the results are obtained by Zhao and Weng [8].

It should be pointed out that in the study undertaken by Zhao et al. [8], the two constituents are considered to have isotropic properties. In the present investigation however, the fibers possess anisotropic material behavior. Using the above approach, one may compute the compliance matrix of a viscoelastic body with orthotropic properties and in turn determine the creep response.

Let us consider a fiber-reinforced composite in which transversely isotropic fibers are uniformly distributed in the matrix material. Here a representative volume element (RVE) of the composite and one of a comparison material (CM) made only of the matrix material are introduced. Let both of the above RVE's be subjected to the same boundary traction $\bar{\sigma}$. Let further denote the elastic coefficients matrix of CM by C_m.

For the real composite however, the mean strain in the matrix is different than that in the CM. Let $\tilde{\varepsilon}$ represent the difference of the two mean values of strain. The mean value of the stress in the CM is $\bar{\sigma}$. On the other hand, in the matrix of the composite, a different mean stress $\tilde{\sigma}$ is present. As a result, following observations can be made:

- in CM, due to the mean strain field ε^0 and the mean stress field $\bar{\sigma}$, the stress-strain relation becomes:

$$\bar{\sigma} = C_m \varepsilon^0 \tag{1}$$

- in the matrix constituent of the RVE composite the mean strain and stress fields are $\varepsilon^m = \varepsilon^0 + \tilde{\varepsilon}$ and $\sigma^m = \bar{\sigma} + \tilde{\sigma}$, respectively. The strain-stress relation is therefore written in the following form:

$$\sigma^m = \bar{\sigma} + \tilde{\sigma} = C_m(\varepsilon^0 + \tilde{\varepsilon}) \tag{2}$$

- in the fiber of the RVE composite the mean strain field differs from that in matrix through an additional term ε^{pt} and hence $\varepsilon^f = \varepsilon^m + \varepsilon^{pt} = \varepsilon^0 + \tilde{\varepsilon} + \varepsilon^{pt}$. In the same manner, the mean stress field is different from that in the matrix by the additional term σ^{pt} and therefore $\sigma^f = \bar{\sigma} + \tilde{\sigma} + \sigma^{pt}$. The stress-strain relation becomes:

$$\sigma^f = \bar{\sigma} + \tilde{\sigma} + \sigma^{pt} = C_f(\varepsilon^0 + \tilde{\varepsilon} + \varepsilon^{pt}) \tag{3}$$

It should be pointed out that C_f is the matrix of elastic coefficients of the fibers. Using Eshelby's equivalence principle one may write the average stress in fiber in terms of the elastic coefficients of the matrix C_m by introducing the term ε^* in the average strain field, i.e.;

$$\sigma^f = \bar{\sigma} + \tilde{\sigma} + \sigma^{pt} = C_f(\varepsilon^0 + \tilde{\varepsilon} + \varepsilon^{pt})$$
$$= C_m(\varepsilon^0 + \tilde{\varepsilon} + \varepsilon^{pt} - \varepsilon^*) \tag{4}$$

where the following relation holds:

$$\varepsilon^{pt} = P\varepsilon^* \tag{5}$$

The four-rank tensor P is Eshelby's transformation tensor and has the symmetry property $P_{ijkl} = P_{jikl} = P_{ijlk}$.

Following the procedure used by Zhao et al. [8], it can be shown that:

$$\frac{E_{11}}{E_m} = \frac{\bar{\varepsilon}_{11}^0}{\bar{\varepsilon}_{11}} = \frac{1}{1 + v_f[a_1 - \nu_m(a_2 + a_3)]} \tag{6}$$

where a_i represent the computed coefficients shown in the above reference. Similar expressions can be obtained for the elastic moduli in other directions.

For the shear moduli one can obtain:

$$\frac{G_{12}}{G_m} = 1 + \frac{v_f}{\dfrac{G_m}{G_{12,f} - G_m} + 2v_m P_{1212}} \tag{7}$$

Following a procedure similar to that just presented, the remaining expressions for the shear moduli are obtained. Finally, in order to determine the expressions for Poisson's ratio, one can use the relation:

$$\nu_{12} = -\frac{\bar{\varepsilon}_{22}}{\bar{\varepsilon}_{11}} = \frac{\nu_m - v_f[a_4 - \nu_m(a_5 + a_6)]}{1 + v_f[a_1 - \nu_m(a_2 + a_3)]} \tag{8}$$

The expressions for the remaining Poisson's ratios ν_{23} and ν_{31} are obtained in a similar manner.

3 EXPERIMENTAL PROCEDURES

Isothermal creep tests have been conducted at room temperature ($23^{\circ}C$) as well as at elevated temperatures. The relative humidity in the laboratory could be kept constant at $50 \pm 5\%$.

The creep testing facility consists of tensile testing stations with a lever/ loading arrangement having a ratio of 10:1 with a maximum working load capacity of 6kN (Fig. 2). The load on the specimen and the strain measured by extensometers were registered electronically by a PC.

A few of the test stations were equipped with a cylindrical heating chambers (Fig. 3) thermostatically controlled. The measured temperature inside the chamber was shown to be kept within about ±1°C. The duration of loading was 10 hours for all tests.

4 MATERIALS AND SPECIMENS

The fiber reinforced materials investigated were commercially available composites. The epoxy resin matrix system was Fibredux 6376C (Ciba-Geigy AG) reinforced with T800 carbon fibers (Toray Industries, Inc.). The matrix of this prepreg system is a toughened epoxy resin with a glass transition temperature of about 180°C. The thermoplastic system used was the APC-2 material (ICI). The semicrystalline PEEK matrix of this system with a glass transition temperature of about 145°C was reinforced with IM6 carbon fibers (Hercules, Inc.). The fiber volume content for both systems was 60% ± 3%. The neat resin plates with a thickness of 3 mm (epoxy resin) and 4 mm

(PEEK) were supplied by the resin manufacturers.

The fiber-reinforced specimens had a nominal length of 150 mm, a width of 10 mm and a thickness of 1 mm. Neat resin specimens had the same length and width as the reinforced specimens, but were 3 mm thick.

5 TESTING PROGRAM

For both materials all the creep tests were performed at temperatures below the glass transition temperature. The maximum test temperature was about 20°C below the glass transition temperature for PEEK and 40°C below the glass transition temperature for the epoxy resin. The test temperatures for PEEK were 23°, 60°, 80°, and 100°C for the neat material and 23°, 80°, 100°, and 120°C for the reinforced one. The test temperatures for the neat and reinforced epoxy samples were 23°, 80°, 120°, and 140°C.

At room temperature five stress levels ranging between 10 and 70% of the ultimate tensile stress were applied. The load levels however, were reduced with increasing test temperature.

6 RESULTS AND DISCUSSION

The Mori-Tanaka approach has been modified in order to take account of the anisotropic behavior of the fiber and the viscoelastic properties of the

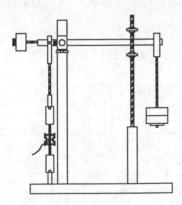

Figure 2. The single loading lever device

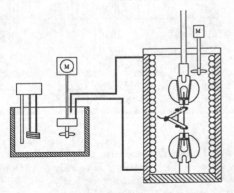

Figure 3: The principle of high temperature testing

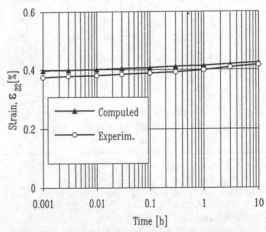

Figure 4. Comparison between experimental and computed results of creep strain ε_{22} for carbon/epoxy subjected to $\sigma_{22} = 30$ MPa at 23°C.

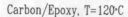

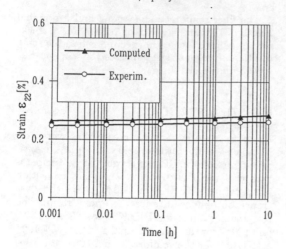

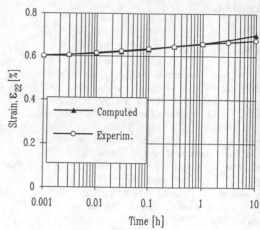

Figure 5. Comparison between experimental and computed results of creep strain ε_{22} for carbon/epoxy subjected to $\sigma_{22} = 19$ MPa at 120°C.

Figure 7. Comparison between experimental and computed results of creep strain ε_{22} for carbon/PEEK subjected to $\sigma_{22} = 47$ MPa at 80°C.

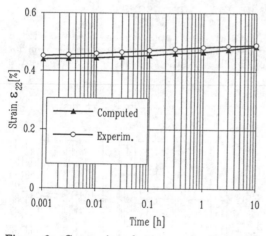

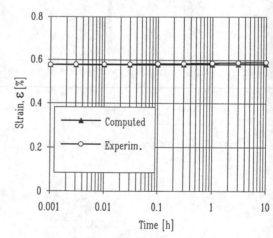

Figure 6. Comparison between experimental and computed results of creep strain ε_{22} for carbon/PEEK subjected to $\sigma_{22} = 36$ MPa at 80°C.

Figure 8. Comparison between experimental and computed results of creep strain $\dot{\varepsilon}$ for neat PEEK resin subjected to $\sigma = 26$ MPa at 23°C.

matrix in a two phase composite material. The viscoelastic behavior of the polymeric matrix is characterized by the nonlinear constitutive equations developed by Schapery [2] which has been successfully used by Brüller [1], [9]. The comparison between the experimentally determined creep curves and those obtained with the presented theory is shown in Figure 4-9. It is readily seen that the theoretical prediction agrees fairly with the experimental results. It should be pointed out that only a part of the results is shown in the presented figures.

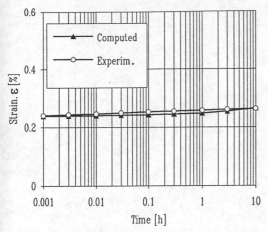

Neat Epoxy Resin, T=80°C

Figure 9. Comparison between experimental and computed results of creep strain ε for neat epoxy resin subjected to $\sigma = 8.6$ MPa at 80°C.

REFERENCES

[1] Brüller, O.S., "Zur Charakterizierung des Lang-zeitverhaltens von Kunststoffen, Arch. Appl. Mech., 63, 363, 1993.

[2] Schapery, R.A., "Stress Analysis of Viscoelastic Composite Materials," J. of Comp. Mat., 1, 228, 1967.

[3] Hashin, Z., "On Elastic Behavior of Fibre Reinforced Materials of Arbitrary Transverse Phase Geometry", J. Mech. Phys. Solids, 13, 119, 1965.

[4] Walpole, L.J., "On the Bounds for the Overall Elastic Moduli of Inhomogeneous Systems-I, II", J. Mech. Phys. Solids, 14, 151 and 289, 1966.

[5] Hill, R., "Theory of Mechanical Properties of Fiber- strengthened Materials: I. Elastic behavior, II. Inelastic behaviour", J. Mech. Phys. Solids, 12, 199, 1964.

[6] Mori, T. and Tanaka, K., "Average Stress in the Matrix and Average Elastic Energy of Materials with Misfitting Inclusions", Acta Metallurgica, 21, 571, 1973.

[7] Eshelby, J.D., "The Determination of the Elastic Field of an Ellipsoidal Inclusion, and Related Problems", Proc. of the Royal Society, London, A241, 376, 1957.

[8] Zhao, Y.H. and Weng, G.J., "Effective Elastic Moduli of Ribbon-Reinforced Composites," J. of Appl. Mech., 57, March 1990.

[9] Brüller, O.S., "Nonlinear Characterization of the Long Term Behavior of Polymeric Materials", Polym. Eng. Sci., 27, 144, 1987.

Recent Advances in Experimental Mechanics, Silva Gomes et al. (eds) © 1994 Balkema, Rotterdam, ISBN 90 5410 395 7

Proposal for a standardised experimental device to study the behavior of laminates composites under low velocity impact loading

V. Martin, F. Collombet, M. Moura & J. L. Lataillade
LAMEF (LAboratoire Matériaux Endommagement Fiabilité), ENSAM-CER de Bordeaux, Talence, France

ABSTRACT: The aim of this work is the development of an experimental set-up and its suitable metrology to obtain with good accuracy the characteristics of low velocity impact loading versus time. Our scope is to define a Standardised Test for the impact of composite plates. It consists of a drop weight set-up with a velocity range between 1 and 10 m/s and for a maximal energy of 50 J. The metrology is based on the Split-Hopkinson Pressure Bars technique. The apparatus is adapted to a parametric study of the impact characteristics, projectile shape, boundary conditions, geometry and stacking of the laminate for example. Experimental observations show two particular kinds of damage : first, on a microscopical scale, there is a transverse matrix cracking, which is a precursor of the second type, delamination on a mesostructural scale.

1 INTRODUCTION

The low resistance of laminated composite structures under low-velocity localised impact (few ten of Joule) is an impediment to their industrial development. Because of the heterogeneity of the composite structures, the damage induced by impacts is localised, multi scale and then corresponds to mechanisms of different response times. The damage zone is not easily visible to the naked eye and depends on the manufacturing conditions. In the range of low-energy impacts, we have studied an impact similar to dropping tools during the maintenance of planes. The accidental aspect of a "low-energy" localised impact hinders a direct investigation of the loading, as well as a measurement of the damage versus time. In the laboratory, we have developed an experimental device adapted to a parametric study of the impact characteristics, projectile shape, boundary conditions, geometry and stacking of the laminate for example. Furthermore we can obtain the overall response of the structure versus time by means of a metrology based on the Split Hopkinson Pressure Bars technique.

The local behaviour of the material in this kind of impact loading acts with the general response of the structure. The interaction is not detectable when the impact velocity leads to the perforation of the structure. In this case, only local phenomena are activated [Ayax 1992].

This duality between material and structure requires a study of multi-scale damage : on the ply scale, the transverse matrix cracking initiates the delamination on a meso-scale. The area of delamination depends on the loading chronology, the geometry and the velocity of the projectile, the local response of the material and the manufacturing conditions.

The behaviour of laminates under low-velocity impact must be studied from the structure itself and not from a reduced specimen without reference to the representative conditions of local dynamic loading on typical structure.

2 EXPERIMENTAL INVESTIGATION

2.1 Material and test specimens

To identify the time of first visible damage and to study their effects on the structure, we define the following specimens.

The material system selected for the first investigation is a E-glass long fibres pre impregnated with epoxy resin (STRUCTIL). The fibre content is 60 %wt. Material plies are precut and laid up by hand to form large panels in three different stackings : $[0_2/90_6/0_2]$, $[0_3/90_4/0_3]$, $[0_4/90_2/0_4]$. The panels are vacuum bagged and cured in an oven or an autoclave in accordance with the manufacturer's recommendations to study the influence of manufacturing conditions on final damage . Following the cure, test specimens are cut with a water jet system to circular 200 mm diameter and 2 mm thick plates.

The three different chosen lay-ups are allowed to have a constant thickness of the plate (ten plies-2 mm), and a constant overall static flexion of the plate.

A study on a reduced specimen is carried out to look at the behaviour of the material and not of the structure.

These samples are manufactured from AMPREG 75 graphite/epoxy prepreg (SP SYSTEMS). The chosen lay-ups are : $[0_2/90_2/0_2/90_2]s$, $[0_4/90_4]s$,

$[0_2/\pm45_2/90_2]$s. After the cure cycle in vacuum bag, small square plates are cut from panels (60 mm side, 2 mm thick).

2.2 Experimental set-up

The drop-weight set-up developed in the laboratory allows localised impacts in the velocity range from 1 to 10 m/s for energies between 1 and 50 J. The projectile slides down a guide rail and impacts on the centre of the plate (Fig.1). It is a cylindrical rod with an hemispherical tip. Its mechanical and geometrical characteristics are defined as follows : steel, length 600 mm, diameter 25 mm for a mass of 2,3 kg. A technical device prevents repeated impacts.

The embedding boundary conditions of the plate are not established but are necessary in order for the experiment to be repeated accurately. The circular plates are clamped on a 180 diameter, the square plates on a 50 mm diameter.

2.3 Metrology

The projectile is instrumented with strain gauges on section A to use the Split Hopkinson Pressure Bars technique (Fig.2). In this way, it is possible to record the propagation of the impact waves in the rod versus time. The time laws of the resultant impact force F(t) applied on the interface rod-plate (1) and the vertical displacement d(t) of the end of the projectile (2) are obtained by numerical treatment of these recorded signals [Bacon 93].

An optical sensor completes the metrology to measure the velocity Vi of the rod just prior to impact.

$$F(t) = E\ S(\ \varepsilon_I\ (t) + \varepsilon_R(t)) \qquad (1)$$

$$d\ (t) = \int_0^t \left[\ c\ (\varepsilon_I(\tau) - \varepsilon_R\ (\tau)) + v_i\right] d\tau \qquad (2)$$

with E Young modulus, S cross section of the rod
ε_I (t) and ε_R (t) incident and reflected wave
c wave velocity in the rod
Vi impact velocity

The use of the Hopkinson metrology in these tests needs a precise measurement of the impact velocity (equation 2). Because of the time-integration of the signal, a minor error on the velocity could have a great influence on the computation of the displacements. Furthermore, during a low velocity impact, the contact time between the impactor and the plate is relatively long (6 ms) in comparison with the contact

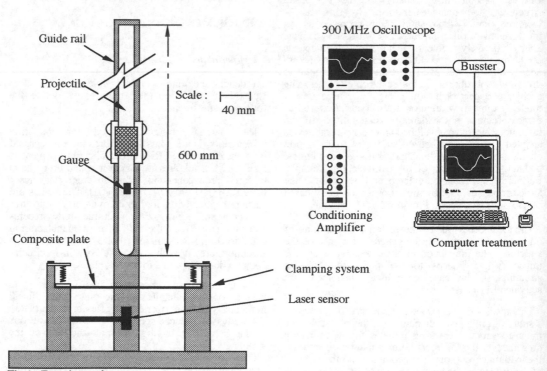

Fig.1 : Experimental set-up

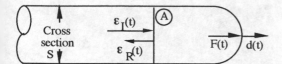

Fig.2 : Representation of the measured and calculated signals on the instrumented rod

time in a "classic" Hopkinson test (0.5 ms).

Many errors can be made in the treatment of signals : on the initial value (geometry of the rod, position of strain gauges...), and on the calibration of the rod.

The first kind of errors is corrected by an optimisation process which determines a new velocity of the wave in the rod. The impactor is calibrated with a standard bar on a dynamic compressive device.

Without contact, a laser sensor measures the displacement of any point of the non-impacted face of the plate. In this study, we consider only the centre of the back face of the plate.

3 STRUCTURAL RESPONSE

The Hopkinson metrology shows the interaction between the impact rod and the plate during the contact.

Impacts on glass epoxy plates with increasing energy (increasing velocity) show an increasing maximal force and a shorter contact time without any change in (Fig.3-4) .

We observe that the general response of the structure doesn't clearly show a change of behaviour during the contact.

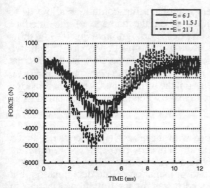

Fig.3 : Example of an impact force at increasing energy levels

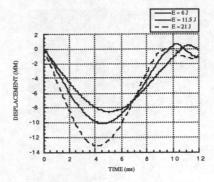

Fig.4 : Example of the displacement of the centre of the plate at increasing energy levels

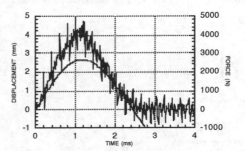

Fig. 5 : Measured displacement and force for a carbon/epoxy plate

The maximal displacement of the centre of the non-impacted side of the plate is very great (about 10 mm) in comparison with the thickness of the specimen (2 mm). Its value increases with the impact energy.

For a square carbon/epoxy plate (Fig.5), we observe a smaller displacement and a shorter contact time for an equal maximal force in comparison with the recorded signals for a glass-epoxy structure.

The Hopkinson metrology allows us to identify precisely the overall response of a structure under a low velocity impact and to show the contact between the impactor and the specimen.

4 LOCAL RESPONSE

In each tested plate, the damage has a multiscale aspect (Fig.6) : on the ply scale, matrix cracking , and on the mesoscale, delamination localised at the interface of two layers with different orientation of fibres.

Fig.6 : Multi-scale aspect of damage

Fig. 9 : First delamination

The macrocracks are in the external non impacted ply in the direction of the fibres. The delamination is at the interface 90°/0° from the impact and propagates in the direction of the fibre of the lower ply which is the direction of the cracks (Fig.7). The wing-shaped delamination is divided on both sides of the contact area.

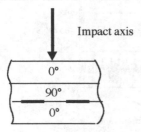

Fig.7 : Location of delamination

Fig.10 : Damage in an autoclave-manufactured plate

Fig.11 : Damage in an oven-manufactured plate

To establish a chronology of damage, impacts with increasing energy are carried out. For each stacking we observe the same energy threshold of damage initiation.

For an impact energy of 1 J, matrix cracking appears without delamination (Fig.8).

For an impact energy of 2.5 J, a delamination area becomes visible in the cracking zone (Fig.9).

We consider for these kinds of structures the matrix cracking sets on the delamination.

The different manufacturing conditions allow us to study the influence on the damage of the interface quality between plies. Photographs (Fig.10-11) show the damage caused by an impact energy of 15 J in two [03/904/03] plates. The macrocracks in the internal ply (90°) are numerous when the plates are cured in an autoclave. But on the other hand, the delamination area is more extensive when the plates are cured in an oven.

In accordance with these results, we consider that a strong adhesion between two plies of different orientation (as in the case of autoclave plates) prevents the initiation and propagation of the delamination, but on the other hand it sets off the initiation of cracks in the internal 90° layer.

Fig.8 : First matrix cracking

The stacking of the plies is one of the main parameters which governs the behaviour of the structure under a low velocity impact. The three different stackings modify the thickness of the two 0° layers. We notice for an impact energy of 15 J that there are macrocracks in the non-impacted 0° layer. The average length of the macrocracks increases with the thickness of the 0° layer (Fig.12), but their number decreases.

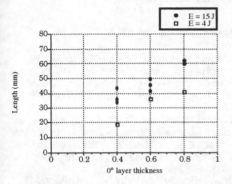

Fig.12 : Average length of macrocracks versus the non impacted 0° layer thickness

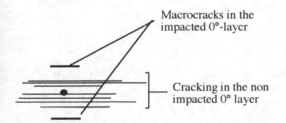

Fig.13 : Schematic representation of the cracking in a $[0_4/90_2/0_4]$ plate

In the $[0_4/90_2/0_4]$ plates, some macrocracks appear in the impacted 0° layer (Fig.13).

This cracking process shows that, when the thickness of the external layers grows, the plate has a behaviour identical to an unidirectional one, with the appearance of few long cracks.

In the same way, the delamination area is greater when the thickness of the 0° layer is more important.

With the aim of studying the propagation of the damage, some tests for increasing impact energy have been carried out. All the $[0_3/90_4/0_3]$ plates are impacted with the same steel projectile. The range of impact energy is from 2.2 J to 22 J. As shown on the photographs (Fig.14), the delamination area extends with the impact energy.

5 STUDY OF RELATIONS : STRUCTURAL RESPONSE VS IMPACT PARAMETERS AND DAMAGE

For the same experimental conditions, the results are shown on Fig. 15-16 for the three different stackings.

The simular results have been obtained for the plates which have the same stacking but are cured under different conditions.

Except in the case of experimental errors, the general response is not modified by the initiation or the propagation of the damage in the different plates. However, the damage under low velocity impact is localised and very small in comparison with the geometry of the plate. In fact, it may not influence the overall response of the structure.

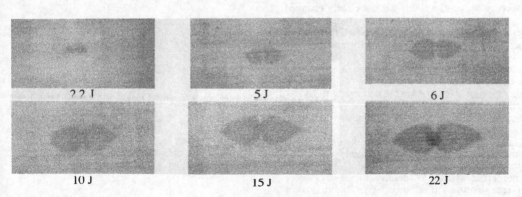

Fig.14 : Photographs of damage area for different impact-energy levels

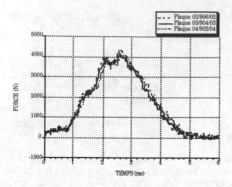

Fig. 15 : Impact force chronology for each stacking

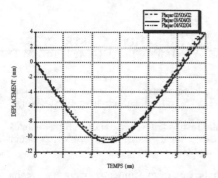

Fig. 16 : Displacement of the centre of the non impacted face for each stacking

6 CONCLUSION

The experimental device developed in our laboratory allows us to obtain impact force histories on plates under low velocity impact. A parametric study based on the influence of the stacking, of the manufacturing conditions and of the impact energy shows that the overall response of the plates has not clearly changed when the damage occurred. In this range of low velocity impact, the damage is so localised that it can't be identified by a change of the overall response.

We define a chronology of damage in a 10-plies glass-epoxy laminate under low velocity impact. The matrix cracking is the first visible damage without any delamination. The nature and the geometry of damage depend on the mesostructure of the plate.

To have a better understanding of the damage process during a low velocity impact test, we develop a hybrid approach mixing experimental, theoretical and numerical aspects to identify the behaviour of composite structures [Collombet 93]. The theoretical and numerical part is based on the development of an original model of damage mixing in a dynamic finite explicit code, a self consistent method to model matrix cracking, and the calculation of contact forces to represent the impact loading and the delamination between plies.

ACKNOWLEDGEMENTS

The authors would like to acknowledge DRET, PIRMAT-CNRS, AQUITAINE Area and CESTA-CEA for their financial support.

REFERENCES

Ayax E., Roux R., Lataillade J.L., "Behaviour of glass polyester laminates to high velocity projectiles", Proceedings of ECCM5, Bordeaux (France), April 7-10 1992, ISBN 2-9506577-02-2, pp 297-311

Bacon C.,1993, "Mesure de la ténacité dynamique de matériaux fragiles en flexion trois-points à haute température - Utilisation des barres de Hopkinson". PhD thesis,Université Bordeaux I n° 840, Janvier 93

Collombet F., Bonini J., Martin V., Lataillade J.L., "Hybrid method to study the behaviour of composite plates under low velocity impact", Euromech 306, Mechanics of Contact Impact, Prague (Czech Republic), September 7-9 1993

On the mechanical behavior of the black cork agglomerate

M.J. Lopes Prates
Instituto Nacional de Engenharia e Tecnologia Industrial, Lisboa, Portugal

J.A. Gil Saraiva
Laboratório Nacional de Engenharia Civil, Lisboa, Portugal

M.J. Moreira de Freitas
Instituto Superior Técnico da Universidade Técnica de Lisboa, Portugal

ABSTRACT: The paper suggests a specific experimental procedure and a specific analytical methodology to identify the physical mechanisms presents in the compression of the black cork agglomerate. It was found that a desordered and frozen polymeric structure is present in the granular interfaces of the material and that this structure can be manipulated by cyclic compression to modify the mechanical behaviour of the material.

1 INTRODUCTION

The black cork agglomerate is a commercial product made exclusively of cork, previously triturated in granules with a typical dimension of about 2 cm, by the industrial process of steambaking (Fernandez 1970). It is usually classified as acoustic, thermic and vibratic corkboard in function of crescent density. The main characteristics were largely described by Fernandez (1972; 1974a,b) and sumarized elsewhere (Andrade 1962; Prates & Saraiva 1990).

From the whole industrial process we emphasize three steps in the autoclave, where the granules are first pressed together, then submitted to an homogenous direct steaming and finally pressed just to the final density. These steps together with the cooling both controlled and natural of the steambaked material, determine the quality of the granular bonding interfaces.

The intergranular interfaces can be seen as rough walls with fragments resulting some sort of cell wall fusion and gluing. Both are originated by polycondensation and polysolidification of part of the material exuded from the walls of the cork granules cells during the steambaking.

The cellular structure of cork is very irregular (Pereira et al. 1987) and the topochemistry of these cellular walls as well as the molecular structure of their chemical components are not well known (Pereira 1988; Graça 1990a,b).

However the effects of hot water treatment and steam heating are known (Rosa & Fortes 1989; Rosa

et al. 1990), as well as some transformations brought about in the structure and chemical composition of the steambaked corkboard (Ferreira & Pereira 1986; Pereira & Ferreira 1989) have been studied.

Among the points to be clarified remains the role played by the core material of the interfaces (Pereira & Ferreira 1989). It is formed by amorphous polymers and, as a result of the polycondensation, it must be considered with a desordered and frozen structure, though no published results showing the existence of this structure via tradicional mechanical tests is known.

Concerning the mechanical properties of the black cork agglomerate, it was seen previously (Prates & Saraiva 1989, 1990a,b, 1991) that i) the coefficient of Poisson can be considered null, ii) the material behaves as isotropic, as theoretically described by Rosa & Fortes (1988), iii) the deformation is allways nonlinear (energy being dissipated), iv) there are four specific steps for the deformation process in which the modulus of elasticity (1st) grows quickly, (2nd) falls quickly, (3rd) stands on a (quasi)patamar and (4th) grows steadily, v) imposing an important predeformation it does never more recovers the 1st and the 2nd stages refered in iv) (a remarkable property to be explored in the field of vibration insulation), vi) the lower values for 3rd stage of deformation are obtained with low density boards and vii) an increasing ambient temperature make the deformation easier.

The aim of this paper is to present an appropriate criteria to identify the real main physical mechanisms present in the deformation process of the black cork agglomerate, for a better understanding of their mechanical behaviour.

2 EXPERIMENTAL DETAILS

The samples used in the tests, all performed at the environmental laboratory conditions (room temperature around 18°C and relative humidity around 65%), were cut in the factory from the commercial product. All have square section A of 150*150mm² but two thicknesses l_0 25 and 50mm were tested. Density ρ_0 of the samples was between 120-170kg/m³.

To access the mechanisms of the deformation process a few series of compression tests were made, varying the experimental procedure. From the whole set of tests (Prates 1993), it will be presented hereby the following series, each one repeated for 3 times: i) 5 cycles for large loading/unloading conditions (50 or 80kN), at a constant cross-head speed of 5mm/min, on 4 classes of samples (2 densities: 120/130 and 160/170kg/m³; 2 thicknesses: 25 and 50mm) and ii) 50 cycles of low and medium loading/unloading conditions (1, 2, 5, 10 or 20kN), at the same cross-head speed of 5 mm/min, on the same class of samples (density: 120/130kg/m³; thickness: 50mm).

3 ANALYTICAL METHODS

One can derive from the experimental pairs of load-(F_i)-displacement(δ_i) pairs of stress(σ_i)-strain(ε_i)

$$\sigma_i = F_i / A; \quad \varepsilon_i = \delta_i / l_0 \tag{1a,b}$$

and density(ρ_i)-cellular gas volume(V_i)

$$\rho_i = \rho_0 /(1-\varepsilon_i); \, V_i = V_0 \times (1-\varepsilon_i) \times (1-\rho_i / \rho_s) \tag{2a,b}$$

ρ_s being the cellular wall density, roughly 1300 kg/m³.

To study the evolution of the elasticity moduli, the tangent($d\sigma/d\varepsilon$) and the chord ($\Delta\sigma/\Delta\varepsilon$) elastic Young's modulus, from the standard ASTM E111, were computed. The first, by the quasi central-difference formula

$$E_i = (\sigma(\varepsilon_i + \Delta\varepsilon_r) - \sigma(\varepsilon_i - \Delta\varepsilon_l))/(\Delta\varepsilon_r + \Delta\varepsilon_l) \tag{3}$$

and, the second, by the linear regression over all the points of each one of the quasi straight line fragments that one can identify along a stress-strain diagram.

The elastic wave velocity can be estimated

$$c_i = \sqrt{E_i / \rho_i} \tag{4}$$

as well as the influence of the cellular gas on the deformation process through the evolution of the thermodynamic coefficient

$$n_i = (\ln p_i - \ln p_{i-1})/(\ln V_{i-1} - \ln V_i) \tag{5}$$

$\Delta V = V_{i-1} - V_i$ being a very small quantity and

$$p_i = \sigma_i + p_0 \tag{6}$$

where p_0 can be roughly taken as the atmosferic pressure.

With these experimental parameters one can construct the diagrams relating the pairs (σ, ε), (E, ε), (E, c) and $(\ln p, \ln V)$ and understand, through the study of their evolution, what kind of mechanism is driving the deformation process. The first allows a comparison with the same type of diagrams for cellular solids, polymers and composites; the second shows the slope values at any point of each curve and at the same equivalent point from curve to curve and so it is not difficult to analyse their gaps; the third, to have an indication how are the changes in the bulk of the material; and, the fourth, to see when the cellular gas takes the command of the deformation process.

4 RESULTS AND DISCUSSION

As can be seen through the diagrams RCTE, RCYE, RCCY, RCLN, RETE and RELN of the serie 111, concerning the first cycle of the large compression tests of each one of the sample classes there are four well defined steps in the deformation process. The first and the last two steps are similar to those of the cellular solids and the second is a typical of composite materials.

The main mechanisms are, i) in the first, the cell wall bending and the cell wall stretching of the granular cellular structure and the bending of the superstructure of the granules interfaces, ii) in the second, a regime of brittle fracture, initiated by crazing, in part of the interfacial superstructure, iii) in the third, the cell wall buckling of the cellular structure of the granules associated with the cell wall stretching and the pressure of the enclosed gas and, iv) in the fourth, the enclosed gas compression.

RCYE111 shows the tangent modulus variation and clearly supports the idea of a deformation process with four distints steps.

RCCY111 shows clearly the existence of two basic kind of wave propagation conditions: the lower branch of each curve is the true superposition of two semibranches, (up and down) in correspondence with the first two deformation steps, which mean two similar physical processes, and the uper branch of each curve is what happens in the fourth step.

RCLN and RELN111 show linear behaviour in the upper branch of the curve (n=constant; fourth deformation step), that a quasi-static politropic process occurs, as can be seen in Suskhov. The mechanism was previously confirmed (Prates & Saraiva 1991)

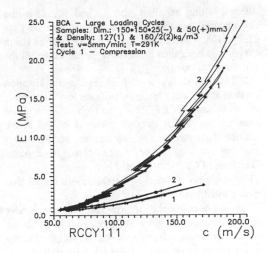

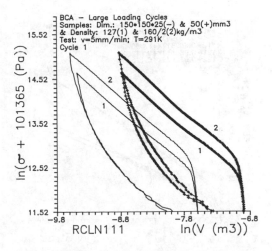

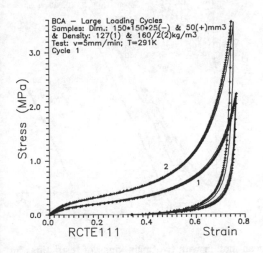

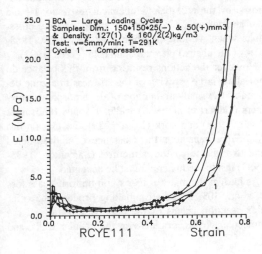

with an internal temperature rising and decreasing in cyclic compression. The changes in the slope of the uper branches are imputed to the destruction caused by the relative motion of the hard carbonized small pieces of wood, which are present in the raw material.

RETE111 shows the traditional step of buckling of the cellular solids with closed cells (Gibson & Ashby, 1988).

The diagrams of the type RCLN and RELN can be used to study the deformation process in compression of all the flexible foams with thin cellular walls.

But, as can be seen through the CCTE, CCYE and CETE of the series 110 and 210 there is something changing at the structural level.

CCTE1101 shows a clear increase in the slope of the compression curve from the first to the second loading cycle, which mean an increase of the material stiffness along the first and second steps of the deformation process, as can better be seen in CETE1105, where the slope of the compression curves was computed by linear regression over all the experimental points.

CCTE2101 shows something like a linearisation process at the same material stiffness level, but now, as can be seen in CETE2105, the high stiffness of the material can extend through the zone of the above defined third step of the deformation process. This growing process can be tracked through diagrams like those of CCYE1101 and 2101.

The question is: what is there in the material that rises its stiffness?

The stiffness of the amorphous polymers under compression is well described as a result of a balance where two kinds of bonding forces (established along and between moleculares chains, each one with brow-

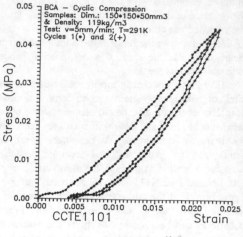

CCTE1101

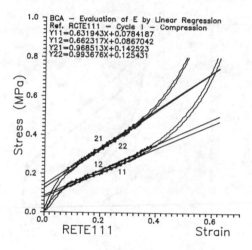

RETE111

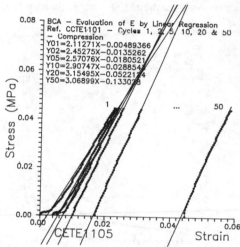

CETE1105

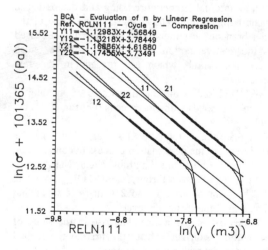

RELN111

nean motion with two main kinds of restrictions imposed by the crosslinks - chemical restrictions - and the entanglements - physical restrictions (like crosslinks when working like hookings (Panyukov 1988-b))- of the chains) play the main role (Perez, 1992).

However the actions resulting from this balance do not allow large growing of the stiffness in a zone defined as the small-strain zone of the amorphous polymers (Gilbert et al. 1986), cellular solids (Gibson & Ashby 1988) and cork (Rosa 1989). To explain this so unique behaviour the existence of a desordered and frozen polymeric structure (Panyukov 1988-1993) inside the material must be accepted.

As known the material had a non uniform mass loss of about 30% in the steambaking process; the complex reaction chains of the depolymerization of the molecular chains, originally in the cellular walls, and their partial polymerization in the granular interfaces must create a new non uniform structure.

This structure must have an high density of defects (Perez 1992) in the glued interfaces, which cause crazing and the nucleation of multiple cracks, giving significant inelastic extension before fracture (Ahmad & Ashby 1988), justifying the shape of the diagram in the defined second step of the deformation process.

In the lower loading cyclic compression the same initial density of crazing can be found but not the same density of nucleation of cracks. Thermal activation, which promote molecular motion (Perez 1992) is present and it can expect to break some chemical bonds and to activate others of the dispersed molecular chains of the same kind of polymers. So, this frozen topological structure (Paniukov 1988a) can be modified, which means to change the quasinetwork of effective hookings (Paniukov 1989).

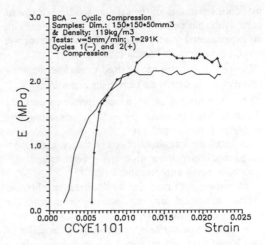

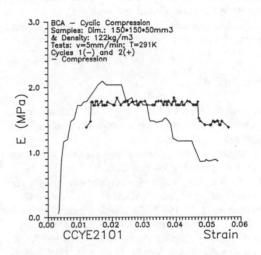

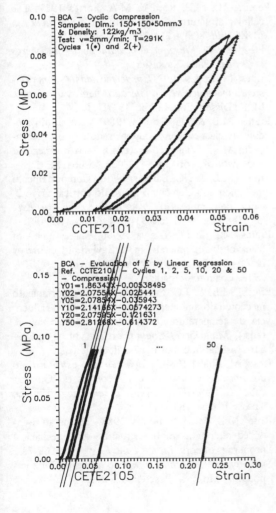

The basic question of the theory - to establish the relationship between the microstructure and the macroscopic elastic characteristics of a polymeric material - remains open (Panyukov 1993, Perez 1992).

However the Panyukov theory can predict the structural changing of the amorphous polymers of some fractions of the material. At the same time the Perez theory of the molecular motion frames quite well the structural changes.

So, we are driven to admit the experimental evidence given by the diagrams.

5 CONCLUSIONS

The (σ, ε), (E, ε), (E, c) and $(\ln p, \ln V)$ diagrams are a powerful tool in the identification of the deformation mechanisms of the cellular solids with closed cells, with thin and soft walls, when associated with an experimental procedure following, in compression tests, a logaritmic increasing loading.

The (E,c) diagram can be used to analyse the structural changes within the composite materials.

The $(\ln p, \ln V)$ diagram is appropriate to evaluate the true importance of the cellular gas along the whole deformation process.

The black cork agglomarate has a frozen and desordered topological inner structure, which can be manipulated in cyclic compression, as done.

The growing stiffness and its linearization obtained in the small-strain zone can be explained through the presence of this structure.

To have a low sitffness, the cokboards must be of good quality, i. e., produced without woody fragments in the raw material and with the glued part of their intergranular superstructure broken. It must work like a cellular solid with closed cells.

REFERENCES

Ahmad, Z. Bin & M.F. Ashby 1988. Failure-mechanism maps for engineering polymers. *Jour. of Mat. Scie.* 23: 2037-2050.

Andrade, A. 1962. *Thermic and acoustic insulation.* Lisboa: Junta Nacional da Cortiça.

Fernandez, L.V. 1970. *Aglomerados negros de corcho (I e II Parte).* C44. Madrid: AITIM-SNMC.

Fernandez, L.V. 1972. *Estudio de las características de los aglomerados expandidos puros de corcho en placas, para aislamento térmico, producidos en España.* SérieC Vol.54. Madrid: AITIM-SNMC.

Fernandez, L.V. 1974a. *Estudio de la calidad de los aglomerados de corcho acústicos y vibráticos.* Série C Vol. 62. Madrid: AITIM-SNMC.

Fernandez, L.V. 1974b. *Aglomerados y discos de corcho. Sus calidades en el aislamiento y en el tapamiento.* Madrid: INIA.

Ferreira, E. & H. Pereira 1986. Algumas alterações anatómicas e químicas da cortiça no fabrico de aglomerados negros. *Cortiça.* 576: 274-278.

Gibson, L. & M. Ashby 1988. *Cellular solids. Structure & properties.* London: Pergamon.

Gilbert. D.G., M.F. Ashby & P. W. R. Beaumont 1986. Modulus-maps for amorphous polymers. *Jour. of Mat. Scie.* 21: 3194-3210.

Graça, J. 1990a. *Constituintes químicos da cortiça e de outros tecidos suberosos.* Assistant dissertation. Lisboa: ISA/UTL.

Graça, J. 1990b. *Topoquímica da parede celular suberosa. Seu reconhecimento por reacções histoquímicas.* Assistant lecture. Lisboa: ISA/UTL.

Panyukov, S.V. 1988a. Theory of polymers with a frozen structure. *Sov. Phys. JETP.* 67(5): 930-939.

Panyukov, S.V. 1988b. Topological interactions in the statistical theory of polymers. *Sov. Phys. JETP.* 67(11): 2274-2284.

Panyukov, S.V. 1990. Theory of structurally disordered polymers. *JETP Lett..* 51(4): 253-255.

Panyukov, S.V. 1993. Theory of elasticity and relaxational dynamics of "soft" solids. *JETP.* 76(5): 808-817.

Pereira, H. 1988. Chemical composition and variability of cork from *Quercus suber* L.. *Wood Sci. Techn..* 22: 211-218.

Pereira, H. & E. Ferreira 1989. Scanning electron microscopy observations of insulation cork. *Mater. Scie. Eng..* A111: 217-225.

Pereira, H.; M.E. Rosa & M.A. Fortes 1987. The cellular structure of cork from *Quercus suber* L.. *IAWA Bull.* 8(3): 213-218.

Perez, J. 1992. *Physique et mécanique des polymères amorphes.* Paris: Lavoisier

Prates, M. Lopes 1993. *Características e comportamento mecânico de aglomerados negros de cortiça.* M. Sc.Dissertation. Lisboa: IST/UTL.

Prates, M.L. & J.G. Saraiva 1989. Variação do módulo de elasticidade do aglomerado negro de cortiça. in M. T. Vieira et al. (eds.), *Novos materiais: que futuro?*, vol 1, p. 413-429. Lisboa: SPM.

Prates, M.L. & J.G. Saraiva 1990a. Comportamento mecânico do aglomerado negro de cortiça. 1° e 2° ciclos de compressão. *Proc. II Enc. Nac. An. Exp. de Tensões,* vol. 2, p.2.31-2.44. Lisboa: APAET.

Prates, M.L. & J.G.Saraiva 1990b. Characteristis of black cork agglomerate. in A. Sayigh (ed.), *Energy and the environment into the 1990s,* vol. 4, p.2709-2713. London: Pergamon.

Prates, M.L. & J.G. Saraiva 1991. Comportamento mecânico do aglomerado negro de cortiça. Influência da temperatura. in A. Pádua Loureiro et al. (eds.), *Materiais91-Comunicações,* vol. 2, p.509-516. Lisboa: Soc. Portuguesa de Materiais.

Rosa, M. Emília 1989.*Relação entre a estrutura e o comportamento mecânico da cortiça do* Quercus suber *L. Contribuição para o seu estudo.* Ph. D. Thesis. Lisboa: IST/UTL.

Rosa, M.E. & M.A. Fortes 1988. Temperature-induced alterations of the structure and mechanical properties of cork. *Mater. Scie. Eng..* 100: 69-78.

Suskhov, V.V.. *Technical thermodynamics.* First impression. Moscow: Peace Publishers.

Recent Advances in Experimental Mechanics, Silva Gomes et al. (eds) © 1994 Balkema, Rotterdam, ISBN 90 5410 395 7

An original approach to stress analysis in thick foams

Laurentiu Neamtu
'Ovidius' University Constantza, Romania

ABSTRACT: The study was made assuming that we're in the elastic range. So we can use superposition if we consider both mechanical and thermal loads. I've elaborated a finite elements method (f.e.m.) original programme using a discrete structure with some in-plane rectangular elements . I've also considered a transversal law of displacements distribution upon Misicu's theory of thick plates ,which I've considered to be an approximation function on one-dimensional elements through the thickness of the plate . So, instead of a 3D element ,I've used a combination of a 2D element and 1D element . Moreover, I've modified the vector of loads in order to introduce the thermic effect as described in Timoshenko's plates and shells theory.

1 INTRODUCTION

Foams, or cellular solids, permit an optimisation of stiffness or of energy absorbtion, for a given weight of material. There are natural foams used for structures (wood) or for thermal insulation (cork) and there are man-made foams which are all functions filled by cellular polymers and which are used for cushions, padding, packaging, insulation. Foams give a way of making very light solids. If combined with stiff skins to make sandwich panels,they give structures which are extremely stiff and light. The engineering potential of foams is considerable and not yet completely realised. The studies on foams mechanics seem also to be neglected, therefore we shall make a study on some thick polystyrene plates under a vertical load p, considering the author's interest in thick laminated composite plates.So,"what's the connection?" one may ask. The answer is that foams are a special class of composite materials, so they might be included- from my point of view- in a sort of a general theory for thick composite plates, even considering laminated plates, but in the special case in which layers have the same properties. This I named a "pseudo"laminated theory, which I used in the analyses of thick foams. The choice of a foam was also dictated by the availability of experimental facilities ,in order to verify the numerical calculus . I also must mention that I was primarily interested in the maximal values of normal stresses σ_{xx} which are given using Misicu's theory (1973) by Kümbetlian & Neamtu (1992) by analytical calculus, and also experimentally determined using strain gages. My f.e.m. programme ,based on the "pseudo"laminated theory ,gave results concerning these stresses which I've compared with the values obtained by the above mentioned means.

2 THERMAL LOADS

As it is shown by Timoshenko (1968),the equation for the bending of a plate following a spherical surface is:

$$\frac{1}{r_x} = \frac{1}{r_y} = \frac{M}{D \cdot (1+v)} \qquad (1)$$

where : $\frac{1}{r_x}$, $\frac{1}{r_y}$ = principal curvatures of

the midplane of the plate ;

D= the flexural stiffness of the

plate = $\frac{E \cdot h^3}{12 \cdot (1-v^2)}$,where E = Young

modulus ; h = thickness of the plate; v = Poisson ratio ; M = bending moment with uniform distribution along the boundary of the plate. Relation (1) may be used in order to determine the efforts due to the variation of the temperature into the plate for some cases of uneven heating. In case that we admit linear variation of temperature between the two faces of the plate but no temperature variation in the planes parallel to the midplane, the situation may be assimilated to that of the simple bending of the plates, following a spherical surface .

If : α = thermal expansion coefficient of the material of the plate ;

t = t1 – t2 , where t1 = temperature of the upper face and t2 = temperature of the lower face, then for

$$\frac{1}{r_x} = \frac{1}{r_y} = \frac{1}{r} \text{ , Timoshenko (1968) gives us}$$

$$\frac{1}{r} = \frac{\alpha \cdot t}{h} \qquad (2)$$

If the plate has built-in bounds, the uneven heating generates bending moments with uniform distribution along the boundary. Those moments are the sequence of the annul of the uneven heating.

Considering eq.(1) and (2) , we may write (Timoshenko, 1968):

$$M = \frac{\alpha \cdot t \cdot D \cdot (1+v)}{h} \qquad (3)$$

so that $\quad \sigma^{max} = \frac{\alpha t E}{2(1-v)} \qquad (4)$

which is a justified relation in case of the plate with all built-in boundary.

In case of a single built-in boundary we may conclude that for that bound, relation (4) is still valid.

3. MECHANICAL LOADS

Misicu (1973) assumed a displacements field given by:

$$u_1 \equiv u = \sum_{i=0,1}^{\infty} \frac{z^{2i+1}}{(2i+1)!} U_i \ ,$$

$$u_2 \equiv v = \sum_{i=0,1}^{\infty} \frac{z^{2i+1}}{(2i+1)!} V_i \ ,$$

$$u_3 \equiv w = \sum_{i=0,1}^{\infty} \frac{z^{2i}}{(2i)!} W_i \qquad (5)$$

and which have to satisfy the differential equations, in tensorial form :

$$\sigma_{ij} = \delta_{ij}\lambda\theta + 2\mu\epsilon_{ij} \quad , \quad \sigma_{ij}, j = 0 \qquad (6)$$

where $\quad \theta = u_{i,i} \ , \ 2\epsilon_{ij} = u_{i,j} + u_{j,i} \quad$ and

$\lambda \ , \ \mu$ – Lamme coefficients.

Considering a bi-harmonica transversal load p, Kümbetlian (1976) obtained analytically, from (5) the functions of displacements in the form :

$$u = zw_{,x} - \frac{1}{1-v} \left[(\frac{h}{2})^2 \cdot z - (2-v) \cdot \frac{z^3}{3!} \right] \Delta w_{,x} -$$

$$\left[\frac{5-2v}{6(1-v)^2} \cdot (\frac{h}{2})^4 \cdot z - \frac{3-2v}{2(1-v)^2} \cdot (\frac{h}{2})^2 \cdot \frac{z^3}{3!} + \right.$$

$$\left. \frac{3-v}{1-v} \cdot \frac{z^5}{5!} \right] \frac{P_{,x}}{D} - \left[\frac{111-147v+91v^2-25v^3}{5!(1-v)^3} \right.$$

$$(\frac{h}{2})^6 \cdot z - \frac{20-31v+19v^2-5v^3}{12(1-v)^3} \cdot (\frac{h}{2})^4 \cdot \frac{z^3}{3!} +$$

$$\left. \frac{2-v}{(1-v)^2} \cdot (\frac{h}{2})^2 \cdot \frac{z^5}{5!} - \frac{4-v}{1-v} \cdot \frac{z^7}{7!} \right] \frac{\Delta p_{,x}}{D} \qquad (7)$$

and analogous for v and w. Considering (7) it was possible to get values for stresses in thick plates, for instance :

$$\sigma_{xx} = -\frac{E}{1-v^2} \left\{ z \cdot (w_{,xx} + vw_{,yy}) + \left[\frac{z}{4h} - \frac{2-v}{6} \right. \right.$$

$$(\frac{z}{h})^3 \right] \cdot h^3 \cdot \Delta w_{,xx} \} - \left[\frac{3vz}{2(1-v)h} - 2v \cdot (\frac{z}{h})^3 \right] \cdot p \cdot$$

$$- \left[\frac{5-2v}{8(1-v)} \frac{z}{h} - \frac{3-2v}{4(1-v)} \cdot (\frac{z}{h})^3 + \frac{3-v}{10} \cdot (\frac{z}{h})^5 \right]$$

$$\cdot h^2 p_{,xx} - \left[\frac{v(5-2v)}{16(1-v)^2} \cdot \frac{z}{h} - \frac{v}{4(1-v)} \cdot (\frac{z}{h})^3 \right.$$

$$\left. + \frac{v}{10} \cdot (\frac{z}{h})^5 \right] \cdot h^2 \cdot \Delta p - \left[\frac{111-147v-91v^2-25v}{640(1-v)^2} \right.$$

$$\cdot \frac{z}{h} - \frac{20-31v+19v^2-5v^3}{96(1-v)^2} \cdot (\frac{z}{h})^3 +$$

$$\left. + \frac{2-v}{40(1-v)} \cdot (\frac{z}{h})^5 - \frac{4-v}{420} \cdot (\frac{z}{h})^7 \right] \cdot h^4 \cdot \Delta p_{,xx} \quad (8)$$

and analogous for all σ_{ij} ,

where i, j = x, y, z , including their maximal values in lateral surfaces, for z = ± h/2. For example, in case of cylindrical bending, for p = P + Qx, the maximal value for σ_{xx}

goes for (Kümbetlian & Neamtu, 1992) :

$$\sigma_{xx}^{max} = -\frac{1}{W_y} Dw_{,xx} - \frac{4-v}{4(1-v)} p \ ; \ W_y = \frac{h^2}{6} \qquad (9)$$

4. THERMO-MECHANICAL LOADS

Using superposition we'll get corresponding values of stresses.

5. CALCULUS BY F.E.M.

The displacements field given by (5) gave me the idea of using a simplified form of a layer-wise laminate theory, where all layers are of similar properties. Moreover, the displacement field (5) offers the possibility of increasing the degrees of freedom of the structure, that is more likewise the physical reality, but simplifies Cosserat medium.

So, if we consider the variational principle of the minimum potential energy, applied to a specific finite element, we may write:

$$\delta\Pi_e = 0 \qquad (10)$$

precisely :

$$\delta\left(\frac{1}{2}\iiint_{V_e}(\{\epsilon\}^T\{\sigma\} - 2\{u\}^T\{X\})\,dV - \right.$$

$$\left.- \iint_{S_1}\{u\}^T\{T\}dS\right) = 0 \qquad (11)$$

where, $\{\epsilon\}^T$ - the transposed matrix with the components of the strain tensor, $\{\sigma\}$ - the matrix with the components of the stress tensor, $\{u\}^T = [\,u\ v\ w\,]$ -vector of displacements, $\{X\}^T = [\,X\ Y\ Z\,]$ -vector of body forces, $\{T\}^T = [\,T_x\ T_y\ T_z\,]$ -traction on the surface of the body.

Now, instead of displacements field (5) we take a finite form, that is possible to be modelled by finite elements. Therefore, we shall consider U_i , V_i , W_i as functions of (x,y) in the plane with z_i and i =1,n, where n will be the number of nodes (planes) that we'll consider to discretize the structure through the thickness of the plate. Thus, instead of (5), we can write :

$$u = [N_i]\{U_i\} \qquad (12)$$

where $[N_i] = [N_1 N_2 \ldots N_n]$ represents the matrix of the interpolation functions on the 1D element which models the plate through its thickness as a vertical rod ; $\{U_i\}^T = \{U_1 U_2 \ldots U_n\}$ is the transposed vector of displacement u components in each (x,y) plane at z_i , i = 1,n. Now, every U_i is approximated by a 2D mesh corresponding to the z_i plane, so we can talk about a new set of shape functions, that we'll note with

$[N_j] = [N_1 N_2 \ldots N_m]$, where m represents the number of nodes per 2D chosen element. Consequently,we may write:

$$U_i = [N_j]\{U_{ij}\} \qquad (13)$$

where $\{U_{ij}\}^T = \{U_{i1} U_{i2} \ldots U_{im}\}$ is the transposed vector of the nodal displacements of the 2D chosen element in the z_i plane.

Now it is obvious that we may write:

$$[U_i] = \begin{bmatrix} [N_j]\,[U_{1j}] \\ [N_j]\,[U_{2j}] \\ \ldots \\ [N_j]\,[U_{nj}] \end{bmatrix} \qquad (14)$$

and so, the displacement u becomes :

$$u = [N_i]\cdot\begin{bmatrix} [N_j]\,[U_{1j}] \\ [N_j]\,[U_{2j}] \\ \ldots \\ [N_j]\,[U_{nj}] \end{bmatrix} \qquad (15)$$

and analogous procedure is followed for v and w. We mention that the first term of the second member of eq.(15) is only z dependent, while the second term is only (x,y) dependent. This is essential when integrating, because it will be no matter which integral will be made first: the surface or the thickness one. Now, if matrix [B] is considered so that we may write ,by respect to the element :

$$\{\epsilon\} = [B]\cdot\{d\} \qquad (16)$$

where : {d} - vector of nodal displacements, then we have the stiffness matrix [k] :

$$[k] = \iiint_{V_e}[B]^T[E]\,[B]\,dV \qquad (17)$$

where, generally,

$$[E] = \begin{bmatrix} 1/E_{11} & -v_{21}/E_{22} & -v_{31}/E_{33} & 0 & 0 & 0 \\ -v_{12}/E_{11} & 1/E_{22} & -v_{32}/E_{33} & 0 & 0 & 0 \\ -v_{13}/E_{11} & -v_{23}/E_{22} & 1/E_{33} & 0 & 0 & 0 \\ 0 & 0 & 0 & G_{xy} & 0 & 0 \\ 0 & 0 & 0 & 0 & G_{yz} & 0 \\ 0 & 0 & 0 & 0 & 0 & G_{zx} \end{bmatrix}$$

Due to space limitations we won't include in this paper the final form of the stiffness matrix, but the idea of the further calculus is from now on obvious.
Integrating by respect to Z axis , eq. (17)

becomes :

$$[k] = \iint_{S_e} [B^*]^T [E^*][B^*] \det[J] \, d\xi \, d\eta \quad (18)$$

where : S_e -surface of the finite element;

   $[J]$ - Jacobi matrix;

   $[B^*]$ - matrix $[B]$ after integration by respect to Z axis;

   $[E^*]$ - matrix of flexural stiffness

and ξ, η - the local coordinate system for the 2D element. I used a simplified application in order to verify my theory. The work was done with some isoparametric quadratic 2D elements, with eight nodes and central node. The global system is (X,Y) and the element has its own axis system (Fig. 1). The transversal interpolation was made with a single 1D element ,but with shape functions of the third order.

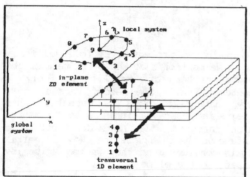

Figure 1

In the system of equations characteristic to f.e.m. :

$$[K]\cdot\{d\} = \{p\} , \quad (19)$$

where $\{p\}$ - vector of loads, we use superposition in order to modify the vector of loads. Therefore, we consider:

$$\{p\} = \{p^m\} + \{p^t\} , \quad (20)$$

where : $\{p^m\}$ - the component due to the mechanical load;

   $\{p^t\}$ - the component due to thermal load;

Hence, a condensed form :

$$\{p^t\} = \iiint_{V_e} [B]^T [E] \{\epsilon_0\} dx \, dy \, dz =$$

$$= \iint_{S_e} [B^*]^T [E^*] \{\epsilon_0^*\} \det[J] \, d\xi \, d\eta \quad (21)$$

where the transposed form of $\{\epsilon_0^*\}$ is :

$$\{\epsilon_0^*\}^T = \frac{(t_2 - t_1)}{h} \{\alpha_x \ \alpha_y \ \alpha_z \ 0 \ 0 \ 0\} , \quad (22)$$

so strains $\{\epsilon_0^*\}$ correspond to the difference of temperature by respect to the thickness of the plate.

6. THE COMPUTING PROGRAMME

It is an original programme written in Turbo-Pascal 6. We can extend the dimension of the analyzed structure using dynamically allocated variables. For the increase of speed we used RAMDRIVE facilities. The programme has a modular conception and it works by the successive run of its three specialised modules : Placa_1.exe , Placa_2.exe , Placa_3.exe . Placa_1.exe receives input data , automatically generates the structure with constant or variable step in case of at least two linear bounds of the structure, and calculates the stiffness matrix. Placa_2.exe calculates nodal displacements and , if request , print them. Placa_3.exe calculates stresses . For the installation of the three modules we need approx. 207 KB of HDD or FDD. The programme has also a various menu and we are allowed to replay and to modify data in any point we wish. To obtain values from tables 1 and 2 ,I used a 2D mesh of 60 elements and the 1D transversal element was taken with a 3rd degree of shape functions.

7. EXPERIMENTAL ASPECTS

There were used thick plates (h/a > 1/10) made of polystyrene ,with a = 315 mm between the constrained margins. There were done strain gage measurements considering all the constrains due to the work with plastic plates (i.e short time measurement in order to avoid over-heating and wrong results etc.). The used adhesive was arabic gumma and it proved to be quite the requested one for the bonding of the gages. The choice of the material was imposed by the necessity of large deflections with reduced loads, in the conditions of large relative thicknesses (h/a) and negligible body forces compared to the applied loads. All the measurements were done at the middle of the built-in margin ,at x = a, in the section with maximal moment. The values for the Young modulus and Poisson coefficient were determined upon the Romanian standard STAS 5874-73-50.

8. RESULTS. CONCLUSIONS

There were analyzed two cases of plate under cylindrical bending, with uniform distributed load P, with different conditions for the two supported margins :
-case 1: plate with one bound built-in and the opposite simply supported (fig. 2a) , which gave the results presented in table 1;
-case 2: plate with both opposite bounds built-in (fig. 2b) , and with results in table 2.

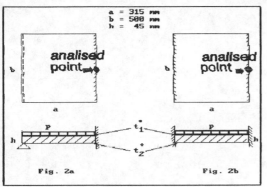

a = 315 mm
b = 500 mm
h = 45 mm

Fig. 2a   Fig. 2b

Figure 2

Material properties were :
E = 51.851 MPa ; ν = 0.2778 ;
$\alpha_x = \alpha_y = \alpha_z = 7.5 \cdot 10^{-5}\ [degr^{-1}]$

Table 1
Plate supported - built-in (Fig. 2a);
VALUES OF MAXIMAL STRESSES σ_{xx} [MPa]

Mechanical load	t °C	p= e-4* 23.648 [MPa]	p= e-4* 29.734 [MPa]	p= e-4* 35.947 [MPa]
Exact theory of thick plate	-2	0.08727	0.11112	0.135468
	0	0.09266	0.11650	0.140853
	+2	0.09804	0.12189	0.146237
Theory of thin plate	-2	0.08152	0.10389	0.126723
	0	0.08690	0.10927	0.132108
	+2	0.09229	0.11465	0.137492
F.E.M for thick plate - My programme	-2	0.08162	0.10437	0.1277
	0	0.08861	0.11141	0.134690
	+2	0.09560	0.11845	0.141780
NONSAP (25 bricks 21 nodes each)	0	0.06550	0.08179	0.09888
Experiment values	0	0.09551	0.12024	0.143277

Table 2
Plate built-in - built-in (Fig. 2b);
VALUES OF MAXIMAL STRESSES σ_{xx} [MPa]

Mechanical load	t °C	p= e-4* 35.947 [MPa]	p= e-4* 42.161 [MPa]	p= e-4* 48.374 [MPa]
Exact theory of thick plate	-2	0.08875	0.10503	0.121302
	0	0.09414	0.11041	0.126687
	+2	0.09952	0.11579	0.132071
Theory of thin plate	-2	0.08268	0.09790	0.113132
	0	0.08807	0.10329	0.118517
	+2	0.09345	0.10867	0.123901
F.E.M for thick plate - My programme	-2	0.08844	0.10445	0.1204
	0	0.09265	0.10866	0.12467
	+2	0.09686	0.11287	0.128890
NONSAP (25 bricks 21 nodes each)	0	0.06568	0.07703	0.08838
Experiment values	0	0.09664	0.112936	0.129712

The errors that occurs - comparing to the exact solutions of thick plates theory - in the f.e.m. calculus are between 1.6 % (in the second case of analyze) and 4.3 % (in the first case of analyze) ,so then very much acceptable .
We recommend the use of this kind of layer-wise approach in the analyze of thick plates . Using space elements difficulties of calculus are amplified and we don't consider as a necessity this compromise, the results being quite satisfactory . Making a more refined mesh, or considering a higher order of interpolation through the 1D element, the results will improve. Of course

there will be a more complicated calculus, so we'll have to decide when to stop with the refinement. Solutions with 3D elements, as we can see, give also good enough results for a simple mesh. But refinements in this case will prove to be more cumbersome than in the case of the combination between the 2D element and 1D element. Moreover, local effects can't be emphasized with 3D elements. For example ,interlaminar stresses that occur in case of laminated composites. So this kind of approach is to be extended to thick laminated composite plates, which are the preoccupation of the author for his Ph.D. research work.

REFERENCES

Ashby M.F. & Jones D.R.H. 1980. Engineering Materials II. Pergamon.
Gârbea D. 1990. Analiza cu elemente finite. Bucharest: Technical Publishers.
Kümbetlian G. 1976. Calculul exact al plăcilor actionate hidrostatic si cu stare complexâ de mobilitate a conturului, pe baza rezolvârii antimediane a teoriei mobilitâtii elastice. Bucharest: Doctorate thesis.
Kümbetlian G. & Neamtu L. 1992. Metode comparative pentru studiul tensiunilor în plăci groase. Craiova, Romania : The Sixth National Conference on Stress Analysis and Material Testing.
Misicu M. 1973. Incovoiere si torsiune. Bucharest: Academy Publishers.
Timoshenko S. 1968. The theory of plates and shells. Bucharest: Technical Publishers.

Recent Advances in Experimental Mechanics, Silva Gomes et al. (eds) © 1994 Balkema, Rotterdam, ISBN 90 5410 395 7

An experimental and theoretical investigation of the mechanical properties of a pre-tensioned unidirectional glass/epoxy composite

J.N.Ashton & A.S.Hadi
Applied Mechanics Division, Department of Mechanical Engineering, UMIST, Manchester, UK

ABSTRACT

The effect of fibre pre-tension on the mechanical properties of E-glass fibre in an epoxy resin has been determined experimentally for different composite pre-stresses (0,25,50,75,100,200MPa) and for each pre-tension at a number of fibre volume fractions. A theoretical model based on the rule of mixtures is used to predict the effect of fibre pre tension on the pre-load strain in the fibre and matrix, the tensile strength and elastic modulus of the composite for a given curing temperature. The results show that pre-tensioning of the fibre increases the tensile strength of the composite. This increase has been found to be a function of fibre pre-tension and fibre volume fraction. Good agreement has been achieved between the theoretical model and the experimental results.

INTRODUCTION

Glass reinforced plastics remain one of the most commonly used of composite materials and are to be found in both structural and non-structural applications.However, glass fibre has a serious disadvantage as a reinforcing material because of its low rigidity which means that the full strength of the fibre cannot be utilised. Pre-tensioning of the glass fibre before constuction of the composite is likely to assist in overcoming the problem of low rigidity. It will also give better fibre alignment in continous unidirectional composites in the direction of the applied pre-tension and enable greater utiilisation of the superior mechanical strength of the glass fibre.

In this work a tensile load has been applied to the fibre bundles during fabrication of a unidirectional glass fibre reinforced composite. The applied tension is released on final curing thus allowing the pre-tensioned fibres to contract inducing pre-stresses in the composite as a whole.

THEORETICAL ANALYSIS

A theoretical model developed by Tuttle[1] has been applied and compared with the experimental results.The model assumes that a perfect bond exists between fibre and matrix,and that both fibre and matrix are isotropic linearly elastic materials. The temperature dependence of the matrix material is ignored. Figures 1(a) and 1(b) show schematically an element of the material before and after the applied tension is released.

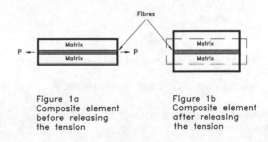

Figure 1a
Composite element
before releasing
the tension

Figure 1b
Composite element
after releasing
the tension

In Figure 1(a) the element consists of a fibre with pre-tension P in an unloaded matrix material. This condition is maintained during the construction and curing process. During this time the fibre and matrix stresses are

$$\delta_p = \frac{P}{A_f} \quad , \quad \delta_{pm} = 0$$

where δ_p and δ_{pm} are initial fibre and matrix stresses respectively

A_f is the fibre cross-sectional area.

Releasing the the applied tension will result in a change in length of the element, as shown in Figure 1(b), which is assumed to be uniform due to the perfect bond between fibre and matrix, and will cause a change in the fibre and matrix stresses to some new values δ_{fp} and δ_{mp} respectively. Since the element remains in equilibrium

$$\delta_{fp} A_f + \delta_{mp} A_m = 0 \qquad (1)$$

where A_m is the matrix cross-sectional area

The stress in the fibre is given by

$$\delta_{fp} = E_f \epsilon_{fp} + \delta_p \qquad (2)$$

The curing process took place at room temperature and it has been assumed that the epoxy resin was sufficiently viscous to allow almost complete relaxation of the thermal residual stresses. Consequently, the curing temperature is taken to be the stress free temperature(2) and the stress in the matrix is given by

$$\delta_{mp} = E_m \epsilon_{mp} \qquad (3)$$

where E_f and E_m are the fibre and matrix elastic moduli and
ϵ_f and ϵ_{mp} are the fibre and matrix strains.

Since perfect bonding between fibre and matrix has been assumed, therefore

$$\epsilon_{mp} = \epsilon_{fp} = \epsilon_p$$

where ϵ_p is the pre-load strain.

noting that

$$V_f = \frac{A_f}{A} \quad , \quad V_m = \frac{A_m}{A}$$

and $\quad E_{11} = E_f V_f + E_m V_m \qquad (4)$

where E_{ij} is the elastic modulus of the composite in the fibre direction and, V_f and V_m are the fibre and matrix volume fractions respectively.

Combining equations (1), (2), (3) and (4)

the pre-load strain is given by

$$\epsilon_p = - \frac{V_f \delta_p}{E_{11}} \qquad (5)$$

THEORETICAL RESULTS

The variation of the pre-load strain with fibre volume fraction for E-glass fibre ($E_f = 72.5$GPa) in an epoxy resin ($E_m = 3.45$GPa) for different fibre pre-tensions is shown in Figure 2. Figures 3 shows the predicted variation of matrix and fibre stress withe fibre volume fraction for a pre-stress of 200MPa. The pre-stressed composite tensile strength is calculated using the Rule of Mixtures, i.e.

$$\delta_c = \delta_f V_f + \delta_m V_m \qquad (6)$$

where δ_c = composite tensile strength

and, δ_f and δ_m are fibre bundle and matrix tensile strengths respectively

For a unidirectional composite

$$\delta_c = \epsilon_{11} E_{11}$$

and for a pre-stressed unidirectional composite

$$\delta_{cp} = E_{11} (\epsilon_{11} - \epsilon_p)$$

where δ_{cp} = tensile strength of pre-stressed unidirectional composite.

noting that

$$\delta_f = E_f \, \epsilon_{11}$$
$$\delta_m = E_m \, \epsilon_{11}$$

and combining equations (4), (5), (6) and (8) gives

$$\delta_{cp} = \delta_f V_f + \delta_m V_m + \delta_p V_f$$

where δ_m and δ_f = matrix and fibre bundle

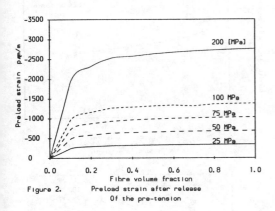

Figure 2.   Preload strain after release
Of the pre-tension

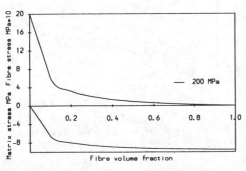

Fibre and matrix stresses after release
of the pre-tension

Figure 3.

COMPOSITE FABRICATION AND TESTING

The filament winding process has been used to produce composites, consisting of unidirectional E-glass fibre (Roving) in an epoxy (My750 & Hy931 manufactured by Cibe-Geigy), using a specially designed pre-tensioning device as shown schematically in Figure 4. Release of the

pre-load was achieved by cutting the composite at the mandrel edge just prior to final cure.

The tensile test specimens have been manufactured and tested in accordance with BS 2782 Part 10 (3) for the testing of unidirectional composites.Testing was carried in an Instron Universal Testing Machine.A total of ten specimens have been tested for each pre-stress and fibre volume fraction.

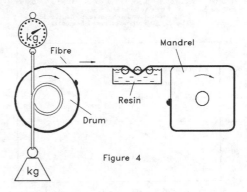

Figure 4

EXPERIMENTAL RESULTS

Figures 5,6,7,8 and 9 show the results of tensile tests to failure for composite pre-stresses 0,25,50,75,100 and 200MPa at fibre volume fractions of 0.3, 0.45 and 0.6. These are plotted as tensile strength against fibre volume fraction. Also plotted are the results of the theoretical analysis together with the predicted tensile strength based on the simple Rule of Mixtures approach. Both the theoretical plots have been extended to zero fibre volume fraction at which the tensile strength is that of the resin only.

Figure 5 also shows a comparison between the experimental results for the composite with zero pre-stress and 200MPa pre-stress. The mechanical properties used in the theoreical model were obtained experimentally from tensile tests on individual fibre bundles and on epoxy resin specimens. Figure 10 shows the variation of fibre bundle tensile strength with fibre gauge length.The variation of modulus of elasticity,obtained from the the tensile tests, with composite pre-stress for different fibre volume fractions is shown in Figure 11. Also plotted on Figure 11 is the predicted elastic modulus calculated using the Rule of Mixtures, equation (4).

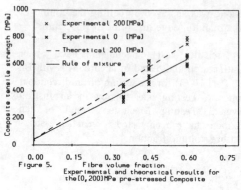

Figure 5. Experimental and theoretical results for the [0,200] MPa pre-stressed Composite

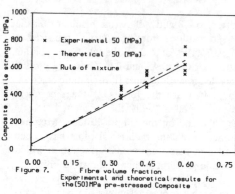

Figure 6. Experimental and theoretical results for the [25] MPa pre-stressed Composite

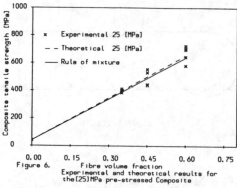

Figure 7. Experimental and theoretical results for the [50] MPa pre-stressed Composite

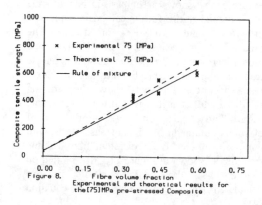

Figure 8. Experimental and theoretical results for the [75] MPa pre-stressed Composite

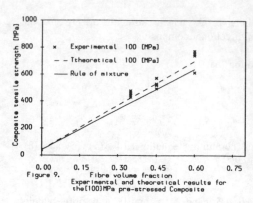

Figure 9. Experimental and theoretical results for the [100] MPa pre-stressed Composite

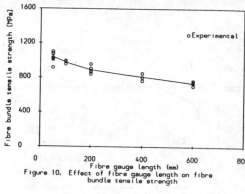

Figure 10. Effect of fibre gauge length on fibre bundle tensile strength

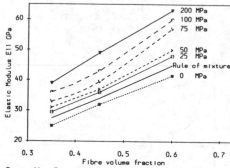

Figure 11. Experimental results for Composite Elastic Modulus for different fibre pre-tension

DISCUSSION

The theoretical results in Figure 2 show that pre-load strain is almost independent of fibre volume fraction for fibre volume fractions greater than 0.1. For example, for a composite pre-stress of 100MPa at $V_f = 0.3$, $\varepsilon_{11} = 1265\mu m/m$ whereas at $V_f = 0.7$, $\varepsilon_p = 1351\mu m/m$ giving only a 6.35% increase. The dependence of pre-load strain on pre-stress can be seen, for example, at $V_f = 0.5$, at pre-stresses of 50 and 200MPa the pre-load strains are $658\mu m/m$ and $2632\mu m/m$ respectively representing a 75% increase. Pre-load strain is primarily a function of fibre pre-load. It can be seen from Figures 5,6,7,8 and 9 that the composite tensile strength increases with fibre volume fraction as expected. These Figures also show that at each fibre volume fraction, the composite tensile strength increases with pre-stress. There is good agreement between the experimental results and theory. Figure 10 shows that fibre bundle tensile strength decreases with increasing gauge length. In Figure 11 it can be seen that at each of the fibre volume fractions tested there is a significant increase in elastic modulus with increase in composite pre-stress. The simple Rule of Mixtures approach overestimates the elastic modulus of the zero pre-stressed composite but significantly under estimates the value for even the 25 MPa pre-stressed composite.

CONCLUSIONS

The theoretical model shows good agreement for each composite pre-stress.Pre-load strain shows little dependence on fibre volume fraction (above 0.1) and a high degree of dependence on composite pre-stress. The elastic modulus in the fibre direction increases with composite pre-stress and with fibre volume fraction.Pre-tensioning of the fibre gives a significant increase in composite tensile strength.

REFERENCES

1. Tuttle,M.E.,'A mechanical/thermal analysis of pre-stressed composite laminates',Journal of Composite Materials,Vol 22,pp780-792, August 1988.

2. Tsai,S.W. and Hahd,H.T.,'Introduction to Composite Materials', Lancaster, PA, Technomic Publishing Co.,Inc, 1980.

3. Anon, BS 2782, Part 10, Glass reinforced Plastics, Method 1003, Determination of Tensile Properties, 1977.

Recent Advances in Experimental Mechanics, Silva Gomes et al. (eds) © 1994 Balkema, Rotterdam, ISBN 90 5410 395 7

Reinforced shells made of hybrid epoxy composites

J.Travassos, R.Prina & A.Bismark
Indústrias e Participações de Defesa, S.A., Lisbon, Portugal

A.Silva, C.Soares & M.de Freitas
Instituto Superior Técnico, Lisbon, Portugal

ABSTRACT: A shell consisting of twenty-two layers of balanced glass fabric, reinforced by six radially arranged plates, consisting of twenty unidirectional glass layers, intended to withstand radial stresses and fourty-four unidirectional carbon layers, to withstand side bending stresses, originated by a non-axi symmetrical load was constructed. For modelling, finite elements for personal computer with composite plate elements were used. In order to compare numerical results with static and dynamic performance of the structure, one shell was instrumented with thirty-six strain-gages and tested with three cases of loading: one axi-symmetrical, other according to one of the six radial reinforcements of the structure and another according to one of the bisectrixes between the reinforcements. In addition dynamic tests were performed with the three cases of loading, in real time conditions. The results achieved make possible to bring forward the concept of structures intended for the pre-established purposes, i. e., to produce composite structures of 300-mm diameter and 1.7-kg weight to withstand a static load of 250 kN and an impact fatigue loading of 82.5 kN (17.5 kJ) with any inclination between 0 and 45° with the symetrical axis of the structure.

1 INTRODUCTION

It may seem strange that a mortar, a weapon from the fifteenth century, though its present shape dates from 1915, continues to be so widely accepted in a time of sophisticated weapons and great technological developments. The specific feature of this weapon, firing at an approximately parabolic trajectory, together with other characteristics, such as low cost, easiness of use, high rate of fire, flexibility, destructive power of ammunition, low maintenance, mobility and relatively low weight, make it optimum for immediate fire support of infantry units.

The mortar is a weapon composed by three component groups to be carried separetely: a barrel, an adjustable mounting provided with a sighting unit and a base-plate. It is normally fired at a high elevation angle (formed between the barrel and the horizontal plane) about 45 to 86 degrees, using an explosive cartridge and several increments, so that the projectile range can vary as a function of the main variables - barrel inclination and number of increments. The traditional mission of the mortar as an infantry weapon is to provide support to

immediate and near fire. According to Machado (1986) its expected missions for the nineties are:
- Supporting combat units, neutralizing and incapacitating the enemy infantry, neutralizing the enemy weapons of medium and long-range direct fire, helping to separate infantry from its associated tanks;
- Helping to combat light armoured vehicles and even tanks;
- Intervening in operations to hinder enemy operations, by throwing smoke shells, mines and jammers;
- Illuminating the battle field.

In the present study, it was found that the mortar can be an excellent device for performing shock tests on structures. Structure resistance to impact in composite materials of polymeric matrix is well present in the literature (Dorey 1985). The purpose of developing new methods for shock testing of materials and structures has been dealt by other researchers (Peraro 1987). It is known that there have been other atempts to use fire-arms for impact tests (BRITE, proposal n°. EOI-541, Development of a shock-machine to test components subjected to very high and long lasting accelerations). So, in view of the study of the variables involved in designing structures made of composite materials and assuming it is better to produce mortars provided with base-plates in this kind of materials (Amaral 1987), a multimaterial structure has been studied, composed by:

- A reinforced shell in composite materials of epoxy matrix with glass and carbon fibres, intended to fulfil the main function of a mortar base-plate, i. e., not to allow the weapon to sink into the ground during the successive impacts, while being as light as possible for being one-man-loaded;

- One aluminum component for distributing stresses resulting from the impact caused by the barrel assembly on the composite structure, without a significant increase of weight considering its natural dimensions;

- One steel component for coupling to the mortar barrel and distributing stresses on the aluminum component;

- Screws and washers for fastening the structure components.

In view of the static and dynamic tests here presented, this twice innovative study is expected to give an important contribution for the aplication of complex-shape epoxy composite materials, intended for the production of light structures of high mechanical performance.

2 The structure

Further to previous works (Silva, Travassos, Freitas e Soares 1993), figure 1 shows a tridimensional sketch of the reinforced shell in composite materials studied within the scope of the present work.

Figure 2, shows the multimaterial structure as described above, subjected to a load according to one of the possible extreme working positions, i. e., with a 45 degrees angle towards the symmetry axis of the structure and along one of the ribs. It shows also a layout concerning:

- One orthogonal axis (H, V) for the ribs and another (R, T) for the shell;

- Six pairs of strain-gages applied on the shell over the ribs, at 83 mm (gage R) and at 106 mm (gage T) from the symmetrical axis of the structure;

- Six pairs of gauges stuck on the shell over the bisectrixes of the ribs, at the same distance of the symmetrical axis of the structure;

- Six pairs of gauges applied on the ribs, at 80 mm from the symmetrical axis of the structure, with one gauge in the direction H at 32.5 mm from the base of

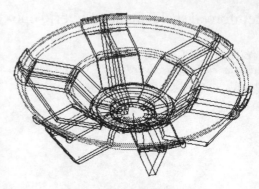

Overall dimensions in mm:
Total height of the reinforced shell = 85
Fixed diameter of the shell = 288
Distance from the base to the lower circ. face = 35
Width of the rib = 61
Height of the shell rim = 11

Figure 1 - Reinforced shell in epoxy composite material

the structure and the other in direction V, at 55 mm from the base;

- Each rib is identified by numbers from 1 to 6 and the bisectrixes between adjacent ribs by letters from A to F;

- Six dashes aligned with the ribs to indicate the face where the gages have been applied on the rib;

- One axis system (0°, 90°) for fiber alignement within the composite material;

- Angle Z to indicate the incidence of loading towards the symmetry axis of the structure.

After the performance of tests for the experimental development of the multimaterial structure produced by compression moulding, three base-plates were manufactured and tested statically and dynamically, two of them intended for strength tests and a third, intended for elastic behaviour tests, according to the following conditions:

I - Composition
Shell: [(0 / 45)$_5$ / 0]$_S$ (Epoxy-Glass)
Nom. thickness = 5.3 mm (50 % fiber vol.)
Rib: [(90$_{C1}$ / 90$_{C2}$)$_{11}$ / (0$_{V1}$ / 0$_{V2}$)$_5$]$_S$
where   C1, C2   (Epoxy-Carbon)
          V1, V2   (Epoxy-Glass)
          0°   Direction H

Nom. thickness = 12.6 mm (60 % fiber vol.)
Note: Rib 1 aligned with 0° of shell
 II - Consolidation
Curing cycle: 140 °C / 1800 s / 0.5 MPa
Weight of the composite structure = 16.6 kN
Weight of the base-plate = 23.6 kN

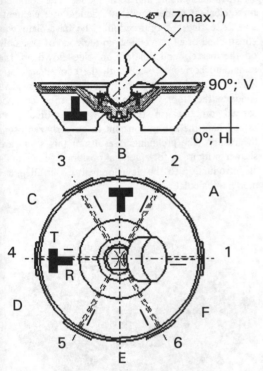

Figure 2 - Layout of the multimaterial structure subjected to a non-axisymmetrical loading

3 Loading conditions

A mortar works like an internal combustion engine. Instead of a piston, the expansion of hot gases in a chamber (the weapon barrel) drives a body which is usually called the projectile. When the propelling charges are ignited, the hot gases from the combustion cause a quick increase in chamber pressure. Then a strong resistance to the initial motion of the projectile, due to inertia and friction forces can be noticed. So, the projectile motion starts usually when the pressure reaches values ranging from 42 to 69 MPa. As the volume of the combustion chamber increases, the pressure tends to decrease.

However, the propellant rate of burning rises, causing a sharp pressure increase until a maximum is reached, which usually occurs near the beginning of the barrel, depending the proper profile from the nature of the propelling charges. Beyond that point, pressure drops till the barrel muzzle, where the pressure is still 10 to 30% of the previous maximum, which means that the projectile still accelerates after leaving the barrel.

Figure 3 shows, as an example, the balistic working conditions of a 60 MM mortar system, where it can be noticed how pressure, velocity and projectile travel along the barrel bore can vary, as a function of time (AMCP 706-107 1963).

The energy developed by the propelling charges is dispersed by several losses in the system, namely, the frictional work between the projectile and the barrel walls, the movement of the base-plate, the translation of propelling gases and mainly the heat losses through the barrel, the ammunition and propelling gases, assuming that translation energy absorbed by the projectile is about 34 % of total energy. This is similar to the performance of internal combustion engines used in the automotive industry.

In a pressure-travel diagram for the projectile motion inside the barrel, this energy is represented by the area under the pressure curve, which is in turn the work performed by the gas expansion by unit of barrel cross-sectional area - which can be easily related with the kinetic energy of the projectile at the barrel muzzle and with the energy absorbed by the base-plate. For the specific case of a 60 MM mortar system, it can be assumed that the total energy (W) absorbed by the base-plate is as follows:

$$W = 0.5 \ m \ v^2 = 0.5 \times 1.40 \times 158^2 = 17487 \ J$$

where:
Mass of the ammunition = 1.401 kg
Muzzle velocity (maximum charge) = 158 m/s
and,
Travel (lenght of the barrel)= 0.511 m
Time of travel inside the barrel = 6.5 s
Maximum pressure = 31.1 MPa

As it can be noticed, the mortars are adjustable devices for shock test, as far as direction and transmitted energy are concerned. They can impart extremely high impact energies to the structures or their models. This is particularly important when comparing the mechanical performance of different structure designs, especially when involving new materials and new technologies. Testing of another kind of structures requires the design of interface fixtures for that particular structure. If it is required to perform impact tests with an inclination angle between 0 and 45° towards the horizontal plane, a special breech mechanism with a movable firing pin can be used.

Whenever more energy is required, alternative mortar systems can be used, such as caliber 81, 120 and 160 MM with different kinds of projectiles, enabling to adjust the energy imparted to the base-plate or to the testing structure. With a 81MM mortar system, the energy imparted to the testing structure can easily rise to a maximum of 100.6 kJ. This flexibility in terms of energy, time of impulse and direction is not easily available in a laboratory environment, thus stimulating the use of mortar systems for structure testing.

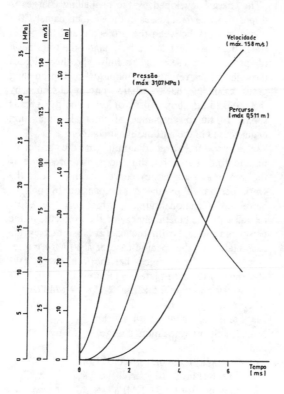

Weapon: 60MM MTR, M2, M19.
Projectile: HE, 49A2.
Charges: M3A1, 4 increments.

Figure 3 - Functioning diagrams of a 60 MM mortar system

4 Results

Three types of analysis were performed:

4.1 Numerical simulation

The base-plate was modeled in a personal computer using the finite element program COSMOS, with the multi-layered elements Shell4L and Shell3L, having 6 degrees of freedom per node. These elements are based on Mindlin theory with linear C° fields and use the concept of decomposing the deformation into well-defined bending and shear modes. The complete finite element model totals 505 nodes, 492 elements and 2826 degrees of freedom. The base-plate was constrained of all movements in the zone of assembly of the barrel, and loaded up side down in the remaining horizontal shell with the equivalent static load. The use of a spring-bed under the base-plate to simulate the floor reaction was also tried but not carried on because there is some difficulty in establishing the spring constant to properly represent the reality. The preliminary results of this study are shown in figure 4, for a general position of loading as an introduction to the devellopment of an adequate model of calculation.

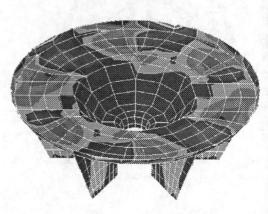

Figure 4 - Reinforced shell FEM grid and stress results

4.2 Experimental static tests

Both base-plates used for the strength test have withstood a load of 250 kN with a 25-degree inclination angle along the rib 1 direction and the bisectrix A; no trace of damage has been found in the structure. It should be emphasized that this load was about three times stronger than the maximum load for 60 MM mortar, according to figure 3 and twice as the maximum possible, using four M3A1 powder increments.

The results obtained at the test of the base-plate intended for load direction Z = 32 degrees, along A →D direction, are as follows:

Strains (ε) in m/m x 10^{-6} obtained on the shell along the direction T, over the ribs:

kN	C1T	C2T	C3T	C4T	C5T	C6T
70	-1861	-2017	-1204	-1007	-1089	-1430

Strains (ε) in m/m x 10^{-6} obtained on the shell along the direction R, over the ribs:

kN	C1L	C2L	C3L	C4L	C5L	C6L
70	+395	+477	+463	+334	+284	+479

Strains (ε) in m/m x 10^{-6} obtained on the shell along the direction T, over the rib bisectrixes:

kN	CAT	CBT	CCT	CDT	CET	CFT
70	+1054	+1019	-346	-620	-290	+264

Strains (ε) in m/m x 10^{-6} obtained on the shell along the direction R, over the rib bisectrixes:

kN	CAL	CBL	CCL	CDL	CEL	CFL
70	-433	-568	-78	+395	-141	-139

Strains (ε) in m/m x 10^{-6} obtained on the ribs along the direction H:

kN	R1T	R2T	R3T	R4T	R5T	R6T
70	+2192	+1166	-1493	-363	-65	+2113

Strains (ε) in m/m x 10^{-6} obtained on the ribs along the direction V:

kN	R1L	R2L	R3L	R4L	R5L	R6L
70	-1429	-431	+1669	+427	+239	-1293

Notes:
Shell:
1 - The tangential strains of the shell are always negative over the ribs
2 - The radial strains are always positive and quite uniform, in spite of load inclination
3 - The strain recorded by the three gages stuck on the bisectrixes of the shell (lower half) is negative, while at three gages on the upper half is positive.
4 - The strain recorded by the radial gage on the shell along the bisectrix of the ribs is positive. The remaining are negative.
From this analysis, it can be found that whenever there are compressions in the tangential direction, tensions are developed in the radial direction and vice-versa.

Ribs:
1 - The results obtained by the horizontal gauges on ribs 3 and 6 are obvious, as they form a perpendicular plane to the applied load (they are placed on opposite sides of the corresponding rib). Due to load inclination, ribs 4 and 5 remain under compression, obviously. Ribs 1 and 2, remain under traction, also due to load inclination and to the face of each rib on which they are stuck.
2 - The vertical gauges always record an extension opposite to the horizontal gages and they also record an opposite extension on opposite ribs, i.e., 1-4, 2-5 and 3-6. These results are perfectly consistent with the Poisson's effect on the ribs, since both horizontal and vertical gages are applied on the same face of the ribs. The rib bends preferably according to the fiber alignement in the composite material.

4.3 Experimental dynamic tests

Both base-plates prepared to the strength test, have withstood several impacts at the maximum charge (average pressure according to figure 3) for various load inclinations, including 45 degrees, 32 impacts in all, with no trace of damage.

Figure 5 shows a layout of the dynamic test performed on a normal proving ground, to assess the possibilities of the available equipment to record the structure response to the mortar impacts, for various loads.

Figure 6 shows the results recorded with the data recording system as represented in the figure, being these results under examination, for comparison with numerical results collected by computer-aided simulation and the static tests performed for the equivalent loading. This will be the scope of a future publication.

5 Final remarks

The static and dynamic tests performed during the present study, lead to the conclusion that it is feasible to produce and to use mortar base-plates from epoxy composite materials.

The reduction in weight, if compared with an identical steel-made base-plate, is doubtless greater than 50%. This steel welded structure weighs about 65 N, against only 23.6 and 28 N of the base-plates subjected to the tests. These base-plates, subjected to the dynamic tests with a 60 MM mortar fired at the maximum charge have evidenced no damages. According to the test results, the expected reduction in weight should be between 57 and 64%, depending on the overall dimensions of the base-plate and the fatigue resistance required.

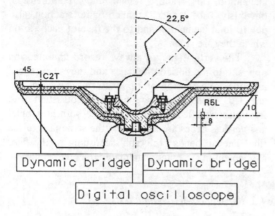

Figure 5 - Layout of the dynamical data system used for data recording

These results will stimulate designers to improve numerical models for an eventual transfer of this technology to larger mortars. This is completely revolutionary and demands new challenges to designers of such a kind of structures in the future.

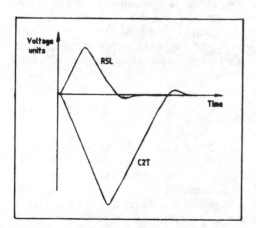

Figure 6 - Dynamical response through two strain-gages applied on the composite structure

References

Amaral, A. P. 1987. *Prato-base para morteiros,* Lisboa: FUTURO, MAR 87, pp. 12-13.

AMCP 706-107 1963. *Elements of armament engineering, part two, ballistics,* engineering design handbook, pp 1-3.

Dorey, G. 1985. *Impact and crashworthiness of composite structures,* Structural impact and crashworthiness, volume 1, keynote lectures, pp. 155-192.

Machado, C. L. 1986. *Memorando relativo aos sistemas de morteiro de 60, 81 e 120 MM.* Lisboa: GED/INDEP.

Peraro, J. S. 1987. *Prediction of end-use impact resistance of composites.* ASTM STP 936, pp. 187-216

Silva, A., J. Travassos, M de Freitas & C. M.M. Soares 1993. *Mechanical bending behaviour of composite T beams.* Composite Structures, 25, 1-4, pp. 579-586.

Recent Advances in Experimental Mechanics, Silva Gomes et al. (eds) © 1994 Balkema, Rotterdam, ISBN 90 5410 395 7

Setting up of biaxial membrane and bending tests in reinforced concrete panels

I. R. Pascu & J. F. Jullien
INSA, Lyon, France

Y. L'Huby
EDF, Lyon, France

ABSTRACT: A new test device designed to allow tests on reinforced concrete panels under both axial forces and bending moments in two orthogonal directions has been developed at the Concrete & Structures Laboratory of INSA Lyon. The specimen is loaded by means of 12 hydraulic jacks. Both the hydraulic system and the data acquisition are controlled by a computer. Some test results are also reported here.

1 INTRODUCTION

The use of reinforced concrete (RC) shells and panels in large structures such as hyperbolic cooling towers, marine platforms, nuclear reactor envelopes etc., has required the development of computer methods and models. Much progress has recently been made in modelling RC. Though these models are more and more refined they cannot take into account all the complexity of RC behaviour. Hence, there is a need for experimental validation.

However, there is a lack of experimental data concerning the behaviour of RC shells and panels under complex loads (axial forces and bending moments in two directions and shear in the plane of the panel). This brought us to carry out an experimental study based on a new test equipment which allows complex loading of RC panels.

This paper presents the new test equipment developed at the Concrete & Structures Laboratory of INSA Lyon. We also present the first results obtained.

2 TEST PRINCIPLE

The test equipment (figure 1) has been designed in order to allow the testing of RC panels under both axial forces and bending moments in two orthogonal directions. The novelty is that we can carry out this type of loading having free edges and a constant bending moment over the whole width of the panel. Axial loads may be tensile or compressive.

Flexural and axial force tests reported in the literature are limited either to biaxial compression and flexure in one direction induced by an eccentric axial load or to uniaxial compression and flexure induced by transversal loads to a slab supported on the edges.

Our test equipment also allow us to choose the ratio between the axial force and the bending moment in the same direction or to choose the ratio between the axial forces (or the bending moments) in two different directions.

The principle of the system is similar to exterior prestressing (see figure 2). There is no need of a reaction frame.

The forces in the loading bars are balanced by the reaction forces in the test specimen. Pinned joints ensure that the loading bars are working only in tension or compression.

Statical equilibrium gives:
- in-plane force $N = N_1 + N_2$
- bending moment $M = (N_1 - N_2)*d$.

3 TEST SET - UP

3.1 Test specimen

The test specimens are square RC panels. The overall dimensions are 2 m by 2 m and the thickness is 10 cm (figure 3). To connect the loading devices on the specimen there are 48 holes, each with a diameter of 32 mm.

Reinforcement is composed of two layers of 6 mm diameter bars spaced at 10 cm on two orthogonal direction parallel to the edges of the panel.

Supplementary reinforcement bars are placed within a distance of 35 cm from the edges in order to

Figure 1. General view of the test equipment

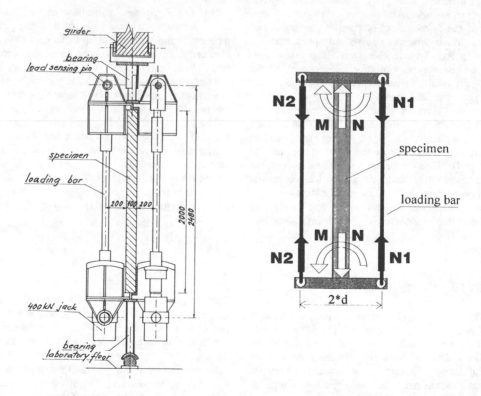

Figure 2. Vertical cross-section and simplified scheme of the test set-up

prevent local failure due to the stress concentrations caused by the loading devices.

Normal concrete, with an average cylinder compressive strength $f_c = 35$ MPa, is used.

The specimen are produced in our laboratory.

3.2 Loading system

The panel to be tested is maintained in a vertical position under a support frame. The bearings are placed at the middle of the upper and of the lower edges of the panel (figure 2).

The lower bearing is a cylindrical joint having the axis parallel to the bottom edge of the panel.

The upper bearing allows three movements: rotation around a horizontal axis parallel to top edge of the panel, rotation around and translation along a vertical axis.

Twelve steel support devices are fixed with high strength bolts on the two faces of the panel. Load bars and hydraulic jacks are fixed to these supports by clevis pinned connections (see figure 1).

Hydraulic jacks introduce in-plane forces and bending moments to the panel.

3.3 Instrumentation and data acquisition

Measures are taken of the forces applied to the specimen, displacements normal to the plane of the panel and strains in the concrete and the reinforcement.

Forces are measured by 12 load sensing clevis pins (Strainsert SPA force transducers) placed in the joints connecting the loading bars to the supports (see figure 2).

Transversal displacements are measured by 8 LVD transducers.

Concrete strains are measured by 20 HBM 50/120 LY41 strain gauges, 10 on one face and 10 on the opposite face of the specimen. The gauges are placed in the central zone of the panel.

Steel strains are measured by 20 Vishay CEA 06-125 UN 120 strain gauges.

There is a total of 60 channels connected to a Centralp data acquisition unit. This unit is connected by the RS232 interface to the same computer which is used to control the test.

4 TEST PROCEDURE

The test procedure depends on the load combination chosen in a particular test.

Loads may be applied in any order. The four hydraulic circuits can be monitored independently.

The only restriction is that load increments must be applied sequentially, on only one circuit at a time.

For example a membrane load in one direction may be applied in the following sequence: a pressure increment $\wp$ is applied by the jacks acting on one face of the panel, then the same increment $\wp$ is applied by the jacks acting on the opposite face of the panel. These two steps may be repeated until the desired load is achieved.

Obviously this procedure induces, at every uneven step, bending moments which are balanced at the next step. If pressure increments are not great, the bending deformations are reversible. We found that 0.5 MPa increments do not affect the final state of the specimen.

The time necessary to apply two load steps is about 40 seconds, including data acquisition.

To apply bending moment, alternate increments of $+ \wp$ and $- \wp$ are applied on the faces of the specimen. The membrane force is maintained constant and bending is increased proportionally to $2 * \wp$ at each step.

5 RESEARCH PROGRAM

The purpose of our research is to determine the influence of in-plane forces on the cracking and on the ultimate moments and curvatures of a RC panel.

Test results may also be used to validate finite element models.

In-plane loads are chosen so that the concrete stresses varies between 5 MPa (about 15% of f_c) in compression and 1 MPa (about 30% of f_t) in tension.

In the first part of our program we will apply in-plane forces in two directions and moment in one direction.

In the second part of our program we will apply in-plane forces and moments in two directions.

6 FIRST TESTS RESULTS

A preliminary test in compression in the y-direction showed a linear elastic behaviour (figure 4).

Note that the strain value in this figure, as in all of the following figures is the average value of the measured strains in the central zone of the specimen.

The "average stress" is the ratio between the applied force and the gross cross-section area.

Two tests were performed in which biaxial in-plane forces and uniaxial bending moments were applied.

Moment-displacement and moment-curvature relationships for the second test are shown in figures 5 and 6.

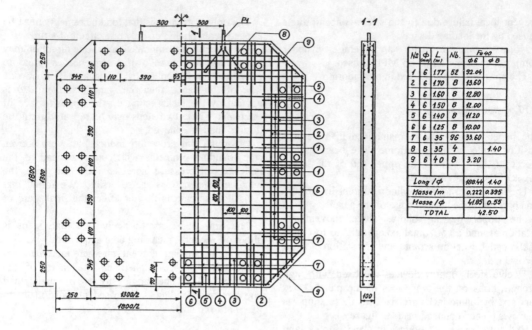

N°	φ (mm)	L (m)	Nb.	Fe 40	
				φ6	φ8
1	6	1.77	52	92.04	
2	6	1.70	8	13.60	
3	6	1.60	8	12.80	
4	6	1.50	8	12.00	
5	6	1.40	8	11.20	
6	6	1.25	8	10.00	
7	6	35	96	33.60	
8	8	35	4		1.40
9	6	40	8	3.20	
Long /φ				188.44	1.40
Masse /m				0.222	0.395
Masse /φ				41.85	0.55
TOTAL				42.50	

Figure 3. Geometry and reinforcement of the test panel

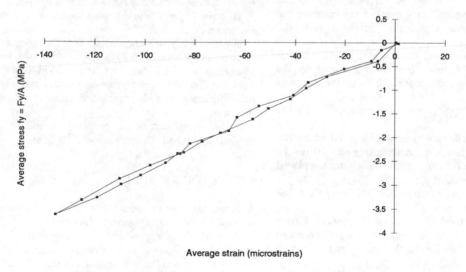

Figure 4. Average stress - average strain curve. Preliminary compression test

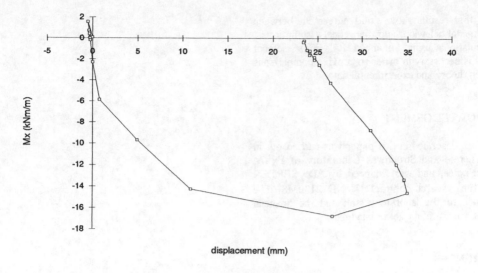

Figure 5. Moment-displacement relationship. Test 02

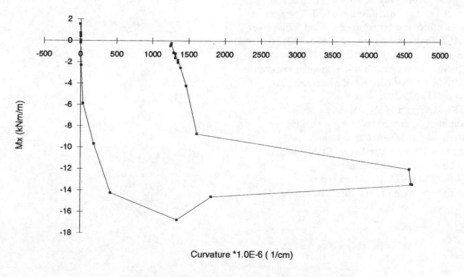

Figure 6. Moment-curvature relationship. Test 02

Curvatures are calculated from the measured strains of the compressed concrete and of the tensile reinforcement, assuming the plane sections remain plane. The strain gauges on the tensile reinforcement are generally in cracked sections.

Cracking of concrete and yielding of tension reinforcement are obvious on these curves.

7 CONCLUDING REMARKS

The new test set-up makes it possible to carry out various tests on RC panels under combined loads.

The instrumentation and the data acquisition system allows us to get a large amount of information about global behaviour (forces and displacements) and local behaviour (strains) of the tested panel.

The first results show good agreement between experimental behaviour and theoretical assumptions.

Calculations using advanced RC finite element models is necessary in order to provide comparisons between theory and experimental data.

ACKNOWLEDGEMENT

The work described in this paper was carried out in the Concrete and Structures Laboratory of INSA, Lyon, France, and was financed by EDF-SEPTEN under the research contract N° ND 2130MS. The assistance to the laboratory staff and the financial support are gratefully acknowledged.

REFERENCES

Aas-Jacobsen, K. 1983. Buckling of walls. *Bulletin d'Information CEB No. 155*:223-250.

Aghayere, A.O. 1988. Stability of concrete walls. *Ph.D Thesis*. University of Alberta, Edmonton.

Aghayere, A.O., J.G. Mac Gregor 1990. Tests on RC plates under combined in-plane and transverse loads. *ACI Structural Journal 87*: 615-622.

Christiansen, K.P., V.T. Frederiksen 1983. Experimental investigation of rectangular concrete slabs with horizontal restraints. *Materials and Structures 93*: 179-191.

Kordina, K., R. Timm 1979. *Experimentelle Untersuchungen zum Stabilitaetsverhalten von bewehrten und unbewehrten Betonwaende*. Institute fuer Baustoffkunde und Stahlbetonbau, Techniche Universitaet Braunschweig.

Swartz, S.E. et al. 1974. Buckling tests on rectangular concrete panels. *ACI Journal 71*: 33-39.

Ductility effects in reinforced concrete beams and slabs

R.H.Scott

School of Engineering and Computer Science, University of Durham, UK

ABSTRACT: An experimental investigation into the relationship between reinforcement ductility and moment redistribution is described involving laboratory tests on two-span reinforced concrete beam specimens. Strain gauged reinforcement was used to obtain very detailed measurements of the reinforcement strain distributions. Results are presented to relate development of moment redistribution to the load and strain histories of the specimens.

1 INTRODUCTION

At the last two Experimental Mechanics conferences, the author reported work undertaken to measure reinforcement strain and bond stress distributions in reinforced concrete structural members (Scott *et al* 1986, 1990). The technique of using internally strain gauged reinforcement was described whereby electric resistance strain gauges (e.r.s.g.'s) were installed within a machined duct running longitudinally through the centre of the bars, thereby avoiding the disturbance to the bond behaviour which would result from installing such gauges around the perimeter. At that time, the smallest bar diameter which had been strain gauged was 12 mm diameter, using a duct size of 4x4 mm in which 100 gauges, each with a gauge length of 3 mm, were installed. However, there later arose a need to strain gauge bars of smaller diameter for use in specimens investigating ductility effects in beams and slabs which prompted a further miniaturisation of the strain gauging technique. These developments, and their use in the ductility investigation, form the subject of this paper.

2 STRAIN GAUGING DEVELOPMENTS

The first new development involved the instrumentation of 10 mm and 8 mm diameter bars. The same basic technique was used, each strain gauged bar being formed by milling two solid bars to a half round and then machining a longitudinal groove in each in which the strain gauges and wiring were installed, the two halves then being glued together to give the appearance of a normal solid bar. However, the duct sizes were reduced to 3.2x3.2 and 2.5x2.5 mm, for the 10 mm and 8 mm bars respectively, in order to maintain the ratio of the cross-sectional area of the machined bar to that of the solid bar at around 88%, a value similar to that achieved for the 12 mm diameter bars. Using e.r.s.g.'s with a 2 mm gauge length, it was possible to install 50 gauges in the 10 mm bars and 30 gauges in the 8 mm bars. Since then further development has enabled bars of only 6 mm diameter to be strain gauged, although with the compromise of using the 2.5x2.5 mm duct (the smallest practicable size) which has reduced the cross-sectional ratio to 78%. The main problem encountered during the manufacture of these latter bars was the considerable distortion which occurred, due to stress relief, during the half-rounding operation, but this was cured by using a specially manufactured spring-loaded jig to keep the bar firmly restrained. It is believed that the gauging of bars as small as 6 mm diameter has not been attempted before.

3 DUCTILITY CONSIDERATIONS

The small diameter bars have been used in tests designed to investigate the associated phenomena of reinforcement ductility and moment redistribution in reinforced concrete beams and slabs, a topic which attracted considerable discussion across Europe during the drafting of EC2, the new Eurocode for concrete structures (EC2 1992).

Current design practice permits the bending

Table 1. Specimen Details

Specimen	Overall Depth (mm)	Support Reinforcement		Span Reinforcement	
		Top	Bottom	Top	Bottom
B3T16A	400	3T16 (0.53%)	3T12	3T12	3T20
B2T20B	250	2T20 (0.93%)	3T12	3T12	3T20
B3T16B	250	3T16 (0.89%)	3T12	3T12	3T20
B3T16BL	250	3T16 (0.89%)	3T12	3T12	3T20
B5T12B	250	5T12 (0.83%)	3T12	3T12	3T20
B2T20C	150	2T20 (1.70%)	3T12	3T12	3T20
B3T16C	150	3T16 (1.63%)	3T12	3T12	3T20
B5T12C	150	5T12 (1.53%)	3T12	3T12	3T20
B5T12CR	150	5T12 (1.53%)	3T12	3T12	3T20
B2T12D	150	2T12 (0.58%)	3T10	3T10	3T12
B3T10D	150	3T10 (0.63%)	3T10	3T10	3T12
B5T8D	150	5T8 (0.68%)	3T10	3T10	3T12
B2T8E	150	2T8 (0.26%)	2T8	2T8	3T8
B4T6E	150	4T6 (0.30%)	2T8	2T8	3T8
W2T12D	150	2T12 (0.58%)	3T10	3T10	3T12
W3T10D	150	3T10 (0.63%)	3T10	3T10	3T12
W5T8D	150	5T8 (0.68%)	3T10	3T10	3T12

moment distributions in the members of a continuous frame at the ultimate limit state to be derived using an elastic analysis. However, the ductility of the tension reinforcement in a reinforced concrete section allows plastic hinges to form, although the amount of rotation which these hinges can undergo is limited by the concrete failing in compression. The development of plastic hinges modifies the elastic moment distribution which is recognised by permitting the elastic moments to be adjusted using a procedure called *moment redistribution*. UK practice (BS8110 1985) is that, subject to maintaining equilibrium between internal and external forces, moments may be reduced by up to 30% in braced frames, or by up to 10% in unbraced frames. EC2 also permits moment redistribution to occur, but distinguishes between what it terms high ductility and low ductility reinforcement, with 30% redistribution being allowed in the former case, but only 15% in the latter. No redistribution at all is permitted in unbraced frames.

During the drafting of EC2, discussion arose concerning the related areas of moment redistribution, reinforcement ductility and allowable rotation of a reinforced concrete section. This led to a programme of tests on two-span beams being undertaken at Durham University using strain gauged reinforcing bars in an attempt to achieve a correlation between the level of strain in the reinforcement and the degree of moment redistribution.

4 SPECIMEN DETAILS AND TESTING

To obtain data pertaining to moment redistribution, it was necessary to test specimens which were statically indeterminate thus two-span beams, the simplest such specimens, were selected for the test programme. Seventeen specimens were tested, each 300 mm wide and 5.2 m long overall and supported to give two equal spans of 2.5 m, each span being loaded at its mid-point. Three beam depths - 150, 250 and 400 mm - were used. The two principal parameters investigated were percentage of tensile reinforcement over the centre support and bar diameter. Table 1 gives specimen details, and shows that six groups (or series) were tested. The first five series, prefixed B and with suffices A to E, all used bar reinforcement, whilst the sixth series, prefixed W, used ribbed wire reinforcement (as used in fabric reinforcement in slabs) but was similar to series D in all other respects. Each series investigated a different nominal percentage of tensile reinforcement over the centre support, using a range of bar diameters (except for the single specimen in Series A) Thus the 0.9% for Series B was achieved using two T20 bars (Specimen B2T20B), three T16 bars (Specimens B3T16B and B3T16BL) and five T12 bars (Specimen B5T12B). Table 1 shows the permutations and also gives the exact reinforcement percentages for each. The full specimen title reflects the bar arrangement used, as indicated in the examples above. Specimens were designed to accommodate 30% moment redistribution at the centre support and Table 1 gives details of the other longitudinal reinforcement used in each specimen. Shear reinforcement was T10 links at 200 mm centres in Series A and B specimens and R8 links at 100 mm centres in all other specimens.

Four bars in each specimen were machined to accommodate electric resistance strain gauges. Fifty such gauges were installed in a duct running longitudinally through the centre of each bar, giving a total of 200 gauges per specimen. One of these bars was provided as part of the top and bottom reinforcement at both the centre support and in one of the spans.

Specimen B5T12CR was similar to B5T12C except that, due to an incident which occurred during its manufacture, its top strain gauged bar over the centre support was replaced - the concrete around this bar was carefully broken-out and a new bar inserted and concreted in - thus providing an opportunity to assess whether moment redistribution behaviour was affected by a repair of this type. Specimen B3T16BL was essentially a repeat of B3T16B.

Specimens were tested by loading both spans incrementally until failure occurred, using manually pumped hydraulic jacks. The applied loads and support reactions, and a full set of strain gauge readings (around 10 000 per test) were recorded at each load stage using a computer controlled data acquisition system.

5 RESULTS

Moment redistribution led, as expected, to support moments being decreased (numerically) and span moments being increased (numerically) when specimens were loaded, as shown in Figure 1 for the last load stage in the test of B3T16C. Three stages in the development of moment redistribution were identified - an initial rapid increase, then an extended period during which there was little change (the serviceability level) followed, at the end of a test (ultimate), by a further rapid increase. This sequence, which occurred at the centre support and in both spans, may be seen in Figure 2, also for specimen B3T16C, the scatter at low load levels

being while the specimen bedded-in onto the three supports. Negative values indicate that moment redistribution led to bending moments being reduced numerically, as at the centre support, whilst positive values indicate that moments were increased numerically, as at the mid-span load positions. This convention is also used in Figure 3 but, for convenience, values in the text and tables will all be quoted as positive.

A full listing of percentage redistribution values at serviceability and ultimate is given in Table 2. For a given section size and reinforcement percentage more redistribution occurred in specimens reinforced with T16 or T20 bars than in those reinforced with T12's or below, illustrated in Figure 3 for the beams in Series B (for clarity, only values for the centre support are shown). By contrast, all beams in Series D followed a similar trend, irrespective of reinforcement diameter (8, 10 or 12 mm) or whether the reinforcement was bar or wire with, as Table 2 indicates, the serviceability level of redistribution at the centre support being in the 12 to 15% range. Specimen B5T12CR, which had a gauged rod replaced, showed levels of percentage redistribution significantly higher than those in B5T12C - by a factor of two for much of the test period - and it also had a reduced failure moment. The difference in behaviour was marked and resulted from only *one* of the *five* bars over the centre support being replaced.

An unexpected feature was the high level of percentage redistribution reached at serviceability, particularly at the centre support where values ranged from 10 to 27%, including seven specimens at, or above, 20%. Peak tensile reinforcement strains for when these levels were first reached are given in Table 3, although there are gaps as precise identification was difficult in some instances. All values were less than 1000 microstrain with two values being under 500 microstrain, significant in that levels of redistribution normally associated with ultimate conditions were occurring while the reinforcement was still *elastic*.

At ultimate, specimens in Series A, B and C failed in shear, while those in Series D and E failed in bending. Bending failures led to very high tensile strains - over 100 000 microstrain in some instances - being developed over the centre support and at the mid-span load positions. These high strains resulted from localised gross yield of the reinforcement and led to plastic hinges being developed. Shear failures, being brittle, occurred at considerably lower strains spread over greater lengths of reinforcement. Peak tensile strains at ultimate for all specimens are shown in Table 3.

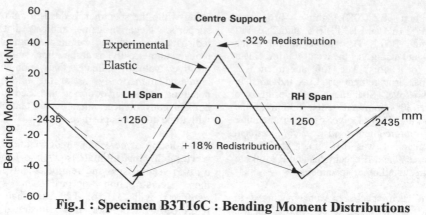

Fig.1 : Specimen B3T16C : Bending Moment Distributions

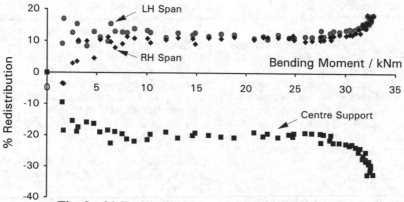

Fig. 2 : % Redistributions for Specimen B3T16C

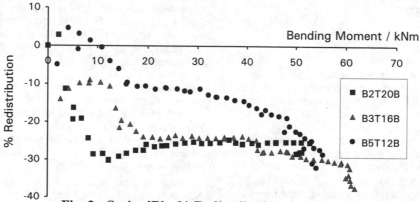

Fig. 3 : Series 'B' : % Redistributions at Centre Support

Table 2. Percentage Redistributions

Specimen	Serviceability		Ultimate	
	Support	Span	Support	Span
B3T16A	27	14	32	17
B2T20B	26	15	28	15
B3T16B	24	14	37	21
B3T16BL	13	7	21	11
B5T12B	11	7	32	18
B2T20C	14	8	24	13
B3T16C	20	12	32	18
B5T12C	10	5	27	15
B5T12CR	20	13	37	22
B2T12D	14	8	27	15
B3T10D	15	10	25	16
B5T8D	12	9	22	13
B2T8E	37	20	41	21
B4T6E	20	11	39	21
W2T12D	15	9	25	14
W3T10D	12	9	32	18
W5T8D	14	8	27	14

Table 3. Peak Strains in Tension Reinforcement (microstrain)

Specimen	Serviceability (Start)		Ultimate	
	Support	Span	Support	Span
B3T16A	710	560	3240	2490
B2T20B	430	520	2250	2390
B3T16B	710	600	18290	5330
B3T16BL	-	-	4150	5440
B5T12B	555	360	4030	2350
B2T20C	885	775	10790	18590
B3T16C	820	635	16380	4720
B5T12C	990	565	12700	2830
B5T12CR	435	340	6430	2930
B2T12D	930	895	31040	33940
B3T10D	835	875	37510	58080
B5T8D	-	-	28220	42290
B2T8E	-	-	>100000	16530
B4T6E	-	-	>100000	19530
W2T12D	-	-	32550	28490
W3T10D	725	560	33020	26570
W5T8D	575	655	27770	25370

Levels of percentage redistribution at ultimate were high (Table 2), with nine specimens showing values of over 30%. High reinforcement strains were not a prerequisite for this, however, as B3T16A and B2T20B managed 32 and 28% respectively with tensile reinforcement strains nowhere exceeding about 3200 microstrain.

The results suggested that the initial rapid development of percentage redistribution was due to reductions in member stiffness caused by flexural cracking. Once the major crack pattern was established, the stiffness of the specimen remained relatively stable and this was reflected by the level of percentage redistribution remaining steady, a situation which prevailed until the reinforcement yielded. Yield of the reinforcement with, in some instances the formation of plastic hinges both at the centre support and at the load positions, led to further reduction in stiffness and thus a renewed increase in percentage redistribution. Investigations using purpose written computer software to model the beam behaviour indicated that, at serviceability, 10 to 15% redistribution at the centre support could be expected purely as a consequence of changes in stiffness as cracks developed. This is consistent with the test results for 8, 10 and 12 mm diameter bars and wire in Series B, C and D, with the larger values for 16 and 20 mm diameter bars coming from increased slip due to their poorer bond performance. The high values for the two Series E specimens require further investigation, but it is clear that variations in flexural stiffness along a member, together with changes in this stiffness distribution as a consequence of cracking, have a more profound influence on moment redistribution behaviour than has previously been appreciated. In particular, the results clearly indicate that high reinforcement ductility is not a prerequisite for a high level of moment redistribution.

Data analysis continues with curvature distributions, changes in the position of the point of contraflexure and the relationship between neutral axis depth and percentage redistribution receiving particular attention A study is being made of the extent over which gross yield occurs in the reinforcement as a consequence of plastic hinge development, these results all being correlated with the modelling software. It is intended that substantial use will be made of the modelling software for further investigations of reinforcement ductility and moment redistribution. Details of this additional work will be given at the conference.

6 CONCLUSIONS

Moment redistribution occurs at all stages in the load history of a reinforced concrete beam or slab. At serviceability, it is strongly influenced by variations in member stiffness. High levels may occur with the reinforcement elastic, exacerbated by slip when bars larger than 12 mm diameter are used.

At ultimate, plastic hinge development further increases moment redistribution levels. However, values of around 30% are also obtainable with shear failures, indicating that high reinforcement ductility is not a prerequisite for a high level of redistribution. Suggestions for revising current design procedures will be proposed in due course.

7 ACKNOWLEDGEMENTS

The assistance of Mr A.R. Azizi, Miss D.M. Dawes and Dr P.A.T. Gill with the experimental work is gratefully acknowledged. Financial support was kindly provided by the Science and Engineering Research Council with additional support from Ove Arup and Partners and Oscar Faber plc.

8 REFERENCES

BS8110 "The Structural Use of Concrete", *British Standards Institution,* London, 1985

Scott R.H. & Gill P.A.T., "Techniques in Experimental Stress Analysis for Reinforced Concrete Structures", *Proc. VIIIth Int. Conf. on Experimental Mechanics,* Amsterdam, Ed. Wieringa H., (Martinus Nijhoff Publishers), May 1986, 87-96 (ISBN 90-247-3346-4)

Scott R.H. & Gill P.A.T., "Stress Analysis of Reinforced Concrete Structural Elements", *Proc. IXth Int. Conf. on Experimental Mechanics,* Copenhagen, August 1990, 1365-1374, (ISBN 87-7740-035-6)

EC2 : Eurocode No. 2, "Design of Concrete Structures", Part 1, General Rules and Rules for Buildings, DD ENV 1992-1-1, 1992

Recent Advances in Experimental Mechanics, Silva Gomes et al. (eds) © 1994 Balkema, Rotterdam, ISBN 90 5410 395 7

Tensile behavior of glass fibre reinforced concrete

J.A.O. Barros
School of Engineering, University of Minho, Guimarães, Portugal

J.A. Figueiras
Civil Engineering Department, Faculty of Engineering, University of Porto, Portugal

C.V.D. Veen
Faculty of Engineering, Delft University of Technology, Netherlands

ABSTRACT: From the results of the research carried out in the last years on fibre reinforced cement based materials, it can be pointed out that, for the fibre contents usually employed in practice, the post-peak tensile behaviour is the most improved material characteristic. However, difficulties in carrying out valid direct tensile tests have limited the research in this field. The scarcity of investigation on the tensile behaviour of glass fibre reinforced concrete (GFRC) is also probably due to the ageing problems of GFRC systems. In order to contribute to a better knowledge of the uniaxial tensile behaviour of GFRC, deformation-controlled uniaxial tensile tests were carried out at Stevin Laboratory (NL). Polymer-modified glass fibre reinforced cement (PGFRC) specimens manufactured by spray up and premix techniques, and GFRC specimens are tested at the age of 28 days. The experimental response of the tested specimens is illustrated and the results are used to validate a computational code developed for the analysis of fibre reinforced concrete (FRC) structures, wherein the most recent concepts of fracture mechanics of brittle materials are included.

1 - BACKGROUND INFORMATION

1.1 - *Glass fibre reinforced concrete*

Glass fibre reinforced concrete manufactured by spray or premix process has proven to be an attractive material for a wide range of applications (PCI 1981, Majumdar 1991). Several factors have contributed for this, namely: manufacture facilities in terms of shape, size and finishing; economy in transport and handling equipment; decrease of superimposed loads on structural components due to the relative low weight of GFRC systems; satisfactory mechanical properties and good physical performances. However, soon it was recognized the long-term decline of the properties of these composites (Majumdar 1974). The chemical attack on glass by the alkaline cement environment and the growth of hydration products, mainly calcium hidroxides, in between the filaments in the strand

have been pointed out as the responsible by the durability of the GFRC mechanical properties (Bijen 1983, Leonard 1984, Bentur 1986, Shah 1988, Li 1991). The ways of improving the retention of the GFRC properties with age have been directed to the following main procedures: reduce the chemical aggressiveness of the cementitious matrix by employing low alkali cements (Akihama 1987); protect the fibres from the corrosive environment through a polymer or admixtures mix addition, which has the capacity to envelop the fibres by a protective film (Bijen 1990); interfer in the fibre chemical composition and fibre surface treatment (Fyles 1986).

The submicron particles of the acrylic polymer applied in the Forton-PGFRC surround the strand filaments, obstructing the fibre engulfment by the cement hydration products and protecting the fibres against a chemical attack. A significative improvement in the durability of these composites under weathering conditions has been claimed (Bijen 1990).

1.2 - *Tensile behaviour*

The deformation-controlled uniaxial tensile test is probably the most suitable procedure to study the brittle behaviour of cement-based materials (Gopalaratnam 1985, Wang 1990, Hordijk 1991). The tensile load-deformation curve schematically represented in Fig. 1 can be regarded as a typical curve of a bar of PGFRC under deformation controlled uniaxial tensile test, as long as the fracture behaviour is governed by the fibre pull-out mechanism. Comparatively to plain concrete (PC), fibre reinforced concrete (FRC) has a less steep softening branch (branch III), a larger ultimate deformation and a more enlarged nonlinear branch before peak load (branch II). The initial part of the pre-peak curve (branch I) appears to be a straight line, indicating that FRC behaves linearly in this region. In this first branch, load does not create new microcracks until the stress level represented by the bending over point (BOP) is reached. Between BOP and peak load, the response is usually described by a nonlinear branch (branch II). This branch is more enlarged when a high percentage of fibres is employed. In this stage a substantial amount of energy is required to propagate the microcracks (process zone). The localization of the process zone is usually associated to a stress concentration which is due to a material heterogeneity or to a structural descontinuity (such as a notch). If the process zone is developed within the measuring length, whose deformation is used as test control parameter, a load-deformation relationship, as indicated by branch III in Fig. 1, can be obtained and used to get the stress-crack opening relationship (Hillerborg 1976). The load that can be transferred decreases with an increasing deformation of the process zone. As a result of this load reduction, the composite material outside the process zone unloads (branch IV). The pull-out mechanism of fibres bridging the fracture surfaces in the process zone is responsible for the energy increase of the FRC as compared with PC.

2 - EXPERIMENTS

2.1 - *Mix composition*

Comparatively to conventional concrete, GFRC mixes are generally characterized by greater cement

factors, higher contents of fine aggregates and smaller sized coarse aggregates.

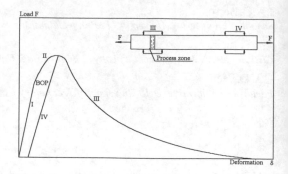

Figure 1 - Load-deformation relation for a (fibrous) concrete bar under uniaxial tensile loading.

Two different mixes were used in the experimental program. The method of mixing the PGFRC compositions determined the mix procedure. In the premix method, it is recomended to mix in the first stage the aggregates, cement, water and admixtures until the mix gets the adequate workability. In a second stage the fibres are added to the mix during the strictly time needed to soak the fibres in the slurry. In the spray-up method, chopped glass fibres and cement slurry are simultaneously sprayed onto a form, usually from separated nozzles, and in fact they are mixed at the moment of placement. The mortar was previously prepared and introduced in a deposit of the spray equipment. The ratio between the amount of glass fibre and cement slurry is controlled by the spray gun operator.

Typical mix compositions of commercial PGFRC systems are given in Table 1.

Table 1 - Mix composition of the PGFRC composites (FORTON 1992)

Constituent (Kg)	Spray-up 5/5	Premix 3/7
Cement PcB	50	50
Sand LG 56	50	50
Forton Compound VF 774	7	10
Water	13	12
AR Glass fibres	6.3	3.75
Characteristics:		
Cement-sand ratio (Weight)	1	1
Water-cement ratio (Weigth)	0.33	0.34
AR-glass fibres (Weight)	5	3
Polymer solids (Volume)	5	7.2
AR glass fibres	Cem-FIL 205/5B roving cut into 31 mm strands during spray-up production	Cem-FIL 60/2 chopped strands 12 mm long

2.2 - *Preparation of the specimens*

The PGFRC tensile specimens were sawed from two 605×755x50 mm panels made by FORTON BV Company. One panel was manufactured by the premix process and the second panel by the spray technique. Until demoulding, the panels were covered with a polyethylene sheet. After demoulding the panels were stored in a climatic room at 20°C and 65% RH. The panels were transferred to the Stevin Lab. at an age of 7 days and stored in a natural environment of the laboratory until testing. Additionally, some GFRC tensile specimens were manufactured in the Stevin Lab. and obtained from sawing cubes of 150 mm side length, in a direction orthogonal to the casting direction.

The dimensions of all tensile specimens were 150x60x50 mm. Two saw cuts reducing the middle cross-sectional area to 50x50 mm were initially used. However, difficulties in glueing the specimens to the machine loading platens forced the reduction of the net area of the specimens to 30x50 mm.

2.3 - *Experimental setup*

Tests were carried out on a closed-loop electro-hydraulic loading machine which has the capacity of 300 kN in tension, see Fig. 2.

Fig 2 - Photo of the testing rig used in a deformation--controlled uniaxial tensile test (Stevin Laboratory).

The averaged signal used for the deformation-controlled tests was given by four LVDT's with a base length of 35 mm (lmeas=35 mm) mounted on the corners of the specimen, see Fig 3.

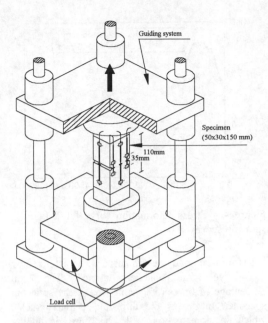

Figure 3 - Schematic representation of the test equipment and the measuring devices.

The deformation rate for GFRC and PC specimens was 0.16 μm/s and 0.08 μm/s respectively. The highest deformation rate was necessary in order to perform a tensile test up to 2 mm of deformation within a working day. Furthermore, doubling the speed of the deformation did not seem to significantly influence the results (Hordijk 1991). The averaged deformation measurements of the four LVDT's with a base length of 110 mm (lmeas=110 mm) was used to calculate the Young's modulus. The specimens were glued to steel platens fixed to the load equipment. More detailed information is given in Barros (1992).

2.4 - *Experimental results*

2.4.1 - PGFRC specimens

Per each Forton' panel three tensile specimens were tested. A typical stress-deformation (σ-δ) curve corresponding to the specimen manufactured by spray up technique is shown in Fig 4. The stress is defined as the load (average result of the four load cells) divided by the net cross-sectional area, while the deformation is the average result of the four corresponding LVDT's (35 and 110 mm base length).

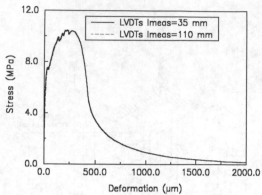

Figure 4 - Stress-deformation curve for a PGFRC spray up specimen.

Before peak load, the σ-δ curve is characterized by a relative large nonlinear branch. The high percentage of fibres (5%) employed in the spray up specimen originates the diffusion of microcracks, which in combination with the fibre debonding mechanism is responsible for this behaviour. At 2 mm of deformation, the specimen had still capacity to transfer some load via crack due to the frictional rigidity mobilized by the fibre pull-out mechanism. After the complete separation of the specimen in two halves, it was observed that the fibres were distributed in parallel layers to the front and rear specimen's face, revealing the process of manufacturing. As in each layer the fibres were 2D randomly distributed, the most part of these fibres offer an effective reinforcement.

The stress-deformation relationship corresponding to a typical premix PGFRC specimen is plotted in Fig. 5.

From comparison between the stress-deformation curves of the spray up and the premix specimens, it can be observed that the strength of the premix specimen is considerably lower, and the nonlinear branch before peak loading is less pronounced. This behaviour can be justified by the lower percentage of fibres (3%) and by the 3D randomly fibre distribution. Since the specimen is loaded in the uniaxial direction, this fibre orientation is not effective as it is in the 2D fibre distribution of the spray up specimens.

2.4.2 - Results of the plain mortar and the GFRC specimens

Additionally, some experiments were performed on plain mortar and GFRC specimens both manufactured at the Stevin Laboratory. Two GFRC and two PC specimens were tested. Typical results of the stress-deformation relationship are shown in Fig. 6.

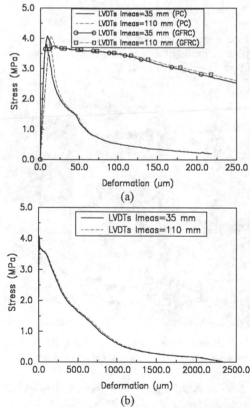

(a)

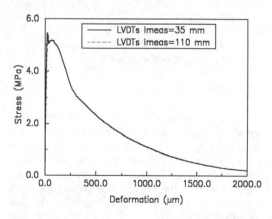

Figure 5 - Stress-deformation curve for a PGFRC premix specimen.

(b)

Figure 6 - Stress-deformation curves for: (a) plain mortar and GFRC specimens, (b) GFRC specimens.

Figure 6(a) clearly illustrates that the glass fibres modify the post-peak behaviour increasing the energy absorption capacity, improving the deformation capacity and increasing the tensile strength marginally as well. A scatter in the deformation at peak load was observed either on the GFRC specimens manufactured in Stevin Lab. as well as on the PGFRC specimens produced with the premix technique. Consequently, it appears that the premix technique tends to scatter the FRC properties, probably due to the difficulty in obtaining a uniform fibre distribution in the mix.

2.4.3 - Comparison of the results

The average values of the most significant properties obtained in the deformation controlled uniaxial tensile tests are summerized in Table 2.

Table 2 - Overview of the main tensile properties

	mortars			
	plain	premix GFRC	premix PGFRC	spray PGFRC
Fracture energy G_f (J/m^2)	230	1912	3017	4286
σ^{peak} (MPa)	3.8	4.4	5.5	10.8
δ^{peak} (μm)	11	55	78	250
δ^{ult} (μm)	$\cong 300$	$\cong 2350$	$\cong 2550$	≈ 2200
(1) w_0 (μm)	$\cong 300$	$\cong 2300$	$\cong 2500$	$\cong 2000$
(2) w_c (μm)	311	2234	2820	2040

(1) estimated ultimate crack opening
(2) critical crack opening, $w_c = 5.14 G_f/f_{ct}$ (Hordijk 1991)

The increase in fracture energy when adding glass fibres in the mix is evident. However, the tensile strength is markedly increased for the PGFRC made with the spray up technique only. Consequently, there is a significative difference between the results of the tensile strength of the premix (P)GFRC and the spray up PGFRC. This property is largely determined by the fibre orientation.

The relations between the normalized stress (stress divided by the tensile strength) and crack opening displacement (σ/f_{ct}-w) are plotted in Fig. 7 for the specimens tested. The relationship proposed by Cornelissen (1986) is also represented in Fig. 7,

$$\frac{\sigma}{f_{ct}} = \left\{1 + \left(c_1 \frac{w}{w_c}\right)^3\right\} \exp\left(-c_2 \frac{w}{w_c}\right) - \frac{w}{w_c}\left(1 + c_1^3\right)\exp\left(-c_2\right) \quad (1)$$

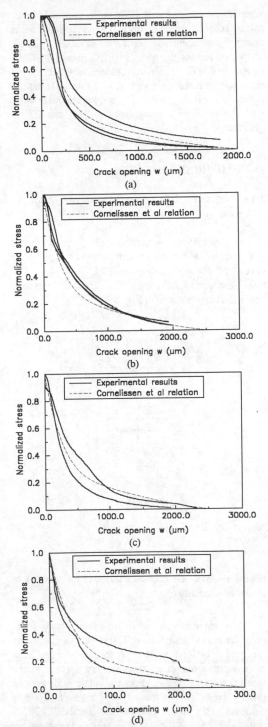

Figure 7 - Normalized stress-crack opening curves: (a) spray PGFRC; (b) premix PGFRC; (c) GFRC; (d) PC.

wherein values of 3 and 6.93 were assumed for c_1 and c_2 respectively. The critical crack opening w_c is material dependent and takes the values 2500, 2000, 2500, 300 μm for the spray PGFRC, premix PGFRC, GFRC and PC specimens respectively, which are close to the results prescribed in Table 2. It can be concluded that expression (1) can approximate fairly well the tensile softening behaviour of the GFRC specimens.

3 - NUMERICAL ANALYSIS

3.1 - *Comparison of experimental and numerical results*

The specimens experimentally tested were analysed by a finite element computational code developed for the analysis of FRC structures. The finite element mesh with the applied boundary conditions is presented in Fig. 8.

Figure 8 - Finite element mesh used in the analysis of the tensile specimens.

A four-node elements were integrated using four-point Gaussian quadracture. The lower boundary was assumed to be fixed and the present analysis had been carried out under direct displacement control at the top of the speciment. For the plain mortar and premix GFRC specimens it was taken a Young's modulus of 25 GPa, while for the premix PGFRC it was assumed a value of 17.5 GPa. These values match fairly well the linear branch of the experimental response. For Poisson's ratio it was

taken a 0.2 value. A linear elastic behaviour was assumed for the material between cracks. The stress-crack opening displacement relationship (expression (1)) was transformed into a crack stress-strain relation, which was used to simulate the softening behaviour (crack constitutive law).

The experimental and numerical response (average stress and deformation as defined in section 2.4.1) are compared in Fig. 9.

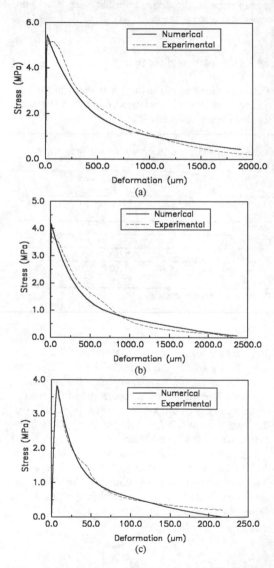

Figure 9 - Numerical and experimental average stress-deformation (lmeas=35mm) relation for: (a) premix PGFRC; (b) GFRC; (c) PC, specimens.

It can be concluded that the present numerical model catch with enough accuracy the experimental tensile behaviour of the specimens analised. In the case of spray PGFRC specimens (this analysis is not included in the present work), the microcracking spreads throughout a large zone of the specimen and the fibre debonding mechanism (interfacial fibre-matrix stress transfer) is improved which can be responsible for the considerable extent of the nonlinear branch before peak load. Therefore, a nonlinear pre-peak tensile constitutive law should be used to simulate this behaviour. In the PC and fibrous specimens reinforced with relative low percentage of fibres, a pre-peak linear elastic tensile behaviour is accurate enough. The nonlinear softening diagram represented by expression (1) is a good relation to model the post-peak behaviour of the cement brittle material.

4 - CONCLUSIONS

The present work is concerned with the tensile behaviour of GFRC specimens with 28 days of age. The following conclusions can be summarised:
- Fracture energy of cement based materials is significantly increased by adding glass fibre to the mix composition.
- The tensile strength is largely determined by the fibre orientation which depends on the mixing method. A tensile strength of about 11 MPa is found when a spray up technique is used for the PGFRC. A tensile strength between 4.5 and 5.5 MPa is found for (P)GFRC mixes made with the premix method.
- Smeared crack models based on finite element techniques wherein softening laws and fracture mechanics concepts are included can capture the experimental response. The softening behaviour of the cement based materials can be adequately represented by expression (1).

ACKNOWLEDGEMENT

The authors wish to acknowledge the Lab. facilities granted by Prof. Walraven of TUDelft. Acknowledgement to the "FORTRON DV" company who offered the PGFRC specimens to the project. Special thanks to mr A. van Rhijn who assisted during the experimental program.

REFERENCES

Akihama, S. & Suenaga, T. & Tanaka, M. & Hayashi, M. 1987. Properties of GFRC with low alkaline cement. Fiber reinforced concrete properties and applications, SP-105, S.P. Shah, G.B. Batson (edts.) 189-210.

Barros, J.A.O. 1992. Behaviour of glass fibre reinforced concrete systems. Faculty of Civil Engineering of Porto University, Technical report.

Bentur, A. 1986. Aging process of glass fibre reinforced cements with different cementitious matrices. RILEM symposium FRC 86, vol. 2, paper 7.3.

Bijen, J. 1983. Durability of some glass fiber reinforced cement composites. ACI journal, July-August 305-311.

Bijen, J. 1990. Improved mechanical properties of glass fibre reinforced cement by polymer modification. Cement & concrete composites 12:95-101.

Cornelissen, H.A.W. & Hordijk, D.A. & Reinhardt, H.W. 1986. Experimental determination of crack softening characteristics of normalweigth and lightweight concrete. HERON 31(2):45-56.

Fyles, K. & Litherland, K.L. & Proctor, B.A. 1986. The effect of glass fibre compositions on the strength retention of GRC. RILEM symposium FRC 86, vol. 2, paper 7.5.

FORTON private communications 1992.

Gopalaratnam, V.S. & Shah, S.P. 1985. Softening response of plain concrete in direct tension. J. Am. Conc. Inst. 82:310-323.

Hillerborg, A. & Modeer, M. & Petersson, P.E. 1976. Analysis of crack formation and crack growth in concrete by means of fracture mechanics and finite elements. Cement and concrete research 6:773-782.

Hordijk, D.A. 1991. Local approach to fatigue of concrete. PhD thesis, Delft University of Tech..

Leonard, S. & Bentur, A. 1984. Improvement of the durability of glass fiber reinforced cement using blended cement matrix. Cement and concrete research 14:717-728.

Li, Z. & Mobasher, B. & Shah, S.P. 1991. Evaluation of interfacial properties in fiber reinforced cementitious composites. Fracture processe in concrete, rock and ceramics 317-326, J.G.M. Van Mier, J.G. Rots, A. Bakker (eds.), RILEM, E. & F.N. Spon (Publ.).

Majumdar, A.J. 1974. The role of the interface in glass fibre reinforced cement. Cement and concrete research 4:247-266.

Majumdar, A.J. 1991. Glass reinforced cement. BSP professional books.

PCI committee on glass fiber reinforced concrete panels. 1981. Recommended practice for glass fiber reinforced concrete panels. PCI Jor., 26(1):25-93.

Shah, S.P. & Ludirdja, D. & Daniel, J.I. & Mobasher, B. 1988. Toughness-durability of glass fiber reinforced concrete systems. ACI materials journal, September-October 352 360.

Wang, Y. & Li, V.C. & Backer, S. 1990. Experimental determination of tensile behaviour of fibre reinforced concrete. ACI materials journal, 87(5):461-468, September-October.

Recent Advances in Experimental Mechanics, Silva Gomes et al. (eds) © 1994 Balkema, Rotterdam, ISBN 90 5410 395 7

FRC strengthened brick masonry walls horizontally loaded

Renato S.Olivito, Giuseppe Spadea & Pierluigi Stumpo
Department of Structural Engineering, University of Calabria, Arcavacata di Rende (CS), Italy

ABSTRACT: The need to recover ancient and historical buildings has contributed to the development of studies about the mechanical behaviour of masonry structures. Several researchers have worked both analytically and experimentally to understand the material as well as possible.

The aim of this work is to furnish an analysis about the carrying capacity of damaged masonry structures, and to suggest an effective repairing technique, which could be used as a practical method to restore the existing building heritage.

1. INTRODUCTION.

Recently, the study of masonry structures has received greater interest from many researchers both from a theoretical and experimental point of view [Del Piero (1989), Page (1983), Como et al. (1985), Maier et al. (1991)]. Moreover, the structural behaviour of masonry is of considerable importance in the renovation and consolidation of monumental and historical buildings.

Masonry is a typical composite material whose components, bricks and mortar, have different mechanical properties, and present a higher compressive than tensile strength [Angelillo et al. (1992)]. Therefore, it seems reasonable to assume that the mechanical properties of the composite material are intermediate between those of the components.

In this work an experimental study about the behaviour of brick masonry walls is carried out by means of small-scale models; this choice was motivated by the possibility of achieving a non-isotropic masonry-like material with a satisfactory respect for the microstructure [Angelillo et al. (1992)]. The models were completed at both the edges by means of reinforced concrete curbs, and were subjected to precise load conditions by using a testing apparatus properly set up.

At first a load condition was imposed on the models to create some damage in the material;

then, a second loading was applied to achieve structural failure of the models.

Successively, the same models were repaired by means of a mortar with a high cement content reinforced by a small volume percentage of polypropylene short fibres. The repaired models were subjected to the second load condition again, and the results were compared with those previously obtained.

The results of this experimental study allow an affirmation that the use of FRM (Fibre Reinforced Mortar) can be a suitable means to ensure a sufficient resistance capacity of the damaged masonry material, also preserving the original appearance as much as possible.

2. MATERIALS.

To obtain the small-scale modellization of the composite, it was chosen to build the wall models adopting a 1:2 length scale and to take care in preparing the component materials, particularly in the cutting of bricks from commercial types and in the sizes of mortar aggregates [Angelillo et al. (1992)].

In figures 1 and 2 the brick cutting and the gravel size percentage of the mortar are shown, respectively.

Two kinds of mortar were prepared, M2 and

M4, by means of 325 Portland cement according to the ANDIL recommendations given in Table 1.

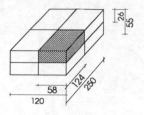

Figure 1. Brick models.

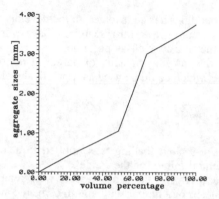

Figure 2. Aggregate sizes curve for the mortar.

Table 1. Mortar compositions and properties.

Type	Volume fractions			Strength [MPa]
	Cement	Hydraulic lime	Gravel	
M1	1	0	3	20
M2	1	0.5	4	10
M3	1	1	5	5
M4	1	2	9	2.5

Cubic specimens with a 100 mm lateral dimension were prepared for both the mortar types, according to UNI 6132-72, and uniaxial compression tests were carried out using a servocontrolled electro-mechanic testing machine with a 0.5% accuracy.

In Figure 3 the results of the tests conducted on the bricks and the mortars are plotted, while in Table 2 the average values of the mechanical properties are given.

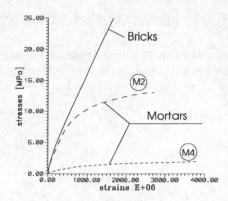

Figure 3. Uniaxial compression tests on bricks and mortars.

Table 2. Mechanical properties of the masonry components.

Property	Mortar type		Bricks
	M2	M4	
E [MPa]	13500	4155	14400
ν	0.186	0.236	0.126

3. MODELS AND TESTING APPARATUS.

Two series of three masonry models were prepared, according to the dimensions shown in mm in Figure 4, using the M2 mortar for the 1st series and the M4 mortar for the 2nd one.

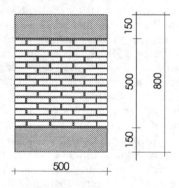

Figure 4. Masonry wall model.

The upper and lower concrete curbs were necessary to obtain the right boundary and load conditions.

The testing apparatus was set up by constructing a steel frame of an adequate stiffness, as shown in Figure 5. Loads and forces were imposed to the model under test by using both hydraulic jacks and tightening nuts, and their values were recorded by means of strain gauge load cells located at the right points.

Linear inductive displacement transducers (LVDT), whose positions were accurately chosen, were adopted to record the displacements at particular points of the model, as shown in Figure 6.

Signals coming from load cells and from transducers were recorded and transformed using a data acquiring unit connected to a personal computer via an IEEE 488 interface card.

4. TESTS AND RESULTS.

To analyze the effects of a previous load condition on the ultimate masonry behaviour, each model was first subjected to a constant vertical load, 13 kN and 10 kN for the 1st and the 2nd series models, and to a horizontal load increasing up to a noticeable changing of the model stiffness.

This technique did not allow for a quantification of the damage, but only ensured that a certain damage in the material had occurred. As an example, in Figures 7 and 8 the horizontal displacements of the 1-3 measuring points vs. horizontal load are plotted for a model of the 1st and one of the 2nd series, respectively.

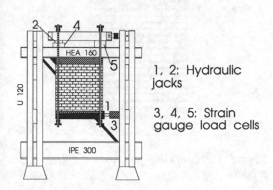

Figure 5. Testing apparatus.

1, 2: Hydraulic jacks

3, 4, 5: Strain gauge load cells

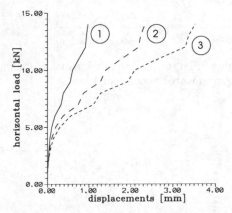

Figure 7. Damage phase - 1st series model. Horizontal displacements vs. horizontal load.

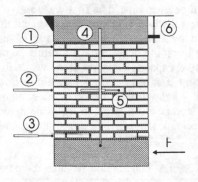

Figure 6. Displacement transducers positions.

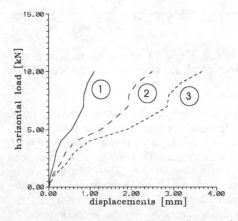

Figure 8. Damage phase - 2nd series model. Horizontal displacements vs. horizontal load.

Successively, each model was subjected to a vertical uniform load and to a horizontal one, both increasing up to the failure of the model. In Figures 9 and 10 the results obtained for the 1st and the 2nd series are plotted.

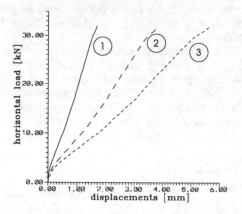

Figure 9. 1st series unrepaired model. Horizontal displacements vs. horizontal load.

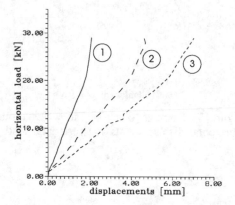

Figure 10. 2nd series unrepaired model. Horizontal displacements vs. horizontal load.

It can be noticed that the mortar composition influenced the masonry behaviour. In fact, the use of a good quality mortar allows one to obtain a masonry material stronger and stiffer than the use of a shoddy one. Moreover, the first model series showed a cracking pattern following the main diagonal (Figure 11), instead of the smeared cracking of the second model series (Figure 12).

Figure 11. Cracking pattern in a 1st series model.

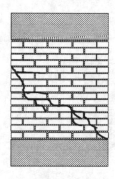

Figure 12. Cracking pattern in a 2nd series model.

Then, the broken models were repaired by using a polypropylene fibre reinforced mortar [Bruno et al. (1990)]. Two main problems should be overcome. The first was to give a ductile capacity to the repaired models, and this was solved by using the polypropylene fibres. The second was the need to have a mortar with high adhesive capacity; this problem was solved by studying a mortar with small-dimension aggregates and with a high cement content. The FRM composition is given in Table 3, whereas its mechanical properties are given in Table 4.

Table 3. FRM composition (volume fractions).

Aggregates [mm]			Cement	Water	Fibres
0.335	0.71	1.4			
0.48	1.92	0.6	0.6	1.5	0.04

Table 4. FRM mechanical properties.

E [MPa]	ν	σ_r [MPa]
21500	0.218	18.65

To test the repaired models, it was considered that they were always made of damaged material; then the models were subjected to a vertical uniform load and to a horizontal one, both increasing up to the failure again. In Figures 13 and 14 the horizontal displacements vs. the horizontal load are plotted.

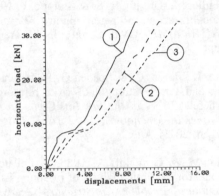

Figure 13. 1st series repaired models. Horizontal displacements vs. horizontal load.

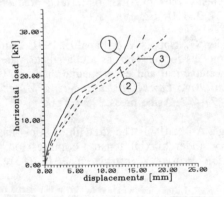

Figure 14. 2nd series repaired models. Horizontal displacements vs. horizontal load.

5. DISCUSSION AND CONCLUSION.

By comparing the results obtained for unrepaired and repaired models the effect of the use of FRM as a repairing material can be observed.

The 1st series of repaired models showed a quasi-linear behaviour, like the unrepaired ones, with an increase in the displacements, mainly due to interface cracks (Figure 15).

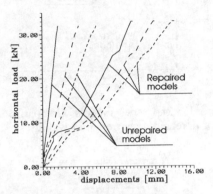

Figure 15. 1st series models. Comparison between unrepaired and repaired walls.

The new cracking zone faithfully follows the previous one apart from a little slippage between interfaces of the bed joints, restrained by FRM (Figure 16).

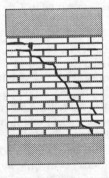

Figure 16. 1st series model. Cracking pattern.

The 2nd series of repaired models showed a more ductile behaviour, with displacements 3 times greater than those obtained for the unrepaired ones (Figure 17); in this case the cracking presents a smeared distribution in the structure of the model (Figure 18).

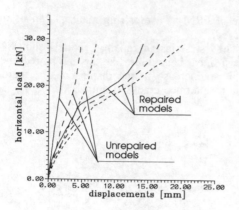

Figure 17. 2nd series models. Comparison between unrepaired and repaired walls.

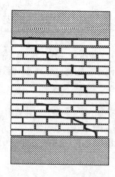

Figure 18. 2nd series model. Cracking pattern.

In conclusion, it can be affirmed that the use of FRM allowed for a good restoration of damaged masonry walls, especially in the case of masonries built using a poor quality mortar.

In fact, for all repaired models the same level of ultimate loading and a great increase of deformability up to failure were obtained, compared with those for the unrepaired models.

Future developments of this work could be a more exact definition of restoration method and analytical support of the results.

ACKNOWLEDGEMENTS.

The financial support of the Ministry of University and Scientific and Technological Research of the Italian Government is gratefully acknowledged.

The Authors also express their acknowledgements to Mr Antonio Pantuso and to the technical staff of the Laboratory of the Department of Structural Engineering of the University of Calabria, which were so helpful in preparing the experimental set up and the tests.

REFERENCES.

Angelillo, M. & Olivito, R. S. 1992. Experimental analysis of horizontally loaded masonry walls, *Rep. of the Civ. Eng. Inst., University of Salerno*, N. 40.

Angelillo, M., Olivito, R. S. & Stumpo, P. 1992. Modellizzazione di pareti murarie. *Proc. of the 11th AIMETA Nat. Congr.*, 35-42, Trento, Sept. 28-Oct. 2.

Bruno, D., Porco, G. & Spadea G. 1990. Cracking and tensile strength of fibre reinforced concrete. *Proc. of the 9th Int. Conf. on Experimental Mechanics*, 1452-1461, Copenhagen, 20-24 Aug.

Como, M. & Grimaldi, A. 1985. A unilateral model for the limit analysis of masonry walls. *Unilateral Problems in Structural Analysis*, Springer, Wien.

Del Piero, G. 1989. Constitutive equations and compatibility of the external loads for linear elastic masonry-like materials. *Meccanica*, 24, 3, 150-162.

Maier, G., Nappi, A. & Papa, E. 1991. Damage models for masonry as a composite material: a numerical and experimental analysis, *Constitutive Laws for Engineering Materials*, 427-432, Asme press, New York.

Page, A. W. 1983. The strength of brick masonry under biaxial tension-compression, *Int. Jou. Masonry Construction*, 3, 1, 26-31.

Recent Advances in Experimental Mechanics, Silva Gomes et al. (eds) © 1994 Balkema, Rotterdam, ISBN 90 5410 395 7

A quality control test for composite masonry/concrete construction

S.W.Garrity
University of Bradford, UK

ABSTRACT: Concrete and masonry are often used compositely in repair work or for new construction. Unless steel ties are used, which is not always desirable, the shear connection between masonry and concrete necessary for composite action is dependent on the bond between the two materials. This paper describes a simple quality control test for measuring the tensile bond strength of the masonry/concrete interface; the specimens can be inspected prior to testing for signs of defects such as shrinkage cracking or voids resulting from poor compaction. The test can also be used for on-site trials, held in advance of construction, to determine the most suitable concrete mix proportions for new or repaired masonry.

1 INTRODUCTION

Many forms of construction are designed where concrete and masonry are assumed to act compositely. In particular, hollow concrete, calcium silicate or clay blocks filled with in-situ concrete are used for low to medium rise building works in many parts of the world. Concrete is also used to repair or strengthen a variety of existing masonry structures ranging from historic monuments and arch bridges to modern buildings damaged by terrorist attack or seismic activity. Various forms of construction in which reinforced concrete and masonry are assumed to act compositely have also been developed; some of these are referred to generically as reinforced masonry (Hendry 1991). This tends to be specified for structures with an exposed masonry finish that are required to resist the large magnitude forces generated by seismic effects, hurricanes, retained earth, vehicle impact or explosions.

1.1 *Concrete/masonry bond*

Unless steel ties are used, which is not always desirable, the shear connection between masonry and concrete necessary for composite action is dependent on the bond between the two materials. In a review of the literature on brick/mortar bond, Goodwin and West (1980) concluded that the rate at which the masonry units absorb moisture from the mortar had the greatest influence on the bond strength. More recently, Lawrence and Cao (1987, 1988) found that the bond between clay bricks and modern cementitious mortars is derived mainly from the mechanical interlocking of the cement hydration products growing on the surface and in the pores of the bricks. In particular, the brick/mortar bond was found to be influenced by the distribution, continuity and degree of microcracking of the hydration products which is dependent on several factors such as the water:cement ratio, free lime content, aggregate grading and curing of the mortar. The size and distribution of surface pores in the bricks and their moisture content at the time of construction can also have a significant effect. Similar factors are also likely to influence the bond between concrete and masonry in composite construction, as are macroscopic defects such as shrinkage cracking and voids in the concrete caused by inadequate compaction.

The importance of achieving good bond between masonry and concrete has been demonstrated in large-scale testing of reinforced brickwork beams where shear failure was immediately preceded by longitudinal cracking at the brickwork/concrete interface (Tellet & Edgell 1983). The premature vertical splitting failure of two laterally loaded full-scale reinforced brickwork walls due to the inadequate shear connection between the concrete and the brickwork, has also been reported (Maurenbrecher et al 1976, Fisher et al. 1989). More recently, in full-scale tests on masonry arch bridges, premature failure occurred as a result of arch ring separation caused by lack of bond (Melbourne & Gilbert 1993) and similar problems are possible when concrete is cast directly onto the extrados of the arch barrel of a bridge as a strengthening measure.

1.2 *Durability*

In reinforced masonry construction, the reinforcing steel is passivated against corrosion by a continuous coating of alkaline cement paste provided by the surrounding mortar or concrete. Research has shown that where concrete is cast against masonry, the wet concrete can be de-watered by the porous masonry leading to plastic shrinkage cracking (Kingsley et al. 1985, de Vekey 1988). Plastic settlement of the wet concrete, that can lead to cracking and reduced bond between the concrete and the steel reinforcement, has also been reported (Garrity 1992). Although there are no reports of corrosion of the steel reinforcement in reinforced masonry structures arising from plastic cracking, there is a concern that plastic shrinkage and plastic settlement effects may reduce the corrosion protection, particularly where exposure to chloride ions from seawater or de-icing salts is likely.

1.3 *Testing of composite concrete/masonry*

Judging from the above, the quality of the masonry/concrete interface is likely to influence the structural performance and, possibly, the durability of composite masonry/concrete construction. It is also clear that the bond between concrete and masonry is highly complex and that testing masonry and concrete as the separate components of composite construction will not reflect how the two materials interact in practice.

Where reinforced masonry construction has been used for many years, as in the seismically active Western United States of America, engineers have devised test procedures that take into account the de-watering of wet concrete (or grout) by porous masonry (Dickey 1987). A typical example is that described by Edgell and de Vekey (1987) for the compressive strength testing of the grout used in reinforced concrete blockwork wall construction. The test is very similar to the conventional cube test for mortar and concrete except that, instead of using steel moulds to form the specimen, the grout is cast in moulds formed from concrete blocks similar to those used in the actual wall. The surfaces of the blocks are covered with absorbent paper which allows water to pass through it but prevents the grout from bonding to the blocks; see Figure 1.

Almost all of the tests that have been developed for monitoring the quality of masonry construction are compressive, flexural or masonry unit/mortar bond strength tests. Although the bond strength of composite masonry/concrete construction has been

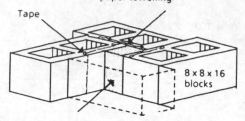

3⁵/₈ in x 3⁵/₈ in x 7⁵/₈ in
(100 x 200 mm) grout test prism

Figure 1. Grout test specimen cast in absorbent mould (Edgell & de Vekey 1987)

measured in the laboratory by Sinha and Foster (1979) and others, as far as the author is aware, no test has been developed that is suitable for routine quality monitoring on site.

As part of a laboratory-based study of composite masonry/reinforced concrete walls, the author devised a simple tensile test which was used to measure the bond strength of the masonry/concrete interface. A futher requirement of the test was that it should permit the identification of any defects within the interfacial zone that might influence structural behaviour and durability such as plastic cracking, cracking due to drying shrinkage or the effects of poor compaction. As the test involved the use of small specimens that could be easily handled and transported to a testing house, it is considered suitable as a site-based quality control test for composite masonry/concrete construction. The main aims of this paper, therefore, are to:-

a) describe the tensile bond strength test for composite masonry/concrete.

b) review the sampling and testing methods.

c) describe how the test might be used in practice.

2 THE TENSILE BOND STRENGTH TEST

In the study referred to above, short lengths of reinforced cellular brickwork wall were constructed; each wall consisted of brickwork built in a cellular bonding pattern around vertical steel reinforcing bars that had been previously cast into a reinforced concrete foundation. The voids or cells in the brickwork were then filled with insitu concrete to form a reinforced masonry wall. When the concrete had gained sufficient strength, each length of wall was rotated into a horizontal position and tested to failure as a beam in four point bending (Garrity 1992).

In order to investigate the bond between the brickwork and the concrete it was decided to carry out direct tensile testing of samples of the brickwork/concrete interface taken from two of the beams tested in the study. The method of sampling and testing is described below.

Both beams were made using the same type of bricks, namely a perforated fired clay brick with a compressive strength of 95.5 N/mm², a water absorption of 6.3% and an initial rate of suction of 0.46 kg/m²/min, laid in a 1:¼:3 cement:lime:sand mortar. The insitu concrete had a compressive strength of 41 N/mm² at the time of testing.

2.1 Sampling and preparation of test specimens

On completion of the bending tests, 44mm diameter cylindrical specimens were cored from the undamaged regions of the beams using rotary coring equipment. In each case, full depth cores yielding two test specimens were taken from the two beams described above. Each full depth core was cut in half and the ends of each half were trimmed with a rotating blade saw to produce a 50mm long test specimen with the brickwork /concrete interface located at its centre as shown in Figure 2. The end faces of each cylindrical specimen were ground parallel to within ± 0.25mm. Two different types of specimen were obtained from the beams, namely those containing a mortar joint, hereafter referred to as *type A* specimens, and those without a mortar joint, referred to as *type B* specimens.

2.2 Test details

Each cylindrical specimen was tested in direct tension in an Instron 4206 universal testing machine at the University of Bradford. The arrangement used for gripping each end of the specimen and locating it in the jaws of the test machine is shown in Figure 2. A double universal joint arrangement comprising a screw-fitted steel end-plate was used to align the specimen in the jaws of the test machine and to ensure that any applied load was purely axial. Any minor variations in the ends of the specimen were accommodated in the thin layer of epoxy-based adhesive that was used to connect the test piece to the steel end-plates. Once set up in the test machine, the specimen was subjected to an increasing tensile force until failure occurred. The force was applied to the specimen via a crosshead which moved at a constant rate of 0.5mm/min.

Figure 2. Test arrangement

2.3 Test results

The results of tests on thirty four cores taken from the two reinforced brickwork beams referred to earlier are presented in Table 1.

3 DISCUSSION

3.1 Review of the sampling and testing

Coring equipment of various sizes can be readily hired at low cost from most plant hire firms and many civil engineering laboratories specialising in concrete or rock testing will be equipped with such machinery. Furthermore many engineers will be familiar with the equipment as it is a commonly used method of sampling hardened concrete from existing

Table 1. Tensile bond strength test results		
Sample No. and Type	Tensile Bond Strength [N/mm²]	Mode of Failure [see below]
Beam 1		
A1	1.39	4
A2	2.74	2
A3	2.14	3
A4	1.13	4
A5	1.5	1
A6	2.06	4
A7	0.71	4
A8	2.09	4
mean tensile bond strength = 1.72 N/mm²		
B1	2.24	2
B2	1.22	1
B3	1.21	1
B4	2.64	2
B5	2.14	1
B6	1.21	1
B7	2.84	2
B8	1.98	2
B9	1.37	1
B10	2.60	1
mean tensile bond strength = 1.95 N/mm²		
Beam 2		
A9	0.54	4
A10	0.59	4
A11	1.23	4
A12	2.80	4
mean tensile bond strength = 1.29 N/mm²		
B11	1.59	1
B12	2.53	2
B13	2.02	1
B14	2.35	3
B15	1.47	1
B16	2.38	1
B17	2.25	1
B18	0.98	1
B19	1.46	1
B20	2.01	1
B21	1.52	1
B22	1.14	1
mean tensile bond strength = 1.81 N/mm²		

Failure modes:

1 Failure at brick/concrete interface.
2 Tensile failure of concrete.
3 Tensile failure of brick.
4 Failure plane passing through brick/mortar interface and along mortar joint.

structures for compressive strength testing.

Obtaining composite brickwork/concrete cores from the reinforced brickwork beams proved to be quick and simple. The only problem encountered during drilling was the wear of the cutting tool caused by steel reinforcing bars. In practice, it is not always possible to avoid cutting through some of the reinforcement even when a covermeter is used to identify its likely position prior to drilling.

The use of 100mm and 150mm diameter cores was investigated in trials, however, the lighter 44mm diameter samples proved to be easier to handle during trimming and when manoeuvring the specimen into the jaws of the testing machine. A common problem when testing comparatively large, low strength specimens in tension is the creation of undesirable torsional and flexural effects by specimen misalignment. Such effects are reduced when using small diameter specimens. Indeed, the use of 44mm diameter cores together with the measures described in 2.2 to ensure axial loading appear to have been successful as none of the specimens tested showed any signs of torsional or flexural failure.

On site, the use of comparatively small cores is preferred for the following reasons:-

i) There is less chance of coring through any steel.
ii) Less material is removed from the structure which is important when thin structural members are being investigated.
iii) There will be less damage to the exposed surface of the structure.
iv) Small samples are easier to transport to a testing house.
v) It is less costly to obtain a representative number of samples for testing.

3.2 *The test results*

The differences in the results presented in Table 1 are very typical of tests on inherently variable materials such as masonry. The results do, however, yield two important points with practical implications. In the first instance, it is clear that the average tensile bond strength of the type A test specimens, that is those containing mortar joints, was less than that of the type B specimens. Hence, when coring samples for testing from composite masonry/concrete construction, it is important to ensure that some of the cores contain mortar joints in order to obtain a representative sample.

The second point demonstrates a possible disadvantage of using small diameter cores, namely that the effects of any minor defects will be magnified. An example of this is seen with six of the cores where failure occurred in the concrete rather than at the brick/concrete interface; in each case, it is very likely that failure was caused by the presence of a large piece of aggregate in the core. Also, with two of the cores, failure occurred in the brick. This is almost certainly due to the presence of intermittent horizontal planes of weakness in the brick due to the extrusion process which was used to form the "green" clay into the required shape prior to firing. Both defects are likely to act as stress raisers in the cores leading to premature failure and underestimates of the interface bond strength. It is not yet known whether cracking would initiate at such locations in a full-scale structure. To draw attention to these potential problems when testing masonry/concrete cores of this type, it is suggested that the mode of failure and location of the failure plane should be recorded in addition to the tensile bond strength.

4 PRACTICAL APPLICATIONS OF THE TEST

4.1 *Bond strength testing before starting construction on site*

With reinforced concrete construction in the U.K. it is common practice for the concrete supplier to produce trial batches of the proposed concrete mixes before starting construction on site. This helps the engineer responsible for the supervision of construction to judge if the materials are suitable for the proposed works and gives adequate time to adjust the mix proportions if necessary. Using the same philosophy, it is suggested that *trial panels* of unreinforced composite masonry/concrete that are representative of the proposed forms of construction, are built before starting work on site. The trial panels should be constructed from the bricks or blocks specified by the architect or engineer and two or three alternative concrete mixes. Cores would be taken from each panel after three, seven and twenty eight days and tested in the manner described in this paper. This approach is recommended for the following reasons:-

i) It gives the engineer or architect on site an opportunity to judge the standard of workmanship of the bricklayers or masons before work is started.

ii) It gives the contractor the opportunity to try different methods of construction.

iii) It shows which of the concrete mixes is likely to suffer from defects such as poor compaction or plastic cracking due to the de-watering effects of the masonry.

iv) It shows which concrete mix is likely to be most compatible with the specified masonry and, in particular, which mix achieves the greatest bond.

v) It allows the engineer to assess the influence of any shrinkage-compensating or plasticising admixtures, proposed by the contractor, on the masonry/concrete bond.

vi) It provides the engineer and contractor with useful information regarding the rate of gain of tensile bond strength. This may influence the rate of construction on site.

vii) It gives *reference tensile bond strengths* when the construction is three, seven and twenty eight days old. These can be used in the quality control process (see 4.2).

4.2 *Quality control testing of composite masonry/concrete on site*

The following procedure is suggested as a means of assessing the quality of composite masonry/concrete during construction :-

a. As described above, *trial panels* of unreinforced composite concrete/masonry should be built before starting construction of the main works on site. The most compatible concrete mix should be identified and the *reference tensile bond strengths* for that mix should be agreed and recorded. Construction of the specified works can then start on site.

b. During construction, when instructed by the supervisory staff, each bricklayer or mason must build a short length of unreinforced wall. This *control wall* must be representative of the actual construction and should be built using the same masonry units, mortar, insitu concrete and masonry bonding pattern as in the specified works. Additional control walls may be built, at the request of the engineer's or architect's representative.

c. Each control wall should be built in similar exposure conditions and be cured in a similar manner to the actual construction.

d. After seven days have elapsed, the control wall should be carefully rotated and lowered onto a flat part of the site. At least ten cores should be drilled from the wall, each core penetrating through the masonry into at least 50mm of the insitu concrete. In order to obtain a representative sample, the cores should be taken from all parts of the control wall.

e. Each core should be carefully inspected for any signs of cracking or poor compaction and the results of the inspection should be given to the supervising engineer or architect. Any premature failures occurring during coring or afterwards must be recorded.

f. The masonry/concrete interface of each core should be indelibly marked with the date of construction and any identifying number. Each core should then be placed in its own polyethylene bag and sealed then taken to a testing house for testing in the manner described in this paper.

g. The results of the tests, including the mode and plane of failure and the age of the specimen when tested, should be reported to the supervising engineer or architect without delay for comparison with the *reference tensile bond strengths*.

4.3 *Testing of masonry strengthened using in-situ concrete*

There are many examples of existing masonry structures that have been repaired or strengthened using insitu concrete. If the engineer or architect needs to measure the bond between the two materials, cores can be taken and tested in direct tension as described earlier. A problem that could arise with historic or listed masonry structures is the need to disguise the core locations. If the end of the core that was originally the exposed face of the masonry structure is retained when the core is prepared for testing, it can be fixed back in its original position with an epoxy resin adhesive once the rest of the void has been filled with mortar or any other suitable material.

5 SUMMARY

The quality of the masonry/concrete interface is likely to influence the structural performance and, possibly, the durability of composite masonry/concrete construction. Testing the masonry and concrete as the separate components of composite construction will not, however, reflect how the two materials interact in practice.

It is proposed to measure the bond strength of the masonry/concrete interface using a simple tensile test. The test is considered suitable for most forms of composite masonry/concrete construction such as new reinforced masonry or insitu concrete repairs to existing masonry structures. A suggested quality control procedure for composite masonry/concrete construction, in which the tensile bond strength test is a key part, is also presented.

ACKNOWLEDGEMENTS

The author wishes to thank Nabil Amoura and Clive Leeming for their assistance with the experimental work and Marshalls Clay Products for providing the bricks for this project.

REFERENCES

de Vekey, R.C. 1988. The shrinkage cracking of concrete infill in reinforced masonry. *Proc. Brit. Mas. Soc.* 2: 85-91. Stoke-on-Trent: British Masonry Society.

Dickey, W.L. 1987. Progress of reinforced masonry. *Proc. 4th N.Amer. Mas. Conf.*: 1.1-1.14. Los Angeles: The Masonry Society.

Edgell, G.J. & R.C.de Vekey 1987. Grouting reinforced brickwork and blockwork: a review of current specifications, practice and experience. *Special Publication 116.* Stoke-on-Trent: British Ceramic Research Limited.

Fisher, K, B.A.Haseltine & W.Templeton 1989. Structural testing of brickwork retaining walls. *Proc. 5th Can. Mas. Symp.*: 827-837. Vancouver: University of British Columbia.

Garrity, S.W. 1992. Reinforced clay brickwork highway structures. *Proc. 6th Can. Mas. Symp.*: 735-749. Saskatoon:University of Saskatchewan.

Goodwin, J.F. & H.W.H.West 1980. A review of the literature on brick/mortar bond. *Proc. Brit. Ceram. Soc.* 30: 23-37. Stoke-on-Trent: British Ceramic Society.

Hendry, A.W. 1991. *Reinforced and prestressed masonry*. Harlow : Longman Scientific & Technical.

Kingsley, G.R., L.G.Tulin & J.L.Noland 1985. Parameters influencing the quality of grout in hollow clay masonry. *Proc. 7th IBMaC*: 1085-1092. Melbourne: Brick Development Research Institute.

Lawrence, S.J. & H.T.Cao 1987. An experimental study of the interface between brick and mortar. *Proc. 4th N.Amer. Mas. Conf.*: 48.2-48.14. Los Angeles: The Masonry Society.

Lawrence, S.J. & H.T.Cao 1988. Microstructure of the interface between brick and mortar. *Proc. 8th IBMaC*: 194-204. Dublin: Elsevier App. Sci.

Maurenbrecher, A.H.P., A.B.Bird, R.J.M.Sutherland & D.Foster 1976. *SCP11 - Reinforced brickwork - vertical cantilevers 2.* St.Neots: Structural Clay Products Limited

Melbourne, C. & M.Gilbert 1993. A study of the effects of ring separation on the load-carrying capacity of masonry arch bridges. In J.E.Harding, G.A.R.Parke & M.J.Ryall (Eds), *Bridge Management 2*: 244-253. London : Thomas Telford.

Sinha, B.P. & D.Foster 1979. Behaviour of reinforced grouted cavity beams. *Proc. 5th IBMaC*: 268-274. Washington D.C.: Brick Institute of America.

Tellet, J & G.J.Edgell. 1983. The structural behaviour of reinforced brickwork pocket-type retaining walls. *Technical Note 353.* Stoke-on-Trent : British Ceramic Research Association.

Recent Advances in Experimental Mechanics, Silva Gomes et al. (eds) © 1994 Balkema, Rotterdam, ISBN 90 5410 395 7

A model for laboratory consolidation of fissured clays

F. Federico
Department of Civil Engineering, University of Roma 'Tor Vergata', Italy

ABSTRACT: Laboratory dissipation tests have been carried out on natural as well as on an artificially fissured clay showing that natural materials may be physically represented by the artificial material, whose meso-structure is controlled according to assigned characteristics. The possibility of a theoretical analysis of the consolidation process was started on the assumption that fissured material is represented as a double porous elastic medium, accounting for fluid exchange between fissures and pores. Experimental outcomes are finally discussed against theoretical forecasts, computed for values of parameters that may reasonably be attributed to the samples, obtaining a significant agreement.

1 INTRODUCTION

During the past two decades, the modelling of mechanical response of fissured or fractured rock masses or highly tectonized clayey soils (scaly clays), in which water contemporary seeps through the net of fissures and through the pores within fragments or blocks, has attracted a progressively increasing deal of interest in geotechnical engineering.

The interest on this subject has recently received a strong impulse due to new problems related to migration of contaminants (Leo, Booker, 1993), as well as related to heat transport in groundwater (Hensley, Saviddou C., 1993).

The available conceptual models may be roughly divided in three categories.

To the first one belong models that represent fractured rocks or soils as equivalent single porous anisotropic continuous media. Main physical quantities (potential, porosity and fluid pressure) have to be considered, in this case, as a gross average over blocks of the material sufficiently large to contain many fractures. Some drawbacks derive from this approach, such as the need for the formulation of an equivalent permeability tensor and its experimental evaluation.

In a second, opposite category fall the models that carefully reproduce the main physico-geometrical characteristics of the set of discontinuities. These models better apply to rock masses constituted by large intact blocks delimited by few, distinct ordered systems of fissures.

Severe difficulties arise, however, if one should try to model flow in complex geotechnical materials: although playing a key role in governing fluid movement, their meso-structure is not easily detectable nor characterizable.

A third line of research aims at characterizing a highly fissured porous medium as a double porous continuous medium; the first system of pores represents the net of fissures; the other, the pores within intact material fragments.

Dual porosity models appear both computationally tractable and sufficiently precise for practical purposes. The main conceptual difficulty lies in the fact that, since the fissure system itself is considered as a porous medium, both pore and fissure flows must be represented as continuous, smooth functions, defined at *each* point of the medium. As a consequence, macroscopic flow requires the overlapping of two different seepage flows, each in the corresponding system of voids, far from actual motion, mainly around the intact, deformable, less permeable blocks.

This matrix-fissure interaction on the scale of the fissure spacing must be carefully modelled. To this purpose, models belonging to this third category may be further distinguished in two fundamental families, the main difference between them being the representation of fluid exchange through the two system of voids.

For models of the first family, the equations that describe the flow in the fissure system contain a source term that represents the exchange of fluid from the pores of the blocks to the fissures; this term, expressed by quantities at the scale of the matrix blocks, is then macroscopically distributed over the entire flow domain, as is the fissure flow itself (Douglas, Arbogast, 1990).

The second family collects dual porosity models (Aifantis, 1980; Wilson, Aifantis, 1982; Ohnishi et al., 1988; Musso, et al.,1990; Valliappan, Khalili-Naghadeh, 1991; Elsworth, Bay, 1992), for which the fluid exchange from one system of voids to the other depends at each point on the difference between the pressures in the pores and in the fissures, through a fluid exchange coefficient: this last one may be in turn related to permeability of both porous systems as well as to shape and volume of single porous blocks (Musso, et al.,1990; Ohnishi et al., 1988).

Beyond the differences among the various double porosity models, a fundamental question, about the possibility that these models could correctly explain the actual mechanism of pore pressure dissipation in complex materials, arises.

This question may find a definite answer only through experiments on *natural* materials. However, difficulties associated with identification and evaluation of parameters describing the meso-structure and the local mechanical behaviour, call for a preliminary set of experiments on *artificially structured* materials, thus eliminating the problems related to variability of the internal structure, always affecting natural materials. This approach has been followed by Sato et al. (1985) and recently by Musso, Federico (1993).

The problem of the representativeness of artificially structured materials is dealt with in the paper. To this purpose, some results of laboratory dissipation tests carried out on artificial as well as on intensely natural fissured clay samples are presented and discussed.

Experimental results clearly outline a feature of the artificial structured material, i.e., the capability of the net of discontinuities to act as a "diffuse" internal drainage, collecting the water seeping from the pores of the fragments, as in the actual behaviour of natural fissured material.

Experimental data obtained from consolidaton tests on artificial materials are then discussed against the corresponding theoretical results (Musso, Federico, 1993), showing a significant agreement.

2 EXPERIMENTAL RESULTS

2.1 Natural stiff fissured clays

Numerous papers on *London clay* (Ward et al., 1959; Garga, 1988) are reported in technical literature. These studies, however, do not sufficiently focus the role of structure on pore pressure dissipation.

Some attention deserve the results of two dissipation tests in a triaxial cell conducted in 1979, within an experimental research programme on the characterization of an intensely fissured, "scaly", clay formation of Sicily. The interest derives from the origin of the undisturbed samples used in the work, that had been obtained in 1962 by means of pressure samplers from the core of Scanzano earth dam (in Sicily) under construction at that time.

The original material is composed by an assembly of flat scales about one millimeter thick and one centimeter long.

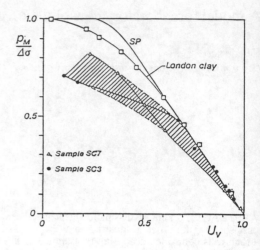

Fig.1 - Measured values of the ratio $p_m/\Delta\sigma$ vs average degre of consolidation $U_v = w(\tau)/w_c$ for natural fissured clays.

After recovery, the samples placed within thick cylindrical containers closed at the extremities by perforated brass plates, were stored under distilled water for a very long period of time, until the date of the tests, in 1979; saturation degree raised from the original value, close to 0.80, up to unity, found at the date of the tests.

Typical values of the Atterberg's limits as well as details of the procedure tests are reported in (Federico, Musso, 1991).

Two specimens were subjected to laboratory consolidation tests in a triaxial chamber. Figure 1 shows, for both samples, non-dimensional pore pressure excess $p_m/\Delta\sigma$ measured at the top cap of the cell test plotted against the average degree of consolidation U_v, expressed as the ratio of the settlements $w(t)$ at time t and w_c (final settlement). Data concerning London clay reported in the cited publication by Garga as well as the solution based on the Terzaghi's theory (curve SP) are also shown for comparison.

The experimental points cover a narrow band; dissipation of measured pore pressures takes place more rapidly than that one expected by theories applying for usual single poro-elastic media (curve SP).

2.2 Artificially structured clays

The material consists of an assembly of numerous small clayey spheres, covered by a thin layer of sand. Space among spheres is filled by fine sand or very small clayey fragments. To regularize the sample before the test, a thin layer of fine sand, working in practice as a porous plate, was been spread over the top of the sample.

The small spheres are firstly placed in layers within a conventional oedometric cell (diameter

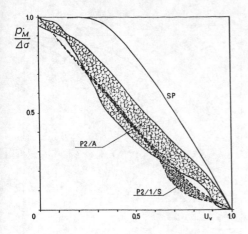

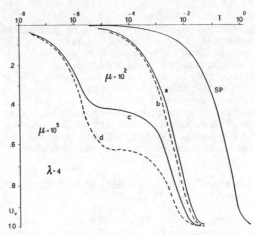

Fig.2 - Measured values of the ratio $p_m/\Delta\sigma$ vs average degre of consolidation $U_v = w(\tau)/w_c$ for two artificially structured clays.

Fig.3 - Theoretical values of the average degree of consolidation $U_v = w(\tau)/w_c$ vs time factor T, for $\lambda=4$; $\mu=10^2$ (curves a,b) or 10^5 (curves c,d); $r_k=0.5$ (curves b,d) or 0.7 (curves a,c).

D=100 mm, height H=39.5 mm); after a preliminary compaction under an initial load $\Delta\sigma'= 152$ KN/m². a new smaller sample (diameter D=63 mm, height H=19 mm) is obtained and placed in a Rowe's cell for consolidation tests.

After saturation, consolidation was started under constant back pressure; the effective consolidation pressures were $\Delta\sigma'= 100$ KN/m², $\Delta\sigma'= 200$ KN/m² and then $\Delta\sigma'= 400$ KN/m².

The results of two sets of consolidation tests (samples P2/A, P2/1/S) are presented and discussed. A detailed description of samples preparation and of experimental results is reported in (Musso, Federico 1993).

It must be emphasized that the samples differ from each other exclusively for the grain size of the sand covering the clay spheres, fine (P2/A) or medium (P2/1/S). The average diameter S of the clay lumps is about 5 mm.

Figure 2 shows, for both samples, non dimensional water pressure $p_m/\Delta\sigma$ recorded at the top cap of the cell vs the degree of consolidation U_v. It may again be observed that the experimental points cover a narrow band, particularly for the sample P2/1/S; moreover, they are located well below to the theoretical curve (SP) obtained by the Terzaghi's theory. In other terms, as in the case of natural fissured clays, dissipation of measured pore pressure takes place more rapidly than that one pertaining to intact, non structured materials.

Consolidation of the sample P2/1/S is in general faster than that one of the sample P2/A; in one case, not reported in the figure, it has not been possible to measure fluid pressure and settlement during the first, almost istantaneous, phase of the phoenomenon. It may therefore be argued that experimental results differ from those corresponding to single porous media as much as more the sample is structured.

3 THEORETICAL MODEL

With reference to a homogeneous, double poro-elastic medium, laterally constrained, resting on an impervious base formation, consolidating under a uniform vertical constant total stress increment $\Delta\sigma$, the set of resolving equations has been solved in closed form (Musso, Federico, 1993), under initial conditions analysed in Musso et al. (1993).

Main non dimensional parameters responsible for the evolution of settlements as well as for the fluid pressure excesses, are $\mu = k_1/k_2$, $\lambda = H/s$, $r_k=K/K*$; coefficients k_1 and k_2 represent the permeability of the set of fissures - the single chunks being imagined as impermeable - and of the intact porous blocks, respectively; H is the thickness of the consolidating sample; s represents a linear dimension characterizing the size of the clay scale; K, $K*$ express the bulk drained moduli characterizing the fissured medium as a whole, being fixed the Poisson's coefficient $\nu = 0.3$, and the porous medium without fissures, respectively.

Fig.3 shows theoretical values of the average degree of consolidation U_v vs time factor $T = C_v \cdot t/H^2$, being C_v the one-dimensional consolidation coefficient of the clay fragments material, expressed, in turn, as $C_v = k_2/(\gamma_w \cdot m_v)$; γ_w and m_v represent the unit weight of the fluid and the oedometric compressibility of the fissured material, respectively.

Differently from the single porosity case, which is always represented by a single curve (SP), double porosity curves are labelled with specific values of the non dimensional parameters. Such curves always lie below the graph SP., i.e., saturated double poro-elastic media consolidate faster than SP media.

The evolution of the process strongly depends upon physico-mechanical parameters describing the meso-structure. The shape of the curves c,d, for example, clearly reveals that in the consolidation process may be distinguished two distinct phases: the first one, extremely fast, regards the net of the fissures; the second phase, delayed in time, the pores of clay fragments. This condition takes place provided the coefficients of permeability pertaining to the porous clay lumps and to the system of fissures are sufficiently different each from other.

4 INTERACTION BETWEEN SOIL AND MEASURING SYSTEM

For both natural or artificial materials, in addition to well known problems affecting experimental data interpretation, related to flexibility of pore-water pressure measuring systems (Gibson, 1963; Viggiani, 1965), a new problem must be taken into account.

In fact, a conventional measuring system feels through the saturated filter, the contemporary contribution of fissures and pores. Thus, the physical conditions associated to the connection of a measuring system at the base of the sample must be modelled by imposing both the equalization of the interstitial fluid pressures with the unknown pressure measured in the system, as well as the continuity of the flow among the two systems of voids and the measuring system (Federico, Musso, 1991).

The analysis of this problem has been carried out by numerical methods. Some theoretical results are shown in fig.4, where non dimensional excess interstitial fluid pressures $p_i/\Delta\sigma$ ($i=1$, fissures; $i=2$, pores) are represented vs the relative depth z/H, for different values of the time factor T (isochrones).

One can observe that, according to the imposed boundary conditions, at the base, where the sample is connected to the measuring system (m.s.), both pressures assume the equal, unknown value p_m recorded through the m.s..

The gradient of the isochrones of the interstitial fluid pressure excess, computed at the base of the sample, assumes opposite values for the two systems of voids: in other terms, an additional fluid exchange, from pores to fissures, takes place through the saturated filter stone.

As a consequence, along the whole sample, the pressure p_1 remains generally higher than the corresponding pressure in absence of the m.s.; on the contrary, especially in proximity of the filter stone, the pressure p_2 assumes values generally lower than the corresponding ones in absence of the m.s.. So, at the same time, the m. s. induces two distinct, but opposite effects. The first one, localized in the filter stone, tends to speed up the consolidation process. The second effect, conversely, is diffused along the whole sample and it generally slows down the process, in reason of the

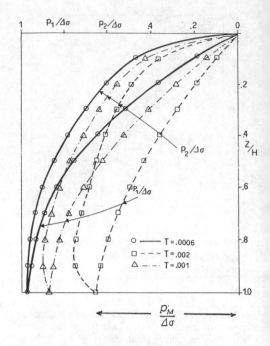

Fig.4 - Non dimensional interstitial fluid pressures $p_i/\Delta\sigma$ ($i=1,2$) vs relative depth z/H, for different values of time factor T. Non dimensional parameters values are: $\lambda=2$, $\mu=10$, $r_k=0.8$.

reduction of the local difference ($p_2 - p_1$).

Due to the composition of these opposite effects, the degree of consolidation U_V remains only slightly modified (Federico, Musso, 1992).

5 EXPERIMENTAL VERSUS THEORETICAL RESULTS

Neither consolidation of London clay nor that of Scanzano dam core are acceptably fitted by the Terzaghi's theory, being the differences larger for Scanzano core data (fig.1). The same observation may be extended to the artificial materials (fig.2).

The unfitting of data by theory does not seem to rely on possible errors of measurement; on the contrary, for a great or less part, it reflects an actual feature of the fissured medium, whose response is governed by the "diffuse", internal drainage represented by the net of fissures.

However, the interpretation of data concerning with natural fissured clays cannot be carried out through double porosity models, owing to the lack of reliable information regarding parameters characterizing the meso-structure.

It is possible to try to interpretate, on the contrary, the experimental results of consolidation tests on artificially structured clays (samples P2/A, P2/1/S).

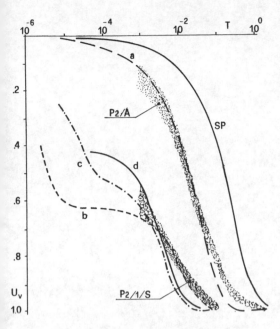

Fig.5 - Comparison between theoretical values and experimental results of laboratory consolidation tests on artificially structured clayey materials.

Taking into account the difference of C_v values for the two samples (Musso, Federico, 1993), experimental data are synthetically represented in fig. 5, to be compared to theoretical curves of the average degree of consolidation U_v vs time factor T.

For the sake of simplicity and in reason of its indifference to the presence of a m. s., computation of U_v has been carried out referring to the case of absence of m.s.. The curves have been obtained for values of parameters r_k, λ, μ, that may reasonably be attributed to the samples: $\lambda = 4$, (curves a,c,d), or $\lambda=2$ (curve b), being $H=19$ mm, $s=5$ mm; $r_k = 0.4$ (a), 0.5 (b), 0.6(c) and 0.7(d); $\mu=10$(a), 10^5(b,d), 10^4 (c), having assigned $k_2=10^{-9}$ cm/sec.

Theoretical curves significantly agree with experimental results, particularly for sample P2/A.

Experimental results, on the contrary, can not be analysed through the theoretical curve corresponding to single poro-elastic media (SP), reported in the same figure.

6 CLOSING REMARKS

Researches specifically undertaken to throw light on the mechanism of speeding-up clay consolidation by an internal net of fissures are rare.

The studies have been generally devoted to determinate strength (Jappelli et al., 1977), compressibility (Bilotta, 1984) and consolidation characteristics through conventional tests. However, the results reported in technical literature generally concern with clay formations affected by a net of fissures not sufficiently continuous to be considered superimposed to the intact porous medium; at the same time, these results are not correlated to the main meso-structural characteristics.

Lacking a reliable set of careful experimental results able to throw light on the role of meso-structure in speeding up consolidation of natural fissured clays, a research program on *artificial* fissured clays has been started.

The experimental results presented and discussed in the paper put into evidence that the artificial material well represents natural fissured clays.

Both kind of materials are affected by a net of discontinuities, that act as a "diffuse" internal drainage, collecting the water seeping from the pores of the intact fragments, thus accelerating the consolidation process. Differently from the case of natural materials, the meso-structure of artificially fissured materials may be easily assigned, controlled and moreover *measured* through peculiar parameters.

A theoretical model for analysing the consolidation process of structured clayey soils has been synthetically presented.

The particular problems related to the interaction between a conventional measuring system of the interstitial fluid pressures and a double porous media have been outlined.

Finally, experimental data of consolidating artificially fissured clays have been compared to theoretical results determined by assigning reasonable values to the meso-structural parameters, showing a significant agreement.

The research work has been financially supported by C.N.R., National Researches Council, contract numbers 90.01426.PF42 and 93.02990.PF42.

REFERENCES

Aifantis E.C. (1980) - On the problem of diffusion in solids. *Acta Mechanica*, 37, 265-296.

Bilotta E. (1984) - Some results from oedometric tests on "scaly" clays: compressibility and swelling (in italian). Rivista Italiana di Geotecnica, XVIII, n. 1, 52-66.

Douglas J., Arbogast Jr and T. (1990) - Dual porosity models for flow in naturally fractured reservoirs. In *Dynamics of Fluids in Hierarchical Porous Media*, Cushman J. H. ed., 177-221.

Elsworth D., Bay M. (1992) - Flow-Deformation Response of Dual Porosity Media. *J. of Geotech. Engrg.*, vol. CXVIII, n.1, January.

Federico F., Musso A. (1991) - Pore pressure dissipation in highly fissured clays. Proc. *X Europ. Conf. on Soil Mech. and Found.Engrg.*, Vol.I, 77-82, Florence.

Federico F., Musso A. (1992) - Soil - pore-water pressure measuring system interaction in consolidating scaly clays (in italian). Research report, Dept. of Civil Engrg., Univ. of Rome "Tor Vergata".

Garga V.K. (1988) - Effect of sample size on consolidation of a fissured clay. *Canad. Geotech. J.*, vol. 25, 76-84

Gibson R.E. (1963) - An analysis of system flexibility and its effect on time-lag in pore-water pressure measurements. *Géotechnique*, vol. 13.

Hensley P. J., Saviddou C. (1993) - Modelling coupled heat and contaminant transport in groundwater. *Int. J. for Num. and Anal. Meth. in Geomech.*, vol.17, 493-527.

Jappelli R., Liguori V., Umiltà G., Valore C. (1977) - A survey of geotechnical properties of stiff highly fissured clays. Proc. Int. Symp."The Geotechnics of Structurally Complex Formations", vol. II, Capri.

Leo C.J., Booker J.R. (1993) - Boundary element analysis of contaminant transport in fractured porous media. *Int. J. for Num. and Anal. Meth. in Geomech.*, vol.17, 471-492.

Musso A., Federico F., Miliziano S. (1990) - One-dimensional consolidation of structured media. Proc. *Int. Conf. on "Mechanics of Jointed and Faulted Rock"*, 673 - 680, Wien.

Musso A., Federico F., Ferlisi S. (1993) -Response of saturated double porosity media to an undrained increase in all-round stress (in italian). *Rivista Italiana di Geotecnica*, n.2, 91-112.

Musso A., Federico F. (1993) - Consolidation of highly fissured clays (in italian). *Rivista taliana di Geotecnica*, n.3, 247-273.

Ohnishi Y., Shiota T., Kobayashi A. (1988) - Finite element double-porosity model for deformable saturated-unsaturated fractured rock mass. In *Num. Meth. in Geomech.*,Swoboda ed., Innsbruck, 765-775.

Sato K., Shimizu T., Ito Y. (1985) - Fundamental Study on permeability and dispersion in double porosity rock masses. *Fifth Int. Conf. on Num. Meth. in Geomech.*, Nagoya, 657-664.

Valliappan S.,Khalili-Naghadeh N. (1991) - Flow through fractured media. *Computer Methods and Advances in Geomechanics*, Beer, Booker & Carter (eds), vol.2,1667- 1672, Balkema.

Viggiani C. (1965) - Interaction between soil and instrument in pore-water pressure measurements (in italian). Proc. *VII Italian Conf. of Geotechincs,*Trieste.

Ward W.H., Samuels S.G., Butler M.E. (1959) - Further studies of the properties of London clay. *Géotechnique*, 15, 33 - 58.

Wilson R.K., Aifantis E.C. (1982) - On the theory of consolidation with double porosity. *Int. J. of Engrg. Science*, vol. 20, n. 9, 1009-1035.

8 Fracture mechanics and fatigue

Recent Advances in Experimental Mechanics, Silva Gomes et al. (eds) © 1994 Balkema, Rotterdam, ISBN 90 5410 395 7

Ratchetting of stainless-steel/steel laminate pipe under cyclic axial-straining or cyclic bending combined with internal pressure

F. Yoshida, T. Okada & T. Izumi
Department of Mechanical Engineering, Hiroshima University, Higashi-Hiroshima, Japan

ABSTRACT: Ratchetting tests were performed by subjecting thin-walled pipes of stainless steel (SUS304), 0.25%-C steel (STPG38) and SUS304/STPG38 laminate to combined steady internal pressure and cyclic axial-straining (or cyclic bending). To discuss the ratchetting behavior quantitatively, the numerical analysis of the ratchetting was also conducted by use of constitutive models with two types of combinations of the nonlinear kinematic hardening rule (NK) and the NK with radial evanescence (NK-RE). The SUS304 pipe has apparent transient ratchetting and the subsequent very small steady ratchetting, which may be called *shake-down*. On the other hand, the STPG38 shows almost no transient behavior, and its rate of ratchetting is much higher than the steady rate of the SUS304. For the SUS304/STPG38 laminate pipes, the steady strain-accumulation rate is much lower than that of the STPG38. This is because the circumferential stress in STPG38 layer decreases during ratchetting with an increasing number of strain cycles.

1 INTRODUCTION

When a pipe is subjected to cyclic axial-load (tension-compression) combined with internal pressure, even if the pressure is rather low, e.g., 10-20% of the yield strength of the pipe, the diameter of the pipe increases with an increasing number of strain cycles. This cyclic-stress-induced strain accumulation is called *ratchetting*. Ratchetting is also observed when a pipe is subjected to cyclic bending, instead of cyclic axial-load, combined with internal pressure. In general, each material has its own characteristics of ratchetting, e.g., for carbon steels (Bright & Harvey 1978a, Hassan & Kyriakides 1992) and 60-40 brass (Yoshida et al. 1978b) ratchetting strain accumulates almost linearliy with an increasing number of strain cycles, however, type 304 stainless steel shows very small steady-ratchetting after a transient period of strain cycles (Yoshida 1991).

Stainless-steel/steel laminate pipes are widely used for chemical industries, petroleum transportation systems, nuclear power plants, etc. If laminate-pipe components are subjected to vibratory bending in case of earthquake or unexpected mechanical vibration, etc., ratchetting will be a serious problem from the viewpoint of structural reliability. However, ratchetting behavior of laminate pipes has never been investigated, although there have been many papers on monolithic pipes.

This paper deals with the ratchetting behavior of stainless-steel/steel laminate pipes. The aim of this work is to examine, both experimentally and theoretically, the ratchetting behavior of laminate pipes consisting of two layers whose ratchetting characteristics are completely different.

By use of thin-walled pipe specimens of type 304 stainless steel (SUS304), 0.25%-C steel (STPG38), and SUS304/STPG38 laminate, two types of ratchetting tests under the conditions: (a) combined internal pressure and cyclic axial-straining and (b) combined internal pressure and cyclic bending, were performed at room temperature.

To discuss the ratchetting behavior quantitatively, the numerical analysis of the ratchetting was also conducted by use of constitutive models with two types of combinations of the nonlinear kinematic hardening rule (NK) and the NK with radial evanescence (NK-RE).

2 EXPERIMENTAL RESULTS

2.1 Experimental procedures

Thin-walled pipe specimens of SUS304 (outer diameter Do=15 mm and wall thickness t=1.0 mm) and STPG38 (Do=24 mm and t=1.0 mm) were machined from a

SUS304 bar of 22-mm diameter and a STPG38 pipe of Do=27.2 mm and t=2.9 mm, respectively. Prior to being tested, the SUS304 specimens were annealed in air after being heated at 1050 °C for 30 min, and the STPG38 specimens were annealed in a furnace after being heated at 750 °C for 30 min. Specimens of SUS304(inside)/STPG38(outside) laminate pipe, as shown in Fig.1, were prepared with the following procedures. A fully annealed SUS304 bar, after being heated at 1050 °C for 30 min, was machined to 0.5- or 1.0-mm thickness tube, and it was annealed again after being heated at 650 °C for 30 min to remove workhardened layer induced by lathe processing. Thus prepared thin-walled SUS304 tube was inserted into an annealed STPG38 pipe specimen, and they were adhesive-bonded together.

Two types of ratchetting tests under the conditions: (a) combined internal pressure and cyclic axial-straining and (b) combined internal pressure and cyclic bending, were performed, as schematically illustrated in Fig 2(a) and (b). In the axial-straining ratchetting test (a), the specimens were subjected to cyclic axial-straining of strain range $\Delta\varepsilon_z$=0.01 or 0.02 at strain rate $\dot{\varepsilon}_z$= 0.001 s^{-1} combined with a constant internal pressure. The axial

and circumferential strains were measured by clip-on extensometers. In the bending-ratchetting test (b), the specimens were subjected to cyclic four-point bending of maximum axial-strain range $\Delta\varepsilon_z$=0.01 or 0.02 at strain rate $\dot{\varepsilon}_z$= 0.001 s^{-1} combined with a constant internal pressure.

The axial and circumferential strains were measured by strain gauges bonded on the outer surface of the specimen, as shown in Fig 2(b). It should be noted that, for laminate pipes, the nominal value (or in other words,

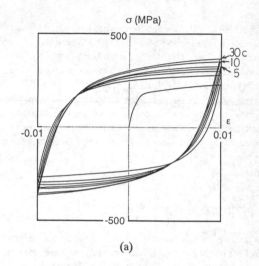

(a)

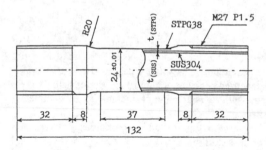

Fig.1 Dimensions (in mm) of SUS304/STPG38 specimen. $t_{(SUS)}$=0.5 or 1.0 mm, $t_{(STPG)}$=1.0 mm.

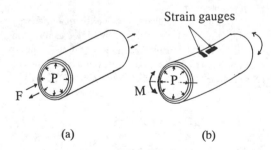

(a)   (b)

Fig.2 Schematic illustration of ratchetting tests under the conditions: (a) combined internal pressure and cyclic axial straining and (b) combined internal pressure and cyclic bending.

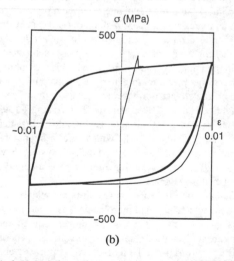

(b)

Fig.3 Uniaxially cyclic stress-strain curves. (a) SUS304; (b) STPG38.

the mean values) of circumferential stress was kept constant throughout a ratchetting test, however, the stresses in the two layers of SUS304/STPG38 were not the same, and as it will be discussed later, the stresses varies with number of strain cycles.

2.2 Cyclic axial-straining combined with internal pressure

Figure 3(a) and (b) shows experimental results of cyclic stress-strain curves of SUS304 and STPG38 under uniaxial cyclic straining. The SUS304 has remarkable cyclic strain-hardening, but the STPG38 has almost no cyclic strain-hardening/softening. The stabilized flow stress level of the SUS304 is almost the same as that of the STPG38.

Figure 4 shows an example of ratchetting strain accumulation in a test of σ_θ=40 MPa, $\Delta\varepsilon_z$=0.02 for a laminate pipe of wall thickness of $t_{(SUS)}/t_{(STPG)}$=0.5-mm/1.0-mm. Circumferential strain accumulates with an increasing number of strain cycles. Throughout the experiments no delamination occured between SUS304/STPG38 layers.

Figure 5(a) and (b) shows the circumferential strain ε_θ (at ε_z=0) versus number of strain cycles N for the SUS304 and the STPG38, respectively. The strain accumulation of the SUS304 was completely different from

that of the STPG38. The SUS304 has apparent transient ratchetting and the subsequent very small steady ratchetting, which may be called *shake-down*. On the other hand, the STPG38 shows almost no transient behavior, and its rate of ratchetting ($d\varepsilon_\theta$ /dN) is much higher than the steady rate of the SUS304.

Figure 6 shows the experimental results for the

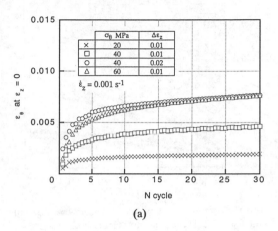

(a)

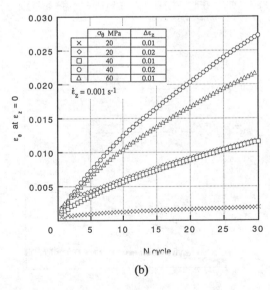

(b)

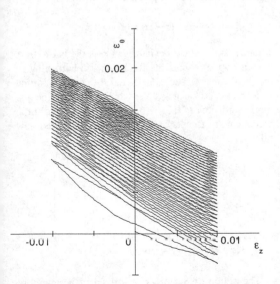

Fig.4 Strain trajectory in a ratchetting test of σ_θ=40 MPa, $\Delta\varepsilon_z$=0.02 for a SUS304/STPG38 laminate pipe of wall thickness of $t_{(SUS)}/t_{(STPG)}$=0.5-mm/1.0-mm.

Fig.5 Experimental results of strain accumulation ε_θ (at ε_z=0) vs number of strain cycles in ratchetting tests under combined cyclic axial-straining and internal pressure. (a) SUS304; (b) STPG38.

SUS304/STPG38 laminate pipes. The transient strain-accumulation rate of the laminates is lower than that of the SUS304, and the steady rate is much lower than that of the STPG38. This is beacause the circumferential stress in each layer of SUS304/STPG38 varies with an increasing number of strain cycles, as it will be discussed later.

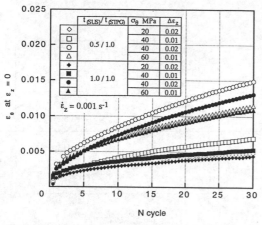

Fig.6 Experimental results of strain accumulation ε_θ (at ε_z=0) vs number of strain cycles in ratchetting tests under combined cyclic axial-straining and internal pressure for SUS304/STPG38 laminate pipes.

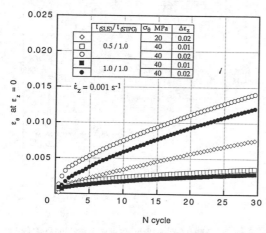

Fig.7 Experimental results of strain accumulation ε_θ (at ε_z=0) vs number of strain cycles in ratchetting tests under combined cyclic bending and internal pressure for SUS304/STPG38 laminate pipes.

When the thickness of SUS304 layer becomes thicker from 0.5 mm to 1.0 mm, less strain accumulation is observed, because the SUS304 layer, which has lower steady rate of ratchetting, prevents the larger ratchetting of the STPG38 layer.

2.3 Cyclic bending combined with internal pressure

Figure 7 shows the experimental results of circumferential strain ε_θ (at ε_z=0) versus number of strain cycles N for SUS304/STPG38 laminates in ratchetting tests under cyclic bending combined with internal pressure. It is found in this figure that the results are almost the same as the corresponding ratchetting tests under cyclic axial-straining combined with internal pressure (see Fig.6), because the stress/strain condition at the measured point in the bending-ratchetting is just the same as the axial-straining ratchetting.

3 ANALYSIS OF RATCHETTING

3.1 Constitutive models

One of the most difficult problems in cyclic plasticity is the accurate description of ratchetting behavior. Here we shall discuss the constitutive modeling for that in the framework of plastic potential theory with isotropic and kinematic hardenings.

A yield function f is given by the equation:

$$f = \frac{3}{2}(s - \alpha) : (s - \alpha) = (Y + R)^2 , \tag{1}$$

where, s is the stress deviator and α is the back-stress deviator which represents the kinematic hardening of the yield surface. The value Y and R denote the initial yield stress and the drag stress which represents the isotropic hardening of the yield surface. The plastic strain increment $d\varepsilon^p$ can be determined by the associated flow rule:

$$\left. \begin{aligned} d\varepsilon^p &= \frac{3}{2} d\bar{\varepsilon} \, \frac{s - \alpha}{J(s - \alpha)} , \\ d\bar{\varepsilon} &= \left(\frac{2}{3} d\varepsilon^p : d\varepsilon^p\right)^{\frac{1}{2}} , \\ J(s - \alpha) &= \left[\frac{3}{2}(s - \alpha) : (s - \alpha)\right]^{\frac{1}{2}} . \end{aligned} \right\} \tag{2}$$

The isotropic hardening rule is given by the equation:

$$dR = b(Q - R) d\bar{\varepsilon} , \tag{3}$$

Q and b : material constants.

For the kinematic hardening, the nonlinear kinematic hardening rule (NK) in Eqn(4) (Armstrong & Frederick, 1966) and the NK with radial evanscence (NK-RE) in Eqn(5) (Burlet and Cailletaud,1987) were employed.

$$d\,\boldsymbol{\alpha}_i = c_i\left(\frac{2}{3}\,a_i\,d\,\boldsymbol{\varepsilon}^p - \boldsymbol{\alpha}_i\,d\bar{\varepsilon}\right),\qquad(4)$$

$i = 1$ and 2, C_i and a_i : material constants,

$$d\,\boldsymbol{\alpha}_A = c_A\left[\frac{2}{3}\,a_A\,d\,\boldsymbol{\varepsilon}^p - \sqrt{\frac{2}{3}}\,(\boldsymbol{\alpha}_A : \boldsymbol{n})\right],\qquad(5)$$

C_A and a_A : material constants,

where n denotes the unit direction of the plastic strain increment.

By using the linear superposition of the NK and the NK-RE, a new kinematic hardening rule is given by Eqn(6) [model A].

$$\boldsymbol{\alpha} = \mu_A\,\boldsymbol{\alpha}_1 + (1 - \mu_A)\,\boldsymbol{\alpha}_A + \boldsymbol{\alpha}_2\,.\qquad(6)$$

With a proper value of μ_A, the constitutive model can describe the transient and the subsequent very small steady ratchetting, as shown for SUS304.

On the other hand, the other type of combination of the kinematic hardening rules of the NK and the NK-RE is given by Eqns(7) and (8) [model B] (Chaboche & Nouailhas, 1989).

$$d\,\boldsymbol{\alpha}_1 = c_B\left[\frac{2}{3}\,a_B\,d\,\boldsymbol{\varepsilon}^p - \mu_B\,\boldsymbol{\alpha}_1\,d\bar{\varepsilon} - \sqrt{\frac{2}{3}}\,(1 - \mu_B)\,(\boldsymbol{\alpha}_1 : \boldsymbol{n})\right]$$

$$(7)$$

C_B, a_B and μ_B : material constants,

$$\boldsymbol{\alpha} = \boldsymbol{\alpha}_1 + \boldsymbol{\alpha}_2\,.\qquad(8)$$

With a proper value of μ_B in Eqn(7), the model can describe the almost steady ratchetting at all over the strain cycles, as shown in the ratchetting of STPG38.

3.2 Analysis of ratchetting for laminate pipes

By using the above two constitutive models, the numerical simulation of the ratchetting were performed for pipes of SUS304, STPG38 and their laminate. For the analysis of the laminates, the condition of isostrain between the two layers of SUS304/STPG38 at the laminate boundary and the linear distribution of strains in the radial direction of a pipe were assumed.

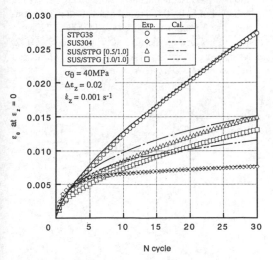

Fig.8 Comparison of experimental and analytrical results of ratchetting behavior for the case of σ_θ=40 MPa and $\Delta\varepsilon_z$=0.02.

4 DISCUSSION

Figure 8 shows the comparison of experimactal and analytical results of ratchetting behavior for the case of σ_θ=40 MPa and $\Delta\varepsilon_z$=0.02. The analytical results for the SUS304 and the STPG38 agree quite well with the experimental results. The present constitutive models can well describe the above-mentioned difference in the ratchetting characteristics between the SUS304 and the STPG38. The analytical results for the laminates agree

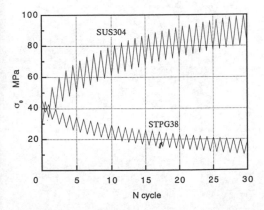

Fig.9 Variation of circumferential stresses in the layers of SUS304/STPG38 in a laminate pipe during ratchetting.

comparatively well with the experimental results.

Both in the experimental results and the analytical results, the transient strain-accumulation rate for the laminates is lower than that of the SUS304, and the steady rate is much lower than that of the STPG38. This is due to the variation of circumferential stress in each layer of SUS304/STPG38 with an increasing number of strain cycles, as illustrated in Fig.9. The circumferential stress in SUS304 layer increases, and the stress in STPG38 decreases with an increasing number of strain cycles. Therefore the rate of strain accumulation of the laminate pipe becomes lower compared with that of the STPG38 pipe. The variation of the circumferential stresses in two layers is caused by the difference in the ratchetting characteristics between the layers, but not the difference in the flow stress of layer metals.

5 CONCLUDING REMARKS

The strain accumulation of SUS304 was completely different from that of STPG38. The SUS304 pipe has apparent transient ratchetting and the subsequent very small steady ratchetting, which may be called *shake-down*. On the other hand, the STPG38 shows almost no transient behavior, and its rate of ratchetting is much higher than the steady rate of the SUS304. The constitutive models with two types of combinations of the NK and the NK-RE can quantitatively well describe these ratchetting behavior.

For the SUS304/STPG38 laminates, the rate of ratchetting in the transient period is lower than that of the SUS304, and the steady rate is much lower than that of the STPG38. In general it can be concluded that the use of a metal of low-ratchetting characteristics as a layer of a laminate pipe is very effective in reducing the ratchetting of the laminate. The reason for that has been revealed by the stress/strain analysis that the circumferential stress in a layer of high-ratchetting characteristics decreases very much during ratchetting with an increasing number of strain cycles.

REFERENCES

Armstrong, P.J. and Frederick, C.O. 1966. A mathematical representation of the multiaxial Bauschinger effect, *GEGB Report* RD/B/N/731, Berkeley Nuclear Laboratories.

Bright, M.R. and Harvey, S.J. 1978a. Cyclic-strain-induced creep under complex stress, in Stanley, P. (ed.), *Nonlinear problems in stress analysis*, Applied Science Publisher, London, p.259-283.

Yoshida, F., Tajima, N., Ikegami, K. and Shiratori, E. 1978b. Plastic theory of the mechanical ratcheting (Analysis by an anisotropic hardening plastic potential and its comparison with experimental results of brass), *Bull. JSME*, 21, p.389-397.

Burlet, H. & Cailletaud, G. 1987. Modeling of cyclic plasticity in finite element codes, *2nd Int. conf. on constitutive laws for engineering materials: theory and application*, p.1157-1165.

Chaboche, J.L. and Nouailhas, D. 1989. Constitutive modeling of ratchetting effects - Part II: Possibilities of some additional kinematic rules, *J. Eng. Mat. Technol.*, *T. ASME Ser.H*, 111, p.409-416.

Yoshida, F. 1991. Uniaxial and biaxial creep-ratchetting behavior of SUS304 stainless steel at room temperature, *Int. J. Pres. Ves. & Piping*, 44, p.207-223.

Hassan, T., Corona, E. and Kyriakides, S. 1992. Ratchetting in cyclic plasticity, Part II: Multiaxial behavior, *Int. J. Plasticity*, 8, p.117-146.

Recent Advances in Experimental Mechanics, Silva Gomes et al. (eds) © 1994 Balkema, Rotterdam, ISBN 90 5410 395 7

Testing and microstructural interpretation of 35CD4 steels in cumulative high cycle fatigue damage

Mirentxu Vivensang & Alexandre Gannier
LAMEF (LAboratoire Matériaux Endommagement Fiabilité), ENSAM-CER de Bordeaux, Talence, France

ABSTRACT: Two-level cumulative damage tests in high cycle fatigue have been carried out in rotative bending loading with normalized as well as quenched and tempered specimens. The important scatter of fatigue life encountered during these tests and the bad correlation of the cumulative damage models, lead us to make surface investigations by Scanning Electron Microscopy (SEM). These observations permit us to visualize and to quantify the damage evolution during the cycles. The behaviour of the two types of steels are underlined, the non-linearity of the high-low cumulative tests can be mostly explained and persistant slip bands (psb) and extrusions can be partly considered as damage parameters.

INTRODUCTION

Recent cumulative tests have been published (Choukairi 1993) on a quenched and tempered 35CD4 (SAE 4135) steel. Trying to verify the cumulative damage laws proposed in literature, they have been mainly confronted to the wide disparity of fatigue life (the distribution of lives were varying from 132400 cycles to 10^6 cycles for a stress amplitude $S=550$ MPa). Moreover, non-linear models were implemented and did not fit the tests performed on the 35CD4 steel. To get round these two problems, surface investigations by Scanning Electron Microscopy have been performed to follow the damage evolution during the cycles. In the same time, we decided to perform a normalized 35CD4 steel. The microstructural confrontation of the two steels might help us to understand their mechanical behaviour.

1. PRESENTATION OF THE STUDY

1.1 *Materials*

The quenched and tempered 35CD4 steel has been austenitised at 850°c during 30 minutes, quenched in oil and tempered at 550°c during one hour. This steel is mainly constituted of quenched martensite we can distinguish in white on figure1.

The darker zones are the lower carbon zones. The grain size is around 10 μm.

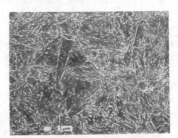

fig.1: microstructure of 35CD4 quenched and tempered

The normalized 35CD4 steel has been austenitised at 850°c during 30 minutes and cooled in the furnace. It is constituted of perlite and ferrite grains in black on figure 2. The grain size is around 20 to 40 μm.

fig.2: microstructure of 35CD4 normalized ×100 0

Table 1: mechanical characteristics of 35CD4 steels

	Re	Rm	A%
35CD4 quen	875 MPa	1068 MPa	11.5
35CD4 norm	370 MPa	650 MPa	18

Their static mechanical characteristics are presented on table 1

1.2 *Tests*

Rotative bending tests are carried out on a prototype resonant machine (frequency = 100 Hz). As long as the crack is not initiated, the operating frequency stays stable. When the crack is initiated, the specimen stiffness decreases and so does the frequency. When the specified discrepancy of frequency is reached, the test is stopped. The specimen of torus diameter 8 mm, is then considered as fractured; a crack of a few millimetres is present on the surface of specimen. The first stage of experiments involves constant amplitude tests on sets of 40 specimens for two different stress levels. The low level is close to the endurance limit and the high level has a fatigue life close to 10^5 cycles. These tests are presented on a cumulative frequency diagram. The second stage involves two-level tests. They are presented on a cumulative damage diagram and confronted to Miner(1945) and Chaboche(1974) damage curves.

1.3 *SEM observations*

To follow the damage evolution on the surface of the specimen, the samples are polished before testing. The undamaged specimens show varying densities of inclusions (silicon oxides) from a sample to another. After a few cycles, persistant slip bands appeared followed by extrusions later in the fatigue life. On figures 3 and 4, we can see these three cues.

fig.3: surface aspect before testing: 35CD4 quenched and tempered **x 1000**

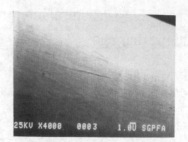

fig.4: surface aspect after 5% of fatigue life at 650 MPa: 35CD4 quenched and tempered

The evolution of psb and extrusions is quantified: for this, tests are stopped and specimens are observed by SEM after 5%, 10%, 30% and 50% of the mean fatigue life, calculated from the above tests.

2. CONSTANT AMPLITUDE TEST - 35CD4 QUENCHED AND TEMPERED

2.1 *Tests*

The fatigue limit of the quenched and tempered steel at 10^6 cycles has been evaluated with a stair-case method: Sd = 520 MPa.

The chosen stress levels are 550 MPa for the low level and 650 MPa for the high level. Test results are shown on figure 5.

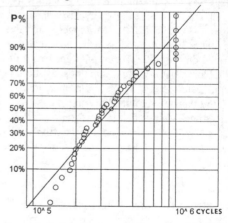

fig.5: cumulative frequency diagram - 550 MPa - 35CD4 normalized

We can notice a wide disparity of life: from 132400 cycles to 10^6 cycles and from 45100 cycles to 112700 cycles for the low and high levels respectively.

2.2 *Observations*

The observations made in parallel on 6 samples, 3 for each stress, gave us the results in figures 7 and 8. The densities of inclusions, psb and extrusions have been censused on each sample.

We have represented on the X-axis the percentage of fatigue life, and on the Y-axis the psb (or extrusion) creating rate per cycle.

fig.9: surface aspect at 5% of fatigue life at 550 MPa - 35CD4 quenched and tempered

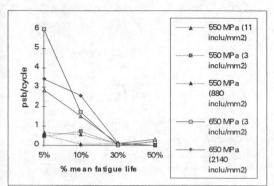

fig.7: psb creating rate per cycle - 35CD4 quenched and tempered

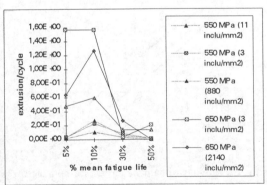

fig.8: extrusion creating rate per cycle - 35CD4 quenched and tempered

Two points can be underlined:

- first, at 5% of fatigue life, we can notice a 10 fold ratio of psb creating rate between the low and the high stress. This difference is visualized on figures 4 and 9.

- second, for the high stress, one of the sample has a very low density of inclusions, and features a very high density of psb at 5% of fatigue life.

3. CONSTANT AMPLITUDE TEST - 35CD4 NORMALIZED

3.1 *Tests*

The endurance limit is Sd = 347 MPa. The chosen stresses are: 347 MPa for the low level and 390 MPa for the high level. The results are as follows: as for the quenched and tempered steel, we have encountered an important scatter of fatigue life: from 369800 cycles to 7 10^6 cycles for 347 MPa and from 110500 cycles to 374100 cycles for 390 MPa.

3.2 *Observations*

As for the preceding material, we have represented alll the results in diagrams (figures 12 and 13):

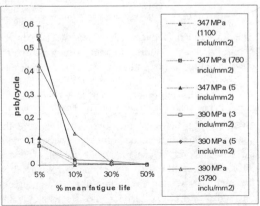

fig.12: psb creating rate per cycle - 35CD4 normalized

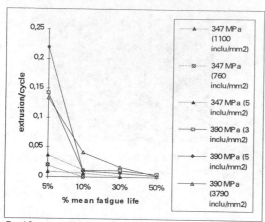

fig.13: extrusion creating rate per cycle - 35CD4 normalized

A few points can be underlined:
   * for a given life ration, the observed damage by SEM is more pronounced for the normalized steel than for the quenched and tempered steel, the size of the psb and extrusions are twice bigger (figure 14).

fig.14: surface damage at 5% of fatigue life at 347 MPa

The difference between the two steels can be explained by the distinct microstructures corresponding to the grain size (Eifler, Macherauch 1990).
   * as for the quenched and tempered steel, at 5% of life, we can see a 5 fold ration of psb creating rate between the two stresses; on the contrary, this difference quickly becomes non-existent and very close to zero between 10% and 30% of fatigue life. This aspect can be explained by the fact that we can encounter a few microcracks (around 60 μm) very early in the life whatever the stress level and the inclusion densities can be.

4 TWO-LEVEL LOADING

4.1 Tests

The following cumulative tests have been performed:
   - For the quenched and tempered steel:
* low-high tests   550 MPa during 20000 cycles
                        650 MPa until fracture
* high-low tests   650 MPa during 15000 cycles
                        550 MPa until fracture
   - For the normalized steel :
* low-high tests   347 MPa during 300000 cycles
                        390 MPa until fracture
* high-low tests   390 MPa during 100000 cycles
                        347 MPa until fracture
All these tests again revealed an important scatter in fatigue life. We have represented these results on the following cumulative damage diagram (figure 15)

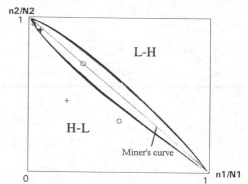

fig.15: cumulative damage curves for both steels
(o: normalized steel)
(+: quenched and tempered steel)

The low-high results fit the linear Miner's rule, which means that the first imposed stress level has no influence on the damage evolution of the second level. But, the main aspect of this curve is the non-linearity of the high-low tests. Both points are far from Miner's curve, which means that the first applied stress leads to an important decrease of the residual life. Although it is a non-linear model, the Chaboche law is far from the reality in our case.

4.2 Observations

SEM observations are made just after the first imposed stress level.

The results of the psb and extrusions creating rates have been added to the diagrams (figures 16, 17, 18, 19).

- quenched and tempered steel

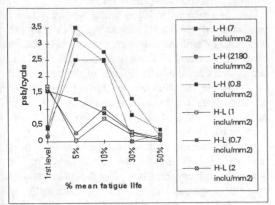

fig.16: psb creating rate per cycle: 35CD4
quenched and tempered

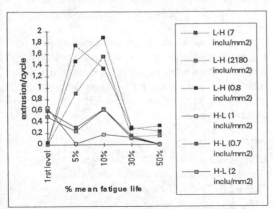

fig.17: extrusion creating rate per cyle: 35CD4
quenched and tempered

* The extrusion diagrams (figures 8 and 17) are in good agreement with the cumulative tests. Actually, for the low-high tests at 5% of the mean fatigue life, we have a greater average extrusion creating rate: 1.3 extrusion/cycle for the low-high test against 0.9 extrusion/cycle for the constant amplitude high stress. The higher rate at the second level gives evidence on a higher damage rate; therefore, the usual gain of fatigue life during the first stress level is annihilated, which is in accordance with the test results.

* For the high-low tests, at 10% of the mean fatigue life the number of extrusions per cycle is around 0.4 extrusion/cycle against 0.2 extrusion/cycle for constant amplitude low stress. In that case, we can see that the first level has an important influence on the second level: there is a ratio higher than 2 between the two extrusion rates. This gives evidence of a higher damage rate, which leads to the usual loss of fatigue life. So, there is a good correlation between the cumulative tests and the observations.

- normalized steel

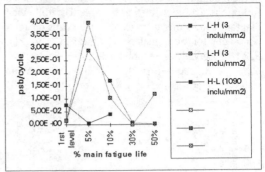

fig.18: psb creating rate per cycle - 35CD4
normalized

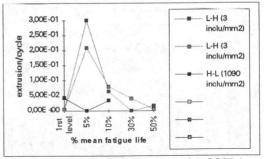

fig.19: extrusion creating rate per cycle - 35CD4
normalized steel

In the case of the low-high tests we can notice the same phenomenon as for the preceding steel: 0.25 extrusion rate against 0.17 extrusion rate for the constant amplitude high test; which means that we have a good correlation with the cumulative tests.

The main characteristic of the high-low tests as shown by SEM observations, is an extrusion creating rate close to zero. This can be explained by the fact that cracks have been already observed just after the first imposed level. In that case, the psb and the extrusions can't be taken into account as damage parameters.

Several authors have already noticed the presence of large micro-cracks, the phenomenon of coalescence observed, leads to non-linear behaviour (Magnin, Bataille, 1991). This damaging process has not been investigated yet.

5 CONCLUSION

Our study brings to light three points:

1- the visualization of the damage on the surface of the specimen during the cycles at a single stress level permitted us to compare the microstructural behaviour of the two steels, and to follow the evolution of the damage. The following diagram shows the average psb and extrusion rates per cycle for both materials (figure 20 and 21). SEM observations enable us to compare the behaviour of the two steels in relation to the main stages of micro-damaging.

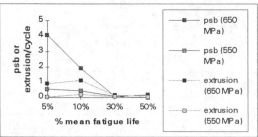

fig.20: average psb and extrusion creating rates/cycle - 35CD4 quenched and tempered

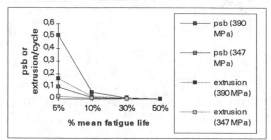

fig.21: average psb and extrusion creating rates/cycle - 35CD4 normalized

* For the quenched and tempered steel, at 5% of the mean fatigue life, persistent slip bands have been encountered in low and high densities respectively for the low and high stress levels. As for the extrusions, they have appeared between 5% and 10% of the mean fatigue life for the low stress and before 5% for the high stress. We can notice the difference of shape between the psb and the extrusion curves.

* For the normalized steel the curves have exactly the same shape. Persistent slip bands and extrusions are present at 5% of the mean fatigue life whatever the stress level can be. Micro-cracks appear earlier in the fatigue life around 10% of the mean fatigue life in some samples of normalized steel. They appeared at 50% of the mean fatigue life for the quenched and tempered steel.

2- The quantification of the damage, mainly presented in this text has shown that the psb creation rate and the extrusion creation rate can be taken into account as damage parameters in all the cases except for the high low test of the normalized steel. In the latter case, the presence of micro-cracks modify the behaviour of the material. The micro-cracks should be taken into account as damage parameters.

3- The influence of the inclusion densities has been shown for the low stress level of the quenched and tempered steel. For the other cases, it doesn't apparently explain the wide disparity of life. The size of the inclusions has not been taken into account here; some authors showed that it could have an important influence on the behaviour of the material (Melander, Larsson, 1993).

The presented models don't explain the non-linearity of the tests. As we are confronted to complex observed phenomena, it is very difficult to take into account all the damage parameters to propose a cumulative model, in the case of industrial metals

REFERENCES
Chaboche,J.L. "Une loi différentielle d'endommagement de fatigue avec cumulation non linéaire." Revue Française de Mécanique n°50-51.1974.
Choukairi, F.Z. Barrault, J "Use of a statistical approach to verify the cumulative damage laws in fatigue" Int. J. Fatigue 15 n°2 1993 pp. 145-149.
Eifler, D. Macherauch, E. "Microstructure and cyclic deformation behaviour of plain carbon and low-alloyed steels " Int. J. Fatigue 12 n°3 1990 pp. 165-174.
Magnin, T. Bataille, A. "Mesure quantitative de l'endommagement en fatigue oligocyclique à partir des mécanismes de croissance en surface des microfissures" Journées de Printemps 1991.
Melander, A. Larsson, M. "The effect of stress amplitude on the cause of fatigue crack initiation in a spring steel" Int. J. Fatigue 15 n°2 1993 pp. 119-131.

Recent Advances in Experimental Mechanics, Silva Gomes et al. (eds) © 1994 Balkema, Rotterdam, ISBN 90 5410 395 7

The inverse crack identification problem: Extension to two simultaneous cracks on free-free beams

Júlio M. Montalvão e Silva & António J. M. Araújo Gomes
Department of Mechanical Engineering, Instituto Superior Técnico, Lisboa, Portugal

ABSTRACT: Most of the published work on the use of changes in the vibration behaviour in order to characterize the development of cracks in structural elements deals with the so-called "direct problem" i.e., models are developed and used to predict the changes in dynamic behaviour. In some cases the theoretical predictions are compared with experimental data. In a recent paper, the authors addressed the so-called "inverse problem" and proposed a method which uses the knowledge of the changes in the values of the experimental natural frequencies in order to predict the location and depth of an "unknown" crack. The method was successfully applied to some examples and results were discussed. In the present paper the authors present their approach and apply it to the case where there are two simultaneous cracks in a free-free beam.

NOMENCLATURE

e	size of edge of cubic additional masses
E	Young's modulus
i	mode number
L	length of beam
L_e	crack location measured from left end of free-free beam
n	number of modes of vibration
p	depth of crack
Q_i	ratio ω_{fi} / ω_i
R	auxiliary variable
ρ	mass density
ω_i	natural frequency (mode i) of an uncracked beam
ω_{fi}	natural frequency (mode i) of a cracked beam
$(\omega_{fi})_e$	expected value of ω_{fi}
$(\omega_{fi})_p$	predicted value of ω_{fi}

1 INTRODUCTION

Structural integrity is of paramount importance due to the possible catastrophic consequences of sudden unexpected failures. Therefore, monitoring through the application of adequate non-destructive methods may be mandatory. The choice of the most appropriate techniques depends on various factors namely size, accessibility, environment and cost.

In particular, when structural elements are subjected to dynamic loads they are prone to develop micro-cracks that propagate and may produce the colapse of the structure. The velocity at which a crack propagates is related with the load spectrum. Though the time interval that mediates the beginning of the propagation of the crack and the final colapse may be a long one, it is difficult or even impossible to detect such a crack, by means of the use of techniques that are based on a visual inspection, when it is located on an inaccessible place.

One way of overcoming the previous difficulty may be the application of techniques that are based on vibration analysis. These techniques seek to correlate the existance and propagation of a crack with variations in the dynamic behaviour of the structure under analysis. The authors are currently involved in the development of such techniques, trying to establish means of detecting cracks in structures through the analysis of the induced variation of the natural frequencies. This work has covered both theoretical and experimental analysis (Gomes & Silva 1990ab, 1991 and 1992ab and Silva & Gomes 1990a) but addresses what is called the "direct problem" i.e., the use of theoretical models in order to predict changes in dynamic behaviour those changes being afterwards verified experimentally.

However, the techniques under discussion will only be useful if one is capable of predicting the existence of a crack based on the previous knowledge of changes in the dynamic behaviour. This is what is called the "inverse problem" and presents a number of difficulties related with the solution of the corresponding system of equations.

In a recent paper (Silva & Gomes 1994), the authors addressed the inverse problem and presented some results obtained for clamped-free beams with only one crack. Theoretical and experimental data was compared and discussed. In this paper, the authors start by making a short reference to some of the work that has been published by the international community in the field of crack identification. Then they take a theoretical model, developed previously, to represent the dynamic behaviour of a simple cracked structural element and apply it into a computer program developed in order to solve the "inverse problem". The examples of application are now free-free beams with two simultaneous cracks.

2 OVERVIEW OF PREVIOUS WORK

As stated previously, most of the authors past work has addressed the "direct problem" in the case of simple structural elements such as beams or thin walled tubes, covering a number of different load and end conditions. The results were encouraging and it seemed natural to extend the analysis to the so-called "inverse problem". This problem was obviously addressed by other investigators. For example, Mayes & Davies (1976) studied the behaviour of a cracked rotor and concluded that the knowledge of the variation of any two natural frequencies would be sufficient to determine the location and depth of the crack.

Other authors have addressed the problem using different types of dynamic analysis. Thus, Achenbach et al (1987) proposed the use of ultra-sounds in order to characterize cracks in solid elements namely in order to determine their location, orientation and thickness. Two years later, Iman et al (1989) proposed the use of an analytical dynamic model in order to detect cracks in large rotors. In the following year Mannan & Richardson (1990) developed a method to detect and characterize cracks based on the determination of the mass, stiffness and damping characteristics, obtained from frequency response functions. At the same time, Stubbs & Osegueda (1990ab) proposed a theoretical model to evaluate damage in beams, plates and shells through the determination of the mass, stiffness and damping matrices. Still in 1990, Rizos et al (1990) developed a method to identify cracks in rectangular cross-section clamped beams, based on modal characteristics namely the amplitudes of vibration in two different points of the beams, for a given mode of vibration.

The probability of detecting multiple cracks using the knowledge of the modes of vibration was addressed by Lu & Ju (1991). In the same year, Samman et al (1991) used a PVC model of a bridge in order to study crack detection based on the comparison of frequency response functions. The analysis of plates imperfections was also carried on by Natke & Cempel (1991) who proposed the use of

numerical methods for the quantification of the influence of the imperfections on the mass, stiffness and damping matrices.

Meanwhile, Nishimura & Kobayashi (1991) discussed the use of boundary element methods for crack identification. The evaluation of the modification of acoustic properties, due to the existence of bidimensional cavities, was proposed by Qunli & Fricke (1991) who used Green functions on the analysis of the variations of the natural frequencies. In the same year, Pandey et al (1991) suggested the use of finite elements assuming that a crack would reduce the structural stiffness and increase the damping. For the same purpose, Shen & Taylor (1991) developed a computational algorithm based on the minimization or maximization of functions directly related with the crack characteristics.

More recently, Swamidas & Chen (1992) studied the changes in the dynamic behaviour of a model of a tower/platform based on the use of electric strain gauges and paying special attention to the changes in the amplitude of the vibration response due to the crack propagation. In the same year, Guigné et al (1992) studied the applicability of acoustic response measurements to the crack detection problem. Meanwhile, the authors extended their work to other load and end conditions including a short analysis of the case where the beams were immersed in water (Silva & Gomes 1992).

The work of Swamidas was recently extended to plates (Swamidas & Budipriyanto 1994 and Chen & Swamidas 1994). In the same year, Klein et al (1994) used vibration and acoustic methods and presented a very interesting experimental study on the changes in modal characteristics due to fatigue cracks developing in a welded cantilevered beam.

The previous overview is meant only as a means of showing the diversity of techniques, in the dynamic domain, that are currently in study. Their discussion would take too much space and might be, by itself, the contents of a state-of-the-art paper. This diversity and the continuously growing number of papers on the subject, published or presented in international conferences, may be taken as the recognition that the practical applicability of the different techniques under study is still far from being achieved, there being a need for further studies.

The solution and applicability of the so-called "inverse problem" is, however, addressed by very few of the interested researchers, as far as damage is concerned in terms of cracks in structural elements. A very recent and interesting study was presented by Meneghetti (1994) who analysed cracked free-free beams detecting the crack locations, though not their depth, by means of a method based on stiffness sensitivity analysis.

In the case of the authors, the results of their previous work encouraged the use of the same principles and theoretical model for the purpose of addressing the "inverse problem". This was performed by applying a specially developed computer program to the prediction of cracks location and depth in simple structural elements where cracks previously produced were assumed as unknown. Clamped-free beams with one crack were initially analysed and results (Silva & Gomes 1994) were very good. The technique was then tested for the case where there exist two simultaneous cracks, with different depths and at different locations. These case studies considered free-free beams with mass loads at the ends and the results are presented and discussed in this paper.

3 THEORETICAL MODEL AND EXPERIMENTAL PROCEDURE

Basically, all the work carried out by the authors was performed by studying transverse (bending) vibrations of simple structural elements such as beams and thin walled tubes. The theoretical models were developed assuming that a cracked beam could be modeled as two shorter beams connected through a torsion spring (Silva & Gomes 1991) that simulates the stiffness of the original beam at the section where the crack exists. The schematic representation in figure 1 shows the model for the case where the beam is free-free and loaded by two cubic masses at the free ends.

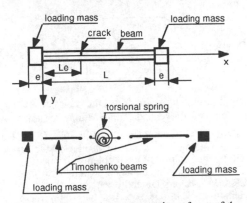

Figure 1 - Schematic representation of one of the modelled beams and cracked beam model

The initial studies considered unloaded free-free straight beams with only one crack. The use of this type of test specimens was decided in order to make it as simple as possible and free from most of the experimental errors due to difficulties in simulating other load and end conditions. The developed theoretical model (Gomes & Silva 1991) was validated by a large amount of experimental data (Gomes & Silva 1990b). Also, at the beginning, cracks were simulated by means of slots. At a later stage a three-point-bending technique was used in order to produce real line cracks in the tested beams.

The encouraging results and experience gained in the previous work led to further analysis now applying it to beams with different cross-sections and also to thin walled tubes, subjected to other load and end conditions. The corresponding theoretical

models were validated experimentally (Gomes & Silva 1992ab). Extension of these models to a beam with more than one crack is a straighforward job and is schematically shown in figure 2.

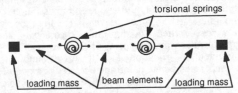

Figure 2 - Schematic representation of the model of a free-free beam with two cracks

The experimental determination of the flexural natural frequencies of the beams under test was performed using a mobility measuring set-up (figure 3) based on a two-channel spectrum analyser. For the case studies presented in this paper, free-free conditions were simulated by means of a very soft suspension.

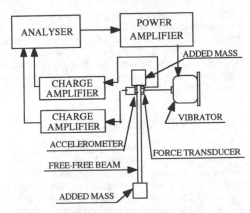

Figure 3 - Schematic representation of the measuring set-up for the case of a free-free beam loaded by two masses at the free ends

Due to the fact that the development of the theoretical models is described in the previously referred papers, the authors do not present it here. Also it would be difficult to do so given the short space available.

4 COMPUTER PROGRAM ALGORITHM

In the above referred work, the authors based the analysis of their results on the frequency ratio

$$Q_i = \frac{\omega_{fi}}{\omega_i} \qquad (1)$$

where ω_{fi} represents the ith natural frequency of a cracked beam and ω_i represents the ith natural frequency of the same beam with no cracks.

Though, in the referred work, the experimental results and theoretical predictions where reasonably similar they did not show a complete agreement. This is a normal situation, even for the simple structures under test, and is due to the assumptions one must take when developing the theory. However, when taking the theoretical and experimental values in terms of the frequency ratio presented in equation (1), it was found that discrepancies where a lot smaller. Further studies showed that a large amount of these discrepancies was due to the influence of the transducers mass and of the rod connecting the vibrator to the test specimen (figure 3). In fact, the use of an impact hammer for the vibration excitation of the test piece and the use of negligible mass strain gauges as a means of measuring the vibration response, allowed a significative improvement of the results (Gomes & Silva 1992b).

The above mentioned facts provided an increase in the confidence on the theoretical model that predicted values of the frequency ratio Q_i almost coincident with the experimental values of the same ratio, despite the fact that the theoretical and experimental absolute values of the natural frequencies did not show such a good agreement. Thus one was led to conclude that for a cracked beam, under different load and end conditions and using the authors' theoretical model, it was possible to state that

Experimental Q_i = Theoretical Q_i

Based on the previous assumption, a computer program named CRACAR (Gomes and Silva 1992c) was developed and implemented on a PC-compatible computer. This program was designed in order to determine the location and depth of one or more cracks, in a structural element of the type under analysis, based on the experimental knowledge of the values of Q_i.

The procedure to evaluate the performance of CRACAR was the following:

i) take the non-cracked structural element and measure some of its natural frequencies ω_i for the desired load conditions;

ii) develop real line cracks (with known location and depth) in the structural element and measure the new values of its natural frequencies ω_{fi};

iii) based on the previously found values of ω_i and ω_{fi}, calculate the corresponding values of Q_i;

iv) feed CRACAR with the experimental values of Q_i, with the geometric and material properties of the structural elements and with the load and end conditions (the program also accepts variations of the geometric and load conditions along the total length of the structural element).

With the previous information, CRACAR starts by generating the theoretical model for the structural element under study, assuming there are no cracks, and calculates the corresponding theoretical values of ω_i. Then, CRACAR takes the experimental values of Q_i and applies them to the theoretical values (for the non-cracked element) of ω_i calculating the corresponding expected values (for the cracked element) of ω_{fi}. These expected values will be designated by $(\omega_{fi})_e$.

The following step is for CRACAR to assume the existence of cracks at different locations L_e, with depths p, and calculate the corresponding values of the natural frequencies $(\omega_{fi})_p$. This step is repeated for different values of the cracks locations (sweeping the structural element) and for different values of the cracks depths. For each situation, CRACAR thus predicts the corresponding values $(\omega_{fi})_p$ of the natural frequencies.

Considering now the quantity

$$R = \sqrt{\frac{1}{n} \sum_{i=1}^{i=n} \left[\frac{(\omega_{fi})_e - (\omega_{fi})_p}{(\omega_{fi})_e} \right]^2} \qquad (2)$$

it will be easy to conclude that the values of $(\omega_{fi})_p$ for one of the cracks combinations assumed by CRACAR will minimize the value of R. The corresponding locations and depths of the cracks will be the solution of the problem at hand.

The sweeping incremental search performed by CRACAR covers the entire length of the structural element, unless one specifies the zone where the search must be done.

The latter case is to be more frequent given the fact that one is usually aware of the structural locations where stresses are larger and therefore where cracks may be expected to develop. Anyway, in order to save computer time, CRACAR starts with a course search and refines it at a later stage, in order to achieve faster results. Also, the operator can specify the increments for the search procedures.

5 EXAMPLES OF APPLICATION

In order to exemplify the performance of CRACAR in solving the "inverse problem", clamped-free straight steel beams, with rectangular cross-sections, were initially tested. The beams had the following geometric and material properties:

Cross-section = 18x32 mm
L = 600 mm
E = 21000 kg/mm^2
ρ = 7850 kg/mm^3

and were loaded, at their free end, by a cubic steel mass. The beams were first tested in order to determine their natural frequencies. Then, real line cracks were produced by means of the three-point-bending technique. The location and depth of these cracks were carefully controlled. New tests yielded the natural frequencies of the cracked beams.

Having previous knowledge of the location and depth of the cracks, allowed an easy determination of the performance of CRACAR. Ten different configurations were considered and the results obtained showed an extremely good agreement as can be seen in Silva & Gomes (1994).

Thus, it was decided to extend the analysis in order to consider the case where two cracks are simultaneously present in the beam. As stated previously, the beams were assumed free-free and mass loaded at their ends by means of cubic steel blocks. In the present study the length of the test specimens was L = 800 mm.

The cross-section size was the same as in the clamped-free examples (18x32 mm). Again, real line cracks were produced by means of the three-point-bending technique though, in these cases, two cracks were produced. The location and depth of these cracks were also carefully controlled.

Three different load conditions were considered:
1 - beam with no cubic masses at the free ends
2 - beam with two cubic masses with edge = 40 mm at the free ends
3 - beam with two cubic masses with edge = 60 mm at the free ends

The geometric properties of the cracks developed in the beams were the following:

Example A - 1st crack location - L$_e$ = 180 mm
   1st crack depth - p = 8 mm (25% of width of cross-section)
   2nd crack location - L$_e$ = 350 mm
   2nd crack depth - p = 12 mm (37.5% of width of cross-section)

Example B - 1st crack location - L$_e$ = 290 mm
   1st crack depth - p = 10 mm (31.25% of width of cross-section)
   2nd crack location - L$_e$ = 360 mm
   2nd crack depth - p = 14 mm (43.75% of width of cross-section)

For the more refined crack location search procedures executed by CRACAR an increment of 5 mm was used. The finer increment for the crack depth search was chosen to be 2.5% of the width of the cross-section. The results that were obtained are presented in tables 1 to 4. Tables 1 and 3 present the experimental and theoretical values of the first three natural frequencies of the uncracked and cracked beams, for the different load conditions.

Table 1. Natural Frequencies for Example A

End condit.	Nat. Freq.	Uncracked beam Exp. (Hz)	Theor. (Hz)	Cracked beam Exp. (Hz)	Theor. (Hz)
No loads	ω_1	263.7	262.3	245.7	244.3
	ω_2	716.4	716.5	695.0	695.0
	ω_3	1385	1386	1313	1313
40x40 mm cube	ω_1	179.9	179.8	168.5	168.4
	ω_2	513.3	515.9	500.2	502.7
	ω_3	1024	1030	977.4	983.5
60x60 mm cube	ω_1	132.5	132.9	124.5	124.9
	ω_2	415.8	422.2	405.9	412.2
	ω_3	830.8	867.5	811.3	847.2

Table 2. Crack Characteristics determined by CRACAR for Example A

Load Conditions used for calculations	No. of Nat. Freqs used for calculations	Crack Location L$_e$ (mm)	Crack depth (% of cross-section width)
1, 2 and 3	3	L$_1$ = 200	p$_1$ - 25
		L$_2$ = 370	p$_2$ - 40
1, 2 and 3	2	L$_1$ = 185	p$_1$ - 25
		L$_2$ = 355	p$_2$ - 37.5
1 and 2	3	L$_1$ = 185	p$_1$ - 25
		L$_2$ = 350	p$_2$ - 37.5
1 and 2	2	L$_1$ = 205	p$_1$ - 22.5
		L$_2$ = 345	p$_2$ - 37.5
1 and 2	1	L$_1$ = 180	p$_1$ - 20
		L$_2$ = 315	p$_2$ - 40
1	3	L$_1$ = 185	p$_1$ - 25
		L$_2$ = 350	p$_2$ - 37.5
2	3	L$_1$ = 195	p$_1$ - 25
		L$_2$ = 355	p$_2$ - 37.5

Table 3. Natural Frequencies for Example B

End condit.	Nat. Freq.	Uncracked beam Exp. (Hz)	Theor. (Hz)	Cracked beam Exp. (Hz)	Theor. (Hz)
No loads	ω_1	263.2	262.3	229.3	228.1
	ω_2	716.1	716.5	688.9	689.3
	ω_3	1385	1386	1312	1312
40x40 mm cube	ω_1	179.7	179.8	159.1	159.2
	ω_2	512.7	515.9	499.3	502.4
	ω_3	1023	1030	970.5	977.8
60x60 mm cube	ω_1	132.3	132.9	118.1	118.6
	ω_2	416.2	422.2	407.2	413.1
	ω_3	837.5	867.5	795.1	823.6

Tables 2 and 4 present the values of the cracks locations and depths as predicted by CRACAR. These predictions were based on both the unloaded and loaded conditions.

Table 4. Crack Characteristics determined by CRACAR for Example B

Load Conditions used for calculations	No. of Nat. Freqs used for calculations	Crack Location L_e (mm)	Crack depth (% of cross-section width)
1, 2 and 3	3	$L_1 = 295$	p_1 - 35
		$L_2 = 365$	p_2 - 45
1, 2 and 3	2	$L_1 = 285$	p_1 - 35
		$L_2 = 365$	p_2 - 45
1 and 2	3	$L_1 = 295$	p_1 - 35
		$L_2 = 365$	p_2 - 45
1 and 2	2	$L_1 = 285$	p_1 - 35
		$L_2 = 385$	p_2 - 45
1 and 2	1	$L_1 = 305$	p_1 - 32.5
		$L_2 = 390$	p_2 - 45
1	3	$L_1 = 295$	p_1 - 35
		$L_2 = 365$	p_2 - 45
2	3	$L_1 = 290$	p_1 - 32.5
		$L_2 = 365$	p_2 - 45

6 FINAL NOTES

The presented results show a good agreement and thus validate the technique used in CRACAR. It is worth emphasysing that the results are good even when one of the cracks is located near the free end of the beam.

Location errors are consistently below 35 mm (4.4% of the beam length). As far as crack depth predictions are concerned, results may be considered very good for all situations. In general, it can be concluded that the use of the knowledge of a larger number of natural frequencies variations leads to more accurate results.

Thus, one may state that the accuracy of the prediction procedure used by CRACAR was confirmed. The only problem that may be an inconvenience is the time the computer takes for the search calculations. However and as stated previously, this problem may be overcome by doing an initial course search ,which determines the zone where a finer search must be done, or by immediately specifying the zone where the search is to be done. The latter is an obvious option given the fact that, in most structures, one knows beforehand the zones where stresses are higher and where stress concentration is bound to be responsable for a crack developing and propagating.

Hence, the accuracy of the results presented in this paper could be inproved if a finer search was performed.

Experimental analysis is not difficult to perform. However, in some cases, it is found that it is not easy to determine the values of some of the natural frequencies. This problem may be overcome by searching values of natural frequencies that correspond to easily determined modes of vibration. This means testing the structure under different loading and/or end conditions, as was performed for the case studies presented in this paper.

REFERENCES

Achenbach, J.D., Sotiropoulos, D.A. & Zhu, H. 1987. Characterization of Cracks from Ultrasonic Scattering Data. *Trans. ASME JAM:* Vol. 54.

Chen, Y. & Swamidas, A.S.J. 1994. Dynamic Characteristics and Modal Parameters of a Plate With a Small Growing Surface Crack. *Proc. 12th IMAC:* Vol. 1, 1155-1161, Honolulu, Hawaii, U.S.A.

Gomes, A.J.M.A. & Silva, J.M.M. 1990. On the Use of Modal Analysis for Crack Identification. *Proc. 8th IMAC:* Vol. 2, 1108-1115, Kissimmee, Florida, U.S.A.

Gomes, A.J.M.A. & Silva, J.M.M. 1990. Experimental Determination of the Influence of the Cross-Section Size in the Dynamic Behaviour of Cracked Beams. *Proc. 2nd IMMDC:* 414-423, Los Angeles, California, U.S.A.

Gomes, A.J.M.A. & Silva, J.M.M. 1991. Theoretical and Experimental Data on Crack Depth Effects in the Dynamic Behaviour of Free-Free Beams. *Proc. 9th IMAC:* Vol. 1, 274-283, Florence, Italy.

Gomes, A.J.M.A. & Silva, J.M.M. 1992. Dynamic Analysis of Clamped-Free Cracked Beams Subjected to Axial Loads. *Proc. 10th IMAC:* Vol. 1, 541-548, San Diego, California, U.S.A.

Gomes, A.J.M.A. & Silva, J.M.M. 1992. Dynamic Analysis of Clamped-Free Cracked Beams with Variable Masses at the Free End. *Proc. 7th ICSEM:* Vol. 2, 1051-1056, Las Vegas, Nevada, U.S.A.

Gomes, A.J.M.A. & Silva, J.M.M. 1992. CRACAR - A Computer Program for Detection of Cracks in Structural Elements Through the Variation of their Natural Frequencies (in Portuguese). *Internal Report:* Faculty of Sciences and Technology, New University of Lisbon.

Guigné, J.Y., Klein, K., Swamidas, A.S.J. & Guzzwell, J. 1992. Modal Information from Acoustic Measurements for Fatigue Crack Detection Applications. *Proc. 11th ICOMAE:* Vol. 1, Part B, Offshore Technology, 585-593, Calgary, Canada.

Iman, I., Azzaro, S.H., Bankert, R.J. & Scheibel, J. 1989. Development of an On-Line Rotor Crack Detection and Monitoring System. *Trans. ASME JVASRD:* Vol. 111.

Klein, K., Guigné, J.Y. & Swamidas, A.S.J. 1994. Monitoring Change in Modal Parameters with Fatigue. *Proc. 12th IMAC:* Vol. 1, 1792-1800, Honolulu, Hawaii, U.S.A.

Lu, Y.M. & Ju, F.D. 1991. Probabilistic Distribution of Multiple Cracks in Structures Due to Random Modal. *IJAEMA:* no. 6(1), 25-34.

Mannan, M.A. & Richardson, M.H. 1990. Detection and Location of Structural Cracks Using FRF Measurements. *Proc. 8th IMAC:* Vol. 2, 1108-1115, Kissimmee, Florida, U.S.A.

Mayes, I.W. & Davies, W.G.R. 1976. The Vibrational Behaviour of a Rotating Shaft System Containing a Transverse Crack. *Vibrations in Rotating Machinery*, C168/76, University of Cambridge.

Meneghetti, U. & Maggiore, A. 1994. Crack Detection by Sensitivity Analysis. *Proc. 12th IMAC:* Vol. 1, 1292-1298, Honolulu, Hawaii, U.S.A.

Natke, H.G. & Cempel, C. 1991. Fault Detection and Location in Structures: A Discussion. *Mechanical Systems and Signal Processing:* no. 5(5), 345-356.

Nishimura, N. & Kobayashi, S. 1991. A Boundary Integral Equation Method for an Inverse Problem Related to Crack Detection. *IJNME:* Vol. 32, 1371-1387.

Pandey, A.K. Biswas, M. & Samman, M.M. 1991. Damage Detection from Changes in Curvature Mode Shapes. *JSV:* no. 145(2), 321-332.

Perchard, D.R. & Swamidas, A.S.J. 1994. Crack Detection in Slender Cantilever Plates Using Modal Analysis. *Proc. 12th IMAC:* Vol. 1, 1769-1777, Honolulu, Hawaii, U.S.A.

Qunli, W. & Fricke, F. 1991. Determination of the Size of an Object and its Location in a Rectangular Cavity by Eigenfrequency Shifts: First Order Approximation. *JSV:* no. 144(1), 131-147.

Rizos, P.F. Aspragathos, N. & Dimarogonas, A.D. 1990. Identification of Crack Location and Magnitude in a Cantilever Beam from the Vibration Modes. *JSV:* no. 138(3), 381-388.

Samman, M.M., Biswas, M. & Pandey, A.K. 1991. Employing Pattern Recognition for Detection of Cracks in a Bridge Model. *IJAEMA:* no. 6(1), 35-44.

Shen, M.H.H. & Taylor, J.E. 1991. An Identification Problem for Vibrating Cracked Beams. *JSV:* no. 150(3), 457-484.

Silva, J.M.M. & Gomes, A.J.M.A. 1990. Experimental Dynamic Analysis of Cracked Free-Free Beams. *Experimental Mechanics*, Vol. 30, no. 1: 20-25.

Silva, J.M.M. & Gomes, A.J.M.A. 1991. Crack Modelling Using Torsional Springs (in Portuguese). *Proc. JFSPM4:* Lisboa, Portugal.

Silva, J.M.M. & Gomes, A.J.M.A. 1992. On the Use of Modal Analysis for Fatigue Crack Detection in Simple Structural Elements. *Proc. 11th ICOMAE:* Vol. 1, Part B, Offshore Technology, 595-600, Calgary, Canada.

Silva, J.M.M. & Gomes, A.J.M.A. 1994. Crack Identification of Simple Structural Elements Through the Use of Natural Frequency Variations: The Inverse Problem. *Proc. 12th IMAC:* Vol. 2, 1728-1735, Honolulu, Hawaii, U.S.A.

Stubbs, N. & Osegueda, R. 1990. Global Non-Destructive Damage Evaluation in Solids. *IJAEMA:* no. 5(2), 67-79.

Stubbs, N. & Osegueda, R. 1990. Global Damage in Solids - Experimental Verification. *IJAEMA:* no. 5(2), 81-97.

Swamidas, A.S.J. & Chen, Y. 1992. Damage Detection in a Tripod Tower Platform (TTP) Using Modal Analysis. *Proc. 11th ICOMAE:* Vol. 1, Part B, Offshore Technology, 577-583, Calgary, Canada.

Swamidas, A.S.J. & Budipriyanto, A. 1994. Experimental and Analytical Verification of Modal Behaviour of Uncracked/Cracked Plates in Air and Water. *Proc. 12th IMAC:* Vol. 1, 745-752, Honolulu, Hawaii, U.S.A.

Recent Advances in Experimental Mechanics, Silva Gomes et al. (eds) © 1994 Balkema, Rotterdam, ISBN 90 5410 395 7

Development of an ultrasonic fatigue device and its application in fatigue behavior studies

Jingang Ni
Department of Jet Propulsion 405, Beijing University of Aeronautics and Astronautics, People's Republic of China

Cloude Bathias
Department of Engineering Materials, Conservatoire National des Arts et Métiers, Paris, France

ABSTRACT: An ultrasonic fatigue device has been developed to study the fatigue life and fatigue crack growth (fcg) behavior of engineering alloys at high frequency (f = 20kHz). In these studies, the operating principle and the layout design of the ultrasonic fatigue machine, specimen design and calculation as well as experimental procedures are presented briefly. Experimental data on the fatigue life and fcg behavior of several alloys are investigated in ultrasonic and conventional fatigue frequencies loading. It is concluded that the ultrasonic the fatigue tests provide an efficient, time-saving, low expensive and widely adaptive means in studying the fatigue properties of metals and alloys.

1 INTRODUCTION

The ultrasonic fatigue testing technique has been introduced to study metals' fracture mechanical characteristics since 1950 when Mason[1] developed the piezo–electric and magnetostrictive type of transducers which permit translating 20kHz electrical voltage signals into 20kHz mechanical displacements. And it was since 1970 that this technique has been used to determine the fcg rates and threshold behavior of metallic materials[2]. An excellent overview expounding the historical and scientific background of these ultrasonic fatigue tests was given in the work of Tien[3]. In reality, the following aspects can be considered as essential reasons to explain why this technique may be developed so rapidly, and in which many scientific establishments and investigators have taken so great interest.

In aircraft and engine engineering, some structures, such as discs and blades of turbine machines, may be subjected not only to viscoelastic–plastic loading of high intensity (regime of mini–maxi rotation), but also to repetitive excitations of small amplitude and at high frequency to the order of kiloheitz, caused essentially by acoustical, aerodynamical, or even mechanical vibrations. This last aspect of fatigue loading, called ultrasonic fatigue, may bring about a particular mechanical damage and rupture mechanism of structures and is deserving of extra attention.

This kind of ultrasonic fatigue loading may attain to 10^9 cycles, it would require 230 days consecutive manipulation by means of the conventional technique with a loading frequency of 50Hz. So it would be time consuming and expensive, even impossible to establish S–N curves for such high cycles fatigue properties of materials.

On the other hand, the determination of fcg behavior or $\triangle K$ vs da / dN relations and threshold stress intensity values by conventional tests (f = 50Hz) will be either strenuous or costly, and furthermore, it becomes even delicate if da / dN is less than 10^{-8} mm / cycle.

By using the ultrasonic fatigue technique, as the exciting frequency is of 20kHz, high cycles fatigue life tests may be fulfilled in a suitable period (10^9 cycles corresponding to 14 hours testing time), and the fcg rates behavior may be determined easily down to the order of 10^{-9} to 10^{-8} mm / cycle.

The ultrasonic fatigue machine with the accessory equipments developed in this study allows of investigating fatigue behavior of engineering materials for a) high cycles fatigue life at 20kHz and at room (T = 30°C), high(T = 800°C) or cryogenic (T = −196°C) temperatures, with cycles to failure being studied up to 10^9 cycles. b) fcg rates and threshold behaviors at room and high temperatures, with da / dN data down to 10^{-8} − 10^{-9} mm / cycle.

In this investigation, theoretical and experimental aspects of the ultrasonic fatigue testing technique are briefly described. Experimental data on fatigue limit and fcg behaviors of several engineering alloys obtained at high (f = 20kHz) and conventional (f = 20 to 50Hz) loading frequencies are explored.

2 SPECIMEN DESIGN

2.1 Fatigue life specimen

The fatigue life specimen designed for the ultrasonic fatigue tests is shown in figure 1, with two quarter–wave length, symmetrical about the center section $(x = 0)$, and a cylindric–toroidal profile at the center part. The dimensions R_0, R_1, R_2, and L_2 are fixed for all materials to facilitate the machining of specimens and the specific length L_1 remains to be determined analytically to have a resonance frequency of the first intrinsic longitudinal vibration $(f = 20\text{kHz})$ and a maximum strain value in the middle section of the specimen $(x = 0)$, which is capable of damaging materials. In this study, we have $R_0 = 31$ mm; $R_1 = 1.5$ mm; $R_2 = 5$ mm; and $L_2 = 14.31$ mm.

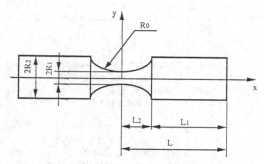

Fig.1 Fatigue life specimen design profile

Resonant length $(f = 20\text{kHz})$:

$$L_1 = \frac{1}{\theta} arctg\{\frac{1}{\theta}[\frac{\beta}{th(\beta L_2)} - \alpha th(\alpha L_2)]\} \quad (1)$$

with , $\quad \theta = 2\pi f_r \sqrt{\frac{\rho}{E_d}}$

$$\alpha = \frac{1}{L_2} arch(\frac{R_2}{R_1}); \quad \beta = \sqrt{\alpha^2 - \theta^2}$$

where $\rho = $ density ; $E_d = $ dynamic Young's modulus;

and $f_r = $ resonant frequency.

Relation between $U_0(x = L)$ and σ_m $(x = 0)$:

$\varepsilon_m = U_0 \cdot \varphi(L_1, L_2) \cdot \beta$

$\sigma_m = E_d \cdot U_0 \cdot \varphi(L_1, L_2) \cdot \beta \quad (2)$

in which, $\varphi(L_1, L_2)$ is a characteristic variable concerning the specimen geometry,

$\varphi(L_1, L_2) = cos(\theta L_1)ch(\alpha L_2) / sh(\beta L_2)$

where, $U_0 = $ maximum displacement amplitude at the extremity of the specimen; $\varepsilon_m = $ maximum strain occurred in the middle section of the specimen; and $\sigma_m = $ maximum stress deduced from Hooke's law.

2.2 Fatigue crack growth specimen

The fcg specimen employed in the ultrasonic fatigue study has been designed to have a geometric profile as shown in figure 2.

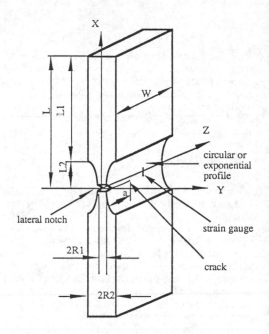

Fig.2 Fcg specimen design profile

Similar to the ultrasonis fatigue life specimen described above in 2.1, the dimensions of this ultrasonic fcg specimen R_0 ,
R_1 , R_2 , L_2 and W are fixed for all materials to facilitate the machining of specimens except L_1 which remains to be determined according to resonance conditions. In this study, we have chosen $R_1 = 1.5$ mm; $R_2 = 4$ mm; $L_2 = 12.2$ mm; and W = 14 mm.

For the given specimen profile, analytic solutions can be summarized as follows (uni–dimensional, exponential profile, notch effect count out):

Resonant length $(f = 20\text{kHz})$:

$$L_1 = \frac{1}{\theta} arctg\{[\frac{\sqrt{\gamma^2 - \theta^2}}{th(\sqrt{\gamma^2 - \theta^2} \cdot L_2)} - \gamma]/\theta\} \quad (3)$$

in which, $\gamma = \dfrac{1}{2L_2} ln(\dfrac{R_2}{R_1})$

Relation between $U_0(x=L)$ and $\sigma_m(x=0)$:

$$\varepsilon_m = U_0 \cdot \psi(L_1, L_2) \cdot \sqrt{\gamma^2 - \theta^2}$$

$$\sigma_m = E_d \cdot U_0 \cdot \psi(L_1, L_2) \cdot \sqrt{\gamma^2 - \theta^2} \qquad (4)$$

in which, $\psi(L_1, L_2)$ is a characteristic function depending on exciting regime and specimen geometric design.

$$\psi(L_1, L_2) = \dfrac{cos(\theta L_1)e^{\gamma L_2}}{sh(\sqrt{\gamma^2 - \theta^2}\, L_2)}$$

Determination of the stress intensity factor ΔK:

There is no analytical solution concerning the calculation of the stress intensity factor in the case of ultrasonic fatigue loading. In this investigation, an analytical – experimental approach is adopted [4] with regard to the peculiarities of the resonance vibration. The following equations may be served to calculate ΔK value in the case of ultrasonic fatigue loading.

$$\Delta K = E_d \cdot \Delta\varepsilon_m \cdot \sqrt{\pi a} \cdot z(\dfrac{a}{w}) \qquad (5)$$

in which, $\Delta\varepsilon_m$ = tensile part of nominal strain amplitude measured in the mid–section of the specimen by meas of dynamic strain gauges, positioned outside of plastic zone; and $z(\dfrac{a}{w})$ = characteristic function, dependent upon the ratio of crack length to specimen geometric size [4].

$$z(\dfrac{a}{w}) = 1.12 - 1.351(\dfrac{a}{w}) + 10.781(\dfrac{a}{w})^2$$

$$- 32.27(\dfrac{a}{w})^3 + 52.11(\dfrac{a}{w})^4 - 30.39(\dfrac{a}{w})^5 \qquad (6)$$

3 EXPERIMENTAL PROCEDURES

The ultrasonic fatigue machine used in this work was developed following the original design of W. Mason [1]. Figure 3 shows the block diagram of the resonant system and apparatus employed in ultrasonic fatigue life and fcg studies.

In principle, a power generator excites a magnetostrictive type of transducer which transforms longitudinal ultrasonic wave to the specimen through a horn in order to obtain an amplification of the strain amplitude in the middle section of the specimen. Each element is designed to have a resonant frequency about 20 kHz and an automatic unit maintains constantly the whole system operating at the resonant frequency.

In operation, a maximum displacement amplitude U_0 is achieved at the end ($x=L$, dis-placement antinode); while the maximum strain value ε_m is obtained in the middle of the specimen ($x=0$, strain antinode). The vibrational amplitude U_0 is measured by means of a dynamic capacitive sensor and the maximum strain or stress values can be evaluated using analytical solutions (Eqs 2,4).

Alternatively, during the ultrasonic fatigue tests, the maximum strain values can be measured directly using miniature strain gauges, suitably positioned on the sample surface. In this study, a system consisting of a Wheatstone bridge amplifier, dynamic strain gauges (0.79 mm by 0.81 mm) and a digital oscilloscope (two channels, 40K memory for each) has been built for direct strain measurements.

Preliminary results indicate that the linear relation between maximum strain and displacement amplitude at the deduced from analytical solutions can be verified for a reasonable deformation extent (cf. Ti–6Al–4V, $\varepsilon_m < 3.0 \times 10^{-3}$), with a relative error of about 5%.

A compressed air cooling system served to eliminate the temperature raise caused by the ultrasonic frequency effect. The temperature at the center section of the specimen was maintained at 35°C, registered by means of a thin thermocouple ($d=0.1$ mm), carefully attached to the specimen surface.

For the ultrasonic fcg testing, a system of video – camera – supervision has been set up to

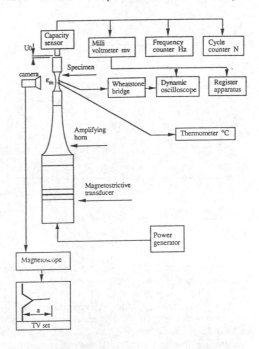

Fig.3 Full resonance ultrasonic system with schematic view of apparatus

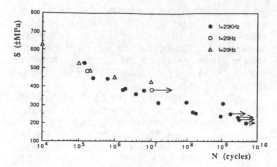

Fig.4　Fatigue life data for Udimet 500

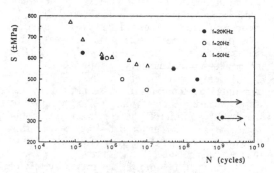

Fig.5　Fatigue life data for 17–4 PH

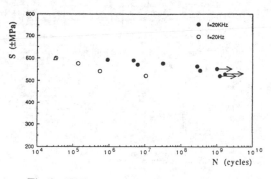

Fig.6　Fatigue life data for Ti–6Al–4V

survey and record crack initiation and propagation processes. The registering information allows to obtain fcg rates da / dN data and to calculate stress intensity factor ΔK values according to the crack length and the nominal strain amplitude obtained in the middle section of the specimen.The registering system refines events to 1 / 25th of a second and magnifies specimen surface till 150 – 200 times.

4　RESULTS AND DISCUSSION

4.1 On the fatigue life behavior

Twenty specimens of Udimet 500, eighteen of Ti–6Al–4V and ten of 17–4PH were tested in ultrasonic fatigue tests (f = 20 kHz, R = −1) with the maximum strain controlled constant in the middle section of the specimen. Some specimens were used for conventional fatigue tests with a loading frequency of 20 to 50 Hz and a push–pull stress alternating cycles imposed. The experimental fatigue life data are shown in figures 4 through 6.

It may be remarked that, for the nickel superalloy Udimet 500, the ultrasonic fatigue data closely matched the conventional fatigue tests results between 10^5 and 10^7 cycles, and the fatigue limit at 20kHz diminishes continuously until 10^9 cycles ; while for 17–4PH and Ti–6Al–4V, the fatigue resistance behavior in ultrasonic fatigue regime seems to be slightly better than that given in the conventional fatigue loading. This effect indicates the alloys respond differently to the ultrasonic fatigue excitation in the resonant regime.

Moreover, our experiments [4] , [5] showed that ultrasonic frequency excitations at lower amplitudes increase effectively the fatigue resistance of materials. And in ultrasonic fatigue tests, while the specimen vibrates longitudinally at the resonant frequency, the middle section of the specimen is excited at constant strain amplitude.

In this investigation, it is considered that the fatigue life behavior of materials depends essentially on both the loading regime and the mechanical characteristics of materials.

4.2 On the fcg behavior

Four different materials, Astroloy (nickel alloy, similar to Udimet 700), 17–4 PH, Ti–6Al–4V and Al–Li 8090, were investigated. The experiments were fulfilled in ultrasonic fatigue loading (f = 20kHz, R = −1) and in conventional fatigue frequencies (f = 20 to 50Hz) with different values of stress ratio applied. The obtained experimental results are presented in figures 7 through 10.

Experimental data indicate that the fcg behavior at ultrasonic fatigue frequency is comparable to which observed in conventional fatigue testing with a more important stress ratio R = 0.7–0.75 imposed. It may be considered that, at ultrasonic fatigue frequency, the closure effect at the crack tip is of no importance and the achieved stress intensity values will be equivalent to the effective stress intensity factors determined in conventional fatigue loading.

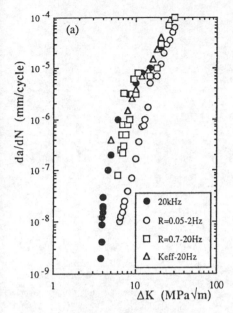

Fig.7　Fcg data for Astroloy

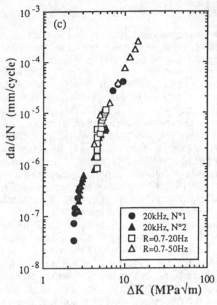

Fig.9　Fcg data for Ti–6Al–4V

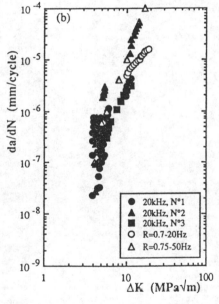

Fig.8　Fcg data for 17–4 PH

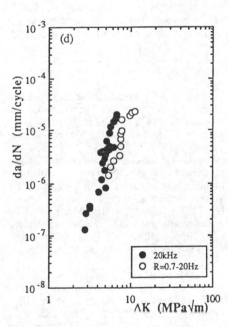

Fig.10　Fcg data for Al–Li 8090

5 CONCLUSIONS

(1) The ultrasonic fatigue testing provides an efficient, time—saving and less expensive experimental means in studying fatigue life (S–N curves) and fcg (ΔK–da / dN) behaviors. This method yields information on the fatigue resistance of materials which is useful when developing both the mechanical and physical aspects of strength theories.

(2) The analytic solutions of the maximum strain value for a given displacement amplitude at the free end of the specimen have been verified experimentally and confirmed to be in good agreement in elastic deformation domain, by means of miniature foil strain gauge measurement.

(3) The experimental results show the different response of each alloy to ultrasonic frequency loading, with Ti–6Al–4V giving the best resistance, followed by 17–4 PH and Udimet 500. Fatigue resistance behavior depends upon the ultrasonic and the mechanical characteristics and microstructure of the material, loading regime, and the surface conditions of the specimen.

(4) The stress intensity factor values ΔK determined in ultrasonic fatigue loading may be comparable with the effective stress intensity values obtained in conventional fatigue tests.

REFERENCES

(1) Mason, W. P., Piezoelectric Crystals and Their Application in Ultrasonics, Van Nostrand, New Youk, 1950, P. 161.

(2) John K. Tien and R. P. gamble, Met. Trans., 2(1971) p. 1933.

(3) John K. Tien, State of Ultrasonic Fatigue. Ultrasonic fatigue, American Institute of Mining, Metallurgical, and Petroleum Engineer, Inc, 1982, pp.1–14.

(4) Ni, J. G., "Study on the Mechanical Behaviors and Damage Mechanisms of Alloys Under Ultrasonic Fatigue Loading", Ph. D. thesis, CNAM de Paris, France, 1991.

(5) C. Bathias and Ni J.G., Study of Lifetime Behavior of Alloys in Ultrasonic Fatigue Testing, ASTM E–9, 2nd. symposium, Pittsburgh, USA 1992.5, pp141–152.

(6) Bathias, C., Ni, J. G., Wu, T. Y., and Lai, D., "Fatigue Threshold of Alloys at High Frequency", Proc., 6th Int. conf. on the Mech. Beh. of Mat., Kyoto, Japan, July 1991, pp463–468.

Recent Advances in Experimental Mechanics, Silva Gomes et al. (eds) © 1994 Balkema, Rotterdam, ISBN 90 5410 395 7

Localized strain behaviors around crack tip during load cycling and fatigue crack extension process

K. Hatanaka & T. Ishikawa
Yamaguchi University, Japan

ABSTRACT: A grid pattern was described in a space of about 30μm with a diamond stylus in the area ahead of the precrack of the specimen. Then the distortion of the grids was measured by means of a photomicroscope. The three dimensional components of the strain were calculated using the Euler's equations, into which the above measurement was introduced. The localized strain in the vicinity of a crack tip was examined under cyclic straining by means of the above grid method, where special attention was given to the strain behavior during one strain cycle in the strain controlled test. The longitudinal strain εy was intensively developed in the directions of $\theta \simeq \pm 60°$ ahead of crack tip, where θ is the angle inclined with respect to the general direction of crack extension.

1 INTRODUCTION

Fatigue crack extension process seems to be controlled by the microstructural factor with the order of dislocations. Then the direct observations of dislocation motion which is accompanied with cyclic loading is very useful for the physical understanding of fatigue crack extension mechanism. This is, however, extremely difficult.

The insitu observations on the fatigue crack extension have been inside a scanning electron microscope, and some useful microscopic information has been obtained (Davidson et al.,1979, Kikukawa et al. ,1974, Kikukawa et al.,1978). Mechanical parameters dominating the fatigue crack extension, however, have not been quantitatively measured around fatigue crack tip in these research works.

In the light of such situations, the localized deformation with the order of grain size was quantitatively measured in the vicinity of fatigue crack tip in the present study, using the grid method which was proposed by one of the authors in the earlier paper(Hatanaka et al.,1992). At that time special attention was paid to the variation of localized strain just ahead of the crack tip during one load cycle.

2 MATERIAL AND TEST PROCEDURES

The tested material is commercially pure copper annealed at 800 °C for 3 hours. Its grain size was 40μm. The specimen has 18mm long parallel part with rectangular crosssection of 16mm × 6mm. A hole of 1.0mm in diameter was drilled at the center of the specimen. Then the slit of 4.0mm in length from the hole center line was produced by spark-machining. Furthermore, the fatigue precracks were introduced at the two slit ends(Hatanaka et al.,1993).

The grid lines were described in a space of about 30μm in both directions of X- and Y- axes with a diamond stylus in the area ahead of precrack, where Y-axis is set up along the loading direction.

Push-pull loading tests were performed using the closed loop type test system, controlling the displacement between the two small dents produced by punching at the points of 1mm distant from the center of the drilled hole above and below, where displacement ratio R_ε was maintained as -1.0. Furthermore, testing machine was operated by hand at several stages in strain cycling process, and stopped from 5 to 9 times during one strain cycle. Then crack opening and closing behavior and localized deformation around crack tip, which were produced by strain reversal were examined in detail.

Figure 1 shows the configurations for measuring the two-dimensional strains which include the drill hole, the slit, the precrack introduced from the slit end and the grid for measuring the strains. The Y-

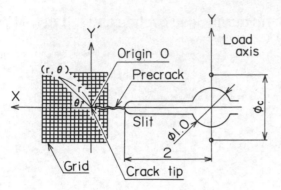

Fig.1. Schema showing grids and defining the coordinate system

and X- axes run parallel to load direction and perpendicular to this through the center of the drill hole. The 1600 grid points formed as intersections of 40 × 40 lines which were described in the Y- and X-directions around crack tip were settled for measurements of the localized strains. The positions of these grid points (X_0, Y_0) are determined with respect to the origin of the above coordinate system before test.

The strain at the position of an aimed grid point $(X_{N(I,J)}, Y_{N(I,J)})$ of the above 1600 grid points at N cycles is calculated as follows:

The displacements u and v of the point $(X_{0(i,j)}, Y_{0(i,j)})$ in the direction of X- and Y- axes are expressed by the following equations,

$$u = (X_{N(i,j)} - X_{0(i,j)}) - (X_{N(I,J)} - X_{0(I,J)}) \quad (2.1)$$

and

$$v = (Y_{N(i,j)} - Y_{0(i,j)}) - (Y_{N(I,J)} - Y_{0(I,J)}) \quad (2.2)$$

where i and j are suffix-notations for expressing the grid points surrounding an aimed grid point $(X_{0(I,J)}, Y_{0(I,J)})$, $(X_{0(I,J)}, Y_{0(I,J)})$ and $(X_{N(I,J)}, Y_{N(I,J)})$ are the coordinates of the aimed points before test and at N cycles, and $(X_{0(i,j)}, Y_{0(i,j)})$ and $(X_{N(i,j)}, Y_{N(i,j)})$ are the ones of the 24 grid points surrounding the aimed point before test and at N cycles. The displacements at the 25 grid points generally form a continuous curved surface, and therefore u and v are approximately expressed by the respective cubic equations with respect to $X_{N(i,j)}$ and $Y_{N(i,j)}$ in the following way.

$$S = A_1 + A_2 X_{N(i,j)} + A_3 X_{N(i,j)}^2 + A_4 X_{N(i,j)}^3$$
$$+ A_5 Y_{N(i,j)} + A_6 Y_{N(i,j)}^2 + A_7 Y_{N(i,j)}^3 \quad (2.3)$$

The constants, $A_1, A_2, \ldots, A_6$ and A_7 included in eq.(2.3) are fixed upon through the least square's regression analysis so that the equations

$$|u - S| \gtrapprox 0 \quad (2.4)$$

and

$$|v - S| \gtrapprox 0 \quad (2.5)$$

might hold. Thus, the equations of u and v which form the continuous curved surface are determined.

The displacement gradients at the point of interest are obtained as follows (Obata et al.,1979); the partial derivatives of the cubic equations of u and v are obtained with respect to $X_{N(i,j)}$ and $Y_{N(i,j)}$, and then the values of $X_{N(I,J)}$ and $Y_{N(I,J)}$ at the point of interest are substituted for $X_{N(i,j)}$ and $Y_{N(i,j)}$ in the equations resulting from partial derivatives.

The displacement gradients estimated in this way are substituted into the Euler's equations expressed by the following,

$$\varepsilon_x = \frac{\partial u}{\partial X} - \frac{1}{2}\left[\left(\frac{\partial u}{\partial X}\right)^2 + \left(\frac{\partial v}{\partial X}\right)^2\right] \quad (2.6)$$

$$\varepsilon_y = \frac{\partial v}{\partial Y} - \frac{1}{2}\left[\left(\frac{\partial u}{\partial Y}\right)^2 + \left(\frac{\partial v}{\partial Y}\right)^2\right] \quad (2.7)$$

$$\varepsilon_z = -\varepsilon_x - \varepsilon_y \quad (2.8)$$

and

$$2r_{xy} = \frac{\partial u}{\partial Y} + \frac{\partial v}{\partial X} - \left[\frac{\partial u}{\partial X}\frac{\partial u}{\partial Y} + \frac{\partial v}{\partial X}\frac{\partial v}{\partial Y}\right] \quad (2.9)$$

where ε_x, ε_y, ε_z and γ_{xy} are nominal strains in the direction of X-, Y- and Z-axes, and shear strain, respectively. In the above analysis, the assumptions of uniformity in displacement in the direction of the plate thickness w, resulting in $\partial u/\partial Z = \partial v/\partial Z = 0$ and $\partial w/\partial Z = \varepsilon_z$, where ε_z is the normal strain in the direction of plate thickness, $\partial w/\partial X = \partial w/\partial Y = 0$ and invariability in volume were introduced (Bossaert et al. ,1968). The other strain components, ε_r, ε_θ and $\gamma_{r\theta}$ were also obtained from transforming the coordinate system from the orthogonal to the polar in which the origin is fixed at the crack tip.

The procedures stated above were repeated for the other grid points, and then the localized strains were estimated in the vicinity of the fatigue crack tip.

3 MEASUREMENTS OF STRAINS

3.1 Strain behaviors of localized strain around crack tip during one strain cycle

Figure 2 is the schema showing the locations on load-displacement hysteresis loop at which the strain and the crack opening displacement were measured during one strain cycle.

Figures 3 (a) to (c) show the distribution of the normal strain in the direction of Y-axis, ε_y and the crack tip opening behaviors at several points on the hysteresis loop during one strain cycle from N=4 to 5 cycles, where the displacement range was controlled at $\Delta\Phi_c$=310μm in the test: (a),(b) and (c) correspond to the points, A, C, and E in Fig.2. The intense positive strain develops in the direction of $\theta\simeq\pm60°$ ahead of crack

tip at the point A diminishes during compressive loading excursion, and then almost vanishes at the compressive tip of the hysteresis loop, the point C, as shown in figs.3 (b). Furthermore, the negative strain is scarcely yielded and the crack tip remains open at this point. In the tensile excursion starting from the point C in Fig.2, the positive strain develops again in the direction of $\theta\simeq\pm60°$ ahead of the crack tip, and the amount of strain and the size of the strain-concentrated area are much greater at the point E than at the point A, as shown in Figs.3 (a) and (c).

3.2 Crack tip opening-closing behaviors during one strain cycle

Figure 4 shows the opening and closing behaviors at the crack tip during one strain cycle at several stages in strain cycling process at $\Delta\phi_c$=310μm . The thin three lines and the bold solid line represent the crack opening displacements at 200μm behind the crack tip and at the center of the crack, ϕ_t and ϕ_c, respectively, which are plotted against the pointed locations on the hysteresis loop shown in Fig.2. The crack tip opening displacement ϕ_t is positive even at the compressive tip of the hysteresis loop, during first one strain cycle, and this becomes large at all the loading points on the hysteresis loop as the cyclic straining process progresses, as shown in Fig 4.

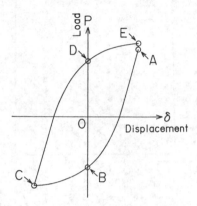

Fig.2. Positions where strain and COD were measured during one cycle

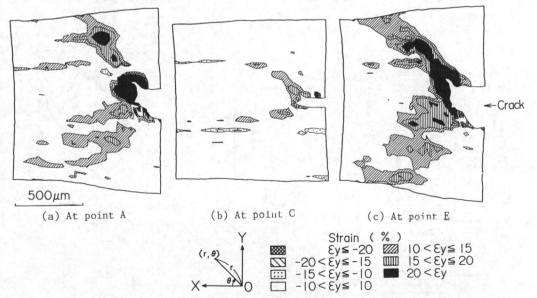

(a) At point A   (b) At point C   (c) At point E

Strain (%)	
$\varepsilon_y \le -20$	$10 < \varepsilon_y \le 15$
$-20 < \varepsilon_y \le -15$	$15 < \varepsilon_y \le 20$
$-15 < \varepsilon_y \le -10$	$20 < \varepsilon_y$
$-10 < \varepsilon_y \le 10$	

Fig.3. Distribution of normal strain ε_y around crack tip and crack tip opening behavior at several loading points in Fig.2, where displacement range was controlled at $\Delta\phi_c$=310μm in the test

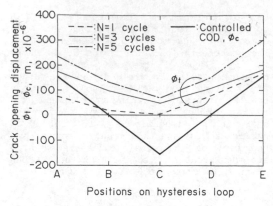

Fig.4. Opening and closing behaviors at the crack tip during one strain cycle

Thus, the crack tip remains open during one strain cycling in the displacement controlled test.

3.3 Fatigue crack extension process

Figure 5 shows the schema on the typical crack extension path in the cyclic straining test at the controlled displacement range, $\Delta\phi_c = 80\mu m$. Generally, the crack branches intermittently into the two at its tip and one of them extends further. Another crack stops growing and closes at its tip during further strain cycling.

Figures 6 (a) and (b) shows the displacement of the grid points in the vicinity of the crack tip during half a cycle which corresponds to the loading excursion from the points C to E through D in Fig.2 at $N=30$ and 60 cycles, respectively, where first and second ones of the two arrows →→ represent the movements of the grid points resulting from loading from the points C to D and from D to E. The locations of the crack tip at the same strain cycles are shown in Fig.5. The

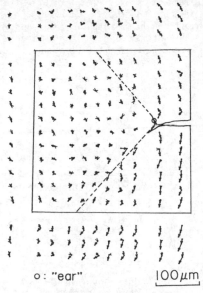

(a) N=30 cycles

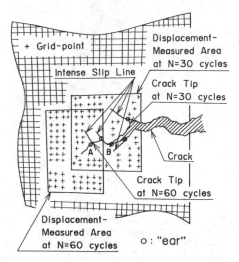

Fig.5. Schema on the typical crack extension path

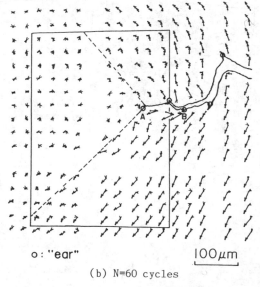

(b) N=60 cycles

Fig.6. Displacement of the grid points around crack tip during tensile half a cycle

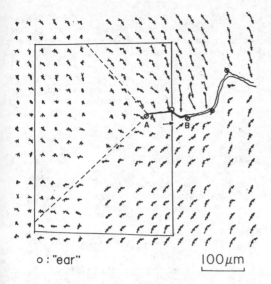

o : "ear" 100μm

Fig.7. Displacement of the grid points around crack tip during compressive half a cycle

crack is in the direction nearly perpendicular to the loading axis at N=30 cycles, as shown in Fig.6(a). This figure also shows that displacement of the grid point is greater in the area of $|\theta|>60°$ and is much smaller in the V-shaped region of $|\theta|<60°$(Neumann et al.,1974). The crack branched into the two directions of $\theta\approx60°$ and $-60°$ at this point and then extended downwards to the left.

Figure 7 shows the displacement of the grid points during half a cycle from the points A to C through B just before half a cycle from the the points C to E under the same test condition. We can observe the localized deformation behavior at the crack tip during one strain cycle at $\Delta\phi_c=80\mu m$ in Fig.6(b) and 7; the reverse displacement occurs in the directions of $\theta\approx\pm60°$ in the regions of $|\theta|>60°$, corresponding to the strain reversal around the crack tip and its amount is much larger in the lower region than in the upper one, where the crack extended downwards to the left. The displacement is still quite small in the V-shaped region at this strain-cycling stage.

The quite large displacement occurs repeatedly in the directions of $\theta\approx\pm60°$ during strain cycling and the crack extends along either of them. As the result, the crack takes a zig-zag route in a short range, heading perpendicularly to the loading axis on the whole. Meanwhile, another one of the branched cracks stop growing and the "ear" is formed at that site along the shear deformation band,

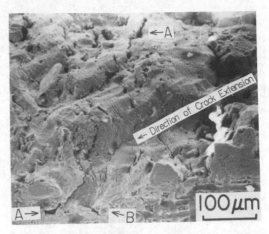

(a) View from direction inclined at 135 degree with respect to specimen surface

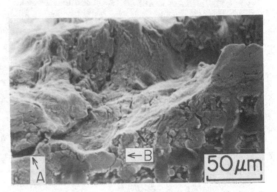

(b) View from direction perpendicular to specimen surface

Fig.8. Fatigue crack extension path and fractographs

which is shown in Figs.5 to 7, being noted as "ear".

Figure 8 (a) shows the fracture and specimen surfaces viewed from the direction inclined at the angle of 135 degree with respect to both the surfaces through scanning electron microscope. The same field is observed from the direction perpendicular to the fracture surface with much larger magnification in Fig.8 (b). The A and B marked in Figs.8 (a) and (b) correspond to the "ears" shown in Figs.5 to 7. It is found from Fig.8 that a furrow remains in the interior region as well as in the specimen surface at the ear. Such a furrow seems to be built up by intensive shear deformation cyclically reversed at the crack tip, judging from the measurements shown in Figs.5 to 6 and discussion associated with them. The chevron type protrusion is also left around the ear in the specimen surface, showing that the direction of the crack extension was changed in this area in Figs.8 (a) and (b). The prominence like this is not so marked in the interior region of the specimen, showing that the fatigue crack extended in quite planar way, and the typical striations are found in Fig.8(a). Such a difference in the fatigue crack extension mode between the surface and the interior region of the specimen is considered to be caused by the intensity of the plastic constraint effect dependent upon the location of the specimen; the intense reversible plastic shear deformation along $\theta \simeq \pm 60°$ is quite easy to occur in the very surface layer of the specimen, while this is fairly limited in its interior region where a plastic constraint effect works. Consequently, the crack takes a zig-zag type path specifically in the surface layer. Such a crack heading for $\theta \simeq \pm 60°$ changes its direction and grows towards the line extending the pre-crack plane at a certain position, since the crack grows nearly perpendicularly to the loading axis in fairly planar manner in the interior region of the specimen.

CONCLUSIONS

The localized strain in the vicinity of the crack tip was measured by means of the grid method in annealed copper strain-cycled under displacement-controlled condition. The special attention was paid to the localized strain and the crack opening-closing behaviors during one strain cycle at that time. The main results obtained are summarized as follows.

1. The intensive normal strain ε_y was developed in the directions of $\theta \simeq \pm 60°$ ahead of the crack tip, where θ is the angle inclined with respect to X-axis direction perpendicular to the loading axis (Y-axis).

2. The crack tip remains open even at the compressive tip of load-displacement hysteresis loop and the crack tip opening displacement becomes large with progress of strain cycling process.

3. Fatigue crack intermittently branches into the two along the intensively shear-deformed zones which are developed in the directions of $\theta \simeq \pm 60°$, and either of them extends in further strain cycling process. A deep furrow is formed at the tip of another crack which stops growing. This extends from the surface to the interior of the specimen.

4. Corresponding to the crack extension process described in (3), the unevenness with a zig-zag shape is left in the surface layer of the specimen. Meanwhile, quite planar fracture surface with typical striation prevails in the interior of the specimen. The difference in the intensity of the plastic constraint effect might cause such a location dependent crack extension mode.

REFERENCES

Bossaert,W.,Dechaene,R. and Vinckier,A. 1968. Computation of Finite Strains from Moire Displacement Patterns. J. Strain Analysis 3,1:65-75

Davidson,D.L. and Lankford,J.1979.Dynamic, Real-Time Fatigue Crack Propagation at High Resolution as Observed in the Scanning Electron Microscope. Am. Soc. Tes. Mat. Spec. Tech. Publ. 675:277-284

Hatanaka,K.,Fujimitsu,T. and Inoue,H.1992. A Measurement of Three-dimensional Strains Around a Creep-crack Tip. J. Soc. Exp. Mech. 32,3:211-217

Hatanaka,K.,Yoshioka,Y. and Ishikawa,T. 1992. Measurements of Localized Strain around Crack Tip during Strain Cycling and Some Considerations on Fatigue Crack Extension in Copper. Trans. Jpn. Soc. Mech. Engng. 59,559:674-681

Kikukawa,M.,Jono,M. and Adachi,M. 1978. Direct Observation of Fatigue Crack by Scanning Electron Microscope and Its Propagation Mechanism. J. Soc. Mat. Sci. 27,300:853-858

Kikukawa,M.,Jono,M.,Yasui,K.,Adachi,M.and Inada,Y. 1974. Microscopic Measurement of Fatigue Damage by Scanning Electron Microscopy. J. Soc. Mat. Sci. 23, 252:708-715

Neumann,P.1974. New Experiments Concerning the Slip Processes at Propagating Fatigue Cracks. Acta Met. 22:1155-1165

Obata,M.,Shimada,H. and Kawasaki,A. 1979. Large Deformation Analysis by the Fine Grating Method Considering the Deformation in the Thickness Direction. J. Japanese Soc. Non-Destructive Inspection 28,8:485-490

Recent Advances in Experimental Mechanics, Silva Gomes et al. (eds) © 1994 Balkema, Rotterdam, ISBN 90 5410 395 7

An experimental study of plastic zone adjustment by stress visualization techniques

K.Taniuchi

Meiji University, Kawasaki, Japan

ABSTRACT: The present report discusses the results of an experiment intended to test the validity of Irwin's plastic zone adjustment. The observation of yield stress patterns was used to make direct experimental measurement of the lengths of plastic zones induced by applying a tensile load to the notch tip of an arc-shaped specimen made from carbon steel. The experimental results show that the experimentally measured values approach the first approximate values more closely than Irwin's Plastic Zone Adjustment values do. It was further discovered that among other things this is due to the redistribution of singular stresses and ductility in the experimental strip. Irwin's Plastic Zone Adjustment should be construed as valid for the adjustment of plastic zone length only under certain conditions.

1 PREFACE

It is widely known that the application of Irwin's plastic zone adjustment to the length of a crack is an important topic in fracture mechanics,[1] but there have been very few instances of experimental investigation of this application. The formula for the loci of the points which satisfy the von Mises yield criterion at a notch tip has been given.[2] However there are indications that the theoretical calculated plastic zone does not coincide with the actual plastic zone configuration.[3] It is thought that one of the reasons why this sort of problem remains is that it is very difficult to make experimental measurements of the plastic zone at the tip of a crack.

By virtue of the yield stress stripes it is possible to make direct observation of the plastic zone with the naked eye.[4] This technique of continuously observing the phenomenon is immensely easy by comparison with etching. Furthermore, when a load is exerted on a notch tip it is possible to make a direct experimental measurement of the size of the plastic zone.[5]

Direct sequential observation was made of the elastic-plastic condition that was brought about at the zone of a notch tip by applying a tensile load to an arc-shaped carbon steel specimen. Measurement was made of the size and shape of the plastic zone that developed from the application of a tensile load. The experimentally obtained values were used to ascertain the validity of drawing the loci of the points which satisfy the von Mises yield criterion and the validity of Irwin's plastic zone adjustment.

2 EXPERIMENTAL PROCEDURE

Figure 1 displays the size of the arc-shaped test specimen[6] used in the investigation. The size of the notch 'a' is shown to be 15.0 mm. The magnified recording of Figure 2 illustrates the shape and dimensions of the zone of the notched tip of the test specimen.

Table 1 displays the chemical components of the test material. The common market variety S50C material was used. A tensile test of the material indicated that its yield stress was 350 MPa. After the test specimen was machine processed it was vacuum annealed. Subsequently the surface of the test specimen was finished off to a fine smoothness with No. 2000 emery paper so that the striped patterns to appear on the surface would be easily observable. The R max value of the surface was measured to be less than 0.1 μm.

Upon using an Instron-type material

testing machine to apply a tensile load to the test specimen, a video camera at a magnification of 15 was used to observe the condition of the yield stress stripes that developed in the zone of the notch tip as the tensile load increased. A crosshead speed of 0.1 mm/min. was chosen. On the stress-strain diagram this value is the straight line parallel to the abscissa and forms a limit of the recording of the yield point elongation.[7] A strain gauge-type extensometer was used to measure the displacement in the test specimen.

3 RESULTS AND DISCUSSION

3.1 The Shape of the Plastic Zone at the Notch Tip

Figure 3 illustrates the yield stress stripes that appear in the vicinity of a notch tip as the tensile load is increased. The recorded data of Figure 3

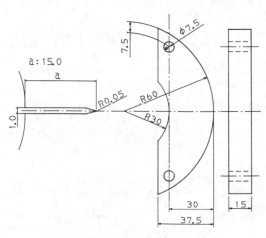

Fig.1 The shape and dimensions of test specimens

Table 1 The chemical composition of test specimens in percentage

C	Si	Mn	P	S
0.46	0.20	0.70	0.027	0.017

Fig.2 Detail of a notch tip

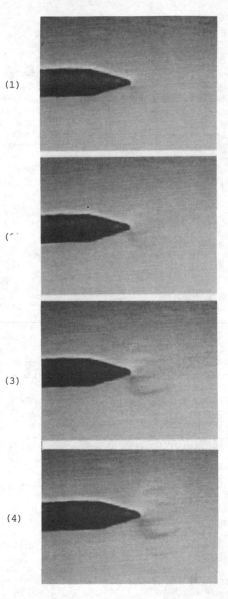

(1)

(2)

(3)

(4)

Fig.3 Propagation of the yield stress Patterns in the notch tip region

changed from (1) to (4) in direct proportion to the increase in tensile load. Inferring from the special properties of yield stress stripes,[4] the zone in Figure 3 on which yield stress stripes can be seen indicates that there is a plastic zone at the notch tip. The plastic zone is shaped like a silk-worm, and as the load is increased the zones grow larger and develop similar shapes.

The plastic zone in Figure 3 appears as a result of the material reaching its yield point. Accordingly it is possible to investigate the experimental results given in Figure 3 on the basis of the yield criterion.

The zone at the top of the crack fulfills the von Mises yield criterion and can be expressed in the following equations.[2]

$$r_y(\theta) = \frac{K_1^2}{4\pi\sigma_y^2} \cdot \left[1 + \cos\theta + \frac{3}{2} \cdot \sin^2\theta \right] \quad (1)$$

$$r_y(\theta) = \frac{K_1^2}{4\pi\sigma_y^2} \left[(1-2\nu)^2 (1+\cos\theta) + \frac{3}{2} \cdot \sin^2\theta \right] \quad (2)$$

The first equation expresses the plane stress condition, while the second equation expresses the plane strain condition.

In order to make a comparison of the experimentally measured shape of the plastic zone with the shape indicated by the loci of the points which satisfy the Von Mises yield criterion, Figure 4 was obtained by drawing the loci of the

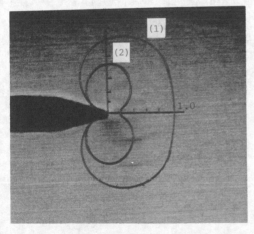

Fig.4 The collation of the experimental measurements of the plastic zone with the loci given by the formula satisfies the von Mises criterion

points of equations (1) and (2) for the plastic zone given in Figure 3. In Figure 4 notation (1) represents equation (1) and notation (2) represents equation (2).

In Figure 4 the shape of the experimentally measured plastic zone and the shape given by the loci derived from the von Mises criterion display an over-all resemblance. The experimentally measured value of the length of the plastic zone is ascertained to be the same as locus (1) of the theoretical values. The experimentally measured value of the height falls in the interval between loci (1) and (2) of the theoretical values.

The shape of the plastic zone in Figure 3 was ascertained to resemble the shape of the plastic zone as derived from etching.[8] It also became clear that the experimentally measured value of the length of the plastic zone was almost the same value as that given by the loci derived from the points satisfying the von Mises yield criterion. In order to accurately obtain the value for the length of the plastic zone is was undertaken to ascertain the relation between the length of the plastic zone and tensile load. To this purpose a diagram was drawn to record the relation between load and displacement.

3.2 The Load-Displacement Diagram

Figure 5 is the load-displacement diagram corresponding to the yield stress stripes in Figure 3. The numbers (1) - (4) that are annotated to the lines drawn in Figure 5 correspond respectively to the numbers (1) - (4) in Figure 3. By using both the data of the plastic zone in Figure 3 and the graph of the load and displacement proportions in Figure 5 it is possible to ascertain the lengths of the plastic zone corresponding to the tensile load.

3.3 The Length of the Plastic Zone

3.3.1 The experimental measurement of the length of the plastic zone

Figure 6 was obtained by collating from Figure 3 and 5 the experimental measurement values r_p corresponding to the tensile load P and arranging them on logarithm coordinates. The experimental measurements of Figure 6 are arranged in relation to the stress intensity factor. The stress intensity factor of the crack of the test specimen employed is given in the following equation:

$$K_1 = \left(\frac{P}{BW^{\frac{1}{2}}}\right)\left[\frac{3X}{W} + 1.9 + 1.1\frac{a}{W}\right]$$
$$\times\left[1 + 0.25\left(1 - \frac{a}{W}\right)^2\left(1 - \frac{r_1}{r_2}\right)\right]f\left(\frac{a}{W}\right) \quad (3)$$

Where P = Load, B = Specimen thickness, X = Loading Hole Offset, W = Specimen Width, a = Crack Length, and r_1/r_2 = Ratio of inner to outer radii.

Figure 7 was formed by correlating the experimentally measured values r_p with K_1 as given from Figure 6.

3.3.2 The equation for the length of the plastic zone[2]

The length r_y of the plastic zone at the tip of the crack is given in the following equation r_y by substituting in the monaxial yield stress value σ_y for the equation of the elastic stress distribution.

$$r_y = \frac{1}{2\pi}\cdot\left(\frac{K_1}{\sigma_y}\right)^2 \quad (4)$$

Equation (4) gives the value for the stress condition of the plane surface.

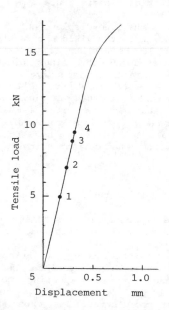

Fig.5 The propagation of yield stress patterns as shown by the load displacement curves

The value given in Equation (4) is called the first approximation. The elastic stress distribution formula σ for the tip of a crack is the yield stress value σ_y of Equation (4). Thus by computing the redistribution of the stress of the part extending beyond the elastic stress distributiobn of σ_y, the following equation results:

$$a_e = a + r_y \quad (5)$$

Irwin considered a_e to be the length of the imaginary crack. This value a_e is called the crack length of Irwin's plastic zone adjustment. By application of Irwin's plastic zone adjustment the length of the plastic zone r_p is made out to be:

$$r_p = 2r_y \quad (6)$$

The relation of the length of the plastic zone to the plane strain condition is expressed by the following equation:

$$r_y = \frac{1}{6\pi}\cdot\left(\frac{K_1}{\sigma_y}\right)^2 \quad (7)$$

3.3.3 The experimentally measured length of the plastic zone and the theoretical value

In order to compare the experimentally measured value r_p of the length of a plastic zone with the theoretical value, the values (b) of Equation (4) as well as those of Equation (7) were collected and recorded as the straight lines (a) and (b) respectively in Figure 7. Line (a) represents the plane strain condition and line (b) represents the plane stress condition. Line (c) is the value for the

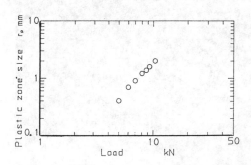

Fig.6 Plastic zone size measurements in relation to load

plane stress condition as obtained by applying Irwin's plastic zone adjustment to the length of the crack.

The experimentally measured values are all regularly located as points along the right side of line (b).

In Figure 7 the positions of the experimentally measured values are aligned along the first approximation line (b), but the results bring up misgivings about the consistency of Irwin's plastic zone adjustment. This is not a problem which can be ignored, so it is appropriate to reflect on the basic difference between the value of the first approximation and Irwin's plastic zone adjustment in order to discover the cause of that difference. If one emends the length according to Irwin's plastic zone adjustment by taking into consideration the stress redistribution for the length of the plastic zone that is brought about by yield stress, hypothetically the crack would lengthen, and the plastic zone according to Irwin's plastic zone adjustment would become twice as long as that of the first approximation. The fact that in Figure 7 the experimentally measured values come closer to the values of the first approximation than the values derived from the application of Irwin's plastic zone adjustment suggests that Irwin's plastic zone adjustment is inappropriate to the circumstances of the present experiment. It is not known whether the results show the stress redistribution. This problem appears to bear some relation to the condition of the yield point elongation in the stress-strain diagram.

3.3.4 The yield point elongation and Irwin's plastic zone adjustment

Figure 8 is a stress-strain diagram for a rectangular tension test specimen which has a thickness of 4 mm and is made of the chemical components stated in Table 1. In Figure 8 the sections a-b are parallel lines with the abscissa. The fact that yield stress σ_y is operating during the time that there is yield point elongation shows that the stress does not exert a higher value than σ_y.

It has been ascertained that when as a result of increasing tensile load there is yield point elongation or, in other words, when a plastic zone develops at the crack tip of a test specimen made of carbon steel or some other ductile material, a yield stress that is higher than σ_y is not operating in that section. This interpretation does not apply for a test specimen which is made of a material which does not undergo yield point elongation.

According to Irwin's plastic zone adjustment, the part within the elastic stress distribution which has a yield stress value less than σ_y forms a plastic zone, while the part which has a value above σ_y becomes the object of stress redistribution, and is supposed to exert stress at the tip of the emergent plastic zone. In the part which has become the object of stress redistribution the part which underwent a stress value of less than σ_y overlaps with the part which underwent a stress value greater than σ_y.

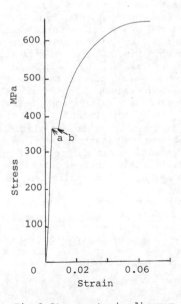

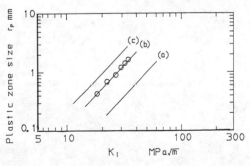

Fig.7 Plastic zone size measurements in relation to K_1

Fig.8 Stress-strain diagram

Obviously it would be irrational to think that the data for the sections a-b in the experimental results of Figure 8 can be similarly elucidated as resulting from the fact that the stress value above the yield stress overlaps in the part where there is yield point elongation. The experimental values of Figure 7 are closer to the first approximation when Irwin's plastic zone adjustment is not applied than when it is applied, showing that stress redistribution is an irrelevant factor.

It has been ascertained that the application of Irwin's plastic zone adjustment was inappropriate for the present experimental conditions.

The experimentally measured value of the length of the plastic zone induced at the notch tip was ascertained to be the same as the value indicated by equation (4) for dimensions of a plastic zone at the crack tip under yield conditions. Furthermore, if one combines the value of the plastic zone induced by tensile load to the value given by the equation (4) for the dimensions of a plastic zone at a crack tip, the aggregate result is the same as that obtained by Irwin's plastic zone adjustment.

4 CONCLUSIONS

A tensile test was applied to an arc-shaped carbon steel test specimen. Measurement by direct observation of yield stress stripes was made of the length of the plastic zone that developed as a result of applying a tensile load to the notch tip. The experimentally measured values of the length of the plastic zone were closer to the first approximation values than the values obtained from the application of Irwin's plastic zone adjustment. From these results it is clear that one should not interpret Irwin's plastic zone adjustment as a compensation for the redistribution of stress. One must instead understand it as compensating for the plastic zone at the tip of the crack.

It was confirmed that there is a partial difference between the shape of the plastic zone given by the loci of the points satisfying the von Mises yield criterion and the shape obtained from experimental measurement.

REFERENCES

1) SEM, Handbook on Experimental Mechanics, (1987), 899, Prentice Hall.
2) Kanninen, M.P. and Popelar, C.H., Advanced Francture Mechanics, (1985), 172-175, Oxford University Press, New York.
3) Broek, D., Elementary Engineering Francture Mechanics, (1986), 109, 110, Martinus Nijhoff Pub.
4) Taniuchi, K., Simple Stress Sensor; Utilizing of Stretcher Strains, ASTM STP 1025, (1989), 217.
5) Taniuchi, K., Visualization of Elastic-Plastic Zones, Proceedings of the 7th International Congress on Experimental Mechanics, (1992), 1005.
6) ASTM E399-90, Standard Test Method for Plane-Strain Fracture Toughness of Metallic Materials, ASTM, Philadelphia.
7) Taniuchi, K., Satoh, T., Results Observed on the Texture of Stretcher-Strains Low-Carbon Steel Plates, Research Reports Faculty of Engineering Meiji University, No.49 (1985), 15.
8) Hahn, G.T. and Rosenfield, A.R., Local Yielding and Extension of a Crack Under Plane Stress, Acta Metallurgica, 13, (1965), 293.

Recent Advances in Experimental Mechanics, Silva Gomes et al. (eds) © 1994 Balkema, Rotterdam, ISBN 90 5410 395 7

Repair of fatigue damage in pre-cracked specimen by laser surface treatment

J. Komotori & M. Shimizu
Department of Mechanical Engineering, Keio University, Yokohama, Japan

T. Nagashima
Production Department, SHOWA DENKO KK, Kawasaki, Japan

ABSTRACT Rotational-bending fatigue tests have been carried out on laser-treated and untreated specimen with special interest in the effect of laser surface treatment on the fatigue strength of pre-cracked specimens. Results show that the laser surface treatment has a beneficial effect on increasing the fatigue strength of the specimen with a crack shallower than the laser-hardening depth. This is because the progress of small surface crack can be suppressed due to the closure at the crack tip, which resulted mainly from the local residual compressive stress generated by laser surface treatment.

1. INTRODUCTION

Laser surface treatment is a process in which a metal surface is rapidly heated by concentrating the laser beam and then rapidly cooled, by the surrounding unaffected area providing self-quenching[1]~[8] and is one of the best methods of hardening the working surface of machine parts and tools which have considerably complex shapes. This is because there is no troublesome process involved, such as that in other types of heat treatment, and no necessity for large scale equipment. There is a possibility of the development of new technique for a prevention of fatigue fracture of structural members in service by making use of the laser surface treatment, since this method has many advantages in modifying the surface structure of the materials.

In this paper, a study has been made using an experimental approach to obtain a basic knowledge concerning the usefulness of laser treatment as a means of repairing the fatigue damage such as the propagation of surface microcracks, with special attention paid on the effect of laser surface treatment on the fatigue strength of specimens with a small pre-crack.

2. EXPERIMENTAL PROCEDURES

A medium carbon steel having carbon content of 0.36% was employed for this study. After full annealing the material at 1200 ℃ for 3 hours, the heat treatments shown in Fig.1 were performed and an hour-glass shape fatigue specimen with a minimum diameter of 8mm was machined. Every specimen was electro-polished and removed by an amount of about 20 μm as to eliminate the residual stress due to machining.

A number of pre-cracked specimens having crack lengths of 200, 500, 800, 1100, 1500 and 2000μm were prepared by fatigue loading at a stress level of 350 MPa to the specimen with artificially introduced micro-pit.

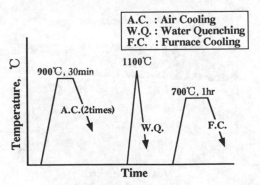

Fig. 1 Heat treatment process

Laser surface treatment was performed by focusing the beam of a 2.4 kw continuous wave YAG laser on a surface of the pre-cracked specimen. Fig.2 illustrates the set-up of laser surface treatment. Process parameters such as laser power, defocus and shutter speed (irradiation time of the beam) were adjusted to produce a martensitic microstructure at the highly localized area on specimen surface, without melting. Working parameters are given in Table 1.

Rotating-bending fatigue tests were carried out on these specimens with a frequency of 48Hz. Crack Opening Displacement (COD) of laser-treated and untreated specimen were examined by measuring the change of the distance of two micro-vickers indentation marks spanning the crack (schematic diagram in Fig.3) when subjecting the static bending stress. The measurements of residual stress were also made on laser-treated specimens in an axial direction by X-ray analysis.

3 RESULTS AND DISCUSSION

3.1 *Microstructure of laser treated region*

Fig.4 shows SEM photographs of the specimen surface and cross section at the laser treated region. It should be noted that the microstructure of the laser treated region has been significantly changed and a dark-etched zone can be observed. To clarify the micro-structural change in these areas, measurements of micro-vickers hardness were made on the specimen after laser surface treatment. Fig.5 (a) and (b) show such a result, both on the surface and cross section of laser-treated specimen, respectively. It can be seen from these figures that (i) the micro-vickers hardness of laser affected zone are about 3 times higher than that of the unaffected region, (ii) these hardened areas are about 2.8mm in diameter on surface and 450μm in depth, and correspond to a dark-etched area observed in Fig.4.

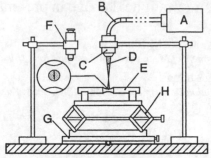

A : YAG Laser	E : Specimen
B : Optical Fiber	F : Microscope
C : Shutter	G : X-Y Stage
D : Laser Beam	H : Jack

Fig. 2 Laser surface treatment set-up

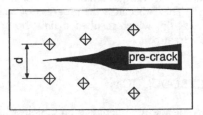

Fig. 3 Micro-vickers indentation marks used for the measurement of Crack Opening Displacement

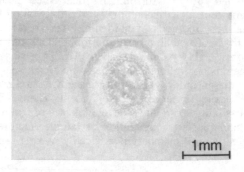

(a) surface

(b) cross section

Fig. 4 Macroscopic view of laser-treated area

Table 1 Laser processing parameters

Power (W)	320
Defocus (mm)	15
Shutter speed (s)	0.4

3.2 Effect of laser surface treatment on fatigue strength of pre-cracked specimen

The results of rotating-bending fatigue test are shown in Fig.6. In this figure, the hollow marks ○ and □ show the fatigue lives of smooth and pre-cracked specimens with a crack length of 200 μm, and the solid marks ■ show the result of laser-treated specimens. A remarkable reduction in fatigue strength is observed in untreated pre-cracked specimens (marks □ in Fig.6). However, the fatigue strength of laser-treated specimen is almost the same as that of smooth specimen. This implies that the laser surface treatment gives a remarkable effect on increasing the fatigue

strength of a specimen with a small crack and suggests that the progress of the small pre-crack can be suppressed by means of laser treatment.

To clarify the result mentioned above, the observation of fracture surface was made using a Scanning Electron Microscope (SEM) on laser-treated specimen having small surface crack length of 200μm. Fig.7 shows the SEM photograph of fracture surface near laser-treated zone. It should be noted that the fracture surface is formed along the edge of laser-hardened area. This means that the small pre-crack remains as a nonpropagating crack and that the final fracture of the specimen is controlled by the crack generating from untreated surface.

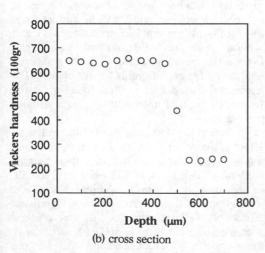

(a) surface

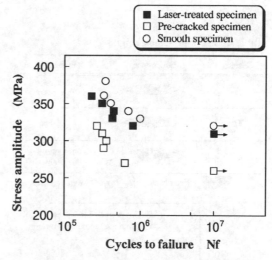

Fig. 6 Effect of laser surface treatment on fatigue life properties

(b) cross section

Fig. 5 Hardness distribution of laser-treated area

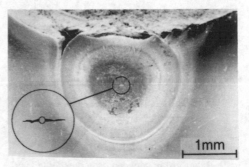

Fig. 7 Typical feature of a fracture surface and laser-treated area (A series specimen)

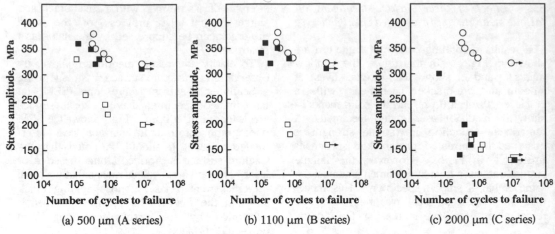

(a) 500 μm (A series) (b) 1100 μm (B series) (c) 2000 μm (C series)

Fig. 8 Results of fatigue test for the specimen having 500, 1100 and 2000 μm pre-crack length
(■ : laser-treated specimen, □○ : untreated specimen)

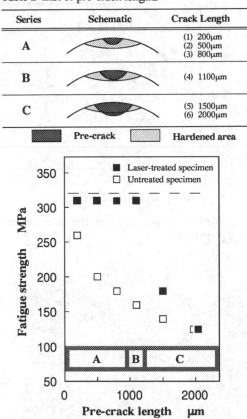

Table 2 List of pre-crack lengths

Series	Schematic	Crack Length
A		(1) 200μm (2) 500μm (3) 800μm
B		(4) 1100μm
C		(5) 1500μm (6) 2000μm

■ Pre-crack   ▒ Hardened area

Fig. 9 Relationship between pre-crack length and
fatigue strength for laser-treated and untreated
specimen.

3.3 Effect of pre-crack length on fatigue strength of laser-treated specimen

To discuss the practical application of laser surface treatment, additional fatigue tests were carried out before and after laser surface treatment for specimens having a wide range of crack lengths.   In Table 2, a list of pre-crack lengths tested in this study is given, with a schematic illustration that represents the relationship between pre-crack depth and laser-hardened depth.

Some examples of fatigue tests are shown in Fig.8 (a)-(c), where marks ○ show the results of smooth specimen and marks ■ and □ show the results of pre-cracked specimen with and without laser surface treatment, respectively. Fig.9 gives the relationship between pre-crack length and fatigue strength for 10^7 cycles. The horizontal dotted line in this figure gives the fatigue strength of untreated specimen without pre-cracks.

It is found that laser surface treatment has an effect of increasing the fatigue strength when the pre-crack depth is shallower than surface hardening depth (A and B series in Table 2). In the case of C series specimen, however, irradiation of laser beam has no significant effect on fatigue strength of pre-cracked specimen (see Fig.8 (c) and Fig.9).   To clarify these results, the observation of fracture surface was made for the specimen having pre-crack length of 2000μm.

Fig.10 shows the typical example of fracture surface near the laser- treated zone in C series specimen. It is clear that the pre-crack introduced before laser surface treatment continuously propagates and leads to the final fracture of the specimen.

Consequently, it can be concluded that small pre-crack, shorter than 1100μm i.e. shallower than surface hardening depth, would not propagate beyond a laser hardened zone. In such a situation, the fatigue strength of specimens with small pre-crack can be increased. In the case of C series specimen, however, the improvement of fatigue strength can not be achieved by laser surface treatment. This is caused by the fact that the propagation of pre-crack occurred beyond the laser hardened zone and it leads to the final fracture of the specimen.

3.4 Mechanism of suppressing the surface crack propagation

To discuss the mechanism of suppressing a small crack propagation, effect of laser surface treatment on the behavior of Crack Opening Displacement (COD) was examined. Fig.11 (a) and (b) show the results, where the value of COD versus distance from crack tip (see schematic illustration in Fig.3) has been plotted. In this figure, solid and hollow marks show the results of laser-treated and untreated specimen, respectively. It is clear that the crack closure occurred after laser surface treatment for A series specimen in which the improvement of fatigue strength had been observed. In C series specimen, however, the

crack tip opening occurs even under stress below the fatigue strength of the smooth specimen.

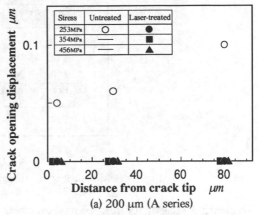

Stress	Untreated	Laser-treated
253MPa	O	●
354MPa	—	■
456MPa	—	▲

(a) 200 μm (A series)

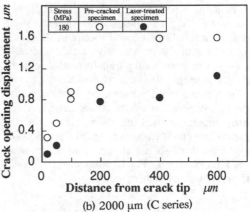

Stress (MPa)	Pre-cracked specimen	Laser-treated specimen
180	O	●

(b) 2000 μm (C series)

Fig. 11 COD for pre-cracked specimen, before and after laser treatment

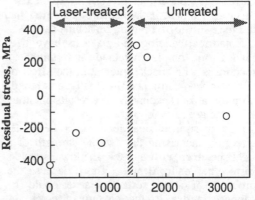

Fig. 12 Distribution of residual stresses

Fig. 10 Typical feature of a fracture surface and laser-treated area (B series specimen)

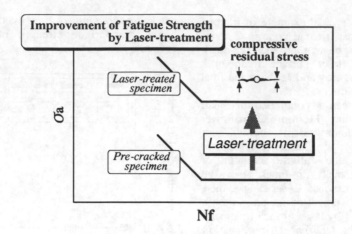

Fig. 13 Schematic diagram illustrating the results of present study

To clarify the reason for this, the measurement of residual stress was made for laser-treated specimen. Fig.12 shows the residual stress distribution in laser-treated specimen. It should be noted that the compressive residual stresses due to the martensitic transformation can be observed within the laser treated area.

From these results it can be concluded that the compressive residual stress generated by laser treatment has a good effect on the improvement of fatigue strength of pre-cracked specimen through the suppression of the small crack propagation. Fig.13 illustrates the results of present study, schematically.

4. CONCLUSION

An attempt was made to obtain a basic knowledge concerning the usefulness of laser surface treatment as a means of repairing the fatigue damage of a running machine part.

Rotational-bending fatigue tests have been carried out on laser-treated and untreated specimens with special interest in the effect of laser surface treatment on the fatigue strength of pre-cracked specimens. The results obtained are summarized as follows:

(1) Laser surface treatment has a beneficial effect on increasing the fatigue strength of a steel specimen with a crack shallower than its surface hardening depth. This is because the progress of small surface cracks could be suppressed due to the closure of crack tip, which resulted mainly from the local residual compressive stress formed by laser surface treatment.

(2) In the case of a specimen having relatively large surface crack, the improvement of fatigue strength could not be observed. In this case, the crack tip opening occurs under stress less than the fatigue strength of the smooth specimen.

ACKNOWLEDGEMENT

The authors wish to thank TOSHIBA Co., Ltd. for cooperation in laser surface treatment.

REFERENCES

(1) M.F. Ashby and K.E. Easterling 1984. The Transformation hardening of steel surface by laser beams-I. hypo-eutectoid steels: *Acta Metall. Vol. 32, No. 11*: 1935-1948.
(2) Akira Kato 1985. Prevention of fracture of cracked steel bars using laser: part I - laser hardening: *Trans. ASME Vol. 107*: 195-199.
(3) P.A. Molian 1987. Fatigue characteristics of laser surface-hardened cast irons: Trans. ASME Vol. 109: 179-187.
(4) J.C. Ion, T. Moisio, T.F. Pedersen, B. Sorensen and C.M. Hansson 1991. Laser surface modification of a 13.5% Cr, 0.6% C Steel: *J. of Mat. Sci. 26*: 43-48.
(5) D.M. Gureyev, A.V. Zolotarevsky, A.E. Zaikin 1991. Laser-arc hardening of aluminium alloys: *J. of Mat. Sci. 26*: 4678-4682.
(6) D.M. Gureyev, V.K. Shuchostanov, S.V. Yamtchikov 1991. Effect of preheat-treatment during laser hardening on residual stress formation in high-chromium irons: *J.of Mat. Sci. 26*: 6023-6026.
(7) W. Kurz and R. Trivedi 1992. Microstructure and phase selection in laser treatment of materials: *Trans. ASME Vol. 114*: 450-458.
(8) P.A. Molian 1987. Fatigue characteristics of laser surface-hardened cast irons: *Trans. ASME Vol. 109*: 179-187.

Recent Advances in Experimental Mechanics, Silva Gomes et al. (eds) © 1994 Balkema, Rotterdam, ISBN 90 5410 395 7

Relation between the scatter in growth rate of microcracks and small-crack growth law

M. Goto & H. Miyagawa
Faculty of Engineering, Oita University, Japan

N. Kawagoishi
Faculty of Engineering, Kagoshima University, Japan

H. Nisitani
Faculty of Engineering, Kyushu University, Fukuoka, Japan

ABSTRACT: In order to study the scatter characteristics of growth rate of microcracks, rotating bending fatigue tests of smooth specimens have been carried out using a normalized 0.21 % carbon steel. The growth data were analysed by assuming the Weibull distribution, and a statistical investigation of the physical basis of scatter in growth rate were performed. Furthermore, the relation between the behavior of microstructurally small cracks and the small-crack growth law, $dl/dN = C_1\sigma_a^n l$, which is effective for determining the growth rate of mechanically small cracks in many ductile materials, was investigated.

1 INTRODUCTION

In the design of machines and structures which must operate at stresses above the fatigue limit, the evaluation of fatigue life is crucial to the safe design of members. The growth law of a small crack must be known in order to estimate the fatigue life, because the fatigue life of members is controlled mainly by the propagation life of a small crack. A small crack whose growth behavior is controlled mainly by the mechanical factor are called a mechanically small crack. Many growth laws for such a small crack have been reported and they have succeeded in describing crack growth behaviors over the limited experimental programs.

On the other hand, it is well known that the fatigue data generally exhibit large scatter. Accordingly, to grasp the scatter of fatigue life is an important task for the safe design of members. The behavior of a microstructurally small crack, a small crack with dimensions of the order of the microstructure, must be studied when examining the physical basis of scatter in fatigue life, because the growth behavior of microstructurally small crack is strongly affected by the inhomogeneity of the microstructure and exhibits anomalously large scatter when compared to the mechanically small cracks (Lankford 1977). Thus, the quantitative evaluation of the scatter in growth rate of a microstructurally small crack is

crucial when we evaluate the scatter in fatigue life of members.

In the present study, using a 0.21 % carbon steel, rotating bending fatigue tests of smooth specimens have been carried out. 15 specimens were fatigued at each stress amplitude and the behavior of microstructurally small crack which led to the fracture of a specimen was examined in all the specimens. The growth data were analysed by assuming the Weibull distribution, and a statistical investigation of the physical basis of scatter in growth rate were performed. Furthermore, the relation between the behavior of microstructurally small cracks and the small-crack growth law, $dl/dN = C_1\sigma_a^n l$ (Nisitani & Goto 1986), which is effective for determining the growth rate of mechanically small cracks in many ductile metals, was studied.

2 EXPERIMENTAL PROCEDURES

The material was a 0.21 % carbon steel in the form of rolled bar about 22 mm in diameter. The specimens were machined from the bars after normalizing for 60 min at 885°C. The chemical composition (wt%) was 0.21 C, 0.21 Si, 0.47 Mn, 0.02 S, 0.21 Cu, 0.06 Ni, 0.09 Cr, remainder ferrite. The mechanical properties are 324 MPa lower yield stress, 498 MPa ultimate tensile strength, 978 MPa true breaking stress and 63.2 % reduction of area.

The shape and dimensions of the specimen are shown in Fig.1. The strength reduction factor of this geometry is close to unity, so that the specimen can be considered as a plain specimen. Before testing, all the specimens were reannealed in vacuum at 600 °C for 60 min to remove residual stresses, and were then electropolished to remove about 25 μm from the surface layer in order to make the changes in surface state easier to observe.

All the tests were carried out using a rotatory bending fatigue machine with a capacity of 14.7 Nm operating at 3000 rev/min. The observations of fatigue damage on the specimen surface and the measurement of crack length were made via plastic replicas using an optical microscope at a magnification of x400. The crack length, l, is the length along the circumferential direction on the specimen surface. The stress value referred to is that of the nominal stress amplitude, σ_a, at the minimum cross section.

3 EXPERIMENTAL RESULTS AND DISCUSSION

In the present study, fatigue tests were carried out at two stress ranges σ_a=260 MPa(=1.06σ_w) and 300 MPa (=1.22σ_w), where σ_w is the fatigue limit, in order to study the probabilistic properties of crack growth rate. That is, a total of 15 specimens were fatigued at each stress range and the behavior of the major crack was examined in all the specimens.

Figure 2 shows the $S-N$ curve. It is found that the number of cycles to failure, N_f, exhibits large scatter for each constant stress range, and this scatter tends to increase and be more significant with a decrease in stress range.

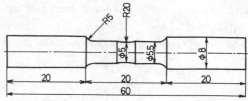

Fig.1 Shape of the specimen.

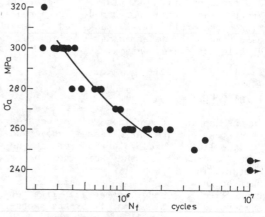

Fig.2 $S-N$ curve.

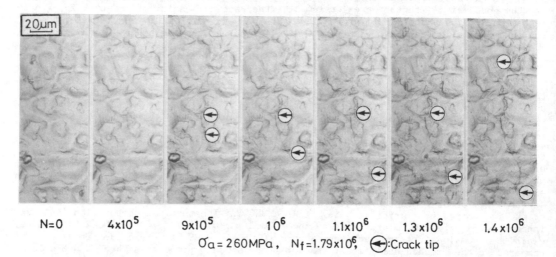

N=0   4x10^5 9x10^5 10^6 1.1x10^6 1.3x10^6 1.4x10^6

σ_a = 260 MPa , N$_f$=1.79x10^6, ⊖:Crack tip

Fig.3 Change in the surface topography around a major crack which led to the final fracture of the specimen.

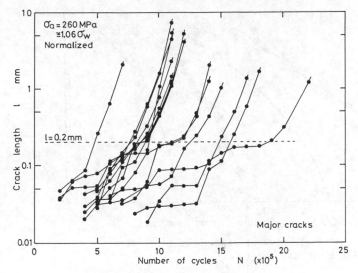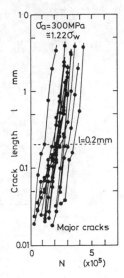

Fig.4 Crack growth curves of major cracks

3.1 Scatter in crack growth behavior

Figure 3 shows the change in surface state around a major crack. Until the initiation of a crack, fatigue damage is accumulated gradually within the same region whose dimensions is closely related to the grain size (about 25 μm in this material). At a later stage a crack is detected in this region (the number of cycles to crack initiation in this case is about 9×10^5 cycles). Figure shows an example of small crack propagation behavior with temporary crack arrest at the ferrite / pearlite grain boundaries. That is, after initiation, the crack grew in an uninterrupted fashion up to 1.1×10^6 cycles. However, negligible propagation is observed from 1.1×10^6 to 1.3×10^6 cycles. At 1.4×10^6 cycles, crack propagation is observed again. This irregular propagation behavior is caused by the inhomogeneity of the microstructure, and has a probabilistic nature. However, the effect of such inhomogeneity on crack propagation declines gradually as the crack grows, and its effect can be disregarded once the crack length exceeds about 0.2 mm. Small cracks with dimensions on the order of the microstructure, e.g. the grain size, are called microstructurally small cracks. It has been shown that these cracks exhibit anomalously high, irregular growth rates when compared to large cracks at the same range of the stress intensity factor.

Figure 4 shows the growth data of major cracks. The scatter for the cracks smaller than 0.2 mm (about 8 times the ferrite

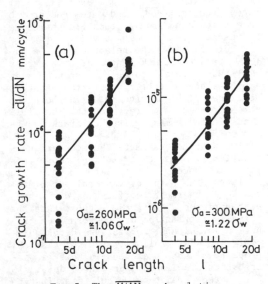

Fig.5 The $\overline{dl/dN}$ vs l relation.

grain size (Goto 1993)) is remarkably large, whereas that for cracks larger than 0.2 mm is not large; certainly in terms of growth rate.

3.2 Statistical investigation of small crack growth rate

It is important to evaluate the scatter in crack growth rate when estimating crack growth life and fatigue life. In what follows, the scatter characteristics of

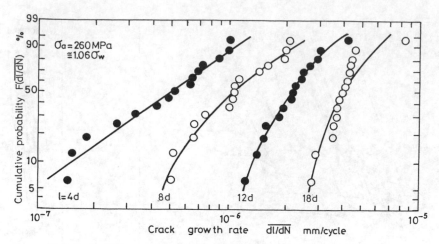

Fig.6　Crack growth rate at each crack length plotted on the Weibull probability paper.

crack growth rate, $\overline{dl/dN}$, are examined. Here, the term $\overline{dl/dN}$ indicates the growth rate calculated from a crack growth curve represented by a smooth curve.

Figure 5 shows the relation between the $\overline{dl/dN}$ and the crack length at each stress amplitude. In all cases, the growth rates exhibit large scatter, but the scatter decreases with an increase in both the stress range and crack length. The full lines indicate mean values of crack growth rate.

If we want to estimate the scatter quantitatively, the distribution properties for each specimen and lifetime must be analyzed. Many studies concerning the statistical treatment of fatigue data have been performed, and these suggest that the Weibull distribution or log-normal distribution is suitable for fatigue data analysis. In what follows, the statistical properties are examined via Weibull distribution studies of each set of data, with the three-parameter Weibull distribution function $F(x)$ expressed by the following equation:

$$F(x) = 1 - \exp\left\{-\left(\frac{x-\gamma}{\eta}\right)^m\right\} \qquad (1)$$

Here the three constants m, η and γ are the shape parameter, scale parameter and location parameter, respectively. Determination of the three Weibull parameters is made by the correlation factor method, i.e. the correlation coefficient is the maximum value determined by a linear regression technique.

Figure 6 shows the distribution of crack growth rates plotted on the Weibull dis-

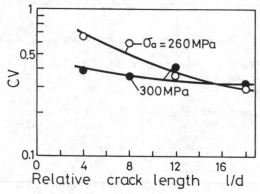

Fig.7　The CV vs l/d relation for crack growth rate distribution.

tribution paper. The crack growth rate is expressed by the Weibull distribution.

Figure 7 shows the CV, coefficient of variance, versus crack length relation for growth rate distributions. The values of CV decrease with an increase in the crack length and it tends to saturate to the value close to 0.3. It is found that the values of CV for cracks smaller than 0.2 mm at stresses close to the fatigue limit are anomalously high. Using Fig.5, we can evaluate the mean value of $\overline{dl/dN}$ at each crack length and stress range. For example, the mean growth rate in which the CV plot exhibits an anomalously high value is nearly equivalent to the threshold level, i.e. $\overline{dl/dN} \approx 10^{-7}$ mm/cycle. When the growth rate sufficiently exceeds 10^{-6} mm/cycle, the influence of stress range and crack

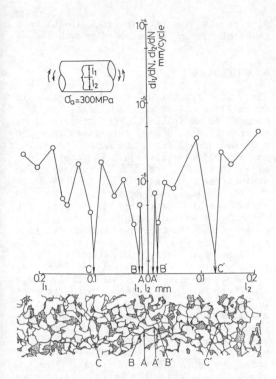

Fig.8 Fluctuation in crack growth rate at grain boundaries and ferrite/pearlite boundaries.

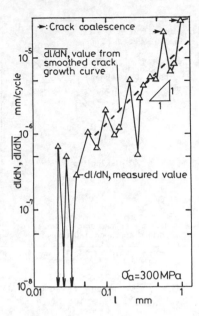

Fig.9 The dl/dN, $\overline{dl/dN}$ vs l relation.

length on the CV vs l relation is small. In the previous crack growth curves (Fig.4), it was shown that the scatter in growth curves can be disregarded once the crack length exceeds about 0.2 mm, even when the stress range is closed to the fatigue limit. The respective mean growth rates for 0.2 mm cracks in such a case exceed 10^{-6} mm/cycle enough. Thus, the crack length associated to a sharp decrease in scatter in the crack growth curve is equivalent to the crack length which correlates to a growth rate in excess to 10^{-6} mm/cycle.

3.3 Fluctuation of crack growth rate and small crack growth law

Figure 8 shows the growth rate of a microstructurally small crack calculated from the raw crack growth data vs crack length relation. Trace of a crack and microstructure is also shown to appreciate the crack propagation path and the position of crack arrest as indicated with characters labelled on the figure. It is found that the crack arrest takes place at the grain boundaries and ferrite/pearlite boundaries. Namely, the fluctuation of crack growth rate is caused by the inhomogeneity of the microstructure.

Figure 9 shows the crack growth rates, dl/dN and $\overline{dl/dN}$, vs l relation including the results of crack length in excess of 0.4 mm. The term dl/dN is the growth rate calculated directly from raw crack growth data and $\overline{dl/dN}$ is from the smoothed crack growth curve. Figure shows the great large fluctuation in dl/dN. On the other hand, $\overline{dl/dN}$ represented by a dotted line is nearly proportional to the crack length when $\overline{dl/dN}$ exceeds 10^{-6} mm/cycle. Moreover, it was shown that the $\overline{dl/dN}$ vs l relation nearly corresponds to the relation extrapolated from a mechanically small crack in a drilled specimen.

Nisitani and Goto have studied the small crack growth behaviour using various kinds of ductile metals and indicated that the growth rate of a small crack in which the condition of small scale yielding does not hold can be uniquely determined by a term $\sigma_a^n l$ not by the stress intensity factor range ΔK. Furthermore, they have proposed a convenient method for predicting the fatigue life based on the small crack growth law, equation (2), in which the effect of material properties is partly considered. The validity of a method have been confirmed by its application to the other researcher's fatigue data (Nisitani, Goto & Kawagoishi 1992).

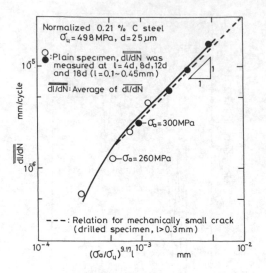

Fig.10 The $\overline{\overline{dl/dN}}$, mean value of $\overline{dl/dN}$, vs $(\sigma_a/\sigma_u)^n l$ relation.

$$dl/dN = C_3\,(\sigma_a/\sigma_u)^n l \qquad (2)$$

Where σ_u is the ultimate tensile strength. Equation (2) holds for a crack whose growth rate exceeds 10^{-6} mm/cycle enough. Equation (2) is experimentally derived from the small crack propagation behaviour which is controlled mainly by the mechanical factor and exhibits little fluctuation in growth rates.

Figure 10 shows the $\overline{\overline{dl/dN}}$, average of $\overline{dl/dN}$ (see Fig.5), vs $(\sigma_a/\sigma_u)^n l$ relation. Here, the constant **n** was inferred by the following equation which holds for the normalized steels;

$$\mathbf{n} = -7.9\times10^{-3}\,\sigma_u + 13.1 \qquad (3)$$

Although there are some scatter in the plots, it is found that the equation (2) is valid when $\overline{\overline{dl/dN}}$ exceeds 10^{-6} mm/cycle enough. A dotted line shows the relation obtained from the mechanically small cracks in plain specimens with a small blind hole. The $\overline{\overline{dl/dN}}$ vs $(\sigma_a/\sigma_u)^n l$ relation nearly overlappes to the relation for mechanically small cracks when $\overline{\overline{dl/dN}}$ exceeds 10^{-6} mm/cycle.

Consequently, the propagation life of microstructurally small cracks can be evaluated approximately based on $\overline{\overline{dl/dN}}$ and the small crack growth law, without using the actual growth rate with remarkably large fluctuation due to microstructural inhomogeneity. Thus, it may be important to clarify the scatter in crack growth rate calculated from the smoothed growth curve.

4. CONCLUSIONS

The main results of statistical analysis on small crack propagation rate may be summarized as follows : The distributions of growth rate, $\overline{dl/dN}$, calculated from the smoothed growth curve can be expressed by the Weibull distribution. The coefficient of variance, CV, for $\overline{dl/dN}$ decreases with an increase in l when $\overline{dl/dN}$ is close to the threshold level. However, for $\overline{dl/dN}$ in excess of 10^{-6} mm/cycle, no significant effect of stress range on the CV is observed and the CV tends to saturate to a value close to 0.3 as the crack length increases. On the other hand, the crack growth rate, dl/dN , calculated from raw crack growth data exhibits extremely large fluctuation, however, the fluctuation of $\overline{dl/dN}$ is not large when compared to dl/dN and $\overline{dl/dN}$ can be determined by the small crack growth law when $\overline{dl/dN}$ exceeds 10^{-6} mm/cycle. Thus, the growth life of microstructurally small cracks may be evaluated based on $\overline{dl/dN}$ and small crack growth law without considering the actual growth rate.

REFERENCES

Goto, M. 1993. Fatigue Fract. Eng. Mater. Struct. 15:953–963.
Lankford J. 1977. Eng. Fract. Mech. 9:617–624.
Nisitani, H. & Goto, M. 1986.The behaviour of short fatigue Cracks. EGF-1:461–478.
Nisitani,H., Goto,M. & Kawagoishi,N. 1992. Eng. Fract. Mech. 41:499–513.

Recent Advances in Experimental Mechanics, Silva Gomes et al. (eds) © 1994 Balkema, Rotterdam, ISBN 90 5410 395 7

Using electrical resistance strain gages as fatigue damage sensors

J.L.F.Freire & C.A.C.Arêas
Catholic University of Rio de Janeiro, Brazil

ABSTRACT: This paper reports experiments where adhesive mounted strain gages were used to indicate fatigue damage in the surface of double cantilever test specimens made of strucutal low carbon steel.

1 INTRODUCTION

Initial fatigue damage can occur very early in the life of a component. Nisitani and Takao (1981) showed that damage (microcracks) can occur as early as 5 to 10% of the expected total fatigue life of rotating beam specimens.

Surface fatigue damage can be monitored in polished specimens by direct observation using microscopy. The initial damage is related to the observation of localized persistent slip-bands and to the generation of microcracks. These cracks are smaller than or of the order of the grain size, depending on the material being studied.

Due to its superficial nature, initial fatigue damage may influence the response of electrical resistance strain gages. Strain gages can monitor the fatigue damage in two ways. They can suffer fatigue by damaging their grids due to strain cycling. In this case, they will indicate how severely was the component surface loaded. They can also be affected by the damage in the component surface caused by the fatigue loading. In this case, the strain gage response will be influenced by slip-band and microcracks.

The present paper reports results of an investigation where strain gages were used in an attempt to indicate surface fatigue damage in low carbon steel specimens.

2 EXPERIMENTAL METHODS

Low carbon steel specimens were used in the fatigue tests. The steel composition and its mechanical properties are given in Table 1.

Table 1. The steel composition and its engineering mechanical properties.

Element	C	Mn	P	S	Si
Percentage	0.20	1.55	0.03	0.07	0.36

Yield strengh (MPa) : 380
Ultimate strength (MPa) : 560
Elongation (%) : 17

The test specimens and the alternating load device were specially designed for the fatigue tests. The specifications to be satisfied were the following :

1. The specimens should be easily mounted.

2. The alternating strains induced at the surface regions prone to develop fatigue damage should be monitored by electrical resistance strain gages.

3. The test frequency should be high enough so that the total fatigue test time could be small.

The double cantilever specimen geometry used in the tests is shown in Figure 1. The tests were carried out in a shaker which excited the specimens in a quasi-ressonant condition. The test frequency was around 400 Hz. This vibrating load apparatus induced fatigue damage in the free surfaces of the thinner sections. These fatigue prone regions are indicated by letters (A to D) in Figure 1. Regions A to D had a milling machine type of finishing except for some tests where a metallographic polishing type of finishing was used.

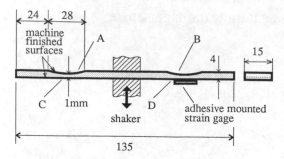

Figure 1. Specimen geometry and fatigue prone surfaces.

Strain gages were mounted in regions C and D with cyanoacrylate cement. Table 2 gives the principal properties of the strain gages used in the tests. Both strain gage types had pre-attached leads to avoid the risks that in-place soldering could bring to the fatigue life of the installations.

The strain gage response was monitored by an oscilloscope and used to controll the strain ranges submitted by the shaker to the specimens. The ranges applied in the tests were equal to +/- 1500 and +/- 1800 $\mu\epsilon$ at the strain gage sites.

While cycling the specimens, the fatigue damage was monitored through the strain gage response. Damage was indicated by a zero-shift in the strain gage response which increased to the point that it became unstable and impossible to control.

Surface damage was observed with a standard optical microscope and with a scanning electron microscope (SEM).

3 RESULTS AND DISCUSSION

After a certain number of cycles the strain gage response started to indicate a positive zero-shift which increased with test time. No change in the natural frequency of the specimens was observed. This indicates that large cracks were not present. Tests results are presented in Table 3. The initial zero-shift of both strain gage types started at lives much lower than those expected for the specimens to endure or present macrocracks. Test lives of 10^5 and 10^4, respectively for strain ranges of +/- 1500 and 1800 $\mu\epsilon$, are cited in the literature as capable of causing initial fatigue damage to the strain gage installations for some strain gage types (Measurements Group).

In order to identify the cause of the zero shift behavior in the present tests, i.e., if it appears

Table 2. Properties of the strain gages used in the tests.

Type I: Kratos (Brazil),   5 mm grid, steel compensation, 120 Ω, epoxy backing, constantan foil

Type II: Kyowa (Japan),   5 mm grid, steel compensation, 120 Ω, phester backing, constantan foil

Table 3. Number of cycles to generate a zero-shift of 500 $\mu\epsilon$ in the strain gages mounted in the milled machine surfaces.

Strain range ($\mu\epsilon$)	Strain gage type	Number of cycles (1000)	Cycles to fracture (1000)
± 1500	I	170 ± 100	650 ± 100
± 1500	II	413 ± 100	650 ± 100
± 1800	I	18 ± 9	140 ± 30
± 1800	II	120 ± 20	140 ± 30

due to fatigue damage in the specimen surface or in the strain gage installation, a number of investigations were develloped and their results will be described in the next paragraphs.

3.1 Low magnification inspection

Visual inspection of the fatigue prone surfaces and of the strain gage installations were made using a stereo-microscope with 16 and 40 times magnification options. There were not evidences of cracking in the strain gage grids, solder dots, lead-wire connections, and in the specimens free surfaces. It should be pointed out that these observation results are valid for the polished and unpolished specimens. Cracks became evident only when the response of the strain gages were lost or completely out of control.

3.2 Static tests

Static tests were carried out with some specimens after they had presented the damaged behavior. These tests consisted in bending statically the double cantilever specimens with increasing loads and in measuring the strains indicated by the gages. The gages showed linear and expected responses under compression but abnormal behavior when the tension strains reached 80% of the alternate test strain amplitude.

3.3 Dynamic tests with new sensors

After presenting the zero-shift behavior, the strain gages were substituted by new ones and the specimens were again submitted to the same dynamic loading. It was verified that the strain response was initially stable. After one thousand cycles the zero-shift started again. This type of check was repeated for specimens that did not suffered the static test and gave the same results. Two specimens had their four test regions (A-D) polished and clean as for metallographic inspection. One strain gage was installed in each specimen at section C. After the first unbalance new strain gages were installed in sections A and B whithout the need for surface preparations and cleaning. Immediate unbalance was verified in all these gages when the tests were resumed.

The results presented above are good indicatives that the fatigue damage is in the surface of the specimens and not in the gages. The one thousand cycles delay to start presenting damage in the case of the unpolished specimens is credited to the surface preparation to install the new gages in substitution to the old ones.

3.4 Strain gage types

Table 2 presents the principal characteristics of the types of strain gages used in this investigation. The principal difference is that the gage type I has a epoxy backing while type II has a phester backing. The test results showed that the gages type I were more sensitive to the surface damage than the gages type II. Gages type I showed a zero-shift of 500 $\mu\epsilon$ as soon as the specimens reached 10 to 20% of their total fracture fatigue life. Gages type II started showing a zero shift of 100 $\mu\epsilon$ only after 50% of the total life. The zero-shift of 500 $\mu\epsilon$ corresponded to 80% of total life for these gages.

3.5 Endurance gage tests

Both strain gage types were submitted to fatigue tests using specimens made of a higher fatigue strength steel. Both gage types endured more than 10^6 cycles without presenting any zero-shift behavior. The alternate test strain range was +/- 1500 $\mu\epsilon$. In these tests, fatigue cracks started in sites far from the strain gage locations and propagated transversally to them, cutting the grid abruptly.

3.6 Polished specimens

The total set of results presented above clearly indicate that the specimens made of low strength steel develop surface fatigue damage which influence the strain gage behavior. It was suspected that very small cracks were generated at the surfaces under the strain gage backings. These cracks open while under tension and close under compression. Plasticity at their roots and plastic strain accumulation in persistent slip-bands create and increase the zero-shift when they are located immediately under the gage grid. Small cracks can also propagate to the botton side of the grid generating changes in the gage resistance.

Specimens with polished surfaces were examined in optical and scanning microscopes before and after presenting strain gage zero-shift due to alternate cycling. After 10^4 cycles and before presenting zero-shift, the specimens revealed sparse and very ligth ondulations on their more strained surfaces. After 10^4 cycles more these specimens showed some zero-shift. Microscope observations of surfaces corresponding to zero-shifts of 100 to 200$\mu\epsilon$ revealed the presence of highly strained regions, persistent slip-bands and small cracks, 2 to 10 μm long. Longer cracks and more damage on the surfaces appeared in the specimens with more cycles. Surface damage increased to the point that the zero-shift became completely out of scale and some small visible cracks started to appear.

Figure 2 shows a SEM photograph of an array of cracks in the damaged region after the specimen had been cycled for 3 x 10^4 and the strain gage was presenting a zero-shift of 100 $\mu\epsilon$.

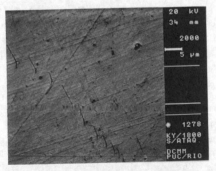

Figure 2. SEM photograph of cracks in the damaged region (zero-shift of 100 $\mu\epsilon$, 3 x 10^4 cycles, $\pm1800\mu\epsilon$).

Figure 3. Surface damage due to fatigue cycling after loosing the response of the strain gage (500 x).

Figure 3 shows a photograph of the same specimen when the zero-shift was out of control and much larger than 1000 $\mu\epsilon$. The larger crack in the figure is aproximatelly 25 μm long and its surface crack opening displacement, COD, is in the order of 1 μm.

Nisitani and Takao discussed some aspects of the initiation of microcracks in ductile materials. They showed that the COD of propagating and non-propagating fatigue cracks differ of one order of magnitude for crack lengths of 100μm. The surface COD of a propagating crack of that size which was subjected to an alternating strain of +/- 1000μm after 2.7 x 10^5 cycles was equal to 0.4 μm.

Fatigue damage in the surface under a strain gage can cause a positive zero-shift in two ways. Small cracks underneath the gages can propagate to the gage and penetrate its backing and a small portion of the thickness of its grid , increasing its resistance. If it penetrates too deep in the foil, or if many small cracks penetrate a little, the gage resistance can reach numbers as high as 180 ohms, as it was measured for one of the gages. Another possibility is that the strain concentration due to plasticity in the slip-bands or a permanent COD of 1 μm can cause a zero-shift strain of 100 $\mu\epsilon$ in a strip of the gage grid sizing 10 mm.

Figure 4 shows SEM photographs of an etched specimen which lost its strain response and showed visible cracks. Figure 4a shows a region where high strains occurred causing heavy deformed slip-bands which degenerated in small microcracks. Figure 4b shows a 100 μm long crack. The neigborhood shows a highly strained slip-bands that did not degenerated in microcracks. It seems that the longer crack, once developed, unloaded its neighborhood and did not allow the slip-bands to open in new small cracks.

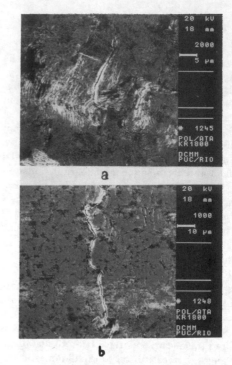

Figure 4. SEM photographs of an etched specimen which lost its strain gage response ($\pm$1800$\mu\epsilon$).

4 CONCLUSIONS

It can be concluded that the electrical resistance strain gages used in the presented experimental sensed the progressive surface fatigue damage generated in the low carbon steel specimens by the alternating strain loading. The fatigue damage was observed to be related to persistent slip-bands and to the initiation of microcracks in the specimen's surfaces. The surface damage induced positive and progressive zero-shift in the strain gage response up to the point that it became uncontrollable and out of the measurement range. The strain gages with epoxy backing indicated the fatigue damage before those which had a phester backing.

REFERENCES

Measurements Group Tech Note 1991. Fatigue characteristics of micro-measurements strain gages. TN-508-1.

Nisitani, H. & Takao, K.I. 1981. Significance of initiation, propagation, and closure of microcracks in high cycle fatigue of ductile metals. *Engineering Fracture Mechanics*. 15, 3-4: 445-456.

Recent Advances in Experimental Mechanics, Silva Gomes et al. (eds) © 1994 Balkema, Rotterdam, ISBN 90 5410 395 7

Fatigue life assessment in notched specimens of 17Mn4 steel

J.D.M.Costa & J.A.M.Ferreira
Department of Mechanical Engineering, University of Coimbra, Portugal

ABSTRACT: This paper presents a study of the fatigue behaviour in notched specimens of 17Mn4 steel. The fatigue include the initiation and the propagation phases. The life of each of these phases is influenced by the materials and by the stress concentration at the root of the notches. In this work is analysed the effect of the position and the radius of circular holes on the fatigue of bending specimens.

The stress concentration factors were calculated for each geometry using the finite element method. The properties of the used materials were obtained by monotonic and cyclic tests in previous work. The fatigue life to crack initiation was obtained experimentally and using theoretical predictions. The fatigue life prediction to crack initiation was obtained using the local strain approach and the Coffin-Manson equation modified by Morrow. The crack propagation prediction was based on fracture mechanics parameters. A good agreement was obtained between the predicted and the experimental results.

1 INTRODUCTION

The traditional approach based on nominal stresses and elastic stress concentration factors has in recent years been supplemented by two more sophisticated methods. These are: the local strain approach, applicable to crack initiation and the fracture mechanics approach, applicable to crack propagation.

In the applications of the 17Mn4 steel, currently used in pressure vessel construction, geometrical descontinuities are frequently responsible for providing the origin of fatigue crack initiation. Many of these defects are better considered as notches than as cracks, providing the introduction of a crack initiation stage, the calculation of which can be made by the application of an appropriate method. This, avoid to underestimate the measured life that result if we used only the fracture mechanics in fatigue life analytical predictions, due to the assumption of the presence of a pre-existing defect acting as a crack.

In most situations the local plastic strain will be sufficiently contained to limit the plastic zone to a small region and it is the behaviour of the material in this region that governs the crack initiation. Therefore, the methods employed in the prediction of fatigue crack initiation life are based in the assumption that only the notch root stresses and strains are responsible for the initiation stage. This is known as the "local strain approach".

At the root of the notch there is a local increase in strain as the notch root stress is limited by yielding.

With this approach the fatigue life to crack initiation at notch root is related to the fatigue life of strain-controlled unnotched laboratory specimens. Thus, the analysis reduces to i) the determination of

the local stresses and strains and ii) the assessment of how those experimental local stresses and strains relate to the known strain-fatigue life curve of the analysed material, taking into account the mean stress effect by application of Morrow's (1965) equation.

Usually, the maximum strain at the notch root is obtained by the application of the Neuber's (1961) rule, which was, lately, extended to cyclic loading situations by Topper et al (1969). It was found (Leis et al 1973) (Glinka 1985) that Neuber's rule often overestimate the local inelastic strains and stresses. The equivalent strain energy density method (Molski et al 1981) is reported (Glinka 1985) (Jones et al 1989) to give better estimates of fatigue life than Neuber's rule

In previous work, (Costa et al 1993) (Ferreira et al 1993) fatigue crack initiation lives were predicted for this material and another similar material using the equivalent strain energy method and were shown to be in good agreement with experimental results obtained in different geometries of specimens from those used in this paper.

This paper assesses the applicability of the equivalent strain energy method to the prediction of fatigue crack initiation life of notched bend specimens in 17Mn4 steel. Two different geometries are considered with very different values of the stress concentration factor.

The fatigue life was predicted as the sum of the initiation life and crack propagation. The crack propagation was computed using the fracture mechanics parameters. The stress concentration factor was computed using the finite element method and the stress intensity factor was computed by the weight functions method. The effect of the hole size

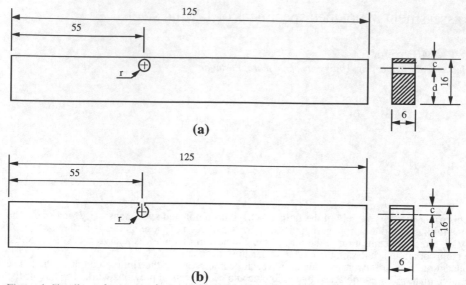

(a)

(b)

Figure 1. Cantilever beam specimens geometries with: a) a circular hole displaced from the center line; b) a key hole.

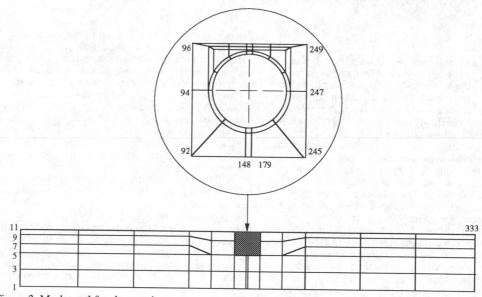

Figure 2. Mesh used for the specimen geometry shown in figure 1a.

on the fatigue life was predicted for both geometries.

2 MATERIAL AND EXPERIMENTAL DETAILS

The specimens used, cantilever beam with a hole near the maximum bending stress, were machined from a plate with 16 mm thickness of 17Mn4 steel, a material widely applied in pressure vessel construction. Specimens details are shown in figure 1. In both geometries, the ratio c/d and r/d are 0.2 and 0.9, respectively.

Chemical composition is given in table 1. Monotonic and cyclic mechanical properties data are given in tables 2 and were obtained in previous work (Costa et al 1993).

The K_t values were determined using a finite element program. The specimen was modeled using eight noded isoparametric elements. One of the meshes used is shown in figure 2. Ranges of c/d between 0.2 and 0.5 and of r/c between 0.2 and 0.9 were used for the determination of the values of K_t.

Fatigue testing of the notched specimens was undertaken in a displacement controlled mechanical

Table 1. Chemical composition of 17Mn4 Steel.

C	Si	Mn	P	S
0,20	0,25	1,20	0,05	0,05

Table 2. Monotonic and cyclic properties of 17Mn4 steel.

Tensile strength, σ_{UTS} (MPa)	596,0
Yield strength, σ_{YS} (MPa)	412,1
Elongation, ε_r (%)	24,1
Cyclic hardening exponent, n'	0,194
Cyclic strength coefficient, K' [MPa]	1120
Fatigue strength exponent, b	0.0504
Fatigue strength coefficient., σ'_f [MPa]	618
Fatigue ductility exponent, c	-0,501
Fatigue ductility coefficient, ε'_f [-]	0,2815

machine at a frequency of 25 Hz using a sinusoidal waveform and a stress ratios R of 0. The load was monitored using a load cell and the crack length initiated at the notch surface was measured using a optical system.

3 FATIGUE LIFE PREDICTION

The fatigue life prediction was computed as the sum of crack initiation and crack propagation life. The crack initiation life N_i was computed taking in account the mean stress effect by using the Morrow's equation (1),

$$\frac{\Delta \varepsilon}{2} = \varepsilon'_f \left(2N_i \right)^c + \frac{\sigma'_f - \sigma_m}{E} \left(2N_i \right)^b \tag{1}$$

where: $\Delta \varepsilon$ is the local elastic-plastic strain at the initiation point, σ_m is the local mean stress, ε'_f and σ'_f represent the fatigue ductility and fatigue strength coefficients respectively, c and b are the fatigue ductility and fatigue strength exponents respectively and E is the modulus of elasticity. $\Delta \varepsilon$ and σ_m were computed from nominal stress and the stress concentration factor using the method described by Glinka (1985), assuming a cyclic stress-strain behaviour described by the Ramberg-Osgood equation

$$\varepsilon = \frac{\sigma}{E} + \left(\frac{\sigma}{K'} \right)^{1/n'} \tag{2}$$

where: σ and ε are the local stress and strain values, K' the cyclic strength coefficient and n' the cyclic strain hardening exponent. The values used for the material parameters in this paper are presented in table 2.

The propagation life was calculated by integration of the Paris law assuming semi-circular flaws with 0.15 mm initial depth located at the root of the notch and propagating in the thickness direction. The stress intensity factor at the root notch region was the basic Raju and Newman (1984) equation multiplied by a magnification factor M_K.

The crack shape during the propagation was assumed by the relationship between the Paris law coefficients, $C_B = 0.9^m C_A$, where: C_B is the Paris law coefficient C for the propagation in surface direction and C_A is he Paris law coefficient C for the propagation in the thickness direction. The values used in the prediction were: $m=3$ and $C_A=1.83 \times 10^{-13}$ (da/dN-[mm/cycle] and ΔK [Mmm$^{-3/2}$]). The final crack length was taken as half of the thickness.

The stress intensity factor was calculated by the weight function method using the general equation:

$$K = \int_0^a \sigma(x) \, m(x,a) \, dx \tag{3}$$

where: $\sigma(x)$ is the stress distribution in the crack plane and $m(x, a)$ is the weight function. The weight function used in this work was derived by Bueckner (1970) for bidimensional edge cracks in plates. The stresses $\sigma(x)$ were computed by finite element method.

The stress intensity factors obtained by equation (3) were compared with a known reference solution K_r introducing the magnification factor M_K, which takes in account the stress concentration, and is calculated by the simple relation:

$$M_K = \frac{K}{K_r} \tag{4}$$

The reference solution used for the stress intensity factor was the Brown (1977) equation applied to a edge crack in a plate in tension.

4 PREDICTED AND EXPERIMENTAL RESULTS

The K_t values were calculated with the stresses distribution computed by a finite elements program using bidimensional isoparametric eight nodes elements. Some of the values obtained for K_t are indicated in the table 3.

From this table we verify an important effect of the geometric parameters c/d and r/c on K_t on both geometries. It is particularly important the effect of r/c on K_t to circular holes.

Using the stress distribution obtained by the finite elements method the magnification factor M_K was computed by equation (4). The figures 3 and 4 present the M_K values for some of the analysed geometries. These figures show the effect of r/c. In the circular hole the M_K factor is very high for all the crack lengths.

Table 3. Stress concentration factor K_t.

c/d	r/c	K_t	
		Figure 1a	Figure 1b
0.5	0.2	1.25	2.63
0.5	0.4	1.78	2.20
0.5	0.6	3.13	1.85
0.5	0.8	6.05	1.70
0.2	0.2	2.10	3.67
0.2	0.4	2.62	2.85
0.2	0.6	4.10	2.40
0.2	0.8	7.50	2.15
0.2	0.9	10.50	-

In this geometry the effect of r/c is very important for all the crack lengths analysed. In the case of key hole the magnification factor is important only for cracks less than $a/(d-r)=0.15$, and the effect of r/c is significant only for cracks less than $a/(d-r)=0.05$.

stress range. As it was expected an important reduction of crack initiation life is observed when the stress range increases. A good agreement was obtained between the predicted and experimental results.

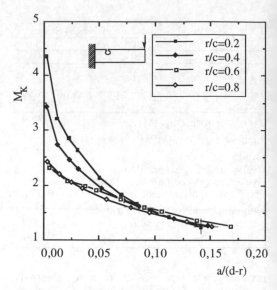

Figure 4. M_K versus a/(d-r). Geometry of figure 1b.

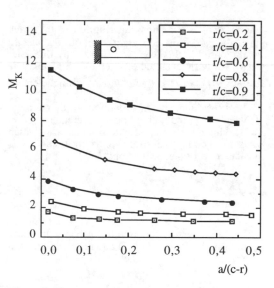

Figure 3. M_K versus a/(c-r). Geometry of figure 1a.

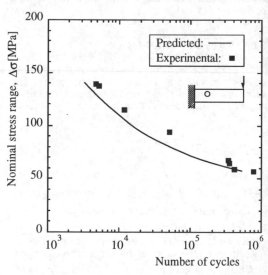

Figure 5. Comparison of experimental and predicted cycles to crack initiation as a function of nominal stress range. Geometry of figure 1a.

In figure 5 are plotted the predicted and experimental results obtained for the initiation life of the plates with circular hole. The number of cycles to crack initiation is plotted as a function of the nominal

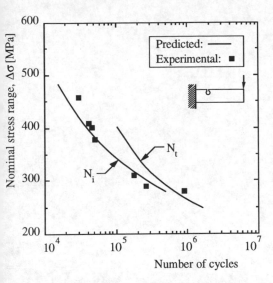

Figure 6. Comparison of experimental and predicted cycles to crack initiation as a function of nominal stress range. Geometry of figure 1b.

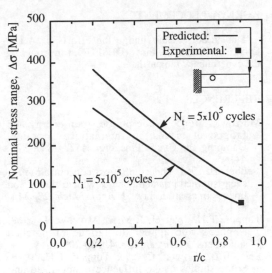

Figure 7. Predicted and experimental results for a fatigue life of 5×10^5 cycles. Geometry of figure 1a.

A similar plot is shown in figure 6 for the plates with key hole. In this figure the predictions of the total life (sum of the crack initiation an crack propagation life) are also plotted. Similary, as in the figure 5, a good agreement between the experimental results and predictions to crack initiation life was observed for this geometry.

In figures 7 and 8 are plotted the predictions of the effect of the geometric parameter r/c on the nominal stress range corresponding to a crack initiation life of 5×10^5 cycles and to a total fatigue life (sum of the crack initiation an crack propagation life) of 5×10^5 cycles. In both figures is superimposed the experimental nominal stress range to a crack initiation life of 5×10^5 cycles for the geometries tested. It was obtained, in both cases, a good agreement between the predicted and the experimental results of crack initiation lives.

For the circular holes the nominal stress range decrease significantly when r/c increases. In opposite the nominal stress range has an important increase when r/c grows from 0.2 to 0.8 for the case of the key hole.

The effect of the geometric parameter r/c on nominal stress range is similar when we consider the crack initiation life or the total life. As it was expected to one imposed life the stress range allowable is higher when we consider the total life instead of the initiation life. In both cases this difference increases when r/c decreases.

5 CONCLUSIONS

1 - For plates with circular holes in cantilever bending the nominal stress range to a fatigue life of 5×10^5

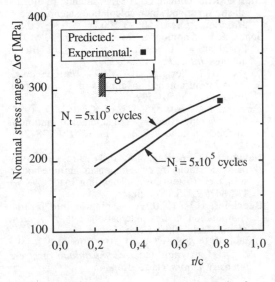

Figure 8. Predicted and experimental results for a fatigue life of 5×10^5 cycles. Geometry of figure 1b.

cycles decreases nearly four times when r/c increases from 0.2 to 0.9.

2 - For plates with key holes in cantilever bending the nominal stress range increases more than fifty per cent when r/c increases from 0.2 to 0.8.

3 - The predicted lives to crack initiation are in good agreement with the experimental results.

ACKNOWLEDGEMENTS

The authors acknowledge the efforts of A.J.P. Oliveira and J.P.T.B. Pires (DEM/UC) for providing the experimental fatigue data.

REFERENCES

Morrow, J. 1965. Cyclic plastic strain energy and fatigue of metals, in International Friction, Damping and Cyclic Plasticity. *ASTM* STP 378: 4-48.

Neuber, H. 1961. Theory of stress concentration for shear strained prismatic bodies with arbitrary non linear stress-strain law. *J. Appl. Mech.* 28: 544-551.

Topper, T.H., Wetzel, R.M. & Morrow, J. 1969. Neuber's rule applied to fatigue of notched specimens. *J. Mat. JMLSA* 4: 200-209.

Leis, B.B., Gowda, C.V. & Topper, T.H. 1973. Some studies of the influence of localized and gross plasticity on the monotonic and cyclic concentration factors. *J. Test. Eval.* 1: 341-348.

Glinka, G. 1985. Energy density approach to calculations of inelastic strain-stress near notches and cracks. *Eng. Fract. Mech.* 22: 485-508.

Molski, K. & Glinka, G. 1981. A method of elastic-plastic stress and strain calculation at a notch root. *Mater. Sci. Engng.* 50: 93-100.

Jones, R.L., Phoplonker, M.A. & Byrne, J. 1989. Local strain approach to fatigue crack life at notches. *Int. J. Fat.* 4: 255-259.

Costa, J.D.& Ferreira, J.A. 1993. Fatigue crack initiation in notched specimens of 17Mn4 steel. *Int. J. Fat.* 15: 501-507.

Ferreira, J.A., Borrego, L.F. & Costa J.D. 1993. Shot peening effect on fatigue crack initiation of surfaces notches. *Proc. 5th ICSP*: 238-245. Oxford.

Newman, J.C. & Raju, I.S. 1984. Stress intensity factor equations fro three dimensional finite bodies subjected to tension and bending loads. *NASA* TM 84793. USA, Virginia.

Bueckner, H.F. 1970. A novel principle for the computation of stress intensity factors. A. Angen. Math. Mech. 50(9): 529-546.

Brown, W.F. & Srawley, J.E. 1977. In D.P. Rooke & D.J. Cartwright (eds), *Compendium of stress intensity factors*, cap. 5. London.

Recent Advances in Experimental Mechanics, Silva Gomes et al. (eds) © 1994 Balkema, Rotterdam, ISBN 90 5410 395 7

Devices for mixed mode propagation: Application to A 42 FP steel

J. Royer & A. Le Van
Laboratoire Mécanique Structures, Ecole Centrale, Nantes, France

ABSTRACT : The aim of this contribution is the effect of mixed modes upon the propagation of a crack. It is shown that using only two specimen geometries and their associated loading device any mixed mode (plane or three-dimensional) can be applied on a crack front by means of a classical testing machine. Experimental results concerned with A 42 FP steel are correlated with numerical values of stress intensity factors. This procedure can be performed both to directly characterize a material behaviour and to validate fatigue criteria under any mixed mode.

1 INTRODUCTION

In the field of linear elastic fracture mechanics, the propagation of fatigue cracks subjected to pure mode I is correctly predicted by Paris law. Nevertheless the behaviour of cracks under mixed mode fatigue is less known. Actually, in this case one has to answer the two following questions :

- what is the direction of the crack propagation ?
- what is the incidence of the mixed mode on the propagation velocity ?

Several criteria are available to solve this problem but first it is convenient to separately consider the two following possibilities.

- Plane mixed modes (modes I + II)

The analysis of the problem is easier because the propagation vector of the crack front remains approximately in a plane perpendicular to the crack front.

- Three-dimensional modes

In the general situation there is a double problem facing us. On one hand it is generally impossible to obtain all along the crack front a mixed mode involving modes I and III only. We have simultaneously a mode II which value is uneasy to cancel. On a second hand the propagation vector of the crack front is no more perpendicular to the front. It results a propagation kinematics which is generally complicated.

Different criteria were proposed by authors such as (Erdogan 1963), (Hussain 1974), (Wu 1978), (Amestoy 1980) in order to predict crack extension according to modes I and II but in the literature we find few propositions relevant to complete three-dimensional modes (Palaniswamy 1974), (Sih 1975). Moreover the validation of these last formulations is rather difficult because it needs correlations between experimental values uneasy to measure and numerical values that can only be obtained after long and expensive three dimensional computations.

We succeeded in solving these different difficulties in order to compare numerical and experimental values about the same specimen. Nevertheless in the present contribution the numerical approach (Le Van 1993) will be omitted for the benefit of the experimentation. Actually we demonstrate that by means of only two simple geometries of specimen with their relevant loading device it is possible to study the behaviour of a crack front under any combination of mixed modes. This can be bring into play using a classical testing machine.

2 LOADING DEVICES

2.1 *Plane mixed mode device*

The plane symmetrical Y specimen (fig. 1) consists of two cracks and two arms which are loaded by means of 20 mm diameter pins. These pins transmit loads $\vec{F}_1$ and $\vec{F}'_1$ at an angle α with

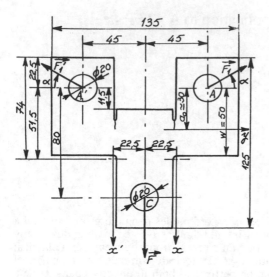

Fig. 1 - Y specimen for plane mixed-mode.

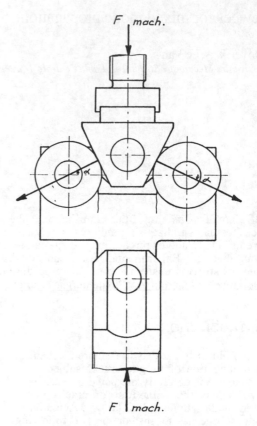

Fig. 2 - Plane mixed-mode loading device

respect to the y axis and the whole equilibrium of the specimen is obtained by a third force $\vec{F}$ parallel to x axis and applied through pin C. Each arm may be considered as a half CT specimen for which the parameter ω is equal to two times the 25 mm thickness.

The crack length a_0 is measured as for a CT specimen according to the distance along the x axis between the center of pin A and the crack front. Under these conditions the loading applied on the cracks is defined by the two independent parameters α and F_1. The mixed mode loading is defined by parameter α whereas the level of load depends on F_1.

The loading device is different with each value of α nevertheless the principle presented on fig. 2 remains the same. At the lower pin C, the load produced by the testing machine is transmitted by means of a clevis. The other loads $\vec{F_1}$ and $\vec{F'_1}$ are applied to pins A and A' through wedges inserted between the needle bearing on both sides of the specimen, thus the α angle is fixed by the angle between the wedge faces. This set up is convenient for static and dynamic loading but the present study is concerned with fatigue loading only. In comparison with (Arcan 1982) specimen for example, our Y specimen allows not only pure mode II on the crack front but also any mixed-mode.

The design of the specimen allows to separately apply a pure mode I on each crack in order to proceed an easy and symmetrical

fatigue precracking of the specimen under the same conditions as a CT specimen. Then the specimens were fatigued under mixed mode loading using the appropriate wedges until we get about 1 mm crack propagation after kinking. Finally a brittle fracture of the remaining ligaments was achieved at low temperature. On the both part of the broken specimens the parameters of the initial crack and those of its propagation were optically measured. These measurements were realized at several points regularly spaced on the crack front in order to take into account the curvature of the crack front. Due to a large scatter the results of the measurements on the faces of the specimens were disregarded. The 25 mm thickness of the Y specimen is enough to assume plane strain conditions on the major part of the crack front. Then the kinking angle and the velocity of the crack propagation were derived from the average of the measurements on the central part of the crack front.

2.2 *Three dimensional mixed-mode device*

This time the specimens have a round section and contain a surface plane crack in the median cross section (fig. 3). The ends were designed so as to apply any type of loading including a torque. A straight sharp notch normal to the specimen axis was machined in the middle cross-section and a crack was then initiated using a classical three-point bending set-up the precracking of the specimens was conducted until the maximum crack depth a_0 reaches 5 mm.

After breaking the specimens it was checked that the geometry of the crack front is quite circular. Then the initial geometry of the crack front can be defined by its two parameters : the maximum crack depth a_0 and the radius R' of crack front.

Cracked specimens were then subjected to mixed loading by using the special device scheme illustrated in figure 4 in which two identical specimens were connected to the loading apparatus. The boundary conditions at the ends were equivalent to simple supports with respect to bending and to fixed ends with respect to torsion. As shown in figure 1, the F action line applied on the lever could be shifted away by a distance from the specimen axis. The two-specimen set and its equipment was equivalent to a beam with a 2ℓ span ($\ell = 125$ mm). Two moments were thus applied on each cross-section including cracks : a torsion moment ($Mx = Fd/2$) and a bending moment ($Mz = F\ell/4$). In these conditions, the cracks of the two loaded specimens were subjected to a mixed mode : mode I was governed by the bending moment and modes II and III by the torsion moment. At any time during the loading, these modes remained proportional to F, and the ratio $\delta = K_{III}/K_I$ kept a steady value depending only on the ratio d/ℓ only.

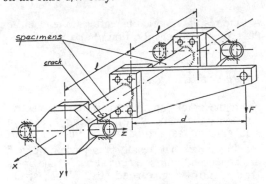

Fig. 4 - Loading device for three-dimensional modes

The testing of specimens was conducted after choosing particular values of the parameter pairs (Fmax, d) of the loading. In order to get an easier interpretation of experimental results the same value Fmax of the load applied on several specimens was kept as long as it was possible to assume a linear elastic behaviour of the crack propagation. The type of mixed mode only results from the length of the lever arm d. Besides, the linear elastic condition another one was retained for the selection of d values, that is to provide about 1 mm propagation of the crack front under a maximum of 150 kcycles.

Under these conditions series of specimens with same mode I and different mode II+III distributions were available to discuss the experimental results.

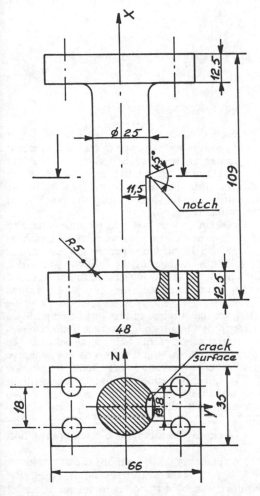

Fig. 3 - Specimen geometry for three-dimensional modes

The control of crack propagation was achieved according to two ways :
- by optic inspection on the lateral surface of the specimen,
- by means of an electric resistivity method, the potential drop being recorded at three points along the crack front.

After fatigue a brittle fracture of the specimens was produced at low temperature to measure the geometry parameters of the crack front before and after fatigue under mixed modes. For the initial crack front the parameters a_0 and R' were measured to derive the stress intensity factor values. The crack propagation was measured at 7 points regularly spaced on the crack front. At these special points the values of crack propagation Δa and kinking angle β were plotted. Nevertheless only the kinking angle in the plane normal to the crack front was measured.

3 RESULTS ON A 42 FP STEEL

All the results reported hereafter are relevant to the ratio of loadings R = Fmin/Fmax = 0.1.

3.1 *Plane mixed-modes*

From the crack length a measured by electric resistivity the velocity of the crack propagation was derived as a function of number N of loading cycles. The results presented on figure 5 are concerned with the central part of the crack front mainly governed by plane strain conditions. For a fixed value of the mixed mode ration K_{II}/K_I the velocity values $\Delta a/\Delta N$ were plotted versus the stress intensity factor (SIF) K_I using a logarithmic scale. An example of the Paris lines thus obtained is represented on figure 5.

We derived the relevant coefficient of Paris law allowing an interpolation as a function of mixed modes K_{II}/K_I. But the shifting of Paris lines according to the mixed mode is obvious on figure 5, the more rapidly as ratio K_{II}/K_I increases.

Beside these predicted results we noticed some distortion at the onset of crack propagation such as a first decreasing of the velocity $\Delta a/\Delta N$ that can be explained by a new crack initiation induced by kinking.

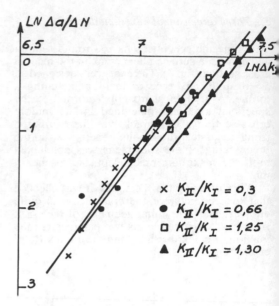

Fig. 5 - Crack velocity under mode I + II

3.2 *Three dimensional modes*

For each considered parameter (Δa or β) of crack propagation we have to take into account the values of the three SIF and this does not allow a simple geometric representation. In addition, this time only discrete values at the measuring points spaced along the crack front are available.

However it was possible to present and to analyse a correlation between the numerical values of SIF and the experimental parameters by taking advantage of the following remarks. On one hand as the initial crack was initiated under pure bending, the computed values confirm that K_I keeps quite a constant value whatever the abscissa of the considered point on crack front. On a second hand the geometries of the initial crack fronts being very similar, the ratio remains the same from one specimen to another one, this ratio is only depending on the abscissa of the considered point. Consequently it was possible to point out the incidence of two among the SIF, the third remaining constant.

- Mode III contriibution on crack propagation

From symmetry considerations it can be shown that K_{II} is always zero at point A on the symmetry axis of the cracked section (deepest point of the crack front). It is then possible to plot curves of crack velocity $\Delta a/\Delta N$ as a

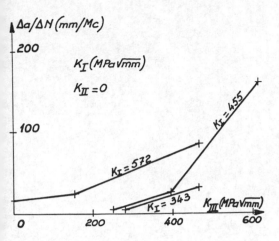

Fig. 6 - Crack velocity under mode I + III

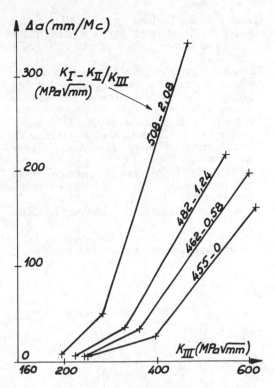

Fig. 7 - Crack velocity under three-dimensional modes

function of ratio K_{III}/K_I for specific values of K_I (fig. 6).

The addition of mode III gives rise to an important increasing of the crack velocity and this is particularly obvious for the value $K_I =$ 455 MPa $\sqrt{mm}$ Nevertheless the other curves demonstrate that K_I is always the main parameter of crack propagation.

- Complete mixed mode

Such loadings occur at any point of the crack front, excepted A. The curves at constant ratio K_{II}/K_{III} on figure 7 point out the incidence of K_{II} upon the propagation. The selected values on this figure characterize especially the incidence of mode II upon the crack propagation.

Similar curves regarding kinking angle β were plotted versus the same parameters.

4 CONCLUSION

The two described specimen geometries and their relevant loading device allow to control the mixed mode applied on a crack front. These geometries assume the major part of the crack front loaded under plane strain conditions and the mixed mode is quite the same all along the crack front of the plane specimen whereas only the mode I remains constant along the crack front of the three-dimensional specimen. One has only to choose the suitable abscissa of the measuring point along the crack front in order to select the required ratio between mode II and mode III. The scale of the different modes can be adjusted by means of independent parameters.

The presented results concerned with a steel can be performed for any other material. They demonstrate it is possible to characterize the behaviour of a crack under any mixed mode defined by the three SIF. This procedure can be bring into play to validate three-dimensional criteria in the field of linear elastic fracture mechanics.

REFERENCES

Amestoy, Bui, Dang Van 1980. Analytic asymptotic solution of the kinked crack problem. *Advances in Fracture Research ICF 5*. Cannes : 107

Arcan, M., Banks-Sills, L. 1982. Mode II fracture specimen, photoelastic analysis and results. *Proc. 7th Int. Conf. on Exp. Stress Analysis-Technion*. Haïfa. Israel. : 187-201 Aug.

Erdogan, F., Sih, G.C. 1963. On the crack extension in plates under plane loading and transverse shear. *J. of Basic Engineering* : 519-527

Hussain, A., Pu, L., Underwood, J. 1974. Strain energy release rate for a crack under combined mode I and mode II. *ASTM-STP 560* : 2-28

Le Van, A., Royer, J.1993. Part-circular surface cracks in round bars under tension bending and twisting. *International Journal of Fracture.* Vol. 61 : 71-99

Palaniswamy, K., Knauss, W.G. 1974. On the problem of a crack extension in brittle solids under general loading. *Mechanics Today Nemat Nasser Edt.*

Sih, G.C. 1975. A three-dimensional strain energy density factor theory of crack propagation. *Mechanics of Fracture 2.* Noordhoff International Publishing Leyden : 15-53

Wu, C.H. 1978. Fracture under combined loads by maximum energy release rate criterion. J. *Appl. Mech.* Sept.

Recent Advances in Experimental Mechanics, Silva Gomes et al. (eds) © 1994 Balkema, Rotterdam, ISBN 90 5410 395 7

Fracture mode transition in high cycle fatigue of high strength steel

Y. Kuroshima
Kyushu Institute of Technology, Kitakyushu, Japan

M. Shimizu
Keio University, Yokohama, Japan

K. Kawasaki
Neturen Co., Ltd, Hiratsuka, Japan

ABSTRACT: The objective of present study is to investigate the problem of fracture mode transition from surface-originated crack to fish eye or vice versa in long life fatigue region. Rotating bending fatigue tests are conducted on quenched and low temperature tempered SAE 9254 spring steels, having either two kinds of artificially introduced defects, i.e., isolated defects (retained ferrite) or connected defects (proeutectoid ferrite). Discussions are made on the relatiohship between the fatigue fracture mode transition and the surface conditions. In addition, the role of microstructure in the formation process of fish eye is examined. A material with isolated defects in its microstructure showed the fracture mode transition. Strengthening of surface layer by mechanical polishing suppressed the initiation of surface crack and resulted in fatigue failure of fish eye pattern. Microscopically this transition was found to be induced by the change of crack formation process from slip band cracking in surface retained ferrite to cleavage cracking in internal retained ferrite.

1 INTRODUCTION

All steels contain non-metallic inclusions and other defects, and they have generally long been regarded as one of the chief cause of fatigue failure in steel.

The inclusions and other defects often act as a crack initiation site for final fatigue failure. This trend is, in general, intensified with increasing material strength. According to the fractographic study of high cycle fatigue in long life region, the origins of fatigue failure are usually classified into two types, i.e., subsurface defect and surface defect. A fracture mode for the fatigue failure where a leading crack initiated at subsurface defect controls the whole fatigue process, is called "fish eye fracture".

Generally, the microscopic fatigue crack originated in surface substrate leads to the final fracture of the specimen. In the case of high strength steel, however, fracture mode transition from the surface substrate-originated type and surface defect type to the fish eye type occurs in long life region.

In addition, the study of high cycle fatigue in very long life region in high strength steels (Asami 1991) has shown that some high strength steel shows no fatigue limit in very long life region over 10^8 cycles. To ensure the reliability of high strength steel it is of importance in engineering practice to examine how the inclusions or defects dominate the fracturing process in long life fatigue region.

The objective of the present study is to investigate why the fracture mode transition from the surface-initiated type to the internal fish eye type occurs in long life fatigue region. Rotating bending fatigue tests were conducted on quenched and low temperature tempered spring steels, having two kinds of artificially introduced defects. The relationship between the fatigue fracture mode transition and surface conditions were discussed. In addition, the role of microstructure in the formation process of fish eye was examined.

2 EXPERIMENTAL PROCEDURE

The material used in this study is a Si-Cr type of spring steel (SAE 9254). In Table 1 is listed the chemical composition of the material. It was machined into an hour-glass shape fatigue specimen with a smallest diameter of 6.5 mm in the gauge length. Two kinds of specimens having different microstructure were prepared by heat treatment as shown in Figure 1. Figure 2 shows examples of microstructure of each specimens. Two types of specimens with different microstructures were employed; A series specimens containing isolated ferrite phases in a

Table 1. Chemical composition. (wt%)

Material	C	Si	Mn	P	S	Cr
SAE 9254	0.55	1.44	0.60	0.028	0.004	0.67

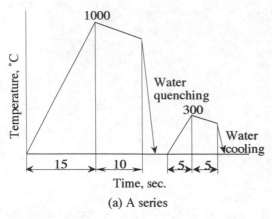

(a) A series

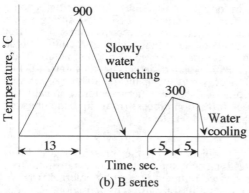

(b) B series

Figure 1. Heat treatment

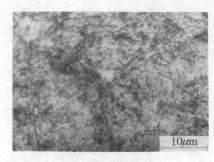

(a) A series

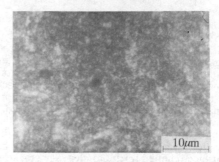

(b) B series

Figure 2. Microstructures of testing materials.

responsible for final failure was identified.

3 EXPERIMENTAL RESULTS

3.1 *Fatigue strength*

Figure 3 shows the results of rotating bending fatigue tests for each series specimen. It is to be noted that the C series specimen having the isolated defects microstructure with work hardened layer exhibited the highest fatigue strength among all series specimens in long life region. The B series specimen having the continuous defects microstructure shows the lowest fatigue strength. With increasing stress amplitudes the test results gradually tend to converge.

3.2 *Fracture mode of fatigue failure*

The Macroscopic observation concerning fracture origin was conducted on the fracture surfaces of all the failed specimens. The results revealed that the difference of long life strength between A and C series is attributed to the transition of fracture mode.

In case of the A series specimens, showing

martensite phase and B series containing continuous ferrite phases filled with a martensite. These ferritic phases consist of very small retained ferrite grains and proeutectoid ferrite grains. These microstructure having small ferrite phases are hereafter referred to as the "isolated defects" or the "continuous defects", because ferrite phase in the microstructure is relatively weak as compared with martensite phase.

After heat treatment, all the specimen surfaces were electropolished and removed by an amount of about 30 ~ 50 μm thick as to exclude the residual stresses. In order to clarify the influence of surface condition on fracture mode transition C series specimen having subsequent surface finish with fine emery paper and alumina powder to the A series specimen, was further used. The C series specimen naturally provides a very thin work hardened surface layer.

Surface residual stress was examined using an X-ray residual stress measurement equipment.

Rotating bending fatigue tests were carried out at a speed of 47.5 Hz at room temperature.

Fractographic observations were done on all the failed specimens with the aid of a scanning electron microscope and the crack initiation site

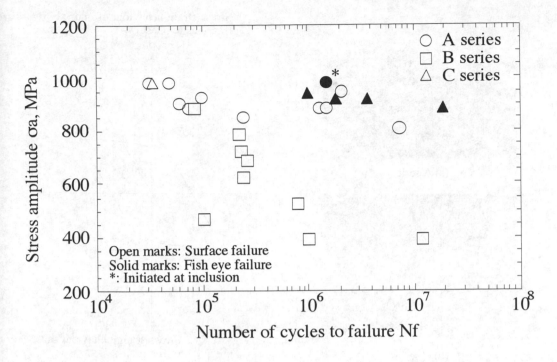

Figure 3. Results of fatigue tests.

lower fatigue strength, almost all the leading cracks were originated in specimen surface except one case (marked with a solid circle in Figure 3). In such an exceptional case, the fatigue crack was formed at a larger non-metallic inclusion below the specimen surface.

In case of the C series specimen, showing the highest fatigue strength, all the cracks were initiated in specimen surface and a typical fracture mode of fish eye pattern, as shown in Figure 4, was observed in each specimen.

In case of the B series specimen, all the specimens were fatigue-failured by the growth of a crack initiated in specimen surface.

3.3 Observation of crack initiation site

Detailed observation of fatigue failure origin was done in long life region in each series specimen. To reveal the microstructures the specimen surfaces were slightly etched after finishing the tests. The slip band cracking formed within larger ferrite grain, as shown with an arrow mark in Figure 5(a), was responsible for crack initiation site in A series specimen.

For the B series specimen, the crack nucleated within the featureless portion of the structure. Through closer observation, it was found that the crack initiation site was located in the chain of ferrite grain.

In the case of C series specimens failed in fish eye pattern, the crack initiated from a featureless facet in specimen subsurface, as exemplified in Figure 5(b). To identify the microstructure of the facet the micro-Vickers hardness tests were done in the fracture surface of several specimens. The results showed that the facet on the fracture surface consisted of the transguranular cracking in a ferrite grain or the intergranular cracking in the ferrite/ferrite boundary in the material.

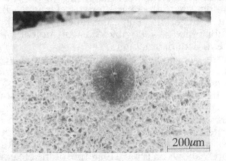

Figure 4. Typical example of fish eye on fracture surface at C series.

(a) A series

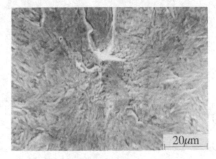

(b) C series

Figure 5. Typical example of crack initiation site on fracture surface.

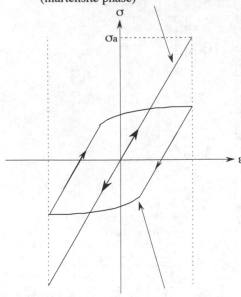

Stress-strain behaviour of matrixl (martensite phase)

Stress-strain behaviour of ferrite grain

Figure 6. Schematics of stress-strain behaviour of ferrite grain and matrix during fatigue test.

4 DISCUSSIONS

4.1 *Fatigue crack initiation*

It is generally accepted that a hard inclusion in material acts as a crack intuition site in high strength steel. However, in this investigation, the cracks nucleated within small ferrite grains formed during heat treatment. That is, the soft ferrite grain surrounded with hard substrate promoted the constraint-free slip band formation and resulted in crack initiation.

The material used can be regarded as one of composite materials, consisting of hard matrix and soft grain. Suppose the fatigue cycling plastic deformation should occur only in the soft grain, due to the difference of yielding level of both compositions, as schematically illustrated in Figure 6. Therefore, the soft ferrite grains in specimen surface would probably behave as a preferential site for fatigue crack initiation. This means that fatigue failure starts from the specimen surface.

On the other hand, the fish eye pattern fracture observed in the C series specimens can be interpreted as being induced by another mechanism. As mentioned above, when the soft ferrite grains are exposed on the specimen surface,

they preferentially operate as a crack initiation site, due to difference of yielding stress between ferrite grain and surrounding matrix. Suppose the ferrite grains existing slightly below the specimen surface. The plastic deformation inside and periphery of the ferrite grain are three dimensionally completely restricted by surrounding hard matrix. This constraints yields stress raiser in the ferrite grains and results in cleavage cracking.

In fact, the metallographic observations revealed that the facets observed at fatigue crack initiation site was consistently formed along {100} planes coincident with the characteristic cleavage plane of BCC materials. Thus, the facet type cracking in subsurface is recognized as an indication of fish eye pattern cracking.

With the increase of internal stress, it is expected that the cleavage cracking of ferrite grain originates at an early stage of loading cycles. In fact, this cleavage cracking was observed in some ferrite grains on the A series specimen surface. However, no cleavage cracking was seen in the C series specimen. The presence of the residual stress might prevent from cleavage cracking. Figure 7 shows an typical example of the cleavage cracking of ferrite grain observed in the A series specimen.

As discussed in the foregoing, the surface ferrite

Figure 7. Cleavage craking of surface ferrite grain at N=1 cycle.

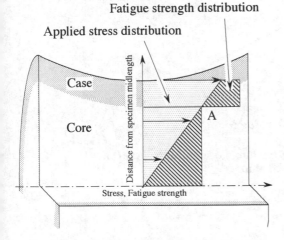

Fatigue strength distribution

Applied stress distribution

Figure 8. Schematically illusration of model of fish eye fracture mechanism for the surface hardened steel.

grains are subjected to higher plastic strain due to the difference of strength between the ferrite grain and the surrounding matrix. However, as the surrounding hard matrix constraints the plastic deformation in ferrite grains, no cleavage cracking occurs in reality in the ferrite grains of the specimen surface. On the other hand, when the ferrite grains are in subsurface, the surrounding matrix three-dimensionally restricts the plastic deformation are elevate the stress in ferrite grains, and results in cleavage cracking of the ferrite grain. This is the reason why the origin of the fish eye pattern cracking works in the internal subsurface.

4.2 Fracture mode transition

In rotating bending, a maximum stress is produced at the specimen surface, due to stress gradient.

In this respect, the fatigue crack was in fact originated at the ferrite grains of the surface in A series specimen. The reason why the C series specimen showed a fracture pattern of fish eye type can be interpreted as fellow.

The surface of A series specimen was finished by electric polishing. In case of the C series specimen, additional polishing using emery paper and alumina powder was given to the surface of the A series specimen. This finishing treatment brought a work hardened surface layer to the C series specimen. Indeed, the Vicker's hardness at the surface of A and C series specimens was 693 HV and 718 Hv, respectively. The distribution of HV hardness quickly reduces in the diametral direction of the C series specimen. The hardness at a point of 30 μm deep in C series specimen was almost the same as A series specimen. In addition, a compressive residual stress of about 400 to 600 MPa was measured at the surface on the C series specimen. It quickly attenuated in the hardened layer and showed a value of 100 MPa at a depth of 30 μm.

Regarding the failure of fish eye pattern, it is generally said that the origin of the fish eye usually locates at the interface between work hardened surface layer and non-hardened core region of the specimen. Those type fracture patterns are common in carburized steel and induction hardened steel. Figure 8 schematically illustrates the relationship between the stress distribution and the fatigue strength distribution. The fish eye is formed at a position where the applied stress coincides with the fatigue strength. In C series specimen, the fish eye was formed at a position of 50 ~ 500 μm deep.

The growth condition of the cleavage crack might be free from the condition of the hardened surface layer. Adding hardened surface layer should have contributed to suppress the growth of the cracks formed in surface layer, then, the crack initiated at deeper position might grow. In consequence, the final fracture of C series specimens in the long life region is induced by the fish eye fracture mode. Figure 9 shows the fracture mode transition in high strength steel can be well understand from the model.

5 CONCLUSIONS

Discussions were made on the relationship between the fatigue fracture mode transition and the surface conditions, and also of the role of the microstructure in the formation process of the fish eye pattern. The results obtained show that (1) the fracture mode transition occurs in the material with isolated defects in the microstructure when a surface fracture mode is prevented by strengthening the surface layer with mechanical polishing and, (2) such a fracture mode transition is associated with the change in

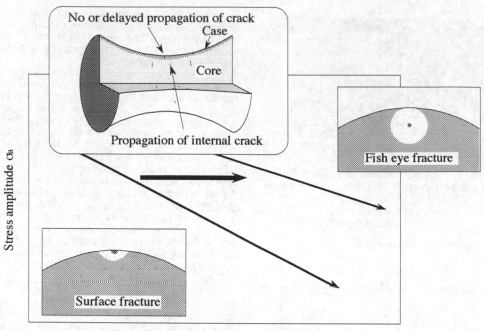

Figure 9. Schematically illustration of fracture mode transition between surface fracture mode and internal fracture mode.

fatigue crack initiation process from slip band cracking of surface retained ferrite to cleavage cracking of internal retained ferrite.

REFERENCES

Asami, K. 1991. Long life fatigue strength of gas-carburized steel. Proc. 6th Int. Conf. Mech. Behav. Mater 2: 499-504. Kyoto.

Recent Advances in Experimental Mechanics, Silva Gomes et al. (eds) © 1994 Balkema, Rotterdam, ISBN 90 5410 395 7

Experimental mechanics on initiation and growth of distributed small creep-fatigue cracks

R. Ohtani, T. Kitamura, N. Tada & W. Zhow
Department of Engineering Science, Kyoto University, Japan

ABSTRACT: High-temperature fracture in heat-resisting alloys subjected to push-pull cyclic loadings is characterized by the initiation and growth of distributed small cracks under creep-fatigue interaction conditions. Emphasis is put on the two subjects: how to evaluate the stochastic nature of small cracks and how to realize the creep-fatigue interaction.

1 NEED OF SMALL-CRACK INVESTIGATION

Facture mechanics approach to the propagation of large and long cracks at high temperatures gives us a definite explanation of the behavior of time-dependent fracture (Ohtani et.al. 1988). However, in the region of small cracks where the half crack length is shorter than about ten times as long as the grain size, the initiation time as well as the growth rate show distinguished scatter (Ohtani et.al. 1990). Figure 1 shows a schematic representation of the growth curves of small cracks in comparison with large cracks. Not only their stochastic resistance and driving force for grain boundary cracking but also their random behavior are still vague. Moreover, striking creep-fatigue interaction takes place in the process of crack initiation and early growth. However, not only the substance of creep-fatigue interaction but also the method of its quantitative evaluation are not precise.

2 METHOD OF SMALL CRACK EXPERIMENTS

First of all, fatigue tests are necessary for observing small cracks. For this purpose, the tests are conducted in air and in a vacuum at high temperatures. They are interrupted at an interval of some fatigue cycles and intermittent observation of the multiple small cracks at the surface and the inside of specimens is made. Smooth bar specimens with 10 mm diameter of a Type 304 stainless steel are used. Figure 2 shows the illustration of the setups of high-temperature fatigue tests.

The number and the length of small cracks are measured on enlarged photographs by manual with the naked eye. In some case, an image analyzing system is adopted for the measurement (Kitamura et.al.1993). The photographs in Fig.2 show the images before and after the analysis. The intergranular cracks are clearly picked out and the measured crack length coincides well with the eye-measured data.

In some tests, observation is done in situ and in a vacuum using a high-temperature microscope combined with a servo-hydraulic fatigue testing

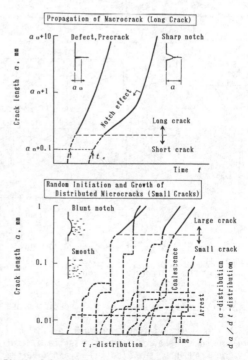

Fig.1 Schematic representation of the random behavior of small crack initiation and growth.

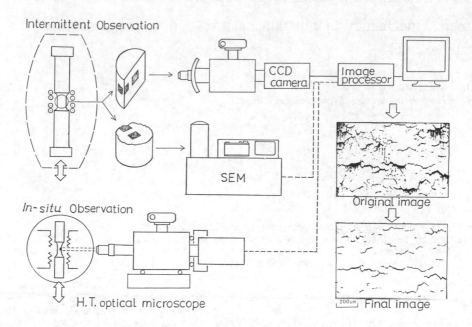

Fig.2 Illustration of the setups of high-temperature fatigue tests and the measurement of small cracks.

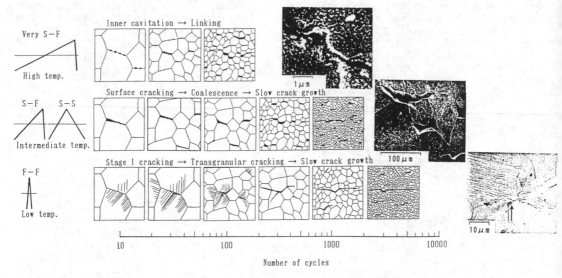

Fig.3 Classification of the type of fatigue fracture in ductile polycrystalline alloys at high temperatures: Type F, pure fatigue type; Type I, creep-dominated surface cracking type; Type C, monotonic-creep inner-cavitation type.

equipment. In this case, miniature specimens are adopted.

High temperature fatigue strength exhibits a time-dependent behavior and is affected by not only the frequency but also the strain waveform (Solomon et.al. 1973, Halford et.al. 1973). The effect is thought to be due to the creep-fatigue interaction. In order to analyze this, tests are conducted under push-pull strain-controlled conditions. The strain waveforms adopted are the triangular type of fast-fast, fast-slow, slow-slow and slow-fast. The fast strain rate in tension or in compression is 1 %/s, and the slow strain rate is 10^{-3} to 10^{-4} %/s.

3 CHARACTERISTICS OF EXPERIMENTAL RESULTS

3.1 Type of cracking

Figure 3 shows the types of cracking during high temperature fatigue in ductile polycrystalline alloys. Type F is the pure fatigue type which shows persistent slip band cracking and/or grain boundary cracking when specimens are subjected to fairly low temperatures or very high tensile strain rate cycles of fast-fast or fast-slow waveforms. Type I is the creep-dominated type with the grain-boundary creep-cracking at the surface when subjected to slow-slow or slow-fast cycles. Type C is the monotonic creep type characterized by grain boundary cavitation inside the material while exposed at higher temperatures or lower strain rates in slow-fast strain cycles (Ohtani et.al. 1991). The observation of inner cavitation and cracking of Type C must be made by cutting the specimen of the interrupted fatigue test and the spatial distribution should be obtained for the three-dimensional analysis.

3.2 Mode of cracking

Figure 4 shows typical views of small surface cracks of Type I in a Type 304 stainless steel.

The cracks or cavities cannot be detected inside the specimen up to the end of fatigue life. The cracks lay nearly perpendicularly to the stress axis. As the oxide film of a few μm-thick covers the specimen surface in air environment, it must be removed by the diamond paste for the observation and measurement. It appears that cracks grow much easier when they are accompanied with the coalescence.

Figure 5 shows one of the results of in situ observation showing the process of initiation and early growth of one of Type I creep-fatigue cracks. It is found that the crack is formed along the grain boundary between adjacent triple points A and B, and that it opens gradually and uniformly. The fatigue crack propagates rather continuously at every cycle, whereas the creep-fatigue crack grows discretely by the unit of a grain boundary facet. Therefore, a definition of the crack initiation is given as a condition in which complete crack-opening takes place along a grain boundary.

3.3 Crack initiation condition

Under creep-fatigue conditions, the crack initiation time becomes shorter and the crack density becomes higher as the strain rate in tension is

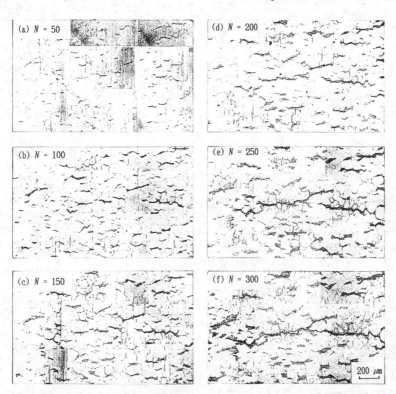

Fig.4 Creep-fatigue cracks distributed on the surface of a Type 304 stainless steel: temperature, 923K; total strain range, 1.0 %; strain rate, 10^{-3} - 1 %/s; number of cycles to failure, 375.

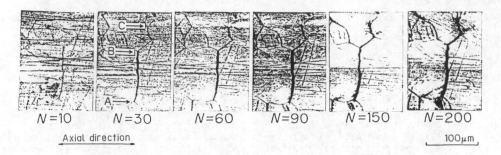

$N=10$ $N=30$ $N=60$ $N=90$ $N=150$ $N=200$

Axial direction $100\mu m$

Fig.5 Process of crack initiation and growth along grain boundaries in a Type 304 stainless steel observed in situ during the creep-fatigue test.

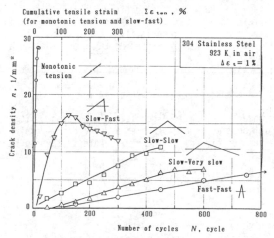

Fig.6 Effect of strain waveform on the increase in the number of surface cracks in a Type 304 stainless steel.

slower. On the contrary, they do as the compressive strain rate is faster. Figure 6 is the test results showing the effect of strain waveforms on the increase in the crack density. Here, the crack density is defined as a number of cracks per unit area of a fixed surface for the cracks of Type I. The results imply that grain boundaries are shielded from cracking by the compressive creep in every unloading process. This might be due to the reversible grain boundary sliding and the extinction of small cavities.

Figure 7 shows cavities and small cracks of Type C generated along grain boundaries on a longitudinal section of the fatigued specimen of a 1Cr-1Mo-1/4V steel subjected to the very low tension and fast compression (10^{-4}-1 %/s) strain waveform at 923K(650°C). Using another specimen being subjected to the same loading condition and fatigued up to the same number of cycles, the test is continued under the different strain waveform of fast-very slow (1- 10^{-4}%/s). It is found that most of the cavities and cracks dis-

appear after a few cycles of the fast-very slow loading.

The increasing rate of crack density with the increase in the number of cycles, dn/dN, is a measure representing the degree of crack initiation. Assuming that the density increases proportionally to the number of cycles at the early stage of each curve, the slope, dn/dN, is constant for each loading condition, and it is given as functions of tensile strain rate, compressive strain rate, strain range and temperature. The magnitude is also given as a function of creep rate, $\dot{\varepsilon}_c$, and stress, σ, at an arbitrary time in a tension half cycle in the form that

$$dn/dN = A_i \left(\int_0^{\tau} \sigma \dot{\varepsilon}_c \, dt \right)^2, \qquad (1)$$

where A_i is a material constant which depends on the compressive strain rate and the temperature and τ is the period of a tension half cycle.

3.4 Crack growth condition

It is difficult to determine the crack growth rate as the small crack does not propagate continuously. The crack growth rate, dc/dN (c is the half length of a surface crack), is then defined as the increment of crack length, Δc, per interval of the interrupted test, ΔN. The interval is about ten per cent of the number of cycles to failure ($\Delta N = N_f/10$).

The dependence of the crack growth rate on the compressive strain rate is not so distinct, being similar to the large crack propagation (Ohtani et.al.1988). Figure 8 shows examples of the test results for a slow-fast and a slow-slow strain cycle. It is also found that the air environment has little effect.

It might be concluded that the reversibility of creep-fatigue damage, or the effect of cyclic loading in low cycle fatigue regime, is more distinguished in the crack initiation process than in the growth process.

It is clear from Fig.8 that the average of the growth rate coincides with the extraporation of large-crack propagation law: da/dN = Cc·ΔJc,

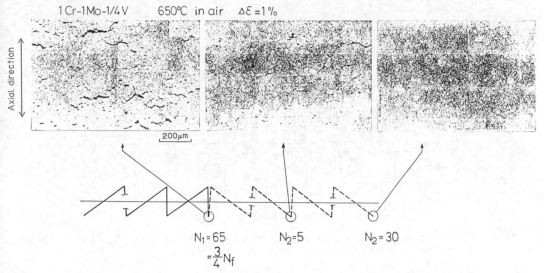

Axial direction

1Cr-1Mo-1/4V   650°C in air   Δε = 1%

200μm

$N_1 = 65$
$= \frac{3}{4} N_f$

$N_2 = 5$

$N_2 = 30$

Fig.7 Cavities and small cracks in a 1Cr-1Mo-1/4V steel when subjected to the very slow-fast strain waveform, and shrinkage or healing of them by the subsequent fast-very slow strain cycles.

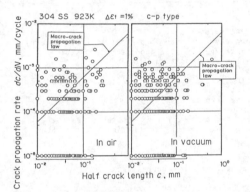

304 SS 923K   Δεt =1%   c-p type

Crack propagation rate dc/dN, mm/cycle

Macro-crack propagation law

Macro-crack propagation law

In air

In vacuum

Half crack length c, mm

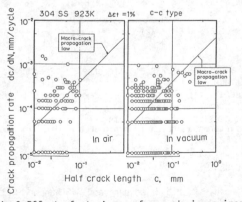

304 SS 923K   Δεt =1%   c-c type

Crack propagation rate dc/dN, mm/cycle

Macro-crack propagation law

Macro-crack propagation law

In air

In vacuum

Half crack length c, mm

Fig.8 Effect of strain waveform and air environment on the growth rate of small cracks in Type 304 stainless steel.

where $\triangle Jc$ is the J-integral range(Ohtani et.al. 1988, Saxena 1993). As $\triangle Jc$ can also be given as a function of creep rate, $\dot{\varepsilon}_c$, and stress, σ, the crack growth rate is expressed as follows:

$$dc/dN = A_g \int_0^\tau \sigma \dot{\varepsilon}_c \, dt, \qquad (2)$$

where A_g is a material constant which depends little on the strain waveform but depends on the crack length and shape, the specimen size and the loading mode. It is also understood from Eq. (2) in comparison with Eq.(1) that the crack growth has a weaker dependence of stress, strain rate and strain range than the crack initiation.

3.5 Creep-fatigue interaction

There are two main factors as the physical substance of creep-fatigue interaction in the process of crack initiation and early growth.
  One is the change in the fracture mode and mechanism according to the loading conditions. As was indicated in Fig.3, the creep fracture yields faster crack initiation and growth and shorter number of cycles to failure than the pure fatigue. Even in the regime of creep dominated fracture, the inner cavitation of Type C, in general, does damage to the material more than the surface cracking of Type I. Figure 9 shows a fracture mode map of the high-temperature fatigue in a Type 304 stainless steel. This map is based on the limited number of the

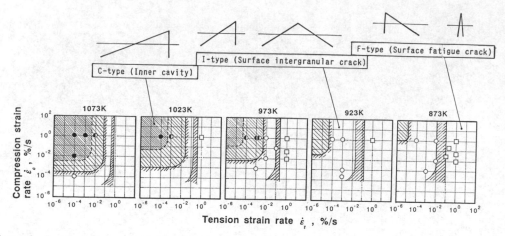

Figure 9 Fracture mode map of the high-temperature fatigue in a Type 304 stainless steel indicated in the diagrams of tensile strain rate versus compressive strain rate.

Table 1 Accelerating and decelerating factors of the irreversible damage accumulation during high temperature fatigue in polycrystalline alloys.

Crack Initiation	Small Crack Growth	Large Crack Growth

Accelerating Factors

(F)(I) Interaction between internal stress and effective stress,
Interaction between strain hardening and recovery (Increase in plastic or creep strain)
         (I) Interaction between elastic and creep deformation
            (Small scale creep condition)
(F)(I)(C) Microstructurally small crack (Stochastic)
(F) Mechanically small crack (Large scale yielding condition)
(F)(I) Chemically small crack (Environmental effect)
(F)(I)(C) Material degradation (Embrittlement for long term use)
(F) Distribution of many cracks   (F) Coalescence of cracks   (F) Cavity-zone cracking
(I)(C) Distribution of many cavities   (I)(C) Cavity-crack coarescence   (C) Unstable fracture
                                                             (Limitation of SCG)

Decelerating Factors

(F) Reversible dislocation gride   (F) Crack arrest by grain boundaries
(I) Reversible grain boundary sliding   (I) Crack arrest by grain boundary junction
(C) Reversible vacancy diffusion   (C) Healing of cavities and cracks
                          (F)(I) Crack closure (Plastic deformation/oxide induced)
                          (F)(I) Crack branching, kinking   (F)(I) Microcrack cloud

(F) Surface fatigue crack,   (I) Surface intergranular crack,   (C) Inner cavitatin crack

authors' experimental results and the boundaries of the mode are not always ascertained. Nevertheless, it is clear that the fracture mode changes according to the compressive strain rate as well as tensile one. Attention should be paid that the inner cavities of Type C is revealed only for very slow tension and fast compression strain cycles.

The other factor is the reversible deformation of every fatigue cycle. Even in creep dominated fatigue,the accumulation of damage is thought to be due to irreversible local deformation at the fracture sites.The change in the amount of irreversibility according the strain waveforms and strain ranges makes a difference between creep-fatigue and pure fatigue or pure (monotonic) creep. This change will also contribute to the difference in the fracture mode. Accelerating and decelerating mechanisms of the damage accumulation are listed in Table 1.

REFERENCES

Halford,G.R.,M.H.Hirschberg and S.S.Manson 1973. ASTM STP 520: 658-669.

Kitamura,T., N.Tada and R.Ohtani 1993. In Behaviour of Defects at High Temperatures, ESIS 15. R.A.Ainsworth & R.P.Skelton(eds): 47-69. Mech. Engg.Publ: London.

Ohtani,R., and T.Kitamura 1988. In High Temperature Creep-Fatigue, JSMS - Current Japanese Materials Research, Vol.3. R.Ohtani,M.Ohnami & T.Inoue(eds): 65-90. Elsevier.

Ohtani,R., and T.KItamura 1990. Proc. 4th Int. Conf. on Creep and Fracture of Engineering Materials and Structures. B.Wilshire & R.W. Evans(eds): 791-802. The Inst. of Metals.

Ohtani,R., T.Kitamura & N.Tada 1991. Materials Science and Engineering-A. 143-1/2: 213-222.

Saxena,A. 1993. JSME Int.J., Ser.A. 36-1: 1-20.

Solomon,H.D. and L.F.Coffin,Jr. 1973. ASTP STP 520: 112-122.

Recent Advances in Experimental Mechanics, Silva Gomes et al. (eds) © 1994 Balkema, Rotterdam, ISBN 90 5410 395 7

Nanoscopic evaluation of corrosion damage by means of scanning tunneling and atomic force microscopes

Kenjiro Komai, Kohji Minoshima, Masahiko Itoh & Takeshi Miyawaki
Kyoto University, Japan

ABSTRACT: This investigation demonstrates that scanning tunneling microscopy (STM) and atomic force microscopy (AFM) are capable of performing in-situ nanoscopic visualization of initiation and growth processes of localized corrosion in aqueous solutions. The nanoscopic initiation and growth mechanisms of localized corrosion of intergranular corrosion and stress corrosion crack of an austenitic stainless steel and 7XXX series aluminum alloys are discussed based upon nanoscopic in-situ visualization by using STM/AFM.

1 INTRODUCTION

Most damage issues in machines and structures are caused by environmentally induced material degradation in an operating environment, including corrosion fatigue and stress corrosion cracking. In order to clarify the fracture and damage mechanisms of these environmentally induced material degradation, serial high-magnification observation of damage initiation and growth processes are necessary.

Scanning tunneling microscopy (STM), that was first developed in 1982 (Binning 1982), is revolutionizing the study of surface physics and electrochemical researches (Masuda 1991a, 1991b). It is capable of imaging nanoscopic topography of surfaces not only in vacuum but also in air or in aqueous solutions. Although STM imaging of nonconducting surfaces is impossible, atomic force microscopy (AFM), that was developed in 1986 (Binning 1986), is capable of imaging topography of nonconducting surfaces (Meyer 1990).

This investigation demonstrates that STM and AFM are capable of performing in-situ nanoscopic visualization of initiation and growth process of intergranular corrosion as well as stress corrosion cracking of an austenitic stainless steel and aluminum alloys. We also discuss the concerned issues in an STM/AFM in-situ visualization in solutions, as well as nanoscopic corrosion damage mechanisms.

2 STM/AFM PROBE MICROSCOPES AND MATERIAL TESTED

The principles of operation of an STM are very simple: an extremely, usually atomically, sharp conducting tip made of Pt-Ir is brought in the proximity of a few nanometers from the sample surface-so close that a small constant voltage applied between the tip and sample produces a current due to the tunneling effect. This tunneling current decreases by an order of magnitude if the distance between the tip and sample increases by an order of about 0.1 nm. It is held constant by adjusting the z-axis of the micropositioner while the tip is raster-scanned. The image is formed from the record of vertical position required to keep the tunneling current constant as a function of x and y positions.

In aqueous solutions, the current due to electrochemical reactions on the tip is superimposed on tunneling current, and therefore, the current must be decreased by one-tenth to one hundredth of tunneling current. For this purpose, an STM operating in solutions uses the tunneling tip, which is glass-insulated for all but the last few microns, in order to minimize the tip surface exposed to a solution, as well as a specially designed bi-potentiostat, and the current associated with electrochemical reaction can thereby be minimized. In order to perform in-situ visualization of growth of stress corrosion cracks, the three-point bending jig was installed in the STM/AFM.

The cantilever used in an AFM is of a micromachined Si_3N_4 type. The sharp tip is positioned in the close proximity of the sample surface, thereby the cantilever being bent by the atomic force between the tip and sample surface. By using the light lever, an image of topography is then performed by keeping the cantilever deflection constant.

In this investigation, the STM unit operating in air, the STM unit in solutions and the AFM unit operating in air and/or solutions were connected

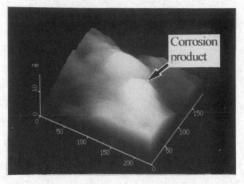 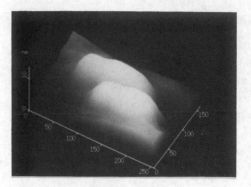

(a) In-situ visualization at an initiating stage
(Duration is 16min).

(b) In-situ visualization
(Duration is 32min).

Fig.1 Three-dimensional visualization of growth process of corrosion products formed
on the suface of 7075 alloy in a 3.5% NaCl solution.

to the probe station manufactured by Seiko Instrument Industries (Japan), and thereby the in-situ visualization were performed in aqueous solutions. The individual unit was mounted on a isolator, and the probe station controlled operation of the units and the necessary image processing of the acquired images.

The materials tested in this investigation were an austenitic stainless steel SUS304 and 7XXX series aluminum alloys, 7N01 and 7075. The stainless steels and 7N01 aluminum alloy were sensitization heat treated, and the 7075 alloy was treated to a peak aging of T6. The samples were ground to #2000 emery paper, and then were polished by 1 μm diamond paste followed by 0.04 μm silica powder.

3 EXPERIMENTAL RESULTS AND DISCUSSION

3.1 *In-situ visualization of growth process of corrosion product in solution by AFM*

Figure 1 shows bird eye view of the results of in-situ visualization of growth process of corrosion products that were forming on a surface of the 7075 aluminum alloy immersed in a 3.5% NaCl solution. From the figures, the corrosion products were observed to be piling up near the corrosion product "hills" shown by an arrow in Fig.1(a), which had existed from the beginning of AFM imaging (immersion duration: 16 min).

In the case of aluminum alloys, passive films form on the surface, giving relatively superior

(a) In-situ visualization
(Duration is 21 min).

(b) In-situ visualization
(Duration is 24 min).

(c) In-situ visualization
(Duration is 29 min).

Fig.2 In-situ visualization of intergranular corrosion of SUS304 in a 3.5% NaCl + HCl solution
(pH = 1.5).

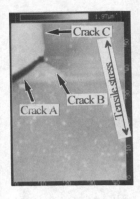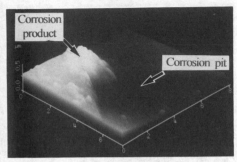

(a) In-situ visualization of crack tip in air.

(b) In-situ visualization of crack tip in a 3.5% NaCl solution (Testing duration is 20min).

(c) In-situ visualization of crack tip (Testing duration is 43min).

Fig.3   In-situ visualization of stress corrosion   crack of 7N01 by AFM   in a 3.5% NaCl solution.

resistance to corrosion. However, once the films are broken down by plastic deformation or the activity of chloride ions, dissolution occurred at that point. In Fig.1, the piling up of corrosion reaction products of aluminum hydroxide of dissolved aluminum and hydroxide ions were visualized by the AFM in an order of nanometer. The contact force between the tip and sample surface is so small that extremely soft surface such as corrosion products was successively topographically imaged.

3.2 In-situ observation of intergranular corrosion by AFM

Figure 2 shows serial imaging of the results of in-situ AFM visualization of intergranular corrosion of a sensitized austenitic stainless steel (sensitization heat treatment: 650° C for two hours) in a 3.5% NaCl + HCl solution (pH: 1.5): the brightness of each position expresses the height, and the most bright point corresponds to the highest, and the darkest one the lowest. A gray scale of each image shows the height difference of the highest and the lowest points. These figures were serial imaging. These demonstrate that three aligned corrosion pits along the grain boundary of 80 nm in depth and 100 nm in diameter were coalescing with each other, through the slender groove shown by an arrow. These pits were growing along the grain boundary, not in the depth (z) direction. This indicates that coalescing of aligned corrosion pits along the grain boundary lead intergranular corrosion.

3.3 In-situ observation of SC crack growth in solution by AFM

In-situ visualization of stress corrosion (SC) crack of the sensitized 7N01 aluminum alloy in a 3.5% NaCl solution was performed. The sample was mounted on the three-point bending jig, and was coated with silicone resin except the observing sample surface, in order to prevent the influence of galvanic corrosion on SC crack growth. An SC precrack was then introduced in a 3.5% NaCl solution at 30°C. After eight hour immersion, it was taken from the solution and was ultrasonically washed in ethanol. In-situ visualization was then performed in a 3.5% NaCl solution at 15°C.

Figure 3(a) shows the topographic image of an SC precrack tip by using the AFM in laboratory air. The SC precrack "A" branched at the triple grain boundary, and it grew into the crack "B" and crack "C". Figure 3(b) shows the in-situ imaging in solution of the crack "B" after 20 minute immersion, where the crack "B" retarded, whereas the crack "C" grew. Figure 3(c) shows the image of the crack "B" after 43 minutes, showing severe corrosion around and ahead of the SC crack tip, with the crack being covered with corrosion products. The crack tip finally changed into pitting corrosion like shape. Clear imaging of crack shape of the growing SC crack "C" also became impossible because it was covered with thick corrosion products.

3.4 In-situ observation of SC crack by AFM in air

It was rather difficult to perform in-situ imaging of SC cracks in this material/environment system, because the crack grew relatively fast and was covered with thick corrosion products. However,

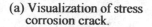
Tensile stress

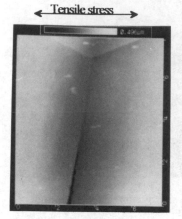

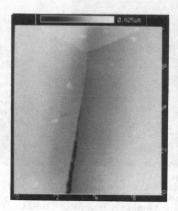

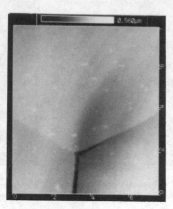

(a) Visualization of stress corrosion crack.

(b) Visualization of stress corrosion crack growth.

(c) Visualization of stress corrosion crack growth.

Fig.4 In-situ visualization of stress corrosion crack of 7N01 by AFM in air.

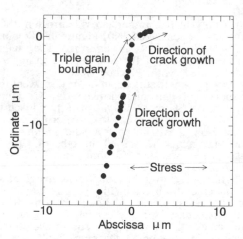

Fig. 5 Loci of crack tip.

it was noteworthy that the SC crack, which was introduced in a 3.5% NaCl solution, grew also in air. This meant that the in-situ, clear AFM visualization of crack shape was possible in air under better condition of no piling up of corrosion products over a crack. An SC precrack was then introduced in a 3.5% NaCl solution, similarly to the experiment discussed above, and in-situ AFM visualization was performed in air.

3.4.1 *SC crack growth along the grain boundary perpendicular to tensile stress*

Figure 4 shows the AFM images of the growing SC intergranular crack (crack length: about 5

mm), where the grain boundary was almost perpendicular to the tensile stress. It is clear that the topographies of deformed crystal surfaces ahead and behind the crack were symmetric against the grain boundary, indicating that the crack grew in pure Mode I. The crack grew successively, and turned to the right at the triple grain boundary, with growing along the grain boundary again.

Figure 5 shows the loci of the crack tip observed by the AFM in the Cartesian coordinate. The loci agreed with the crack shape after the test. In Fig.5, the time interval of the imaging was almost constant, indicating that the crack growth rate decreased with approaching the triple grain boundary. The average growth rate was about 3.3 nm/s, and the crack grew continuously in the order of microns.

3.4.2 *SC crack growth along the grain boundary inclined to tensile stress*

Figure 6 shows the AFM images of the growing SC crack (crack length: about 7 mm) along the grain boundary which inclined to the tensile stress. Before the Mode I crack opening of the growing crack from the upper right hand corner approached the triple grain boundary, the crack opening was observed at the triple grain boundary (see Fig. 6(a)). This indicates that the initially existing crack (crack ①) grew in the mixed mode of Modes I, II and III as far as the triple grain boundary. In fact, Fig.6 shows that not only Mode I displacement but also Modes II and III displacement was observed for the crack ①. The crack ② was also intergranular crack.

Figure 7 shows the loci of the observed crack tip by the AFM. The direction of tensile stress

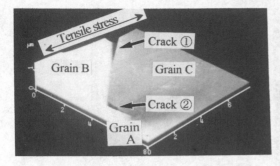

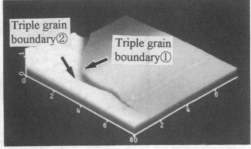

(a) Visualization of stress corrosion crack.

(b) Visualization of stress corrosion crack growth.

Fig.6 In-situ visualization of stress corrosion crack of 7N01 by AFM in air.

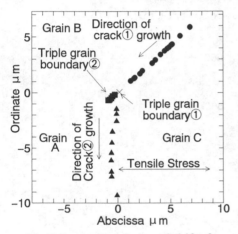

Fig.7 Loci of crack tip and shift of triple grain boundary.

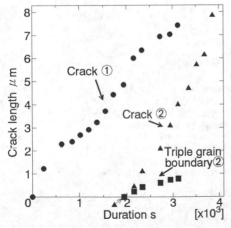

Fig.8 Growth curve of stress corrosion crack and shift of triple grain boundary with testing duration.

parallels the x-axis (abscissa). Figure 8 shows the relationship between time and crack length, which was measured along the individual growing direction. From these figures the grain "C" was detaching from the grains "A" and "B" with crack extension, and the average growth rate of the crack ① was about 2.4 nm/s, whereas the rate was 3.9 nm/s for the crack ② . The average Mode II crack opening rate at the triple grain boundary was 0.65 nm/s. These indicate that the growth rate of the crack ② was higher than that of the crack ①, because the crack ② grew along the grain boundary almost perpendicular to the tensile stress, and therefore the crack grew in Mode I loading, whereas the crack ① grew in the mixed mode of Modes I, II and III. In fact, no Mode II and III displacement was observed for the crack ②.

Figure 9 shows the AFM images of the growing SC intergranular crack (crack length: about 5mm) of 7075 in laboratory air. Cross sections of A-A' and B-B' lines in Fig.9, which respectively corresponded to pre- and post-cracking section, are shown in Fig.10. At pre-cracking section, difference of height of grain, i.e., mode III displacement, increased with crack tip approaching (Fig.10(a)) because of the concentration of stress. However, at post-cracking section, the difference decreased (Fig.10(b)) because of the relaxation.

4 CONCLUSIONS

This investigation demonstrates that STM and AFM are capable of performing in-situ, nanoscopic and topographic visualization of initiation and growth processes of localized corrosion. This investigations yielded the following conclusions.

1. STM and AFM were capable of performing in-situ nanoscopic visualization of corrosion damage including corrosion product forming, intergranular corrosion and stress corrosion

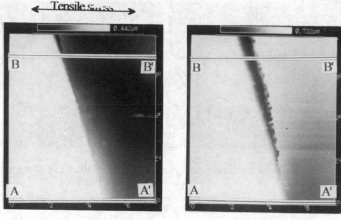

← Tensile stress →

(a) In-situ visualization of crack tip.　　(b) After 90 min.

Fig.9 In-situ visualization of stress corrosion crack of 7075 in air.

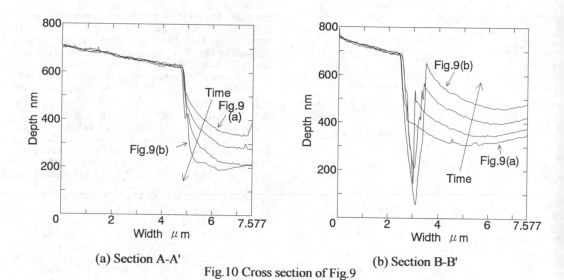

(a) Section A-A'　　　　　　　　　　(b) Section B-B'

Fig.10 Cross section of Fig.9

cracking. Compared with the two, an AFM was more suitable for the in-situ visualization from the standpoint of the capability of observing non- conductiong surface, stability, and so on.

2. Intergranular corrosion was formed by coalescence of aligned pitting corrosion which initiates along grain boundaries.

3. The crack growth rate of intergranular SC crack of an aluminum alloy decreased with approaching the triple grain boundary. The growth rate under Mode I loading was higher than that under a mixed loading mode.

REFERENCES

Binning, G., Rohrer, H., Gerber, Ch., and Weibel, E., *Appl. Phys. Lett.*, Vol.40, 1982, 178.
Binning, G. and Quate, C.H., *Phys. Rev. Lett.*, 56, 1986, 930.
Masuda, H., Matsuoka, S. and Nagashima, N., *Jour. Jap. Corros. Eng.*, Vol.40, 1991, 754.
Masuda, H., Nagashima, N. and Matsuoka, S., *Trans. JSME*, Series A, Vol.57, 1991, 2270.
Meyer, G. and Amer, N.M., *Appl. Phys. Lett.*, Vol.56, 1990, 2100.

Recent Advances in Experimental Mechanics, Silva Gomes et al. (eds) © 1994 Balkema, Rotterdam, ISBN 90 5410 395 7

Determination of the threshold for fatigue crack propagation using different experimental methods

K. Golos
Warsaw University of Technology, Poland

R. Pippan & H. P. Stüwe
Erich-Schmid-Institut für Festkörperphysik, Österreichische Akademie der Wissenschaften, Leoben, Austria

ABSTRACT: Pre-cracks generated in purely compressive cycles permit a determination of the threshold of stress intensity range and the crack growth curves with increasing load amplitude only. They also permit to measure the effective threshold of stress intensity range. This new method is compared with the load shedding technique according to ASTM and the decreasing of the stress intensity range at a constant maximum stress intensity.

1 INTRODUCTION

The mostly used method to determine the threshold of stress intensity range ΔK_{th} is the so-called load-shedding method (ASTM 1993). According to this method, a pre-fatigue crack is initiated at an intermediate value of stress intensity range ΔK ($K_{IC} \cdot (1 - R) > \Delta K > \Delta K_{th}$, where R is the stress ratio and K_{IC} is the fracture toughness). Then the load amplitude, and hence ΔK are successively reduced in steps until the growth rate is virtually zero, where two conditions should be fulfilled, the reduction rate for the stress intensity range ΔK

$$\frac{d\Delta K}{\Delta K \, da} \leq -0.08 \, [\text{mm}^{-1}] \qquad (1.1)$$

(a is the crack length) and the maximum value for the load reduction per step

$$\Delta(\Delta P) \geq 0.1\Delta P \qquad (1.2)$$

(ΔP is the load amplitude). The condition (1.1) causes a very time-consuming experiment, especially as one approaches the low growth rates near the threshold. Therefore, it is frequently violated which may lead to erronous values for the threshold of stress intensity range ΔK_{th}. Cadman et al. (1981) show that reduction rates smaller than the ASTM-limit can influence also the measured ΔK_{th} values.

An alternative technique is the stepwise increase of the load amplitude on specimens pre-cracked in cyclic compression (Suresh 1985, Pippan 1987a, Novack and Marissen 1987). This technique allows to determine the threshold ΔK_{th} and the growth rates "from below". In addition it permits to determine the effective threshold $\Delta K_{eff \, th}$ in a convenient way (Pippan 1987b and Pippan et al. 1993).

Another technique was proposed by Döker et al. 1981, which was extensively performed by Herman et al. 1988. In this test the maximum stress intensity factor is kept constant and the stress intensity range is decreased by increasing the minimum stress intensity factor. It is assumed (Döker and Herman) that the threshold determined by this way should be $\Delta K_{eff \, th}$. The load history of these three methods are shown in Fig. 1. The purpose of this paper is to compare

- the measured ΔK_{th} values of method I and II

- and $\Delta K_{eff \, th}$ of method II and III.

2 EXPERIMENTS

The material used was the alloy 7020/T5. After heat treatment its yield strength was 400 MPa and its ultimate strength 430 MPa. CT specimens ($B = 6$ mm, $W = 50$ mm and notch depth ≈ 15 mm) were machined in the LT-direction.

A standard electromechanical resonance machine was used (frequencey ≈ 150 Hz). Tests were conducted in air at room temperature. Crack length was monitored continuously by the direct current potential method (Fig. 2). The setup was calibrated with specimens that where cut to different depths.

- Method I (Fig. 1a):

  According to ASTM the rate of the load shedding with increasing crack length was chosen to

$$c = \frac{1}{\Delta K} \frac{d(\Delta K)}{da} = -0.07 \, \text{mm}^{-1}, \qquad (2.1)$$

where the amplitude of the load ΔP and the mean load was decreased after each extension of about 0.5 mm.

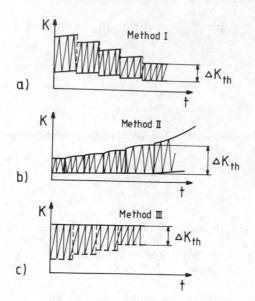

Figure 1: Schematic illustration of the loading procedure.

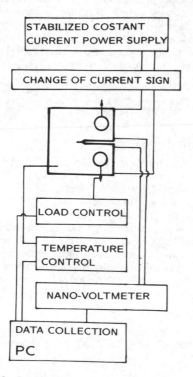

Figure 2: Set up for crack growth measurements by the potential drop technique.

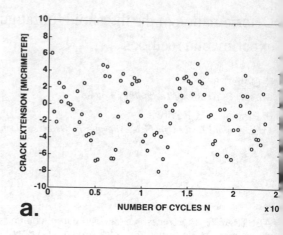

a.

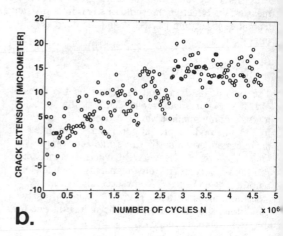

b.

Figure 3: Crack extension as a function of cycles at $R = 0.1$, at $\Delta K = 0.8$ (a) and $0.95\,\text{MPa}\sqrt{\text{m}}$ (b).

- Method II (Fig. 1b):

  The pre-cracked specimens were produced with a constant load amplitude in full compression ($R = 20$). The initial length measured from the notch root was about 0.3 mm. The advantage of pre-cracking the specimen in cyclic compression is that the stress intensity where the crack closes is below zero at the beginning of a tension-tension fatigue test. This permits to measure two threshold values with increasing load amplitudes, ΔK_{th} and $\Delta K_{\text{eff th}}$. The typical result of such test is shown in Fig. 3 and 4.

  Fig. 3 presents the results near $\Delta K_{\text{eff th}}$. At the load amplitude which corresponds to a ΔK of $0.8\,\text{MPa}\sqrt{\text{m}}$ no increase in the crack length was observed (the seemingly large scatter reflects the fine scale used for presentation). After an in-

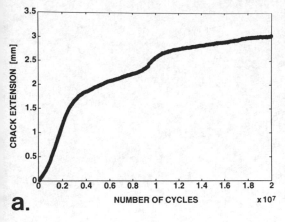

a.

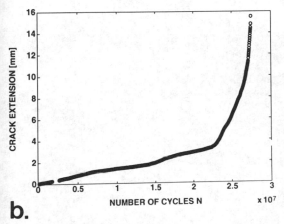

b.

Figure 4: Crack extension as a function of cycles at $R = 0.1$ where ΔK rises from 2.5 to 3.1 MPa$\sqrt{\text{m}}$ (a) and the crack growth test where ΔK increases from 3.5 MPa$\sqrt{\text{m}}$ (b).

crease of ΔK to 0.95 MPa$\sqrt{\text{m}}$ the crack grows and then stops. Since there is no closure of the crack at the beginning of the test we now have an upper and a lower bound for $\Delta K_{\text{eff th}}$. With further increments of the load amplitude the crack starts to grow and stops as soon as the closure stress intensity factor $K_{\text{cl}} = K_{\text{max}} - K_{\text{eff th}}$.

Finally there is an increment where the crack does not stop (Fig. 4) and we obtain an upper and a lower bound for ΔK_{th}.

- Method III (Fig. 1c):

This test involves decreasing load amplitude with constant K_{max} with a linear increase in K_{min} until crack growth stops. The K_{max} value was chosen as 7 MPa$\sqrt{\text{m}}$ (the same value was used to pro-

duce the precrack) and the load was changed after extensions of the crack of about 0.5 mm and the reduction rate of the stress intensity range with increasing crack length was

$$\frac{\Delta(\Delta K)}{\Delta(a)} = -0.2 \frac{\text{MPa}\sqrt{\text{m}}}{\text{mm}}. \qquad (2.2)$$

3 RESULTS AND DISCUSSION

The experiment shown in Fig. 3 and 4 yields the following results (at $R = 0.1$ with method II):

1. $\Delta K_{\text{eff th}}$, the effective threshold is between 0.8 and 0.95 MPa$\sqrt{\text{m}}$ and

2. ΔK_{th} for a "long" crack is between 3.1 and 3.5 MPa$\sqrt{\text{m}}$.

At $R = 0.7$ we obtained with method II $\Delta K_{\text{eff th}}$ between 0.85 and 1 MPa$\sqrt{\text{m}}$ and ΔK_{th} between 1.3 and 1.5 MPa$\sqrt{\text{m}}$.

The threshold determined with method I at $R = 0.7$ was also between 1.4 and 1.5 MPa$\sqrt{\text{m}}$, but at $R = 0.1$ the difference of the threshold values obtained with method I and II was significant. Therefore, different experiments were performed at this stress ratio. The first test with method I was started with a length of the pre-crack measured form the notch root of $\approx$ 5 mm and an initial $\Delta K = 8$ MPa$\sqrt{\text{m}}$. The measured threshold ΔK_{th} in this test was 4.4 MPa$\sqrt{\text{m}}$. In the second test the length of the pre-crack was 1 mm and the initial $\Delta K = 6$ MPa$\sqrt{\text{m}}$. The reduction rate in both tests was equal. The obtained threshold ΔK_{th} in the second test was 3.8 MPa$\sqrt{\text{m}}$. The length of the crack where we reached the threshold in these two tests differed significantly (this length measured from the notch root was 13 mm and 7.5 mm, respectively). Therefore, it seems that in this case the closure stress intensity at the threshold is influenced by the length of the crack, although one usually assumes a "long crack" behaviour.

Two experiments were performed also with method II where the length of the pre-crack measured from the notch root was 0.1 and 0.3 mm which were produced with $\Delta K \approx 10$ and $\Delta K \approx 18$ MPa$\sqrt{\text{m}}$ (at $R = 20$), respectively. The load steps in both tests are equal. No difference in the measured ΔK_{th} values was observed.

The stress ratio in the test with method III increases from 0.1 to 0.8, and the measured ΔK_{th} value was 1.4 MPa$\sqrt{\text{m}}$.

Therefore, we come to the following conclusions:

- The K_{max} constant test (method III) did not lead to the correct value of $\Delta K_{\text{eff th}}$ as determined by method II.

- No difference between threshold values obtained with method I and II was observed at high R-ratios.

- At low stress ratios the ASTM technique leads to higher values of ΔK_{th} – this may cause a nonconservative design.

- The method II permits to measure $\Delta K_{eff\,th}$ in a very simple way and the lowest ΔK_{th} values. In addition it should be noted that this method allows to determine a R-curve for the threshold (Pippan et al. 1993) which can be used to estimate the behaviour of short cracks.

Acknowledgement

This work was supported by the "Forschungsförderungsfonds für die gewerbliche Wirtschaft", Projekt Z 1-6/585 in cooperation with AMAG and Forschungszentrum Seibersdorf, K. Golos received a fellowship from the Austrian Federal Ministry of Science and Research.

REFERENCES

[1] Annual Book of ASTM Standards Vol.03.01 1993. E647,93, ASTM Philadelphia:679-706.

[2] Cadman, A.J., R. Book & E. Nicholson 1981. Effect of test technique on the fatigue threshold ΔK_{th}. In J. Bäcklund, A.F. Blom & C.J. Beevers (eds), *Proc. Fatigue Threshold*:59–75. Warley, U.K., EMAS.

[3] Nowack, H. & R. Marissen 1987. Fatigue crack propagation of short and long cracks: physical basis, prediction methods and engineering significance. In R.O. Ritchie & E.A. Starke (eds), *Proc. Fatigue*87:207–230. Warley, U.K., EMAS.

[4] Pippan, R. 1987a). The growth of short cracks under cyclic compression. *Fatigue Fracture of Engineering Materials and Structures* 9:319–328.

[5] Pippan, R. 1987b). Threshold and crack growth tests on pre-cracked specimens produced in cyclic compression. In R.O. Ritchie & E.A. Starke (eds), *Proc. Fatigue* 87:933–940. Warley, U.K., EMAS.

[6] Pippan, R., L. Plöchl & F. Klanner 1993. Schwellwert für den Ermüdungsrißfortschritt bestimmen. *Materialprüfung* 35:333–338.

[7] Suresh, S. 1985. Crack initiation in cyclic compression and its application. *Engineering Fracture Mechanics* 21:453–463.

Recent Advances in Experimental Mechanics, Silva Gomes et al. (eds) © 1994 Balkema, Rotterdam, ISBN 90 5410 395 7

Evolution of the internal microdeformations during appearance and development of fatigue crack

C. Gheorghies
Physics Department, University of Galati, Romania

L. Palaghian
Mechanics Department, University of Galati, Romania

ABSTRACT: Some experimental results regarding the evolution of the structural changes at crystalline lattice level during occurrence and propagation of fatigue crack are presented. The measurements have been achieved beginning with the moment of introducing the load and finishing with the damage moment of the tested sample. The structural changes, evinced by X-ray diffraction, have a cyclic character and in each stage of damage actuate a specific mechanism.

1 INTRODUCTION

Analysis of the occurrence and development mechanism of cracks is, at present, one of the main concerns in the study of the fatigue damaging process. Within the limited durability range, the fatigue damaging has a cumulative character and occurs in many distinct stages namely: the incubation process; the appearance and the attainment of critical dimensions with a few microcracks and the final breaking.

According to P.I.E. Forsyth (1962) the occurrence and development of microcracks (first stage of damage) are checked by local tension field (second order internal microtensions). The occurrence and the development of the macrocracks (second stage of damage) are determined by the presence and dimensions of plastic deformation enclaves at their peaks.

According to A.G. Pineau and R.M. Pelloux (1974) the dimensions of enclaves must outrun the dimension of the structural elements. With small tension J.L. Robinson and C.I. Beevers (1973) show that the plastically deformed area at the crack peak has a dimension which is of the same size order as the dimension of the structural elements. This leads to the growth of damage process sensibility as compared with the structural modifications.

In order to study the fatigue damage mechanism of OL52 steel, in this paper some structural changes, which occur in the superficial layer of certain standard samples subjected to the fatigue process within a limited durability range, are analysed.

2 METHODOLOGY FOR ESTIMATING STRUCTURAL CHANGES

The OL52 steel standard samples have been subjected to plane bending in symetric alternative cycle fatigue test at a tension level of $\sigma = 175.10^6$ N/m^2 (limited durability range). For the second stage of damage the structural changes in the plastic zone at the crack peak was evaluated.

By means of X-ray diffractometry the analysis was made of the evolution of certain fine structure parameters in the same area when the test was interrupted at certain intervals. The superficial layer of the zone subjected to fatigue was irradiated by X-rays on a DRON-3 equipment. The diffraction spectrum of X-radiation ($\lambda = 0.70926$ Å) was traced within the angular interval $2\theta \in [39°-42°]$ and the evolution of the profile of diffraction line(220) corresponding to the ferritic-perlitic stage of steel, was studied depending on the stress duration.

The level of second order internal microdeformation was evaluated by determining the width of the diffraction line (220). This is directly proportional to the lack of homogeneousness of the interplane distance inside the mosaic block, η, which can be determined by the relation:

$$\eta = (\frac{\Delta d}{d})_{[220]} = \frac{B_{[220]}}{4.tg\ \theta_{[220]}} \qquad (1)$$

where $B_{[220]}$ = the physical width of the line(220); and $\theta_{[220]}$ = Bragg diffraction angle associated to the gravity center of the same line.

The level of dislocation density in the crystalline lattice, ρ, was estimated by the ratio $(I_{fon}/I_{max})_{[220]}$ because,

$$\rho \sim \left(\frac{I_{fon}}{I_{max}} \right)_{[220]} \qquad (2)$$

where I_{fon} = the minimum intensity of the diffraction line (220); and I_{max} = the maximum intensity of the diffraction line (220).

The level of the first order internal tension, σ_I, can be estimated by relation

$$\sigma_I = - \frac{E}{\mu} (ctg\ \theta_{[220]})(\Delta\theta)_{[220]} \qquad (3)$$

where E = Young's modulus; μ = Poisson's coefficient; and $(\Delta\theta)_{[220]}$ is given by the relation:

$$(\Delta\theta)_{[220]} = \theta_{[220]} - \theta_{[220]st} \qquad (4)$$

where $\theta_{[220]st}$ = Bragg diffraction angle associated to the line gravity center of the standard sample which is characterized by the absence of the structural defects.

The fatigue macrocrack length was evaluated during the fatigue test by means of an optical microscope (magnification, G=100).

3 EXPERIMENTAL RESULTS AND DISSCUTION

Experimental determinations show that the most intense structural modifications occur both in the first thousand stress cycles(figure 1, range A, curves 1 and 2) and in occurence stage of the fatigue macrocrack(figure 2, curve 2 and 3).Within such intervals strains and releases of the crystalline lattice occur with high amplitudes but with different frequencies (higher frequencies in the first thousand stress cycles and lower ones after the occurrence of the fatigue macrocrack, $3.1.10^4$ cycles). In the first thousand cycles(the incubation period)an important role is played by the dislocations which accumulate and outrun the obstacles of the crystalline lattice. After the occurrence of the macrocrack,the dislocation density maintains at the relatively constant level, their influence on the damage process is reduced, the role of second order microtensions being prevailing. In range B of figure 1 the second order internal tensions oscillate with a lower amplitude as compared with those of range A. In this range, the most important is the dislocation accumulation, thus prepearing the beginning of the second stage of damage. The curve 4 in figure 2 shows that the

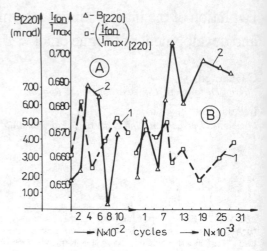

Fig.1.

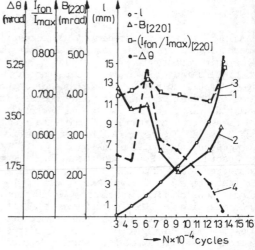

Fig.2.

first order internal tensions decrease according to the crack length (figure 2, curve 1) increase as a consequence of relaxations.

Some our researches(L.Palaghian and C.Gheorghies(1993))pointed out that if the second order internal tensions have a lower level, then the period until the occurrence of the macrocrack is longer.

4 CONCLUSION

4.1 The fatigue damage process develops in the three stages when some structural changes at the crystalline lattice level occur.

4.2 The structural modifications occur by steps with different amplitudes and frequencies at the level of each stage, the role of the dislocation respective of the second order internal tension being, in turn, prevailing.

REFERENCES

Forsyth,P.I.E.1962.Proc.Crack Prop. Symp., Cranfield, College of Aeronautics:76-94.
Palaghian,L. & C.Gheorghies 1993.Eigth Int. Conf.on.Fract.Col.of.Abstr.I.170.Kiev.
Pineau A.G.& R.M.Pelloux 1974.Metal.Trans. 5:1103-1112.
Robinson J.L. & C.I.Beevers 1973.Metal.Sci. J.9:153-159.

Recent Advances in Experimental Mechanics, Silva Gomes et al. (eds) © 1994 Balkema, Rotterdam, ISBN 90 5410 395 7

Experimental investigation of fatigue crack propagation in 7475-T761 aluminium alloy

C.P.M. Pereira, M.P. Peres & V.A. Pastukhov
FEG/UNESP, Guaratinguetá, SP, Brazil

ABSTRACT: The retardation effect of overloads on crack propagation under cyclic loading is investigated. The dependence of both intensity of retardation and its extension at crack line from number of overload cycles and some acceleration of crack propagation after retardation caused by extended overload block are observed. The possibilities to describe these phenomena using known models of crack propagation are discussed.

1. INTRODUCTION

The lifetime of structural elements containing strong stress concentrators is determined by the rate of subcritical crack growth. In the case of crack propagation under cyclic loading with variable parameters, the shape of maximal and minimal load functions effect significantly on crack behaviour. In particular, the overloads can provoke retardation of crack growth, delay or complete crack arrest after return to reference loading. The detailed investigation of such phenomena under different materials and loading conditions and its adequate description is important for optimization of structural design and evaluation of residual lifetime of cracked elements. The retardation effect of overloads is sensitive to various parameters such as overload ratio, number of cycles in overload block, current crack length, etc., and its particular properties related to different structural materials is a subject of experimental investigations.

As regards theoretical description of retardation, it's based on hypothesis of residual stress arising in near crack tip zone plastically affected during period of overload. The known models (Elber 1971; Willenborg et al. 1972; Wheeler 1972) are formulated in terms of crack closure or of effective stress intensity factor range and use quasi-static evaluation of plastically affected zone size and some empirical parameters. The development of this approach (Matsuoka & Tanaka 1979) is based on experimental observation of plastically affected segment of crack propagation which results in corrective factor for monotonic evaluation of plastic zone size. The recently formulated and developed so-called "local approach of fracture" in the framework of continuum damage mechanics (Lemaitre 1986) opens new opportunities for prediction of subcritical crack behaviour. For this concept an adequate approximation of stress field in total net-section is vital and the behaviour of plastically affected zone should be carefully described.

The main objective of present work is experimental investigation of crack growth retardation caused by overload in 7475-T761 aluminium alloy widely used in aircraft industry. The significant retardation effect for overload ratio 1.2 and more and the important role of number of overload peaks for crack growth retardation in close high-strength aluminium alloys has been observed (Torres et. al 1992). Other parameter affecting on retardation behaviour is current crack size at themoment of unloading or partial unloading (end of overload block) because of growth of plastically affected zone with crack length.

In present investigation the special attention is paid at influence of number of overload peaks on intensity of retardation and on the size of overload affected segment of crack propagation under different overload ratio and for overloads applied at different crack length.

The other aim is to analyze the known theoretical models of fatigue crack propagation under variable amplitude loading with respect to experimentally observed properties of retardation.

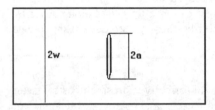

Figure 1. Center cracked specimen

2. EXPERIMENTAL PROGRAM

2.1 Material and Specimens

The specimens are cutted in longitudinal direction from sheet of aluminium alloy 7475-T761 (yielding strength σ_{ys} = 393 MPa; ultimate strength σ_{us} = 462 MPa; modulus of elasticity E = 68900 MPa; sheet thickness 4.16 mm). The gage length of center cracked specimens is 300 mm and width is 75 mm, the initial crack length 2a = 14 mm (figure 1).

2.2. Loading Conditions

The fatigue crack propagation is investigated at loading frequency 50 Hz and fixed amplitude of load ±3.15KN. The tests with four different values of constant mean stress (9.45, 14.7, 19.2 and 36.9KN) are fulfilled. First two regimes are used as reference in investigation of overload effect. The overload blocks of 1, 10, 100 or 1000 cycles with mean stress 19.2KN are applied at crack length a = 10, 15 and 20 mm.

3. RESULTS

The transition of crack growth process from subcritical to critical phase is observed at crack length 25mm < a < 32mm. Generally, the retardation caused by overloads prolongates significantly fatigue life and effect of overload increase with overload ratio.

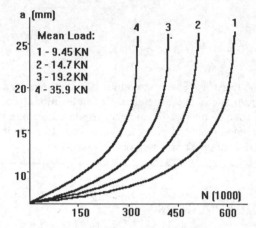

Figure 2 Crack growth under constant amplitude ±3.15KN and constant mean stress.

3.1 Constant mean stress

The significant dependence of crack growth rate from mean stress is observed in tests with fixed loading parameters. For fixed amplitude of loading the increasing of mean stress is resulted in decreasing of fatigue life.

The experimental curves "crack length versus number of load cycles" are characterized by strong increasing of crack growth rate during crack propagation (figure 2).

3.2 Retardation at reference load 9.45±3.15KN

The tests with overloads 19.2±3.15KN and reference load 9.45±3.15KN correspond to the overload ratio 1.77. For high strength aluminium alloys it's a range of strong retardation effect and close to the threshold of crack arrest. The overload affected segment of crack propagation is determined by comparison of curves "a-a_0 versus N-N_0" (a_0 is a crack length at the end of overload block; N_0 is corresponding number of cycles) for overload and reference curves (for

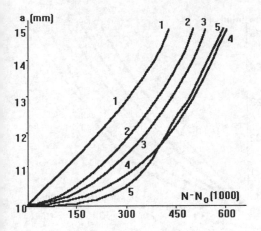

Figure 3. Retardation after overload (ratio 1.77) at crack length 10 mm. 1 - reference curve; 2 - 1; 3 - 10; 4 - 100; 5 - 1000 overload peaks.

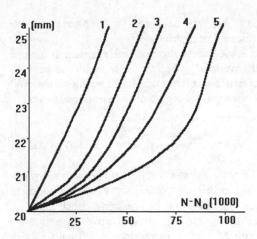

Figure 5. Retardation after overload (ratio 1.77) at crack length 20 mm. 1 - reference curve; 2 - 1; 3 - 10; 4 - 100; 5 - 1000 overload peaks.

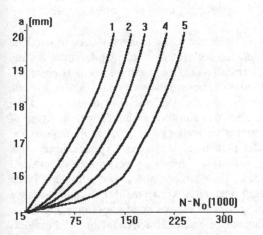

Figure 4. Retardation after overload (ratio 1.77) at crack length 15 mm. 1 - reference curve; 2 - 1; 3 - 10; 4 - 100; 5 - 1000 overload peaks.

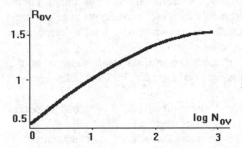

Figure 6. Normalized overload affected zone size (R_{OV}) versus number of overload peaks (N_{OV}).

constant loading parameters tests). This comparison given by figures 3, 4 and 5 for crack length 10, 15 and 20mm respectively shows an essential influence of number of overload peaks

From the analysis of data presented at figures 3, 4 and 5 can be concluded:

1. Both extension of overload affected segment and intensity of retardation increase significantly with number of peaks in overload block, however, the tendency to saturation of this influence is observed.

2. The extension of zone where crack behaviour after overload is different from one in loading program without overloads is less than quasi-static evaluation (size of monotonic plastic zone under overload minus size of monotonic plastic zone under reference load) in the cases of 1 or 10 overload peaks.

3. This extension is more than quasi-static evaluation in the cases of 100 or 1000 overload peaks.

4. For all three values of crack length the growth of overload affected zone normalized by quasi-static evaluation with number of overload peaks

may be approximated by common curve presented at figure 6.

5. Some acceleration of crack growth at the end of overload affected zone with respect to reference test (with constant loading parameters) is observed for big number of overload peaks.

3.3 Retardation at reference load 14.7±3.15KN

The maximal stress ratio (overload to reference load) in this case (1.25) usually corresponds to less intensity of retardation effect than in previous case. In present investigation it's resulted in increasing difficulties in determination of overload affected zone because of increasing similarity of compared curves. However, the intensity of overload effect may be analyzed the total number of cycles of crack propagation from initial length to 25mm. Figure 7 represents the crack retardation after overload at length 20mm where overload affected zone also can be observed.

The main conclusions from analysis of the results for overload ratio 1.25 are following:

1. The retardation effect of overload increases with number of overload peaks.
2. This influence has a tendency to saturation.
3. For 1000 cycles in overload block the acceleration of crack growth at the end of overload affected zone is observed.

The comparison with results for overload ratio 1.77 shows that:

1. For the fixed amplitude of loading and maximal stress of overload the intensity of retardation decreases with increasing of maximal stress of reference load.
2. Normalized overload affected zone size at crack length 20mm for both cases corresponds to approximation presented at figure 6.

4. DISCUSSION

Let's analyze the experimentally observed properties of retardation caused by overload with respect to main approaches in theoretical modelling of subcritical fatigue crack growth.

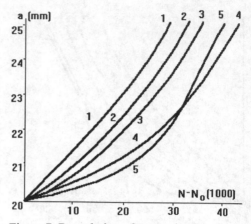

Figure 7. Retardation after overload (ratio 1.25) at crack length 20 mm. 1 - reference curve; 2 - 1; 3 - 10; 4 - 100; 5 - 1000 overload peaks.

4.1. Influence of number of overload peaks

The influence of number of cycles in overload block on both intensity of retardation effect and extension of overload affected zone is observed. It should be mentioned that for the lifetime predictions the last one isn't decisive, but the correct description of retardation in terms of number of load cycles is of most importance. The models of traditional type, relating crack growth rate directly to parameters of external load, successively solve this problem using empirical approximations for N_{OV}-dependence of corrective factors or of stress intensity factor thresholds. Such corrective factors or thresholds are applied exclusively in overload affected zone determined by quasi-static evaluation which also can be supplied by empirica corrective factor like in model of Matsuoka & Tanaka (1979).

The main advantage of this approach is neglecting of other, more complicated, aspects of near crack tip stress distribution under cyclic loading like N_{OV}-dependence of plastic affected zone size or equations "stress versus distance from crack tip". The other side of this advance is a necessity to analyze a big array of experimental data on crack propagation to determine all used parameters. In other words, this approach represents mainly an approximation of experimentally obtained points by curves of expected shape and has limited possibilities to prediction of crack behaviour under new combination of materials, geometric and loading conditions.

Contrary, the promising "local approach of fracture" gives a predictions of subcritical crack behaviour using only experimental parameters obtained in lifetime tests of smooth specimens. This advantage is also balanced, in this case by necessity of careful analysis of stress evolution in net-section of element with crack growing under cyclic loading. The direct application of quasi-static approximations of overload affected zone size following the analysis of Rice (1967) can give an adequate results for extended overload blocks when the steady state behaviour may be considered (Pastukhov et al. 1993). However, the N_{ov}-dependence of plastic affected zone size can't be neglected like in approach analyzed above. The description of plastic zone growth with number of overload cycles is the only tool for description of properly dependence of fatigue lifetime from number of overload peaks. It should note that the application of empirical approximations based on crack growth tests like presented at figure 6 is contrary to philosophy of considered approach and, thus, the detailed theoretical investigation of plastic zone behaviour based on energy balance is necessary for correct application to short overload blocks.

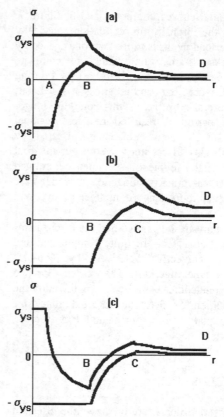

Figure 8. Approximation of near crack tip maximal/minimal stress fields under reference load (a), overload (b) and immediately after overload (c).

4.2. Acceleration after retardation

In the framework of traditional type models the acceleration of crack propagation at the end of overload affected zone may be explained only by dispersion of experimental data. However, the "local approach to fracture" based on damage mechanics concept naturally describes this phenomenon and its sensitivity to number of overload peaks.

Let's consider for simplicity the quasi-static analysis of stress evolution near crack tip under cyclic zero-to-tension loading with one overload block (Rice 1967; Matsuoka & Tanaka 1979) presented at figure 8. Such schemes are widely used for explication of retardation phenomenon by residual stress arising near crack tip after overload. However, from this analysis it's clear that during period of overload (fig 8, b) the

material at the part of plastic zone (region BC) and out of this zone (region CD) is operating under more severe loading condition than in the case of reference load (fig. 8, a). From the point view of damage mechanics it means the faster reduction of materials local lifetime which is resulted in accelerated crack propagation throughout this zone. In the region BC the contrary tendency exists: after overload the local loading conditions are softer than under reference load. In the region CD the overload influence may be resulted only in accelerated degradation of material directly related to extension of overload block. This theoretical conclusions are in good correspondence with experimental observations. In particular, the "local approach of fracture" formulated in terms of damage mechanics explains the behaviour of experimental curves number 4 and 5, figure 7. Initially, the intensity of retardation caused by 1000 overload peaks is more than one caused by 100 peaks. It may be related to complete development of zone, plastically affected during overload, where the residual stress arises. Later, when the crack tip passes this zone, the longer overload period is resulted in less residual lifetime in the zone elastically affected by overload. This effect may overweight the first one and total lifetime of specimen determined by subcritical crack propagation may decrease with further increasing of number of overload peaks.

This analysis shows also that the mechanisms of retardation effect are quite complicated and can't be correctly described by "on-off" Heaviside-type functions. The existence of two contrary tendencies complicates significantly the determination of overload affected zone size from the analysis of experimental data on crack propagation.

5. CONCLUSIONS

(1) The experimental study shows the effect of overload ratio and number of overload peaks on retardation of fatigue crack propagation in 7475-T651 aluminium alloy.
(2) The growth of overload affected zone with number of cycles in overload block is observed. This phenomenon should be carefully studied and described to develop the damage mechanics models of crack growth aimed at short overload blocks.
(3) Some acceleration of crack growth after retardation caused by extended overload block is observed. This property of process is completely described by "local approach of fracture" based on damage mechanics concept and neglected by models of traditional type.

Acknowledgements

The support of this work by FAPESP (Research Foundation of Sao Paulo State, Brazil) through Process N.92/2770-0 is gratefully acknowledged.

REFERENCES

Elber, W. 1971. The significance of fatigue crack closure. *ASTM STP 486*: 230-242.
Lemaitre, J. 1986. Local approach of fracture, *Eng. Fracture Mechanics*, 25: 523-537.
Matsuoka, S., & K.Tanaka, Influence of stress ratio at baseline loading on delayed retardation phenomena. *Eng. Fracture Mechanics*, 11: 703-715.
Pastukhov, V.A., H.J.C.Voorwald, & M.Barbosa. 1993. Simulation of subcritical crack growth under cyclic load and/or creep conditions. *Proc.14 CILAMCE, Sao Paulo, Brazil*, 2: 840-849 (in Portuguese).
Rice, J.R. 1967, Mechanics of crack tip deformation and extension by fatigue. *ASTM STP 415*: 247-311.
Torres, M.A.S., C.A.R.P.Baptista, H.J.C.Voorwald & J.A.M. de Camargo 1992. Effect of the overload cycles on fatigue crack growth retardation. *Processing, Properties and Applications of Materials. Proc. Int. Conf., Birmingham, UK.*: 983-987.
Wheeler, O.E., 1972 Spectrum loading and crack growth. *J. of Basic Engineering, Trans. ASME*, 94.
Willenborg, J., R.M. Engle & H.A.Wood 1971. *Technical Memorandum 71-1-FBR.*

Recent Advances in Experimental Mechanics, Silva Gomes et al. (eds) © 1994 Balkema, Rotterdam, ISBN 90 5410 395 7

Anisotropic fatigue crack propagation of Ti-6A1-4V alloy in 3% NaCl solution

M. Nakajima & T. Shimizu
Toyota College of Technology, Japan

K. Tokaji & T. Ogawa
Gifu University, Japan

ABSTRACT: Fatigue crack propagation (FCP) has been investigated on a Ti-6Al-4V alloy in 3%NaCl solution. Experiments were conducted on three different microstructures prepared with annealing at 705°C (AN705) and 850°C (AN850) and solution-treatment and aging (STA), and on two orientations. In T-L orientation for AN705 and AN850 and both orientations for STA, an abrupt increase in FCP rates took place at a certain ΔK value, ΔK_{SCC}, which was strongly related to SCC under cyclic loading. Fractographic examination revealed that fracture surfaces consisted of extensive cleavage when $\Delta K \geq \Delta K_{SCC}$. A considerable anisotropy in FCP was observed for the annealed microstructures. Their crack paths in L-T orientation suddenly changed toward the rolling direction, suggesting the strong sensitivity to corrosion fatigue crack propagation (CFCP) in T-L orientation. The pole figures showed that the annealed microstructures had a transverse texture in which (0002) planes were parallel to the rolling direction. This indicated that the largest susceptibility to aqueous NaCl was found for loads acting perpendicular to (0002) planes.

1 INTRODUCTION

It has been found so far that high strength titanium alloys were susceptible to sustained load cracking (SLC) due to internal hydrogen (Meyn 1974, Boyer and Spurr 1978) and stress corrosion cracking (SCC) (Meyn and Brooks 1981). In corrosion fatigue crack propagation (CFCP), several studies have shown that enhanced FCP rates were observed in aqueous environments, in particular contained Cl⁻, when compared to laboratory air (Meyn 1971, Dawson and Pelloux 1974, Wanhill 1976, Ryder *et al.* 1978, Dawson 1981, Yoder *et al.* 1983). However, the role of microstructure in CFCP has not always been clarified. From this point of view, the present investigation has been undertaken. In the preliminary tests in 3%NaCl solution, it was observed that the crack paths in L-T orientation for annealed microstructures suddenly changed toward the rolling direction as shown

in Fig.1, suggesting that the CFCP was strongly dependent on both microstructure and orientation.

In the present paper, FCP characteristics were studied on different microstructures and orientations of a Ti-6Al-4V alloy in 3%NaCl solution and the effects of microstructure and crystallographic texture on CFCP were discussed based on the pole figure analysis and fractographic examination.

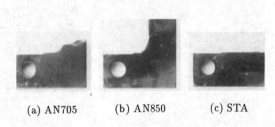

(a) AN705   (b) AN850   (c) STA

Fig.1. Fatigue crack paths in 3%NaCl solution (L-T orientation).

2 EXPERIMENTAL PROCEDURES

The material used is a Ti-6Al-4V alloy. The chemical composition (wt.%) is 0.008N, 0.16O, 0.004H, 0.19Fe, 0.006C, 6.29Al, 4.13V, <0.001Y, balance Ti. Three microstructures were prepared with heat treatments, *i.e.* annealing at 705°C (AN705) and 850°C (AN850), and solution treatment and aging (STA). The annealed microstructures consist of recrystallized α-phase and β-phase. Since the annealing temperature of AN850 is higher than AN705, the grain sizes of α -phase for the former are somewhat larger than the latter. The microstructure of STA consists of primary α and (α+ β) mixture resolved from α'-phase. The mechanical properties are summarized in Table 1. Note

the considerable anisotropy.

Experiments were performed at a cyclic frequency of 1Hz and a stress ratio of R=0.05 using CT specimens with two orientations, *i.e.* L-T and T-L orientations; L-T refers to loading in the longitudinal (rolling) direction and crack propagation in the transverse (width) direction. The environment was a 3%NaCl solution, in which the hydrogen-ion concentration, pH, and the dissolved oxygen content were approximately 5.0 and 7.82ppm, respectively. Fracture surfaces were examined by a scanning electron microscope (SEM), and the crystallographic texture of the alloy after heat treatments was characterized by determining the distribution of (0002) poles using a X-ray diffraction equipment.

Table 1. Mechanical properties.

Meterial	Specimen orientation	0.2 % proof stress	Tensile strength	Breaking strength on final area	Elongation	Reduction of area
		$\sigma_{0.2}$ (MPa)	σ_B (MPa)	σ_T (MPa)	δ (%)	ϕ (%)
AN705	L	880	933	1143	14	26
	T	1032	1063	1256	11	26
AN850	L	848	944	1165	14	27
	T	997	1067	1272	12	25
STA	L	1108	1176	1407	11	26
	T	1189	1249	1480	9	26

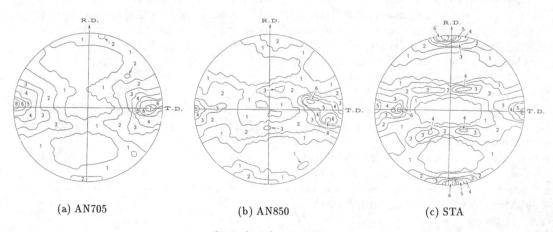

(a) AN705   (b) AN850   (c) STA

Fig.2. (0002) pole figures.

3 EXPERIMENTAL RESULTS

3.1 *Texture*

(0002) pole figures are shown in Fig.2; where RD and TD represent the rolling direction and transverse direction, respectively. AN705 has a transverse texture in which (0002) planes are oriented parallel to the rolling direction (Fig.2(a)). AN850 also has a similar texture, but the anisotropy of the texture is weaker than AN705 (Fig.2(b)), because of the higher annealing temperature. In both annealed materials, the influence of processing can not be eliminated, since the heat treatments were performed at temperatures below β-transus. On the other hand, STA is characterized by a random crystallographic texture (Fig.2(c)), indicating considerably less anisotropy than the annealed materials.

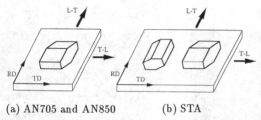

(a) AN705 and AN850   (b) STA

Fig.3. Schematic illustration of orientation of hexagonal unit cells and loading direction.

Figure 3 shows a schematic illustration of the relationship between orientation of hexagonal unit cells and loading direction. It should be noted that for the annealed materials the load acts perpendicular to (0002) planes in T-L orientation.

3.2 *FCP behaviour*

Figure 4 shows the FCP characteristics for all the materials in room air and in 3%NaCl solution. For AN705 and AN850, the FCP rates of both orientations are similar in room air, indicating that the effect of orientation is not observed. On the other hand, FCP behaviour in 3%NaCl solution exhibits a considerable anisotropy. The crack paths in specimens with L-T orientation suddenly changed toward the rolling direction (see Fig.1). In T-L orientation, an abrupt increase in FCP rates takes place at a certain ΔK value, 14MPa$\sqrt{\text{m}}$ for AN705 and 15MPa$\sqrt{\text{m}}$ for AN850, which is considered to be related to cyclic SCC.

For STA, the FCP rates in 3%NaCl solution are enhanced at a certain ΔK value in both orientations. The extent of acceleration is more remarkable in T-L orientation than in L-T orientation.

3.3 *Fractography*

The fracture morphology in room air was mostly transgranular, accompanied with striations at high

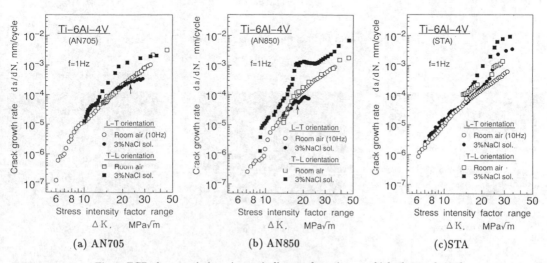

(a) AN705   (b) AN850   (c)STA

Fig.4. FCP characteristics. Arrow indicates the point at which the crack path began to deflect toward the loading direction.

ΔK levels, regardless of microstructure and orientation (Ogawa *et al.* 1992).

SEM micrographs of the fracture surfaces in 3%NaCl solution are shown in Figs.5 to 7. The appearances of fracture surfaces for the annealed materials, AN705 and AN850, in T-L orientation (Fig.5 and Fig.6) are similar, and the fracture surfaces at $\Delta K = 11\mathrm{MPa}\sqrt{\mathrm{m}}$ where FCP rates were not enhanced consist of transgranular with a small fraction of cleavage. On the other hand, at $\Delta K = 27\mathrm{MPa}\sqrt{\mathrm{m}}$ where the enhanced FCP was observed, extensive cleavage facets can be seen.

For STA, the features of the fracture surfaces are similar in both orientations (Fig.7(a) and (b)). In the fracture surfaces at $\Delta K = 15\mathrm{MPa}\sqrt{\mathrm{m}}$ where the FCP was not enhanced, transgranular fracture accompanied with secondary cracks and facets which correspond to primary α phase is observed. On the other hand, at $\Delta K = 27\mathrm{MPa}\sqrt{\mathrm{m}}$ where the FCP was enhanced, brittle fracture including cleavage facets and tubular voids is seen, which is very similar to SCC fracture surfaces in titanium alloys (Meyn and Bayles 1987). Therefore, the enhanced FCP in 3%NaCl solution may be attributed to cyclic SCC independent of microstructure and orientation.

4 DISCUSSION

A considerable anisotropy in FCP behaviour in 3%NaCl solution was observed for the annealed microstructures. The deflection of crack path in L-T orientation indicates the strong sensitivity to CFCP in the rolling direction. Blue and brown stains were observed more clearly on the specimens' sides and fracture surfaces of the annealed materials. The macroscopic fracture surfaces are shown in Fig.8 for all the materials and both orientations. Such stains can be seen as black area on

Crack growth direction $\longrightarrow$

 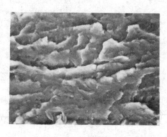 

Microstructure $\Delta K = 11\mathrm{MPa}\sqrt{\mathrm{m}}$ $\Delta K = 27\mathrm{MPa}\sqrt{\mathrm{m}}$

$\underset{\text{10 }\mu m}{\llcorner\quad\lrcorner}$

Fig.5. SEM micrographs of fracture surfaces for AN705 (T-L orientation).

Crack growth direction $\longrightarrow$

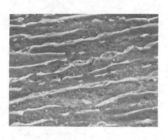 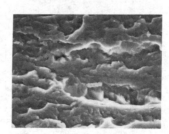 

Microstructure $\Delta K = 11\mathrm{MPa}\sqrt{\mathrm{m}}$ $\Delta K = 27\mathrm{MPa}\sqrt{\mathrm{m}}$

$\underset{\text{10 }\mu m}{\llcorner\quad\lrcorner}$

Fig.6. SEM micrographs of fracture surfaces for AN850 (T-L orientation).

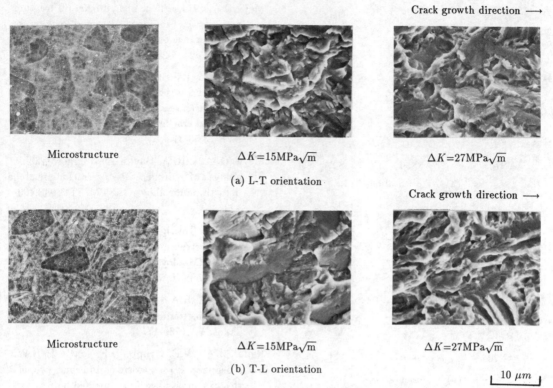

Crack growth direction ⟶

Microstructure $\Delta K=15\mathrm{MPa}\sqrt{\mathrm{m}}$ $\Delta K=27\mathrm{MPa}\sqrt{\mathrm{m}}$

(a) L-T orientation

Crack growth direction ⟶

Microstructure $\Delta K=15\mathrm{MPa}\sqrt{\mathrm{m}}$ $\Delta K=27\mathrm{MPa}\sqrt{\mathrm{m}}$

(b) T-L orientation

|⎯ 10 μm ⎯|

Fig.7. SEM micrographs of fracture surfaces for STA.

the fracture surfaces. In addition, bubbles could be seen streaming from cracks. These observations suggest that an electrochemical reaction had taken place at the crack tip, generating hydrogen. Produced hydrogen can enter into the metal and hydrides would precipitate on or near (0002) planes (Paton and Spurling 1976, Hall 1978). Furthermore, a cleavage fracture was often observed on or near (0002) planes (Meyn 1971).

As shown in Fig.2, the annealed microstructures had a transverse texture in which (0002) planes were oriented parallel to the rolling direction. Thus, the cleavage fracture can take place more easily in T-L orientation, loads being applied perpendicular to (0002) planes, than in L-T orientation, leading to anisotropy in CFCP behaviour.

two annealed microstructures, AN705 and AN850, and a solution-treated and aged microstructure, STA, and on two orientations, L-T and T-L. In T-L orientation for AN705 and AN850 and both orientations for STA, an abrupt increase in FCP rate took place at a certain ΔK value, which was related to SCC under cyclic loading. Fractographic examination revealed that the fracture surfaces at which the enhanced FCP was observed consisted of extensive cleavage. A considerable anisotropy in FCP behaviour was observed for the annealed microstructures which showed a transverse texture, in which (0002) planes were oriented parallel to the rolling direction.

5 CONCLUSIONS

FCP behaviour was investigated on a Ti-6Al-4V alloy in 3%NaCl solution. Tests were conducted on

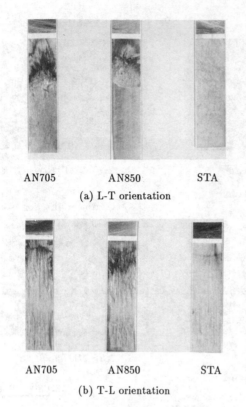

AN705   AN850   STA

(a) L-T orientation

AN705   AN850   STA

(b) T-L orientation

Fig.8. Macroscopic appearance of fracture surfaces.

REFERENCES

Boyer,R.R. & W.F.Spurr 1978. Characteristics of sustained-load cracking and hydrogen effects in Ti-6Al-4V. Metal.Trans. 9A:23-29.

Dawson,D.B. & R.M.Pelloux 1974. Corrosion fatigue crack growth of titanium alloys in aqueous environments. Metal. Trans. 5:723-731.

Dawson,D.B. 1981. Fatigue crack growth behavior of Ti-6Al-6V-2Sn in methanol and Methanol-water solutions. Metal. Trans. 12A:791-800.

Hall,I.W. 1978. Basal and near-basal hydrides in Ti-5Al-2.5Sn. Metal. Trans. 9A:815-820.

Meyn,D.A. 1971. An analysis of frequency and amplitude effects on corrosion fatigue crack propagation in Ti-8Al-1Mo-1V. Metal.Trans. 2:853-865.

Meyn,D.A. 1974. Effect of hydrogen on fracture and inert-environment sustained load cracking resistance of α-β titanium alloys. Metal. Trans. 5:2405-2414.

Meyn,D.A. & E.J.Brooks 1981. Microstructural origin of flutes and their use in distinguishing striationless fatigue cleavage from stress-corrosion cracking in titanium alloys. ASTM STP 733:5-31.

Meyn, D. A. & R. A. Bayles 1987. Fractographic analysis of hydrogen-assisted cracking in alpha-beta titanium alloys. ASTM STP 948:400-423.

Ogawa,T., K.Tokaji & K.Ohya 1992. The effect of microstructure on fatigue crack growth in Ti-6Al-4V alloy. J. Soc. Mater. Sci., Jpn. 41:502- 508 (in Japanese).

Paton,N.E. & R.A.Spurling 1976. Hydride habit planes in titanium-aluminum alloys. Metal. Trans. 7A:1769-1774.

Ryder, J.T., W.E. Krupp, D.E. Pettit & D.W. Hoeppner 1978. Corrosion-fatigue properties of recrystallization annealed Ti-6Al-4V. ASTM STP 642:202-222.

Wanhill,R.J.H. 1976. Environmental fatigue crack propagation in Ti-6Al-4V sheet. Metal.Trans. 7A:1365-1373.

Yoder, G. R., L. A. Cooley & T. W. Crooker 1983. Effects of microstructure and frequency on corrosion-fatigue crack growth in Ti-8Al-1Mo-1V and Ti-6Al-4V. ASTM STP 801:159-174.

Recent Advances in Experimental Mechanics, Silva Gomes et al. (eds) © 1994 Balkema, Rotterdam, ISBN 90 5410 395 7

Assessment of low-cycle fatigue strength of ductile cast iron on the basis of quantitative analysis of spheroidal graphite morphology and matrix structure

S. Harada
Kyushu Institute of Technology, Kitakyushu, Japan

T. Ueda
Hitachi Metals Co., Ltd, Fukuoka, Japan

M. Yano
Oita University, Japan

ABSTRACT : The LCF tests were conducted on ductile cast irons(DI) to examine the effect of matrix structure and nodule count on LCF strength. Materials with three kinds of matrix, structures, i.e., ferritic(FDI), pearlitic(PDI) and austempered(ADI) matrix and materials FDI with low nodule count(LNC) and high nodule count(HNC) were prepared In case of FDI,PDI,ADI, the fatigue life trend differed depending on the testing condition, i.e., cyclic strain- or stress controlled cycling. However, all the fatigue data were well evaluated by means of the Manson universal slope method. In case of the materials LNC,HNC, the fatigue life trend reversed at $N_f=10^3$ and the difference of the nodule count brought the difference in fatigue process. Fatigue crack propagation tests were also carried out and the growth rates of small crack and large crack were compared, by taking into account the crack closure effect. The fatigue crack growth rate showed the material dependency and the growth rate of the small crack in unnotched specimen was faster than that in large crack.

1 INTRODUCTION

Ductile cast iron (DI) has a history over 40 years. In spite of development of various kinds of new materials, it is in current reevaluated as one of the low cost new materials for its advantageous mechanical properties. On the other hand, it is widely believed that DI is a low strength-reliability material as well as gray cast iron. Although much efforts have been done to ensure the strength reliability so far, they have mostly been concerned with qualitative analysis. To ensure higher strength reliability DI should first be recognized as one of the composite materials, consisting of matrix structure and spheroidal graphite (SG). In fact, the SG-morphology-related parameters such as nodule count ng, mean diameter Dm, nodularity hg and area fraction fg and the matrix structure-related parameters such as area fraction of ferrite or pearlite and content of retained austenite in austempered DI, are at present widely used to do the microstructure-oriented strength evaluation of DI. In this respect, the present authors have already conducted a of studies (Harada et al., 1989, 1992, 1993).

The main objectives of the present study are to estimate quantitatively the effects of the SG morphology and the matrix structure on the fatigue life and the crack growth rate under low-cycle fatigue(LCF) condition. In addition, crack growth rate between a small and a large crack in each material were compared through measurement of crack closure behavior.

2 MATERIALS AND EXPERIMENTAL PROCEDURES

The materials tested are ferritic, pearlitic and bainitic ductile cast irons, designated as FDI, PDI and ADI, respectively. They were fabricated using a Y-shaped mold of 25mm wide(bottom size of Y-shaped block) x 135mm high x 235mm long. All the specimens were collected at the bottom of each Y-shaped block where the materials are supposed to be cleanest. As seen in Table 1, all the materials have almost same chemical composition. ADI was prepared by heat treatment; heating at 875°C for two hours, then reheating at 375°C for two hours in salt bath. The results of the SG-morphology-related parameters measured using a image processing analyser with a cut off of $8\mu m$ in SG diameter showed no difference among three materials. Therefore, these materials were used to extract only the effect of the matrix structure on the fatigue strength of the smooth specimen and on the growth rate of small and large crack. In addition, the effect of nodule count on the fatigue strength was also examined using a ferritic ductile cast iron with low and high nodule count, designated as LNC and HNC, respectively. Their chemical compositions are also listed in Table 1. The static mechanical properties and the SG morphology-related parameters of each material are tabulated in Table 2 and Table 3, respectively. The nodularity is defined as a ratio of the area of SG to the area of the tangential circle. It should be noted that the

Table 1 Mechanical properties of the material tested

	C	Si	Mn	P	S	Cr	Cu	Mg
FDI	3.76	2.21	0.19	0.017	0.016	——	——	0.038
PDI	3.70	2.08	0.34	0.014	0.008	——	0.52	0.035
ADI	3.65	2.14	0.35	0.025	0.010	0.04	0.50	0.038
HNC	3.61	2.61	0.41	0.02	0.010	0.03	0.08	0.030
LNC	3.74	2.58	0.45	0.02	0.010	0.02	0.09	0.030

Table 2 Mechanical properties of the material tested

	Tensile strength σ_B (MPa)	Elongation ϕ (%)	Young's modulus E (GPa)	Hardness H_B
FDI	442	21.2	157	146
PDI	863	5.2	163	252
ADI	930	9.8	167	350
HNC	438	22.2	172	152
LNC	455	15.7	171	149

Table 3 SG morphology-related parameters

	Nodule count n (/mm^2)	Area fraction f (%)	Mean diameter D_m (μm)	Nodularity h (%)
FDI	226.7	13.1	23.0	61.1
PDI	163.3	11.4	34.3	68.0
ADI	172.1	11.2	27.6	65.3
HNC	229.6	12.6	21.4	64.3
LNC	84.7	12.8	36.3	63.9

tensile strength becomes higher in the order FDI, PDI, ADI while the elongation becomes higher in the order PDI, ADI, FDI. In the case of the ferritic materials LNC and HNC, the tensile strength is almost same and the elongation is rather higher in HNC than LNC. An hour-glass-shaped cylindrical specimen with 8mm in minimum diameter was used for unnotched fatigue. Either cyclic stress- or strain-controlled fatigue tests were done for smooth specimens in a closed loop electrohydraulic servo-controlled fatigue testing machine. In the case of cyclic strain-controlled testing, diametral change at the minimum cross section of specimen was detected using a pair of linear differential transformers attached to the specimen surface.

On the other hand, the fatigue tests for the large crack growth were done under pulsating loading. Each specimen surface was finished to make observation of crack initiation and propagation easier.

The measurement of crack closure in smooth specimen was done with the aid of an interferometric displacement gauge (Akiniwa et al 1991).

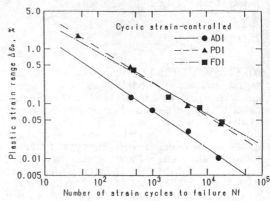

(a) Results under cyclic strain-controlled loading

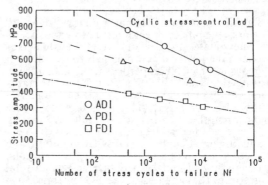

(b) Results under cyclic stress-controlled loading

Figure 1. LCF lives of the materials FDI, PDI, ADI under strain- and stress-controlled cycling

3 EXPERIMENTAL RESULTS AND DISCUSSIONS

3.1 Effects of matrix structure and nodule count on the fatigue life of smooth specimen

Figs.1(a)(b) show the diagrams of the fatigue lives in material FDI,PDI,ADI under strain- and stress-controlled cycling, respectively. It is clear in both figures that the fatigue cycling conditions yields the opposite trend of the fatigue lives of each material. Considering the fact that each material showed different behavior of cyclic hardening and converting σ (cyclic stress amplitude) versus Nf (number of stress or strain cycles to failure) into $\Delta\varepsilon_p$ (cyclic plastic strain range) by taking cyclic s versus $\Delta\varepsilon_p$ relationship measured at the half of the fatigue life, the results of both cycling conditions can be represented on the same diagram, as shown in Fig. 2. Obviously both results almost coincides in each material. Furthermore, the results of FDI,PDI are almost identical and ADI still implies the fatigue lives shorter than FDI,PDI. The fatigue life difference still observed in Fig.2 is attributed to the difference of the tensile strength and elongation in each material. To take this factor into account in fatigue live estimation the Manson universal slope method is applied and the results are shown in Fig.3. Three different lines and solid marks in Fig.3 represent the predicted and the measured fatigue lives of each material. Evidently the fatigue lives of each material can be well evaluated by the Manson method.

Fig.4 shows the diagram of the fatigue lives of the HNC,LNC. The fatigue life trends of both material reverse at Nf=10³. This trend in longer fatigue life region coincides with the result that increasing nodule count enhances the fatigue life in high cycle fatigue region.

3.2 The effect of matrix structure and nodule count on the fatigue process of smooth specimen

In cases of FDI, PDI, ADI, the fatigue process was dominated by the growth of microcracks mainly originated at SG. The microcrack was propagated by mending SG-originated subcracks. According to the results of the fractographic observation, each material showed different structure sensitivity. In case of FDI, having higher ductility, debonding between SG and matrix was often observed on the fractured surface. On the contrary, PDI and ADI, having higher strength and lower ductility, showed the structure sensitivity higher than that of FDI. That is, no delamination at SG-matrix interface was observed and a leading crack was in some cases originated at a microshrinkage.

Fig. 5 shows the relationship between the microcrack growth rate da/dN and the cyclic J integral range ΔJ measured at different stress level in each material. Obviously the crack growth rate is stress-level dependent in each case. This is due to the effect of crack closure and will be discussed later.

Regarding to the fatigue process of HNC,LNC, the difference in nodule count brought microscopic fatigue process. Fig. 6 shows the relationship between the density of subcracks d and the cycle ratio N/Nf. In case of HNC, in spite of nodule count higher than that of LNC, the forma-

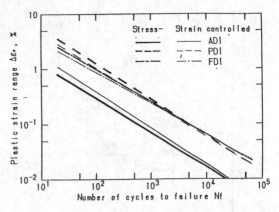

Figure 2. Comparison of both LCF lives by converting σ to Δep using cyclic stress-strain relation

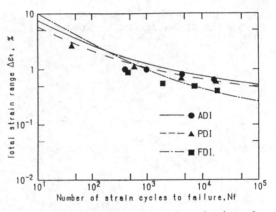

Figure 3. LCF fatigue lives plotted using the Manson universal slope method

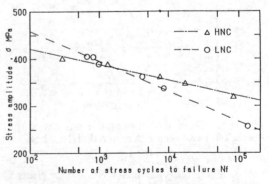

Figure 4. LCF lives of the materials HNC, LNC

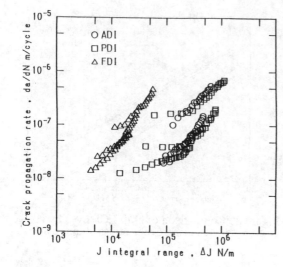

Figure 5. Stress-level dependent microcrack growth rate plotted against ΔJ in materials FDI, PDI, ADI

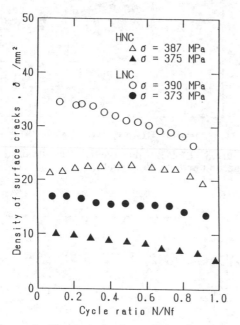

Figure 6. Change of subcrack density with increase of stress cycles in materials HNC, LNC

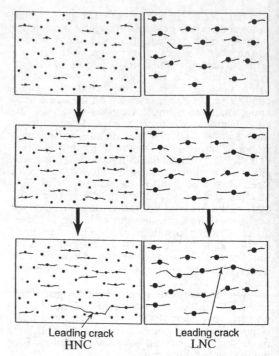

Leading crack
HNC

Leading crack
LNC

Figure 7. Schematical illustrations of the difference in dominant fatigue process in materials HNC, LNC

ter formation of a leading crack, no subcracks are formed and the fatigue process is completely dominated by the growth of the leading crack. On the contrary, large sized subcracks were often coalesced each other and no leading crack was formed until the later stage of the fatigue process in LNC. It can be said that the fatigue process HNC is controlled by the growth of a single crack, and the coalescence of subcracks dominates the fatigue process in LNC.

3.3 Comparison of crack growth rate between a small and a large crack in each material.

Figs. 8 (a)(b) shows the relationship between the small crack growth rate da/dN and the effective cyclic J integral range ΔJ eff in FDI and ADI, respectively. The broken line in the figure represents the results of FDI. It should be noted that da/dN shows no stress-level dependency when plotted against dJeff and is faster in FDI than in ADI. The metallograhic observation of the crack growth revealed that the crack growth in FDI was rather straightforward, while crack tip bifurcation, retarding the crack growth rate, was often observed in ADI.

Fig.9 shows the results of the large crack growth rate in each material, being plotted by da/dN versus ΔK eff (effective cyclic stress intensity fac-

tion of subcracks is less than HNC and the change of the subcrack density in total fatigue process is weaker than LNC. These differences can be understood by the difference of the fatigue process in both materials, as schematically shown in Fig. 7. In case of HNC, formation of subcracks is few and each subcrack grows independently. Af-

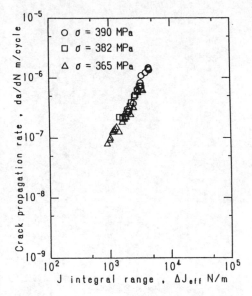 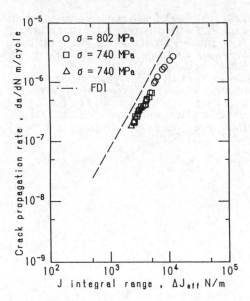

Figure 8. Small (micro-) crack growth rate plotted
against ΔJ_{eff} in materials FDI, ADI

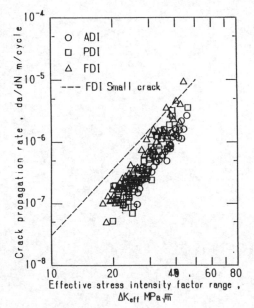

Figure 9. Comparison of crack growth rate between small (micro-) crack and large crack in materials FDI, PDI, ADI

tor range) relationship. The broken line in the figure represents the results of the small (micro-) crack growth rate in FDI. It is to be noted that the large crack growth rate shows material dependency, i.e., it is faster in the order of ADI,PDI,FDI. Furthermore, it should be noted that da/dN is faster in small crack than in large

crack in FDI. Thsese trends in da/dN are to be suggested as being induced by the following reasons. Martensitic transformation of retained austenite and crack tip bifurcation in ADI are responsible for da/dN. In addition, cyclic ceep phenomenon, which accelerates the crack growth, whould have affected the crack growth rate in FDI where the fracture ductility is highest.

4 CONCLUSIONS

Summrizing the results obtained so far, it is concluded as follows.
(1)The unnotched LCF strengths of the materials FDI,PDI,ADI showed an opposite trend, depending on the testing. This difference, however, could be well estimated when the results were plotted using the Manson universal slope method.
(2) The difference of nodule count in FDI brought the difference in fatigue life and in fatigue mechanism.
(3) The large crack growth rate showed material dependency. The crack growth rate was fastest in FDI. The growth rate of the small crack was faster than that of the large crack in FDI.

REFERENCES

Harada, S., et al. 1989. Microstructure-oriented assessment of the static tensile properties of a spheroidal graphite cast iron. *PVP-89*. Vol.174:219-224. Honolulu.
Harada, S. et al. 1992. Low-cycle fatigue of ductile cast iron. *LCF-3*: 124-129. Berlin.

Harada et al. 1993. Small crack growth in ductile cast iron with different microstructure. H. Nisitani, S. Harada & T.Kobayashi(eds), *Strength of Ductile Cast Iron 93*:91-96. Kitakyushu: JSME-MMD.

Akiniwa, Y. et al. 1991. Dynamic measurement of crack closure behaviour of small fatigue cracks by an interferometric strain/displacement gauge with a laser diode. *Fati. Frac. Engng. Materi. Struct* Vo.14. No.2/3: 317-328.

Creep behaviors of notched and unnotched specimens under stress change

A. M. Abdel El-Mageed & G. Abouelmagd
Faculty of Engineering & Technology, Minia University, Egypt

ABSTRACT: Under creep conditions, materials are usually subjected to time-dependent stresses, but most of available creep data are concerned with constant stress. More accurate description of creep deformation during exposure to time-varying stresses is an essential demand for many applications. The response of 0.4 carbon steel to a rectangular cyclic stress – time profil. varying between the stresses σ_1 and σ_2 ($\sigma_1 > \sigma_2$) have been investigated at 550 °C. A stress drop represents a recovery period causing a slow down in the growth rate of intercrystalline voids or a sintering out. Also, the internal stress developed in the previous high-stress period is decreased causing an increase in the strain rate. Also, circumferential 60 °V-notch specimens are tested mainly, to qualitatively study the material suitability for accidental stress concentrations and locally deformation under multiaxial stresses. Based on the present investigation, the exist of notch sharply decreases the creep rate under constant stress consequently increases the creep life.

The present work aiming to (a) interpret a new phenomenon which has been observed during the monotonic creep of notched specimens and named 'self-stress change', (b) a better understanding of the relation between creep rate and time also, the notch strengthened under constant and and cyclic stresses.

1 INTRODUCTION

Current use of material frequently entails simultaneous exposure to more than one hostile environment. At elevated temperature, for example, performance in critical elements may be determined by fatigue, creep, creep under cyclical stress–strain conditions, corrosion, erosion or wear, in addition to a general degradation of properties arising from microstructural deterioration.

The cyclic creep leads to complex behaviours and it should be clear that, the simple constant load tests will prove inadequate in attempts to establish a comprehensive model of material (*Boyle and Spence* 1983). Most of of cyclic loading studies, are focused on describing the immediate creep response after stress change (*Urcola and Sellars* 1987), and (*Goldhoff* 1971). The fatigue–creep interaction behaviours of 316 stainless steel has been investigated by *Manson et al* (1971) and *Plumbridge* (1993). They concluded that the creep plays a dominant role in high temperature cyclic life and the fatigue creep interaction behaviours can be enhanced

according to the type of material processing.

The rupture life of notched specimens is an indication of the ability of a material to deform locally without damage –cracking– under multiaxial stresses (*Whittenberger* 1985). Notched specimens are used principally as a qualitative material selection tool for comparing its suitability for components that may contains deliberate or accidental stress concentrations. *Siegfried* (1971) studied the effect of notch on the emprittlement of Cr-Ni-Ti steel and stress distribution which shown schematically in Fig.(1).

The present investigation, is concerned with creep behaviours under constant and cyclic stress conditions for unnotched and notched specimens. This an important engineering problem for a number of high temperature structures, such as, high temperature turbine blades and materials operating in nuclear reactors. The response of the materials to stress change during creep is complex and therefore, the attention is focused on creep strain, strain rate and creep life behaviours for notched and unnotched specimens.

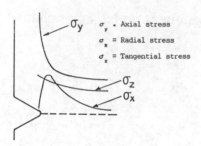

σ_y = Axial stress

σ_x = Radial stress

σ_z = Tangential stress

Fig. 1 Stress distribution in a notch specimen under tensile stress

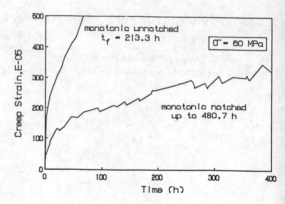

Fig. 2 Typical monotonic creep curves for notched and unnotched specimens

2 EXPERIMENTAL PROGRAMME

The tested carbon steel has the following chemical compositions (weight %) : 0.40 C, 0.023 S, 0.44 Mn, 0.028 P, and 0.13 Si. The tensile properties of unnotched and notched specimens are given in Table 1.

Table 1. Tensile properties of tested steel

Property Specimen	Ultimate strength σ_u, (MPa)	Proof strength $\sigma_{P_{0.2}}$,(MPa)	Ductility δ, (%)
Unnotched	950	690	6.0
Notched	1590	1050	2.4

The creep tests are conducted on A Zst 3/3 creep machine at a temperature of 550°C and stress changes between 60 and 45 MPa. The creep strain is measured with a resolution of 1.E-05, which allows a sensitive measurement of creep strain response after stress changes.

A circumferential 60°V-notch specimens are used, with a cross-sectional area at the base of the notch one half that of the unnotched section. The notch is carefully machined with tip radius (ρ) of 0.35 mm.

The creep stress for both unnotched and notched specimens is pulsating between 60 and 45 MPa according to a rectangular time-fraction with periods of 12, 24, 36 and 48 h.

3 RESULTS AND DISCUSSIONS

3.1 Monotonic creep

under the tested creep conditions, (T = 550°C and σ = 60 MPa), the normal creep curve for notched and unnotched specimens is shown in Fig.2. On the present work, the symbols have been excluded from most of the figures for cyclic strain response appearance. The creep strain response for notched specimen is not monotonically increased with time as unnotched one (Fig.2). The sequential increasing and decreasing of strain is similar to the creep strain behaviour under stress changes, then this phenomenon named, in present work, 'self-stress change'.

A trial is performed, based on the elastic stress field solution (*Efits and Liebowiez* 1972),to interpret the "self-stress changes" phenomenon. The axial stress, σ_y (Fig.1) shows a high stress peak in the notch root, which depends on the notch radius of curvature,ρ. In the course of creep time, the radius of curvature increases due to local deformation at the notch region, and thus results in a large reduction of the axial stress, which cause the first decreasing of creep strain. The decreasing of axial stress causes a contraction of notch region, then the radius of curvature decreases (resharpness) and consequently σ_y increases again and thus causes increasing of the following creep strain.

In Fig (3), the creep strain rate, $\dot{\varepsilon}$ for notched specimen is fluctuated between positive and negative values due to self-stress change phenomenon. During the self-stress change, the maximum axial stress is instantaneously balanced the internal back stress, σ_i = f(T,σ), therefore, the creep strain rate ,$\dot{\varepsilon}$ reaches zero values.

Both of creep strain and creep strain rate have lower values for notched specimen relative to unnotched one, consequently, the creep life,t_f is much longer in case of notched specimen.

3.2 Cyclic creep

The stress reduction acts as an intermediate heat treatment of the material (recovery),

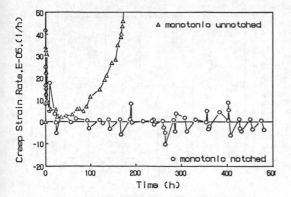

Fig. 3 Creep strain rate of notched specimens under monotonic testing

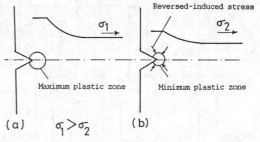

Fig. 5 Schematic illustration of plasticity induced closure

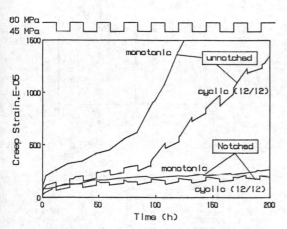

Fig. 4 Creep strain Vs. creep time for monotonic and cyclic loading

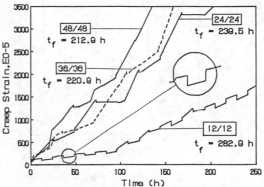

Fig. 6 Creep strain of unnotched specimens under different cycle periods

3.2.1 Unnotched specimens

The creep behaviours of unnotched specimens under different rectangular periodic loading are illustrated in Fig.6. After each stress change the internal back stress, representing the material creep resistance, changes gradually so that a new transient creep stage is initiated as shown in the inset of Fig.6. The, periods of negative creep rate occurrence, after stress reduction (Fig.7), will imply that the stress had been reduced to a level below that of the average internal stress, so that the specimen experienced a net compressive stress $(\sigma_i > \sigma)$. Also, the reason of accelerated creep strain rate, $\dot{\varepsilon}_s$ after the sudden stress increase, is the gradually increasing of the internal stress to adapts the high stress conditions, thus leading to relatively high effective stress.

To clarify the above interpretation, Fig.8 is made, which represent a schematic illustration of the internal stress role on creep strain response under cyclic loading. The given numerical values of internal stresses, σ_i are taken from *Abouelmagd and Abd El-Mageed* (1993)

therefore, the creep strain under cyclic loading is lower for notched and unnotched specimens comparing to monotonic loading as shown in Fig.4.

In addition to recovery, in case of notched specimens, notch region contraction take place, which decreases the creep strain. The contraction of notch region is explained based on the fracture mechanics concept using the phenomenon 'Plasticity-Induced Closure' (*Elber* 1971). As a result of stress decreasing, the plastic zone size around the notch root is contracted. This contraction induced a reversed stress, which reduce the axial stress, as shown schematically in Fig 5. Therefore, the creep strain is lower for notched specimens and consequently the creep life is longer.

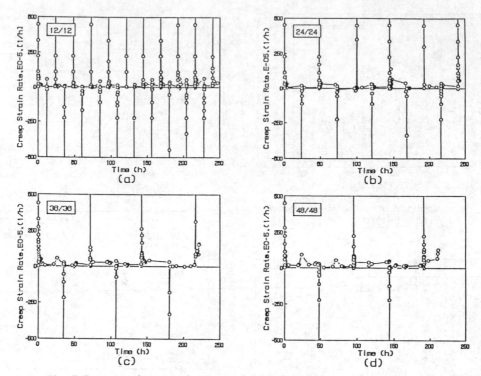

Fig. 7 Creep strain rate of unnotched specimens under different cycle periods

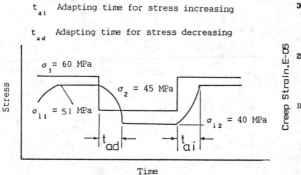

Fig. 8 Schematic representation of internal stress change under cyclic loading

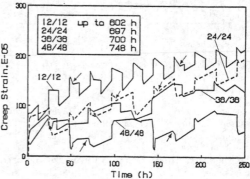

Fig. 9 Creep strain of notched specimens under different cycle periods

3.2.2 Notched specimens

The dependence of creep strain on cycle period for notched specimens is shown in Fig. 9. The intermediate strain decreasing during high and low stress levels, which indicated by the arrows, can be attributed to the 'self-stress change' phenomenon. The same effect of this phenomenon on creep strain rate, $\dot{\varepsilon}$ is shown in Fig. 10.

3.3 Comparison between notched and unnotched specimens

Generally, the cyclic loading and specimen notching reduce the creep strain but the notch has the appreciable effect (Fig. 4). Also, the influence of cycle periods on creep strain is not similar for notched and unnotched specimens. The cycle period (12/12) shows lower values of creep strain compared to

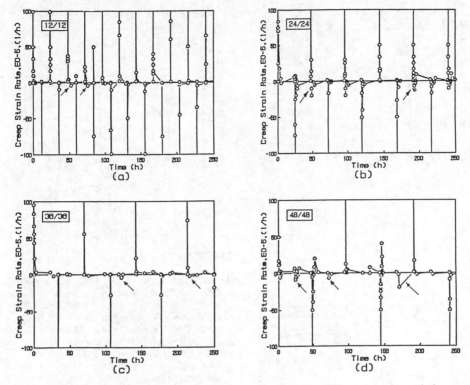

Fig. 10 Creep strain rate of notched specimens under different cycle periods

other cycle periods in unnotched specimens, while in case of notched specimens, the strain of period (12/12) has the higher values.

The above results may be explained based on the two phenomenons, recovery and self-stress change. In case of unnotched specimens, the lower strain of period (12/12) caused by the higher number of recovery (transient creep) compared to other periods (Fig. 6). On the other hand, the notched specimens are influenced by the two phenomenons, but the self-stress change has the dominant role. therefore, the period (48/48) gives the lower creep strain where it has longer period time for self-stress change effect as shown in Fig.9.

An enlightening picture emerges when creep strain at 200 hour is plotted as a function of cycle period, in Fig. 11, for notched and unnotched cases. The notch existence has a remarkable effect on creep strain rate under cyclic loading. The notched specimens show lower values of steady state creep rate, $\dot{\varepsilon}_s$ compared to the unnotched specimens as shown in Fig. 12.

The cyclic loading of unnotched specimens has a negligible effect on creep life compared to the life of monotonic creep under the high stress, σ=60 MPa, (Fig 2 and Fig 6) while it has a great influence for notched specimens

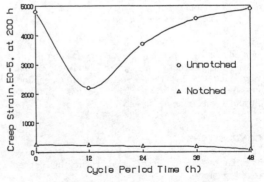

Fig. 11 Dependence of creep strain on cycle period

(Fig.2 and Fig 9). Since the elapsed creep time of notched case is relatively long, the experiments are interrupted before fracture. As well known, based on *Monkman-Grant* relation, that the creep life varies linearly with steady state creep rate, $\dot{\varepsilon}_s$. Therefore, the creep life is weighted by the average creep rate to study the effect of cycle period on creep life.

For the tested steel, the creep life of unnotched specimen is enhanced at a cycle

1217

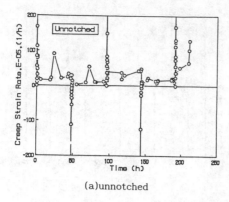

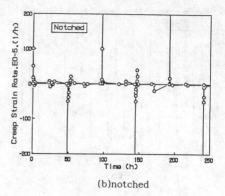

(a)unnotched   (b)notched

Fig. 12 Creep strain rate at period of 48 h

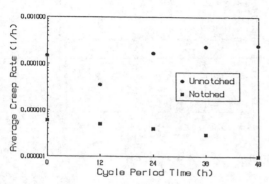

Fig. 13 Dependence of average creep strain
rate on cycle period

period of 12 h, but with increasing period
time the effect of cyclic loading becomes
negative. While the cyclic loading enhances
creep life regardless the period time for
notched specimens (Fig. 13).

CONCLUSIONS

The following are concluded from the present
investigation:

1. The creep life is appreciably enhanced by
notch existence, under monotonic loading,
which attributed to 'self-stress change'
phenomenon.

2. For both monotonic and cyclic loading,
the specimen notching decreases creep strain
and creep strain rate , which is benefit for
most of the applications.

3. The improvement of creep life under
cyclic loading is negligible and observed only
at short cycle periods for unnotched
specimens, while the improvement has a
remarkable levels ,always, and specially a'
long periods time for notched case.

REFERENCES

Abouelmagd,G & A.M Abd EL-Mageed. 1993.
Internal and friction stresses of carbon
steel under creep conditions. Proc. 4th
PEDD:164-174.Cairo:Ain shams university.

Boyle,J.T.& J.Spence 1983. Stress analysis for
creep. England: Butterworth.

Efits,J. & H.Liebowitz. 1972. On the modified
westergaard equations for plane crack
problems. Int.J. Fracture Mech.8: 383- 392.

Elber,W. 1971. In damage tolerance in aircraft
structures. STP 486.p.230-242. ASTM.

Goldhoff,R.M. 1971. Creep recovery in heat
resistant steels. The A.E. Johnson memorial
volume. A.I.Smith & A.M.Nicolson (eds),
p.81-108. London: Applied science publishers.

Manson,S.S., G.R.Halford & D.A.Spera. 1971.
The role of creep in high temperature low
cycle fatigue. The A.E. Johnson memorial
volume. A.I.Smith & A.M.Nicolson (eds),
p.229-248.London:Applied science publishers.

Plumbridge,W.J.1993. The influence of
processing on the fatigue and fatigue- creep
behaviours of stainless steels. Proc.1st
AMPT'93:875-882. Dublin city university.

Siegfried,W.1971. Determination of factors
causing embrittlement in time-to- rupture
tests. The A.E.Johnson memorial volume.
A.I.Smith & A.M.Nicolson (eds), p. 181-224.
London: Applied science publishers.

Urcola,J.J.& C.M.Sellars 1987. A model for a
mechanical equation of state under
continuously changing conditions of hot
deformation: Acta Metall. 35: 2659-2669.

Whittenberger,J.D. 1985. Creep, stress-
rupture, and stress-relaxation testing.
J.R.Newby (ed.), p.315-318. Vol.8. Ohio:ASM.

Recent Advances in Experimental Mechanics, Silva Gomes et al. (eds) © 1994 Balkema, Rotterdam, ISBN 90 5410 395 7

Fatigue properties of Al-Si-Cu-Ni-Mg casting alloys at elevated temperatures

N. Hasagawa
Department of Mechanical Engineering, Gifu University, Japan

T. Miyabe
Tochigi R&D, Honda R&D Co., Ltd, Hagamachi, Haga-gun, Japan

T. Yamada
Asahi Tec Cop., Kikukawamachi, Shizuoka, Japan

ABSTRACT: Rotating bending fatigue test of modified eutectic casting alloys (AC8A) and hypereutectic casting alloys (AC9A and AC9B) were carried out in a temperature range between room temperature and 673 K. The fatigue strength of sodium modified AC8A is superior to that of phosphorus modified AC8A. The fatigue strength of AC9B with lower silicon content is greater than that of AC9A. The decrease in fatigue strength is due to coarse primary silicon. It is concluded that the fatigue properties of Al-Si-Cu-Ni-Mg casting alloys are influenced by primary silicon grains.

1 INTRODUCTION

There is an ever increasing need for better automotive fuel economy and improved engine performance. To meet this need, one of the most effective means is to reduce the weight of engine parts. With this in mind, aluminum casting alloys are widely used for automotive parts. Especially, Al-Si-Cu-Ni-Mg casting alloys (AC8A and AC9A) are used for piston parts, because of their heat resistant properties. But with these alloys, primary silicon grains reduce cast properties, toughness and mechanical properties. So sodium is ordinarily added to eutectic casting alloys (AC8A) to obtain dendrite structure, and phosphorus is added to hypereutectic casting alloys (AC9A and AC9B) to yield fine primary silicon grains.

However, fatigue properties of these alloys at elevated temperatures have not been fully clarified in spite of their use at that temperature. The present study has been carried out to elucidate the fatigue properties of these modified Al-Si-Cu-Ni-Mg casting alloys by means of rotating bending fatigue tests in a temperature range between room temperature and 673 K.

2 EXPERIMENTAL PROCEDURE

The alloys used in this study are eutectic Al-Si-Cu-Ni-Mg casting aluminum (JIS-AC8A) and hypereutectic Al-Si-Cu-Ni-Mg casting aluminum (JIS-AC9A and JIS-AC9B). AC8A was modified by sodium or phosphorus, designated AC8A-Na and AC8A-P, respectively. AC9A and AC9B were modified by only phosphorus. The chemical composition, mechanical properties and heat treatment conditions of these alloys are listed in Table 1, Table 2 and Table 3, respectively. The values of tensile strength at elevated temperature exposure for 100 h at testing temperature are also listed in Table 2. The microstructures of alloys are shown in Fig. 1. AC8A-Na exhibits a dendrite structure, but the other alloys, AC8A-P, AC9A and AC9B, show coarse primary silicon and fine eutectic silicon. The primary silicon grain sizes of AC8A-P, AC-9A and AC9B are 16 μm, 28 μm and 23 μm, respectively. The order of primary silicon grain size is on the same order as silicon content.

Figure 2 shows the shape of the fatigue specimen. The surface of the specimen was polished by emery paper and then finished by buffing. The fatigue tests were carried out by rotating beam fatigue machines of the uniform bending type in laboratory air. Machine speed was about 60 Hz.

The fatigue testing temperatures were chosen room temperature, 473 K, 573 K and 673 K. Specimens were held at the testing temperature for 30 min before loading. The specimens of exposed for 100 h at 473 K or 573 K were also fatigue tested.

A replicating procedure was used to investigate surface crack propagation behavior.

Table 1　Chemical composition of materials(wt%)

Material	Cu	Si	Mg	Zn	Fe	Mn	Ni	Ti	Ca	Na	P	Al
AC8A-Na	0.98	12.3	1.08	0.006	0.14	0.003	1.03	0.15	0.001	0.005	-	bal.
AC8A-P	1.04	11.9	1.08	0.007	0.14	0.003	1.14	0.15	0.001	0.001	0.009	bal.
AC9A	1.02	23.5	1.15	0.007	0.17	0.003	1.02	0.14	0.001	0.001	0.010	bal.
AC9B	1.01	19.5	1.07	0.007	0.17	0.003	1.03	0.14	0.001	0.001	0.016	bal.

Table 2　Mechanical properties of materials

	Tensile strength σ_B (MPa)				Young's modulus(GPa)
	Room temp.	473K	573K	673K	
AC8A-Na	209	186	77	31	79
AC8A-P	203	203	85	34	79
AC9A	183	146	79	43	87
AC9B	193	155	80	40	87

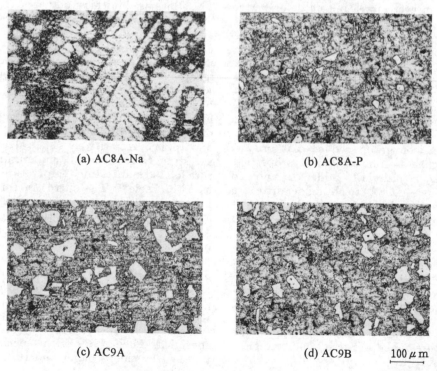

(a) AC8A-Na　　　　　　　(b) AC8A-P

(c) AC9A　　　　　　　(d) AC9B　　100 μm

Fig. 1　Microstrucutres of alloys

Table 3 Conditions of heat treatments

	Solution treatment	Aging treatment
AC8A	758K,3h	468K,5h
AC9A AC9B	758K,3h	528K,5h

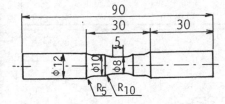

Fig. 2 Geometry of fatigue specimen

3 RESULTS AND DISCUSSION

3.1 σ-N_f curves and σ/σ_B-N_f curves

σ-N_f Curves of alloys are shown in Fig. 3(a) ~ (d), where σ is stress amplitude and N_f is the number of cycles to failure. The fatigue strength of all alloys over the whole cycle region decreases with increasing temperature.

In these figures at 473 K and 573 K, the results (solid symbol) for specimens that exposed to test temperature for 100 h, are also shown.

At 473 K, the fatigue strength which exposed for 100 h almost equals that for 30 min. Since 473 K is almost the same as the aging temperature of T7 treatment as shown in Table 3, the exposure to 473 K does not affect the fatigue strength.

On the other hand, fatigue strength of alloys exposed at 573 K for 100 h is remarkable lower compared with that for 30 min. But the difference in fatigue strength between the two exposure times decreases with stress cycles, and the difference disappears at 10^7 cycles.

From comparison of the fatigue strength at room temperature between AC8A-Na and AC8A-P, although the difference of tensile properties is small, the fatigue strength of AC8A-Na with dendrite structure is greater than that of AC8A-P over the whole stress cycle. This result indicates that in eutectic casting alloy (AC8A) the sodium mod-

ification is more successful than the phosphorus modification in improving the fatigue strength. On the other hand, in hypereutectic casting alloys (AC9A and AC9B), the fatigue strength of AC9B with lower content silicon is greater than that of AC9A.

Since the size of silicon grain and the number of that increase with increasing silicon content, the coarse silicon grain lowers fatigue strength.

At high temperature (573 K and 673 K), the difference in fatigue strength of these alloys disappeared and the fatigue strength of all alloys became the same. No effect of the microstructure on fatigue strength at high temperature was recognized.

Fig. 4(a) ~ (d) are σ/σ_B-N_f curves at room temperature, 473 K and 573 K, where σ_B is tensile strength. In these figure, here, at 473 K and 573 K, the results of 100 h exposure specimen at each temperature were used. In all alloys, σ/σ_B-N_f curves of each temperature are the same, and there is no difference in alloys except AC8A-P. Thus, the fatigue strength can be estimated from the tensile strength at each temperature and the difference in fatigue strength is obtainable from comparison of tensile strength of each alloy.

3.2 Fatigue crack propagation behavior

Fig. 5 and Fig. 6 show the crack propagation rate da/dN vs. the stress intensity factor range ΔK plots at room temperature and 573 K, respectively. The crack propagation tests at 573 K used specimens exposed for 100 h. The crack propagation behaviors of each alloy at both temperatures show anomalous behavior. This behavior is characteristic of the short crack.

The da/dN of all alloys at 573 K is faster than that at room temperature, though da/dN vs. ΔK plots of both temperatures show a wide scatter.

At room temperature, da/dN of AC9A with higher silicon content is highest and the alloy of the lowest da/dN is AC8A-Na with dendrite structure. At 573 K, da/dN of AC9A is also highest.

The crack propagation rate of AC9B at 573K is lowest in these alloys, and the difference of da/dN between room temperature and 573K is small. AC9B show the best crack propagation resistance at 573 K.

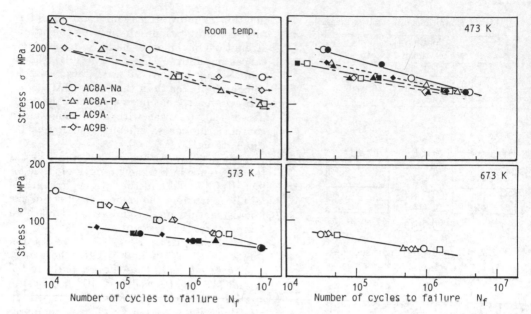

Fig. 3 σ-N$_f$ curves curves of alloys. (a) Room temperature, (b) 473 K, (c) 573 K, (d) 673K
Solid symbol: Specimen exposed for 100 h at the test temperature

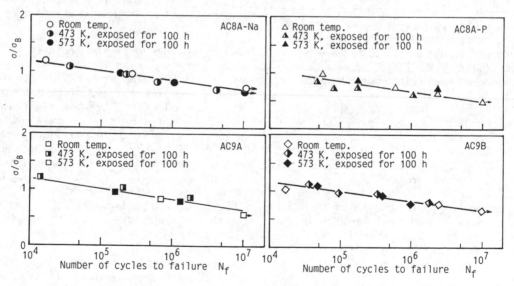

Fig. 4 σ/σ$_B$-N$_f$ curves of alloys.   (a) AC8A-Na, (b) AC8A-P, (c) AC9A, (d) AC9B

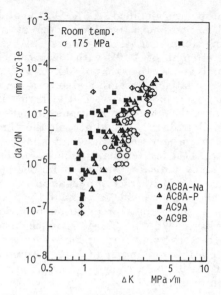

Fig. 5 Relationship between da/dN and ΔK
at room temperature

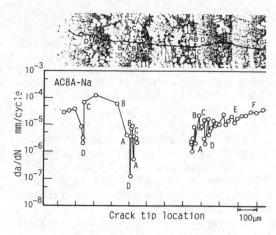

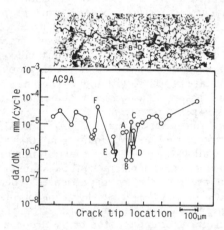

Fig. 7 Relationship between da/dN and crack
length of AC8A-Na at room temperature

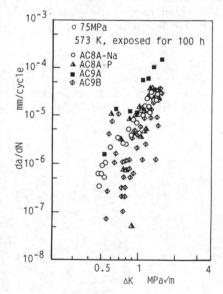

Fig. 6 Relationship between da/dN and ΔK
at 573 K

Fig. 8 Relationship between da/dN and crack
length of AC9A at room temperatrure

The crack propagation resistance of AC9A is lowest at both room temperature and 573 K due to larger silicon content.

The effect of microstructure on crack propagation behavior of all alloys were observed.

For instance, the relation between da/dN and crack length at room temperature at stress amplitude 175 MPa of AC8A-Na with dendrite structure and that of AC9A with lowest propagation resistance are shown in Fig. 7 and Fig. 8, respectively.

In order to confirm the effect of microstructure on crack propagation behavior, an optical micrograph of crack is also shown above each figure. The letters on the micrograph show the position of the crack tip at different cycles. Corresponding letters are shown on the da/dN vs. crack length plot in these figures.

1223

In AC8A-Na (Fig. 7), when cracks propagate along the grain boundary of primary aluminum (A→B), the crack propagation rate increases. However, when the crack propagates across the grain boundary (C→D), the crack propagation rate markedly decreases because grain boundaries becomes the barrier against crack propagation.

In AC9A (Fig. 8), when a crack propagates in the aluminum matrix and reaches the primary silicon grain boundary (A→B), the crack propagation rate decreases. On the other hand, when cracks propagate in silicon grains (B→C), the crack propagation rate increases dramatically.

Crack propagation rate also increases along the silicon boundary. However, when a crack propagates from silicon grain to aluminum matrix across the silicon grain boundary (C→D), the crack propagation rate decreases. As mentioned above, the crack propagation has a tendency to accelerate when a crack propagates across the primary silicon grain or along the boundary of primary silicon. Since the size and the number of primary silicon increases with silicon content, crack propagation rate of AC9A with higher content silicon is fastest. The origin of the cracks in AC9A is primary silicon.

The same observations of AC8A-P and AC-9B were carried out. At room temperature, in each alloy, the effects of primary aluminum and primary silicon on crack propagation behavior are remarkable; in particular, when a crack propagates from the primary silicon grain to the aluminum matrix across the silicon grain boundary, the crack propagation rate markedly decreases. The influence of microstructure disappears above a crack length of about 1 mm.

Finally, the sodium modification of the eutectic casting alloy (AC8A) is also more successful than the phosphorus modification from a crack propagation standpoint, because this treatment does not crystallize the primary silicon which is the origin of the crack. In hypereutectic casting alloys (AC9A and AC-9B), the lower silicon content alloy is superior from the same standpoint.

4 CONCLUSIONS

It was confirmed that the fatigue strength of sodium modified AC8A is superior to that of phosphorus modified AC8A. On the other hand, in phosphorus modified hypereutectic casting aluminum (AC9A and AC9B), fatigue strength of AC9B with lower content silicon is greater than that of AC9A. The decrease in fatigue strength is due to a coarse primary silicon grain size. But at high temperature, the difference in fatigue strength of these alloys decreased and became the same. And the fatigue strength of all alloys decreases as the test temperature increase. Further study indicates that at room temperature and 573 K crack propagation rate of each alloys show a wide scatter and the properties of short crack propagation behavior. Moreover, da/dN at 573 K is much faster than that at room temperature. At room temperature, sodium modified AC8A and at 573 K AC9B both show good crack propagation resistance.

5 REFERENCES

Adachi, M. 1984. Modification of eutectic Al-Si system casting alloys (in Japanese): Journal of Japan Institute of Light Metals 34,361-366.

Adachi, M. 1984. Modification of hypereutectic Al-Si system casting alloys (in Japanese): Journal of Japan Institute of Light Metals 34,430-436.

Recent Advances in Experimental Mechanics, Silva Gomes et al. (eds) © 1994 Balkema, Rotterdam, ISBN 90 5410 395 7

Influence of mean loading on fatigue crack growth velocity in 10HNAP steel under tension

G.Gasiak & J.Grzelak
Technical University of Opole, Poland

ABSTRACT: The paper shows results of investigations on development of fatigue crack growth under tension. Seventeen levels of mean stress were tested in specimens made of 10HNAP steel of higher-strength and higher-corrosion resistance. From the tests it results that a mean loading level strongly influences velocity of fatigue slot propagation. The greater mean loading is, the lower crack growth velocity is when the mean loading level is low. For mean levels the crack growth velocity stabilizes and if the mean loading level becomes high, intensity of crack growth velocity increases.

1 INTRODUCTION

Velocity of fatigue crack growth is dependent on many various factors, such as mean loading, asymmetry of a cycle, spectrum of loadings, geometry of a specimen, microstructure, a kind of material machining (Geary 1992, Kocańda and Szala 1985, Petit et al.1988).

Influence of mean loading P_m and stress ratio R on crack growth velocity is dependent on their values, the velocity range and a kind of material (Gasiak and Grzelak 1993, Kocańda and Śnieżek 1992, Kocańda 1980, Radon 1982, Wnuk 1977). Increase of mean loading or stress ratio usually causes increase of crack growth velocity (Szczepiński 1984). Investigations on crack growth velocities in low-alloy steels for various combinations of stress ratios were discussed by Paris et al. (1972).

This paper shows the results of investigations on influence of mean loading on fatigue crack growth velocity in 10HNAP steel of higher--strength and higher-corrosion resistance under tension for a wide range of changes of the stress ratio R.

2 TESTS AND THEIR RESULTS

Flat specimens made of 10HNAP steel (R_e = 402 MPa, R_m = 494 MPa) were tested. The specimens were made of sheets 10 mm in thickness. The specimen axis was parallel to the direction of rolling and its dimensions, thickness g, width B and length L, were 8 x 35 x 250 mm respectively. The specimens applied for cracking velocity tests had a central circular slot, 4 mm in diameter, with side notches 0.8 - 1 mm in length (2ℓ = 6 mm). The specimen surfaces were polished after grinding.

The specimens were subjected to variable tension at constant amplitude of loading P_a = 15kN (it met a stress amplitude up to the crack propagation σ_a = 65 MPa).

Range of the stress ratio R was changed from R = -0.5 to R = 0.7. The tests were made for seventeen levels of mean loading (i = 17): $(P_m)_i$ = (5, 10, 15, 20, 25, 30, 35, 40, 45, 50, 55, 60, 65, 70, 75, 80, 85) kN. The stress ratios were R = -0.5, -0.2, 0, 0.143, 0.25, 0.333, 0.4, 0.455, 0.5, 0.538, 0.571, 0.6,

0.625, 0.647, 0.666, 0.684, 0.7, respectively.

The tests were carried on a hydraulic pulsator at frequency of 13 Hz. Increments of crack length were measured with a microscope with an accuracy of 0.01 mm; numbers of loading cycles N_j were registered.

The results obtained were shown on graphs $2\ell_j = f(N_j)$, where $2\ell_j$ is an instantaneous crack length measured from the slot top after N_j cycles.

The slot length was measured at its both sides. The total slot increment was a sum of the present slot lengths on the left and right sides, ℓ_j^L and ℓ_j^P, i.e. $2\ell_j = \ell_j^P + \ell_j^L$. The results of measurements and the graphs $2\ell = f(N)$ were the base for calculations and description of the fatigue crack growth velocity. The fatigue crack growth velocity was determined by a ratio of the crack length increment, $\Delta(2\ell_j)$, to the corresponding increment of a number of loading cycles ΔN

$$\frac{\Delta(2\ell)_j}{\Delta N_j} \approx \left[\frac{d(2\ell)}{dN} \right]_j \qquad (1)$$

where

$$\Delta(2\ell) = 2\ell_j - 2\ell_{j-1}, \quad \Delta N_j = N_j - N_{j-1}$$

While tests a uniform increment of crack length was observed at both ends of the slot. We do not mention all the results obtained but those presented here show phenomena occurring while the fatigue crack growth in 10HNAP steel very well.

Courses of fatigue crack growth velocities are shown in Figs.1 and 2 as functions $d(2\ell)/dN = f(\Delta K)$ and $d(2\ell)/dN = f(K_m)$ respectively for two mean loading levels, $P_{m1} = 25kN$ and $P_{m2} = 85kN$.

The sets of measuring points shown in Figs.1 and 2 are described by Paris equation

$$\frac{d(2\ell)}{dN} = C_1(\Delta K)^{m_1} \qquad (2)$$

$$\frac{d(2l)}{dN} = C_2(K_m)^{m_2} \qquad (3)$$

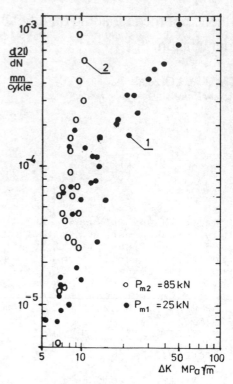

Fig.1   Velocity of fatigue crack growth depending on range of stress intensity coefficient

where
$\Delta K = M_k \Delta\sigma \sqrt{\pi\ell}$ is a range of stress intensity factor,
$K_m = M_k \sigma_m \sqrt{\pi\ell}$ is the mean stress intensity factor,
$\Delta\sigma = 2\sigma_a$ is the stress range equal to the double stress amplitude.
$P_a = const$, $P_{mi} = const$.

The constants C_1, C_2 and exponents m_1, m_2 are dependent on the material. For the specimens tested a correction coefficient, M_k, was introduced. The coefficient takes into account finiteness of the specimen dimensions and a shape of the initial slot (Kocańda and Szala 1985)

$$M_k = \ell + 0.1\left(\frac{2\ell}{B}\right) + \left(\frac{2\ell}{B}\right)^2$$

1226

After logarithmizing the both sides of (2) and (3) we obtain equations

$$\ln \left[\frac{d(2\ell)}{dN}\right] = m_1 \ln (\Delta K) + \ln C_1,$$

$$\ln \left[\frac{d(2\ell)}{dN}\right] = m_2 \ln (K_m) + \ln C_2$$

Values of C_1, C_2 and m_1, m_2 were determined with the method of linear regression (Freund 1967).

Using equations (2), (3) it was possible to make graphs of relationships $d(2l)/dN = f(\Delta K)$ and $d(2l)/dN = f(K_m)$ for all the specimens tested. There is no place for all the results obtained, so we publish only exemplary courses for five ranges of mean loading (Figs.3 and 4): $1-P_{m1} = 10kN$, $2-P_{m2} = 15kM$, $3-P_{m3} = 25kN$, $4-P_{m4} = 40kN$, $5-P_{m5} = 85kN$.

Parameters C_1, C_2, m_1, m_2 in equations (2) and (3) describing graphs 1-5 in Figs.3 and 4 are:

$1 - C_1 = 4.79 \; 10^{-7}$, $\quad m_1 = 1.584$,
$2 - C_1 = 1.98 \; 10^{-7}$, $\quad m_1 = 2.155$,
$3 - C_1 = 3.98 \; 10^{-7}$, $\quad m_1 = 2.065$,
$4 - C_1 = 1.56 \; 10^{-7}$, $\quad m_1 = 1.560$,
$5 - C_1 = 1.95 \; 10^{-10}$, $\quad m_1 = 6.240$,
$1 - C_2 = 1.08 \; 10^{-6}$, $\quad m_2 = 1.502$,
$2 - C_2 = 1.98 \; 10^{-6}$, $\quad m_2 = 2.155$,
$3 - C_2 = 1.41 \; 10^{-7}$, $\quad m_2 = 2.055$,
$4 - C_2 = 3.47 \; 10^{-8}$, $\quad m_2 = 2.211$,
$5 - C_2 = 1.26 \; 10^{-12}$, $\quad m_2 = 4.700$.

From Figs.3 and 4 it results that velocity of the fatigue crack growth is strongly influenced by the mean loading level under tension of specimens of 10HNAP steel. It has been found that as the mean loading increased, the

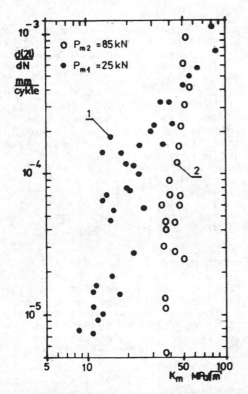

Fig.2 Velocity of fatigue crack growth depending on the mean stress intensity coefficient

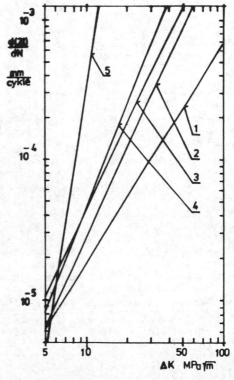

Fig.3 Plots of crack growth velocity depending on the range of stress intensity coefficient

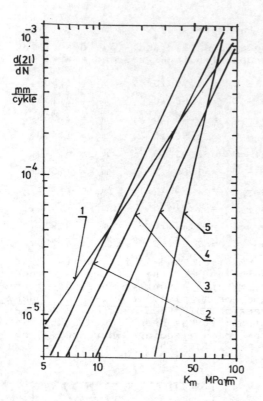

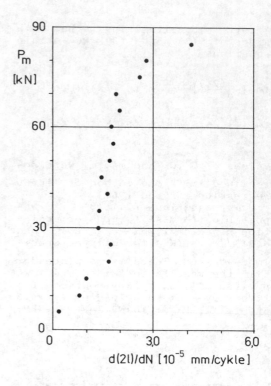

Fig.4 Velocity of fatigue crack growth depending on the mean stress intensity coefficient

Fig.5 Velocity of fatigue crack growth depending on mean loading for the range of stress intensity coefficient ΔK = 7 MPa $\sqrt{m}$

fatigue crack growth velocity increased too (Fig.3). The slot growth intensity changed together with a change of mean loading level P_m (Fig.5).

Within low levels of mean loading velocity intensity decreases together with its increase; for mean levels (20kN < P_m < 70kN) crack growth velocity stabilizes and for high levels of the mean loading intensity of crack growth velocity increases together with its increase.

Fig.6 shows fatigue fracture in a specimen occured under the mean loading P_m = 15 kN (low). The fatigue fracture surface is smooth and it means that the slot propagation was uniform. Fatigue fracture (without any distinct plastic strains) dominates here.

Fig.7 shows a fatigue fracture under P_m = 55 kN and P_a = 15 kN (high mean loading). It can be seen

that the fatigue slot propagation is graded. It is connected with a magnitude of the plastic zone in front of the slot. Owing to the plastic zone development the slot becomes more obtuse and the material consolidates. In this case participation of immediate fracture (zones with visible plastic strains) is greater than the fatigue fracture surface.

3 CONCLUSIONS

From the tests carried on the following conclusions can be drawn:

1. The mean stress level strongly influences velocity of the fatigue crack growth in the specimens of 10HNAP steel loaded with the constant amplitude P_a = 15 kN.

2. For low levels of mean loading intensity of the fatigue crack growth velocity decreases as the mean loading increases. For mean

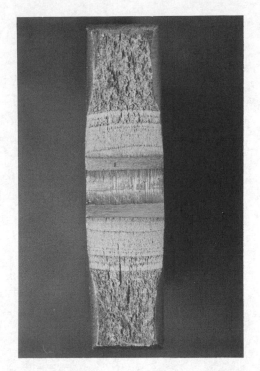

Fig.6 Fatigue fracture occurred under mean loading P_m = 15 kN and amplitude P_a = 15 kN.

Fig.7 Fatigue fracture occurred under mean loading P_m = 55 kN and amplitude of loading P_a = 15 kN.

levels the crack growth velocity stabilizes and for high levels the fatigue crack velocity intensity increases with the loading increase.

3. In case of high mean loadings the lines on the fatigue fracture surface suggest a graded propagation of the slot.

REFERENCES

Freund, J.E. 1967. Modern elementary statistics, p.323-342. New Jersey, USA, Prentice-Hall Inc.

Gasiak, G. & J.Grzelak 1993. Badania wpływu naprężenia średniego na prędkość pękania zmęczeniowego przy rozciąganiu stali 10HNAP. Zeszyty Naukowe Politechniki Świętokrzyskiej. Mechanika 50:141 -148. Kielce (in Polish).

Geary, W. 1992. A review of some aspects of fatigue crack growth under variable amplitude loading. Int.J.Fatigue 6:377-386.

Kocańda, S. 1980. Wpływ asymetrii cyklu obciążenia na prędkość rozwoju pęknięć zmęczeniowych w stali o podwyższonej wytrzymałości 15G1ANb i jej złączach spawanych. IX Symp. Doświadczalnych Badań w Mechanice Ciała Stałego. Warszawa 1:183-186 (in Polish).

Kocańda, S. & J.Szala 1985.Podstawy obliczeń zmęczeniowych, p.256 - 266. Warszawa.PWN (in Polish).

Kocańda, S. & L.Śnieżek 1992.Rozwój pęknięć zmęczeniowych w laserowo wzmacnianych elementach ze stali niskowęglowej. XV Symp.Doświadczalnych Badań w Mech. Ciała Stałego. Warszawa p.149-152 (in Polish).

Paris, P.C., R.J.Bucci, E.T.Wessel, W.G. Clark, T.R. Mager 1972. Extensive study of fatigue crack growth rate in A 533 and A 508 steels. ASTM STP pp.141-176

Petit, J., D.L.Davidson, S.Suresh & P.Rabbe 1988. Growth under variable amplitude loading. Elsevier Applied Science, p.397, London.

Randon, J.C.   1982. Fatigue crack
growth in the treshold region.
In: J.Backland, A.F.Blom and C.J.
Beevers   (eds.).   Fatigue
Thresholds, II:911-930.U.K. EMAS.
Szczepiński, W.   1984.   Metody
doświadczalne   mechaniki   ciała
stałego. X:301-316, Warszawa. PWN
(in Polish).
Wnuk, M.P. 1977. Podstawy mechaniki
pękania, p.183-192. Kraków. Wyd.
AGH (in Polish).

Recent Advances in Experimental Mechanics, Silva Gomes et al. (eds) © 1994 Balkema, Rotterdam, ISBN 90 5410 395 7

Low cycle fatigue strength of Hi-Mn austenitic steel and its transformation properties

S. Nishida & N. Hattori
Saga University, Japan

T. Shimada
Nippon Steel Corporation, Japan

S. Iwasaki
Saga University (Graduate School), Japan

ABSTRACT : This paper refers to the low–cycle fatigue properties of a high manganese steel (HM steel) and typical stainless steel (SUS304), which are the non–magnetic and austenitic ones. The fatigue tests had been performed not only under the strain–controlled conditions but also stress–controlled ones in order to investigate their fatigue strength and transformation properties, respectively. Then, it is confirmed that there is no significant difference on fatigue crack initiation behavior between both materials. In addition, HM steel shows stable behavior after initial rapid hardening region in the curve of the plastic strain to the cyclic number to failure under the strain–controlled condition, while SUS304 becomes gradually increase with increasing cyclic numbers. Though HM steel does not change for its magnetic properties with increasing the plastic strain amplitude, SUS304 remarkably changes for them due to the strain–induced martensite.

1. INTRODUCTION

It is said that high manganese non–magnetic austenitic steel (HM steel) has been expected to be a promising component material of the track for a linear induction motor train, nuclear fusion electric power systems etc., because it shows extra–stable non–magnetic properties, high tensile strength and ductility even in low temperature. This alloy steel contains much manganese, which exists in large quantities in the ocean depths, instead of the expensive nickel composition. Therefore, there would be no problems for it as being rich in natural resource.

As HM steel is kind of new material and does not have come into wide use yet, there are few reports (Maekawa 1990, Nishida 1994) about its mechanical properties but also its fatigue properties.

The authors had performed the low cycle fatigue tests under the strain–controlled conditions. It is well known that the relation between the plastic strain range and number of cycles to failure is generally expressed by Manson(1954)–Coffin(1954) law $\Delta \varepsilon_p N_f^\alpha = C$ about most of metals except high strength–low ductility steel, Ti alloys, etc.,

where $\Delta \varepsilon_p$ is plastic strain range, α and C are material constants. In addition, considering the practical use, the fatigue tests had been also performed under the stress–controlled conditions, in which the cyclic stress was determined at the levels corresponding to the middle life in the strain–controlled diagrams.

The object of this paper refer to the fatigue strength snd transformation properties of HM steel based of above two kinds of conditions comparing with those of a typical stainless steel SUS304, which is also well–known as a non–magnetic one. Especially, fatigue bahaviors of both materials had been investigated in view point of fatigue crack initiation by the successive–taken replica method.

2. EXPERIMENTAL PROCEDURE

The materials used in this test are HM steel and SUS304, which are both non–magnetic austenitic steels of rolled thick plates for a market. Tables 1 and 2 list their chemical composition and mechanical properties, respectively. Figure 1

Table 1. Chemical composition of the materials.   mass%

Materials	C	Si	Mn	P	S	Ni	Cr	Mo	Nb	N	Al
HM steel	0.178	0.23	23.77	0.024	0.0012	0.23	2.05	–	0.04	0.0428	0.003
SUS304	0.048	0.50	0.81	0.026	0.007	8.47	18.49	0.12	–	0.0465	–

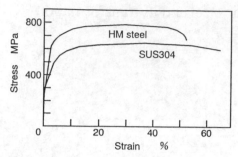

Fig. 1. Stress–strain curves.

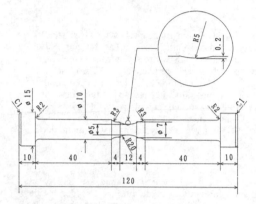

Fig. 2. Shape and dimensions of hourglass type specimen.

Table 2. Mechanical properties of the materials.

Materials	$\sigma_{0.2}$ MPa	σ_B MPa	El %	RA %
HM steel	251.9	751.7	56.0	65.0
SUS304	253.8	613.5	66.0	72.0

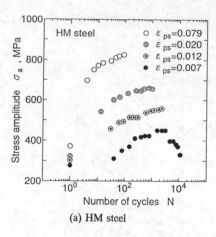

(a) HM steel

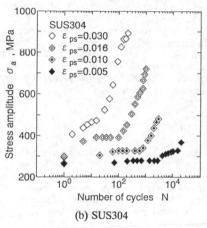

(b) SUS304

Fig. 3. Change of stress amplitude with number of cycles.

shows stress–strain curves of HM steel and SUS304.

Fatigue 2 shows the shape and dimensions of hourglass type fatigue specimen, which was cut out by coinciding the rolling direction with the specimen axis and by making a partial shallow notch at its rolling surface side. This shallow notch exists for limiting fatigue–damaged part and does not affect for its fatigue strength at all (Nisitani 1978). All specimens were annealed in vacuum for one hour at the temperature of 600 °C after polishing with fine emery paper to the longitudinal direction , and thereafter electro–polished to the depth of about 50 μ m.

The fatigue tests had been performed using an electro–hydraulic servo type fatigue testing machine with 98kN load capacity. The low–cycle fatigue tests had been devided into the next two kinds; one was the diametral strain–controlled test using a diametral extensometer, i.e., $R_{\varepsilon t}$ (= ε_{min} / ε_{max})=−1, the other was thestress–controlled test at the stress level corresponding to the middle life in cyclic stress–strain diagrams obtained by the strain–controlled condition， i.e., R_{σ} =(σ_{min} / σ_{max})=−1.

Fatigue crack initiation and crack growth rate were observed by the successive–taken replica method in the circumferential direction at the specimen's surface.

3. RESULTS AND DISCUSSION

3.1 Cyclic stress–strain response

Figure 3 shows the change of stress range with number of cycles at the each strain range, respectively. It can be seen that the both materials shows the remarkable cyclic strain hardening, which caused an increase in the stress

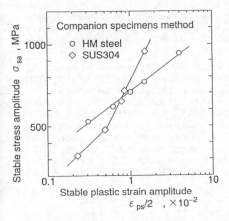

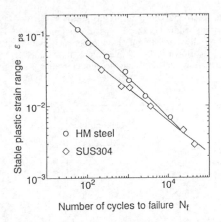

Fig. 4. Cyclic stress–strain curves.

Fig. 5. Relation between stable plastic strain range ε_{ps} and number of cycles to failure N_f.

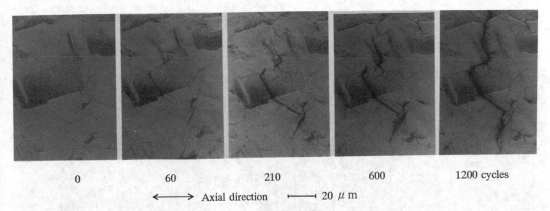

Fig. 6. Successive observation of fatigue crack initiation (HM Steel ε_{ps} =1.24 $\times$ 10^{2}).

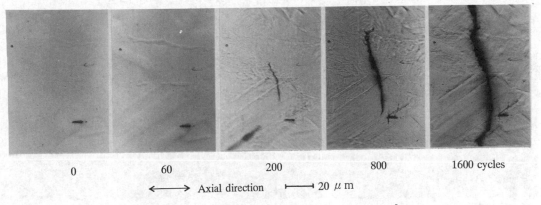

Fig. 7. Successive observation of fatigue crack initiation (SUS304 ε_{ps} =0.93 $\times$ 10^{2}).

required to keep the constant plastic strain amplitude. In addition, there was the significant difference on the hardening behavior between both materials. As shown in the above figure, the results for HM steel shows relatively stable behavior after

the initial rapid hardening region. In contrast, those for SUS304 exhibits the gradually incease with incresing number of cycles corresponding to plastic strain amplitude.

Figure 4 shows the cyclic stress–strain curves, which are determined from the companion specimens method. Each data in this figure was used at middle of the life of the stress–strain curves under the former strain–controlled tests. As shown in this figure, it must be noted that the datum for SUS304 are on the two linear lines. Similar observations have been reported by the other authors(e.g. Hatanaka 1984). On the other hand, the datum for HM steel are expressed by a linear relation.

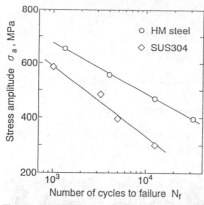

Fig. 8. S–N curves.

3.2 Strain–controlled test

Figure 5 shows the relation between stable plastic strain range and number of cycles to failure. In general, the controlled variable is usually the total strain range, while the fatigue life in low cycle range is successfully expressed by the plastic strain

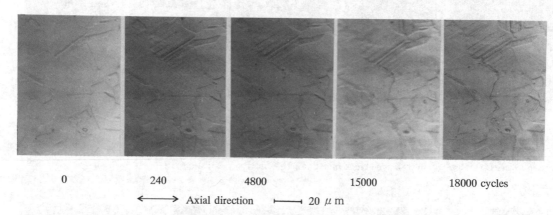

| 0 | 240 | 4800 | 15000 | 18000 cycles |

←——→ Axial direction ├——┤ 20 μ m

Fig. 9. Successive observation of fatigue crack initiation (HM Steel σ_a =397MPa).

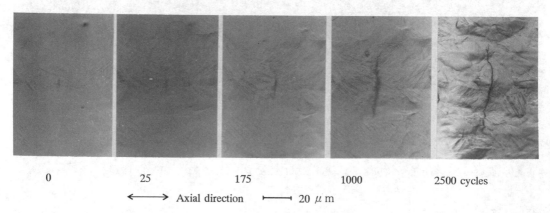

| 0 | 25 | 175 | 1000 | 2500 cycles |

←——→ Axial direction ├——┤ 20 μ m

Fig. 10. Successive observation of fatigue crack initiation (SUS304 σ_a =396MPa).

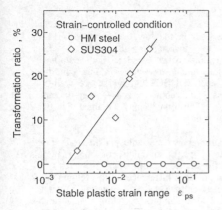

Fig. 11. Change of transformation properties.

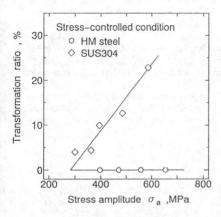

Fig. 12. Change of transformation properties.

range. The elastic strain must be subtracted from the total strain in order to calculate the plastic strain. However, the elastic and plastic strain values change during the constant total strain-controlled tests, because the materials, in most cases, show the cyclic hardening or softening. As shown in this figure, there is scarcely difference between HM steel and SUS304. On the other hand, SUS 304 shows slightly shorter life relative to HM steel. The datum are on the linear relation for both materials, respectively. When the above expression by Manson-Coffine law is used to describe these relationships, the constants for HM steel are α =1.79 and C=1.27, and those for SUS304 are α =2.38 and C=0.051.

Figure 6 shows the representative successive observation results for fatigue crack initiation behavior of HM steel obtained by replica method tested at a constant strain amplitude, ε_{ps} =1.24 $\times 10^2$. The slip bands are at first generated in grains at early stage. Then, fatigue cracks initiate in grain boundary or in its neighbourhood, and

finally propagate mainly as a transcrystalline crack.

Figure 7 shows the result of SUS304 at a constant strain amplitude, ε_{ps} =0.93 $\times 10^2$. The crack initiation behavior for SUS304 is very similar to that for HM steel. In addition, the micro-cracks for both materials initiate within 10 % of the total fatigue life. Thus, the significant difference on fatigue crack initiation behavior between both materials was not observed.

3.3 Stress-controlled test

The stress-controlled condition tests in this papers were carried out using the stress values at middle of the life under the constant strain amplitude condition, as shown in Figure 3. Figure 8 shows the S-N curves of the two kinds of materials under the stress-controlled condition. It can be seen from this figure that the curve for each material has a linear relation, respectively. However, SUS304 shows shorter life relative to the HM steel as compared with that of strain-controlled condition. This is reasonable from the fact that the hystetesis loop of HM steel exhibits more elastic behavior as compared with that of SUS304 at the same stress amplitude controlled condition.

Figures 9 and 10 show the representative successive observation results for fatigue crack initiation behavior of HM steel and SUS304 using the replica method tested at a constant stress amplitude, respectively. As shown in these figures, there was no large difference on the fatigue crack initiation behavior between both materials under the stress-controlled condition. In addition, the fatigue crack initiation behavior under the stress-controlled condition was essentially the same as compared with that under the strain-controlled condition, and the micro-crack also initiates within 10 % of the total fatigue life.

3.4 Transformation property

The magnetic property had been measured using the ferrite meter. The measured point was 1.5 mm apart from each fracture surface. Figure 11 shows the relation between transformation ratio and plastic strain range after tests under the strain-controlled conditon. It can be seen in this figure, the curve of SUS304 approximately shows a linear relation. On the other hand, the transformation ratio of HM steel keeps essentially zero. Considering above the cyclic stress-strain curve behavior, it is considered that the trasformation will be caused by the strain-induced martensitic deformation in SUS304.

Figure 12 shows the relation between trans-formation ratio and stress amplitude after tests under the stress-controlled condition. SUS304 shows the magnetic properties as well as the result under strain-controlled condition. There is the liner

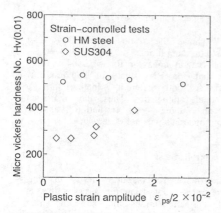

Fig. 13. The relation between plastic strain amplitude and micro vickers hardness number.

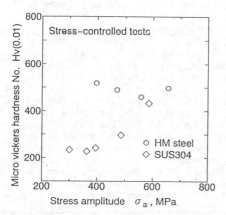

Fig. 14. The relation between stress amplitude and micro vickers hardness number.

relation between the transformation and stress amplitude. On the other hand, HM steel shows the non-magnetic properties as well as the result under strain-controlled condition. It is noted that SUS304 also exhibits the magnetic properties after the tests.

Figure 13 shows the relation between the micro vickers hardness number and plastic strain amplitude after tests under the strain-controlled condition. The variation of hardness number of HM steel was hardly observed over the all strain range. On the other hand, that of SUS304 gradually becomes increased with increasing plastic strain amplitude.

Figure 14 shows the relation between the micro vickers hardness number and stress amplitude after stress-controlled tests. As shown in this figure, the hardness number of SUS304 also becomes increased with increasing stress amplitude. In contrast, that of HM steel is nearly constant over the test stress range.

4. CONCLUSIONS

The fatigue tests in the low cycle range had been performed to investigate the fatigue crack initiation characteristics under the completely pull-push load under low-cycle fatigue conditions using HM steel and SUS304, which are well-known as austenitic and non-magnetic, respectively. The conclusions obtained in these tests were as follows :

(1) The relation between stable plastic strain range and number of cycles to failure for both materials depends on the Manson-Coffin law.
(2) There is no large difference on the fatigue crack initiation behavior between both materials under the strain-controlled condition. In addition, the fatigue crack initiation behavior under the stress-controlled condition is essentially the same as compared with that under the strain-controlled condition.
(3) Both materials shows the remarkable cyclic strain hardening unter the constant strain controlled conditions. SUS304 exhibits the gradually increase with increasing number of cycles corresponding to plastic strain amplitude. On the other hand, HM steel shows relatively stable behavior after the initial rapid hardening region.
(4) SUS304 exhibits the magnetic property after tests under the stress-controlled and strain-controlled condition. On the other hand, HM steel keeps non-magnetic one.

REFERENCES

Coffin, L.F. 1954. Trans. ASME, Vol.76,931
Hatanaka, K. & T.Fujimitsu 1984. Trans. Jpn. Soc. Mech. Eng.,(in Japanese) Vol.50, No.451, 291
Maekawa, I. & S.Nishida et al.3 1990. Trans. Jpn. Soc. Mech. Eng.,(in Japanese) Vol.56, No.225, 1051
Manson, S.S. 1954. Technical Note, Nat. Advis. Comm. on Aeronaut.
Nishida, S. & N.Hattori et al.2 1994. J.Soc.Mater. Sci.,Jpn.,(in Japanese) Vol.43, No.486
Nisitani, N. & Y.Hasuo 1978. Trans. Jpn. Soc. Mech. Eng.,(in Japanese) Vol.44, No.377, 1

Recent Advances in Experimental Mechanics, Silva Gomes et al. (eds) © 1994 Balkema, Rotterdam, ISBN 90 5410 395 7

Ultrasonic evaluation of material degradation due to hydrogen attack of 1/2-Mo steel

Naotake Ohtsuka & Yasunori Shindo
Ryukoku University, Ohtsu, Japan

ABSTRACT: 1/2-Mo steel was hydrogen attacked in an autoclave at 773K and 14.7 MPa for 5 to 300 hours. Ultrasonic characteristics of the steel was measured by the pulse echo digital-overlap (PEDO) method at 25, 10 and 2 MHz. The longitudinal wave velocity and attenuation of ultrasonic wave showed high precision more than 0.1 % and good correlation with the degradation due to hydrogen attack. The RMS ratio of back surface echo to scattering back ground noise signal and the fictitious fractal dimension of wave form of wave amplitude of back ground noise signal also proved apparent relation to the degree of hydrogen attack. These are considered to be caused by the increase of scattering of ultrasonic wave due to the formation of voids or micro-fissures.

1 INTRODUCTION

Pressure vessel steel used in hydrogen environment under high pressure at elevated temperature are subjected to hydrogen attack (HA). Therefore non-destructive evaluation of material degradation due to HA is regarded as one of key technologies to assure the integrity of hydrogen plants. In this paper ultrasonic technique was applied to low alloy steel and the relation between the ultrasonic characteristics and the degradation due to HA was discussed.

2 EXPERIMENTAL PROCEDURE

2.1 *Test specimen*

1/2 Mo steel was exposed in an autoclave at 773 K under hydrogen pressure of 14.7 MPa for 5, 20 and 300 hours. The chemical composition of the material are shown in Table 1. After taking out from the autoclave, the specimens with the thickness of 30, 10 and 5 mm cut down out of the steel were ground and the thickness was measured by a micrometer. The Charpy impact test was conducted so as to measure the degree of degradation of the specimens due to HA.

2.2 *Ultrasonic measurement*

The pulse echo overlap (PEO) method is to measure the sound velocity with relatively high precision by overlapping the phase of two back surface echoes of ultrasonic wave on an oscilloscope (Matec 1988). However, as the time interval of oscillating ultrasonic pulses is adjusted by observing the CRT display of the oscilloscope, flickering of the indicated wave forms yields error. Therefore pulse echo digital-overlap (PEDO) method was used (Ohtsuka & Shindoh 1992). In the method about 10 cycles of 25 to 5 MHz pulses were excited by Matec Model 7700 Pulse Modulator and the detected wave form of the pulse echo was once recorded in Lecroy 9400 Dual Digital Oscilloscope with 8 bit 32 kB memories. Magnifying the time axis 100 times in high resolution mode and overlapping the recorded two back surface echoes on the CRT display, the zero crossings of individual cycles of the echoes were made to coincide by adjusting the time difference of the echoes. The time difference, Δt, and the mean reduction ratio of the peak-to-peak amplitudes, r, of 10 pulses between the

Table 1. Chemical composition of test material (wt %).

C	Si	Mn	P	S	Cr	Mo
0.20	0.27	0.66	.013	.019	0.07	0.57

two echoes were measured. The longitudinal wave velocity, V, and attenuation, α, of ultrasonic wave in the specimen were determined by the following equations.

$$V = L/\Delta t \qquad (1)$$
$$\alpha = (20/L) \cdot \log(1/r) \qquad (2)$$

where L is a round-trip distance of the specimen. In order to improve the reliability, the averages of 10 measured values of V and α were used in the same condition. The sampling time base of the digital oscilloscope was 2 or 5 μs/Div.

In order to investigate the characteristics of the scattering back ground noise signal of ultrasonic wave, the root-mean-square (RMS) and the fictitious fractal dimension of the noise signal during 500 ns between the 1st and 2nd back surface echoes were also measured. The value of RMS of the noise signal was reversely divided by RMS of the 1st back surface echo so as to compensate the difference of setting condition of ultrasonic measurement. The fictitious fractal dimension was determined by the following equation.

$$D = 1 - s \qquad (3)$$

where s is the slope in the logarithmic diagram on the relation between an arbitrary scale, d, and the summation of the scale, l, in a constant time interval.

2.3 *Hydrogen attack parameter*

In order to represent the degradation due to HA quantitatively, the following HA parameter was proposed from the equation on the thermal activation process (Ohtsuka & Shindoh 1993).

$$P_n = T(C + \log t + n \log P_H)/1000 \qquad (4)$$

where T and t are the absolute temperature (K) and exposed duration (hour) in hydrogen and P_H is the partial pressure of hydrogen (MPa). C and n are constants and the optimum values of 10 and 0.5 were used based on the literature relations of void diameters along grain boundaries as a function of T, t and P_H in hydrogen.

3 EXPERIMENTAL RESULTS AND DISCUSSION

3.1 *Charpy impact energy*

The absorbed energy in the specimens in the Charpy impact test at room temperature is shown in Fig.1 as a function of the HA parameter, P_n. The impact energy decreases as the increase of P_n. This indicates that the damage due to HA increases as the increase of exposure time in hydrogen. The result on the higher impact energy for the shortest exposure time in the test than for the virgin specimen is considered to show the effect of heating history.

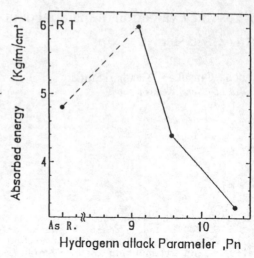

Fig.1 Relation between absorbed energy in the Charpy impact test at room temperature and HA parameter.

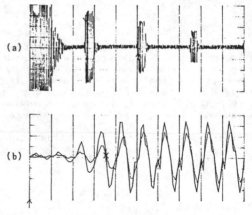

Fig.2 Typical longitudinal waveforms of the virgin specimen.

3.2 *Waveform*

Examples of CRT display of measured ultrasonic waveform at 10 MHz are shown in Fig.2 for the virgin specimen and in Fig.3 for the specimen exposed for 300 hours in hydrogen. In the two figures, (a) and (b) indicate the whole waveform with sampling time base of 5 μs/Div and the overlapped waveform of the 1st and 2nd back surface echoes in high resolution mode, respectively.

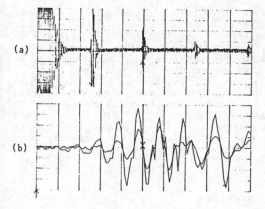

(a)

(b)

Fig.3 Typical longitudinal waveforms of the specimen exposed for 300 hours in hydrogen at 773 K and 14.7 MPa.

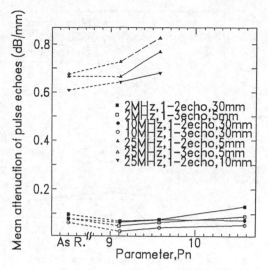

Fig.5 Relation between attenuation of back surface echo and HA parameter.

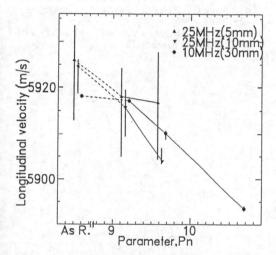

Fig.4 Relation between longitudinal wave velocity and HA parameter.

3.3 *Wave velocity*

The relation of the longitudinal wave velocities for 10 and 25 MHz with the HA parameter is shown in Fig.4. The result of 10 MHz shows that the maximum change in the wave velocity in the test is as few as 0.4 % and that the fewer change less than 0.1 % is measurable by the PEDO method. In the case of 25 MHz the scattering band of the measurement shown as the vertical bars in the figure is larger than 10 MHz and the wave velocity can not be measured as the exposure time becomes longer. The reason is supposed to be the insufficient time precision of the

digital oscilloscope at high frequency and the excessive attenuation with severe HA.

3.4 *Attenuation*

Figure 5 indicates the change of attenuation of back surface echo due to HA. The attenuation increases with the progress of HA regardless of frequency. The value of the attenuation in 25 MHz is considerably larger than the other frequencies. This is considered to be caused by the scattering of ultrasonic wave with shorter wave length at the microscopic voids or fissures.

3.5 *RMS ratio*

The relation between the RMS ratio and the HA parameter is shown in Fig.6. The figure indicates that the RMS ratio decreases with the increase of degradation due to HA and the tendency becomes more obvious with lowering of frequency. This is suspected to be caused by the relative increase of amplitude of scattering background noise signal to that of the back surface echo with the advance of HA.

3.6 *Fictitious fractal dimension*

The measured fictitious fractal dimension of the wave form of scattering background noise signal at 10 and 25 MHz is

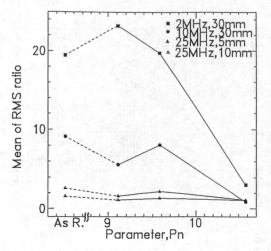

Fig.6 Relation between RMS ratio and HA parameter.

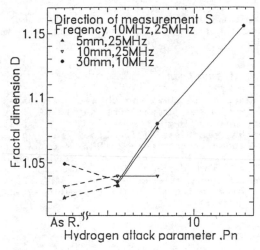

Fig.7 Relation between fictitious fractal dimension of the wave form of scattering back ground noise signal and HA parameter.

indicated in Fig.7. The fictitious fractal dimension increases as the HA damage increases in the figure. The same reason as for the RMS ratio is expected that the increase of the damage accelerates the scattering of ultrasonic wave and distortion of wave form.

4 CONCLUSIONS

(1) It was confirmed that longitudinal wave velocity and attenuation are measurable with sufficient precision for nondestructive evaluation of degradation due to HA by the PEDO method. However, attenuation of high frequency as 25 MHz is too large to measure the velocity.
(2) Apparent relation of the RMS ratio and fictitious fractal dimension of scattering back ground noise signal to the degradation due to HA was found. These are considered to be caused by the increase of scattering of ultrasonic wave with HA damage and the distortion of wave form.

REFERENCES

Matec Instruments 1988. *Product Catalog*. Rhode Island: Matec.

Ohtsuka,N. & Shindoh,Y. 1992. *Serviceability of Petroleum, Process, and Power Equipment*. PVP-Vol.239/MPC-Vol.33: 139-143. ASME.
Ohtsuka,N. & Shindoh,Y. 1993. *Proc. 70th JSME Conference* (I). No.930-9: 539-541. JSME (in Japanese).

Recent Advances in Experimental Mechanics, Silva Gomes et al. (eds) © 1994 Balkema, Rotterdam, ISBN 90 5410 395 7

Use of the Villari effect in fatigue of sintered steel

J. Kaleta & J. Żebracki
Institute of Materials Science and Applied Mechanics, TU Wrocław, Poland

ABSTRACT: The paper shows how a coupled magneto–mechanical effect can be used to monitor fatigue behaviour of selected sintered steel denoted as KA. The specific effect employed in this study is the reverse magnetostriction (or the Villari effect). The quantities measured were: magnetic induction B, magnetic field intensity H and magnetic loop energy ΔM (in the B-H coordinate system). Along with the above magnetic quantities the mechanical ones such as hysteresis energy ΔW (in the $\sigma - \varepsilon$ coordinate system), plastic deformation ε_{pl} and stress σ were recorded. As a result a variety of relationships linking magnetic-mechanical and magnetic quantities were obtained. In particular the measured quantities were determined at the cyclic yield point σ_{cpl} and fatigue limit levels σ_f. Additionally the $\Delta W/\Delta M$ ratio was evaluated at a stress value $\sigma_a = \sigma_f$ (σ_f = fatigue limit in symmetric tension-compression) . Material tested was sintered steel with alloyed components (Ni, Mo) subjected to uniaxial tension-compression. The explored range of lives covered the HCF and the fatigue limit regions. Presented are also some selected comparative results obtained from a conventional steel E355-CC.

1 INTRODUCTION

Sintered steels have become a commonplace choice in designing components for fatigue service. Some characteristics features of these materials such as porosity (Fig. 1.) resulting from the fabrication

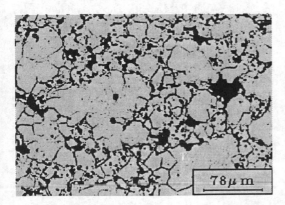

Figure 1. Typical structure of P/M steel

process have a well pronounced effect on their fatigue strength. A deeper insight into the process of fatigue in this class of materials and reliable methods of assessing component service lives have recently become a challenge facing investigation in the field.

Fatigue and cyclic properties of sintered steels differ appreciably from those of conventional steels [Dudziński 1993, Kaleta 1990, 1991a, b, 1992a]. The useful range of lives is usually limited to the HCF and the fatigue limit regions. The stress levels which are considerably higher than the fatigue limit usually result in brittle fracture following a small number of loading cycles. Investigating the LCF region is therefore hardly an easy task and conventional methods of presenting fatigue data in the form of S-N curves are of little value.

The HCF region in sintered steel is characterized by small plastic deformations when compared with conventional steels and the plastic deformation energy ΔW or plastic deformation ε_{pl} are difficult to measure, to give but these two examples. The difficulties are even greater after the fatigue limit has been attained because the ΔW value are then by two orders of magnitude smaller than that recorded in conventional steel [Kaleta 1990, 1991a].

Differences between the attainable plastic strain ε_{pl} values for sintered and conventional steel are shown in Fig. 2. It should be noted that there are two independent axes of strain, each for one material. As can be seen the measured values of ε_{pl} differed by one order of magnitude.

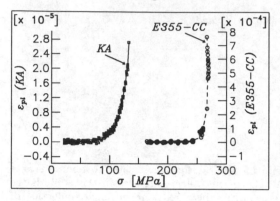

Figure 2. Plot of $\varepsilon_{pl}(\sigma)$ for sintered steel (left) and conventional steel (right)

The circumstances outlined above make the determination of typical characteristics of cyclic deformation and cumulation of plastic deformation energy a difficult task for an experimenter and it is generally acknowledged that only a narrow HCF region can be successfully investigated in this way.

If conventional experimental methods are of limited use in this case then various physical phenomenon-related techniques are even of greater interest than in conventional materials. By these techniques we mean measurements of various forms of energy, temperature changes, acoustic emission, electric resistivity and others. It is hoped that great potential can be expected from the so called coupled phenomena in which different physical fields interact to produce valuable information pertaining to material's behaviour. The coupled thermomechanical field may serve a good example of this experimental trend.

Recent years have seen successful theoretical attempts to explain some magnetic-mechanical effects associated with the process of fatigue in ferromagnetic materials. It is to be noted that the majority of sintered steels can be placed in this class. Of a few known coupled magnetic-mechanical effects [Jiles 1988, Kaleta 1992b, c, Schroder 1989, Titto 1989] the reverse magnetostriction (also called the Villari effect) has been chosen by the present authors as an especially promising.

2 THE AIM OF THE STUDY

The present study was generally aimed at presenting fatigue behaviour of sintered steel in terms of magnetic-mechanical or purely magnetic quantities. The results were compared with those expressed in conventional terms of hysteresis energy ΔW or plastic deformation ε_{pl}.

A specific goal consisted in answering a question whether it is possible to reliably determine the cyclic yield and fatigue limit in an accelerated test.

An additional comparative test on a conventional steel E355-CC was performed to set a background against which coupled phenomena in the region of hardly discernible plastic deformation typical of sintered steel could be evaluated.

3 EXPERIMENTAL DETAILS

The material tested was sintered steel with a addition of $1.2\%Cu$ and alloyed components. The tested P/M steel is denoted as KA. The specimens were (fabricated by Krebsöge Co., Germany) pressed and sintered to obtain flat specimens with a rectangular cross-section 6×8 mm. The sintering conditions chemical and mechanical properties of the material are shown below.

Table 1. Sintering conditions, chemical and static properties of P/M steel investigated.

Temp/time ($°C/min$)	Cu (%)	Ni (%)	Mo (%)	UTS (MPa)	E (MPa)
1120/30	1.5	5.0	0.5	380	141 826

A series of fatigue tests was performed beforehand. Figure 3 and Table 2 show the cyclic properties of tested P/M steel. The N_t value denotes the numbers of cycles at which the two curves representing, respectively, plastic and elastic strain cross each other.

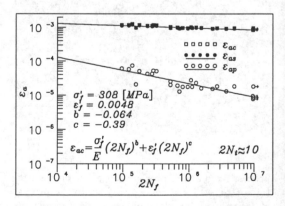

Figure 3. $\varepsilon_a(2N_f)$; sintered steel KA

It should be noted that the cyclic properties characterized by N_t, σ'_f, ε'_f, b and c are essentially different from citied in literature for conventional steels.

Table 2. Cyclic properties of KA sintered steel

$2N_t$ cykle	σ'_f [MPa]	ε'_f [-]	b [-]	c [-]
10	308	0.0048	-0.064	-0.391

The tests were made using both a multiple step test and a constant amplitude of the stress ($\sigma_a = const$) test, in direct tension–compression ($R_\sigma = -1$). The sinusoidal input signal had a frequency of 25 Hz. The stress amplitude was varied over a wide range of values from $\sigma_a = 20MPa$ to $\sigma_a = 135MPa$ (with step equal $5MPa$ or in higher region with step equal $2MPa$). Each stress level of the incremental step test was applied for one hour which resulted in about 90 thousand cycles. The machine used was hydraulic pulser MTS 810 connected with a PC AT via a set of A/D converters. The measurement set–up is shown in Fig. 4.

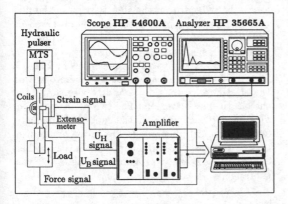

Figure 4. Measurement set–up

In the on–line fatigue tests the hysteresis energy ΔW and plastic strain ε_{pl} were determined. Alongside recorded was voltage U_B induced in a coil placed within the magnetic field produced by a cyclically loaded specimen. In contrast to earlier investigations [Kaleta 1992b, c] an additional coil wound around a crescent–shaped core was mounted which was used for measuring voltage U_H. A coil and measuring channel for determining voltage U_H were calibrated using a standard solenoid producing field of a known intensity H. It can be easily demonstrated that voltages U_B and U_H so measured are proportional to the first

derivative of magnetic induction dB/dt and magnetizing force dH/dt, respectively. The values of B and H were determined by integrating and scaling signals U_B and U_H. Integration of the dB/dt and dH/dt signals was carried out in a purely analytic way using coefficients of the Fourier series representations of signals U_B and U_H. A custom-designed software package APTL [Kasprzak 1989] was used to expand the all mechanical and magnetical signals into Fourier series. Next, all the quantities mentioned above such as ΔW, ε_{pl}, B, H were expressed analytically as functions of respective harmonic components $\sigma(t)$, $\varepsilon(t)$, $U_B(t)$ and $U_H(t)$. A similar list of selected experimental results for the conventional E355-CC steel is also shown. The E355-CC results are not analysed in the report and given for comparative purposes only.

4 ANALYSIS OF RESULTS; DISCUSSION

Having recorded both mechanical ($\sigma(t)$ and $\varepsilon(t)$) and magnetic signals ($U_B(t)$ and $U_H(t)$) we were able to get not only typical $\sigma - \varepsilon$ plots but also a large variety of combined mechanical–magnetic characteristics sensitive to the load level and number of cycles. Below presented are selected relationship resulting from the Villari effect.

Fig. 5 shows a single cycle of the $\sigma(t)$, $U_B(t)$, $B(t)$ for a selected load level.

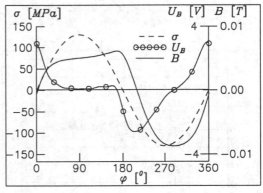

Figure 5. Plot of $\sigma(\varphi)$, $U_B(\varphi)$ and $B(\varphi)$

In a similar way were presented signals U_H and H (Fig. 6). In analysis of the magnetic signals U_B and U_H the harmonic analysis was employed.

A key problem is the ability to determine in an accelerated test the cyclic yield point σ_{cpl} and the fatigue limit σ_f. It must be emphasized that if σ_{cpl} can be found for each single specimen then σ_f becomes a sensible quantity only for a population of specimens. In sintered steel a small degree of scatter is observed, so it was assumed finally that $\sigma_{cpl} \sim 80MPa$ and $\sigma_f \sim 120MPa$. For the load $\sigma_a = \sigma_f$ the other mechanical and magnetic quantities were found to be $\Delta W = 3 \times 10^{-3} MJ/m^3$, $\varepsilon_{pl} = 7 \times 10^{-6}$ and $\Delta M = 3 \times 10^{-8} MJ/m^3$. Further, it is worth noticing the dependence between the magnetic and magneto–mechanical quantities on the one hand and stress on the other. Especially their variation around the σ_{cpl} and σ_f values is of considerable interest.

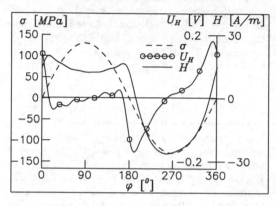

Figure 6. Plot of $\sigma(\varphi)$, $U_H(\varphi)$ and $H(\varphi)$

From earlier investigations [Kaleta 1993] it follows that all the loops, namely $\sigma - \varepsilon$, U_B-ε, B-ε and B–H exhibit high sensivity to the stress amplitude level. The magneto–mechanical loop $U_B - \sigma$ (Fig. 7) and the magnetic loop $B - H$ (Fig. 8) are

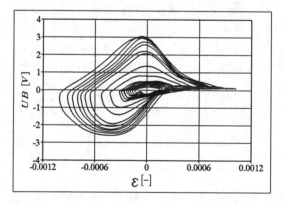

Figure 7. Loop $U_B - \varepsilon$; sintered steel KA

strongly affected by the load level within the elastic region. Once a marked plastic deformation has occurred ($\sigma_a > \sigma_{cpl}$) the $U_B - \sigma$ loop in the first quadrant gets heavily elongated. In the magnetic loop $B - H$ in turn a characteristic sharp projection

arises. It should be noted that this behaviour has been also found in other ferromagnetic materials.

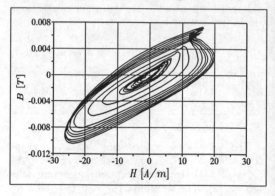

Figure 8. Loop $B - H$; sintered steel KA

Another way of presenting how the Villari phenomenon is affected by the load level is relationship $U_B(\sigma_a)$ (Fig. 9), where $U_B=(U_{Bmax} + U_{Bmin})/2$. The plot is apparently linear up to a knee corresponding to the fatigue limit ($\sigma_f = 120MPa$) and preceded by well pronounced plastic deformation. The magnetic loop energy ΔM is a highly sensitive characteristics in the elastic range ($\sigma \leq \sigma_{cpl}$) as can be judged from the step portion of the plot in Fig. 10. A region of saturation is observed on the plot on attaining some marked plastic deformation in a specimen.

A plot of $\Delta W(\sigma)$ is also presented in Fig. 10. The ΔW value is clearly seen to be zero within the elastic range ($\sigma \leq \sigma_{cpl}$) and to rise sharply in the elastic–plastic region. The two plots are a convincing demonstration of the fact that magnetic and mechanical phenomena are fully complementary in this case. It is further seen that the Villari effect may be of great prognostic value in the elastic range.

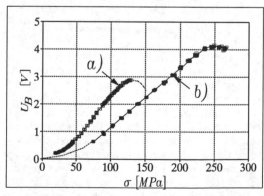

Figure 9. $U_B(\sigma_a)$: a) sintered steel KA, b) E355–CC steel

Both the plot of $U_B(\sigma)$ (Fig. 9) and $\Delta M(\sigma)$ (Fig. 10) are characterized by three distinct intervals. The first one cores the range $0 < P_I \leq \sim 40 MPa$, in the second one good linearity is observed up to the value $\sigma_a = \sigma_f$. In the third interval $P_{III} > \sigma_f$ a saturated condition is attained. For large plastic deformations (not an easy task in sintered steel) even a drop in values of U_B and ΔM in the third interval is observed. This is shown by the dashed line in Fig. 9. For many other ferromagnetic materials a saturated condition has been observed at this stage or even a decrease in the signals U_B and ΔM values.

The Fourier analysis of signals U_B and U_H can be used for testing the applicability of particular harmonic components in predicting the chance of σ_{cpl} and σ_f to occur. The role of the first three odd components has been already stressed elsewhere. For the sake of illustration the variation in the fifth component of signal U_B, $5compU_B = U_B * \sqrt{(sin^2(5\omega t) * cos^2(5\omega t))}$, is shown in Fig. 12 . The plot changes its course for $\sigma_{cpl} \sim 80 MPa$ and gets saturated in the vicinity of $\sigma_f \sim 120 MPa$.

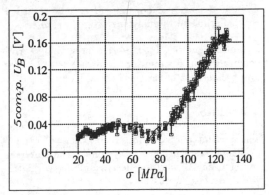

Figure 12. Fifth component of signal U_B for sintered steel KA

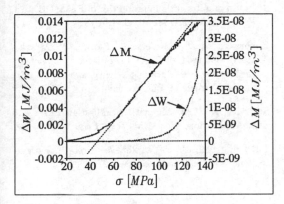

Figure 10. $\Delta M(\sigma_a)$ and $\Delta W(\sigma_a)$ for sintered steel KA

It is to be noted that the ratio $\Delta W / \Delta M$ for $\sigma_a = \sigma_f$ is equal to $1 \cdot 10^5$. A similar value was obtained for conventional ferritic–pearlitic steel. Another characteristic sensitive to the presence of a cyclic yield point is the arithmetic peak values $X_{0H}(\sigma)$, where $X_{0H} = (H_{max} + H_{min})/2$. The quantity so defined shows an abrupt change for $\sigma = \sigma_{cpl}$ (Fig. 11).

The above review of possibilities offered by the Villari effect, though a bit cursory out of necessity, inspires hope that extensive use of magnetic - mechanical characteristics in fatigue testing is not very far ahead. At the same time a new vast field of fundamental research is opening as there will be a great need for physical models accounting for coupled mechanical - magnetic effects.

5   CONCLUSIONS

1. The Villari effect was found especially suitable for investigating fatigue-related degradation of sintered steel.This was due to a high accuracy which could be achieved in determining such magnetic-mechanical or magnetic characteristics and their linear behaviour in the elastic range of loading. The onset of marked plastic deformation makes these characteristics to undergo an appreciable qualitative change. They may therefore serve a useful indicator of a material's condition and, more specifically, can be used for determining the cyclic yield point σ_{cpl} and fatigue limit σ_f in an accelerated test,

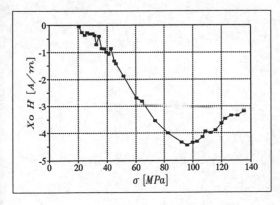

Figure 11. $X_O H(\sigma_a)$; sintered steel KA

2. The Villari effect is a sensitive measure of changes occurring in cyclically loaded materials. A broad range of available magnetic–mechanical

characteristics allows to detect the influence of load level and number of cycles,

3. The selected components of signals U_B and U_H are a clear manifestation of the Villari effect. They show strong linearity in the elastic region and a sharp bend corresponding to the onset of plastic deformation,

4. Such characteristics $\Delta U_B(\varepsilon)$, $5compU_B(\sigma)$ as well as such mechanical - magnetic loops as $U_B - \varepsilon$, $B - \varepsilon$ and magnetic loops $B - H$ may be regarded especially useful in analyzing fatigue processes,

5. The magnetic loop $B - H$ energy value corresponding to the endurance limit is 1×10^5 times lower than that of the mechanical loop $\sigma - \varepsilon$.

REMARKS

The presented results are an outcome of a joint Polish-German project: " Systematische Untersuchungen zur Bewertung des Ermüdungsverhalten von Sintereisenwerkstoffen".

ACKNOWLEDGEMENTS

Financial support under grant 7 7037 92 03 is gratefully acknowledged.

REFERENCES

Dudziński, W. & J. Kaleta, P. Kotowski, L. Zbroniec, J. Żebracki 1993. Fatigue properties and microstructure of alloyed sintered steel KA: *Report No. 7*: Inst. of Materials Science and Appl. Mechanics. Wrocław: TU Wrocław.

Jiles, D.C. 1988. Review of Magnetic Methods for Nondestructive Evaluation. *NDT Int. 21, No 5*: 311–319.

Kaleta, J. & A. Piotrowski, H. Harig 1990. Cyclic Stress–Strain Response and Fatigue Limit of Selected Sintered Steels: *Proc. 8th ECF*: 445–448. Torino: EMAS.

Kaleta, J. & A. Piotrowski, H. Harig 1991a. Cyclic Properties and Hysteresis Energy Accumulation in Selected Sintered Steels: *Proc. FEFG/ICF*: 570–575. Singapore: Elsevier.

Kaleta, J. & A. Piotrowski 1991b. Das Hysteresis–Meßverfahren zur Charakterisierung des Ermüdungsverhaltens von Sinterstählen: *Hauptversammlung der DGM*: 110. Graz: DGM.

Kaleta, J. & A. Piotrowski 1992a. Hysteresis Energy and Fatigue Life of Selected Sintered Steels: *Proc. 3rd ICLCFEPBM*: 473–478. Berlin: Elsevier.

Kaleta, J. & J. Żebracki 1992b. Villari Effect to Determine the Fatigue Limit in Steel. *Proc. 4th ICSF-PLTI*: Vienna: TU Wien.

Kaleta, J. & J. Żebracki 1992c. Villari's Effect in Fatigue Tested Medium-Carbon Steel. *XV Symp. on Exp. Mechanics of Solid State*: 128–131. Jachranka: TU Warsaw.

Kaleta, J. & J. Żebracki 1993. Use of the Villari effect to monitor fatigue properties of E355–CC steel. *Proc. 5th ICFFT*: Montreal: EMAS.

Kasprzak, W. & J. Kaleta, R. Błotny, W. Myszka, M. Niżankowski 1989. APTL–Software, *Report No. 25–26*: Inst. of Materials Science and Appl. Mechanics. Wrocław: TU Wrocław.

Schroder, K. 1989. Magnetic Barkhausen Effect and its Application. *Nondestr. Test. Eval. 5 Iss 1*: 3–8.

Titto, K. 1989. Use of Barkhausen Noise in Fatigue. *Nondestr. Test. Eval. 5 Iss 1*: 27–37.

Recent Advances in Experimental Mechanics, Silva Gomes et al. (eds) © 1994 Balkema, Rotterdam, ISBN 90 5410 395 7

Investigation of cadmium deposition effect on fatigue resistance of ABNT-4340 steel

M. P. Peres, C. P. M. Pereira & H. J. C. Voorwald
Faculdade de Engenharia, Guaratinguetá, Brazil

ABSTRACT: The influence of technologic parameters of cadmium electrodeposition on cyclic fatigue properties of ABNT-4340 steel is investigated. Fatigue tests for maximal stress from 0.5 to 0.9 yielding strength are performed. The optimal combination of working temperature and time of process has been determined.

1. INTRODUCTION

The increasingly design requirements for advanced aerospace structures have driven the development of new materials with improved mechanical properties and fatigue resistance (Wadsworth & Froes 1989; Frazier et al. 1989).

High temperature capability, low density and hydrogen embrittlement resistance area properties required for conventional aircraft and advanced space propulsion system (Paton 1991). To improve and design metals and alloys for aerospace applications, investigations are aimed at strengthening mechanisms, phase transformations, plasticity, creep, fatigue, environmental effects and dynamic and static fracture (Rosenstein 1991).

Fatigue failures are mostly the result of initiation and growth of cracks caused by the application of cyclic loading (Manson 1971). Because high strength materials are used in structures that should be able to carry a predetermined load, the critical crack size to fracture is very small. Under these conditions, in some components the designing against the occurrence of fatigue damage use the safe-life approach (Best 1986).

The increasing need for information on material behaviour has led to a number of test programs for materials. The categories of variables that can influence fatigue behaviour may be classified into mechanical, metallurgical and environmental (Sudarshan et al. 1990). The mechanical variables include the maximum stress, stress amplitude, cyclic frequency and waveform, interaction effects in variable amplitude loading, state of stress, residual stress, crack size, crack orientation and shape. The metallurgical variables include: alloy composition, distribution of alloying elements, microstructure, heat treatment, mechanical working and texture, and the environmental variables that contribute to corrosion fatigue are: temperature, pH, electrochemical potential, the concentrations of damaging species and the nature of the films formed.

The fatigue failure process has been divided into three main stages, namely: precrack deformation, fatigue crack nucleation and fatigue crack propagation. Interest in fatigue crack propagation has been increasing steadily with the development of improved methods for detecting cracks and for predicting their growth using linear-elastic and elastic-plastic fracture mechanics approaches (Vesier & Antolovich 1990; Rai et al. 1990; Smith 1990; Miyamoto et al. 1990; Voorwald et al. 1991).

Fatigue crack initiation occurs at the surface although sub surface nucleation has also been reported. It is possible to obtain empirical equations in which the endurance limit is, for steels, correlated to tensile strength of the

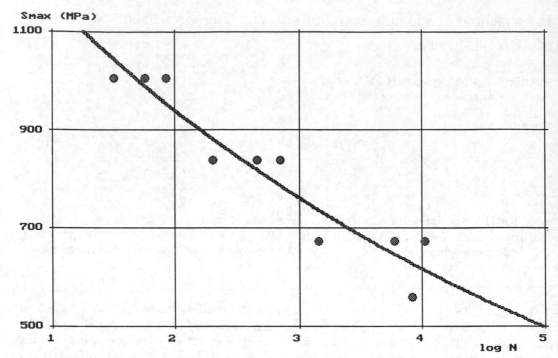

Figure 1. S-N curve for basic material

material (Brand & Sutterlin 1980).

Material soundness and cleanliness are recognized as major factors on fatigue resistance. Localized imperfections in the form of inclusions are highly instrumental in producing strain concentrations resulting in the fatigue crack nucleation (Lankford & Kusenberger 1973). In high strength steel, fatigue sub-surface nucleation may occur and produce the fish-eye condition (Murakami et al. 1989) . The geometric nature of the descontinuity, descohesion at the interface between matrix and particle fracture itself may be the cause of the great range of scatter of fatigue data for high strength steel (Murakami & Usuki 1989).

The aim of this investigation is to analyze the effects of cadmium deposition on the fatigue strength of a 4340 steel in a mechanical condition of (1100-1200) MPa, used in structural components subjected to high stress levels.

2. EXPERIMENTAL

2.1. *Material*

The experimental program was performed on tests samples quenched from (815-845) °C and tempered in the range (520 51) °C for 2 hours. Mechanical properties of the alloy are: (38-42) HRC, yield strength σ_{ys}= 1118 MPa and rupture strength σ_{us}=1210 MPa. The chemical composition of 4340 steel is listed in table 1.

Table 1. Chemical composition of alloy 4340.

	C	S	Mn	Cr	Ni	Mo
Stan-	0,38	0,025	0,60	0,70	1,65	0,20
dard	0,43	max.	0,80	0,90	2,00	0,30
4340	0,39	<0,01	0,69	0,74	1,70	0,23

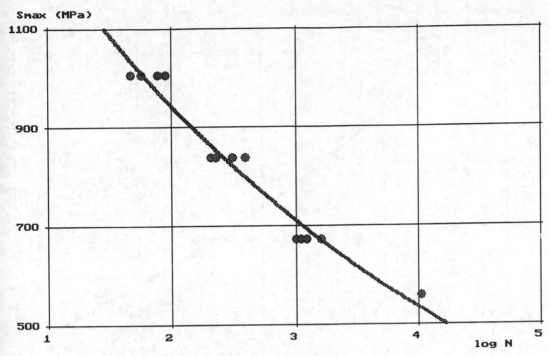

Figure 2. S-N curve for basic material with cadmium deposition

2.2 *Specimens and loading conditions*

Rotating bend fatigue specimens were machined from the bar, polished and inspected by magnetic particles and dimensional. Three groups of test specimens were prepared and tested: 1) smooth specimens of base material; 2) specimens with cadmium deposition; 3) specimens with cadmium deposition, heat treated. The S-N behaviour was established by using a sinusoidal loading at frequency of 50 Hz without mean stress (R=-I) at constant load and room temperature.

3. RESULTS AND DISCUSSIONS

The S-N behaviour of 4340 steel with different conditions is shown in figures 1, 2, 3 and 4 which represent, respectively, experimental data for base material, base material with cadmium deposition and base material with cadmium deposition treated at 190 °C for 3 and 8 hours.

It can be observed from figures 1 to 4, that in case $\sigma=0,90\sigma_{ys}$, no significant difference in test results occurs. From the experimental data it is possible to conclude that in case $\sigma=0,75\sigma_{ys}$ the same tendency was observed for fatigue life as for higher stress levels figure 2 shows a reduction in the fatigue life of the 4340 steel, caused by cadmium deposition, in the case $\sigma=0,65\sigma_{ys}$ due, probably, to hydrogen embrittlement. After heat treatment (190 °C/3 hours), better results were obtained for stress levels equal or lower than $0,6\sigma_{ys}$. Small difference in results occurred for 190 °C/3 hours and 190 °C/8 hours.

It was also observed less dispersion in experimental data of cadmium without heat treatment when compared with conditions indicated in figures 3 and 4 for low stress levels ($\sigma < 0,6 \sigma_{ys}$).

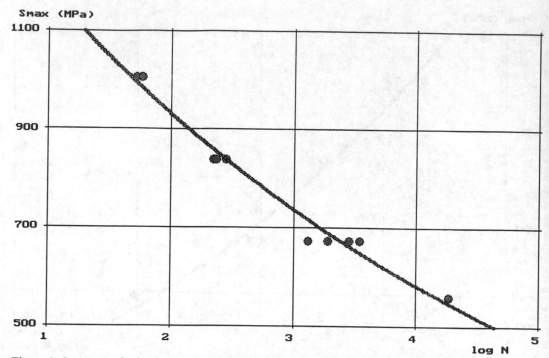

Figure 3. S-N curve for basic material with cadmium deposition. Heat treatment at 190ºC / 3h

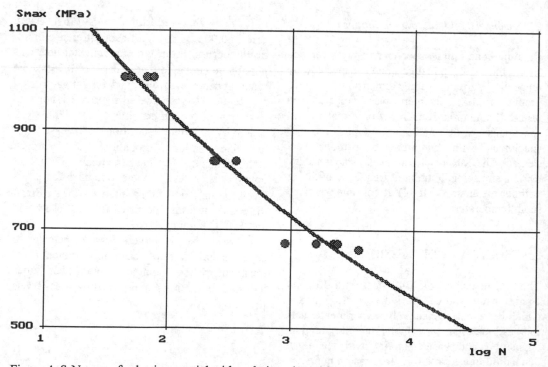

Figure 4. S-N curve for basic material with cadmium deposition. Heat treatment at 190ºC / 8h

4. CONCLUSIONS

1. It can be observed that for $\sigma=0,90\sigma_{ys}$ and $\sigma = 0,75\sigma_{ys}$, no significant differences in test results occurs in conditions studied.

2. The experimental data show a reduction in the fatigue life of the 4340 steel caused by cadmium deposition, in the case $\sigma=0,6\sigma_{ys}$.

3. After heat treatment (190 °C/3 hours), higher fatigue life was obtained for stress levels equal or lower than $0,6\sigma_{ys}$.

4. No significant differences occurred for 190 °C/3 hours and 190 °C/8 hours.

REFERENCES

Best, K.F. 1986. *Aircraft Engineering*, July:. 14-24.

Brand, A. & R. Sutterlin 1980. *Limited d'Endurance des aciers en fonction de la charge de ruptura*. Cetim. France.

Frazier, E.W.; Lee, E.W. ; Donnellan, M.E. & J. J. Thompsom 1989. *JOM* : 22-26.

Lankford, J. & F. N. Kusenberger 1973. *Metallurgical Transactions*. 4:553-559.

Manson, S.S. 1971. *American society of testing and materials*. 254-346.

Miyamoto, H.; Kikuchi, M. & T. Kawazoe. 1990. A study on the ductile fracture of al alloys 7075 and 2017. *Int. Journal of Fracture*. 42:389-404.

Murakami, Y. Kodama, S. & S. Konuma. 1989. *Int. J. Fatigue* , 5:291-298.

Murakami, Y. & H. Usuki 1989. *Int. J. Fatigue*, 11:299-307.

Sudarshan, T.S.; Srivatsan & D.P. Harvey II. 1990. Fatigue process in metals - role of aqueous environments. *Engineering fracture mechanics*. 36::827-852

Paton, N.E. 1991. Materials for advanced space propulsion systems. *Materials Science and Engineering*, A 143. 21-29.

Rai, R.E.; Sundaram, P. ; Pandey, R.K. ;et al 1990. Crack initiation and growth resistance in pressure vessels materials. *Engineering fracture mechanics*. 37:163-173.

Rosenstein, A.H. 1991. Overview of research on aerospace metallic structural materials *Materials Science and Engineering*, A 143. 31- 41.

Smith, E. 1990. Predicting crack arrest in a nuclear reactor pressure vessel during a hypothetical pressurized thermal shock event. *Int. J. Pres. Ves. & Piping* 42:217-235.

Voorwald, H.J.C.; Torres, M.A.S. & C.C.E.Pinto Junior. 1991. Modelling of fatigue crack growth following overloads. *Int. J. Fatigue*. 423-427.

Vesier & S.D. Antolovich 1990. Fatigue crack propagation in Ti-6242 as a function of temperature and waveform. *Engineering fracture mechanics*. 37:753-775.

Wadsworth, F.H. & F. H. Froes 1989. *JOM*. May: 12-19.

Recent Advances in Experimental Mechanics, Silva Gomes et al. (eds) © 1994 Balkema, Rotterdam, ISBN 90 5410 395 7

Effect of negative stress ratio on fatigue crack propagation in polymers

T.Shiraishi, H.Ogiyama & H.Tsukuda
Ehime University, Matsuyama, Japan

ABSTRACT: The effects of negative stress ratio on the fatigue crack propagation rate and crack closure behavior in several polymers were investigated. Fatigue cracks were propagated under cyclic axial loading in various negative stress ratios. Crack closure measurements were made by a compliance method with a strain gauge glued near the crack tip on the specimen surface. The fatigue crack propagation rate, da/dN, increased with decreasing stress ratio in the relationship of da/dN versus the maximum stress intensity, K_{max}, in all polymers tested. This result implies that the compressive stress range in the cyclic loading causes the acceleration in fatigue crack propagation. The degree of this acceleration depended both on the stress ratio and on the maximum applied stress level. The crack closure behavior suggests that such an acceleration for these polymers can be explained on the basis of the plasticity-induced crack closure concept just as for metals.

1 INTRODUCTION

A number of studies have been made on the effect of positive stress ratio on the fatigue crack propagation for polymers (Hertzberg 1980). However, there are very few studies on the effect of negative one on the crack propagation behavior (Shiraishi 1988).

For metallic alloys, some studies on this subject have been reported (Okazaki 1985, Kurihara 1986, Ogiyama 1990, 1991), and it is well known that the fatigue crack propagation rate at the same maximum stress intensity factor increases with decreasing negative stress ratio. This phenomenon can be explained on the basis of the fatigue crack closure concept (Elber 1971).

At present, it is ambiguous whether or not the results obtained for metals are applicable for polymers.

The purpose of the present study is to clarify the effects of negative stress ratio on the fatigue crack propagation rate and the crack closure behavior in several polymers.

2 EXPERIMENTAL PROCEDURE

Commercially available plates of polyvinyl chloride (PVC, 5mm thick), polymethyl methacrylate (PMMA, 5mm thick) and polyamide (PA, 6mm thick) were machined to single edge notch tensile specimens of the size of 180mm in length and 50mm in width. A 3mm long pre-crack was formed by pushing a razor blade to the bottom of a sawcut on a side of the specimen.

The tensile properties of these materials were shown in Table 1. Both PVC and PA have clear yield points, and the former has higher yield stress than the latter. In the case of PMMA, however, there existed no yield point because of brittleness.

The fatigue tests were carried out with an electro-hydraulic servo-type testing machine (push-pull loading type). The cyclic loading frequency was 1 Hz and the test temperature was about 25°C. Fatigue cracks were propagated under various conditions of negative stress ratios with a constant maximum applied stress level for each

Table 1 Tensile properties of materials.

Material	Young's modulus (GPa)	Yield stress (MPa)	Tensile strength (MPa)
PVC	3.4	65.7	65.7
PMMA	3.1	–	69.0
PA	1.3	45.8	45.8

Table 2 Test conditions.

Specimen	σ_{max}, MPa	R		
PVC	9.8	-1.5	-1	0
	6.9	-1		
PMMA	4.9	-2	-1	0
	2.9	-1		
PA	11.8	-1	-0.5	0
	8.8	-1		

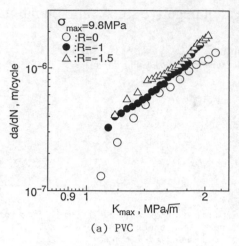

(a) PVC

polymer. The effect of maximum stress level on the crack propagation at a constant negative stress ratio was also examined. Table 2 shows the stress conditions used in this experiment, where R ($=\sigma_{min}/\sigma_{max}$) is the stress ratio, σ_{min} the minimum applied stress and σ_{max} the maximum applied stress.

The fatigue crack length was measured on the specimen surface by a traveling microscope.

The crack closure measurements were made at the frequency of 1 Hz by a compliance method with a strain gauge (gauge length is 0.2mm) glued at a position about 1mm above the crack plane and slightly behind the crack tip (Shiraishi 1984, 1988), which is one of the methods used for metals (Tokaji 1980).

The fatigue fracture surfaces were observed with an optical microscope.

3 EXPERIMENTAL RESULTS AND DISCUSSION

3.1 Effect of negative stress ratio on da/dN at a constant maximum applied stress

Figure 1 shows the relation between the fatigue crack propagation rate, da/dN, and the maximum stress intensity factor, K_{max}, for different stress ratios at a constant maximum stress for each polymer. Here, K values were calculated using the formula for single edge notch tensile specimens (Pook 1968). It is clear that da/dN increased with decreasing stress ratio for all polymers tested. The authors, in the previous study (Shiraishi 1988), have found the same tendency for polycarbonate (PC). This tendency was marked in PA, namely, such a tendency was clearly seen even at R=-0.5 in this material as shown in Fig. 1(c).

If the compressive stress range in the cyclic loading has no effects on the fatigue crack propagation, da/dN at the

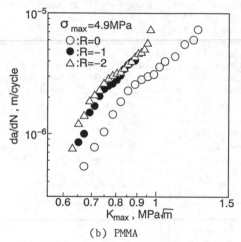

(b) PMMA

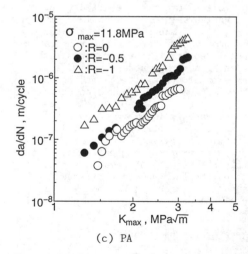

(c) PA

Fig. 1 Effect of negative stress ratio on da/dN at a constant maximum stress.

1254

same K_{max} must be equal for different stress ratios. Therefore, the results shown in Fig. 1 imply that the fatigue crack propagation was accelerated by the cycling of the compressive stress range in the fatigue loading with a negative stress ratio. This tendency of fatigue crack acceleration in the case of negative stress ratios was similar to that reported for various metals (Okazaki 1985, Ogiyama 1990, 1991).

3.2 Appearance of fracture surface

The typical examples of the photographs of the fatigue fracture surfaces for negative stress ratios for each polymer are shown in Fig. 2.

For PVC, the so-called discontinuous growth bands (Hertzberg 1979, 1980) were observed on the fracture surfaces for negative stress ratios as well as for R=0 (see Fig. 2 (a)). These bands are known to be formed by the failure of the craze zone initiated ahead of the crack tip. For PMMA and PA, on the other hand, the striations were observed (see Fig. 2 (b) and (c)), which were also similar to those for R=0, respectively.

These results suggest that the fatigue crack propagation mechanisms for these polymers were not appreciably affected by the compressive stress range in the cyclic loading with negative stress ratios in this experiment.

3.3 Crack closure behavior

In metals, the importance of the crack closure behavior in many phenomena of fatigue crack propagation has been confirmed by many workers, and it is generally accepted that the effective stress intensity factor range, ΔK_{eff}, based on the crack closure concept was an important factor controlling the fatigue crack propagation rate.

The fatigue crack acceleration described above has also been explained on the basis of the concept of the so-called plasticity-induced crack closure (Elber 1971), which is caused by the residual compressive stresses left in the wake of the crack. Namely, the cycling of the compressive stress range in the fatigue loading with a negative stress ratio causes the reduction of the residual compressive stresses left in the wake of the crack. Consequently, the crack opening stress level is reduced and then the effective stress range based on the crack closure concept is increased, which results in the crack acceleration

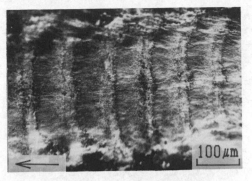

(a) PVC (R=-1.5, K_{max}=1.2MPa$\sqrt{m}$)

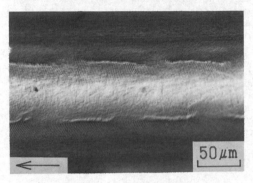

(b) PMMA (R=-2, K_{max}=0.7MPa$\sqrt{m}$)

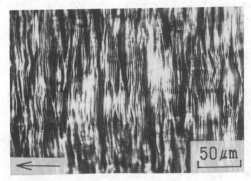

(c) PA (R=-1, K_{max}=2.8MPa$\sqrt{m}$)

Fig. 2 The appearance of fatigue fracture surfaces in the case of negative stress ratios. The arrows indicate the crack propagation direction.

(Okazaki 1985, Ogiyama 1990, 1991).
This explanation has been supported by the results that the fatigue crack propagation rates for negative stress ratios were well evaluated by ΔK_{eff} for various

metals (Okazaki 1985, Kurihara 1986, Ogi-yama 1990, 1991).

In this experiment, in order to clarify the applicability of this explanation to the case of polymers, the crack closure behaviors in these polymers were examined. The crack opening levels were firstly determined to be the inflection points on the load-subtracted strain curves obtained by the method described above under various stress ratios. By using the determined values of crack opening levels, the values of σ_{op}/σ_{max} were calculated for each polymer, where σ_{op} is the crack opening stress (see Fig. 3). For PA, the crack closure at R=0 could not be detected, probably because the considerable blunting of the crack tip took place due to the low yield strength and the high maximum applied stress level. This blunting behavior was clearly observed by the traveling microscope. But, even in this polymer, the crack closure was detected at a certain compressive site in the case of the negative stress ratios. From Fig. 3, it should be noted that the crack opening stress level decreased with decreasing stress ratio for each polymer. This tendency was similar to that reported for metals.

Therefore, it was expected that the crack closure behavior might account for the crack acceleration in these polymers in the same way as for metals.

Then, the data of da/dN for various stress ratios were plotted against ΔK_{eff} (see Fig. 4). As expected, a good correlation was found between da/dN and ΔK_{eff} for each polymer.

In these polymers, the situation near the crack tip is considered to be somewhat different from that in metals. Especially, in the glassy polymers such as PVC and PMMA, a craze zone which is peculiar to these polymers is formed ahead of the crack tip. However, even in these polymers, the residual compressive stresses may be left to some degree in the wake of the fatigue crack, because the crack will propagate through the tensile deformation zone formed ahead of the crack tip. If so, it is reasonable that the occurrence of crack closure and its role in the fatigue crack propagation may be similar to the case of metals. While there are not so many studies on the crack closure for polymers, its occurrence has been detected for PMMA (Pitoniak 1974) and PC (Shiraishi 1984, 1988, Murakami 1987).

The results obtained in this experiment suggest that the explanation on the basis of the plasticity-induced crack closure concept described above will be applicable for the dependence of the fatigue crack

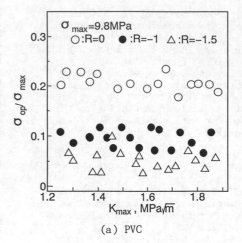

(a) PVC

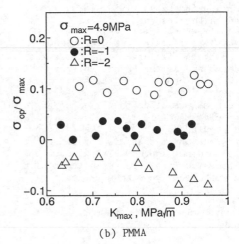

(b) PMMA

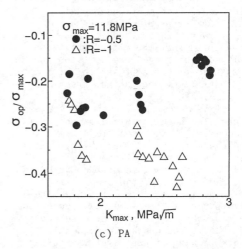

(c) PA

Fig. 3 Effect of negative stress ratio on σ_{op}/σ_{max}.

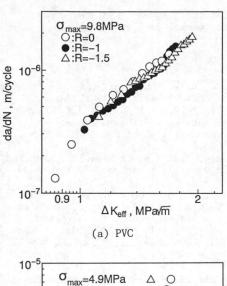

(a) PVC

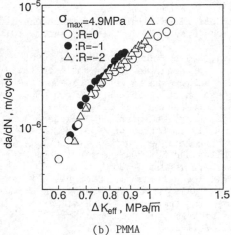

(b) PMMA

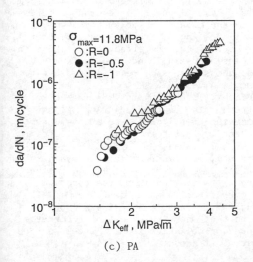

(c) PA

Fig. 4 Relation between da/dN and ΔK_{eff}.

propagation rate on negative stress ratio for polymers.

3.4 Effect of maximum stress level on da/dN at a constant negative stress ratio

For metallic alloys, it has also been pointed out that the maximum applied stress level significantly affected the fatigue crack propagation rate in the case of negative stress ratio, namely, the higher maximum stress level caused the higher da/dN at a constant stress ratio (Ogiyama 1990, 1991).

Figure 5 shows the effect of the maximum applied stress level on da/dN at a constant negative stress ratio (R=-1) for each polymer. It is clearly seen that da/dN increased with increasing maximum stress level for these polymers. This tendency is similar to that for metals.

Such a tendency was considered to be probably because the higher maximum stress level, consequently the larger compressive stress range, would cause the larger reduction of the crack opening stress level as reported for metals (Ogiyama 1990, 1991). As expected, the crack opening stress level decreased with increasing maximum stress level for each polymer (see Table 3).

Furthermore, as seen from Fig. 6, da/dN in this case was also well evaluated in terms of ΔK_{eff}.

These results suggest that the compressive stress range in the cyclic load with a negative stress ratio is an impor-

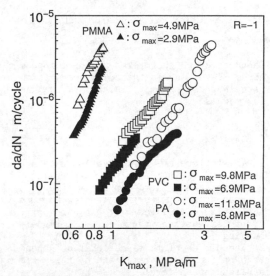

Fig. 5 Effect of maximum applied stress level on da/dN at R=-1.

Table 3 Effect of maximum applied stress level on the average value of σ_{op}/σ_{max} at R=-1.

Specimen	σ_{max}, MPa	σ_{op}/σ_{max}
PVC	9.8	0.09
	6.9	0.18
PMMA	4.9	0.02
	2.9	0.10
PA	11.8	-0.37
	8.8	-0.24

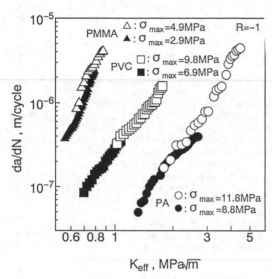

Fig. 6 Relation between da/dN and ΔK_{eff}.

tant factor which governs the crack opening stress level and so the crack propagation rate in these polymers in the same way as for metals.

4 CONCLUSIONS

1. The fatigue crack propagation rate at the same K_{max} increases with decreasing negative stress ratio for all polymers tested. This implies that the crack propagation is accelerated by the cycling of the compressive stress range in the fatigue loading.
2. The degree of this acceleration depends both on the stress ratio and on the maximum applied stress level.
3. These results can be explained on the basis of the plasticity-induced crack closure concept just as for metals.

REFERENCES

Elber, W. 1971. The significance of fatigue crack closure. ASTM STP486:230-242.

Hertzberg, R.W., M.D.Skibo & J.A.Manson 1979. Fatigue fracture micromechanisms in engineering plastics. ASTM STP675: 471-500.

Hertzberg, R.W. & J.A.Manson 1980. Fatigue of engineering plastics. New York:Academic Press.

Kurihara, M., A.Katoh & M.Kawahara 1986. Analysis on fatigue crack growth rates under a wide range of stress ratios. Trans. ASME. PVT, 108:209-213.

Murakami, R., S.Noguchi, K.Akizono & W.G. Ferguson 1987. Fatigue crack propagation and crack closure behavior in polycarbonate and fibre reinforced polycarbonate. Fatigue Fract. Engng Mater. Struct., 10: 461-470.

Ogiyama, H., H.Tsukuda & Y.Soyama 1990. Effect of negative stress ratio on fatigue crack growth in medium carbon steel and two phase stainless steel. J. Soc. Mat. Sci. Japan, 39:406-411.

Ogiyama, H., T.Shiraishi, H.Tsukuda & Y. Soyama 1991. Effect of negative stress ratio on fatigue crack growth in 2017-T3 Al alloy and SUS 304 steel. J. Soc. Mat. Sci. Japan, 40:575-580.

Okazaki, Y., A.Fukushima, H.Misawa & S. Kodama 1985. Effect of compressive part of cyclic stress range on crack propagation behavior. J. Soc. Mat. Sci. Japan, 34:1167-1173.

Pitoniak, F.J., A.F.Grandt, L.T.Montulli & P.F.Packman 1974. Fatigue crack retardation and closure in polymethylmethacrylate. Engng. Fract. Mech., 6:663-670.

Pook, L.P. 1968. The effect of friction on pin jointed single edge notch fracture toughness test specimens. Int. J. Frac. Mech., 4:295-297.

Shiraishi, T. & Y.Soyama 1984. Effect of stress ratio on fatigue crack propagation behavior of polycarbonate at low stress intensity factor range level. J. Soc. Mat. Sci. Japan, 33:1311-1316.

Shiraishi, T., T.Kuroshima & Y.Soyama 1988. Effect of compressive stress on fatigue crack propagation in polycarbonate. J. Soc. Mat. Sci. Japan, 37:795-800.

Tokaji, K., Z.Ando & K.Morikawa 1980. Effect of sheet thickness on retardation behavior of a high tensile steel. J. Soc. Mat. Sci. Japan, 29:808-814.

Recent Advances in Experimental Mechanics, Silva Gomes et al. (eds) © 1994 Balkema, Rotterdam, ISBN 90 5410 395 7

Local crack-tip strain behavior during fatigue crack initiation and propagation in polymer

A. Shimamoto
Saitama Institute of Technology, Japan

E. Umezaki
Nippon Institute of Technology, Saitama, Japan

ABSTRACT: In order to investigate whether the two fatigue processes, crack initiation and propagation, can be combined, the change of local notch–root strain and its history are measured as well as the change of local crack–tip strain and the local strain history of a fatigued element ahead of the propagating crack tip up to failure in a polycarbonate subjected to low–cycle fatigue tests by the fine grid method. As a result, the existence of a unified local strain field where the two fatigue processes can be substantially combined is experimentally confirmed. Therefore the local crack–tip strain may possibly be examined a simpler, one–parameter approach for fatigue life estimation.

1 INTRODUCTION

Recently, polymers with high elastic moduli and strengths have been widely used as machine parts and structural members, in tandem with the rapid development of industry, and their conditions of use are becoming severe. Therefore understanding of deformation and fracture phenomena characteristics of polymers is important for attaining better reliability of machines and structures, and preventing accidents caused by fracture.

Many theoretical and experimental studies have indicated that local strains near the notch root and the crack tip strongly affect initiation and propagation of fatigue cracks, respectively. However in spite of such studies, fatigue crack initiation and propagation at the notch root are still considered to be very complicated because these phenomena are thought to be strongly influenced by the states of plastic deformation at the notch root through the effects of notch radius, loading conditions, the mechanical properties, cyclic prestrained conditions and histories of the materials. The difficulties and limitations in evaluating small–crack propagation behaviors with continuum analysis of the fracture mechanics parameter ΔK are also pointed out. At present the fatigue crack initiation process is considered to be different from the propagation process. However there is the possibility that crack propagation can be treated as a succession of crack initiation events.

To clarify the above-mentioned problems, we believe that it is essential to obtain information on fatigue small–crack initiation and propagation behaviors at the notch root, and on local deformation at the tip of a propagating small crack.

Majumdar (1974) have reported the correlation between fatigue crack propagation and low–cycle fatigue properties. Shimada (1987) have proposed the concept of local strain at the crack tip for fatigue crack initiation and propagation. Lin (1991) have predicted crack propagation by an extended ASME method. They, however, studied the fracture mechanics of metallic materials, not polymers.

Our group (Shimamoto 1989,1990,1991,1992) has previously reported fracture and strain behavior near the notch root and the crack tip in polymers under low–cycle fatigue employing the fine grid method, and their life evaluation. In this study, we measure the change of local notch–root strain and its history (see the left part of Fig.1), as well as the change of local crack–tip strain and the local strain history of a fatigued element ahead of the propagating crack tip up to failure (see the right part of Fig.1) in a polycarbonate subjected to low–cycle fatigue tests by the fine grid method. Subsequently, we discuss whether or not a unified local strain field in which the two fatigue processes, crack initiation and propagation, can be combined exists.

2 SPECIMENS AND EXPERIMENTAL METHOD

The specimens used were a polycarbonate (Lexan

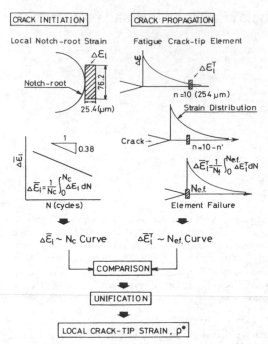

Fig.1 Local crack–tip strain approach for fatigue crack initiation and propagation, and its concept.

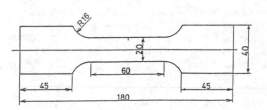

Fig.2 Shape and dimensions of specimen.

Table 1 Mechanical properties of polycarbonate.

Tensile strength	(MPa)	60.80
Elastic limit	(MPa)	40.21
Modulus of longitudinal elasticity	(MPa)	2099
Photoelastic sensitivity	(mm/ N)	0.146

9030) commercially available. Its mechanical properties are given in Table 1, and specimen dimensions, in Fig.2. A round notch with a depth of 1.5mm and a radius of 0.4mm was created on one side by machining. In order to measure local strain at the notch root and the crack tip, fine dot grids with 25.4μm pitch and 4μm diameter were printed on the surface of the specimen using a

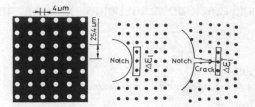

Fig.3 Fine dot grids and schematic view of local strain measurement area.

photographic printing technique, as shown in Fig.3.

The specimens were dried naturally by keeping them in a thermostatic room under conditions of temperature 20°C and humidity around 65% for a sufficiently long period of time before use, and fatigue tests were conducted in the same room. Fatigue tests were carried out using a hydraulic servo–controlleduniaxial–tension–compression–type fatigue machine of 9807N capacity with a digital servo–controller under the conditions of constant load amplitudes of σ=8.0MPa, 10.6MPa, 13.3MPa and 15.9MPa at a stress ratio $R=\sigma_{min}/\sigma_{max}$ of 0. Cyclic frequency was 0.02Hz with a sine waveform.

The changes of local strain at the notch root during fatigue crack initiation, $\Delta\varepsilon_l$, and of local strain of a fatigued element ahead of the propagating crack tip, $\Delta\varepsilon_l^T$, were measured by the fine grid method. Photographs were taken of grids in the area of the notch root and the local zone of the crack tip using a camera equipped with automatic exposure and automatic–film advancing functions, through a relay lens and an optical microscope of 100–fold magnification, at regular intervals. Local strain was obtained by directly measuring the deformation of grids on the negative with reference to the grid in the pretest state, through an enlarging profile projector of 10–fold magnification. For strain calculation, the gauge length was 76.2μm (three grid lengths). Local regions selected for measuring notch–root and crack–tip strains are shown in Fig.3. The error in strain measurement by this method was within ±0.4 percent.

3 EXPERIMENTAL RESULTS AND DISCUSSIONS

3.1 Correlation between local–strain damage accumulation curves for crack initiation and propagation

In order to analyze the crack propagation process, the change and history of local strain in a small fatigued element, A (10th grid element ahead of the crack tip, see Fig.4) (Shimada 1987), were investigated until the element was broken by the advancing crack tip. An example of the change of $\Delta\varepsilon_l^T$ up to the number of $N_{e.f.}=2913$ is shown in Fig.4. The distance from the location of the crack tip to point A (i.e., $N_{e.f.}$ point) is also shown in the horizontal axis below the number of cycles, N. As seen from this figure, the value at point A increased rapidly with the approach of the crack tip.

In order to analyze the crack initiation process, an average local strain accumulation value was proposed as follows (Shimada 1987):

$$\Delta\bar{\varepsilon}_l = \frac{1}{N_c}\int_0^{N_c}\Delta\varepsilon_l\, dN . \qquad (1)$$

The relationship between average local strain accumulation value, $\Delta\bar{\varepsilon}_l$, and crack initiation cycle, N_c, is shown in Fig.5. Each data point fell on a line of slope -0.38 in a log–log coordinate graph, regardless of the values of applied stress. We call this line the "local strain damage accumulation curve" (Shimada 1987).

In order to discuss whether any correlation exists between crack initiation and propagation processes from the viewpoint of the local strain value, the average local crack–tip strain accumulation value, $\Delta\bar{\varepsilon}_l^T$, was also calculated in a similar manner as $\Delta\bar{\varepsilon}_l$. Then, the curve of $\Delta\bar{\varepsilon}_l^T$ versus $N_{e.f.}$ during crack propagation was compared with the $\Delta\bar{\varepsilon}_l$ versus N_c curve during crack initiation.

As a result, very similar lines were obtained, as shown in Fig.5. Based on this result, it is thought that the fatigue crack might also propagate under failure conditions or by a mechanism similar to the crack initiation process. In an attempt to unify the two processes of crack initiation and propagation, we focussed attention on the local strain history within a small area of a (1) "heavily deformed region, R_h," or (2) "effective crack–tip strain element, ρ^*," adjacent to the crack tip.

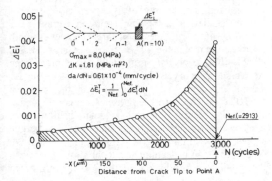

Fig.4 The changes of $\Delta\varepsilon_l^T$ at point A as the crack tip advances.

3.2 *Heavily deformed region, R_h*

The heavily deformed region, R_h, is a cyclically strained area which suffers strain of a wide range over a threshold value, $\Delta\varepsilon_l^T \geq \Delta\varepsilon_{l\,(th)}^T$, wherein the fatigue damage caused in R_h substantially controls the state of crack propagation. Figure 6 shows an illustration of the idea that one may generalize the local crack–tip strain accumulation curve ($\Delta\varepsilon_l^T$ versus $N_{e.f.}$ curve) by introducing the threshold value, $\Delta\varepsilon_{l\,(th)}^T$. As a result of deleting the strain

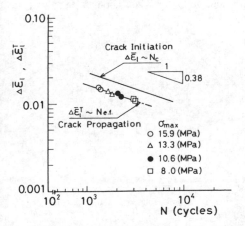

Fig.5 Correlation between the local strain damage accumulation curve ($\Delta\bar{\varepsilon}_l$ versus N_c) and the crack–tip strain damage accumulation curve ($\Delta\bar{\varepsilon}_l^T$ versus $N_{e.f.}$) on crack propagation.

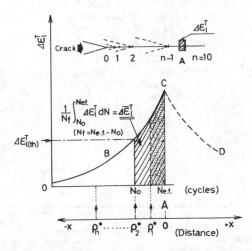

Fig.6 Generalization of local crack–tip strain accumulation curve ($\Delta\varepsilon_l^T$ versus $N_{e.f.}$) for crack propagation, using the threshold value of local crack–tip strain, $\Delta\varepsilon_{l\,(th)}^T$ or the material elemental size, ρ^* from the crack tip.

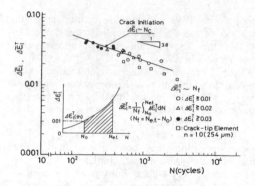

Fig.7 Generalization of local crack tip strain damage accumulation curve ($\Delta\bar{\varepsilon}_l^T$ versus $N_{e.f.}$) for crack propagation by deleting the strain range below the various threshold values of $\Delta\varepsilon_{l\,(th)}^T$ and its fit to $\Delta\bar{\varepsilon}_l$ versus N_c curve.

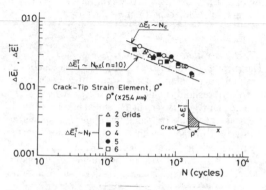

Fig.8 Generalization of $\Delta\bar{\varepsilon}_l^T$ versus $N_{e.f.}$ curve by the concept of material elemental size, ρ^*, from the crack tip, and its fit to $\Delta\bar{\varepsilon}_l$ versus N_c curve.

range below the various $\Delta\varepsilon_{l\,(th)}^T$ values on the $\Delta\varepsilon_l^T$ versus $N_{e.f.}$ curve, the best agreement between the generalized $\Delta\bar{\varepsilon}_l^T$ versus N_f ($=N_{e.f.}-N_0$) plots and $\Delta\bar{\varepsilon}_l$ versus N_c curve could be obtained in the cases of $\Delta\varepsilon_{l\,(th)}^T \geq 0.01$, as shown in Fig.7.

3.3 Effective crack–tip strain element, ρ^*

The curve of $\Delta\varepsilon_l^T$ versus $N_{e.f.}$ was generalized by applying the concept of ρ^*, i.e., the effective crack–tip strain field for fatigue damage in crack propagation. The length of ρ^* was set as the pitch of grids (25.4μm) toward the $-X$ direction from point A, and the fit of the data points to the $\Delta\bar{\varepsilon}_l$ versus N_c curve for each case of ρ^* is shown in Fig.8. As shown in the illustration of Fig.6, the relationship between the strain history of $\Delta\varepsilon_l^T$ until the $N_{e.f.}$ point (i.e., BC curve in $-X$ direction) and the change in strain at the front region of point A (CD curve in $+X$ direction) is substantially similar versus the distance, X. ρ^* in Fig.8 is then described for the front of the crack tip. Generalized data points of $\Delta\bar{\varepsilon}_l^T$ versus N_f ($=N_{e.f.}-N_0$) are plotted in a log–log coordinate graph. The variation of ρ^* is not very large; however, with decrease in the value of ρ^* from six grid lengths (6×25.4μm), each data point tends to approach the $\Delta\bar{\varepsilon}_l$ versus N_c curve. The best agreement to the $\Delta\bar{\varepsilon}_l$ versus N_c curve was obtained for the case of $\rho^*=101.6$μm (four grid lengths). As a consequence, the similar results obtained in generalization by ρ^* and by R_h strongly suggest that the two fatigue stages, crack initiation and propagation, are unified, when taking into account the local strain history in the small region adjacent to the propagating crack tip.

Next, the general crack propagation rate, da/dN, as estimated by correlating the local crack–tip strain value in ρ^* for each applied stress with the

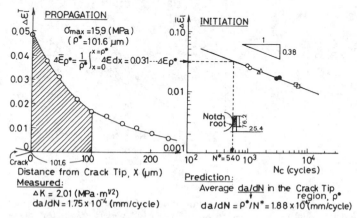

Fig.9 Procedure for estimating the crack propagation rate by combining the average crack–tip strain value in ρ^* with local strain damage accumulation curve for crack initiation.

Table 2 Comparison between calculated and measured da/dN.

Stress (MPa)	Calculated da/dN (mm/cycle)	Measured·da/dN (mm/cycle)	Error (%)
15.9	1.88×10^{-4}	1.75×10^{-4}	7.4
13.3	1.34×10^{-4}	1.29×10^{-4}	3.9
10.6	1.20×10^{-4}	1.02×10^{-4}	17.6
8.0	0.64×10^{-4}	0.61×10^{-4}	4.9

$\bar{\varepsilon}_l$ versus N_c curve, and an example is shown in Fig.9, where it can be seen that the average strain range, $\Delta\bar{\varepsilon}_{\rho^*}$ in ρ^* ahead of the propagating fatigue crack tip can be calculated by dividing the integrated local strain history (hatched part) by the distance of ρ^*. N^* represents the estimated number of cycles at which the material in ρ^* failed due to advance of the crack tip. We can determine N^* by transposing the value of $\Delta\bar{\varepsilon}_{\rho^*}$ to the local strain damage accumulation curve ($\Delta\bar{\varepsilon}_l$ versus N_c). Using the linear cumulative damage law of local strain (Shimada 1987), the average crack propagation rate in the distance of ρ^* is calculated by

$$\frac{da}{dN} = \frac{\rho^*}{N^*} . \qquad (2)$$

da/dN calculated for each applied stress using $\rho^* = 101.6\mu m$ is given in Table 2 with measured values. As seen in Fig.9 and Table 2, the calculated crack propagation rate is in good agreement with the measured value.

According to these results, the fatigue crack initiation and propagation processes can be unified, and da/dN can be estimated by using the cyclic strain distribution just ahead of the propagating crack tip and the local strain damage accumulation curve for crack initiation. The effective crack–tip strain element, ρ^* presents the possibility for an easier one–parameter approach in fatigue life analysis.

4 CONCLUSIONS

Two fatigue processes, crack initiation from the notch root and crack propagation, in a polycarbonate were investigated under four applied stresses at a stress ratio of 0 by the fine grid method. Results were as follows.
(1) The existence of a unified local strain field where the two fatigue processes, crack initiation and propagation, can be combined was confirmed.
(2) The two processes were similar when by taking into account the local strain history in a small region, ρ^*, adjacent to the propagating crack tip.
(3) The crack propagation rate could be estimated by combining the average crack–tip strain range in ρ^* with the local strain damage accumulation curve for crack initiation.
(4) ρ^* presents the possibility for a simpler, one–parameter approach in fatigue life analysis.

REFERENCES

Lin, Y. & E.Lin 1991. Prediction of crack propagation by extended ASME method. Inter .J. Frac. 48(1):71–79.
Majumdar, S. & J.Morrow 1974. Correlation between fatigue crack propagation and low–cycle fatigue properties. ASTM STP 559:159–182.
Shimada, H. & Y.Furuya 1987. Local crack–tip strain concept for fatigue crack initiation and propagation. J. Eng. Mater. Technol. 109(4):101–106.
Shimamoto, A., S.Takahashi & A.Yokota 1991. Fundamental study on rupture by low–cycle fatigue of polymers applying the fine–grid method. Exp.Mech. 31(1):65–69.
Shimamoto, A., E.Umezaki, F.Nogata & S.Takahashi 1990. Strain behavior near fatigue crack tip in polymers and their life evaluation. In T.H.Hyde & E.Ollerto (eds.), Applied Stress Analysis:51–59, London:Elsevier Applied Science Publishers.
Shimamoto, A., E.Umezaki, F.Nogata & S.Takahashi 1992. Strain behavior near fatigue crack tip in polycarbonate and its life evaluation. Trans.Jpn.Soc.Mech.Eng. (in Japanese) 58(555)A:2023–2027.
Shimamoto, A., E.Umezaki, & S.Takahashi 1989. A fundamental study on rupture by low–cycle fatigue of polymers employing the fine–grid method. J.Jpn.Soc.NDT (in Japanese) 38(12):1101–1106.

Recent Advances in Experimental Mechanics, Silva Gomes et al. (eds) © 1994 Balkema, Rotterdam, ISBN 90 5410 395 7

Temperature dependent morphology of fracture surface and crack growth mechanism in a viscoelastic epoxy resin

Kazuo Ogawa & Masahisa Takashi
Aoyama Gakuin University, Tokyo, Japan

Akihiro Misawa
Kanagawa Institute of Technology, Japan

Takeshi Kunio
Keio University, Kanagawa, Japan

ABSTRACT: In this study, the authors will discuss on sub-microscopic mechanism for slow and stable crack growth under the constant strain rate in a viscoelastic epoxy strip, paying attention to the morphology of fracture surface from semi-micro characteristic feature. It is well known that the mechanical properties and fracture behaviors of polymeric material reflect some aspects of molecular chain motions which are definitely dominated by their chemical structures, crosslink density and temperature. In addition, the formation of fracture surface would be understood from a view point of microscopic heterogeneity such as chain entanglement which appears in front of growing tip of crack. Taking the discussions mentioned into account, the authors would like to propose a model for micro-mechanism of crack growth and crack surface formation in a polymeric material, in which the time and temperature dependent mobility of molecular entanglement is well related to the configuration of fracture surface and the crack growth resistance.

1 INTRODUCTION

It is well known that the micro-structure of polymeric material which consists of 3D crosslink and entanglement of long molecular chain is quite different from that of metal. The crack growth behavior as well as the mechanical properties of polymer such as epoxy resin shows, in general, remarkable dependence on time and temperature. Since so called amorphous polymer has no intermediate structures, such as grains in usual metal or crystalline materials, it becomes more difficult to proceed quantitative discussions on fracture mechanisms. Then, there are various difficulties in the establishment of unified understanding on the mechanism and behavior in materials of this type, even if one could collect and integrate diverging information obtained from the basis of different sciences and with the wide range of scale in observation.

The authors will , thus, discuss in this paper the fracture and crack growth mechanisms in an intermediate structure in a scale of molecular entanglement through the qualitative and quantitative evaluation of fracture surface morphology of the material adopted. Reflecting the remarkable dependence of molecular mobility on time and temperature, fracture surface configuration varies widely with the velocity of crack growth. The morphological characteristics remained on fracture surface could be expected to tell us some aspects of submicroscopic structure related with heterogeneous local deformation, i.e. strain and

stress, around a growing crack tip. Namely, attention is paid to river pattern on fracture surface in this study, which consists of submicroscopic lotus-root structure of molecular entanglement.

Taking the discussions mentioned into account, the authors propose a model for micro-mechanism of crack growth and crack surface formation in a polymeric material, in which the time and temperature dependent mobility of molecular entanglement is well related to the morphological characteristics of fracture surface and crack growth resistance.

2 EXPERIMENTS

2.1 *Mechanical properties of the material*

The materials adopted in this experiment was a type of hard epoxy prepared by mixing Bisphenol-A type resin (Epikote 828, Shell Chem.) with an amine type hardener (Triethylene-Tetramine). Fig.1 shows the master curves of relaxation modulus $E_r(t')$ in tension measured under several constant strain rates and temperatures, then composed using the W.L.F. shift factor. It is obvious from the figure that the mechanical property, $E_r(t')$, of the material shows a remarkable dependence on the reduced time t', thus both on time and temperature. The material is, then, considered a typical example of viscoelastic material. The value of $E_r(t')$ varies more than two hundred times from rubbery to glassy state over a wide range of the reduced time more than ten figures. Also, the master curve of $Er(t')$ is approximated by a Prony series to apply to the computation of $J'(t)$-integral, i.e. the extended J-integral for viscoelastic materials.

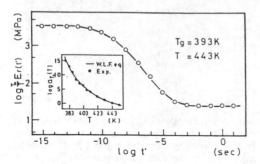

Fig.1 Master curve of the relaxation modulus $Er(t')$

2.2 *Geometry of specimen and loading condition*

The specimen adopted is a strip of 80mm width and 40mm length having an initial crack C_0 (20mm) from a side edge. Since it has been already pointed out by authors in a previous paper (Ogawa 1990) that the quality, sharpness and smoothness, of initial crack influences definitely the subsequent crack growth behaviors, the initial crack is carefully prepared utilizing the natural growth of crack under a controlled temperature and loading rate condition.

A constant rate of displacement loading, 8.33×10^{-3}mm/sec, was applied both at upper and lower grip fixed rigidly not to change loading angle with increase of crack length. According to this method, a crack travels straight forward without swerve from the transverse center line of specimen and the crack growth behavior shows good reproducibility. Four constant temperatures were adopted for tests to cover the viscoelastic and rubbery regions of mechanical behavior of the material.

3 CRACK GROWTH BEHAVIOR AND CRACK GROWTH RESISTANCE

The good reproducibility of crack growth behavior measured with the current experimental techniques used to be applied to polymeric materials could not be expected because of lack in effort for the preparation of sharp and smooth crack tip. Fig.2 compares two examples of microphotograph at the root of growing crack from (a) an initial crack made by a knife edge and (b) a natural initial crack prepared by the method of natural growth under a controlled loading. In the case of (a), many sub-cracks in each ligament are seen around the tip of main crack in stead of a single tip in the case of (b).

The reason for poor reproducibility of crack growth behavior from an initial crack of type (a) is placed at this point. Moreover, it should be pointed out that the crack growth data obtained with an initial crack of type (a) are apt to give us an over estimation of crack growth resistance because of large energy absorption due to tearing between adjacent sub-cracks. On the other hand, a crack started from the initial crack of type (b) travels straight forward without any swerving from the center line of specimen and shows good reproducibility. Also, we can evaluate exactly the minimum resistance for crack growth in this case.

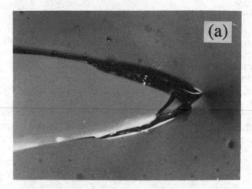

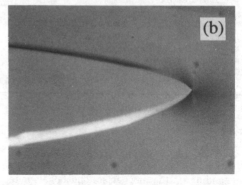

Fig.2 A typical example of crack tip just ahead of initial crack front, (a) a knife initial crack, and (b) a natural initial crack

Fig.3 shows some examples of crack growth curve obtained at different temperature conditions. The curves are constructed measuring the interval of marking formed with the slight cyclic impact superposition technique (Takashi 1983) during the constant rate of displacement loading. Each curve show a good linearity in a double logarithmic scale of crack length increment, C, and time elapsed, t, i.e. a crack growth curve of $C = \alpha t^\beta$, from the beginning of crack growth threshold in the order of magnitude of micro-meter. Also, the temperature dependent properties of the material, namely rubbery and viscoelastic behaviors, are reflected in variation of the slope, β, of the line.

The facts observed in crack growth behavior imply that there might not be the criticality in crack growth threshold in a sense of macroscopic discontinuous phenomena, although it is required to measure the incubation time t^* for calculation of the crack growth resistance. Thus, discussions on the criticality for crack growth threshold has to be complemented with more detailed observations from the other points of view.

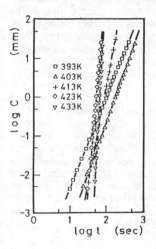

Fig.3 Several examples of crack extension curve

The authors have already pointed out in one of the papers (Ogawa 1991) that there could be a time and temperature independent resistance J'_{Ic} to crack growth threshold over a wide range of test conditions. Selecting an incubation time corresponding to the crack growth increment of 10 μm, a material unique property for crack growth which is independent on time and temperature is successfully obtained from accurate experiments on crack growth and by use of $J'(t)$-integral for a linearly viscoelastic material within a restriction under monotonically increasing loading.

4 MORPHOLOGICAL CHARACTERISTICS OF FRACTURE SURFACE

A great attention should be, in particular, paid for the roughness measurement of fracture surface of polymeric material because of their extreme softness and fragility, thus any kind of roughness measurement have to be performed without touching them. Fig.4 shows typical examples of fracture surface photograph in the vicinity of initial crack front taken with (a) a polarized light microscope in low magnification and (b) a laser microscope in high magnification. While the fracture surface of (a) is covered with fine stream line marks similar to river pattern over the whole area, that of (b) shows some microscopic structure which looks like a lotus root lined up in the direction of crack growth.

The characteristic feature of fracture surface is, however, seen in the neighborhood just ahead the initial crack front, reflecting the temperature and time dependent mobility of molecular entanglement around moving crack tip. The variation of fracture surface roughness along crack growth increment is studied by use of a double beam interference microscope and an image processing system. Fig.5 shows a typical example of maximum roughness, R_{max}, of the fractured surface along the distance from initial crack front. It can be pointed out for each fracture surface obtained under a wide range of test condition that its roughness distribution has a peak value R_{max}^* of the maximum roughness immediately forward the initial crack within a few tens of micrometer. The value of R_{max}^* also varies more than three times over wide ranges of the temperature. The lower the temperature, the rougher the surface becomes.

T = 403K
Crack Propagation

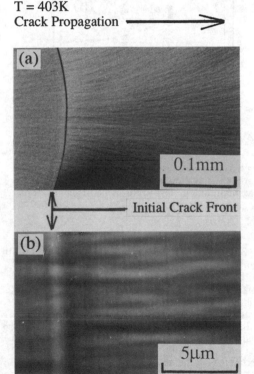

Fig.4 A typical example of fracture surface at neighborhood of an initial crack front, (a)polarized light microscope (b)Laser microscope.

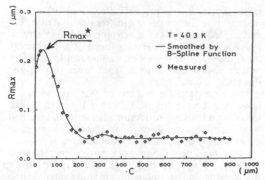

Fig.5 Variation of the maximum surface roughness, Rmax, along the distance from the initial crack front.

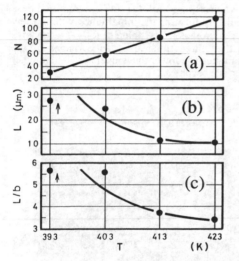

Fig.6 Temperature dependence of the lotus-root like mark, (a) the number of lotus-root mark, (b) the maximum length and (c) the ratio of the maximum length to width.

Since the region in which the characteristic morphology of fracture surface appears is restricted in the vicinity of the tip of initial crack, a limited small area of 80 * 80μm is observed with a laser microscope and analyzed with a image processing technique.

In order to evaluate the characteristics of this louts-root mark quantitatively, the statistical parameters such as the number(N), the perimeter(D), the maximum length(L), the width(b), are extracted. It is noteworthy that the feature parameters regarding the lotus-root mark are remarkably dependent on temperature as shown in

Fig.6. With lowering temperature, D and L increase while N decreases, then the aspect ratio L/b of the lotus-root mark also increases. This fact may implies that a submicro structure around crack tip, which takes part in the growth of crack and the formation of fracture surface morphology, has at least remarkable temperature dependence in changing the size of molecular entanglement. Some clues for understanding the relation between chemical parameters and morphological parameters could be, then, discussed.

The material adopted in this experiment has three dimensional network structure as shown in Fig.7. The crosslink density $\rho=1.06[kmol/m^3]$ is estimated from the rubbery modulus. Since the molar volume, Vm, of the material is easily obtained using molecular formula as $56.63nm^3$, the number, $UN = V/Vm$, of unit molecular structure in a lotus-root structure could be roughly estimated, where V is the volume of lotus-root structure. The relative variation of a ratio, $UN_{(Ti)}/UN_{(T0=423)}$, is shown in Table 1, where $UN_{(T0=423)}$ is the smallest number obtained in this experiment. The ratio varies about 15 times with decrease of temperature. On the other hand, the variation of the maximum roughness, R_{max}, is seen in Fig.4 as about 4 times. This would come from the change in flatness of the aspect ratio of the lotus-root structure. The fact mentioned above could be a useful information for better understanding of the relation between some parameters in the highpolymer chemistry and the submicroscopic observation.

Epikote 828

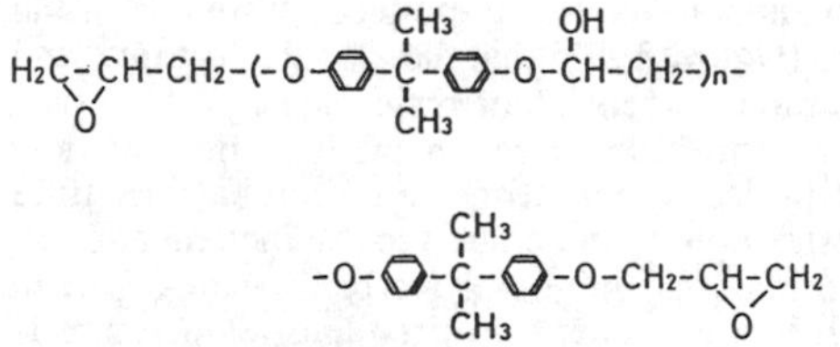

$$H_2C-CH-CH_2-(-O-\bigcirc-\underset{\underset{CH_3}{|}}{\overset{\overset{CH_3}{|}}{C}}-\bigcirc-\overset{OH}{O-CH-CH_2-)_n-}$$

$$-O-\bigcirc-\underset{\underset{CH_3}{|}}{\overset{\overset{CH_3}{|}}{C}}-\bigcirc-O-CH_2-CH-CH_2$$

Triethylene Tetramine

$$H_2N-(-CH_2-CH_2-NH-)_m-CH_2-CH_2-NH_2$$

Fig.7 Molecular structure of epoxy resin and hardener adopted.

Table 1. Temperature dependence of chemical parameter

T[K]	393	403	413	423
$UN=V/V_m[\times10^7]$	58.9	29.4	5.82	3.94
$UN_{Ti}/UN_{T=423}$	14.9	7.43	1.47	1.0

5 CRACK GROWTH MECHANISM

Taking the characteristic features of fracture surface configuration and its variation along crack growth into account, we will be able to consider a model for the variation of stress and strain around the moving crack tip. The time and temperature dependent mobility of entangled long molecular chain could reasonably be related to that of fracture surface roughness in the material used.

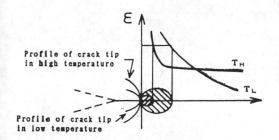

Fig.8 Different sizes of the region affected by high strain just ahead of accelerating crack

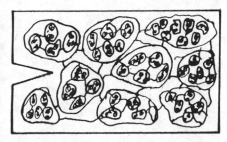

Fig.9 Model of the distribution of several step of size and mass of chain entanglement

The authors would like to discuss a model which can explain the fracture surface roughness variation along crack growth. Let us now consider the variation of stress distribution at the tip of crack before and after the threshold of crack growth. The distribution of crack tip stresses of the starting crack could be affected by two different factors, i.e., one is the acceleration of crack growth and the other is the relaxation of stress particularly around the tip of crack. It is not difficult to assume that the region under a certain level of stress or strain environment could become broader and bigger by the former effect, instead of opposite effect by the latter. In Fig.8, strain distributions considered in the both cases of high and low temperature are compared.

The strain distribution at high temperature is quite different from that at low temperature. At high temperature, the higher strain can occur in a small region near the tip of crack, namely the concentration of strain is restricted just at the root of crack, then it levels out quickly over the width of specimen. Accordingly, the fracture surface roughness could be smoother at high temperature.

Since the strain distribution becomes, moreover, steeper at high temperature, the severe strain region would behave similar to the case of stress. From the view point of molecular chain structure of polymeric material, it will be easily imagined that the size and mass of chain entanglement can not always be constant and uniform over the whole volume. It would be reasonable to consider that there exist inevitably several uneven sizes of entanglement which would be dependent on temperature and crack growth velocity. Fig.9 shows a model of the distribution of several steps of size in molecular chain entanglement. Thus, we can also discuss the relation between the fracture surface roughness, the stress or strain distribution at the tip of crack and the size variation of molecular chain entanglement. At low temperature and under slow crack growth conditions, the stress and strain distribution near the crack tip will be dull and a fairly large region could be involved to create new surface, then large lumps of entanglement would included. At high temperature, another combination of processes could be occur.

6 CONCLUDING REMARKS

In order to supplement lack in the knowledge of intermediate scale fracture mechanism for crack growth behavior in a thermoset epoxy resin, morphological aspect of submicro structure remained on fracture surface such as a louts-root mark and the relation to macroscopic crack growth behavior are carefully investigated involving their temperature dependence.

The results obtained are briefly summarized as follows;
1. The louts-root mark evaluated with a laser maicroscope and computer aided image processing shows remarkable dependence on temperature.
2. Some clues for better understanding of the relation between the chemical parameter of molar volume and the morphological parameter of lotus-root mark are obtained
3. Standing upon the results obtained, a qualitative model for the temperature dependent mechanism of crack growth is proposed.

ACKNOWLEDGMENT

The authors appreciate a partial financial supports of the Center for Science and Engineering Research, Research Institute of Aoyama Gakuin University.

REFERENCES

Ogawa, K. and M. Takashi (1990). The quantitative evaluation of fracture surface roughness and crack propagation resistance in epoxy resin. Trans. JSME, ser.-A, 56, 1133-1139.

Ogawa, K., A. Misawa, M. Takashi, and T. Kunio (1991). Evaluation of crack growth resistance and fracture surface characteristics in several in epoxy resins. Proc. 6th Int. conf. Vol. 4 105-110.

Takashi, M., H. Ohtsuka, A., Misawa and T. Kunio (1983), Crack growth behavior and crack opening profile in viscoelastic strips. Proc. 26th Japan cong. Mat. Res., 247-252.

Recent Advances in Experimental Mechanics, Silva Gomes et al. (eds) © 1994 Balkema, Rotterdam, ISBN 90 5410 395 7

A fatigue accelerated test method to rank plastic piping materials

K.Chaoui
Mechanical Engineering Institute, University of Annaba, Algeria

A.Moet
Macromolecular Science and Engineering Department, Case Western Reserve University, Cleveland, Ohio, USA

ABSTRACT: A new testing procedure to study the resistance of polyethylene pipe material to brittle cracking is achieved using a modified ASTM C-specimen. Low frequency tension-tension fatigue is employed as the acceleration agent. On the mechanistic level, it is found that an initial stage of brittle failure dominates the fracture process which becomes gradually ductile at about halfway the specimen width. The damage zone is characterized by a set of crazes ahead of the crack-tip. The model of crack layer developed by A. Chudnovsky and A. Moet is used to estimate the specific energy of fracture to brittle crack propagation. For PE2306-IIC and PE2306-IA, this parameter is found to be 351 and 264 J/g respectively at a load level of 25% of the yield stress. This result is in agreement with field performance of both materials.

1 INTRODUCTION

Although plastic pipes are designed for several decades of continuous use (ASTM D-1598 1981), unpredicted brittle failure may occur much before the design life is reached (Palermo and Deblieu 1985). Such failures commonly initiate from stress concentration regions and propagate in stable fashion through the structure's wall thus terminating its service life. As a result, the resistance of plastic pipes to brittle fracture, which is also known as slow crack propagation, became a main issue in the assessment of their long-term performance.

At the present time, the most favored approach is the sustained pressure test (ASTM D-1598 1981). In view of the unacceptable length of time required to observe brittle fracture and also unfounded extrapolation especially in the presence of a knee-type stress-time relationship (Raske 1987), active research seeks the development of an alternative test procedure which should accelerate slow crack propagation to a reasonable length of time. Furthermore, a model is sought to obtain from accelerated tests fundamental parameters responsible for the material's resistance to slow crack growth.

In this report, we describe a proposed procedure for such a test. In addition, the crack layer model, which is derived from irreversible thermodynamics (Moet & Chudnovsky 1986), is employed to extract the specific energy of fracture.

2 EXPERIMENTAL PROCEDURE

Polyethylene pipes are obtained from the National Bureau of Standards, Washington, DC, following the reference standard polyethylene resins and piping materials (Crissman 1987). The pipes are manufactured by Plexco Inc., USA in accordance with ASTM D-2513 and are known as PE2306-IIC. The minimum pipe wall thickness is 11 mm and the average outside diameter is 115 mm. The density is 939 kg/m³ and the melt flow index is in the range 0.4 to 1.5.

The choice of specimens geometry is dictated by two specific considerations: (1) the influence of the specific state of residual stress distributions on the crack propagation behavior and (2) the associated microstructural variances. Subsequently, the criteria set for the test must emphasize the design of a specimen geometry made from a pipe in order to include the thermomechanical history imparted by the processing operations. The conditions are: (1) test specimens must be prepared from a pipe so that to include geometrical constraints represented by residual stresses and microstructural gradients, (2) multiple specimens must be prepared from the same pipe portion to avoid variances among lots and extruders so that acceptable test reproducibility is ensured, (3) brittle or brittle-like fracture must be produced and (4) test duration must be short as compared to the sustained pressure test.

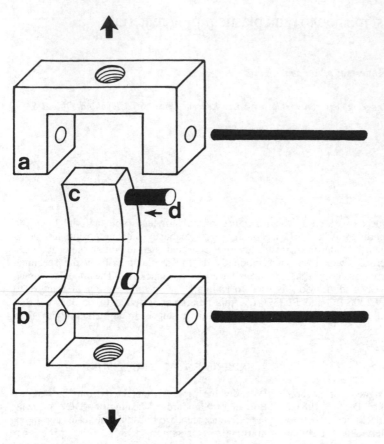

Figure 1. Schematic of test specimen and its gripping fixture, (a) and (b) upper and lower clevis grips, (c) specimen and (d) reinforcing steel sleeves.

The arc-shaped specimen for metallic pipes and cylinders described in (ASTM E399 1981) is used as a basis for this approach. The modifications concerned the specimen size and the loading points. The ratio of the specimen thickness to width is increased from 0.5 to 2.5 to ensure a higher rigidity at the loading lines. Also, reinforcing sleeve are pressed into the holes to reduce to a minimum the friction polymer-metal. Figure 1 shows the specimen geometry and its gripping fixture.

As compared to the ASTM test, the distance separating the two holes is reduced to a minimum of 25 mm to keep the inherent bending stresses at the lowest value. The distance identified as X in ASTM E399 is negative in this case and the axis of the holes is no longuer tangent to the inner radius.

A 2.5 mm deep notch is introduced at the pipe bore using a razor blade mounted on a manually driven press. Tension-tension crack propagation experiments are carried out on an MTS servohydraulic machine. The maximum stress is set at 5.62 MN/m² which corresponds to 25 % of the yield stress measured from compression molded sheets of the same resin. The minimum to maximum load ratio is 0.1 and the test frequency is kept constant at 0.5 Hz to avoid hysteric heating of the material. The crack length is followed at a magnification of 200X with the aid of a travelling optical microscope interfaced with a closed loop video system.

3 RESULTS AND DISCUSSION

3.1 Crack propagation mechanisms

Figure 2 is a composite micrograph illustrating the fracture surface of a failed C-shaped specimen

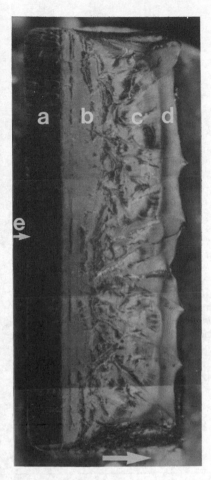

Figure 2. Optical photomicrograph of fracture surface of PE2306-IIC material (X7).

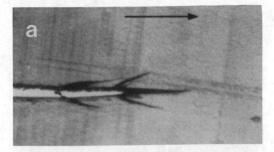

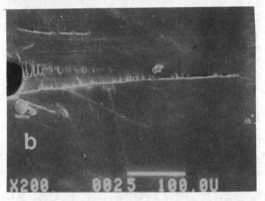

Figure 3. OM (a) and SEM (b) examination of the damage zone in the brittle regime.

under fatigue acceleration as described above. The 2.5 mm notch is represented by region «a». It is observed that two main regimes dominate the propagation span. Up to a crack length of 5 mm, the surface displayed a relatively flat and less fibrillated texture with evident discontinuous fracture bands. This regime is identified as brittle (region b) in contrast to the regime dominating the second half of the fracture surface (region c). The latter is mostly ductile-type failure. Drawn fibrils and yielded features at the surface are visually observable. As the crack advanced, the brittle contribution is lessened at the expense of increased ductility towards ultimate specimen separation. In addition, the specimen thickness measured at the fracture surface decreased up to 87% of the initial value. Past the point of maximum thinning, ultimate failure occured by tearing of the remaining ligaments (region d). It should be noted that this description from low magnification microscopy tends to oversimplify a very complex mix of hierarchical events.

3.2 Damage zone characterization

It is known that the rate of damage accumulation in the crack-tip region (active zone) provides resistance to crack propagation. In this case, damage growth is studied using interrupted tests at preset crack lengths. Subsequently, the specimen is sectioned at its middle (Figure 2, arrow e) using a diamond wafering blade cooled with water. A thin section of this position is examined under transmitted light.

In the brittle regime, the crack is preceeded by a main craze coupled with a pair of secondary crazes (Figure 3a). SEM examination shows the voids forming the craze (Figure 3b).

Beyond a 5 mm crack length, the fracture process involved more material yielding and drawing. The crazes of the ductile regime are very long since some of them initiated in the brittle regime. SEM analysis shows that the craze zone is constituted of multiple crazes which appeared mixed and

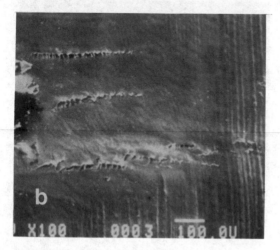

Figure 4. OM (a) and SEM (b) examination of the damage zone in the ductile regime.

intersecting at high magnification (Figure 4a). Two damage machanisms are deduced: (1) crazing is dominant in the brittle regime and (2) in competition with the first, material yielding is observed (Figure 4b) which extends up to 4 folds between Figures 3 and 4.

3.3 Crack propagation analysis

Figure 5 exhibits a typical logarithmic crack propagation rate as a function of crack length. Acceleration is reduced at about 3.5 mm and around 5 mm, there is an inflexion point which, in fact, correponds to the transition shown in fractographic analysis. Commonly, stage II is analyzed in terms of Paris-Erdogan equation. Unfortunately, this equation does not reflect any information about the mechanisms though they control the fracture process.

Since long-term failure is brittle in nature, the analysis is limited to the brittle regime (up to 5

mm crack length). The crack layer model is advanced as a tool to account for damage growth and its law is derived as (Chudnovsky and Moet 1985):

$$t_0 \, \frac{da}{dN} = \frac{\beta \, \dot{w}_i}{\gamma \, R_1 - J_1} \tag{1}$$

The quantity t_0 (da/dN) is the rate of crack propagation in m²/cycle with t_0 the initial specimen thickness. $\dot{W}_i$ is the irreversible work calculated from hysterisis loops evolution in J/cycle. The resistance moment R_1 is an integral quantity representing the amount of damage necessary to cause crack excursion (Chudnovsky 1984). It is evaluated as the change in volume of the damaged material per unit surface created and multiplying the result by the material density which gives R_1 in kg/m². The energy release rate (J_1) is evaluated experimentally from the evolution of load displacement curves (Chaoui 1989). Finally, β is a proportionality constant and γ is the specific energy of fracture in (J/g). Equation (1) is linearized so that β and γ are obtained simultaneously as a slope and an intercept respectively. The results are summarized in Table 1.

Table 1. Comparison of the specific energy of fracture of PE2306-IIC and PE2306-IA.

Pipe	γ (J/g)	β	r
PE2306-IIC	351	4.46 10^{-5}	0.94
PE2306-IA	264	2.36 10^{-5}	0.95

The same analysis is extented to a less tenacious polyethylene. It is found that the model predicts such difference since a lower γ is obtained for PE2306-IA. In order to confirm this finding, lifetime comparisons are made and summarized in Table 2. Concerning times, it is concluded that PE2306-IIC is tougher and more crack resistant than PE2306-IA.

Table 2. Comparison of lifetime data for PE2306-IIC and PE2306-IA.

Pipe	N_i 10^3 cycles	N_b 10^3 cycles	N_t 10^3 cycles
PE2306-IIC	140	65	230
PE2306-IA	10	10	30

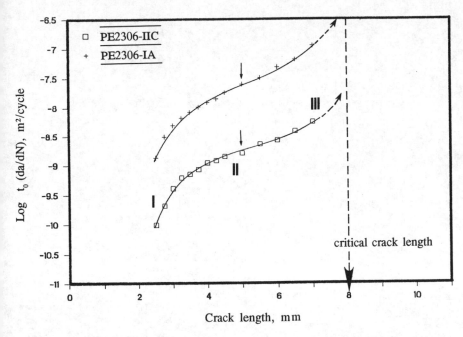

Figure 5. Logarithmic crack growth rate as a function of crack length.

N_i, N_b and N_t represent the number of cycles to initiation, brittle propagation and the total lifetime respectively.

4 CONCLUSIONS

An accelerated testing procedure is advanced to produce brittle fracture in different grades of polyethylene pipes in remarkably short times. The test permits to obtain crack propagation mechanisms and damage growth. The crack layer model provides an analytical framework to study the brittle regime and extract the specific energy of fracture. The last parameter can be employed to rank polyethylene pipes and assess their resistance to brittle fracture.

REFERENCES

ASTM D-1598 1981. *Annual book of ASTM standards*. Part 34.

ASTM E-399 1981. *Annual book of ASTM standards*. Part 10.

Chaoui,K. 1989. PhD Thesis. *A Theory for accelerated slow crack growth in MDPE fuel gas pipes*. CWRU, Cleveland, Ohio.

Chudnovsky,A. 1984. *NASA contractor report*. N° 174634.

Chudnovsky,A. & A.Moet 1986. A theory of crack layer propagation in polymers. *J. Elastomers and Plastics*. 18:50-55.

Crissman,J.M. 1987. *GRI final report*. 87-0236

Palermo E. & I.K.Deblieu 1985. Rate process concept applied to hydrostatically rating PE pipe. *Proc. 9th Plastic fuel gas pipe symp.*: 215-240, New Orleans, LA.

Raske,D.T. 1987. Analysis and application of the constant tensile load for PE gas pipe materials. *Proc. 11th plastic fuel gas pipe symp.*: 102-116, New Orleans, LA.

Recent Advances in Experimental Mechanics, Silva Gomes et al. (eds) © 1994 Balkema, Rotterdam, ISBN 90 5410 395 7

Crack tip plastic zones in a chromoplastic material

C. Atanasiu, D. M. Constantinescu & St. Pastramă
'Politehnica' University, Bucharest, Romania

ABSTRACT: Single edge notch (SEN) tensile panels made from an elastic-perfectly plastic chromoplastic material are tested in order to study the influence of the crack length to width of specimen (a/W) ratios on the plastic zone shape and size ahead the crack tip. The initially Dugdale-type plastic zone shape extends to a "butterfly" shape for short crack ($a/W = 0.1$ and $a/W = 0.2$) specimens, and to an approximately elliptical one for moderate deep ($a/W = 0.4$) and deep ($a/W = 0.7$) crack specimens. The plastic zone extends self similar in small-scale yielding and not self similar in large-scale yielding. Under small-scale yielding the plastic zone extent on the crack axis is plane strain for $a/W = 0.7$, and between plane strain and plane stress for $a/W = 0.4$ and $a/W = 0.2$.

1 INTRODUCTION

Chromoplasticity is an original experimental method developed in Romania by Bălan et al. (1963). The chromoplastic material, a vinylic compound, is elastic and linear for stresses smaller than the yield limit, and perfectly plastic when plastic deformation appears. At this moment, the chromoplastic phenomenon produces a change in the material's colour, turning white when flow takes place under tensile stresses, and black when flow is produced by compressive stresses.

The plastic behaviour near the tip of a straight through-thickness crack has been examined by several investigators. Dugdale (1960) assumed, for a thin sheet of mild steel, that the plastic zone develops as a narrow strip ahead of the crack tip, and under the assumption of plane stress obtained an estimate of the plastic zone length. Considering small geometry changes and small-scale yielding, Hutchinson (1968a, b) determined the near crack tip fields for plane stress and a strain hardening material. Rice and Johnson (1970) underlined the importance of large crack tip geometry changes, and for plane strain and perfect plasticity analyzed crack tip blunting and its influence on the stress field ahead the crack tip. Nishimura and Achenbach (1986) have studied the crack tip zone, from the crack tip to the elastic-plastic boundary, for a blunting crack in plane stress and an elastic-perfectly plastic material, but without taking into account the differential thinning of the plate. Large deformation theory and three-dimensional finite elements were used by Hom and McMeeking (1990) to analyze the near tip stress and deformation fields, and the extension and shape of the plastic zone. Their analysis considered mode I loading applied to a thin sheet, a perfect plastic and strain hardening response of the material, by taking into account crack tip blunting and the differential thinning of the plate. Narasimhan and Rosakis (1988) performed a finite element analysis of a stationary crack under mode I plane stress and small-scale yielding conditions. Among other results, they studied the shape and extension of the plastic zone showing that it is less rounded and spreads more ahead of the crack tip with decreasing hardening. The determined numerical shape and size of the plastic zone agreed well with the experimental (caustics) one as long as no boundary interaction effects were present (contained yielding).

The influence of specimen configurations, different crack length to width of specimen ratios, and material properties was studied for small-scale yielding to large-scale yielding by using the finite element method. McMeeking and Parks (1979) analyzed the crack tip fields in large-scale yielding and established the size requirements for specimens used in fracture toughness testing in order to ensure J-dominance. Shih and German (1981) established the minimum size requirements essential to a single parameter fracture criterion (J-dominance or crack tip opening displacement). For different specimens, they also established the size of the region at the crack tip dominated by the HRR (Hutchinson-Rice-Rosengren) singularity. Gdoutos and Papakaliatakis (1986) studied the influence of the plate geometry and crack to width ratios of a center cracked panel on the elastic-plastic boundary. Sorem et al. (1991) and Dodds et al. (1991) analyzed three-point-bend specimens in order to establish the effects of crack depth on elastic-plastic fracture toughness, demonstrating the limitations of a single parameter fracture criterion because for the same specimen at different crack depths the stress fields ahead of the crack tip reveal large differences at identical crack tip opening displacements.

In this paper, we present the results of the experimental analysis in which the crack tip plastic zone shape and size, the boundary interaction effects, and the yielding mechanism at the crack tip are studied. Single edge notch (SEN) tensile panels, with different crack length to width of specimen (a/W) ratios, are tested from small-to-large scale yielding, and finally to the failure of the specimen.

2. EXPERIMENTAL TESTS

2.1 Experimental procedure

The experimentally tested chromoplastic material is elastic-perfectly plastic, and its properties are:

modulus of elasticity in tension $E = 3400$ MPa
Poisson's ratio $\nu = 0.41$
yield stress in tension $\sigma_0 = 36$ MPa.

A single edge notch (SEN) panel of width W, thickness h, height H, and crack length a is considered. The panel is subjected to an uniform uniaxial tensile stress, σ, perpendicular to the crack axis. Four panels have a width of $W = 70$ mm, a thickness $h = 10$ mm, a height of $H = 300$ mm, and different crack-depth to specimen width ratios (a/W): $a/W = 0.1$, $a/W = 0.2$, $a/W = 0.4$, and $a/W = 0.7$. Another panel has the same height and thickness and a width $W = 100$ mm, and $a/W = 0.7$. The edge through-thickness notch is saw cut and has a width of 0.2 mm. The stress is monotonically increased until is observed that the material ahead the crack tip is changing its colour, turning white. Photographs are taken in order to analyze the shape and the extension of the plastic zone from small-scale yielding to large-scale yielding. When there are observed significant changes, the plastic zone is marked with a pen directly on the specimen while the tensile stress is kept constant. For each specimen it is therefore possible to obtain a complete understanding of the fracture processes that took place. Thus, the influence of different a/W ratios, and the boundary interaction effects are studied. While the stress σ is increased, for each panel, the plastic zone in small-scale yielding, the blunting of the crack tip, further increase of the plastic zone, stable crack growth, and the final extension of the plastic zone at the edge of the specimen can be observed . Some remarks are done during each test, and afterwards, on the photographs is measured the extent of the plastic zone through the specimen width, and is analyzed the mechanism by which yielding is produced during stable crack growth.

2.2 Experimental results

We are going to consider a specimen with a short crack if a/W ratios are 0.1 and 0.2, a moderate deep crack if a/W ratio is 0.4, and a deep crack if a/W is 0.7.

For all the cracked specimens the shape of the crack tip plastic zone is about the same as long as the small-scale yielding conditions are valid and there is no crack tip blunting. One can observe a Dugdale-type strip yield zone ahead of the crack tip, and as mentioned by Anderson (1991) this indicates that yielding occurs by crazing which is a mechanism that is more likely to be produced in polymers because of the triaxial stresses ahead the crack tip. As the tensile stress σ is increased, and on the macroscopic level is observed the crack tip blunting, the shape and extent of the plastic zone depend on the crack to width specimen ratios, a/W. For short crack specimens ($a/W = 0.1$ and $a/W = 0.2$) one can not observe that the plastic zone extends "backward" to the free surface of the specimen. On the contrary, it extends only ahead the crack tip, and from the initial extent along the crack axis it spreads more on the height of the specimen in a "butterfly" shape. Finally, the crack extends more along the direction normal to the crack axis than along the crack axis. For specimens with a moderate deep crack ($a/W = 0.4$) and a deep crack ($a/W = 0.7$) the plastic zone spreads more along the crack axis than transverse to the crack, somehow in an elliptic shape.

We measure the extent of the plastic zone along the crack axis, X (from the crack tip), and normal to the crack axis, Y. For the same load level (tensile stress σ) X and Y are the maximum measured lengths. The X extent of the plastic zone is shown in Fig. 1, and the Y extent in Fig. 2, for the four values of a/W with specimen width $W = 70$ mm, and $a/W = 0.7$ for specimen width $W = 100$ mm.

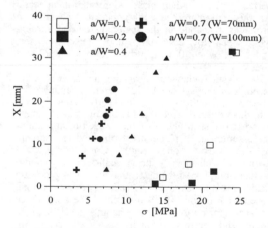

Fig. 1 Extent of the plastic zone along the crack axis.

If a/W is equal to 0.1 or 0.2 the maximum X or Y extent is up to 10 mm before stable crack growth occurs. For $a/W = 0.1$, the "butterfly" plastic zone shape extends at an angle $\alpha \approx 20°$ and $\alpha \approx 17°$ measured from the crack axis, with a ratio of the extent at angle α to the one along crack axis of 1.24 and 1.3. After a crack growth of about 11 mm, in large-scale yielding when the plastic zone has already reached the edge of the specimen, a sudden

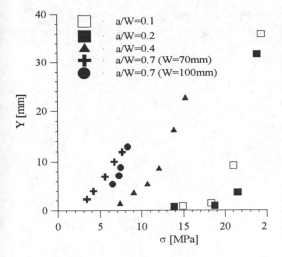

Fig. 2 Extent of the plastic zone transverse to the crack axis.

For $a/W = 0.1$ and $a/W = 0.2$ the previously discussed plastic zone shapes are illustrated in Fig. 3, respectively Fig. 4.

The plastic zone extends only ahead the crack tip and there is no turn "backward". For $a/W = 0.4$ the plastic zone spreads in a completely different manner (Fig. 5 and Fig. 6).

The initially Dugdale-type plastic zone is changing its shape as the tensile stress is increased, and becomes more rounded. Such a shape was predicted as a result of a finite element analysis by Narasimhan and Rosakis (1988), under small-scale yielding and plane stress, for a perfectly-plastic material.

An interesting investigation of the plastic zone extent is obtained in dimensionless coordinates for all the tested specimens. One can normalize the X and Y extents to the thickness of the specimen. In Fig. 7 the maximum extent transverse to the crack axis is represented as a function of the maximum extent along the crack axis.

unstable propagation is produced with a change of the crack path at an angle $\beta \approx 53°$. As it was a singular situation, this unstable propagation may be considered an exception. For $a/W = 0.2$ the angles α are approximately 7.5° and 18° (the stress levels are about the same as for $a/W = 0.1$), and the ratios of the aforementioned extensions are 1.03 and 1.3. As it results from Fig. 1 and Fig. 2 the X and Y extents of the plastic zone are about the same, a little bit smaller for $a/W = 0.2$ than for $a/W = 0.1$. For the moderate deep crack, $a/W = 0.4$, and for the deep crack, $a/W = 0.7$, the plastic zone spreads ahead of the crack tip, the extent X being greater than the extent Y. The "butterfly" plastic zone does not appear any more, and the initially Dugdale-type shape changes, as the tensile stress is increased, to a more rounded one.

Fig. 4 Plastic zone shape when stable crack growth occurs for $a/w = 0.2$.

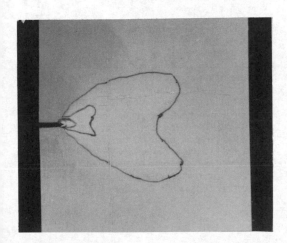

Fig. 3 Plastic zone shape when stable crack growth occurs for $a/W = 0.1$.

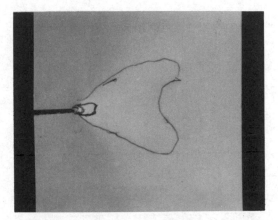

Fig. 5 Plastic zone shape in small-scale yielding for $a/W = 0.4$.

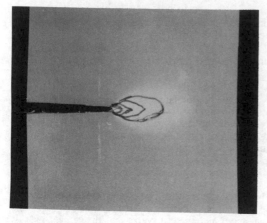

Fig. 6 Plastic zone shape in large-scale yielding for $a/W = 0.4$.

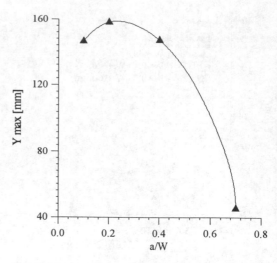

Fig. 8 Maximum extent of the plastic zone at the edge of the specimen.

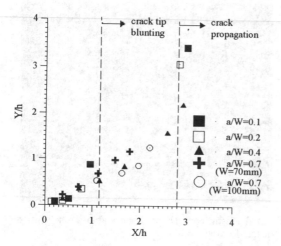

Fig. 7 Normalized extension of the plastic zone.

Fig. 9 Developement of plastic zone for $a/W = 0.2$.

Up to the blunting of the crack tip on macroscopic level, the extension of the plastic zone is self similar for all the specimens regardless the a/W ratios. The X extent of the plastic zone is in this case about the same as the thickness of the specimen, so we may consider that small-scale yielding conditions are valid. In this region we can estimate an approximate dependence of the extents of the plastic zones as $Y = 0.63X$. When crack tip blunting is already evident the influence of the a/W ratio changes the self similar X and Y extents of the plastic zone. But, what is more important, and was shown before, is that the shape of the plastic zone is completely different. For $a/W = 0.7$, because of the strong boundary interaction effects, the plastic zone already reaches the edge of the specimen before stable crack growth

occurs. For this ratio, the shape of the plastic zone is the same as for $a/W = 0.4$.

After the failure of all specimens the boundary interaction effects can be also analyzed by studying the Y maximum extent of the plastic zone measured at the edge of the specimen, transverse to the crack axis. The influence of a/W ratios is shown in Fig. 8.

As one can expect, for longer cracks the plastic zone spreads less. However, the maximum extent of the plastic zone is obtained for $a/W = 0.2$, so, a transitional behaviour between $a/W = 0.1$ and $a/W = 0.2$ is possible from this point of view. In Fig. 9 is shown the final plastic deformation that appears in the specimen with $a/W = 0.2$ after its failure.

3. DISCUSSION

3.1 *Plastic zones*

The chromoplastic material, a viscoelastic one, exhibits rate dependent mechanical properties. As a consequence, the framework of linear elastic fracture mechanics, and the concept of stress intensity factor are not guaranteed. However, one can calculate for different load levels and a/W ratios the value of the instantaneous mode I stress intensity factor, K_I. Under small-scale yielding conditions, the extent of the plastic zone on the crack axis can be made dimensionless by the parameter $(K_I/\sigma_o)^2$. In this way one can study the influence of a/W ratios on the plastic zone extent, X. We use the already known relation

$$X = m\left(\frac{K_I}{\sigma_o}\right)^2 \qquad (3.1)$$

where K_I = mode I stress intensity factor; σ_o = yield stress; and m = dimensionless parameter.

The parameter $m = 0.053$ according to the model of Irwin for plane strain, and $m = 0.393$ according to the model of Dugdale for plane stress. A qualitative interpretation of the experimentally obtained plastic zone extents on the crack axis can be made for all the tested specimens up to the limit of crack tip blunting on macroscopic level (Fig. 10).

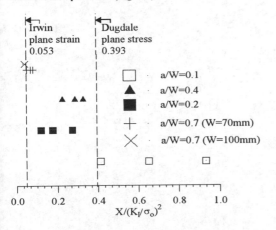

Fig. 10 Dimensionless extent of the plastic zone on the crack axis.

As it was shown before, for $a/W = 0.1$ and $a/W = 0.2$ the plastic zone has a "butterfly" shape, so the X extent is not completely relevant. Nevertheless, a different extent is produced at these two ratios. If $a/W = 0.1$ the plastic zone extends more than the plane stress model predicts. If $a/W = 0.2$ the extent of the plastic zone is between plane strain and plane stress models. The same happens if $a/W = 0.4$. It is remarkable to underline that for $a/W = 0.7$ the X

extent of the plastic zone is almost identical to the one predicted by the plane strain model. For the two specimens tested at this crack to width ratio the ligament size between the crack and free edge of the panel is about twice and three times greater than the thickness of the specimen.

Finite element analyses made under the assumptions of small-scale yielding and plane stress predict for a perfectly-plastic material: $m = 0.28$ (Hom and McMeeking 1990) and $m = 0.29$ (Narasimhan and Rosakis 1988). The experimentally established plastic zone extent along the crack axis is influenced by the value of a/W ratio. The shape of the plastic zone determined by Narasimhan and Rosakis (1988) is close to our results only for moderate deep ($a/W = 0.4$) and deep ($a/W = 0.7$) cracks.

We may reasonably suggest that the shape and size of the plastic zone is influenced by the crack to width of specimen ratio, and as the load is increased the extent of the plastic zone is self similar in small-scale yielding and not self similar in large-scale yielding.

3.2 *Yielding mechanism*

A polymer yields by shear yielding or crazing. The chromoplastic material loaded in tension yields by turning white, and loaded in compression yields by turning black. In our experiments the stresses ahead the crack tip were only tensile, and the yielded white zones have initially a Dugdale-type shape indicating craze yielding (Anderson 1991).

When stable crack growth occurs one can clearly notice in Fig. 11 the craze zones that appear ahead the crack tip.

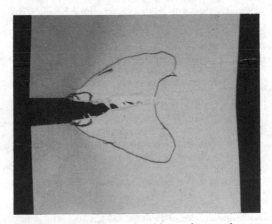

Fig. 11 Craze yielding and stable crack growth.

4 CONCLUSIONS

Specimens with different a/W ratios are studied in order to analyze to ratio effect on crack tip plastic zone shape and size. Initially, in small-scale yielding, a Dugdale-type strip yield zone is observed ahead the crack tip. As the load is increased, for $a/W = 0.1$ and $a/W = 0.2$, the plastic zone does not turn to the free surface of the specimen, and it spreads ahead the crack tip in a "butterfly" shape. For $a/W = 0.4$ and $a/W = 0.7$ the plastic zone extends mainly ahead the crack tip (Fig. 6).

If the extents along the crack axis and transverse to the crack axis are made dimensionless by dividing to the thickness of the specimen (Fig. 7), they are self similar in small-scale yielding regardless the a/W ratio, up to the moment when crack tip blunting is evident on macroscopic level. In large scale yielding the extent of the plastic zone is not self similar, and is influenced by the a/W ratios.

Under small-scale yielding, distinctive features between short and deep cracks appear when referring to the extent of the plastic zone along the crack axis for plane stress (Dugdale) and plane strain (Irwin). For deep cracks ($a/W = 0.7$) the plane strain condition is obtained. For a moderate deep crack ($a/W = 0.4$) and a short crack ($a/W = 0.2$) the extent is in between the one predicted by the plane strain or plane stress.

It seems that a transition region results between $a/W = 0.1$ and $a/W = 0.2$ when the plastic zone length is analyzed. As is shown in Fig. 10, in small-scale yielding the extent of the plastic zone along the crack axis for $a/W = 0.1$ is greater than the one obtained for $a/W = 0.2$. In large-scale yielding (Fig. 8), after the failure of the specimen, the extent of the plastic zone transverse to the crack axis is smaller for $a/W = 0.1$ than the one obtained for $a/W = 0.2$.

REFERENCES

Anderson, T.L. 1991. *Fracture Mechanics - Fundamentals and Applications*. Boca Raton: CRC Press.

Bălan, St., S.Răutu & V.Petcu 1963. *Cromoplasticitatea*. Bucuresti: Editura Academiei.

Dodds, R.H., T.L.Anderson & M.T. Kirk 1991. A framework to correlate a/W ratio effects on elastic-plastic fracture toughness (J_c). *Int. J. Fracture* 48: 1-22.

Dugdale, D.S. 1960. Yielding of steel sheets containing slits. *J. Mech. Phys. Solids* 8: 100-104.

Gdoutos, E.E. & G.Papakaliatakis 1986. The influence of plate geometry and crack material properties on crack growth. *Engng Fracture Mech.* 25: 141-156.

Hutchinson, J.W. 1968a. Singular behaviour at the end of a tensile crack in a hardening material. *J. Mech. Phys. Solids* 16: 13-31.

Hutchinson, J.W. 1968b. Plastic stress and strain fields at a crack tip. *J. Mech. Phys. Solids* 16: 337-347.

Hom, C.L. & R.M.McMeeking 1990. Large crack tip opening in thin elastic-plastic sheets. *Int. J. Fracture* 45: 103-122.

McMeeking, R.M. & D.M.Parks 1979. On criteria for J-dominance of crack-tip fields in large-scale yielding. In J.D.Landes et al. (eds.), *Elastic-PlasticFracture*, ASTM STP 668: 175-194.

Philadelphia: American Society for Testing and Materials.

Narasimhan, R. & A.J.Rosakis 1988. A finite element analysis of small-scale yielding near a stationary crack under plane stress. *J. Mech. Phys. Solids* 36: 77-117.

Nishimura, N. & J.D.Achenbach 1986. Finite deformation crack-line fields in a thin elasto-plastic sheet. *J. Mech. Phys. Solids* 34: 147-165.

Rice J.R. & M.A.Johnson 1970. The role of large crack tip geometry changes in plane strain fracture. In M.F.Kanninen et al. (eds.), *Inelastic Behavior of Solids*: 641-672. New-York: McGraw-Hill.

Shih, C.F. & M.D.German 1981. Requirements for a one parameter characterization of crack tip fields by the HRR singularity. *Int. J. Fracture* 17: 27-43.

Sorem, W.A., R.H.Dodds Jr. & S.T.Rolfe 1991. Effects of crack depth on elastic-plastic fracture toughness. *Int. J. Fracture* 47:105-126.

Recent Advances in Experimental Mechanics, Silva Gomes et al. (eds) © 1994 Balkema, Rotterdam, ISBN 90 5410 395 7

Crack-tip caustics at bimaterial interfaces and in mechanically anisotropic materials

H. P. Rossmanith & R. E. Knasmillner
Institute of Mechanics, Technical University of Vienna, Austria

D. Semenski
Faculty of Mechanical Engineering and Naval Architecture, University of Zagreb, Croatia

ABSTRACT: In this contribution, the method of caustics will be applied to cracks at bimaterial interfaces and to cracks in mechanically anisotropic materials. The theoretical solution for the light deflection and the parametric equation for the singular caustic curve has been derived. It has been found that the shape of the caustic curve departs from the classical form and becomes strongly dependent on the mechanical characteristics of the materials.

Caustics at crack tips in materials with strong mechanical anisotropy are seriously distorted and data assessment on the basis of this caustic curves becomes a serious problem. Since only sections of caustics can be used for data evaluation, a general method for the determination of stress intensity factors from caustics has been developed.

1 INTRODUCTION

The method of caustics, or the shadow-spot technique, has become one of the favorite tools for the experimental determination of stress intensity factors [1].

Several versions of the method of caustics are successfully employed in studying the behavior of stress and strain around crack tips. The most common methods are the transmission method and the reflection method [2, 3].

The physical principle behind the transmission method of caustic is the inhomogeneous deflection of light rays during their passage through a plate specimen due to two effects: the reduction of the thickness of the specimen and the change of the refractive index of the material as a consequence of stress intensification. In the reflection method of caustics, the surface distortion effect alone is operative.

In fracture mechanics, caustics have been extensively used for the determination of stress intensity factors. The result of a caustic experiment is normally one single shadow spot or a time sequence of shadow spots, the shape and the size of which are indicative of the level of stress intensity at the crack tip.

2 INTERFACE CAUSTICS

Consider a given field of principle stresses $\sigma_{1l}(x, y)$ and $\sigma_{2l}(x, y)$ in a plane dissimilar plate specimen with an interface crack as shown in Figure 1, the subscript $l = 1, 2$ holds for the upper and lower halfplane, respectively, separated by the interface.

For a light ray impinging at point $P(x, y)$ in the model plane, geometrical optics in conjunction with the theory of elasticity yield the deflection (i.e. the shift $\vec{w}$ on the screen of the image) of point $P(x, y)$ from its undisturbed position $P'(X, Y)$ to the position $P^*(X, Y)$.

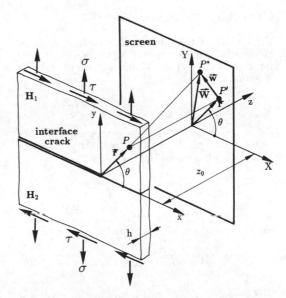

Figure 1: Semi-infinite crack at the interface of two dissimilar regimes

1283

For the following theoretical considerations, it is essential to distinguish between mechanical anisotropy and optical anisotropy. For optically anisotropic materials, the mapping equation is given by [1]:

$$\vec{W}_l^{\pm} = \vec{W}_l^{\pm}(X, Y) = \vec{r}(x, y) + \vec{w}_l^{\pm}(x, y) \quad (1)$$

For mechanically isotropic materials, the deflection vector $\vec{w}_l^{\pm}(x, y)$ becomes (no summation)

$$\vec{w}_l^{\pm} = -z_0 h_l c_l \nabla \left[(\sigma_{1l} + \sigma_{2l}) \pm \varepsilon_l (\sigma_{1l} - \sigma_{2l}) \right] \quad (2)$$

Here, z_0 denotes the distance between the model and the screen, h_l is the thickness of the model, c_l is the elasto-optical constant and ε_l is the factor for the optical anisotropy of the materials. The superscript $\pm$ holds for the two caustic branches obtained for optically anisotropic materials ($\varepsilon_l \neq 0$). For optically isotropic materials and for the front face reflection method of caustics ($\varepsilon_l = 0$), the superscripts $\pm$ become meaningless.

The solutions of

$$J_l = \frac{\partial(X, Y)}{\partial(x, y)} = \frac{\partial X}{\partial x} \frac{\partial Y}{\partial y} - \frac{\partial Y}{\partial x} \frac{\partial X}{\partial y} = 0 \quad (3)$$

define the initial curves in the x, y-plane (model plane) for the caustic curves on the screen (X, Y-plane).

In the general case of a model containing a crack at the interface between two different optically anisotropic materials and employing the transmission method of caustics ($z_0 < 0$), four caustic branches will be obtained [4].

2.1 Mechanically isotropic materials

For a crack at the interface of two dissimilar mechanically isotropic materials, the leading singular term of the complex Goursat stress potential is given by [5]:

$$\Phi_l(z) = K_l z^{-1/2 - i\beta} \quad \text{where} \quad \beta = \ln(g/2\pi) \quad (4)$$

and $z = x + iy = r e^{i\theta}$ are the complex coordinates in the model plane. The complex stress intensity factors K_l are defined by

$$K_l = K_{Il} - i K_{IIl}, \quad K_2 = g K_1 \quad (5)$$

The interrelation of the stress intensity factors K_l is defined by the factor g which depends on Poisson's ratios ν_l and the shear moduli μ_l:

$$g = \frac{\kappa_1 \mu_2 + \mu_1}{\kappa_2 \mu_1 + \mu_2} \quad \text{where} \quad \kappa_l = \frac{3 - \nu_l}{1 + \nu_l} \quad (6)$$

The branches of the caustic curves in complex notation are obtained by combining equations (1)

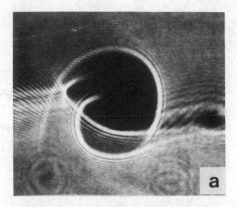

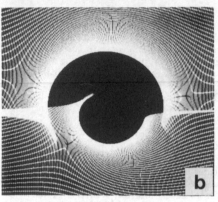

Figure 2: Caustic at an interface crack,
a) experimentally recorded caustic,
b) numerically generated caustic

and (2) and by the use of Kolosov's formula (see eg. [6]):

$$W_l = R_l(\theta) e^{i\theta} -$$
$$- z_0 h_l c_l \left(\tfrac{1}{2} - i\beta \right) \overline{K}_l R_l(\theta)^{-3/2 + i\beta} e^{i\theta(3/2 - i\beta)} \quad (7)$$

where the branches of the initial curve $R_l(\theta)$ are determined by the solutions of equation (3):

$$R_l(\theta) = R_l^* e^{2\theta\beta/5} \quad \text{with}$$

$$R_l^* = \left| z_0 h_l c_l \left(\tfrac{1}{2} + i\beta \right) \left(\tfrac{3}{2} + i\beta \right) K_l \right|^{2/5} \quad (8)$$

The limits of the values of the angle θ (see Figure 1) are given by $0 \leq \theta \leq \pi$ for $l = 1$ (material 1) and $-\pi \leq \theta \leq 0$ for $l = 2$ (material 2).

An experimentally recorded caustic at an interface crack tip and the related numerically generated computer image is shown in Figure 2.

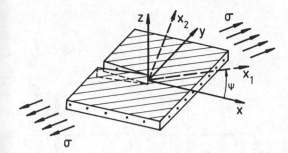

Figure 3: Geometry and loading of a mechanically anisotropic plate specimen (fiber-reinforced material) with a crack

3 CAUSTICS IN MECHANICALLY ANISOTROPIC MATERIALS

In the following, small strains are assumed and, therefore, the linearized general constitutive equation for anisotropic materials in Voigt's notation is given by

$$\varepsilon_j = M_{jk}\sigma_k \quad (j,k = 1,2,\ldots 6) \quad (9)$$

where M_{jk} denote the elastic constants for plane stress and σ_k and ε_j are the stress and the strain vector, respectively.

Fiber-reinforced materials form a very important class of light-weight – high-strength composites frequently used in structural engineering. Since most of these materials are fabricated as thin plates or thin shells of nontransparent materials plain stress conditions ($\sigma_3 = \sigma_{zz} = 0$) are assumed and the reflection technique of caustics will be appropriate.

Consider a thin mechanically anisotropic plate fabricated of fibre-reinforced material where one or more families of fibers form either an obliquely oriented grid under conditions of plane stress as shown in Fig. 3.

In the case of the reflection technique, the general mapping equation for caustics (1) reduces to

$$\vec{W} = \vec{W}(X,Y) = \vec{r}(x,y) + \vec{w}(x,y) \quad (10)$$

and the deflection vector $\vec{w}$ for the reflection method is determined by

$$\vec{w}(x,y) = z_0\nabla(\Delta s(x,y)) = -z_0 h\nabla(M_{3k}\sigma_k) \quad (11)$$

where $\Delta s(x,y)$ is the change of the light path and the mapping equation equ(10) finally becomes

$$\vec{W} = \vec{r} - z_0 h\nabla(M_{3k}\sigma_k) \quad (k = 1,2,\ldots 6) \quad (12)$$

In two-dimensional theory of elasticity, the state of stress and strain in the solid body can be cast in terms of complex stress functions [7]:

$$F(x,y) = 2\mathbf{Re}\sum_{i=1}^{3} F_i(z_i)$$
$$\Psi(x,y) = 2\mathbf{Re}\sum_{i=1}^{3} \lambda_i F_i'(z_i) \quad (13)$$

where $z_i = x + \mu_i y$ and the stresses are given by

$$\sigma_{pq} = -F_{,pq} + \delta_{pq}F_{,rr} \quad \sigma_{3p} = e_{pq}\Psi_{,q} \quad (14)$$

where the subscripts p,q,r run over the range $(1,2)$ and e is the alternating tensor. The complex values μ_i follow from the compatibility equations and for the following calculations three distinct roots of $l_6(\mu) = l_2 l_4 - l_3^2 = 0$ are choosen such, that $\mathbf{Im}(\mu_i) > 0$. The polynomials l_i are given by

$$l_2(\mu) = M_{22}\mu^2 - 2M_{45}\mu + M_{44}$$
$$l_3(\mu) = M_{15}\mu^3 - (M_{14}+M_{56})\mu_2 + \\ + (M_{25}+M_{46})\mu - M_{24} \quad (15)$$
$$l_4(\mu) = M_{11}\mu^4 - 2M_{16}\mu^3 + (2M_{12}+M_{66})\mu^2 - \\ - 2M_{26}\mu + M_{22}$$

equation and $\lambda_i = -l_3(\mu_i)/l_2(\mu_i)$. equation

For a straight, semi-infinite crack, the analysis yields the mapping equation

$$\vec{W} = \vec{r} - z_0 h\left\{\left[2\mathbf{Re}\sum_{i=1}^{3}\Omega_i F_i''''(z_i)\right]\vec{e}_x + \right. \\ \left. + \left[2\mathbf{Re}\sum_{i=1}^{3}\Omega_i\mu_i F_i''''(z_i)\right]\vec{e}_y\right\} \quad (16)$$

where

$$\Omega_i = M_{31}\mu_i^2 + M_{32} - M_{34}\mu_i - M_{35}\lambda_i + M_{36}\lambda_i\mu_i \quad (17)$$

and the complex function F'''' is given by

$$F''''(z_i) = -\frac{1}{4}\sqrt{\frac{1}{2\pi}}\sum_{j=1}^{3}N_{ij}^{-1}K_j z_i^{-3/2} \quad (18)$$

and N_{ij}^{-1} are the components of the inverse matrix of N_{ij} (see Ref.[7]):

$$N = \begin{pmatrix} 1 & 1 & 1 \\ -\mu_1 & -\mu_2 & -\mu_3 \\ \lambda_1 & -\lambda_2 & -\lambda_3 \end{pmatrix} \quad (19)$$

For static plane elasticity problems in mechanically isotropic materials, these equations reduce to those given in Ref [1].

One numerically generated caustic of a crack tip in an anisotropic material is shown in Fig. 4. In the case of orthotropy, the influence of a mode-2 contribution on the shape of the caustic is more

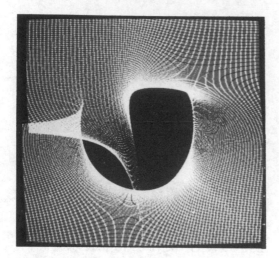

Figure 4: Numerically generated caustic at a crack tip in anisotropic material

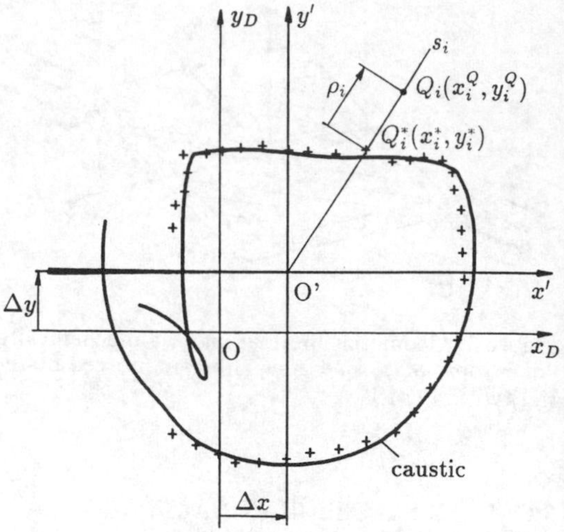

Figure 5: Coordinate systems, experimental data points and best fit caustic

pronounced than in the isotropic case and introduces folds and swallow-tails as new characteristic features of catastrophe theory into caustics. Detailed analysis shows that there is a one-to-one correspondence between the inflexion points on the direction surfaces of the elastic constants and the "caustic elements" within the caustic contour. This degeneracy of the caustic causes additional complexities for data reduction and evaluation procedures [8].

4 DATA PROCESSING TECHNIQUES

The traditional diameter measurement based method for crack-tip caustics is not suitable anymore in the case of mechanically anisotropic materials where the caustic boundary is inflicted by elements of catastrophe theory such as swallow tail formations, etc. We suggest a multi-point overdeterministic data reduction method on the basis of interactive computer graphics which has previously been developed for isotropic materials [9]. This method works very quickly and reliable even when crack tip caustics at interfaces or in anisotropic materials are seriously distorted.

The development of methods for the determination of stress intensity factors is closely associated with the development of the experimental techniques for recording crack tip caustics or shadow spots.

The hardware requirement consists of an image scanner (e.g. a CCD-camera with A/D-converter) and an image storage device provided with a monitor and a "mouse". The digitized and stored caustic pattern displayed on the monitor is required for the interactive selection of certain points on the screen.

For data selection in this novel and improved technique, an essentially arbitrary coordinate system (x_D, y_D) is placed parallel to the crack with the origin O within the caustic area, as shown in Figure 5, the knowledge of the exact position of the crack tip is not necessary.

If the location of the crack tip was known, the coordinate system could be placed appropriately as $\Delta x = \Delta y = 0$. In general, however, the exact site of the crack tip is not known a priori and the distances Δx and Δy will be non-zero. Next, the region of the experimental caustic is selected where data points can easily be identified. In general, the part of the caustic adjacent to the faces of the crack is blurred and obscured (blurr zone **B**) and should therefore be discarded (Figure 6).

The transformation $x_D = x' + \Delta x$ and $y_D = y' + \Delta y$ transforms the image equation to the following expressions:

$$
\begin{aligned}
X_D &= f(K_1, \mu; \Delta x, \Delta y; \varphi) \\
Y_D &= g(K_1, \mu; \Delta x, \Delta y; \varphi)
\end{aligned}
\qquad (20)
$$

where the as yet unknown quantities K_1, μ, Δx and Δy will be determined by a least squares method by minimizing the sum of the squares S of the differences between the experimentally recorded and selected data points $(*)$ and the numerically generated data points (Q):

$$
S = \sum_{i=1}^{n} \left\{ \left(x_i^{*2} + y_i^{*2} \right)^{1/2} - \left(x_i^{Q2} + y_i^{Q2} \right)^{1/2} \right\}^2 \quad (21)
$$

In the interactive data reduction method, a set of m radials s_i, homogeneously distributed within the zone of the clearly visible region of the caustic, **A**, is automatically chosen for point identification

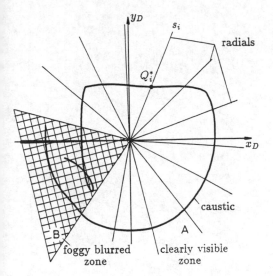

Figure 6: Data acquisition regions **A** and **B**

(Figure 6). This selection leaves open the proper position of the experimental data points on the finite width of the experimentally recorded light intensity distribution which makes up for the caustic. Their proper positions are determined from an evaluation of the density distribution along the radials s_i within the caustic range.

The last step in the K-determination procedure is the visual inspection of the result by plotting the theoretical caustic line and the associated experimental input points together with the stored image of the analyzed caustic on the monitor. If the coincidence of the analytically generated caustic with the experimental recorded caustic is unsatisfactory, the filter characteristics, i.e. the relative grey level for the experimental caustic line, can be set to new values or some of the data points can be repositioned or deleted for a new analysis.

The method presented here converges very rapidly and, even in cases where an appreciable part of the caustic is obscured or not fit for data analysis, very reliable results have been obtained during an extensive test program for caustics at interfaces and caustics in anisotropic materials, where caustics have been numerically generated, fuzzied, data-reduced, and reconstructed on the basis of the calculated data.

ACKNOWLEDGEMENT

The authors would like to acknowledge the financial support granted by the Austrian National Science Foundation under project number *FWF P 6632*.

REFERENCES

[1] A.J. Rosakis (Editor). Special issues on: *The Optical Method of Caustics I, II*. Optics and Lasers in Engineering. Elsevier Applied Science, 13, 1990 and 14, 1991.

[2] P. Manogg. Investigation of the rupture of a Plexiglas plate by means of an optical method involving high speed filming of the shadows originating around holes drilling in the plate. *Int.J. Fracture Mechanics*, 2:604–613, 1966.

[3] P.S. Theocaris and E. Gdoutos. An optical method for determining opening mode and edge sliding mode stress-intensity factors. *Journal of Applied Mechanics*, 39:91–97, 1972.

[4] H.P. Rossmanith, R.E. Knasmillner, and J. Zhang. An interactive approach for K-determination from caustics at interface cracks. In: *Proc.Int.Conf. MMFF-91, Vienna, Austria*, pages 33–47. MEP, London, 1993.

[5] P.S. Theocaris and C.A. Stassinakis. Complex stress intensity factors at tips of cracks along interfaces of dissimilar media. *Engg.Fract.Mech.*, 14:363–372, 1981.

[6] H.P. Rossmanith. The method of caustics for static plane elasticity problems. *J. of Elasticity*, 12(2):193–200, 1982.

[7] A. Hoenig. Near-tip behavior of a crack in a plane anisotropic elastic body. *Engg.Fract.Mech.*, 16(3):393–403, 1982.

[8] H.P. Rossmanith and R.E. Knasmillner. Recent advances in data processing from caustics. In: Proc. of Int. Seminar on *Dynamic Failure of Materials*, Vienna, pages 260–272. Elsevier, 1991.

[9] R.E. Knasmillner. Vielpunktmethode zur Bestimmung des Spannungsintensitätsfaktors mit Hilfe der Methode der Kaustik. *ÖIAZ*, 131:318–320, 1986.

Recent Advances in Experimental Mechanics, Silva Gomes et al. (eds) © 1994 Balkema, Rotterdam, ISBN 90 5410 395 7

On the use of crack opening interferometry for determining long-range cohesive forces at interfaces

K.M. Liechti & Y.M. Liang
Engineering Mechanics Research Laboratory, Department of Aerospace Engineering and Engineering Mechanics, The University of Texas at Austin, Tex., USA

ABSTRACT: Measurements of the normal crack opening displacements (NCOD) of interfacial cracks between glass and epoxy were consistently lower than even linearly elastic predictions. Such results suggest that relatively long-range cohesive tractions were active. A series of experiments and finite element analyses were used to determine the traction/separation law which was then found to be in reasonable agreement with the distribution due to electrostatic effects. Estimates on the shielding provided by such cohesive forces were made for several fracture mode-mixes.

1 INTRODUCTION

A number of interfacial fracture studies have revealed that toughness values increased with larger amounts of mode II loading (Liechti and Hanson, 1988; Cao and Evans, 1989; Thouless, 1990; Wang and Suo, 1990; Liechti and Chai, 1992; Liechti and Liang, 1992; O'Dowd, Stout and Shih, 1991; Thurston and Zehnder, 1993). Several explanations of the toughening effect have been offered including asperity shielding (Evans and Hutchinson, 1989) and plasticity effects (Tvergaard and Hutchinson, 1992). However, neither mechanism appeared to be active in the case of a glass epoxy interface (Liechti and Chai, 1992) in spite of the fact that plastic zone sizes differed significantly, although remaining small-scale, from mode I to mode II. Recent observations of electrostatic charging during interfacial fracture (Zimmerman et al., 1991) and contact (Horn and Smith, 1992), which could potentially act as shielding mechanisms, motivated the present work.

The first objective of this study was to increase the resolution of crack opening interferometry so that NCOD could be measured closer to the crack front. This then allowed measured and predicted NCOD to be compared so that the traction/separation law associated with the long-range cohesive forces could be extracted and compared with distributions due to electrostatic effects.

2 EXPERIMENTS AND ANALYSIS

Optical interferometry (Liechti, 1993) was used to measure the normal crack opening displacements (NCOD). A monochromatic light beam was introduced through the glass to the crack surfaces at normal incidence and the reflected beams were brought together by a microscope to form interference fringe patterns. The NCOD, δ, for a bright fringe of order m were determined through

$$\delta = m\lambda / 2 \qquad (1)$$

where λ was the wavelength of the incident light.

For the wavelength of light (546 nm) used in the experiments reported here, the resolution in NCOD from equation (1) is 273 nm. This can be halved by considering dark and bright fringes. Further increases in resolution can be obtained by measuring the light intensity between fringes, as was done by Voloshin and Burger (1983) for photoelastic fringes. Although the analog version of the digital expression (1) can be applied to interpolate between fringes that are well removed from the crack front, modifications must be considered for interpolations near the crack front in order to account for refraction effects and larger crack opening angles.

The refraction effect essentially shifts the location of fringes and was first examined by Fowlkes (1975) who considered a crack in a homogeneous material under mode I loading. Liang (1993) applied the same approach to the epoxy glass configuration being considered here and found that refraction effects could contribute up to 5% error for distances from the crack front up to 10µm from the crack front, the spatial resolution of the apparatus used in this study.

Another effect, which can significantly influence the measured NCOD near the crack front, is the reflection of light from crack faces with a large crack opening angle. Assuming normal incidence, the actual NCOD, $\delta(\xi)$, are related to the light path difference, δ_{LP}, by

$$\delta_{LP} = \frac{2\delta}{1 - \tan^2 \gamma} \qquad (2)$$

where $\gamma = \tan^{-1}(d\delta / d\xi)$. In order to obtain the actual NCOD, we rewrite (2) as

$$\frac{d\delta}{d\xi} = \sqrt{1 - \frac{2\delta}{\delta_{LP}(\xi)}} \qquad (3)$$

and solve it numerically. Far from the crack front, the slopes are on the order of 0.01 radians or smaller, and the true NCOD can be obtained from (1). However, if we are interested in finding the NCOD ahead of the first order fringe, as will be discussed next, then the crack opening angle may be large and equation (3) should be adopted.

An additional consideration is that equation (1) has been used to obtain NCOD in an essentially digital manner from one fringe to another. The resolution in NCOD can be increased by measuring the variation in light intensity between fringes as has been suggested by Voloshin and Burger (1983) for photoelastic fringes. Considering normally incident light and using a wave representation, we can find (Liang, 1993), that

$$\sqrt{\frac{I}{I_p}} = \sin\left[\frac{\pi}{\lambda} \frac{2\delta}{1 - (\delta')^2}\right] \qquad (4)$$

where $(I_p)I$ is the (maximum) intensity of the fringe patterns, and $\delta = \delta(\xi)$ is the actual distribution of the NCOD. The main implication of equation (4) is that the NCOD ahead of the first order fringe can be determined, their resolution being limited only by that of the intensity measurement device. The scheme outlined above was applied to a cracked specimen under bond-normal loading (Fig. 1). Interpolations (Fig. 2) were made between the crack front and the first and second order bright fringes. The consistency of the interpolations between the first and second order bright fringes established confidence in the results obtained close to the crack front where the NCOD decrease more rapidly than is suggested by $\sqrt{r}$ behavior. The measurements indicated that K-dominance lies between the first and second dark fringes. The additional drop in NCOD closer to the crack suggests that some relatively long-range cohesive tractions are active.

The stress analyses that were used to examine the cohesive stresses were made with the finite element code ABAQUS*. Several stress/strain laws were considered for the epoxy, including linearly elastic,

rate-independent and rate-dependent plasticity responses. The parameters for these models were obtained from uniaxial tension, plane strain compression, and combined normal and shear stress experiments at various rates (Liang, 1993). The rate-independent plasticity analyses that were conducted were mainly J_2-deformation theory analyses. The results from such analyses were designated as EPRI in Figures 3 and 4. The rate-dependent flow rule that was incorporated in the finite element analyses had the form:

$$\dot{\bar{\varepsilon}}^p = D\left(\frac{\sigma_y}{\sigma_o} - 1\right)^q \qquad (5)$$

where $\dot{\bar{\varepsilon}}^p$ is the rate of equivalent plastic strain, σ_y is current yield stress, σ_o is the quasi-static yield stress, and D and q are material properties that were obtained from various constant strain rate, pure shear experiments (Liang, 1993). The results obtained from these analyses were designated EPRD in Figures 3 and 4. The measured NCOD values are now compared with predictions based on the different constitutive behaviors described above for cases involving pure bond-normal and shear dominant (positive and negative) loadings.

3 RESULTS

The first comparisons are made (Fig. 3) for two levels of bond-normal loading, with the NCOD values between the crack front and the first bright fringe having been interpolated from (4). For the lower load level ($v_o = 1.78$ μm), the nonlinear (EPRI, EPRD) and linear solutions were the same in the domain being examined and reasonably matched the measured values far from the crack front. However, the measured values were noticeably lower near the crack front. At the higher load level ($v_o = 4.37$ μm), there were signs of blunting when the EPRI and linear solutions were compared. The EPRD results indicated that no blunting should occur, as they were in agreement with the NCOD from the elastic prediction. Neither solution really matched the measured NCOD near the crack front which were noticeably smaller.

The differences noted above for bond-normal loading were greater when a bond normal loading was added sequentially to a shear load. The differences were greatest for negative shears (Fig. 4) where the EPRI analysis predicted a large degree of blunting. A smaller degree of crack tip blunting was predicted by the EPRD model. Again, it was evident that the measured NCOD were even lower than the values obtained from the elastic analysis. Similar but smaller differences were apparent for positive shear . The consistently higher values of predicted NCOD suggested that relatively long-range cohesive stresses could be acting on the crack faces.

* We are grateful to Hibbit, Karlsson and Sorensen, Inc. for making ABAQUS available under an academic license.

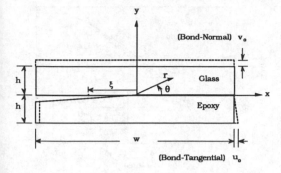

h = 0.95 cm
w = 17.8 cm
2b = 0.20 cm (specimen thickness)

Fig. 1 Geometry of the edge-cracked bimaterial strip specimen

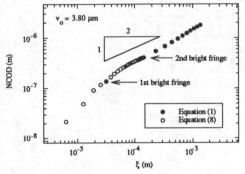

Fig. 2 NCOD ahead of the first and second order fringes

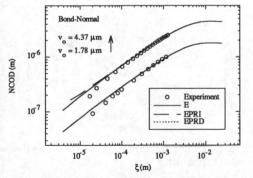

Fig. 3 Comparison of measured NCOD with elastic (E), rate-independent plasticity (EPRI) and rate-dependent plasticity (EPRD) predictions for bond-normal loading

The measured NCOD in Figures 3 and 4, showed a clear departure from elastic and elastoplastic analysis close to the crack front. In addition, the departure was not due to blunting, as might be expected, but in the opposite direction; near the crack front the NCOD were always lower than even the elastic predictions. These results and the observations of Zimmerman et al. (1991) and Horn and Smith (1992) motivated an examination of electrostatic effects on NCOD. The results of the examination are preliminary but should encourage further work. The first step that was taken was to grow a crack, arrest it (without crack face contact) and then apply a cyclic bond-normal loading and unloading that produced crack face contact during unloading and produced no further crack growth during the loading portion. The measured NCOD from two series of experiments are shown in Figures 6a,b where the NCOD, normalized by the applied bond-normal displacement, are plotted as a function of distance from the crack front, for zero numbers of subsequent contacts. In the first case (Fig. 6a), there was no effect of subsequent crack face contact. In the second case (Fig. 6b), there was a difference between the NCOD profile after zero contact and all subsequent contacts, which did not, however, differ from one another. It is not entirely clear why the responses shown in Figures 6a and 6b differ. It is possible that discharging occurred in the first case but not the second, due to differences in environmental conditions. These will have to be more carefully controlled in the future.

The possibility that electrostatic effects were causing the lower NCOD near the crack front was explored by conducting two analyses. First an electrostatic traction vs. NCOD law was developed by integrating Coulomb's law for the force between point charges. The charge density was left as a free parameter. The second (finite element) analysis used the difference between measured and predicted NCOD to extract the tractions along the crack faces. The two traction vs. NCOD laws were then compared in order to establish the charge density that was needed. The details now follow.

The electrostatic force between two point charges is governed by Coulomb's law, which has the following form

$$F = \frac{ke_1e_2}{d^2} \qquad (6)$$

where F is the electrostatic force, e_1 and e_2 are the electrical charges, d is the distance separating the point charges, and $k = 1/4\pi\varepsilon_o$ (8.99×10^9 m/Farad). The analysis was extended to an elemental section of the crack faces and integrated to yield the tractions on the crack faces (Liang 1993)

$$T_y = 2k\eta_1\eta_2 \tan^{-1}\left(\frac{B}{\delta}\right) \text{ and } T_x \approx 0 \qquad (7)$$

1291

where η_1 and η_2 are the charge densities on the top and bottom crack faces. The equation indicates that the electrostatic stresses are strongly mode I dominant.

The distribution of surface tractions was also determined by using the differences between the measured NCOD and the values predicted by the rate-dependent plasticity analyses. The rate-dependent analysis was first conducted, subject to the applied displacements that were recorded during the crack initiation experiments. This resulted in the differences in NCOD such as those that are shown in Figures 3 and 4. The NCOD were brought into agreement in a separate analysis where the globally applied displacements and the difference in NCOD along the crack faces were applied at the same rate. In spite of the nonlinear material behavior and the coupling between crack face tractions and NCOD, this simple superposition resulted in agreement between measured and predicted NCOD without any iteration. The required tractions were then obtained from the reactions along the portions of the crack faces where NCOD had originally differed.

The tractions derived from Coulomb's law and the finite element analyses are compared in Fig. 7, in a plot of tractions vs. NCOD. The broken lines are from the finite element analysis for two cases where no contact had occurred (Fig. 6a,b) and one case where one contact had occurred (Fig. 6b). The results of a Coulomb analysis (7) with an equal charge density of 5×10^{-5} C/m^2 on the glass and epoxy crack faces are shown as the full line, which is quite close to the single contact case. The distribution for the zero contact cases in Fig. 7 could be matched by using a charge density of 3.4×10^{-5} C/m^2 (Liang, 1993). Such charge densities appear to be quite reasonable for polymers. For example, the typical charge density for a polyethylene/metal interface in air is 5×10^{-5} C/m^2 (Lowell and Rose-Innes, 1980). It is also interesting to note that the tractions, for a given NCOD, were higher following contact, suggesting that it had produced additional charging, as noted by Horn and Smith (1992). Of course, the association of the tractions with electrostatic effects in this study is indirect in the sense that no measurements of charging were made. Nonetheless, the NCOD measurements definitely indicate that long-range tractions were active along the crack faces, whatever their origin may have been.

The question then arises as to how much shielding was provided by the long-range tractions. The calculation was made by making use of a judicial choice of J-integral contours. With the J-integral defined as

$$J = \int_{\Gamma} (W dy - \mathbf{T} \cdot \frac{\partial \mathbf{u}}{\partial x} ds) \qquad (8)$$

the contribution associated with the tractions was obtained by choosing two contours; one outside the

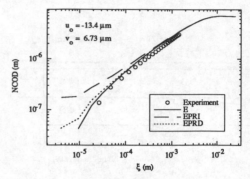

Fig. 4 Comparison of measured NCOD with elastic (E), rate-independent plasticity (EPRI) and rate-dependent plasticity (EPRD) predictions for negative shear loading

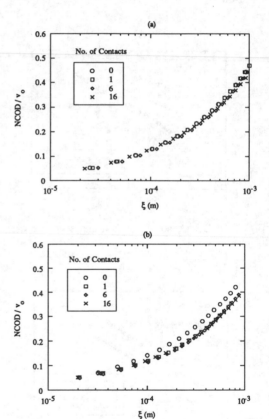

Fig. 5 NCOD/v_0 for various contact times in three series of experiments

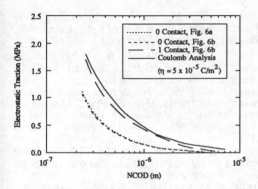

Fig. 6 Electrostatic traction as a function of NCOD

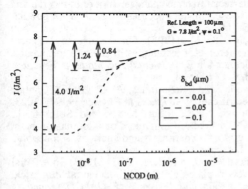

Fig. 7 The contribution of crack face tractions to the J-integral

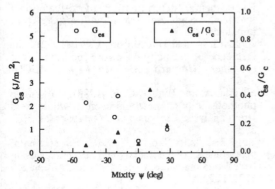

Fig. 8 Electrostatic shielding as a function of fracture mode-mix

close to the crack tip. The presence of nonzero tractions on the crack faces gives rise to a nonzero contribution to J which is the shielding provided by the tractions. Thus, the shielding value of J is

$$J_s = \int_b^a T_y(\xi) \frac{\partial \delta}{\partial \xi} d\xi \qquad (9)$$

where a is the distance from the crack where the tractions become zero for large NCOD (~10^{-5} m in Fig. 7) and b is a location close to the crack front where breakdown must occur due to charge neutralization. For convenience at this stage, we will consider that b is greater than any process zone, such as a Barenblatt cohesive zone that governs the intrinsic adhesion of the interface. For fixed applied displacements, the NCOD are some function of the distance from the crack front so that the variation of shielding within the shielding zone is

$$\tilde{J}_s(\delta) = \int_{\delta(\xi=a)}^{\delta} T_y(\hat{\delta}) d\hat{\delta} \qquad (10)$$

where δ is the NCOD at some point in $b \leq \xi \leq a$. The integration (10) was carried out for a bond-normal loading ($\psi = 0.1°$) with G=7.8 J/m². The separation law, $T_y(\delta)$, that was used was the zero contact case in Fig. 7 extrapolated to δ=0.1, 0.05, 0.01 μm, and then taken to be zero thereafter, effectively making δ=0.1, 0.05, 0.01 μm the respective NCOD at breakdown. The result is shown in Figure 8 where J drops from the global G level (7.8 J/m²) to values of 6.96, 6.56, 3.8 J/m² after charge neutralization. The shielding J_s (G_{es}) provided by the tractions is the difference between the two levels and can be seen in Fig. 8. The importance of determining the breakdown distance is clear as can be seen from the resulting difference in shielding. If the breakdown NCOD were greater than 0.05 μm, then there was not much sensitivity to them. However, breakdown NCOD of the order of 0.01 μm led to significant shielding levels. Clearly, higher resolution measurements of NCOD are required to resolve this issue.

The procedure that was used to generate the traction/separation laws (Fig. 7) was used to generate similar laws for a number of the crack initiation experiments. The differences between the rate-dependent plasticity analyses and the measured NCOD (such as those appearing in Fig. 3 and 4) gave rise to a series of traction/separation laws similar to the one shown in Figure 7. The distributions were function of mode-mix, with a negative mode-mix generating the highest tractions for a given NCOD, perhaps due to the sliding frictional contact that is generated near the crack front during a portion of the loading.

The corresponding shielding values were determined by applying equation (10) with a breakdown NCOD of 0.1 µm. The results exhibited a large degree of scatter when plotted as a function of mode-mix (Fig. 9). The normalized values of G_{ss} exhibited a gradual rise from negative shear to positive shear, which did not follow the U-shaped distribution of toughness vs. mode-mix (Liechti and Chai, 1992; Liang 1993). There could be several reasons for the scatter, such as variation in humidity and surface roughness (on the order of nanometers). These certainly call for a series of experiments where humidity is carefully controlled such as in vacuo. However, nanoscale fracture surface roughness may be hard to avoid and could cause problems if breakdown NCOD are of the same order. The choice of 0.1 µm as the NCOD at which breakdown occurred led to shielding levels that ranged from 10 to 40% and, as discussed earlier (Fig. 7), represent a lower bound. Thus if electrostatic effects were indeed contributing to the long-range cohesive tractions that were observed, they could provide a significant degree of shielding.

Although the measured NCOD and all predictions were in agreement far from the crack front, the measured NCOD consistently fell below even the linear analyses near the crack front. The lower NCOD were ascribed to the presence of long-range cohesive tractions acting on the crack faces, possibly due to electrostatic effects. The long-range tractions provided a larger degree of shielding than either plastic dissipation or asperity shielding for the conditions and materials that applied in this study. Closer control of environmental effects are expected to reduce the scatter in the examination of electrostatic effects while the development of a separation law for the glass epoxy interphase region will allow the viscoplastic dissipation associated with initiation be extracted in a more rigorous fashion.

ACKNOWLEDGEMENTS

The authors would like to acknowledge the financial support that was provided by the Office of Naval Research (N00014-90-J-4024) and the National Science Foundation (MSS-9201929). Thanks also go to Ms K. Hinders for type setting.

REFERENCES

Cao, H. C. and Evans, A. G. (1989). An experimental study of the fracture resistance of bimaterial interfaces. *Mechanics of Materials* 7, 295-304.

Evans, A. G. and Hutchinson, J. W. (1989). Effects of non-planarity on the mixed mode fracture resistance of bimaterial interfaces. *Acta Metall. Mater.* 37, 909-916

Fowlkes, C. W. (1975). Crack opening interferometry - the effects of optical refraction. *Engineering Fracture Mechanics* 7, 689-692.

Horn, R. G. and Smith, D. T. (1992). Contact electrification and adhesion between dissimilar materials. *Science* 256, 362-364.

Liang, Y.-M. (1993). Toughening mechanisms in mixed-mode interfacial fracture. Ph.D. Dissertation and *Engineering Mechanics Research Laboratory Report EMRL #93/12,* , The University of Texas at Austin.

Liechti, K. M. and Hanson, E. C. (1988). Nonlinear effects in mixed-mode interfacial delaminations. *Int. J. Fracture* 36, 199-217.

Liechti, K. M. and Chai, Y.-S. (1992). Asymmetric shielding in interfacial fracture under in-plane shear. *J. Appl. Mech.* 59, 295-304.

Liechti, K. M. and Liang, Y.-M. (1992). The interfacial fracture characteristics of bimaterial and sandwich blister specimens. *Int. J. Fracture* 55, 95-114.

Liechti, K. M. (1993). On the use of classical interferometry techniques in fracture mechanics. Chapter 4 in *Experimental Techniques in Fracture,* ed. J. S. Epstein.

Lowell, J. and Rose-Innes, A. C. (1980). Contact electrification. *Advances in Physics* 29, 947-1023.

O'Dowd, N. P., Stout, M. G. and Shih, C. F. (1991). Fracture toughness of alumina- niobium interfaces: experiments and analyses. *Philosophic Magazine A* 66, 1037-1064.

Rosakis, A. J. and Lee, Y. J. (1993) Interfacial cracks in plates : a three-dimensional numerical investigation. *Cal. Inst. of Tech. GALCIT report.*

Thouless, M. D. (1990). Fracture of a model interface under mixed-mode loading. *Acta. Metall. Mater.* 38, 1135-1140.

Thurston, M. E. and Zehnder, A. T. (1993). Experimental determination of silica/copper interfacial toughness. *Acta Metall. Mater.* To be published.

Tvergaard, V. and Hutchinson, J. W. (1992). The influence of plasticity on mixed mode interface toughness. *Harvard University Report* MECH-193.

Voloshin, A. and Burger, C. P. (1983). Half fringe photoelasticity : a new approach to whole field stress analysis. *Experimental Mechanics* 23, 304-313.

Wang, J.-S. and Suo, Z. (1990). Experimental determination of interfacial toughness curves using Brazil-nut-sandwiches. *Acta. Metall. Mater.* 38, 1279-1290.

Zimmerman, K. A., Langford, S. C. and Dickinson, J. T. (1991). Electrical transients during interfacial debonding and pullout of a metal rod from an epoxy matrix. *J. Appl. Phys.* 70, 4808-4815.

Recent Advances in Experimental Mechanics, Silva Gomes et al. (eds) © 1994 Balkema, Rotterdam, ISBN 90 5410 395 7

Impact resistance of bonded joints

F.Cayssials, J.L.Lataillade, D.Crapotte & C.Keisler
LAMEF (LAboratoire Matériaux Endommagement Fiabilité), ENSAM-CER de Bordeaux, Talence, France

ABSTRACT: The introduction of adhesives in the car industry will be dependent gradually on the increase of experience and confidence in such a method. The behaviour of adhesive joints under high strain rate needs to be clearly understood.
The technique of tensile Hopkinson bar is well adapted to determine the energie release rate Gc as well as the rupture strength of adhesive joints at high strain rates (10^3 to 10^4 s^{-1}). The results show that Gc is highly sensitive to the loading rate , the failure modes and the low temperatures. In the same way, this investigation has highlighted the effect of loading rate and the influence of mechanical and surface characteristics on the joint strength.
A new experimental set-up (the inertia wheel) has been studied and proved to be useful to finely define a bonded joint.

1 INTRODUCTION

The impact behaviour of a new, lightweight bonded structure requires careful examination. Adhesives being polymeric materials, have properties that are in general rate dependant. There is, therefore a need to give designers some confidence that under impact conditions, bonded joints will maintain their structural integrity. Unfortunatly, little amount of work has been carried out to establish properties of bonded joints at high strain rate. It is in this context that the tensile Hopkinson bar apparatus were used to assess, at high loading rate (10^3 to 10^4 s^{-1}), the energie release rate of a rubber modified epoxy/steel sheet joint.and the shear strength of an epoxy joint bonded with different steel sheets. At the present time, further tests are being carried out on a new experimental set up (the inertia wheel) based on the inertia technique, in order to characterize the impact resistance of bonded joints.

The inherent characteristics of the two instrumented impact test, the tensile Hopkinson bar apparatus and the inertia wheel will be analysed and compared through these results.

2 THE FRACTURE MECHANICAL APPROACH AND THE TENSILE HOPKINSON BARS.

Since the strength of adhesive joints is governed by the presence of flaws, the application of the fracture mechanics to adhesive joint failure is well justified.

The particularity of the study undertaken by D.Grapotte and J.L.Lataillade (Grapotte, Lataillade 1992) was not only to initiate an interfaciale crack in the specimen but also to use the Hopkinson tensile bars technique. This technique allows us to reach high loading rates ($\dot{G}c=10^5$ Kj.m^2.s^{-1}) while knowing perfectly the history of the loading.

2.1 *Adhesive joint specimen*

The specimen was made of two steel sheet squares (40X40 mm^2) bonded with a central slip of epoxy resin (10X40 mm^2). A PTFE-covered adhesive film was stuck on one plate in order to create a defect of known initial lenght (Fig.1). The hardening of the adhesive was obtained by curing (the temperature was raised from 20°C to 150°C for 30mn., maintained at 150°C for 60mn.and then brought down to 20°C in 120 mn.). The substrate used was a 0.67 mm thick steel sheet, simply degreased using trichlorothylene. The specimen was then fixed onto the supports with a cyanocrylate adhesive and subjected to different modes (Fig.2).

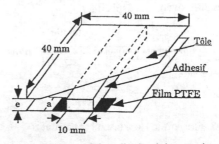

Figure 1: Schema of the adhesive joint specimen.

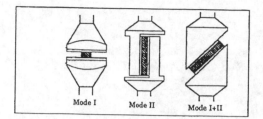

Figure 2: Specimen and the three different modes.

2.2 *Experimental devices*

The dynamic tests were carried out on a tensile Hopkinson bar apparatus; the principle is based on the theory of one dimensional elastic wave propagation.

The projectile is propelled by an air gun against the input bar's anvil. A stress wave is generated in the input bar; the presence of the specimen induces an impedance mismatch so that incident wave is transformed into a reflected wave and a transmitted wave (Fig.3).

The waves are recorded by two strain gauge bridges and stored in a DRAM numerical oscilloscope (Lecroy 9400). A Zimmer OHG optical extensometer is used to follow the relative displacement $\Delta(t)$ of the two substrates. For the low temperature tests, a nitrogen cooling cabinet is set up. According to the "classic Hopkinson analysis" The force $F(t)$ acting in the specimen is deduced from the transmitted wave $\varepsilon_T(t)$:

$$F(t) = A_B E_B \varepsilon_T(t)$$

$$U = \int_0^{Tf} F(t)\, d\Delta(t) \quad (1)$$

A_B bar section, E_B bar modulus,
U fracture energy, Tf time of fracture

Two different loading conditions have been defined as following:

Dyn 1: impact speed $\approx$ 8m/s.

Dyn 2: impact speed $\approx$ 12m/s.

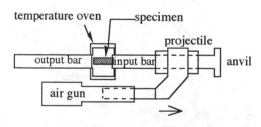

Figure 3: The tensile Hopkinson bar apparatus

2.3 *The derivation of the fracture energy release rate, Gc*

The method used to determine the fracture energy release rate is the Williams and Kinloch 's method (Williams 1984):

The curve U (the fracture energy) versus the initial crack length (a) is plotted. As soon as the curve exibits a linear zone, it is possible to determine the fracture energy release rate Gc by mesuring the slope of the line (Fig. 4). U has been calculated experimentally for six different values of "a" (5, 7.5, 10, 12.5, 15 and 20 mm) according to the relation (1).

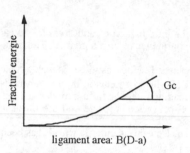

Figure 4: Typical curve U versus B(D-a)

2.4 *Results*

In addition to the dynamic results, some static results (500 mm/mn) obtained on a conventional tensile machine (Instron 1186) are presented (Table 1 et 2).

Like polymers, Gc is highly sensitive to the loading rate, the failure mode and the low temperatures. Whatever the loading rate or the temperature may be, the most brittle failure mode is the mode I+II.

The tensile Hopkinson bar apparatus is well adapted to the fracture mechanic of bonded joints at high loading rate ($\dot{G}c=10^5$ Kj.m^2.s^{-1}). However, the dispersion of the fracture energy values are quite important and the method used to determine the energy release rate Gc requires numerous tests. But since Gc is calculated by linear regression from the curve U versus B(D-a), no important dispersion of Gc arise.

Table 1: Results of Gc (Kj.m^2) & $\dot{G}$c (Kj.m^2.s^{-1}) for different loading rates.

Mode I		Mode II		Mode I+II	
Gc	$\dot{G}$c	Gc	$\dot{G}$c	Gc	$\dot{G}$c
0.41	2.9	5.6	25.4	0.75	4.1
4.6	$1.1\ 10^5$	7.7	$1.9\ 10^5$	2.7	$6.3\ 10^4$
4.1	$1.2\ 10^5$	10.8	$3.3\ 10^5$	4.7	$1.3\ 10^5$

Table 2: Results of Gc (Kj.m^2) & $\dot{G}c$ (Kj.m^2.s^{-1}) for different temperatures.

	Mode I		Mode II		Mode I+II	
	Gc	$\dot{G}c$	Gc	$\dot{G}c$	Gc	$\dot{G}c$
20°C	4.6	1.1 10^5	7.7	1.9 10^5	2.7	6.3 10^4
0°C	2.7	6.5 10^4	4.7	1.1 10^5	2.1	4.9 10^4
-20°C	2.3	5.0 10^4	4.3	9.7 10^4	1.7	3.9 10^4
-40°C	2.2	5.8 10^4	4.1	9.1 10^4	1.6	3.5 10^4

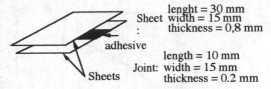

Sheet : lenght = 30 mm
width = 15 mm
thickness = 0,8 mm

adhesive

Joint : length = 10 mm
width = 15 mm
thickness = 0.2 mm

Sheets

Figure 5: the adhesive joint specimen.

2.5 Conclusion

The tensile Hopkinson bar apparatus is well adapted to the fracture mechanic of bonded joints at high loading rates and different temperatures. The dispersion of the fracture energy values is likely to arise from the actual treatment of the Hopkinson waves. Since the fracture time of the adhesive joint is very short and the geometry of the supports is complex, some initial hypothesis inherent to the Hopkinson bar treatment might not be verified. In particular, the equilibrium of the forces at the interfaces, the propagation of an uniaxial wave through the specimen (Hauser 1965).

3 EFFECT OF THE SUBSTRATE ON THE ADHESION IN ADHESIVE BONDS LOADING AT HIGH STRAIN RATE

Since the steel sheets used in the car industry present different surface properties, it is necessary to make sure that the adhesive will be compatible with the substrate. The adherent's nature, the surface properties, the selected treatment become the main parameters to evaluate impact resistance of adhesive joints.

A study undertaken by C. Keisler and J.L.Lataillade (Keisler, Lataillade 1992) was devoted to the evaluation of the strength of adhesive joints (with different steels) in order to identify the parameters likely to influence joints properties when fractured .

To reach a characteriscal value of adhesion, the impact resistance of bonded joints has been determined in shearing by the Hopkinson bar technique (Fig. 3).

3.1 Adhesive joint specimen

The specimen was made of two steel sheets bonded with an epoxy resin filled with mineral particles (Fig.5). The thickness of the joint was either 0.2 mm or 0.5 mm. The hardening of the adhesive was obtained by curing (30 mn at 180°C). Different steels were used, each having its own mechanical and surface properties. The sheets were degreased with

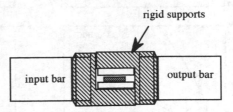

rigid supports

input bar   output bar

Figure 6: Rigid supports.

acetate ethyl. Then, the specimen is bonded on rigid supports with a cyanocrylate adhesive, the whole thing being inserted between the two bars of the apparatus (Fig. 6).

3.2 Experimental set up

The Hokinson bar technique has previously been presented (paragraph 2.2).

The shear stress and the strain rate are related according to the following law:

$$\tau_{rup} = A \, \dot{\gamma}^{\alpha}$$

3.3 Results

The different surface properties of A, B, C and D substrates are given in the table 3; Ra represents the average arithmetical roughness and the Rt factor, the difference between the highest and lowest points of the topography.

As Kinloch and Vallat have already suggested (Kinloch 1987), the results of this experience reveal that the strength of the adhesive joints goes with two parameters: one linked to the interfacial properties and the second to the energy dissipation in the joint. Roughness is a parameter likely to modify adherence (Fig.7, 8).

Table 3: 3D roughness parameters.

Acier	A	B	C	D
	XES	Solphor	IFHR	Soldur
Ra (μm)	1.674	1.396	1.121	1.897
Rt (μm)	13.397	9.257	8.945	11.716

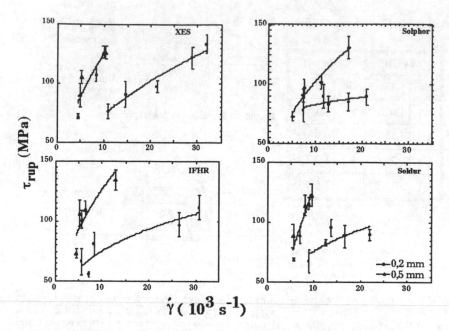

Fig. 7: Variation of rupture strength versus strain rate for the A, B, C and D steels

Table 4: The coefficients of the equations $\tau_{rup} = A \, \dot{\gamma}^{\alpha}$

Acier	thickness	A	α	R
XES	0.2 mm	23.6	0.49	0.98
	0.5 mm	40.1	0.49	0.86
IFHR	0.2 mm	36	0.31	0.92
	0.5 mm	44.8	0.45	0.87
Solphor	0.2 mm	65.3	0.10	0.71
	0.5 mm	36.7	0.44	0.97
Soldur	0.2 mm	38.2	0.30	0.72
	0.5 mm	17.3	0.87	0.92

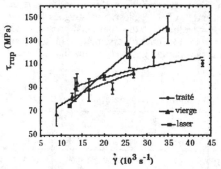

Fig. 8: Influence of D sheet surface state on the joint behaviour.

3.4 Conclusion

From the tensile Hopkinson bar apparatus, the experimental methodology set up proved to be useful to characterize in shearing, a single lap joint.at high strain rate. It is very possible, with this technique to apprehend the interface metal/resin behaviour.

A correlation between the interface microstructure and the fracture joint behaviour was established.

The remarks, already noted in the first study and concerning the experimental method could be emphasized here as well (paragraph 2.5). Somehow, since only the rupture strength was assessed no important disparite in the results arise. The principle difficulty was to keep the same loading rate range for the two different geometries of the joint (Fig. 7).

4 STUDY OF THE IMPACT RESISTANCE OF A BONDED JOINT WITH THE INERTIA WHEEL.

The conception of bonded structure in the car industry must fulfill some demanding criteria linked to performance and reliability. What is the role of parameters of the interface in the behaviour of bonded joint under impact load and its durability ?

In order to answer this question and finally to formulate a prediction model of the behaviour of bonded joints, a new experimental set up (the inertia wheel) has been studied and compared to the tensile Hopkinson bar apparatus by shearing an epoxy bonded with two galvanized steel sheets (of laser roughness type).

4.1 Adhesive joint specimen

The specimen geometry used is the former specimen geometry (paragraph 3.1). The thickness of the joint is 0.2 mm. the two adhesive joints tested are: a flexibilized epoxy (XEP91717) bonded with two galvanised steel sheet (soldur), and a brittle epoxy (NA 84) bonded with two steel sheet no galvanised (soldur).

4.2 Experimental set up

This new experimental set up based on the inertia technique is very suitable; the velocity of the wheel can be adjusted in a very precise manner (2 m/s to 25 m/s). This dynamical tensile machine is made up of a wheel which is able to accumulate a high kinetic energy. At the lower speed, That energy is already five times more important than the required energy to break the joint. When the wheel reaches the right velocity, the specimen positions itself just before a hammer thanks to a pendular mechanism. The pendulum rotation is due to a pneumatic jack with an adjustable pressure (Lopez 1992) (Fig. 8).

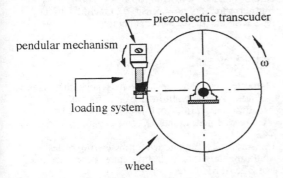

Figure 8: The inertia wheel

4.3 The measuring system

A charge amplifier converts the electric charge suppied by the piezoelectric transcuder into a voltage signal proportional to a force. The signal is stored in a DRAM numerical oscilloscope (LECROY 9400). A Zimmer OHG is also used to follow the relative displacement of the two substrates.

The high energy accumulated by the wheel allows us to get a strain rate nearly constant throughout the test, in contrast to the tensile Hopkinson bar technique. This fact has been confirmed by the direct displacement measurement (Fig.9). So, the deformation can readily be obtained by the following equation:

$$\overline{\gamma_r} = \frac{V_{wheel} \cdot t_f}{h}$$

γ_r, the mean yield strain
V_{wheel}, velocity of the wheel
t_f, the failure time
h, the thickness of the specimen

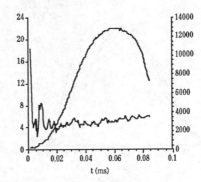

Figure 9: the force and the displacement speed versus time for a test at 4m/s with a XEP 91717.

4.4 Results`

the results obtained on the inertia wheel with the two adhesives are given in table 5.

Table 5: Some results obtained with the two adhesive joints tested on the inertia wheel. S is the standard deviation.

	$\dot{\gamma}$ $(10^3 s^{-1})$	τ_r (Mpa)	Er (kj/m^2)	γ_r (%)
XEP galvanised steel	20	83 s= 7	6 s = 1.1	45 s = 7.2
	35	110 s = 11	10 s= 1.6	40 s = 2.5
NA84 simple steel	20	113	1 s = 0.07	10 s = 1

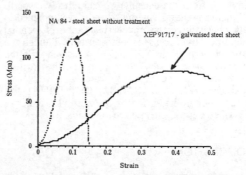

Figure 10: Behaviour of the XEP91717 and NA84 at 4 m/s

1299

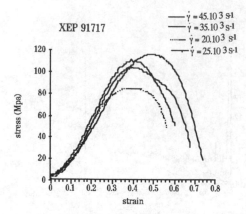

Figure 11: Behaviour of the XEP 91717 at different strain rates.

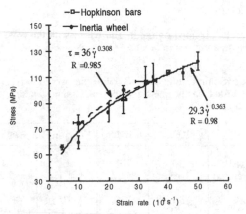

Figure 12: The rupture stress of the XEP 91717 versus the strain rate.

The yield strains of the XEP 91717 are very important whilst the ones of the NA 84 are those expected (Fig.10 and 11). The reasons are not yet clearly understood, but some new experiments are in preparation to attempt to further our understanding.

Anyway, from these results we can already come to the conclusion that the inertia wheel is well adapted to finely characterize the impact resistance of bonded joints, in contrast to the Hopkinson bars. Though the rupture strength of the XEP 91717 measured with the two apparatus are very close (Fig.12), the fracture energy can not be precisely calculated from the Hopkinson bar technique (Fig.13). On the other hand, the dispersion in the the inertia wheel results, either in terms of energy or strength are low (table 5).

5 CONCLUSION

From both devices, different experimental methods have been developed in order to characterize the impact resistance of bonded joints. Both have their advantages. The Hopkinson bars technique has proved to be useful to assess the energy release rate Gc at different temperatures as well as the rupture strength. But since the fracture time of bonded joints are very short and the specimen geometry complex, the inertia wheel is much more adequate to finely define bonded joints in term of fracture energy. Furthermore, it is now possible to realize temperature tests with the inertia wheel apparatus.

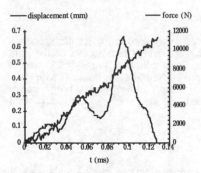

Figure 13: The stress and the displacement versus time, measured on the Hopkinson bars with the XEP .

REFERENCES

Grapotte, Lataillade 1991. Mechanical characterization of adhesive joints. Influence of the rate of loading. Sixth Int. Conf. Mech. Beh. of Mat. - Congrès International, Kyoto, Japan, ICM6, 437-442.

Hauser, 1965. Techniques for measuring Stress-Strain relations at high strain rates. Paper presented at SESA, 395-399.

Keisler, Lataillade, Charbonnet 1992. Effect of the substrate on the adhesion in adhesive bonds loading at high strain rate. Euradh'92, Karsluhe, Germany, 584-589.

Kinloch, Chapman and Hall 1987. Adhesion and adhesives, science and technology. Chap.III.

Lopez, mémoire, 1992. Conception et réalisation d'une machine à choc pour l'étude de la déchirure à grande vitesse de tôles composites", CNAM.

Williams, 1984. Fracture. Mechanics of polymers. Ellis Horwood Limited

ACKNOWLEDGEMENTS

The authors would like to thank SOLLAC, RENAULT, PEUGEOT CECA, the Ministry of Industry supporting this research.program and the "Agence Nationale de la Recherche et de la Technologie" for the grant associated to the Ph.D Thesis of F. Cayssials, as well CNRS and Region Aquitaine for their financial support connected with the doctorate of D. Grapotte.

Recent Advances in Experimental Mechanics, Silva Gomes et al. (eds) © 1994 Balkema, Rotterdam, ISBN 90 5410 395 7

Luminance measurement to evaluate the fatigue damage of notched FRP plates

H. Hyakutake & T. Yamamoto
Fukuoka University, Japan

ABSTRACT: Fatigue tests were carried out on notched specimens of a glass–fiber/epoxy laminate for a wide range of notch–root radii and stress amplitudes. Our attention was focussed on the fatigue–damage zone near the notch root. The process of fatigue damage initiation and growth was studied by measuring the luminance distributions near the notch root. The experiment shows that the number of cycles to failure is determined by both the maximum elastic stress at the notch root and the notch–root radius. On the basis of the concept of severity near the notch root, the experimental results can be clarified. Applying the fatigue failure criterion derived here, we can make an accurate estimate of the fatigue life of notched specimens of FRP.

1 INTRODUCTION

Because of their importance in design applications, the fatigue fracture for notched specimen of fiber–reinforced plastics has been the subject of considerable works. The attention in these studies is mostly on revealing the fatigue damage and the mechanism of fatigue damage development at the notch root.

Our goal is to elucidate the fracture behavior of FRP containing stress concentrations in various notch geometries and to develop a limiting condition for predicting the fatigue strength of notched bars of FRP.

Studying elastoplastic stress near the notch root, we have obtained a fracture criterion for notched bars under static load (Nisitani & Hyakutake 1985). The criterion is based on the concept of severity near the notch root. Several experiments have shown that the criterion is applicable to notched FRP plates over a wide range of notch geometries and dimensions of specimens (Hyakutake et al. 1990).

The aims of the present research are to provide experimental evidence of the validity of the fracture criterion based on the severity near the notch root for the fatigue failure of notched FRP plates. This is accomplished by obtaining experimental data of fatigue tests of pulsating tension on a glass–fiber/epoxy laminate containing notches for a wide range of notch–root radii and stress amplitudes.

To evaluate the fatigue–damage near the notch root, we measured the luminance distributions with CCD camera.

2 EXPERIMENTAL PROCEDURE

The material used was a commercial sheet of FRP, which was a glass–fiber/epoxy laminate. The FRP sheet is designated Japanese Industry Standards (JIS) K 6912, and the material symbol is EL–GEM. The dimensions of the FRP sheet were 1 m width, 2 m length, and 2 mm thickness. Photomicrographs of a transverse section of the FRP sheet are shown in Fig. 1. There are eight layers of glass cloth, and the diameter of the glass fiber is about 0.01 mm. All specimens were cut from the sheet, so that the principal direction of specimen coincided with longitudinal direction of the sheet.

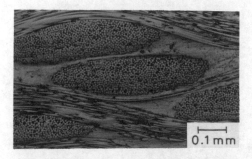

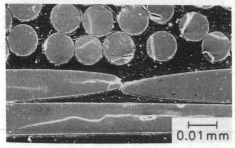

Fig. 1. Photomicrographs showing a transverse section of the FRP sheet (JIS: EL–GEM).

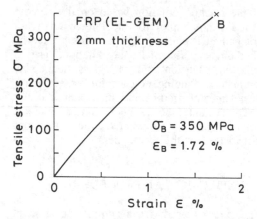

Fig. 2. Tensile stress–strain curve of the FRP sheet.

Figure 2 shows the tensile stress–strain curve of the smooth specimen, and the values of stress and strain at fracture (point B). The specimen fails in a brittle manner.

The shapes and dimensions of specimens of the notched plate are shown in Fig. 3. The specimens having a width of 20 mm were notched in a U–shape on both sides at the midpoint of their length. The

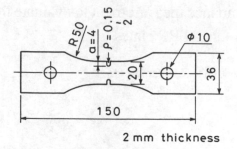

Fig. 3. Test–specimen dimensions (mm).

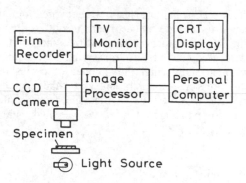

Fig. 4. Luminance–measuring system.

notch–root radius ρ had the following four values: 0.15, 0.5, 1 and 2 mm. The notch depth was 4 mm in all specimens. The geometrical stress concentration factor K_t was taken from the calculations of Neuber. The range of the value of K_t is 2.21 to 6.83.

Fatigue tests of pulsating tension were made on a servohydraulic material test system at frequency 5 Hz with an R value of 0.1.

To evaluate the fatigue damage, we measured successively the luminance distributions near the notch root during fatigue testing. The luminance–measuring system with CCD camera is shown in Fig. 4.

3   RESULTS AND DISCUSSION

Figure 5 shows the S–N curve for notched FRP plates in fatigue tests of pulsating tension. The coordinate σ_n is the maximum nominal stress and N_f is the number of cycles to fracture. It can be seen from Fig. 5 that there is little effect of the notch–root radius ρ on the fatigue life N_f. It seems likely that the notch sensitivity decreases in fatigue tests of notched FRP plates (Maier et al. 1987).

An example of the process of fatigue damage growth on the surface of the specimen is shown in Fig. 6. Fatigue damage appeared near the notch root in an early stage in all specimens; it occurs within the first 10 to 20 % of the fatigue life.

Most of the fatigue life is therefore the process of crack growth. This is the reason why there is little effect of the notch–root radius ρ on the fatigue life N_f, as shown in Fig. 5. On the other hand, the experiment shows that the number of cycles to fatigue damage initiation is determined predominantly by the notch–root radius ρ .

Attempts to determine the fatigue damage for fiber–reinforced composite materials have used, for example, the delaminated area near the notch root (Jen et al. 1993) and X–ray radiographs (Spearing et al. 1991). We observed the decrease of the luminance near the notch root. It is evident that the decrease of luminance near the notch root was associated with irreversible damage and microfracture of composite. Figure 7 shows an example of the change of luminance distributions near the notch root during fatigue testing. The specimen shown in Fig. 7 is the same one in Fig. 6. The striped patterns with light and shade correspond to the value of luminance. The area of fatigue–damage zone increased with increasing number of loading cycles, as shown in Fig. 7.

Figure 8 shows the growth of the area of fatigue–damage zone of the specimen shown in Fig. 7. The relative luminance (R.L.) is the ratio of the luminance at a number of loading cycles to the luminance before testing. It is evident that the fatigue damage accumulated severely at the region where the value of R.L. is small. Figure 9 shows the growth curves of the area of fatigue–damage zone where the value of R.L. is 70 %. There is the point of the rapid increase of the area of fatigue–damage zone in all specimens.

Closer observation near the notch root was made by means of a scanning electron microscope. When the area of fatigue–damage zone where the value of R.L. is 70 % increased rapidly, several fine cracks appeared near the notch root. Figure 10 shows the cracks on the surface of specimen near the notch root.

From the experimental results mentioned above, we can determine the point of fatigue damage initiation N_d, which is the number of loading cycles when the growth curve of the area of fatigue–damage zone where the value of R.L. is 70 % increased rapidly. It should be noted that the value of N_d is determined by the maximum nominal stress σ_n and notch–root radius ρ , as shown in Fig. 9.

The severity near the notch of notched bar is determined by both the maximum elastic stress σ_{max} and the notch–root radius ρ (Nisitani & Hyakutake 1985). It is suggested that the elastoplastic stress distributions near the notch root after small–scale yielding are the same in all specimens, for which both the maximum elastic stress and the notch–root radius are equal in all cases.

On the basis the evidences mentioned above, the fracture criterion for a notched bar is expressed as (Nisitani & Hyakutake 1985)

$$\sigma_{max} = \sigma_{max,c}(\rho) \qquad (1)$$

where σ_{max} is the maximum elastic stress at fracture and is determined as the product of the nominal stress σ_n and the geometrical stress concentration factor K_t. The parameter $\sigma_{max,c}$ on the right–hand side of Eq. (1) is the material constant, which is governed by the notch–root radius ρ only and is independent of notch geometry and specimen size. The validity of the fracture criterion Eq. (1) for

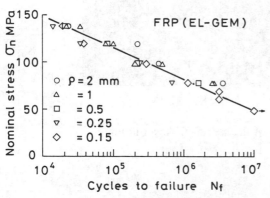

Fig. 5. S–N curve for notched FRP plates.

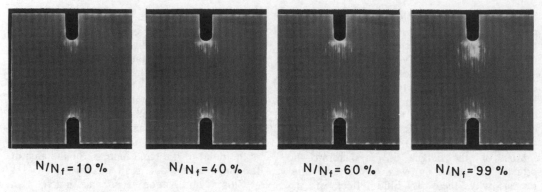

N/N$_f$ = 10 %   N/N$_f$ = 40 %   N/N$_f$ = 60 %   N/N$_f$ = 99 %

Fig. 6. Fatigue damage on the surface of specimen near the notch root (notch–root radius ρ = 1 mm, σ_n = 98.1 MPa, N_f = 5.08 $\times$ 10^5).

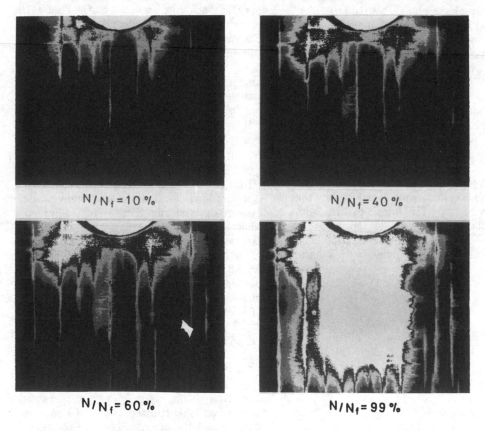

N/N$_f$ = 10 %   N/N$_f$ = 40 %

N/N$_f$ = 60 %   N/N$_f$ = 99 %

Fig. 7. Luminance distributions near the notch root of the specimen shown in Fig. 6.

notched FRP plates was confirmed previously (Hyakutake et al. 1990). In view of the concept of severity mentioned above, it is appropriate to discuss the fatigue failure criterion for notched bars in terms of a combination of the maximum elastic stress σ_{max}, the notch–root radius ρ and the number of cycles to fatigue damage initiation N_d (Hyakutake et al. 1993). It is reasonable to assume that the fatigue failure criterion for a notched bar in cyclic load is expressed as

$$\sigma_{max} \cdot (N_d)^m = C(\rho) \qquad (2)$$

where m is the material constant. The parameter C on the right–hand side of Eq. (2) is the material constant, which is governed by the notch–root radius ρ only and is independent of notch geometry and specimen size. For $N_d = 1$ under static load, Eq. (2) reduced to Eq. (1).

Figure 11 shows the relationship between σ_{max} and N_d for a constant notch–root radius ρ. There are four kinds of ρ. As seen in Fig. 11, each experimental point fell in close proximity to a characteristic straight line for which ρ is constant and the four characteristic straight lines are parallel. From the experimental results mentioned above, we will confirm the validity of the fatigue failure criterion of Eq. (2) derived from the concept of severity near the notch root. The value of material constant m is 0.12 for the FRP used in our research.

It is likely that the parameter C on the right–hand side of Eq. (2) corresponds to the area of fatigue–damage zone near the notch root. Figure 12 shows the relationship between the area of fatigue–damage zone where R.L. = 75 % at N_d and the maximum elastic stress σ_{max}. It can be seen that the area of fatigue–damage zone is governed predominantly by the notch–root radius ρ and is independent of the maximum elastic stress σ_{max}.

It is evident from the present studies that for any specimen, the number of cycles to fatigue damage initiation N_d can be determined. The value of N_d is determined by both the maximum elastic stress σ_{max} ($= K_t \cdot \sigma_n$) and the notch–root radius ρ, as can be seen from Fig. 11. It should be noted that the value of N_d is from 10 to 20 % of the fatigue life N_f.

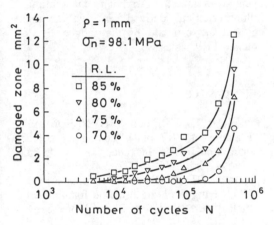

Fig. 8. Area of fatigue–damage zone versus the number of cycles.

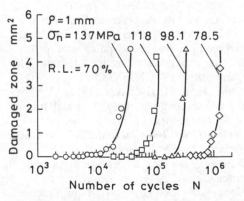

Fig. 9. Growth of the area of fatigue–damage zone where R.L. = 70%.

Fig. 10. Cracks near the notch root at N_d (notch–root radius ρ = 1 mm, σ_n= 98.1 MPa).

4 CONCLUSIONS

The fatigue fracture behavior of notched FRP plates was studied over a wide range of notch–root radii and stress amplitudes. The luminance distribution near the notch root was measured successively

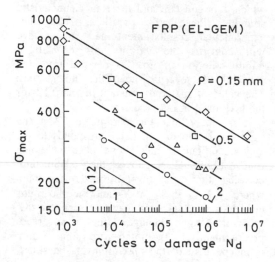

Fig. 11. Effect of the notch–root radius ρ on the relationship between the maximum elastic stress σ_{max} and the number of cycles to fatigue damage initiation N_d.

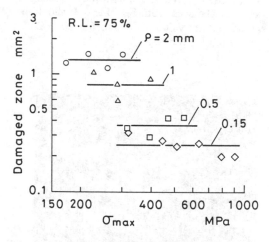

Fig. 12. The maximum elastic stress σ_{max} versus the area of fatigue–damage zone where R.L. = 75 % at N_d.

during fatigue testing to evaluate the fatigue damage.

It was found that fine cracks appeared at the notch root at an early stage: it occurs within the first 10 to 20 % of the fatigue life. The luminance decreased rapidly when the cracks were initiated near the notch root. We determined the point of fatigue damage initiation, which was the number of loading cycles when the area of fatigue–damage zone near the notch root increased rapidly.

On the basis of the concept of severity near the notch root, the experimental results can be clarified and a fatigue failure criterion is determined in terms of a combination of the maximum elastic stress, notch–root radius and the number of cycles to fatigue damage initiation. Applying the fatigue failure criterion derived here, we can make an accurate estimate of the fatigue life for notched FRP plates.

REFERENCES

Hyakutake, H., T. Hagio & H. Nisitani 1990. Fracture of FRP plates containing notches or a circular hole under tension. Int. J. Pres. Ves. & Piping 44–3: 277–290.

Hyakutake, H., T. Hagio & T. Yamamoto 1993. Fatigue failure criterion for notched FRP plates. JSME Int. J., Ser.A 36–2: 215–219.

Jen, M.H.R., Y.S. Kau & J.M. Hsu 1993. Initiation and propagation of delamination in a centrally notched composite laminate. J. Compos. Mater. 27–3: 272–302.

Maier, G., H. Ott, A. Protzner & B. Protz 1987. Notch sensitivity of multidirectional carbon fibre–reinforced polyimides in fatigue loading as a function of stress ratio. Composites. 18–5: 375–380.

Nisitani, H. & H. Hyakutake 1985. Condition for determining the static yield and fracture of a polycarbonate plate specimen with notches. Eng. Fract. Mech. 22–3: 359–368.

Spearing, M., P.W.R. Beaumont & M.F. Ashby 1991. Fatigue damage mechanics of notched graphite–epoxy laminate. Composite Materials: Fatigue and Fracture. T.K. O'Brien(ed.). ASTM STP 1110, p.617–637. Philadelphia, ASTM.

Recent Advances in Experimental Mechanics, Silva Gomes et al. (eds) © 1994 Balkema, Rotterdam, ISBN 90 5410 395 7

Fracture phenomena and strain distribution at the vicinity of crack tip in short fibre composites

Leszek Gołaski
Kielce Technological University, Poland

Jerzy Schmidt
Foundry Research Institute, Cracow, Poland

Małgorzata Kujawińska & Leszek Sałbut
Warsaw University of Technology, Optical Engineering Division, Poland

ABSTRACT: A study on the failure process of short glass fibre reinforced phenolic resin composite is presented. A fracture mechanics approach has been adopted and Mode I tests have been carried out. The damage accumulation was analyzed during loading. For these purposes acoustic emission with amplitude analysis and high-sensivity grating /moire/ interferometry have been applied. Three groups of acoustic events with different amplitude range have been distinquished. The crack opening was measured by the clip gauge technics. The effect of the specimen orientation /T-L, L-T/ has been investigated. The initial investigation to form the hybrid experimental technique applying simultaneously the grating interferometry for full-field displacement and strain analysis and acoustic emission for critical fracture points determination is described for characterizing the fracture mechanics in short fibre composites.

1. INTRODUCTION

Short fibre reinforced plastics are a new generation of materials, which exhibit high strength accompanied by low plasticity. For these materials fracture toughness is one of the most important mechanical properties. However, the fracture toughness determination meet a number of problems which are not encountered when testing metals. These problems result from the structure of composite prepared by injection moulding technology. During injection the fibers in the skin layers are situated near in parallel to the direction of mould filling while in the core of sample the fibres are aligned near perpendicularly to the filling direction. Thus the short fibre composites exhibit a layered structure and the skin to core layer thicknesses ratio depends on the thickness of the composite sheet.

The main problem in fracture testing of short fibre reinforced composites concerns the identification. The linear load displacement characteristic with violent fracture where the beginning of fracture can be found under the maximum load appear in selected samples only. For most of them there is some nonlinearity which appears as a result of different failure processes which occur at the vicinity of the crack tip and the real initiation load is not defined exactly.

None of these problems are solved satisfactorily and up till now there exist no standard method for fracture tests on short fibre composites. The protocol for fracture testing of these materials is under development by ESIS Task Group on Polymers and Composites. The aim of this paper was to undertake an investigation into the influence of crack orientation on fracture toughness and to study the failure processes of short fibre reinforced composites.

2. MATERIAL AND SPECIMEN GEOMETRY

The composite material employed was a phenolic short glass fibre reinforced composite IXEF 1022 supplied by Solvay-Belgium.The samples were machined from 5 mm thick sheets prepared by injection moulding technology. Compact specimens were machined with L-T and T-L crack orientation. The letter L denotes the longitudinal direction i.e. the mould filling direction while the letter T the transverse direction. The first letter denotes the notch direction while the second one the applied load direction.

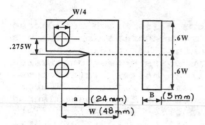

Fig.1. Compact tension configuration /CT/

3. EXPERIMENTAL PROCEDURE

There is no standardized method for fracture testing of short fibre composites and the tests were performed in agreement with "Testing protocol to characterize the toughness of plastics" developed by Williams /1991/ and "Appendix 2" to this protocol by Moore /1992/.

The critical load at crack growth initiation which was used for fracture toughness estimation was taken in accordance with the approach by Williams /1991/. For samples with linear load-displacement response up to fracture the maximum load is taken as a critical one.In the case of some nonlinearity just before the crack growth the critical load corresponding to the crack initiation was determined by the intersection of load displacement diagram with a straight line drown from the origin of diagram with the slope 5 % less than the slope of linear part of load displacement diagram.

Herein two experimental methods for the fracture testing of the composite were employed,specificly, the acoustic emission and grating interferometry /historically, moire interferometry/ methods.

3.1. Acoustic emission, AE, method

The samples with acoustic emission sensor attached were loaded using computer controled Schenck PSB 100.

The clip gauge 10 mm was mounted in special way near the loading pins. The GACEK processor for acoustic emission measurement and amplitude analysis has been used.

The details about AE method applied for short fibre composites are given by Gołaski and Schmidt /1994/.

Stress intensity factor, K_Q, was calculated from the relationship:

$$K_Q = f \frac{P_Q}{BW^{\frac{1}{2}}} \qquad (1)$$

where f depends on a sample geometry in Kapp et al /1985/.

3.2. Grating interferometry, GI, method

Grating interferometry is a technique for high-sensitivity studies of in-plane displacements on flat surface /Post 1987/.The system used for the tests is a four-beam optical arrangement which has been described by Post /1987/, Czarnek /1991/ and modified for automatic fringe pattern analysis by Sałbud et al. /1992/, Kujawińska et al. /1993/; it is illustrated in Fig. 2.

A laser light is expanded through the pin hole and then is formed into a collimated beam through the collimating lens, iluminating the specimen and an interferometer consisting of a light frame and three adjustable mirrors, M_1, M_2, and M_3. In Fig.2/a/, half the incident beam B_1 impinges directly on the specimen surface while the other half B_2 impinges on the mirror M_3 and then reflects to the specimen in a symmetrical direction. In Fig.2/b/,two incident beams A_1 and A_2, impinge on the mirrors, M_1 and M_2, and then illuminate the set of grating lines perppendicular to the y axis to produce the v displacement fringes. In this four-beam optical arrangement, coherent beams A_1 and A_2, and B_1 and B_2 are produced to illuminate the specimen simultaneously. Therefore, when taking the u displacement field, a filter F_x which is an

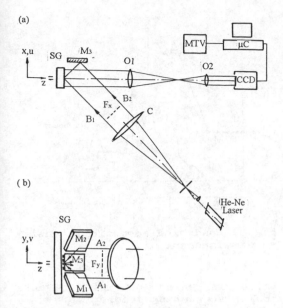

(a)

(b)

Fig.2.Optical arrangement for u /a/ and v /b/ displacement measurement using moire interferometry; M_1, M_2, M_3 - mirrors; F_x, F_y - filters; C - collimator lens; O1-O2 - imaging objectives; SG - specimen grating.

opaque object with a slot allows only the coherent beams B_1 and B_2 to illuminate the specimen. Similarly, when taking the v displacement, a filter F_y with different position of slots allows only coherent beams A_1 and A_2 to illuminate the specimen. The interferogram of the u or v displacement contour is collected by a charged coupled device /CCD/ camera using an image objective lens O1 and O2.

All of the light which emerges essentially normal to the specimen is collected and the specimen surface is focused onto the CCD image sensor consisting of a scanned array /512x512 pixels/ of light sensitive silicon, with processing hardware to convert the scanned output to an analoque video signal format. The signals are recorded and analysed using an IBM or compatibile PC.

Herein, the computer program DZF of the automated fringe pattern analysis /copyright by Institute of Design of Precise and Optical Instruments/ based on the spatial-carrier phase shifting method

/Kujawińska 1992, Poon et al. 1993/ is used. After obtaining the u and v displacement fields, the strains can be calculated by numerical differentiation i.c.

$$\varepsilon_x = \frac{\partial u}{\partial x} \; ; \quad \varepsilon_y = \frac{\partial v}{\partial y} \; ; \quad \gamma_{xy} = \frac{\partial u}{\partial y} + \frac{\partial v}{\partial x} \quad (2)$$

4. RESULTS

The results for measurement of stress intensity factor of fibre composites, in dependence on sample orientation, are
sample L-T $K_c = 8,50$ MPam$^{\frac{1}{2}}$
sample T-L $K_c = 8,31$ MPam$^{\frac{1}{2}}$
The values of K_c for both orientations are similar for these sample dimensions. However different loading-displacement pathes were observed for each of these orientations. They are for:
- linear up to failure and unstable crack propagation for T-L orientation,
- linear with a number of "pop in" events for L-T orientation.
The term "pop in" means discrete crack extension.

The failure processes in this type of composites involve a number of mechanisms including matrix plastic deformation and failure, fibre pull out and/or breakage and bunches of fibres pull out. These elementary events are combined in a way very complicated for the description of damage growth and fracture. At a sufficient stress level some of these mechanisms may result in critical events leading to crack initiation. In composites with visco elastic matrix it is probable that the critical event and crack initiation do not coincide and take place not at the same time.

To follow the failure processes the acoustic emission /AE/ method, including amplitude analysis of AE signals, was employed. This method was applied with good results to analysis of interlaminar fracture mechanisms in long fibre composites /cf Gołaski, 1992/. For this purposes GACEK processor together with the broad band sensor was used. The AE signal was evaluated as a number of events. The events were counted versus time in 15 levels with the difference of 6dB between adjacent

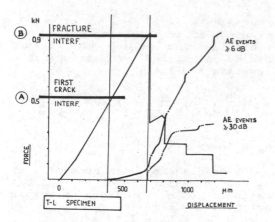

Fig.3. Load and acoustic events versus displacement

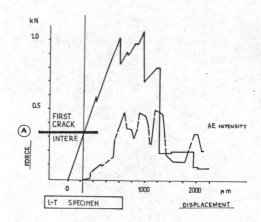

Fig.4. Load and acoustic intensity vs displacement

levels. Thus AE amplitude analysis was made in the range 6-90dB. Next different low and high level filters were used to select proper AE events for further analysis. AE events, loading and displacement were recorded at the same time.

The diagrams of load and AE events vs. load-point displacement are presented in Figs. 3-5. The amplitudes are separated in three groups /LA-low amplitude $\geqslant$ 6dB/; /MA-medium amplitude $\geqslant$ 30dB/; /HA-high amplitude $\geqslant$ 54dB/. The amplitude distribution suggests that two mechanisms dominate in the failure process of the tested composite.

Additionally the loads at the critical fracture points obtained by grating interferometry method are marked on the diagrams.

For all tests and both samples orientations the high agreement between AE and GI results is obtained. LA and MA amplitudes up to critical failure obtained for T-L samples orientation with linear up to failure load are shown in Fig.3. Some LA amplitudes have started and first burst was observed under a load equal to ∿ 0,55 Pmax /point A, Fig.3/. Critical cracking /point B/ has been observed near the maximum load point. From the maximum load point all three LA, MA and HA amplitude groups can be observed. Every next burst is accompanied by increasing in AE events in three selected ranges of amplitude.

Fig.4 and Fig.5 present the results from samples with L-T orien-

tation in which "pop in" events took place during loading.

The first small cracking was observed by grating interferometry method, and it agrees with the starting point of the AE events curve.

LA events took place when the burst intensity increased and the second crack tip appeared.

The fracture of sample was preceded by a significant increase of MA acoustic events.

The presented LA, MA and HA amplitude distributions suggest that the three presented mechanisms dominate in the failure process of the tested composite. The failure process observation and the similarity of the critical values of stress intensity

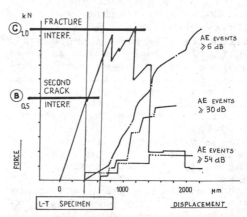

Fig.5. Load and acoustic events versus displacement

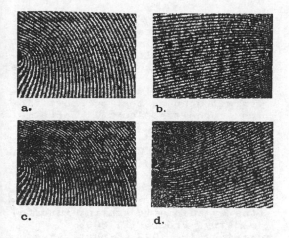

a. b.

c. d.

Fig.6. The interferograms representing u and v in-plane displacement fields before /a;b/ and after /c;d/ first crack, respectively

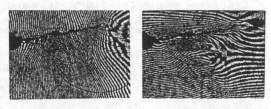

a. b.

Fig.7. The interferograms representing u displacement field before /a/ and after /b/ the second crack

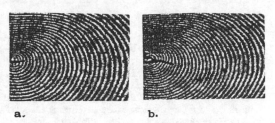

Fig.8. The interferograms representing u and v displacement fields after the critical crack and unloading of the sample

factor for L-T, T-L specimens serve to indicate the importance of matrix in the composites tested.

GI method proved the proper indication of the critical points during the cracking process as shown in Fig. 6-8.

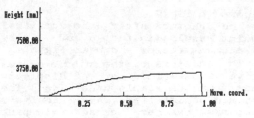

a.

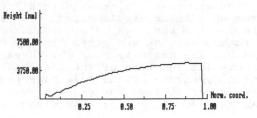

b.

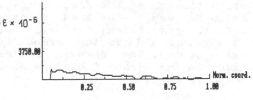

c.

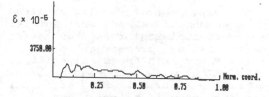

d.

Fig.9. The x-crossections of the displacement fields in the L-T specimen before /a/ and after /b/ the crack, and the respective strain distributions /c/ and /d/

The fringe patterns in Fig.6 represent the in-plane displacement for u- and v-directions before and after the first crack apears.

Figures 6b and 6d represent the sample behaviour in point A /see Fig.4/. The displacement fields before and after the second crack occuring, point B /see Fig.5/, are

1311

shown in Fig.7.

The fracture after the critical point results in the u and v displacement fields shown in Fig.8.

The analysis of the interferograms gives whole-field map of displacements in the sample under load. The method of interferogram analysis /spatial carrier phase stepping method/ restricts the displacement determination to the non-linear term only.

The exemplary y-crossections of the non-linear displacement fields in the L-T specimen before and after crack occurring are given in Fig.9 /a,b/. The comparison of the curves shows the increase of the displacement gradients in the neighbourhood of the crack tip. The overall non-linear term increases significantly.

At loading /F = 210 N/ the strain distribution in y direction has a uniformly decreasing character. After initiation of the first crack /F=340 N/, due to the load relaxation, the significant distortion of the strain field in the crack tip region occurs. The changes of the values in the strain field are presented in Fig.9.

5. CONCLUSIONS

The experimental results obtained from AE and GI methods are initial tries to implement them for study of the failure process of short glass fibre reinforced composites.

AE method with different amplitude ranges is useful for critical fracture points determination.

GI method gives full-field displacement and strain determination around a crack tip.

The methods give good agreement in the estimation of the critical levels in the failure process. We predict that the simultaneous application of AE and GI methods will lead to the proper quantitative description of the phenomenon of the initiation of the crack in short fibre reinforced composites.

REFERENCES

Czarnek R. 1991. Three mirror,four-beam moire interferometer and its capabilities, Opt.Lasers Eng.,15, 93-101

Gołaski L. 1992. Pękanie i odporność na pękanie kompozytów obciążonych zgodnie z drugim sposobem. Zeszyty Naukowe Politechniki Świętokrzyskiej, Mechanika M.48, Kielce 93-112

Gołaski L., Schmidt J. 1994. Fracture toughness test of short fibre composites. Journal Theoretical and Applied Mechanics 1.32, /PAN/, 260-272

Kapp J.A., Leger G.S., Gross B., 1985.Fracture Mechanics Sixteenth Symposium ASTM, STP 868, 27-44

Kujawińska M.,Sałbut L.,Czarnocki P. 1993. Material studies of composites by automatic grating interferometry. Proc.SPIE, 2004

Kujawińska M. 1992. Expert system for analysis of complicated fringe patterns. Proc.SPIE 1755

Moore D.R. 1992. Testing Injection Moulded Discontinuons Fibre Reinforced Composite, Appendix 2,ESIS Polymers and Composites Task Group, London

Post D. 1978. Moire Interferometry. SEM Handbook on Experimental Mechanics, A.S.Kobayashi ed,Prentice Hall

Poon C.Y., Kujawińska M., Ruiz C. 1993. Spatial-carrier phase shifting method of fringe analysis for moire interferometry. Journal of Strain Analysis, 28, 79-88

Sałbut L.,Kujawińska M.,Patorski K. 1992. Polarization approach to high-sensitivity moire interferometry. Opt.Eng., 31, 434-439

Williams J.G. 1991. A Linear Elastic Fracture Mechanics /LEFM/ Standard for Determining K_c and G_c Plastics. ESIS Polymers and Composites Task Group, London

Recent Advances in Experimental Mechanics, Silva Gomes et al. (eds) © 1994 Balkema, Rotterdam, ISBN 90 5410 395 7

Study on fracture mechanisms of AFRP: [0 /±45 /90]s with observation methods and AE methods

Satoshi Somiya
Keio University, Japan

Taku Sugiyama
Graduate School of Keio University, Japan

Abstract: Tensile tests for AFRP [0 /±45 /90]s laminate were conducted in order to analyze the microscopic fracture mechanisms by use of an ultrasonic flaw detector, and at the same time, to study a relation between these fracture phenomena and the power spectrum characteristic of fracture sound. As a result, it was cleared that cracks occurred in 90 layer in the center grow and progress outwardly to cause delamination between layers in order, and the delamination was evaluated quantitatively. Also, in the analysis by the AE methods, we studied on the power spectra of some AE waves. As a result, we reached conclusions that various fracture phenomena and the features of power spectra correspond to each other, and that it is possible to specify the fracture mechanisms by analyzing the power spectra.

1 INTRODUCTION

A large number of researches on the relation between the mechanical properties and fracture phenomena of FRP including ones by the acoustic emission method have been performed, like M. Faudree (1988), J. Block (1988) and S.M.Bleay (1991). S. Somiya (1991) also tried to investigate the fracture phenomena on the AFRP by using parameters of AE energy, and peak-amplitude distribution as well as through surface observations, but it was difficult to allow the result based on the AE method to correspond to the fracture phenomena.

Through the research using these AE methods, however, the usefulness of the AE methods was showed by finding that a large number of fracture sounds are generated from the low stress level in unidirectional AFRP, and by finding, by investigating the reason, the occurrence of fine debonding around fiber which is orientated in the tensile direction. It is an object of this study to clarify the fracture mechanisms of AFRP [0 / ±45/ 90]s laminate at each level of low stress to final fracture by means of the optical methods and ultrasonic equipment, and specify the fracture mechanisms using the AE methods and studying the power spectra of fracture sounds.

2. EXPERIMENTAL METHODS

Eight sheets of unidirectional prepregs made of KEVLER 49/#2500 and epoxy resin P005-07 were laminated on fiber, and formed by using the autoclave. Laminate configuration was of [0/±45 /90]s, and the volume fraction was 55%. The specimen is a strip in accordance with the shape of a tensile specimen type II for CFRP according to the standard in JIS, and its gauge length, width and thickness are 120, 25, 1.2(mm) respectively. The both ends of specimen were bonded with small pieces made of GFRP. The test conditions were temperatures of $23 \pm 2^{\circ}$ C and humidities of 50% ± 5%. The elastic modulus and tensile strength of the material were 19.8 GPa and 1.3 Gpa respectively.

For the AE test, NF900 (NF (Co.)) was used. The AE wave was amplified to 50 dB by use of a pre- and a main-amplifier. A data recorder DM-100(ONO Co.) was used to sample at 1 MHz. A low-pass filter was used in order to input only signals 1 MHz or less, which are Nyquist frequencies during FFT analysis. For the thresholds, Vh = 170 mV and Vl = 120 mV were used for measurement. The sensors were coated with grease and fixed with tape. For the ultrasonic flaw detector, UH Pulse 100 made by Olympus Optical using pulse waves having a frequency of 30 MHz was used.

3 FRACTURE MECHANISMS FOR AFRP[0/+45/90]s MATERIAL

3.1 Observation of fracture mechanisms on surface of specimen

Fig. 1 shows a stress - strain curve and variations in event rate of the test material. Although the AE wave was generated at an exceedingly low stress and increased, it was observed by a optical microscope that fiber debonding occurs along the fiber and continuously grows until it is fractured on the surface of 0 layer as shown in Fig.2. This phenomenon is quite the same as the result of the unidirectional AFRP made of the same prepreg, presented by S. Somiya (1992).

In order to observe the fracture phenomena within the laminate and between the layers, one side of the specimen under load was observed using a CCD video microscope. For this purpose, the plane was smoothly cut by a diamond cutter.

As the cause for delamination between layers, the occurrence of a shearing stress between layers to be calculated on the basis of the theory of lamination for an anisotropic body is considered. According to this theory, since a difference in angle between $-45°/+45°$ layers is a maximum, it was predicted that delamination would occur between these layers.

Fig.3 schematically shows an aspect of occurrence of microscopic fracture resulting from an increase in the stress. As shown in Fig.3b, the microscopic fracture started as debonding around the fiber within $90°$ layer, and grew as a crack perpendicular to the tensile direction. When this crack arrived at the boundary between $90°$ and $-45°$ layers, delamination occurred as shown in Fig.3c. The crack which further developed passed through the $-45°$ layer, and caused delamination between $+45°/-45°$ layers as shown in Fig. 3d. This crack further developed to reach $0°$ layer, and caused delamination between $+45°/0°$

layers, one of which furthered fracture in 0 layer to cause final fracture. This mode of fracture is different from the mode of fracture in the unidirectional AFRP.

3.2 Observation of fracture mechanisms within specimen

It is well known that delamination easily occurs on the side of cross-plied lami-

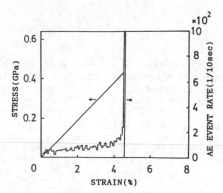

Fig.1 Typical Stress-Strain Curve and AE event rate

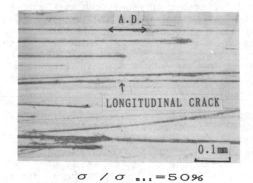

$$\sigma / \sigma_{max} = 50\%$$

Fig.2 Longitudinal Crack observed on surface on unidirectional AFRP

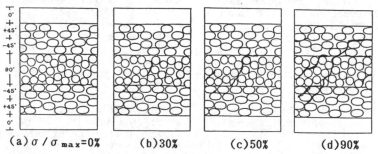

(a) $\sigma / \sigma_{max} = 0\%$ (b)30%   (c)50%   (d)90%

Fig.3(a)-(d) Schematic drawings of Crack growth behavior within AFRP[0/±45/90]s

nate, and it is considered as the cause that the delamination starts from the damage caused when the specimen was cut, and that the shear deformation based on the shearing stress mentioned above is concentrated on the free interface. Therefore, the information obtainable from observation of the side of the specimen does not always indicate the phenomenon within the specimen correctly. Thus, we investigated the aspect of occurrence of delamination within the specimen during the deformation using an ultrasonic flaw detector.

Specimen which after loads of 30%, 50%, 80% and 95% immediately before breaking had been applied against breaking strain, and the loads were removed were prepared. These specimens were put into the ultrasonic flaw detector to select the depth of the plane between laminates as shown in Fig.4 By scanning the pulse wave along the plane, the sliced image was obtained. Fig.5 shows the aspect of occurrence of debonding in an interface of $-45^\circ/+45^\circ$ at 95% of the breaking strain, and the debonding grew long and narrow in $+45^\circ$ direction chiefly along the fiber. The width being hundreds of micron, the debonding was clearly different from the debonding around the fiber which was observed in such 0° layer as shown in Fig.2. Further the amount of debonding is almost the same in the width direction although slightly more on both sides A and B of the specimen, and it was confirmed that the effect of edge was exceedingly small in the material under test.

Next, occurrences of debonding at the interfaces at $90^\circ/-45^\circ$, $-45^\circ/+45^\circ$ and $+45^\circ/0^\circ$ were investigated. In order to quantitatively evaluate the debonded area, a value obtained by dividing the debonded area by the total measured area was defined as " Degree of Release". Fig.6 shows the aspect of an increase in the deformation process of "Degree of Release" on each laminated surface. From this figure, it can be seen that the debonding within 90° layer starts the occurrence at 30% in breaking strain, and increases as the stress increases. Also, it was confirmed from Fig.6 that it was at an interface between $90^\circ/-45^\circ$ layers that delamination occurred most. The delamination between $+45^\circ/-45^\circ$ layers suddenly increased as the final fracture was approached. These facts agree quite well with the result obtained from the observation of the side of specimens. We obtained the result that the occurrence of delamination does not result from the shearing stress, which is an influence from crossplied laminate, but depends upon the cracks occurred within 90 layers.

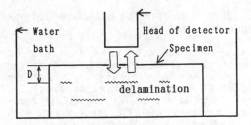

D: Depth to Delamination from Surface

Fig.4 Illustration of Ultrasonic measurement Method

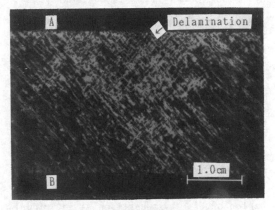

Fig.5 Image photograph of delamination at +45/-45 interface in AFRP By UH 100

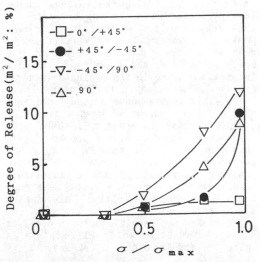

Fig.6 "Degree of Release" dependence on applied load

4. ANALYSIS OF FRACTURE MECHANISMS OF AFRP BASED ON AE METHODS

A research was conducted in order to specify the fracture mechanisms from the features of the power spectrum. In this research, we did not take the position that there exist, in the individual fracture phenomena, specific frequencies, each of which corresponds to them respectively, but thought that the features would appear in the power spectrum. On analyzing the frequency, the resonance characteristics of the AE sensor directly affects the power spectrum, and it is necessary to correct the sensitivity for each frequency. Since, however, it is not easy generally to obtain a sensor having an uniform resonance characteristic, we decided to use an AE sensors popularly used and to compare the patters of the power spectra.

At first, we analyzed the individual AE waves, and tried to obtain the average feature of a large quantity of AE waves by taking it into consideration that the material under test would produce hundreds of thousands of AE waves before the final fracture. As a method of processing a large quantity of AE waves, a "Evaluation method using weighted power spectrum" proposed by the author et al was used here. It is an object of this method to grasp the features of spectra within a period of time by superimposing individual power spectra of a large quantity of fracture sounds to prepare power spectra for entire AE waves which are generated in the specific period.

Fig.7 shows a power spectrum for a typical frequency of one AE wave with the power ratio, in which the maximum number is represented as 1, in the ordinate and the frequency in the abscissa, and many peaks are observed. Further, a large quantity of power spectra of such AE wave are superimposed, and a new index in which the maximum value in the ordinate is newly set to 1, is shown in Fig.8 using Power Ratio. This figure shows a result obtained by analyzing about 80 AE waves obtained in a certain time interval, in which peak values are connected by a line. The figure shows the result up to 30% in fracture stress ratio, power spectra for 30% to 60%, and the result for 60% to immediately before the fracture. The peaks seem commonly in these three figure are called A,B,C and D.

The peak C noticeably starts to occur at a low stress as shown in Fig.8a, and is quite the same as a peak of AE wave 130 to 160 KHz caused by debonding between fiber of unidirectional material pointed out in the previous report by S. Somiya (1992).

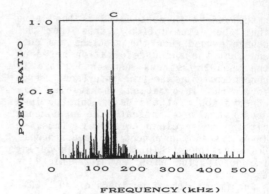

Fig.7 Power spectrum for typical frequency of one AE wave.

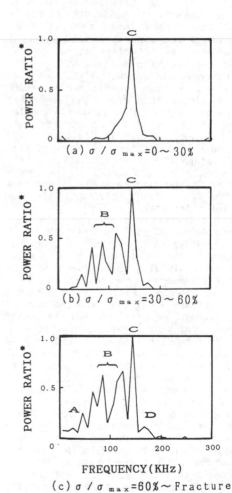

(a) $\sigma / \sigma_{max} = 0 \sim 30\%$

(b) $\sigma / \sigma_{max} = 30 \sim 60\%$

(c) $\sigma / \sigma_{max} = 60\% \sim$ Fracture

Fig.8 Weighted Power spectrum dependence on applied load

1316

It was confirmed that this peak is dependent upon the occurrence of debonding around aramid fiber within O layer, which is the outermost layer. The peak B starts to occur when the stress ratio exceeds 30%, and changes the frequency value little by little depending upon the stress level. However, the peak B consists of tree peaks. The reason for distinguishing the peak B from the peak C is that while the peak B is exceedingly low in the AE wave in which the peak C noticeably occurs as the result of analysis of a large number of AE waves grasped at 30% or more in breaking stress ratio, it was confirmed that there coexist two types of AE waves having no peak C in AE waves in which the peak B occurs. Moreover, since as the fracture is approached, peak A grows, and fiber decomposes and is cut, and further peak D increases, it was distinguished from peak C. When peak A occurred, the resin around the fiber was cracked.

Summarizing the above results, the peak B suggests that it indicates the occurrence of delamination. The height of the peak increases together with the progress of the fracture, and this tendency agrees with Fig.6, thus reaching a conclusion that the peak B indicates the occurrence of delamination. However, this research could not analyze the difference in power spectrum of fracture sound generated from between layers. It was clarified that the power spectrum which occurs in the course of fracture of unidirectional material and the material under test are clearly different from those. It turned out that the variation in power spectrum well corresponds to that in fracture phenomenon. The foregoing shows that it is possible to measure the degree of fracture from an inspection by the AE methods.

5. CONCLUSIONS

We observed the fracture mechanisms within the AFRP [0/±45/90]s material by use of a metallurgical microscope, a CCD video microscope and an ultrasonic flaw detector, and tried to analyze fracture sound by the AE method.

As the microscopic fracture mechanisms of AFRP, it was confirmed that the crack which occurred in 90° layer progresses toward the surface layer, and causes delamination whenever it reaches the laminated interface, and that a great difference between in the interior and in the cut surface in the width direction of the specimen cannot be seen in the occurrence of delamination.

The power spectrum for fracture sound was analyzed, and it was found that the features of power spectrum correspond to the fracture mechanisms. Also, it was found that it is effective for analyzing the frequency when a large quantity of AE waves are generated to use the " Evaluation method using "Weight power spectrum", and that it is possible to specify the cause for fracture by using the distribution shape of the power spectrum.

ACKNOWLEDGMENT

We thanks Mr. J. Mastui and Mr. S Matsuda in TORAY Co. for providing the materials. This work was sponsored by scientific research fund of Iketani.

REFERENCE

Faudree, F. Baer, E. & A. Hiltner 1988. Characterization of damage and Fracture processes in short fiber BMC composite by acoustic emission. J. of Composite materials. 22:1170-1196.

Block, J. 1988. Characterization of damage progression in fiber reinforced composites by acoustic emission. Engineering Application of New composites. ed. S.A. Paipotis: Omega Scientific.

Bleay, S. M. & V. D. Scott. 1991. Microfracture and micromechanics of the interface in carbon fiber reinforced Pyrex glass. J. Mat. Scie. 26:3544-3552.

Somiya, S. Menges, G. & M. Effing 1991. Analysis of peak amplitude distribution of unidirection and quasi-isotropic AFRP on tensile loading. 6th Int. Conf. of Mechanical behavior of materials. 4:411-416.

Somiya, S. 1992 Monitoring of microscopic damage progression in uniderectional AFRP on tensile test by acoustic emission. 7th Int. Cong. on Experimental Mechnics, 12:1560-1565.

Recent Advances in Experimental Mechanics, Silva Gomes et al. (eds) © 1994 Balkema, Rotterdam, ISBN 90 5410 395 7

Effect of fiber orientation on fracture toughness of CFRP

H. Ishikawa
University of Electro-Communications, Chofu, Tokyo, Japan

T. Koimai
Japan Airlines, Haneda Airport, Ota, Tokyo, Japan

T. Natsumura
Ishikawajima-Harima Heavy Industry Co., Ltd, Tanashi, Tokyo, Japan

ABSTRACT: In order to understand the effect of the fiber orientation of carbon fiber reinforced epoxy laminate on Mode I interlaminar fracture toughness, some experiments are carried out by using the specimens with several kinds of stacking sequences. In the case of no bridging fibers, there is no effect of the stacking sequences on the fracture toughness and the value seems to be the mimimum one for the material system.

1 INTRODUCTION

Carbon fiber reinforced epoxy (CFRP) laminates are being used in stiffness critical structural components. In addition to the applicatiom, because of the demand for reduced structural mass, CFRP laminates are tried to be used for some strength critical applications. The sensitivity of strength critical components to damage can affect their inservice conditions. Some research works (for example, (Whitney, 1982 and Ishikawa, 1991)) indicate that one of the strength problems of CFRP lamonates was less fracture toughness.

In this paper, some experimental investigations were carried out to consider the effect of the stacking sequence on the interlaminar fracture toughness of CFRP laminates. Especially, we made focus on the effect of the fiber bridging. As a parameter of the fracture toughness, the strain energy release rate was used.

2 EXPERIMENTS

The laminate specimens with the stacking sequences of [0/0]$_9$, [15/-15]$_9$, [30/-30]$_9$, [45/-45]$_9$, [0$_8$/30//0$_9$], [0$_8$/45/0$_9$], and [0$_8$/60//0$_9$] were fabricated using the graphite / epoxy (T300 / #2500) unidirectional prepreg. In the present study, the

Table 1 Material properties.

Longitudinal Young's modulus	106.1	GPa
Transverse Young's modulus	6.3	GPa
Longitudinal Poisson's ratio	0.34	
Transverse tensile strength	0.06	GPa
Interlaminar shear strength	0.1	GPa
Fiber volume fraction	60	%

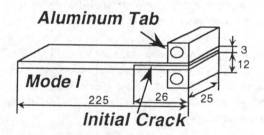

Fig.1 Specimen geometries (mm).

former four laminate specimens are called as cross-ply laminates (CPL), and the latter three laminate specimens as laminates with an oblique-ply (LOP). In the interface of the center in thickness of specimen, a pre-crack (initial crack) was inserted, shown in Fig.1. Then, the double slash "//" , also, indicates the position of the interface of the pre-crack. The length of the pre-crack is 20 mm. The mechanical properties of the unidirectional prepreg are presented in Table 1. The geometry of the specimen

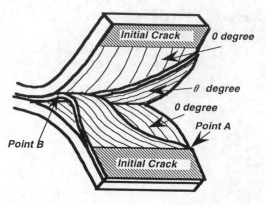

Fig.2 Schematic representation of mid layer bridging.

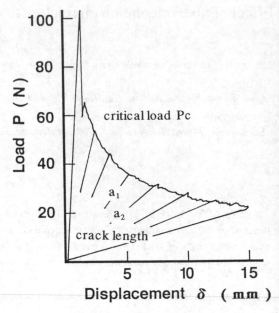

Fig.3 P- δ curve for the specimen with stacking sequence of $[0/0]_9$.

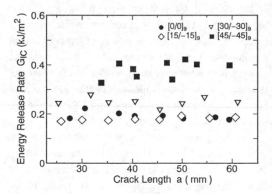

Fig.4 Energy release rate for CPL specimen.

with a pre-crack in the mid-thickness plane is shown in Fig.1.

By using the aluminum tabs joined to the end of the specimen by adhesive, the load was applied under the displacement control at room temperature. The crosshead speed of the testing machine was 0.5 mm/min. On both side surfaces of each specimen, the white color paint was pasted in order to measure the crack length accurately. To obtain the compliance C of the specimen at each crack length, static cycle of loading and unloading was performed in the process of the crack extension, and the relationship between the load P and the loading point displacement δ was recorded. Also, the critical load Pc which corresponds to the onset of crack growth at each crack length was measured. The critical strain energy release rate Gc can be given by the following expression;

$$G = \frac{P^2}{2B} \frac{\partial C}{\partial a} \qquad (1)$$

where B is the width of specimen.

3 EXPERIMENTAL RESULTS AND DISCUSSIONS

In the crack extension process, two kinds of bridging in the debonded interface were observed. One is fiber bridging that is often observed in general specimens of CFRP laminate. Another is layer bridging that is originally observed in the present specimen. The layer bridging is schematically

represented in Fig. 2. The former is occurred in the cases of CPL specimen, except the stacking sequencies of $[0/0]_9$, and $[15/-15]_9$. The latter is occured in the cases of LOP specimen. In CLP specimens, the amount of the bridging fibers is increased with the increase of the difference of the relative angle between the fiber directions of the cross-ply. The undulation and translaminar of the debonded crack caused the fiber bridging.

First, the experimental results for CPL specimen are shown in Figs. 3 and 4. Fig. 3 show the critical

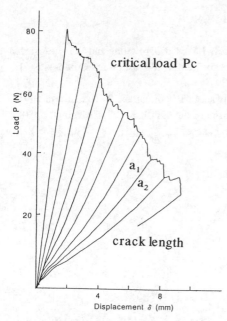

Fig.5 P- δ curve for the specimen with bridging
layer ($[0_9//45/0_8]$).

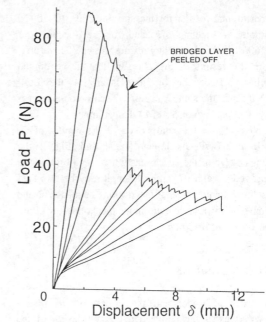

Fig.6 P- δ curve for the specimen in which
the bridging layer is peeled off ($[0_9//45/0_8]$).

load Pc and the relationship between the load Pc and
the loading point displacement δ in the process of the
crack extension. Introducing the compliance C of the
specimen for each crack length a and critical load into
Eq.(1), the strain energy release rate Gc is obtained
and the results of are shown in Fig. 4. The values of
the Gc for the specimens with $[0/0]_9$, and $[15/-15]_9$.
which have no bridging are the same level. The Gc
for the other CPL specimens has the higher value than
the level. The reason seems to be the occurence of the
fiber bridging, as mentioned above. However, it is
necessary to understand the quantitative relation
between the amount of the bridging fibers and the
energy release rate, and this will be investigated.

Next, in the LOP specimen, the bridging of the mid
layer between its upper and lower half layers of the
specimen occured. In order to consider the effect of
the bridging on the Gc, the bridging layer was peeled
off artificially for several specimens of $[0_8/45//0_9]$ and
$[0_8/60//0_9]$. The P- δ curve for the $[0_8/45//0_9]$
specimen without the artificial peeling is shown in
Fig. 5. The P- δ curve for the specimen with the
same stacking sequence, where the bridging layer was
peeled off, is shown in Fig. 6. The value of the load

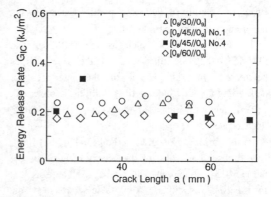

Fig.7 Energy release rate for LOP specimen.

decreases down suddenly and the crack extended
stably in the interface between the 0° and 45° layers
without translaminar.

The experimental results of Gc for the specimens
are shown in Fig.7. For the $[0_8/45//0_9]$ specimen, two
kinds of data, No.1 and No.4, are shown. The data of
No.1 are the results of the specimen without peeling
off. In the specimen of No.4, the data with the crack
length that is larger than about 50 mm are the results
of the specimen with the peeling off. In Fig.7, the

experimental results for the specimen of $[0_8/30//0_9]$ without the peeling off and for the specimen of $[0_8/60//0_9]$ with the peeling off are shown. Therefore, from the results, it may be said that if we do the peeling off of the bridging layer for the LOP specimens, the same value of Gc can be obtained for any stacking sequences of LOP specimens.

Next, From the comparison of the results of the CPL and LOP specimens (Fig.4 and Fig.7), the values of Gc of the specimens where each bridging is removed are almost same. Then, it may be also said that there are no effect of the stacking sequence on the strain energy release rate under the condition of no occurrence of bridging.

4 CONCLUSIONS

In order to understand the effect of the fiber orientations in the above and beneath layers of the interface on Mode I interlaminar fracture toughness, some experiments are carried out by using the specimens with several kinds of stacking suquences. From the results of the experiment, the following conclusions can be remarked:

(1). For the cases, such as the $[30/-30]_9$ and $[45/-45]_9$, that the difference of the fiber angles between both side layers of the interfaces of debonding is increased, the fiber bridging is observed.

The behavior increased the values of the energy release rate.

(2). No bridging occured in the $[15/-15]_9$ and $[0/0]_9$ specimens. The value of the energy release rate for the $[15/-15]_9$ specimen was almost the same as that for the $[0/0]_9$ specimen. This value seems to be the mimimum value for this material system.

(3). In the $[0_8 30//0_9]$, $[0_8/45//0_9]$ and $[0_8/60//0_9]$ specimens, the mid layer bridging occured. When the mid layer bridging was peeled off, the value of fracture toughness was also similar to the mimimum value.

REFERENCE

Whitney, J.M., C.E.Browning and W. Hoogsteden 1982 J. Reinforced Plastics and Composites 1: pp.297.
Ishikawa, H., T. Koimai and T. Natsumura 1991 Preprint of the 69th JSME Fall Meeting, pp.385 (in Japanese).

Recent Advances in Experimental Mechanics, Silva Gomes et al. (eds) © 1994 Balkema, Rotterdam, ISBN 90 5410 395 7

Erosion damage of ceramics due to hot water jet impact

U. Iwata & T. Sakuma
Central Research Institute of Electric Power Industry, Tokyo, Japan

ABSTRACT: In order to apply ceramics in much wider fields, it is necessary to estimate the mechanical properties under high temperature. This paper describes the experimental investigations of erosion damage of ceramics due to hot water jet impact. Erosion damage depends heavily upon the velocity and temperature of jet. Damaged depth increases according to the increase of the velocity and the temperature of jet. Thus, cavitation erosion tests are performed in order to clarify the influence of temperature of jet on erosion damages. As a result, it is clarified that making a comparison between damage due to hot water jet impact and that of cavitation in relation to the hardness Hv and fracture toughness KIC of ceramics, both damages are found to be proportional to $H_v^{-1/4} \cdot K_{IC}^{-4/3}$. Thus, it is clarified that erosion damage due to hot water jet impact is caused by the interaction between cavitation and shear stress due to high speed flow.

1 INTRODUCTION

Structural ceramics such as *SiC* and *Si₃N₄* are superior in heat resistance, erosion, corrosion and high temperature strength to metal materials. Some ceramic parts of automobile are in practical use(Katayama, 1986; Ito,1988 ; Hempel, 1986). The application of ceramics to diesel engine (Kawamoto,1989) and gas turbine(Sasa, 1982) is also expected and some parts of them have been developed.

Some components in the power plants are exposed to high temperature and pressure flow conditions. Especially, drain valve is subject to erosion damage due to the impact of high temperature-speed mixture of water and steam. For the extension of their lives, use of ceramics and other metals as new material is widely studied.

Concerning the erosion resulted from droplets' impact, many works have been reported. They are performed by striking droplets on materials under room temperature. Influences of velocity and size of droplets on damage have been investigated (Hancox,1966; Heymann,1967). It is important to evaluate the damages under the actual conditions such as high temperature and high pressure. However, studies under these conditions, especially in relation to temperature, are very few.

In the case of damage due to hot water jet, authors discussed damages due to high temperature water jet impact and showed the influence of temperature on damage using sintered silicon carbide, silicon nitride and alumina as specimen (Iwata,1992,1993).

This paper describes damage due to hot water jet impact under high temperature-speed conditions using various kinds of ceramics and metals as specimens. The influence of velocity and temperature of jet on damaged depth has been discussed, and has also been discussed in relation to the hardness and the fracture toughness of ceramics by making a comparison between damage due to hot water jet impact and that of cavitation.

2 EXPERIMENTAL PROCEDURE

2.1 *Jet erosion tests*

When the compressed water is gushing from the nozzle to the atmosphere, the compressed water is dispersed in droplets in time. But within a short

Fig.1. Jet patterns due to the change of inlet pressure P_{in} (T_{in} =448K); (a):P_{in} = 2.9 MPa, (b):P_{in} = 7.4 MPa, (c):P_{in} = 16.7 MPa.

distance from nozzle, water is not dispersed but forms water jet as shown in Fig.1. The dispersing point is dependent upon the inlet temperature T_{in} and pressure P_{in}.

When high temperature compressed water is gushing from the nozzle to the atmosphere, water jet is formed within a short distance from nozzle and in the farther distance, droplets jet is formed as mentioned above. Erosion damages are affected by these different type of jet(Iwata,1993). Therefore, tests of water jet impact are carried out. Compressed water is jetted vertically toward the specimen set at the distance L=10mm from the nozzle as shown in Fig.2. Velocity of jet is measured by Phase/Doppler anemometry(Argon-Ion laser). And the temperature of water jet is measured by radiation pyrometer. Dimensions of nozzle are 0.3mm in diameter and 1mm in thickness, and nozzle is made by diamond.

Specimens used are sintered ceramics, type 304 stainless steel(SUS304) and cobalt base alloy(stellite 6B). Mechanical properties of ce-ramics are shown in Table 1. The specimen is the disk with the size of 16 mm and 3 mm in diameter and thickness respectively. And the surface of specimen is finished by abrasive cloth. The water with dissolved oxygen content of 8.36 ppm at room temperature, pH 7.48 and SiO_2 content less than 0.02 ppm is used for tests.

The amount of damage is estimated by maximum damaged depth. And maximum depth is measured by the instrument of surface roughness by stylus method.

2.2 Cavitation tests

In order to clarify the influence of the temperature of jet on damage due to hot water jet impact, cavi-tation tests induced by ultrasonic vibration are carried out.

The temperature and the pressure of water bath are 293 K and 0.2 MPa respectively. Shape and dimension of Specimens used are same as jet ero-sion tests. The specimen is attached directry to the horn. And the amplitude of vibration is 40 μm.

The amount of damaged mass is measured by a electric balance and is converted into damaged volume.

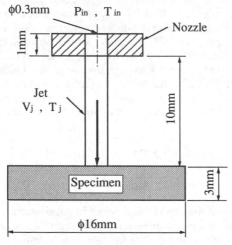

Fig.2. Schematic of jet erosion test.

Table 1. Mechanical properties of ceramics.

Ceramics	Hardness (GPa)	Fracture toughness (MPa·m$^{1/2}$)
Al$_2$O$_3$-A	17.2	3.5
Al$_2$O$_3$-B	15.4	5.5
SiC	27.5	4.6
Si$_3$N$_4$-A	14.7	6.5
Si$_3$N$_4$-B	16.7	5.5
Sialon-A	17.7	7.7
Sialon-B	13.7	6.0
PSZ	11.8	9.5
Mg-PSZ	10.0	15

3 DAMAGE DUE TO THE JET IMPACT

Amount of damage is mostly estimated by the damaged weight or volume. However, these estimations have many errors of measurement and amount of damage is not proportional to test duration in the acceleration period.

In case amount of damage is estimated by the maximum damaged depth, the maximum damaged depth is proportional to the test duration for a period except incubation and deceleration.

Fig.3 shows the maximum damaged depth D_{max} for various materials due to the hot water jet impact in relation to the test duration T_d under the condition that inlet temperature T_{in} and pressure P_{in} are 423 K and 14.7 MPa respectively. The maximum damaged depth D_{max} increases proportionally to the increase of test duration for various materials. This relation between the maximum damaged depth and the test duration is same on other temperature and pressure conditions.

3.1 Influence of the velocity of jet

Fig.4 shows the average damaged depth D_v in relarion to the velocity V_j of jet under the condition of the inlet temperature T_{in} =423 K constant and the inlet pressure P_{in} =2.9-14.7 MPa. The average damaged depth D_v is obtained by equation(1),

$$D_v = \frac{D_{max}}{Q_j T_d} \qquad (1)$$

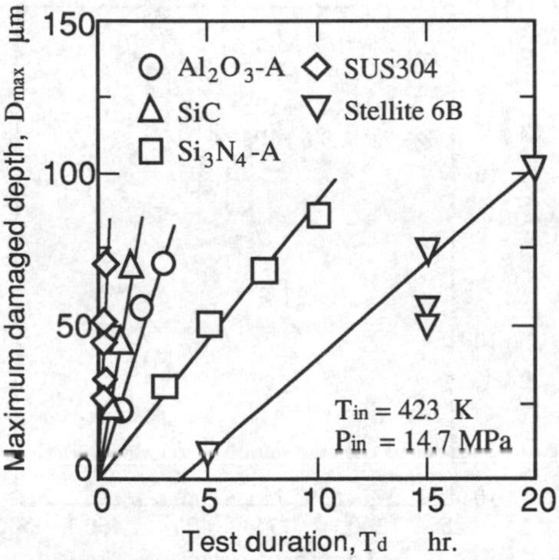

where D_{max} is maximum damaged depth, Q_j is flow rate of jet and T_d is test duration.

The average damaged depth D_v increases according to the increase of the velocity V_j of jet regardless of any kind of ceramics or metal. And the average damaged depth D_v increases proportionally to the nth power of the velocity of jet. For Al_2O_3-A, Si_3N_4-A and $SUS304$, the amount of damage is different according to the materials,

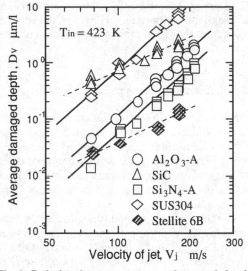

Fig.4 Relation between average damaged depth D_v and the velocity V_j of water jet for various materials

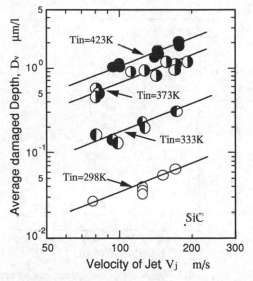

Fig.5 Average damaged depth D_v of silicon carbide in relation to the velocity V_j of jet

Fig.3 Maximum damaged D_{max} depth with time: Inlet temperature T_{in} and pressure P_{in} are 423 K and 14.7 MPa

1325

but the values of n are same.

Furthermore, the values of n of silicon carbide and stellite are different from them. Thus, the cause of damage for silicon carbide and stellite is considered to be different from alumina, silicon nitride and stainless steel.

Damaged depth also depends upon the temperature of jet. Fig.5 shows the average damaged depth D_v of silicon carbide in relation to the velocity V_j of jet as a parameter of inlet temperature T_{in}. Damaged depth increases according to the increase of inlet temperature. However, the value of n does not change in spite of the variation of inlet temperature.

3.2 Influence of the temperature T_j of jet

Fig.6 shows the relation between the average damaged depth D_v and the temperature T_j of jet for various ceramic materials at inlet pressure P_{in} =14.7 MPa constant. The average damaged depth D_v depends strongly upon the temperature T_j of jet and increases according to the increase of the temperature.

For the variation of inlet pressure P_{in}, the temperature T_j of jet affects the average damaged depth D_v. Fig.7 shows the average damaged depth D_v of Si_3N_4-A in relation to the temperature T_j of jet as a parameter of inlet pressure P_{in}.

The average damaged depth becomes large in comparison with that of room temperature of jet.

Accordingly, within a low range of the velocity (within a low inlet pressure range), the average damaged depth D_v is increased by raising the temperature of jet. The cause of damage dependant strongly upon the temperature T_j of jet is considered to be due to the cavitation generated at the collision of jet with specimen because the generation of bubbles of steam is increased accoring to the increase of temperature and the amount of damage is increased according to the increase of bubbles.

3.3 Observation of damaged surface

As shown in Fig.4, the rates of damage of SiC and Stellite 6B are different from other ceramics and metal. Thus, damaged surface is observed.

Fig.8 shows the micro-sketch of the damaged cross section of Al_2O_3-A, SiC and Stellite 6B. Inlet temperature T_{in} and pressure P_{in} are 423 K and 14.7 MPa respectively. And test durations of Al_2O_3-A, SiC and Stellite 6B are 2 hr., 1 hr. and 10 hr. respectively.

Damaged depth D becomes maximum in the neighbourhood of the part of collision of jet with diameter of 0.3 mm. Jet after colliding the specimen flows in high speed on the damaged surface. Accordingly, the shape of damaged cross section becomes wide owing to the shear stress due to high speed flow.

In case of Al_2O_3-A shown in Fig.8(a), the damaged depth D becomes large in the colliding part

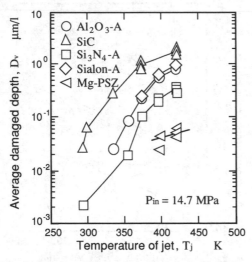

Fig.6 Relation between damaged depth D_v and the temperature T_j of water jet for various materials.

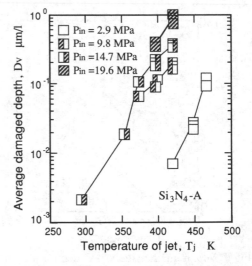

Fig.7 Damaged depth D_v of silicon nitride in relation to the temperature T_j of water jet.

of jet, but the damaged width at the surface of specimen is relatively small. On the contrary, in case of *SiC* shown in Fig.8(b), the damaged depth and width become large in spite of short duration time compared with that of *Al2O3-A*. This result shows that resistance against the shear stress of *SiC* is small. In case of *Stellite 6B* shown in Fig.8(c), erosion resistance is large in comparison with other materials. Thus, damaged depth D is very small even after 10 hr. test duration. But, damaged part is observed in the region of high speed flow and resistance against shear stress is relatively small same as *SiC*.

For these various damaged shapes according to materials, the damaged cross sectional area S_c at

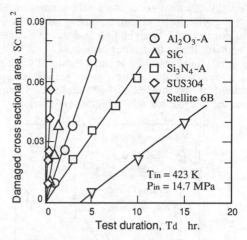

Fig.9 Damaged cross sectional area S_c at maximum depth with time: Inlet temperature T_{in} and pressure P_{in} are 423 K and 14.7 MPa

maximum depth is proportional to test duration. Fig.9 shows the relation between the damaged cross sectional area S_c and the test duration T_d under the condition of T_{in} =423 K and P_{in} =14.7 MPa.

4 DAMAGE DUE TO THE CAVITATION

Damaged volume of ceramics due to the collision of solid particles has reference to hardness H_v and fracture toughness K_{IC}(Evans, 1978). Suppose that the mechanism of damage due to cavitation is same as that of the collision of solid particles, relation between the damaged volume V_{cav} due to cavitation and hardness H_v and fracture toughness K_{IC} is expressed as followimg equation.

$$V_{cav} \propto H_v^{-1/4} K_{IC}^{-4/3} \qquad (1)$$

Fig.10 shows the damaged volume for various kinds of ceramics due to cavitation in relation to hardness and fracture toughness of materials. On the log-log diagram, alumina-A, silicon carbide, silicon nitride and sialon-A lie on a straight line. The mechanism of damages of these ceramics due to cavitation are considered to be same as that of the collision of solid particles.

Jet erosion tests are conducted for these ceramics. Fig.11 shows the relationship between the damaged depth and the hardness and fracture toughness under various inlet temperatures of jet. Alumina-A, silicon nitride and Mg-PSZ lie on a

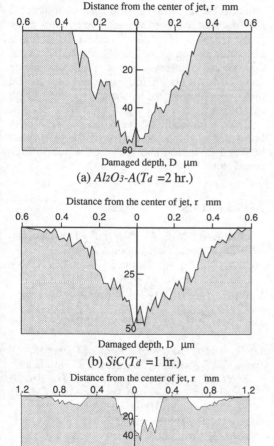

(a) *Al2O3-A(Td =2 hr.)*

(b) *SiC(Td =1 hr.)*

(c) *Stellite 6B(Td=10 hr.)*

Fig.8 Micro-sketch of damaged cross section: Inlet temperature T_{in} and pressure P_{in} are 423 K and 14.7 MPa

straight line for various inlet temperatures of jet. On the other hand, silicon carbide-A and sialon-A do not lie on a straight line and their damaged depth are larger than other ceramics.

From these results, it is considered that the erosion damages due to hot water jet impact are caused by the interaction between the cavitation and the shear force due to high speed flow.

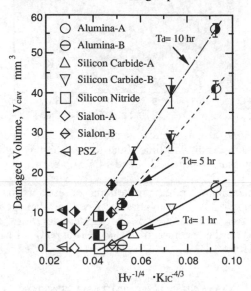

Fig.10 Damaged volume due to cavitation in relation to hardness and fracture toughness.

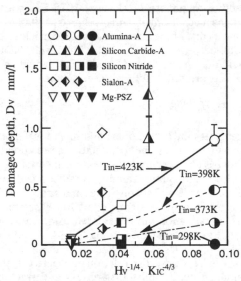

Fig.11 Damaged volume due to hot water jet in relation to hardness and fracture toughness.

5 CONCLUSION

In order to apply structural ceramics to some components exposed to high temperature and speed flow conditions such as drain valve of power plant, it is necessary to estimate the erosion resistance due to the cavitation and the abrasion resistance due to high speed flow.

In this paper, hot water jet impact tests are carried out for various kinds of ceramics and metals. Erosion damage depends heavily upon the velocity and the temperature of jet. In order to clarify the influence of temperature on damage, cavitation tests are conducted.

Making a comparison between damages due to hot water jet impact and those of cavitation in relation to the hardness H_v and fracture toughness K_{IC} of ceramics, both damages are found to be proportional to $H_v^{-1/4} \cdot K_{IC}^{-4/3}$ and damages due to hot water jet impact are mainly caused by cavitation erosion. However, in case of silicon carbide and stellite 6B, both damages due to hot water jet impact are also caused by shear force due to high speed flow.

Thus, it is clarified that erosion damage due to hot water jet impact is caused by the interaction between cavitation and shear stress due to high speed flow.

REFERENCES

Evans, A.G., M.E. Gulden and M. Rosenblatt 1978. *Proc. R. Soc.* London, A. p. 361.

Hancox, N.L. & J.H. Brunton 1966. *Phil. Trans. Soc.* London, A260, p. 121.

Hempel, H. & H. Wiest 1986. *ASME*, 86-GT-199, p. 1.

Heymann, F.J. 1967. *Proc. 2nd. Int. Conf. Rain Erosion,* p. 683.

Ito, T. 1988. *Ceramics Japan*,23[7], p. 638.

Iwata, U., T. Sakuma, H. Takaku, T. Honda and F. Yoshida 1992. *Proc. Int. Conf. The Processing, Properties and Applications of Metallic and Ceramic Materials*, Vol.1, p. 499.

Iwata, U., T. Sakuma and H. Takaku 1993. JSME-ASME Int. Conf. on Power Engineering-93, Vol.1, p. 77.

Katayama, K. & S. Yamazaki 1986. *Trans. Jpn. Soc. Mech. Eng.*, 52[481], p. 2182.

Kawamoto, H. 1989. *Text of Course, Jpn. Soc. Mech. Eng.*,890-77, p. 29.

Sasa, T., A. Koga and M. Kurita 1982. *Bull. Ceram. Soc. Jpn.*, 90[8], pp447.

Recent Advances in Experimental Mechanics, Silva Gomes et al. (eds) © 1994 Balkema, Rotterdam, ISBN 90 5410 395 7

On the dynamic fracture behavior of the structural ceramics at various temperatures by caustics

K. Shimizu & S. Takahashi
Kanto Gakuin University, Yokohama, Japan

M. Suetsugu
Suzuka College of Technology, Japan

ABSTRACT: Fracture behavior of the partially stabilized zirconia (PSZ) and the silicon-nitride ceramics (Si_3N_4) is investigated under dynamic loading at various temperatures by using the method of caustics combined with the ultrahigh-speed camera. The values of dynamic fracture toughness K_{Id} and crack-propagation fracture toughness K_{ID} are obtained, and it is revealed that there is a dynamic effect on these values in PSZ and no dynamic effect is seen in Si_3N_4 up to the temperature of 1000°C. The dynamic crack arrest toughness K_{Ia} is found to exist for PSZ. Finally, a clearly visualized pattern based on AE-wave is shown and some discussions on this pattern are given.

1 INTRODUCTION

It is important to evaluate the strength of the ceramic materials under dynamic loading at high temperature. There are, however, only a few studies on such a problem due to the difficulties in performing the experiment under such severe conditions. Sakata et al. (1988) measured the value K_{Id} of dynamic fracture toughness at high temperature and Kobayashi et al. (1989) tried to obtain the value K_{ID} of stress intensity factor (SIF) for a fast running crack by using a hybrid experimental-numerical procedure. They employed strain gage technique and/or finite element method in their studies.

On the other hand, the method of caustics is a very powerful technique for such a problem. There are many papers concerning the dynamic crack behavior of various materials by caustics (For example, Kalthoff 1986). However, there is no work on application of this method to the ceramics. The authors have also been studying about the method of caustics (Shimizu et al. 1978 and Shimizu et al. 1985) and already presented a paper on the fundamental application techniques of caustics to the ceramics (Shimizu et al. 1990).

In the present paper, we first, describe a technique to apply the caustic method on measuring the dynamic SIF of the ceramics under dynamic loading at various temperatures. Subsequently, some results on the fracture behavior of partially stabilized zirconia (PSZ) and silicon-nitride ceramics (Si_3N_4) are presented. And Finally, visualized patterns based on AE-wave which is emitted during fracture are illustrated.

2 TEST SPECIMEN

We used the plates of partially stabilized zirconia (PSZ) and silicon-nitride ceramics (Si_3N_4). Figure 1 shows the fundamental configuration of the specimen. An artificial notch is cut as shown in Fig.1 by using a diamond wheel. The notch tip has a semicircular shape with various radius ρ. The specimen surface is highly polished. In some specimens, a natural crack is introduced by tapping a wedge into the plate of PSZ and by using BI (Bridge Indentation)

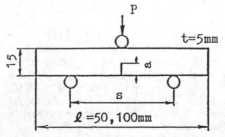

Fig.1 Configuration of the specimen

method for the plate of Si_3N_4. By using these specimens with a natural crack, the fracture toughness values K_{1c} of PSZ and Si_3N_4 are found to be 4.9MN/m$^{3/2}$ and 6.8MN/m$^{3/2}$, respectively.

3 EXPERIMETAL APPARATUS AND TEST PROCEDURE

The experimental equipment for the dynamic loading test at high temperature, constructed at our laboratory, is employed, details of which is described in the previous study (Shimizu et al. 1990). This setup includes the ultrahigh-speed camera (Imacon 790), He-Ne gas laser light source (2mW or 35mW) and the dynamic loading frame with an electric-resistance-type heating furnace. The framing speed of the ultrahigh-speed camera is from 10^4FPS (Frames per second) to 10^7FPS. The virtual image of the caustic pattern which is formed behind the specimen is recorded by using the collimated light and the high-sensitive Polaroid film 667. The specimen is dynamically loaded under three-point bending through a rod which is impacted by a falling weight. The heating furnace with a window of silica glass is prepared at our laboratory, by which the specimen can be heated up to 1200°C.

The equations for determining the stress intensity factor K_1 from the diameter of the caustic pattern, and radius r_0 of the initial curve are given as follows,

$$K_1 = \frac{1.671}{Z_0 t C_0} \times \frac{1}{\lambda^{3/2}} \times \left(\frac{D}{\delta}\right)^{5/2} \quad (1)$$

$$r_0 = D/(\delta \lambda) \quad (2)$$

where, c_0 is a constant of caustics and λ is a magnification factor of the optical system. c_0 is given by Eq.(3) for surface reflection technique.

$$c_0 = \mu /E \quad (3)$$

In these equations, D is the diameter of the caustic pattern, t is the specimen thickness, and z_0 is the distance between the specimen and the image plane. E and μ in Eq.(3) are Young's modulus and Poisson's ratio, respectively. The value of δ depends on the crack speed for a fast running crack and is determined referring to the theory by Nishioka et al. (1990).

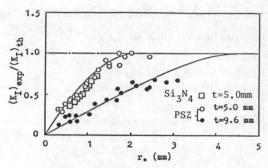

Fig.2 Effect of radius r_0 of the initial curve on SIF

In the present study, the convergent light is used for the static test and the collimated light is used for the dynamic test. Though the constant c_0 defined by Eq.(3), generally, depends on the strain rate, a value for the static condition is employed here for dynamic test too, considering that the change of its value is small. Furthermore, it is already revealed that radius r_0 of the initial curve exerts influence on K-value in caustics (Shimizu et al. 1978 and Shimizu et al. 1985) and we carried out an experiment to study about this effect in the ceramic materials of PSZ and Si_3N_4. Figure 2 illustrates the effect of radius of the initial curve on SIF. The ordinate means the ratio of the experimental value K_1 obtained by caustics and the theoretical one. For the ceramic materials used in this study, there is seen a similar result to the other materials as it is shown in Fig.2. We applied the correction method for the effect of r_0, the procedure of which is described in the other paper (Shimizu et al. 1990)

4 EXPERIMENTAL RESULTS AT ROOM TEMPERATURE

The specimens shown in Fig.1 with widths of 15, 20 and 30mm are used. The constant c_0 in Eq.(3) is determined to be 1.46x10^{-12}m^2/N for PSZ and 0.97x10^{-12}m^2/N for Si_3N_4 from the experiment using the strain gages.

Figure 3 shows examples of the caustic pattern under dynamic loading, the numbers in which mean the order of frames and the interval of these frames is 1μs. In Fig.3, we can observe another visualized pattern outside the caustics. These patterns are based on AE-wave which is generated due to

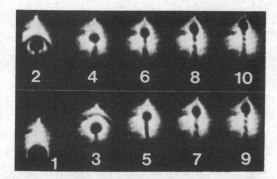

(a)PSZ, ρ =0.2mm, z_o=1380mm, 10^6FPS

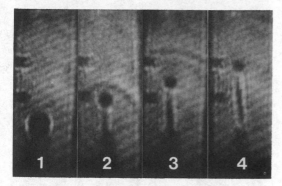

(b)Si$_3$N$_4$, ρ =1.0mm, z_o=1460mm, 10^6FPS

Fig.3 Examples of the caustic pattern under dynamic loading at room temperature

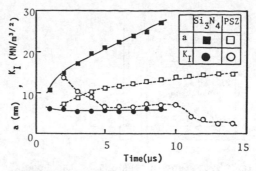

Fig.4 Variations of K_I and the crack length a with time

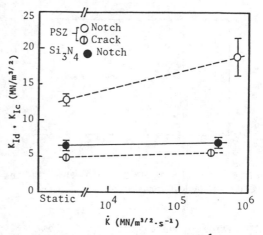

Fig.5 Relationship between K_{Id} and $\dot{K}$

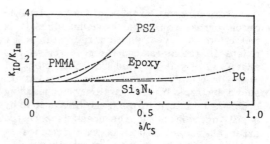

Fig.6 Relationship between K_{ID}/K_{Im} and $\dot{a}/C_s$ for various materials

the initiation of fracture and some discussions are given in later section. Figure 4 shows the variations of the dynamic SIF K_I and the crack length a with time, which are obtained from the caustic patterns as illustrated in Fig.3. In most specimens of PSZ, K_I-value showed oscillating phenomenon just before the initiation of fracture and its value is abruptly decreased at the beginning of fracture. The dynamic behavior of Si$_3$N$_4$ is considerably different from that of PSZ, showing that K_I-value is not decreaed at the initiation of fracture and remains constant.

Figure 5 shows the relationship between the dynamic fracture toughness K_{Id} and the stress intensity factor rate $\dot{K}$. The K_{Id}-value for PSZ has a tendency to increase with $\dot{K}$, particularly in the specimen with notch of ρ =0.2mm, but there is no effect of $\dot{K}$ on K_{Id} for the specimen of Si$_3$N$_4$ even with notch of ρ =0.2mm. The relationship between the crack-propagation fracture toughness K_{ID} for a fast running crack and the crack velocity a is shown in Fig.6, where

experimental results for the other materials of Plexiglas (PMMA), Polycarbonate (PC) and epoxy resin obtained at our laboratory are also shown for reference. The K_{ID}-value is normalized by K_{Im} which is the value of SIF at the crack velocity $\dot{a}\dot{=}0$ and a is normalized by C_s which is the theoretical value of the

1331

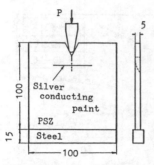

Fig.7 Shape and dimensions of the specimen for the test on dynamic crack arrest toughness of PSZ

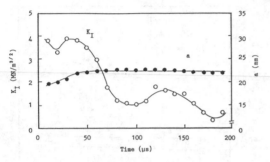

Fig.8 Variations of K_I with time in the dynamic crack arresting test

distortional wave velocity. From this figure, a tendency is seen that more brittle materials have a stronger dependency of K_{ID} on $\dot{a}$.

Furthermore, an experiment on the dynamic crack arrest toughness is tried, since it is important to reveal whether the value of K_{Ia} exists in the ceramics. The specimen of PSZ shown in Fig.7 is used for this purpose, where a slant artificial notch is cut as shown by broken line. A slowly propagating crack is initiated through the impact load by a falling weight and the crack is arrested after it propagated a certain distance as shown in Fig.8. After the crack propagated slowly at a speed of about 100m/s, it is arrested at $K_{Ia}=4MN/m^{3/2}$ and thereafter K_I-value is decreased with oscillation. Thus, the arresting phenomenon exists in PSZ and the value of K_{Ia} is approximately the same as K_{IC}.

5 EXPERIMENTAL RESULTS AT HIGH TEMPERATURE

The specimen with w=15mm shown in Fig.1 is used and the maximum tested temperature is

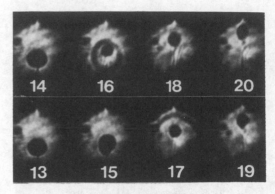

(a) PSZ, T=800℃, ρ =0.2mm, z_0=1710mm, 10⁶FPS

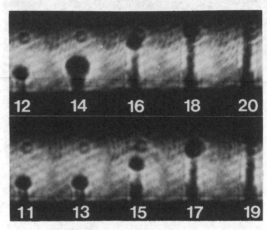

(b)Si₃N₄, T=1000℃, ρ =0.2mm, z_0=1830mm, 10⁶FPS

Fig.9 Examples of the caustic pattern under dynamic loading at high temperature

around 1000℃. The caustic pattern was observed up to this temperature for both the ceramic materials used. The constant c_0 in Eq.(3) is determined by substituting the theoretical value of K_I and the diameter D of the caustic pattern obtained under the static bending test into Eq.(1). The values of c_0, thus obtained, are employed both in static and dynamic tests.

Examples of the caustic pattern are shown in Fig.9, the numbers in which are the order of frames and the interval of each frame is 1μs. Despite such a severe circumstance as under dynamic loading at high temperature, the clear caustic patterns are obtained. Figure 10 shows the variatioins of K_I and the crack length a with time for the specimens in Fig.9. For PSZ, K_I-value is gradually increased until the crack begins to extend and is sharply decreased just at the

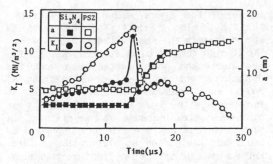

Fig.10 Variations of K_I and the crack length a with time

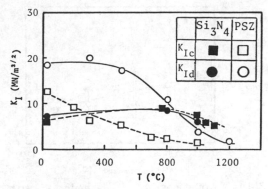

Fig.11 The values of K_{Id} and K_{Ic} versus temperature

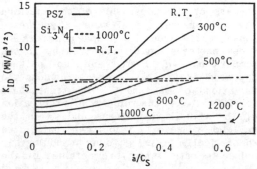

Fig.12 K_{ID} versus $\dot{a}/C_s$ at various temperatures

initiation of fracture. On the other hand, in Si_3N_4, a sudden increase in K_I-value is seen right after the initiation of fracture. This is caused probably by an effect of non-linear property of this material at this temperature.

The values of K_{Id} and K_{Ic} versus temperature relations are shown in Fig.11. In PSZ, the value of K_{Id} is larger than K_{Ic} and begins to decrease at around 500℃. In this material, the effect of temperature on K_{Id} is different from that on K_{Ic} and hence, the value of K_{Id}/K_{Ic} , which means the dynamic effect on fracture toughness, variates with temperature. From Fig.11, it can be seen that this dynamic effect becomes maximum at 800℃ and vanishes at 1200℃. On the other hand, for the material of Si_3N_4, there is not so large difference between K_{Id} and K_{Ic}, reaching maximum at 800℃, and no dynamic effect on the fracture toughness is seen.

Figure 12 illustrates the relationship between K_{ID} and $\dot{a}/C_s$ at various temperatures. As temperature is raised, K_{ID}-value is generally decreased and the dependency of K_{ID} on $\dot{a}$ is diminished in PSZ. For Si_3N_4, Fig.12 reveals that the value of K_{ID} is not dependent on $\dot{a}$ and moreover, there is no effect of temperature on K_{ID}-$\dot{a}$ relation up to 1000℃.

6 VISUALIZED PATTERN OF AE-WAVE

As it is already revealed in Figs.3 and 9, a clear circular ring pattern is observed outside the caustics in the plate of PSZ when the crack begins to extend. The formation of this pattern is based on AE-wave which is generated by fracture at the crack tip, and it is found that the propagating speed of this pattern coincides with the Rayleigh wave

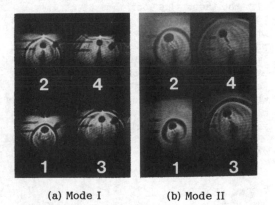

(a) Mode I   (b) Mode II

Fig.13 Examples of the visualized pattern of AE-wave at room temperature (PSZ, ρ =1.0mm, 10^6FPS)

velocity of the material. Figure 13 shows examples of such a visualized pattern of the specimen with a large notch tip radius of 1mm. In this figure, we can see a reflected

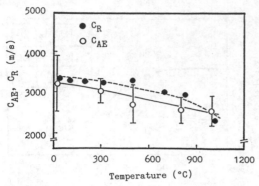

Fig.14 Variations of C_{AE} and C_R with temperature for PSZ

AE-wave at the upper side of the specimen and moreover, AE-wave generated subsequently from the propagating crack. These patterns can be observed up to the temperature of 1000℃ and no pattern appeared at 1200℃. Figure 14 depicts the variations of C_{AE} and C_R with temperature for PSZ. C_{AE} is the velocity of propagation of the visualized patterns and C_R is the Rayleigh wave velocity calculated theoretically by using Young's modulus and Poisson's ratio, which are obtained by the technique proposed by the authors (Shimizu et al. 1990). As shown in Fig.14, the experimental value C_{AE} is in good agreement with the theoretical one, C_R.

In Si₃N₄, such a clear visualized pattern as in PSZ cannot be observed, but a similar pattern is obtained for a specimen with a large notch tip radius of 1mm. Recent experiment at our laboratory revealed that this visualized pattern can be observed only when the crack is propagating at a speed more than 80 percent of the limiting crack velocity v_o. v_o is given by the relation $0.38C_b$, where C_b is the longitudinal wave velocity.

7 CONCLUSIONS

(1) The validity of the method of caustics together with the ultrahigh-speed camera is verified in studying the dynamic fracture behavior of ceramics up to the temperature of about 1000℃.
(2) At room temperature, K_{Id} is slightly increased with $\acute{K}$ and K_{ID} is increased with $\mathring{a}$ for PSZ. In Si₃N₄, no dependency of K_{Id} on $\acute{K}$ is seen and K_{ID} is constant over the wide range of the crack velocity. It is shown that K_{Ia}-value exists for PSZ and is approximately the same value as K_{IC}.
(3) At high temperature, for PSZ, K_{Id} and K_{IC} are decreased with temperature and the dynamic effect K_{Id}/K_{IC} becomes maximum at 800℃ and vanishes at 1200℃. K_{ID} is increased with $\mathring{a}$ and this dependency is diminished with increase of temperature. In Si₃N₄, K_{IC} and K_{Id} are slightly increased at 800℃ and diminished at higher temperature, and no dynamic effect for fracture toughness is observed up to 1000℃. There is no dependency of K_{ID} on $\mathring{a}$ up to 1000℃ for this material.
(4) A visualized pattern based on AE-wave is obtained. It is shown that the velocity of this pattern for PSZ coincides well with the Rayleigh wave velocity of the material at each temperature. The pattern can be observed only when the crack speed exceeds 80 percent of the limiting crack velocity.

The work was partly supported by the Ministry of Education, Science and Culture of Japan.

REFERENCES

Kalthoff, J.F. 1986. Fracture behavior under high rates of loading. Eng. Fract. Mech. 23:289-298.
Kobayashi, A.S. & K.H.Yang 1989. Dynamic stress intensity factors of brittle materials. Adv. Fract. Res. 1:621-632.
Nishioka, T. & H.Kittaka 1990. A theory of caustics for mixed-mode fast running cracks. Eng. Fract. Mech. 36:987-998.
Sakata, M. S.Aoki, K.Kishimoto, Y.Fujino & T.Akiba 1988. Measurement of dynamic fracture toughness of ceramic meterial at elevated temperatures by impact test with free end bend specimen. J. Mater. Sci. Japan. 37:910-915.
Shimizu, K. & H.Shimada 1978. Determination of stress intensity factor by the method of caustics. J. of Japanese Soc. for Non-destructive Inspection. 27:399-406.
Shimizu, K., S.Takahashi & H.Shimada 1985. Some propositions on caustics and an application to the biaxial-fracture problem. Exp. Mech. 25:154-160.
Shimizu, K., K.Otani & S.Takahashi 1990. On the strength evaluation of the zirconium ceramic plates by the direct surface reflection caustics. Proc. 9th Int. Conf. on Exp. Mech. 1462-1471. Copenhagen:Denmark

Recent Advances in Experimental Mechanics, Silva Gomes et al. (eds) © 1994 Balkema, Rotterdam, ISBN 90 5410 395 7

Probabilitical estimation of thermal shock fatigue strength of ceramics quenched in water

T. Sakuma & U. Iwata
Central Research Institute of Electric Power Industry, Tokyo, Japan

N. Okabe
Heavy Apparatus Engineering Laboratory, Toshiba Corporation, Yokohama, Japan

ABSTRACT:The retained strength after repeated thermal shock was analyzed theoretically and the probability of crack occurrence of ceramics was predicted. In this analysis, the Weibull statistical theory of fracture was applied to repeated thermal shock of ceramics on the assumption that thermal shock fracture of ceramics was due to crack growth from one of initial flaws. For the purpose of verification of this proposed analysis, tests for alumina, silicon carbide and silicon nitride with various diameters were conducted under various repeated thermal shocks conditions by means of water quenching. As a result, analytical results were in good agreement with results of repeated thermal shock tests. Thus, the probability of crack occurrence can be evaluated with the normalized fracture strength using effective volume and effective stresses hold time.

1 INTRODUCTION

Brittle ceramics are easily fractured by thermal stress due to rapid changes of temperature. Thus, it is essential to estimate the thermal shock resistance in application as the structural materials.

Thermal shock resistance is generally estimated by the quenching temperature difference that the retained strength is reduced rapidly (Hasselman, 1970; Niihara, 1982; Becher, 1981).

However, the temperature difference can not always be clarified owing to the dispersion of fracture strength. This is mainly caused by the quenching method or flaws that are generated at manufacturing and processing. Furthermore, ceramic fracture strength is subjected to loading conditions such as unloading or loading stress rate and stress hold time.

Different fracture strength is also caused by the existing probability of initial flaws. Authors, therefore, have proposed the novel quench method(Sakuma, 1991) and have proposed the evaluation method of the probability of crack occurrence(Sakuma, 1992,1993).

On the thermal shock fatigue, crack growth behavior has been investigated using notched ceramic specimens(Maekawa, 1990;Akiyama, 1990). The bending strength change after repeating thermak shock tests was investigated using reinforced ceramics(Kurushima, 1993). However, probabilitical estimation of thermal shock fatigue resistance is not studied thoroughly.

In this paper, probability of crack occurrence due to repeated thermal shock is predicted theoretically and repeated thermal shock tests for various kinds of ceramics are carried out by means of water quenching.

This paper also discusses the probabilitical estimation of crack occurrence after repeated thermal shock tests in relation to the effective stresses hold time and effective volume at water quenching.

2 ANALYSIS OF STRENGTH DAMAGE AFTER REPEATED THERMAL SHOCK

2.1 *Normalized Strength Evaluation Method*

Fracture strength of ceramics is subjected to a slow crack growth rate from an initial flaw with the size of a expressed as eq.(1).

$$\frac{da}{dt} = A K_{Imax}^n = B \left(\sigma_{max} \sqrt{a} \right)^n \tag{1}$$

By integrating eq.(1) allowing the stress to vary with time, and suppose that ceramics can be devided into a large number of elements and stress is constant in each element, then

$$\sigma_{imax}^n \, t_{eff} = C \left(\frac{1}{\sqrt{a_i}} \right)^{n-2} \tag{2}$$

where σ_{imax} and a_i are the maximum principal stress and the maximum flaw size of element of i respectively. The effective stress hold time is then

$$t_{eff} = \int_0^{t_f} \left\{ \frac{\sigma_i(t)}{\sigma_{imax}} \right\}^n dt \tag{3}$$

Suppose that the inert strength of ceramics follows two-parameter Weibull distribution, the distribution of a can be expressed as

$$F(a_i) = exp \left\{ - \left(\frac{a_i}{a_o} \right)^{-m/2} \right\} \tag{4}$$

By using eq.(2) and eq.(4), and applying the weakest link model, and suppose that t_{eff} is same in every element, the non-fracture probability R of ceramics as a whole under uniaxial stress is

$$R = exp \left\{ -b_o \left(\sigma_{max}^n \, t_{eff} \right)^{m/(n-2)} \int_V \left(\frac{\sigma_{imax}}{\sigma_{max}} \right)^{mn/(n-2)} dV \right\} \tag{5}$$

Subsequently, considering the value of n is 20-100, the fracture probability P_f is given by

$$P_f = 1 - exp \left\{ - \left(\frac{\sigma_{max} \, t_{eff}^{1/n} \, V_{eff}^{1/m}}{\sigma_o} \right)^m \right\} \tag{6}$$

where m and σ_o are the shape and scale parameter of two-parameter Weibull distribution of the delayed fracture strength. And the effective volume is expressed as

$$V_{eff} = \int_V \left(\frac{\sigma}{\sigma_{max}} \right)^m dV \tag{7}$$

Therefore, the fracture strength characteristics of ceramics are given by eq.(8) from which it is observed that the fracture strength is subjected to V_{eff} and t_{eff} and is scattered according to the Weibull distribution.

$$\sigma_f \, t_{eff}^{1/n} \, V_{eff}^{1/m} = \sigma_o \left\{ ln \left(1 - P_f \right)^{-1} \right\}^{1/m} \tag{8}$$

2.2 Analysis of strength damage due to repeated thermal shock

The fracture strength of ceramics due to thermal stress at quenching is also assumed to be subjected to a slow crack growth rate from an initial flaw. Employing the maximum principal stresses of thermal stress generated by thermal shock, the effective stress hold time t^*_{eff} is obtained according to eq.(3) based on the change in load stress with time σ_{max} (t) , and the effective volume V^*_{eff} is obtained according to eq.(7) based on the maximum principal stresses distribution σ_{max} (r,θ,z). Thus, the reduction of strength $\Delta\sigma$ due to repeated thermal shock is given by

$$\Delta\sigma = \sigma_{max} \left(N \, t^*_{eff} \right)^{1/n} V^{*\,1/m}_{eff} \tag{9}$$

where N is number of cycles and the effective volume is obtained by

$$V^*_{eff} = \int_V \left\{ \varphi \frac{\sigma(r, \theta, z)}{\sigma_{max}} \right\}^m dV \tag{10}$$

where σ_{max} is the maximum principal stress, and $\varphi=1$ for $\sigma \geq 0$, and $\varphi=0$ for $\sigma<0$. Therefore, the retained strength of ceramics after thermal shock σ_R is expressed as

$$\sigma_R \left(N \, t_{eff} \right)^{1/n} V_{eff}^{1/m} = \sigma_o \left\{ ln \left(1 - P_f \right)^{-1} \right\}^{1/m} - \sigma_{max} \left(N \, t^*_{eff} \right)^{1/n} V^{*\,1/m}_{eff} \tag{11}$$

Furthermore, the probability of crack occurrence at thermal shock is when $\sigma_R =0$ according to eq.(11). Therefore, the relation between the thermal shock stress σ_{max} and the fracture probability P_c at thermal shock is obtained by following equation.

$$P_c = 1 - exp \left\{ - \left(\frac{\sigma_{max} \left(N \, t^*_{eff} \right)^{1/n} V^{*\,1/m}_{eff}}{\sigma_o} \right)^m \right\} \tag{12}$$

3 REPEATED THERMAL SHOCK TESTS

3.1 Experimental procedure

Repeated thermal shock tests are performed by quenching in water. Specimen is heated by electric furnace and is dropped freely into water with supporting holder. Dropping height from furnace to water is 600 mm. Distance of falling through water is 600 mm.

Specimens used are sintered alumina, silicon carbide and silicon nitride. The geometric figure of the specimen is a cylindrical body with conical nose that does not introduce ambient gas such as air(Sakuma, 1991). Diameter of specimen is

7.5mm for alumina, 7.5 and 10mm for silicon carbide and 3.5, 5.0 and 7.5mm for silicon nitride. The schematic of thermal shock test by quenching in water is shown in Fig.1 and the shape and dimensions of specimen used are shown in Fig.2 respectively.

The temperature of specimen is measured by thermocouple that is installed in the dummy similar to the specimen. The temperature of quenching water T_o is controlled at 293K constant.

Crack occurrences are recognized by liquid penetrant tests.

3.2 Weibull parameters of specimens

In order to predict the probability of crack occur-

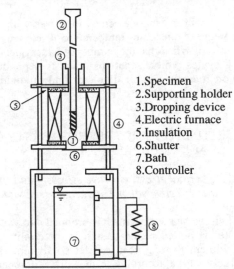

1.Specimen
2.Supporting holder
3.Dropping device
4.Electric furnace
5.Insulation
6.Shutter
7.Bath
8.Controller

Fig.1 Schematic of thermal shock test by quenching in water.

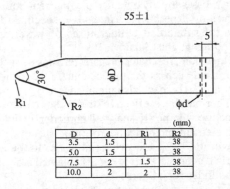

			(mm)
D	d	R1	R2
3.5	1.5	1	38
5.0	1.5	1	38
7.5	2	1.5	38
10.0	2	2	38

Fig.2 Shape and dimensions of specimen used.

rence due to repeated thermal shock, it is necessary to obtain the shape and scale parameter of ceramics. These parameters are obtained by four point bending tests.

Fig.3 shows the Weibull plots of fracture stresses of silicon carbide before quenching.

From results of bending tests for three kinds of ceramics before quenching, normalized fracture strengths are obtained by following equation,

$$\tilde{\sigma}_f = \sigma_f t_{eff}^{1/n} V_{eff}^{1/m} \qquad (13)$$

where σ_f is fracture strength, t_{eff} is effective stresses hold time and V_{eff} is effective volume obtained respectively by four point bending tests,

The shape parameter m and the scale parameter σ_o of two-parameter Weibull distribution are obtained by Weibull statistical analysis using normalized fracture strengths. Values of crack growth rate index n of alumina, silicon carbide and silicon nitride are 44, 37 and 55 respectively(Sakuma, 1992). The shape parameter m and the scale parameter σ_o of specimens used are shown in Table 1.

Table 1 Shape and scale paramete of specimens used.

Specimen	D(mm)	m	σ_o(MPa)
Al2O3	7.5	18.7	332
SiC	7.5	8.3	387
	10	7.7	309
Si3N4	3.5	9.4	766
	5.0	11.6	712
	7.5	7.1	1108

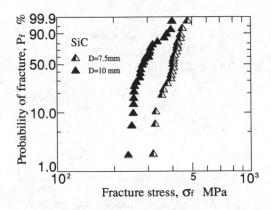

Fig.3 Weibull plots of fracture stresses of silicon carbide before quenching.

3.3 Crack occurrence ratio after repeated thermal shock tests

From the liquid penetrant tersts after repeated water quenching, it is observed that crack is growing from one of flaws and its starting point is at any location on the surface of cylindrical body.

The relationship between crack occurrence ratio P_o and number of water quenching cycles N is show in Fig.4-6. Crack occurrence ratio P_o is defined as following equation,

$$P_o = \frac{k_c}{k} \times 100 \quad (\%) \tag{14}$$

where k_c is the number of specimens cracked and k is the number of specimens used under same quenching conditions.

Crack occurrence ratio P_o increases according to the increase of number of cycles. Number of

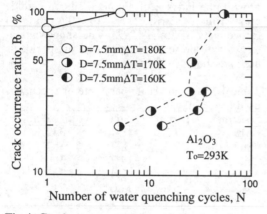

Fig.4 Crack occurrence ratio of alumina after repeated water quenching tests.

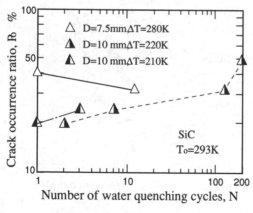

Fig.5 Crack occurrence ratio of silicon carbide after repeated water quenching tests.

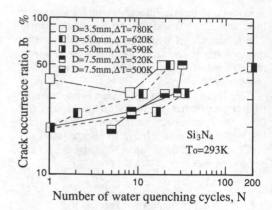

Fig.6 Crack occurrence ratio of silicon nitride after repeated water quenching tests.

cycles which generates crack increases with the decrease of quenching temperature difference ΔT.

It is clarified that the number of water quenching cycles which generates crack is dependent upon materials, their diameter and quenching temperature difference ΔT.

4 ESTIMATION OF THE PROBABILITY OF CRACK OCCURRENCE

4.1 Thermal stress, effective stresses hold time and effective volume at quenching in water

When the heated specimen is dropped into water, water is boiled in the course of the cooling of specimen. Boiling heat transfer coefficients are largely varied according to the surface temperature of specimen. These are also varied according to the subcooled temperature of water and the falling speed of specimen. Authors, in previous paper, have already obtained the transient boiling curves by experiments using silver specimen for several subcooled temperatures of water and dropping heights(Sakuma, 1992).
Furthermore, mechanical characteristics of specimen used such as thermal conductivity, specific heat and so on have also obtained.

Thermal stress, effective stresses hold time and effective volume are analyzed numerically by finite-element method when ceramic specimen as shown in Fig.2 is quenching in water. In the numerical analysis, boiling curves obtained are used as boundary conditions.

Fig.7 shows the relation between the dimensionless maximum thermal stress σ^*_{max} and Biot number B_i. Dimensionless maximum thermal stress and Biot number are defined as following equations respectively,

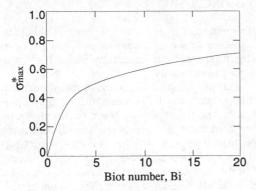

Fig.7 Dimensionless maximum thermal stress at quenching in water in relation to Biot number.

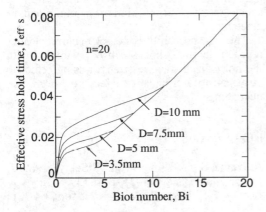

Fig.8 Effective stresses hold time at quenching in water in relation to Biot number.

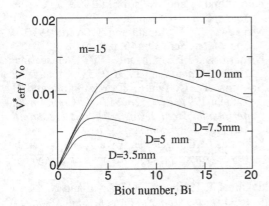

Fig.9 Effective volume at quenching in water in relation to Biot number.

$$\sigma_{max}^* = \frac{(1 - \mu_c) \, \sigma_{max}}{a_c \, E_c \, \Delta T} \qquad (15)$$

$$B_i = \frac{\alpha_m \, D}{2 \, \lambda_c} \qquad (16)$$

where μ_c, λ_c, a_c and E_c are Poisson's ratio, thermal conductivity, coefficient of thermal expansion and Young's modulus of ceramics, σ_{max}, ΔT and α_m are maximum thermal stress at quenching, quenching temperature difference and average boiling heat transfer coefficient respectively.

Fig.8 shows the variations of effective stresses hold time t^*_{eff} with Biot number B_i for various diameters at crack growth rate index $n = 20$. Effective stresses hold time varies with diameter and increases according to the increase of diameter within the range of small Biot number. Effective stresses hold time also varies with crack growth rate index n. Effective stresses hold time decreases with the increase of n.

Fig.9 shows the variation of effective volume for various diameters at shape parameter $m = 15$. V_o is a geometric volume of specimen. Effective volume V^*_{eff} depends upon the diameter of specimen D. Effective volume also depends upon the shape pameter m. Effective vcolume decreases according to the increase of m.

4.2 Probability of crack occurrence

The probability of crack occurrence due to repeated thermal shocks is calculated by eq.(12).

On the other hand, the maximum thermal stress, the effective stresses hold time and the effective volume after thermal shock can be obtained in relation to Biot number by results of numerical analyses. Biot number can be calculated in relation to quenching temperature difference (Sakuma, 1992). Accordingly, the normalized fracture strength after repeated thermal shocks expressed as following equation can be obtained.

$$\tilde{\sigma}_f^* = \sigma_{max} \left(N \, t_{eff}^* \right)^{1/n} \left(V_{eff}^* \right)^{1/m} \qquad (17)$$

Fig.10, 11 and 12 show Weibull plots of the probability of crack occurrence in relation to the normalized fracture strength after repeated thermal shocks of alumina, silicon carbide and silicon nitride respectively. Each line is obtained by calculation according to eq.(12). Each mark is plotted applying median rank method to results of Fig.4, 5 and 6.

From these results, it is concluded that the estimation method of the probability of crack occurrence after repeated thermal shocks almost satisfies experimental results. Thus, the probability of crack occurrence can be evaluated with the normalized fracture strength using effective stresses

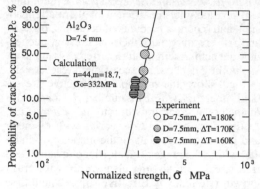

Fig.10 Probability of crack occurrence of *Al2O3* after repeated thermal shocks.

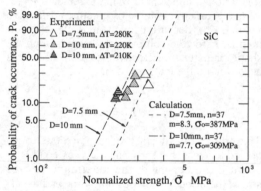

Fig.11 Probability of crack occurrence of *SiC* after repeated thermal shocks.

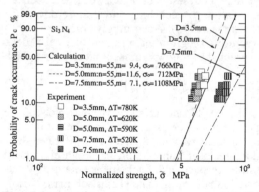

Fig.12 Probability of crack occurrence of *Si3N4* after repeated thermal shocks.

hold time and effective volume.

5 CONCLUSION

It is essential to estimate the thermal shock fatigue resistance of ceramics in applications as structural materials especially under high temperature conditions.

This paper has proposed the probabilitical estimation method of thermal shock fatigue resistance. The pobability of crack occurrence after repeated thermal shock is analyzed theoretically on the assumption that the strength of ceramics follows the two-parameter Weibull distribution. For the verification of proposed analysis, tests for three kinds of ceramics are carried out under various thermal shocks conditions by water quenching. As a result, analytical results were in good agreement with results of repeated thermal shocks tests for various kinds of ceramics and diameters.

Thus, the probability of crack occurrence after repeated thermal shocks can be evaluated with the normalized fracture strength using effective stresses hold time and effective volume at quenching.

REFERENCES

Akiyama, S., Y. Kimura and M. Sekiya 1989. *J. Soc. Mater. Sci.,* 38-435, p. 1415.

Becher, P.F. 1981. *J. Am. Ceram. Soc.,* 64-1, p. 37.

Hasselman, D.P.H. 1970. *J. Am. Ceram. Soc.,* 53-9, p. 490.

Kurushima, T,and K. Ishizaki 1993. *J. Ceram. Soc. of Jpn.,* 101-5, p. 596.

Maekawa, I., H. Shibata and T. Wada 1990. *J. Soc. Mater. Sci.,* 39-445, p. 1380.

Niihara, K., J.P.Singh and D.P.H. Hasselman 1982. *J. Mater. Sci.,* 17, p. 2553.

Sakuma, T., U. Iwata, T. Mizukami and H. Takaku 1991. *Proc. 2nd. World Conf. on Experimental Heat Transfer, Fluid Mechanics and Thermodynamics,* p. 537.

Sakuma, T., U. Iwata, H. Takaku and N. Okabe 1992. *Trans. Jpn. Soc. Mech. Eng.,* 58-547, A p. 476.

Sakuma, T., U. Iwata, H. Takaku and N. Okabe 1993. *Trans. Jpn. Soc. Mech. Eng.,* 59-557, A p. 131.

Recent Advances in Experimental Mechanics, Silva Gomes et al. (eds) © 1994 Balkema, Rotterdam, ISBN 90 5410 395 7

Initiation and growth of interface crack and evaluation of fracture criterion for ceramic-metal joints

Yoshio Arai, Eiichiro Tsuchida & Masakatsu Naitoh
Saitama University, Urawa, Japan

ABSTRACT : Fracture initiation and stable crack extension on the interface were measured by the ultrasonic method for the Si_3N_4-carbon steel joints. Based on the results, the fracture mechanisms of the ceramic-metal joints were studied. Analytical studies on the joining residual stresses for the Si_3N_4-carbon steel joint specimen were performed by the finite element method.

1 INTRODUCTION

The introduction of ceramic-metal joining is important to apply the ceramics to the structural components. It could make the best use of the merit and make up for the demerit. The fracture strength of the ceramic-metal joints is, however, influenced by the residual stresses and the stress concentration(the singular stress field) caused by the difference of the thermal expansion coefficients and the elastic moduli [1] [2] [3] [4] [5].

In this study, fracture initiation and stable crack extension on the interface were measured by the ultrasonic method for the Si_3N_4-carbon steel joints. Based on the results, the fracture mechanisms of the ceramic-metal joints were studied. Analytical studies on the joining residual stresses for the Si_3N_4-carbon steel joint specimen were performed by the finite element method.

2 EXPERIMENTAL PROCEDURE

Si_3N_4 and Japan Industrial Standards (JIS) S45C (carbon steel) were joined by the activation metal, vacuum brazing method. A copper(Cu) sheet was used as the interlayer and a Ti-Ag-Cu alloy was used as the brazing filler metal. Material properties and conditions of joining are listed in Tables 1 and 2, respectively. The configuration and dimensions of specimens and specimen numbers(T. P. No.) are shown in Fig. 1 and Tables 3. The

thickness of specimens were varied from 2.5 mm to 14 mm. Four point bending test procedures in JIS were adhered to [6]. The ultrasonic echo voltage reflected on the ceramic-metal interface were measured continuously during the bend test. The central frequency of the ultrasonic pulse were 4 MHz and the angle beam method were used. The strains near the interface were measured by the wire strain gauge with gauge length 2 mm.

The relation between the interface crack length and the ultrasonic echo gain were calibrated using the notched ceramic-metal joints. The ultrasonic echo height, A, are calculated by the following equation.

$$A = 20 \log \frac{H - H_{c0}}{H_{c1} - H_{c0}} \qquad (1)$$

where, H is the ultrasonic echo voltage reflected on the interface notches with given length, H_{c0} is the ultrasonic echo voltage reflected on the interface before introduction of the notches, H_{c1} is the ultrasonic echo voltage reflected on the interface notch with 1 mm long.

The ultrasonic echo height, A_0, for the specimen before the bending test are defined by the following equation.

$$A_0 = A_c - A_E + 20 \log \frac{H_0}{H_{c0}} \qquad (2)$$

where, A_c is the amplifier gain at the calibration test, A_E is the amplifier gain at the bending test, H_0 is the ultrasonic echo voltage before the bend-

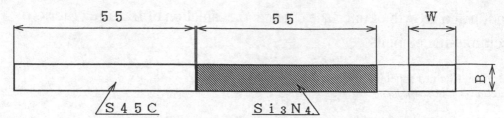

Fig. 1 Specimen configuration.

Table 1: Material properties

	Si_3N_4	Cu	S45C
E (GPa)	304	108	206
ν	0.27	0.33	0.3
$\alpha(\times 10^{-6})$	3.0	17.7	12.0
σ_B(MPa)	980	-	-

Table 2: Condition of joining

Brazing filler	:	Ti-Ag-Cu
Temperature	:	1073~1123 K
Atmosphere	:	Vacuum, 1×10^{-5}torr
Interlayer	:	Cu(thickness 0.2mm)

Table 3: Specimen size(mm)

T. P. No.	W	B
1425	14	2.5
144	14	4
1566A	15	6.6
1566B	15	6.6
148	14	8
1014A	10	14
1014B	10	14
305A	30	5
305B	30	5
3066A	30	6.6
3066B	30	6.6
1514	15	14

ing test. A_0 decreased with increasing the reflection times.

The ultrasonic echo height, A_{ex}, for the specimen during the bending test are defined by the following equation.

$$A_{ex} = A_c - A_E + A_0 + 20 \log \frac{H - H_0}{H_{c1} - H_{c0}} \qquad (3)$$

where, H is the ultrasonic echo voltage reflected on the interface or the interface crack during the bending test.

3 EXPERIMENTAL RESULTS

The relations between the nominal bending stress, σ, and the measured strain, ε, obtained by the four-point bending tests for the Si_3N_4-carbon steel joints are shown in Fig. 2. The linear relations were obtained in T. P. 148. The non-linearity were observed in T. P. 1566B at about 80MPa.

The relations between the ultrasonic echo height, A, reflected on the interface and the nominal bending stress are shown in Fig. 3. The ultrasonic echo heights increase immediatly after the starting the bend loading in T. P. 1425 and 148. The increases of the ultrasonic echo height can be interpreted by the increases of the cracked area on the interface. In T. P. 1566A and 1566B, the ultrasonic echo heights increase at the nominal bending stresses at which the nonlinearity of $\sigma - \varepsilon$ relation occured. The stresses are judged to be the initiation stresses of the interface fracture.

The results of the calibration test are shown in Fig. 4. The relation between the notched area, s, and the ultrasonic echo height, A, is given by the following equation:

In $-32.98 \leq A \leq 0$,

$$s = 0.015A + 0.5 \qquad (4)$$

In $0 \leq A \leq 30$,

$$s = 0.0032A^2 + 0.0932A + 0.4692 \qquad (5)$$

Using the above best fit relations, the interface crack length were calculated. The relations between the interface crack length, a, and the nominal bending stress, σ, are shown in Fig. 5. The interface crack open immediately after the starting the bend loading in T. P. 1425 and 148. In

T. P. 1566A and 1566B, the onset of the interface fracture are observed at about half the bending strength. In all specimen, the stable crack growth occured before the final fracture.

The illustration of the fracture surface is shown in Fig. 6. The fracture origin in the fracture surface is the cramic-interlayer interface. The mechanical parameter at the edge of the interface and the interface fracture toughness shouldbe used to describe the onset of the interface fracture. The interface cracks grow stably on the interface before the final fracture. The strength of the ceramic-metal joints are given by the nominal bending stress at the final fracture. So, the strength of the ceramic-metal joints are controlled by the characteristics of the interface crack growth. The final fracture occured with the deflection of the crack from the interface into the ceramic. To describe the final fracture, the elastic-plastic mechanical parameter for the interface crack and the criteria of the crack deflection should be used.

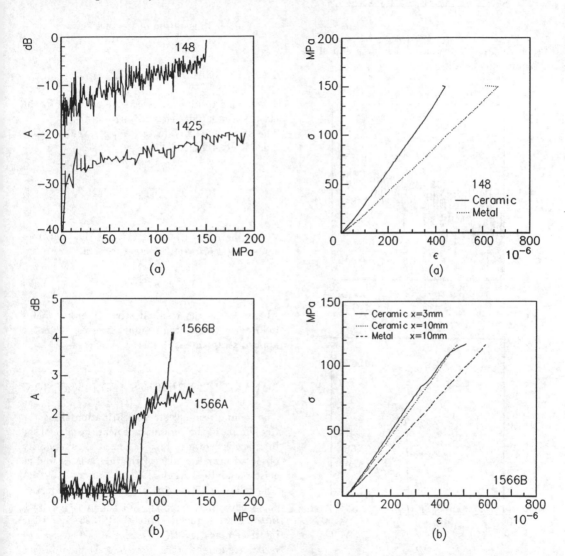

Fig. 2 Relations between nominal bending stress and strain: (a) T. P. 148; (b) T. P. 1566B.

Fig. 3 Relations between ultrasonic echo height reflected on interface and nominal bending stress: (a) T. P. 148 and 1425 ; (b) T. P. 1566A and 1566B.

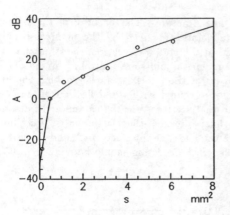

Fig. 4 Relation between notched area and ultrasonic echo height.

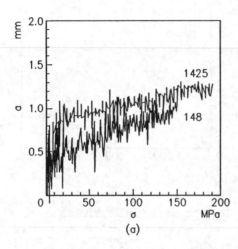

(a)

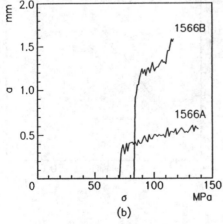

(b)

Fig. 5 Relations between interface crack length and nominal bending stress: (a) T. P. 148 and 1425 ; (b) T. P. 1566A and 1566B.

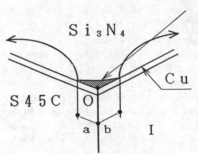

Fig. 6 Illustration of fracture surface.

4 ANALYTICAL PROCEDURE

The joining residual stresses were analysed by the elastic-plastic finite element method [7].
The stress intensity factor, K, for the interface crack is given by Rice [8].

$$(\sigma_y + i\tau_{xy})_{\theta=0} = \frac{K r^{i\varepsilon}}{\sqrt{2\pi r}} \qquad (6)$$

We estimate the stress intensity at the final fracture using a principle of superposition.

$$K = K^R + K^L \qquad (7)$$

where, K^R is the residual stress intensity factor induced by the joining residual stresses, K^L is the applied stress intensity factor.

5 ANALYTICAL RESULTS AND DISCUSSIONS

The joining residual stress distribution along the free surface at the ceramic side near the edge of the interface is shown in Fig. 7. The stress singularity observed near the edge of the interface as well as elastic analytical results [9] [10].
The relation between the stress intensity factor, K_1, where $K = K_1 + iK_2$, of the interface crack in the ceramic-metal joints(T. P. 1566A) and the interface crack length, a, is shown in Fig. 8. The resistance for the crack growth along the interface increase with crack advance. The unstable fracture occured when the stress intensity factor reached the fracture toughness of the ceramic(Si_3N_4).

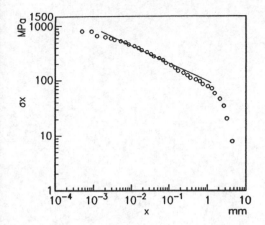

Fig. 7 Analytical results of joining residual stress(elastic-plastic finite element method).

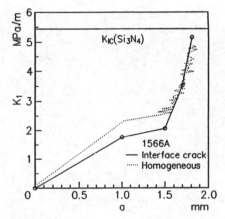

Fig. 8 Interface fracture resistance curve measured using ultrasonic method.

6 CONCLUSIONS

Fracture initiation and stable crack extension on the interface were measured by the ultrasonic method for the Si_3N_4-carbon steel joints. Based on the results, the fracture mechanisms of the ceramic-metal joints were studied. Analytical studies on the joining residual stresses for the Si_3N_4-carbon steel joint specimen were performed by the finite element method. The results obtained are as follows:

(1) The fracture initiation and the stable crack extension on the interface can be measured by the ultrasonic method. The ultrasonic echo height increases with the fracture ini-

tiation. The stable crack extension can be also estimated through the ultrasonic echo height. The fracture initiation occures before the maximum load at the ceramic-interlayer interface.

(2) The interface cracks grow stably on the interface before the final fracture. The strength of the ceramic-metal joints are given by the nominal bending stress at the final fracture. The strength of the ceramic-metal joints are controlled by the characteristics of the interface crack growth.

(3) The final fracture occures when the interface crack deflect into the ceramic. Using applied and residual stress intensity factor for the interface crack and the stress distribution around the interface crack the final fracture can be successfully predicted.

References

[1] Dalgleish, B. J., Lu, M. C. and Evans, A. G., Acta Metall., 36-8 (1988), 2029.

[2] Cao, H. C., Thouless M. D. and Evans, A. G., Acta Metall., 36-8 (1988), 2037.

[3] Evans, Lu M. C., Schmauder S. and Ruhle, M., Acta Metall., 34-8 (1986), 1643.

[4] He, M. Y. and Evans, A. G., Acta Metall., to be published.

[5] Kobayashi, H., Arai, Y., Nakamura, H. and Nakamura, M., Proc. KSME/JSME Joint Conf. Fracture and Strength '90, (1990), 236.

[6] JIS-R1601-1981.

[7] Kobayashi, H., Arai, Y., Nakamura, H. and Sato, T., Mater. Sci. and Engng, A143, (1992), 91.

[8] Rice, J. R., Trans. ASME, J. Appl. Mech., 55 (1988), 98.

[9] Bogy, D. B., Trans. ASME, J. Appl. Mech., 35 (1968), 460.

[10] Blanchard, J. P. and Ghoniem, N. M., Trans. ASME, J. Appl. Mech., 56 (1989), 756.

Recent Advances in Experimental Mechanics, Silva Gomes et al. (eds) © 1994 Balkema, Rotterdam, ISBN 90 5410 395 7

Experimental determination of fatigue crack growth in fiber reinforced concrete

Henrik Stang
Department of Structural Engineering, Technical University of Denmark, Lyngby, Denmark

Zhang Jun
Beijing Building Construction Research Institute, People's Republic of China

Abstract *Fatigue fracture has been studied extensively in the past for metals and ceramics and is now relatively well understood. For concrete, however, the knowledge of fatigue fracture is still rather restricted, and for fiber reinforced concrete knowledge is even more limited. The objective of the present study is to obtain basic crack growth data for fiber reinforced concrete under cyclic flexural loading.*

In this paper, a number of steel fiber reinforced concrete beams were tested in three point fatigue bending. Testing was stopped at different stages in the load history for different specimens and the crack length was measured directly on polished sections by means of fluorescence microscopy. At the same time, the cracking process was investigated indirectly by means of acoustic emission. In this way both quantitative and qualitative results describing the crack initiation and propagation in fiber reinforced concrete under fatigue load were obtained as a function of the load cycles and the crack growth rate $\left(\frac{da}{dN}\right)$ was calculated and analyzed.

The test results show that the damage process of fiber reinforced concrete under cyclic three point bending (using deflection, crack length or accumulated acoustic emission counts as damage measure) is similar to that of plain concrete. The damage process can be divided into three stages, $(0-0.05)N_f$, $(0.05-0.90)N_f$, $(0.90-1.00)N_f$, where N_f is the number of cycles to failure. The crack growth rate in stage 2 can be divided into two parts. Up to about 40 percent of the fatigue life, the rate decreased with cycles and after that it increased. The Paris Law (which was introduced in the analysis of the fatigue crack growth in metals) is not applicable to the tests conducted in the present investigation.

1 Introduction

Much attention has been paid in recent years to fatigue problems by researchers and structural designers. This increasing interest is due to several reasons:

- The use of new types of structures such as marine structures subjected to wind and wave loading.

- The use of new types of materials such as high strength concrete and fibre reinforced concrete of which little is known on their long term behavior.

- The growing recognition that the effects of repeated loading on the characteristics of the materials (static strength, stiffness, durability etc.) may be significant under service loading even if the loading does not cause a fatigue failure.

A number of studies have been made in the past to evaluate the fatigue performances of plain concrete. Fiber reinforced concrete is a relatively new structural material developed through extensive research and development during the last two decades. It has already found a wide range of practical applications and has proved a reliable structural materials having superior performance characteristics compared to conventional concrete. Incorporation of steel or other types of fibers in concrete has been found to improve several of its properties, primarily cracking resistance, impact and wear resistance and ductility. For this reason fiber reinforced concrete is now being using in increasing

amounts in structures such as airport pavements, highway pavements, bridge decks, machine foundations and storage tanks. However, the fatigue performance of fiber reinforced cement and concrete has not been paid much attention to by researchers and users. At present little data is available on fatigue behavior of fiber reinforced cement and concrete. A review of literature on fatigue of fiber concrete (10 references) can be found in [1]. A database must be developed based on well designed experiments of cyclic loading on such materials. Furthermore, theoretical models should be developed to provide the understanding of the factors controlling fatigue response and for future design of materials with modified cracking behavior and with superior fatigue properties.

The primary objective of this investigation is to provide basic experimental data for fatigue crack growth through determination of the behavior of concrete beams reinforced with steel and polypropylene fibers when subjected to constant amplitude fatigue flexural loading.

The methods which can be applied for measuring the fatigue crack length are limited in number. For plain concrete, the crack mouth opening displacement (CMOD) compliance method and the concept of equivalent crack length [2] has been used successfully by some authors [3], [4], [5]. For fiber reinforced concrete this method can not be used because the fiber bridging effect.

In this paper, direct observations of crack length using a fluorescent epoxy impregnation technique is combined with overall compliance measurements to allow for translation of deflection versus cycle ratio plots to real crack length versus cycle ratio plots thus making direct determination of crack growth rate possible.

2 Test Setup

2.1 Materials, mix and specimens

A rapid hardening cement was used for all specimens. The fine aggregate used was natural sand with a maximum particle size of 4 mm. The coarse aggregate used was natural stone, having a maximum particle size of 8 mm. In our investigation, two different fibers were used: a 12 mm fibrillated polypropylene fiber and a hooked-end steel fiber, 30 mm long, 0.5 mm in diameter. The recipe for the fiber concrete is outlined in table 1 while the basic fiber properties are outlined in table 2.

The concrete mixing procedure used was as follows: The polypropylene fibers are first mixed with the fine and coarse aggregate together with $\frac{1}{3}$ of the required water. The fibers and aggregate were mixed for 5 minutes in a pan mixer with a special activator to ensure complete wetting of the aggregates and even distribution of the fibers. Next the cement was added, followed by the remaining $\frac{2}{3}$ of the water with the super plastiziser mixed in. All the ingredients were mixed for 3 minutes. Finally, the steel fibers were sprinkled in by hand to insure even distribution and all ingredient were then mixed for another 5 minutes.

The test specimens were beams with dimensions 50 mm width, 100 mm depth and 350 mm length. This type of specimen was used in all flexural tests, static as well as cyclic. After casting and finishing the surface, the specimens were covered with a plastic sheet and cured for 24 hours at room temperature. They were then removed from the mould and placed in water maintained at 23° C until beginning of the tests. Because of the large number of specimens and time needed for fatigue testing, all of the specimens were water cured for at least 60 days. At that age the specimens were removed from the water and the surfaces were coated with paraffin wax to minimize the effects of both age and moisture loss on the test results.

2.2 Testing equipment

The beams were loaded in three point bending with a span of 345 mm. Deflections at the middle were measured using two standard dynamic Instron extensometers type 2620-602 with 12.5 mm gauge length fitted on each side of the specimen on a special measuring device recording the relative deflection between the loading point and a point located between the two supports of the standard Instron bending rig which form the basis of the setup. The average of the two extensometer signals were used as a measure for the deflection. All of the flexural tests were carried out in an 10 kN, 6022 Instron testing machine.

The damage development was measured using acoustic emission. The acoustic emission was recorded using the Locan 320 equipment from Physical Acoustics Corporation with two sensors each located on the surface at opposite ends of the specimen. The sensors used were general purpose sensors type R15 also from Physical Acoustics Corporation.

2.3 Loading history

For static flexural tests, the loading rate was deformation controlled using the deflection of the specimen at the midspan as feed back signal. The deflection was increased at a constant rate of 0.10 mm/min according to ASTM C1018.

For fatigue tests, one stage constant amplitude fatigue loading was used. The tests were carried out in load control using a sinusoidal wave form with a frequency of 2 Hz. A constant ratio R between the minimum and maximum load levels

$$R = \frac{P_{min}}{P_{max}} = 0.2 \tag{1}$$

and a constant maximum load level S_{max}

$$S_{max} = \frac{P_{max}^{fatigue}}{P_{max}^{static}} = 0.9 \tag{2}$$

was maintained throughout the testing.

3 Crack Length Measurement

For crack length measurement the following destructive method was used at different stages in the loading history. The constant amplitude fatigue test of different specimens is stopped at different stages in the fatigue life. The beams are then vacuum impregnated with epoxy containing a fluorescent dye. Next, sections of size 100x100x10 mm are cut from the middle part of the specimens for further fluorescent epoxy impregnation. Finally, the surface is polished in order to facilitate observation. The crack length corresponding to different stages in the loading history can then be determined by visual inspection under ultra violet illumination.

Note, that this type of crack length measurement is different from the CMOD/ equivalent crack length measurement usually adopted in concrete fatigue [3], [4], [5] and [2]. It is believed that the equivalent crack length approach is not applicable to fiber reinforced concrete because of the very large difference between real and equivalent crack length which is due to the fiber bridging effect.

4 Test Results and Interpretation

4.1 Static properties

The quasi-static load-deflection curves from 6 specimens along with the constructed mean curve are shown in fig. 1. From this the mean first crack load level and the mean peak load can be determined as 7.35 kN and 9.46 kN, respectively.

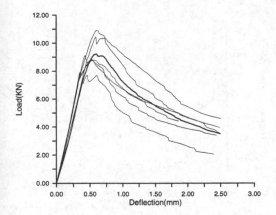

Figure 1. The quasi-static load – deflection curves showing all specimens and the constructed mean load – delection curve.

4.2 Fatigue test results

Seven specimens were used for crack length measurement tests as described above. All tests were carried out as constant amplitude fatigue tests with $R = 0.2$ and $S_{max} = 0.9$.

The test results are shown in table 3. As an example the crack pattern observed in specimen 1 is shown in fig. 2.

As shown a distributed type of cracking can be observed around the main crack. In the present study the crack length is measured as the longest distance between the bottom of the beam and a crack tip connected to the main crack.

The relationship between the crack length and the corresponding maximum deflection in the fatigue test, is shown in fig. 3. A linear fit of the experimental results give the following relationship between crack length a and maximum deflection u_{max}: .

$$a = Au_{max} + B \qquad (3)$$

where $A = 110.1$ and $B = 13.1$ mm under a fatigue load with $R = 0.2$ and $S_{max} = 0.9$.

Care should be taken in the use of the above equation for crack lengths larger than or smaller than the crack lengths which form the basis for the regression analysis. Furthermore, due to the non-linearity of the material the relationship should not be formulated as a crack length/ compliance curve.

A number of beams were also tested to failure under the same load loading conditions. The relationships between maximum deflection and the cycle ratio are shown in fig. 4. Using the relationship (3) the maximum deflection can easily be translated to crack length resulting in estimates for crack lengths in the range from 60 to 95 mm as a function of cycle ratio. The crack length/ cycle ratio relationships are shown in the same figure using the right y-axis.

Figure 2. The crack pattern observed in specimen 1 after 194 cycles. The visualization is due to fluorecent epoxy impregnation and ultra violet lightening.

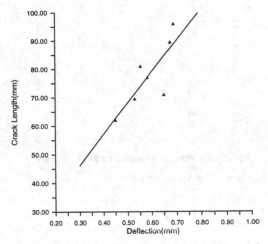

Figure 3. The relationship between crack length and deflection observed in the present investigation with specimens loaded in fatigue with $R = 0.2$ and $S_{max} = 0.9$.

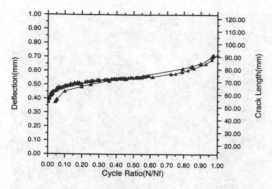

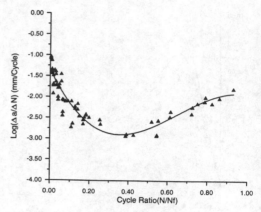

Figure 4. The relationship between maximum deflection or crack length versus cycle ratio.

Figure 6. Crack growth rate as function of cycle ratio during stage 2 crack growth.

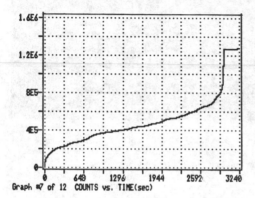

Graph #7 of 12 COUNTS vs. TIME(sec)

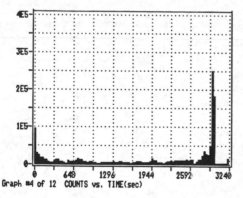

Graph #4 of 12 COUNTS vs. TIME(sec)

Figure 5. AE characteristic during the entire load history: AE count rate and accumulated AE activity as functions of time crack growth.

Damage development was also measured by means of acoustic emission (AE) technique. Interpretation of the results from AE recordings on composite materials is not an easy task since many different mechanisms (matrix cracking, fiber debonding as well as fiber pull-out or fiber breakage) all contribute to the AE signal generation. Neverthe-

less, acoustic emission is a powerful tool in non-destructive testing and is useful in detection of microcracks. The AE activity of the specimens was monitored as described above through the entire load history. All of the AE data are stored, displayed and analyzed by the LOCAN 320 system. In fig. 5 typical results for accumulated counts versus time and count rate versus time are shown. Clearly, the damage evolution can be divided into three stages. The first stage is ranging from 0 to about 5 percent of total life and is characterized by a rapid increase in accumulated AE signals. The second stage is ranging from 5 to 90 percent of the fatigue life and is characterized by an almost constant AE count rate. Finally, a rapid increase in AE count rate takes place from 90 to 100 percent of the fatigue life. The fatigue life primarily depends on the second stage where a more or less uniform cracking process takes place.

From these results and from fig. 4 we can see that the crack length in the second stage is growing from 65 mm to 85 mm . The crack growth rate can also be calculated. The results are shown in fig. **??** . From these results it can be seen that the crack growth rate shows a decreasing tendency up to about 40 percent of the fatigue life while it increases in the last 60 percent of the fatigue life.

For metals and ceramics the well know Paris-Erdogan law [6], [7] has been shown to describe crack growth satisfactory:

$$\frac{da}{dN} = c(\Delta K)^n \qquad (4)$$

where a is crack length, N is number of cycles, ΔK is stress intensity factor range, and c and n are empirical material constants.

The stress intensity factor range can be calculated according to

$$\Delta K = (P_{max} - P_{min})\frac{f(\frac{a}{d})}{bd^{0.5}} \qquad (5)$$

where b is specimen width and d is specimen depth. The $f(\frac{a}{d})$ -function depends on specimen geometry , and can be obtained by elastic finite element analysis. For the present three-point bending specimens we have:

$$f(\frac{a}{d}) = (1 - \frac{a}{d})^{-\frac{3}{2}}(1 - 2.5(\frac{a}{d}) + 4.49(\frac{a}{d})^2$$
$$-3.98(\frac{a}{d})^3 + 1.33(\frac{a}{d})^4) \quad (6)$$

According to the crack length data based on the present method, the relationship between crack growth rate and stress intensity factor range is shown in fig. ??. Two linear plots are also shown together with the data points given by:

$$\frac{da}{dN} = 3.56\,[\text{mm}]\,(\Delta K\,[\text{Nmm}^{-\frac{3}{2}}])^{-3.12} \quad \text{for} \ \ N < 0.4N_f \quad (7)$$

and

$$\frac{da}{dN} = 0.15\,[\text{mm}]\,(\Delta K\,[\text{Nmm}^{-\frac{3}{2}}])^{1.84} \quad \text{for} \ \ N > 0.4N_f \quad (8)$$

It follows from the figure that the crack growth rate is reduced with increasing stress intensity factor range up to about 40% of the fatigue life while a more "normal" relationship between crack growth rate and stress intensity factor range is observed in the last 60% of the fatigue life. Thus, contrary to what is found in testing conventional concrete, see eg. [3] and [4] Paris' law does not seem to be applicable to fiber reinforced concrete. This is probably due to the major role that fiber bridging plays in the crack growth mechanisms, while energy dissipation at the crack tip only plays a minor rôle.

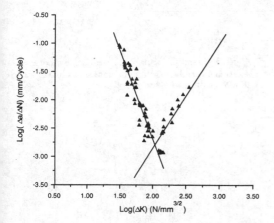

Figure 7. The relationship between crack length increments per cycle versus stress intensity factor range an two linear fits corresponding to crack growth during the first 40% and the last 60% of the fatigue life.

5 Conclusions

The following conclusions may be drawn from this study

1. A reliable method for the determination of crack growth rates based on observation of actual crack length has been proposed. Observation of crack length is based on a fluorescent epoxy impregnation technique and determination of crack growth rate is based on a proposed relationship between crack length and maximum deflection under the same loading conditions as the actual fatigue tests.

2. The crack length and accumulated acoustic emission counts show the same development patterns with time or cycles. A rapid increase from 0 to about 5% of the total life (stage 1) , an uniform increase from 5 to 90% of fatigue life (stage 2) and then a rapid increase until failure (stage 3). The fatigue life depends on stage 2 cracking. This corresponds well with similar observations in conventional concrete.

3. With the loading conditions used here ($S_{max} = 0.9$ and $R = 0.2$) - i.e. a maximum fatigue load well above the static first crack strength - most of the fatigue life took place under fairly long crack lengths: $\frac{a}{d} \in [0.65; 0.85]$.

4. With the loading conditions and specimens used here the crack growth of fiber reinforced concrete under cyclic loading can be divided into two parts. In the first 40% of the fatigue life the crack growth rate reduces with number of cycles and in the last 60% it increases.

5. The Paris' law is not applicable to the tested specimens. This is probably due to the major influence from the fibre bridging effect on the crack growth. For modelling purposes it is therefore recommended that the fatigue crack growth is directly related to the fiber bridging effect.

6 Acknowledgements

This investigation was part of a project on fatigue of fiber reinforced concrete carried out during the period January 1993 - December 1993 at Department of Structural Engineering, Technical University of Denmark. The project is part of the scientific and technological corporation between Denmark and People's Republic of China, project 15, Fiber Reinforced Cement-Based Composites. The authors wish to acknowledge the financial support from Danida. Zhang Jun would like to express his appreciation and gratitude to Dr. Henrik Stang and Professor Emeritus Herbert Krenchel for their steady guidance and support throughout this work. The authors also want to thank the staff at the Department of Structural Engineering for their enthusiastic help during the work.

Table 1: Fiber concrete recipes.

Component	Amount
cement (kg/m^3)	500
sand (kg/m^3)	810
gravel (kg/m^3)	810
water (kg/m^3)	238.5
superplastizier. (kg/m^3)	20
hooked steel fiber (Vol. %)	1
polypropylene fiber (Vol. %)	1

Table 2: Fiber properties, length l, diameter d, geometry (shape, cross section), tensile strength σ_t^u, Young's modulus E_f.

Fiber type	l (mm)	d (mm)	geometry	σ_t^u (MPa)	E_f (GPa)
Hooked steel fiber	30	0.5	hooked, circular	1350	210
Polypropylene	12	0.048	fibrilated, rectangular	400	12

Table 3: Results for crack length measuments during constant amplitude fatigue loading with load level $S_{max} = 0.9$

Specimen nr	Max. deflection mm	Crack length mm	Cycles N	Cycle ratio $\frac{N}{N_f}$
1	0.651	71	194	0.022
2	0.451	62	301	0.033
3	0.582	77	505	0.056
4	0.553	81	615	0.068
5	0.675	89.5	5399	0.600
6	0.529	69.5	5999	0.667
7	0.689	96.0	9000	1

References

[1] Balaguru, P.N., and Shah, S.P. *Fiber Reinforced Cement Composites* McGraw-Hill, Inc. 1992, p. 530.

[2] Swartz, S.E., and Go, C.G. Validity of Compliance Calibration to Cracked Concrete Beams in Bending. *Journal of Experimental Mechanics* 24, No. 2, June 1984, pp. 129-134.

[3] Bazant, Z.P. and Xu, K. Size Effect in Fatigue Fracture of Concrete *ACI Materials Journal* July-August, 1991, pp. 390-399.

[4] Bazant, Z.P. and Schell, W.F. Fatigue of High-Strength Concrete and Size Effect. *ACI Materials Journal* 90, No. 5, September-October 1993. pp. 472-478.

[5] Philip, C.P. and Calomino, A.M. Effect of Fatigue on Fracture Toughness of Concrete. *Journal of Engineering Mechanics* 112, No.8, August, 1986, pp. 776-791.

[6] Paris, P.C., Gomez, M.P. and Anderson, W.E. Rational Analytic Theory of Fatigue. *Trend in Engineering* 13, No.1, January. 1961.

[7] Paris, P.C. and Erdogan, F. Critical Analysis of Propagation Laws. *Transactions of ASME, Journal of Basic Engineering* 85, 1963, pp. 528-534.

Recent Advances in Experimental Mechanics, Silva Gomes et al. (eds) © 1994 Balkema, Rotterdam, ISBN 90 5410 395 7

Damage growth in concrete specimens via ultrasonic analysis in the frequency domain

R.S.Olivito & L.Surace
Department of Structures, University of Calabria, Cosenza, Italy

ABSTRACT: This work deals with the experimental and numerical study of the propagation of ultrasonic waves in concrete specimens subjected to uniaxial compression test. Firstly the frequency-domain analysis and the experimental programme are described, then the numerical analysis is tackled by carrying out a F.E.M. procedure. The comparison between numerical results and experimental ones confirms that the ultrasonic analysis in the frequency-domain is an effective tool for the microcracking study.

1 INTRODUCTION

The ultrasonic testing techniques are widely used in determining the mechanical properties of structural materials (concrete, masonry, wood, steel) .

In recent years non destructive techniques have been employed in providing an average measure of crack density rather than information on individual cracks [Daponte et al (1990), Shah et al (1970),Suaris et al (1987),Thompson et al (1983)]. The structural material contains microdefects even before the application of any load. Therefore, ultrasonic waves, travelling through concrete are modulated and attenuated as a consequence both of the material's dishomogeneity and of the different acoustic impedance of the components.

Thus, the travelling waveform undergoes a decrease in amplitude due to the scattering process occuring at each matrix-aggregate interface. From the ultrasonic signal it is possible to obtain information by measuring the pulse velocity or the pulse attenuation of the ultrasonic waves, or finaly by analyzing the ultrasonic signal in the frequecy domain. In particular, the frequency-domain analysis is characterized by high sensitivity with respect to the pulse velocity or attenuation technique [Daponte et al (1990)].

This technique analyzes the spectrum of the ultrasonic signal coming from concrete specimens under test. From the spectrum variation it is possible to detect the beginning or the growth of microcracking in the sample tested.

In the present work the behaviour of an ultrasonic wave propagating in a cubic concrete specimen subjected to loading is simulated via a F.E.M. procedure.

This F.E.M. program is developed to take into account the variation of the Young's modulus along the horizontal and vertical direction induced by loading.

Numerical results obtained by using this method are compared with recent experimental results. The study emphasizes that ultrasonic analysis in the frequency domain provides quantitative information about the beginning and development of microcracking in concrete and thus provides a useful tool for ultrasonic inspection.

2 FREQUENCY-DOMAIN ANALYSIS

The propagation of ultrasonic waves is influenced by the occurrence of heterogeinities: air-filled discrepancies such as cracks wich affect pulse velocity, attenuation and scatter of waves in the material [Wiberg (1993)]. The ultrasonic techniques for studying crack growth in concrete under a uniaxial compression test are various; their applicability and usefulness depend on the sizes and locations of the defects. Laboratory studies have shown that, in the time-domain analysis, pulse attenuation is a more sensitive indicator of the formation of cracks than pulse velocity. Recently a new laboratory ultrasonic technique has been employed with success using the frequency-domain analysis of ultrasonic signals. A pulse, travelling through concrete, encounters dishomogeneities which increase as the load applied increases, and this causes a spreading of the trasmitted energy from the frequencies of the transducer used to a wider spectrum. From this spectrum variation it is possible to detect the beginning or the growth of microcracking in concrete. To carry out the pulse spectral analysis the experimental measuring apparatus must be supplied with a universal waveform analyzer in addition to the usual ultrasonic equipment. The waveform analyzer can perform a wide set of signal processing functions, which are executed by a single key-stroke, and can be automatically re-executed for each new data acquisition. In order to obtain the signal spectrum between two loading steps it is necessary that the waveform analyzer calculates the spectrum with an adequate number of points in the shortest time. To reach this goal it is necessary to analyze the transient signal representation.

In this study recourse was made to a chirp representation by using the segmented CHIRPZ transform (SCZT) rather than the ordinary FFT tecnique. This provides a method whose resolution can be increased without increasing the transform size, while a corresponding small part within the original frequency range can be analyzed at the same time. The SCZT allows us to increase the resolution in the range under examination and to obtain the spectrum of the ultrasonic signal between two loading steps, and consequently to follow the development of the cracking in the specimen.

3 MATERIALS AND TESTING APPARATUS

The experimental investigation was conducted on two series (A and B) of cubic concrete specimens with 150 mm sides, prepared respectively by using type 325 and 525 Ultracem cement, river sand of 6 mm max grain size, gravel of 7-12 mm max and coarse aggregate of 13-28 mm max. The mix proportions for the two series were respectively: 1:2.666:1.333:1.333 and 1:2.078:1.342:0.974. The mechanical properties of the concrete are given in Table 1.

The concrete cubic specimens were prepared

conforming to Italian standard specifications; the compression tests were performed by using a testing machine equipped with electro-hydraulic regulation and closed-loop circuit. All tests were conducted at a constant strain rate and the measurements were made at 50 KN load intervals up to maximum load.

Table 1. Mechanical properties of concrete.

Series	γ KN/m^3	W/C Ratio	σ_r (MPa)	N Samples
A	0.240	0.45	350	40
B	0.245	0.35	550	40

The measuring apparatus (Figure 1) consists of a commercially available ultrasonic unit, a piezoelectric transducer set of nominal frequency of 54 KHz, a universal waveform analyzer, a personal computer and a plotter.

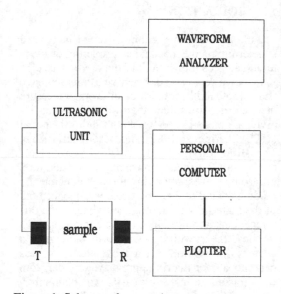

Figure 1. Scheme of measuring apparatus.

The transducers were mounted on a plexiglass frame with spring loading to accomodate lateral expansion of the specimen under test. The ultrasonic measurements were performed conforming to the EN-ISO 8047. Initially, in order to have a high frequency resolution, the optimal number of points to be acquired was defined (4096 points). At each loading step a signal of 4096 points was acquired by means of the universal waveform analyzer at 1 MHz sampling frequency and the corresponding spectrum was obtained by processing the data stored by computer. The σ-ε relationships of the concrete was obtained by experimental tests.

4 NUMERICAL ANALYSIS

4.1 *Formulation of the problem*

The dynamic behaviour of the concrete specimen subjected to vibrations imposed by the transmitted transducer (50 mm diameter) of 54 KHz nominal frequency, is described by the discretized equation

$$[\mathbf{M}]\ddot{\mathbf{q}} + [\mathbf{D}]\dot{\mathbf{q}} + [\mathbf{K}]\mathbf{q} = \mathbf{f}(t) \qquad (1)$$

where [**D**] is the damping matrix of the structure, [**K**] the stiffness matrix, [**M**] the mass matrix, **f**(t) the nodal loading vector and **q** the displacement vector.

The solution of the eq. (1) in closed form is only reached in some particular cases. Therefore, the analytical study of the problem is carried out by discretizating the structure in a number of finite elements of appropriate shape and size, nodal equilibrium conditions of which allow us to solve the problem. For the single finite element it is necessary to define an adequate displacement function which must satisfy continuity along the boundary surface between two adjacent elements to reach the convergence of the procedure.

The particular cubic shape of the element requires continuity not only for the displacements, but also for the rotations with respect to two hortogonal axes placed on the boundary plane. The definition of displacement functions which satisfy the requirements is highly complex.

On the contrary, it is simple to furnish displacement functions which ensure the continuity of displacement, violating the condition of continuity at the rotation, except for the nodal points where this latter is imposed.

In this spirit, if x y z is a rectangular cartesian coordinate system (Figure 2), the displacement components along the axes are given by equations with 24 constants defined as a function of 8 nodal displacements and 16 rotations.

In matricial form these displacement components can be put :

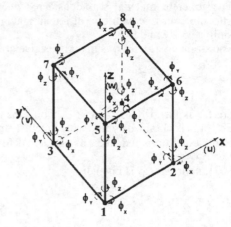

Figure 2. Tridimensional finite element

$$\mathbf{u}(x,y,z) = \mathbf{X}\,\mathbf{a}^{T} \qquad (2)$$

$$\mathbf{u}(x,y,z) = \mathbf{Y}\,\mathbf{a}^{T} \qquad (3)$$

$$\mathbf{u}(x,y,z) = \mathbf{Z}\,\mathbf{a}^{T} \qquad (4)$$

where

$$\mathbf{X} = (\ 1, x, y, \dots z^{3}{}_{xy}\ )$$

$$\mathbf{a}^{T} = (\ a_0, a_1, \dots a_{23}\ )$$

By denoting with (δ) the node displacement vector

$$\delta^{T} = (\ u_1,\dots,u_8,\ \varphi_{y1},\dots,\varphi_{y8},\ \varphi_{z1},\dots,\ \varphi_{z8}\ ) \qquad (5)$$

we obtain the equation :

$$\mathbf{C}\,\mathbf{a} = \delta \qquad (6)$$

where **C** is a 24x24 square matrix.

The inversion of eq.(6) allows us to obtain the displacement function along the x axis of a generic point P of the element as function of 24 node displacements:

$$\mathbf{u}(x,y,z) = \mathbf{X}\,\mathbf{C}^{-1}\delta \qquad (7)$$

Analogous relationships are valid for y and z axes.

In order to put the kinematic components in a single vector, the eq. (7) can be modified as:

$$\mathbf{u}(x,y,z) = \mathbf{M}\,\mathbf{q}_e \qquad (8)$$

$$\mathbf{v}(x,y,z) = \mathbf{N}\,\mathbf{q}_e \qquad (9)$$

$$\mathbf{w}(x,y,z) = \mathbf{P}\,\mathbf{q}_e \qquad (10)$$

were **M**, **N** and **P** are the displacement coefficient vectors.

The concrete material was considered linear elastic and isotropic with two different Young's moduli: Ex=Ey and Ez.

The strain energy for this material is given by

$$\pi\,(\mathbf{u}_e) = \frac{1}{2}\,\mathbf{q}_e^T\,\mathbf{K}_e\,\mathbf{q}e \tag{11}$$

where Ke is the stiffness matrix of the element. The eq.(11) is obtained by imposing kinetic compatibility conditions and takes into account deformation by shearing.

The kinetic energy is given by

$$\tau\,(\mathbf{u}_e) = \frac{1}{2}\,\dot{\mathbf{q}}_e^T\,\mathbf{M}_e\,\mathbf{q}e$$

where Me is the mass matrix of the element.

4.2 Numerical modelling

The numerical analysis was carried out by utilizing a finite element procedure. The discretization adopted consists of subdividing the concrete specimen of 150 mm sides into 27 small 8-node cubic isoparametric elements.

This discretization allowed us to simplify and schematize the load and constraint conditions so as to respect the response of the structure and to reduce the processing time.

In order to explain the calculation algorithm in Figure 3 the flow-chart is given.

Having once assigned the data input regarding the concrete specimen sizes, the discretization of finite elements, the mechanical properties, the mix proportions and the first loading step, the "Randomize" procedure then distributes the coarse aggregate at random laws among the discretized finite elements, by associating a fixed mass to each of them .

Local stiffness, the elements of which depend on the variation law of tangent modulus taken from either the compression or the tensile reading, and the mass matrix are successively determined .

The "Solve" procedure calculates the eigenvalues and eigenvectors of the problem by using the inverse iteration method.

The "Nodal Force" procedure determines the number of finite elements on which the pulsating stress of the trasmitted transducer acts by associating the corresponding quantity of force to element node.

The "Spectrum" procedure provides to give the frequency spectrum as a function of the vibration modes previously obtained and by determining the coefficient of participation for each mode. This procedure allows us to obtain the displacement values at the receiving transducer.

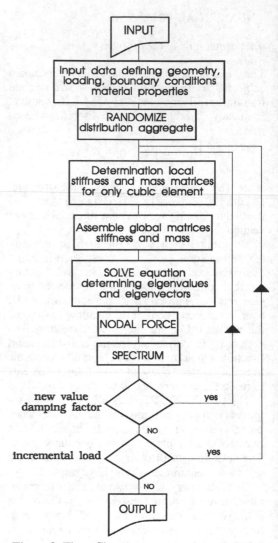

Figure 3. Flow-Chart

5. RESULTS AND CONCLUSIONS

Figures (4) show the comparison of experimental and numerical results. We can observe that there is a satisfactory agreement between the experimental and numerical results for frequencies close to those of the nominal transducer. Whereas at low frequencies the numerical values stray from the experimental ones.

Finally Figure (5) shows the comparison of the displacement of the points of the concrete surface adjacent to the receiving transducer as a function of time.

The numerical analysis carried out on the propagation of ultrasonic waves in a cubic concrete specimen has confirmed the physical validity of the approximated procedure set up and

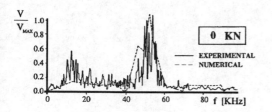

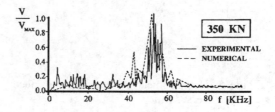

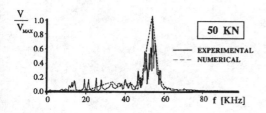

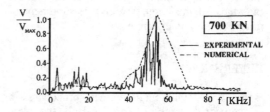

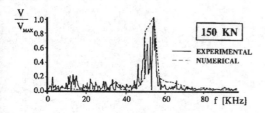

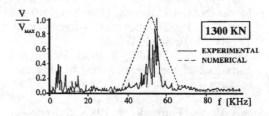

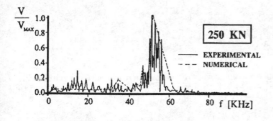

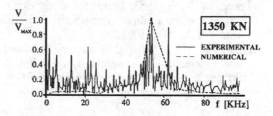

Figures 4. Frequency spectra

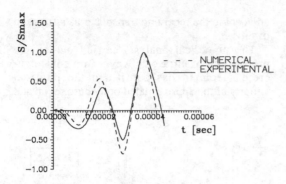

Figure 5. Displacement of receiving transducer versus time

it has emphasized that the ultrasonic analysis in the frequency domain furnishes a useful tool for the study of the beginning and development of microcracking in concrete.

REFERENCES

Bathe, K. J. 1982. Finite element procedures in engineering analysis. Prentice-Hall, Inc. New Jersey.

Daponte, P. Maceri, F. and Olivito, R. S. 1990. Frequency domain analysis of ultrasonic pulses for the measure of damage growth in structural materials. 1990 Ultrasonics Symposium, 2: 1113-1118. Honolulu:U.S.A.

Shah, S. P. and Chandra, S. 1970. Mechanical behaviour of concrete examined by ultrasonic measurements. Journal of Materials. 3:550-563.

Suaris, W. and Fernando, V. 1987. Detection of crack growth in concrete from ultrasonic intensity measuraments. Materials and Structures. 20:214-220.

Thompson, R. B. 1983. Quantitative ultrasonic nondestructive evaluation methods. Jour. of Appl. Mech. 50:1192-1201.

Wiberg, U. 1993. Material characterization and defect detection in concrete by quantitative ultrasonics. Doctoral Thesis. Royal Inst. of Tecnology, Dep. of Structural Eng., Stockholm.

9 Biomechanics

Recent Advances in Experimental Mechanics, Silva Gomes et al. (eds) © 1994 Balkema, Rotterdam, ISBN 90 5410 395 7

A preliminary 3D photoelastic analysis of the stress distribution on a proximal femur

J.A.O. Simões, J.A.G. Chousal, M.A.P.Vaz & A.T. Marques
Department of Mechanical Engineering and Industrial Management, University of Porto, Portugal

A.N.V. Costa
Institute of Mechanical Engineering and Industrial Management, Porto, Portugal

ABSTRACT: Several experimental and numerical techniques can be used to study stresses in bones. The frozen stress photoelasticity is one of the techniques applied to measure elastic stresses. The present work describes a three-dimensional photoelastic analysis using the stress-freezing and slicing technique applied to study the stress distribution on a proximal femur. An image processing system was used for the analysis of the photoelastic fringe patterns. A real shape and size casting technique was developed to obtain our femur models. A simple loading device was designed to simulate the biomechanics of the hip joint. From this study, it was seen that the principal type of stresses imposed in a femur are in the form of compression and flexure. Out-of-plane forces causing both bending and torsion also exist. The existence of these forces implies an important limitation inherent to this preliminary study. It is seen that the bending stresses are the most predominant.

1 INTRODUCTION

The stress analysis of structural elements of the human body has historically generated scientific interest. Borelli and Galileo in the 1600's were active in this field, known nowadays as biomechanics. Following Koeneman [1], engineering methods developed during the industrial revolution were applied to the analysis of the femur by Meyer(1867), Wolff(1870) and Koch(1917).

The quantification of stresses in structures is a requirement in most fields of engineering. An important load bearing structure of the bio-engineering is the skeleton. It is important to understand the dynamic biomechanics of this structure in order to design suitable methods for the repair of fractured bones and prostheses for replacement of skeleton components. To develop these methods, one needs to understand the phenomena of bone remodelling and long-term reliability of skeleton joint implants by predicting the stress distribution in these structures. Through a correct understanding of the stress distribution and bone response, it is possible to develop prosthetic designs and design features which are likely to lead to problems due to stress shielding.

2 THE FEMUR BONE

The femur is the longest and strongest bone of the human skeleton. By analysing this structure, we find a certain order in the bone formation, in particular the arrangement of internal bone, trabecular, in the proximal femur, figure 1 [2].

These trabeculae form arcades that lie in a position similar to the stress trajectories calculated for a structure of like shape and loaded in the same manner, the Fairbairn crane [3].

Figure 1. Diagrammatic representation of load-bearing trabeculae of femoral head and neck.

A femur is mainly composed of cortical and cancellous bone. The cortical bone is a mineralised tissue in the body and carries the majority of the loads. It is characterised by lamellar sheets of tissue, 3-4 μm thick [4]. Cancellous bone is anisotropic in mechanical properties. It is stiffer, with high failure tensile and compressive moduli, but can fail at a lower strain when loaded parallel to the predominant spicular direction than that when loaded in other directions.

3 BIOMECHANICS OF THE HIP JOINT

The principal question of the biomechanics of the hip joint is to know how the loads are transferred from the pelvis into the femur. Analysis of the force actions transmitted at the hip joint has been a problem thereby investigated during all these years. The study of the resultant loading actions at the hip joint is very complex due to the dynamic-force system to which the joint is subjected. The complex anatomical configuration of the femur, the anisotropy and rate-dependent mechanical properties of the bone material are some characteristics which make this analysis very difficult.

A scattered literature exists on the analysis of these forces. This analysis has been performed by many researchers, such as Pauwels (1935) pioneer work [5], Inman (1947), Blant (1956), Strange (1963), and many others, assuming a planar-force system [6]. A three-dimensional analysis of this typical configuration was performed by Williams [7]. A direct determination of hip joint force was made by Rydell (1966), in which strain gauges were fitted to femoral head prostheses implanted in two patients who had suffered fractures of the neck of the femur [8].

Due to the complex anatomical system of twenty two muscles acting at the hip and fourteen muscles and six ligaments at the knee, it is extremely difficult to calculate the stress distribution on the femur. It is necessary to have a correct knowledge of the physical activity of the muscles and their anatomical location, the tension in relevant muscles and ligaments may then be inferred, and the joint forces obtained [6]. The activity of walking, running, climbing, etc., is a result of the combinations of hip joint motions. These activities lead to stresses in the hip joint.

A simple planar force system can be used to analyse the loading conditions associated with the hip joint, figure 2 [9].

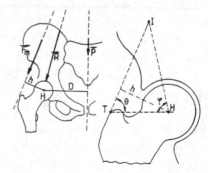

Figure 2. Forces and geometric characteristics of the hip joint.

From the analysis of figure 2, one can easily obtain the static equilibrium relationship of Pauwels as

$$|P| D = |Fm| h \qquad (1)$$

were the abducting force, Fm, is given as

$$|Fm| = \frac{|P|\, D}{IT\ IL \sin\varphi} \sqrt{IT^2 + IL^2 - 2\ IT\ IL \cos\varphi} \quad (2)$$

and the hip force, R, is

$$R = \sqrt{P^2 + Fm^2 + 2\,|P|\,|Fm| \sin\theta} \quad\quad (3)$$

It is then possible to obtain the relationship between the hip and abducting force. The hip force is 3 to 4 times the body's weight and makes an angle of approximately 20° with a vertical line. By considering a planar system instead of a three-dimensional one, the distortion is of 15 to 20°, so this simplification seems acceptable [9].

4 PHOTOELASTICITY IN BIOMECHANICS

Photoelastic methods have been applied to biomechanical studies since late 1930's. Most of these studies have been done to investigate stresses in bone and implants in the orthopaedic and dental fields. In orthopaedic engineering the first application of photoelastic methods was described by Milch (1940) with reference to stresses in the upper end of the femur [3].

The stress analysis of joint replacement components has not attracted many photoelastic investigations. However some experiments have been reported concerning stresses in femoral stems of hip prostheses and in the surrounding cement mantle (Kennedy et al., 1979; Orr et al., 1985, 1986) [3].

The application of photoelastic methods to the study of biomechanics requires assumptions, such as anisotropy and non-homogeneity, that must be taken into account in this type of study. Such considerations are addressed by many researchers when doing experiments and do not need to prevent useful results being obtained within the limitations of the assumptions.

With a photoelastic study the maximum shear stress distribution and the principal stress directions can be assessed [10]. The map of the principal stress directions and shear stress distribution of the proximal femur is one of the aims of the present study.

5 PHOTOELASTIC MODEL

To obtain the photoelastic model we had to fabricate a mould, process the resin mixture, stress-freeze the model and finally cut the model in thin slices to be analysed. The analysis consists on the observation of patterns of dark and light lines, known as fringes. These fringes, known as isoclinics and isochromatics, give us information about directions and magnitudes of principal stresses. The photoelastic fringe pattern of the models were analysed by an image processing technique developed at our Department (DEMEGI).

6 MOULD FABRICATION

The models were produced to real shape and size by precision casting. Since the geometry of the proximal femur used is irregular, we made the mould with a flexible material to easily extract the embedded prototype. The mould was obtained with two half moulds. The material used was a Silicone RTV-M533 rubber of the Walker-Chemic GmbH, Germany. After obtaining the two half moulds, both where put together with screws and nuts. On one of the half moulds we welded pins to obtain a correct positioning of both half moulds and therefore a perfect cavity of the femur. The half moulds were obtained by pouring the silicon rubber into containers made of sheet steel for this purpose. Figure 3 shows a photo of the two half moulds used.

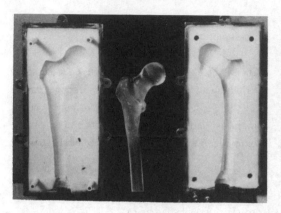

Figure 3. Mould.

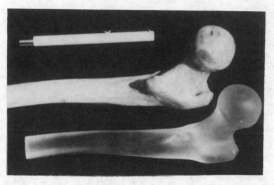

Figure 4. Proximal femur and photoelastic model.

Figure 3 shows a photo of the two half moulds used in our work.

A postcure of the model was then carried out at 60°C during 2hours. With this very simplified method we obtained very good photoelastic models, with no air bubbles inside and no residual stresses.

7 MIXTURE PREPARATION

The best materials for three-dimensional photoelastic studies are the epoxy resins. The epoxy resins are optically sensitive, chemically stable, and can be cast into thick pieces without significant residual stresses.

We used the ARALDITE LY 554 epoxy resin of CIBA-GEIGY and the LY 554 hardener. A very clean container was used to hold the total amount of mixture needed. We placed the resin and the hardener in the container and mixed it at room temperature. The hardener was slowly added to the resin while the mixture was mechanically mixed. The stirrer speed was kept low to avoid the introduction of air bubbles. After thoroughly mixed, the mixture was introduced into an oven and vacuum was done for about thirty minutes to take out some micro bubbles. The mixture was then poured into the cavity of the mould by gravity in contact with the runner wall to prevent air being carried into the mould. We did not pre-heat the mould. Once poured, the cast liquid resin was kept at room temperature during 48 hours. The casting was then removed from the mould. Figure 4 shows the proximal femur used and the obtained photoelastic model.

8 STRESS-FREEZING

To stress-freeze the model it was necessary to find the softening point of the same casting material. For this purpose we made a Dynamic Mechanical Thermal Analysis (DMTA) test. We found that the softening point was around 60°C, figure 5.

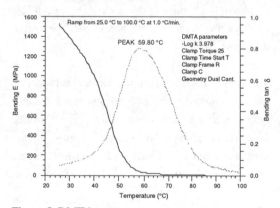

Figure 5. DMTA test.

The model was then heated up to a temperature of 65°C, 5°C above the softening point of the resin during 7 hours. Figure 6 shows the stress-freezing cycle done to the model.

1364

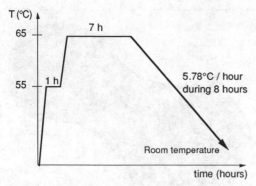

Figure 6. Stress-freezing cycle.

To find the value of the stress-optic coefficient at the stress-freezing temperature we made a calibration specimen from the same batch of plastic material subjected to the same stress-freezing cycle as the model. The stress-optic coefficient obtained was approximately 8.15 N/fringe/mm.

9 LOADING DEVICE

To load the model we designed a very simple loading device, figure 7. The structure of this device was made with U profiles and with sufficient rigidity to avoid the influence of other kind of loading.

The model was loaded with a 82N hip force and 41N abducting force. The direction of the abducting force was of 15° with a vertical line. The direction of the hip force on the head of the femur is rather difficult to state due to the loading configuration. The femoral head is not very spherical which implies some difficulty to find the loading direction.

In this preliminary work the loading conditions acting on the hip joint were simplified. We were more concerned with the three-dimensional stress-freezing method, than simulating the accurate loading condition on the hip joint. Since the model obtained is made of an isotropic material, we can only analyse the stress distribution.

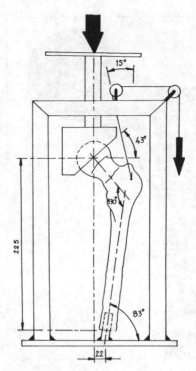

Figure 7 - Loading device.

10 SLICE CUTTING

A Struers Exotom saw was used to cut thin slices (2mm thick) from our stress-freezing models. The cutting disc had a diameter of 350 mm and a thickness of 2.5 mm, which is not the suitable disc to cut the slices. We cutted the slices by using a low blade speed and oil bath to minimise heating in the slice. A test was performed to determine if machining cutting stresses where induced in the slice, and we verified that no cutting stresses were found in the slices because no fringes were found on a slice cutted from a model where the stress-freezing cycle was not applied. The slices were then dried at room temperature.

11 RESULTS AND DISCUSSION

A photoelastic oriented image processing package[11]

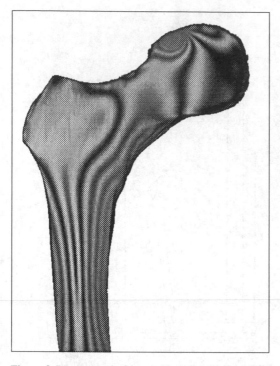

Figure 8. Isochromatic fringes (dark field) of the first model.

Figure 9. Isochromatic fringes (dark field) of the second model.

with phase shift techniques was used to map the fringe pattern and principal directions in the plane of the model slices.

One model was loaded with a 82N hip force, so that no abducting forced was applied. Experimental results presented at figure 8 illustrate the stress pattern for this model.

A second model was loaded with a more realistic loading condition. It was loaded with a 82N hip force and a 41N abducting force. The experimental results presented at figure 9 illustrates the stress pattern.

The principal stress directions for both models obtained can be seen in figures 10 and 11.

It can be observed that the predominant stresses imposed in a femur are mainly in the form of compression and flexure. The influence on the stress

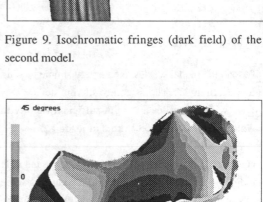

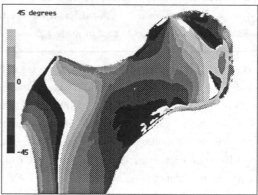

Figure 10. Principal stress directions (first model).

pattern and principal directions by the abducting force can be seen by comparison of figure 8 with 9 and 10 with 11 respectively. The principal stress directions of our models are similar to the directions of the trabecular of the typical femur bone structure.

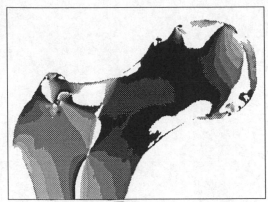

Figure 11. Principal stress directions (second model).

12 CONCLUSIONS

As the start of an extensive research work to be developed, a three-dimensional photoelastic study was done. Since the main equipment required is commonly found in mechanical engineering departments, like ours, it was possible to evaluate our method with minimal expense.

Frozen stress photoelasticity has been proved to be a suitable tool to analyse stresses in bones, in this case on femurs. This method is somehow limited to the analysis of elastic stresses. Its great advantage, over other experimental and numerical techniques, is the determination of the maximum stress and directions for any geometry irrespective of the curvature of the surface or stress gradients.

With this study it is possible to evaluate the correlation between the principal stress directions and the orientation of the trabecular structure of the proximal femur. Also an accurate stress pattern can be achieved by using a three-dimensional photoelasticity analysis.

ACKNOWLEDGMENTS

The authors gratefully acknowledge all the facilities and support given by Prof. Silva Gomes and to the technical assistance staff of CEMACOM(Centre of Composite Materials) and CETECOFF(Centre of New Technologies for Casting of Metals) of INEGI(Institute of Mechanical Engineering and Industrial Management).

REFERENCES

1. Koeneman, J. B. 1989. *Finite Element Analysis in Orthopaedics*. Biomedical Materials and Devices Materials Research Society Symposium. Proceedings Vol. 10: 533-538.
2. Morrey, B. F. 1991. *Joint replacement arthroplasty*. Churchill Livingstone.
3. Miles, A. W. & Tanner, K. E. 1992. *Strain Measurement in Biomechanics*. Chapman and Hall.
4. Black, J. 1988. *Orthopaedic Biomaterials in Research and Practice*. Churchill Livingstone.
5. Pauwels, F. 1935. *Der Schenkelhalsbruch ein Mechanischen Problem*. Ferdinand Enke. Stuttgart, Germany.
6. Paul, J. P. 1971. *Load Actions on the Human Femur in Walking and Some Resultant Stresses.* Experimental Mechanics, Vol. 28, Nº2: 121-125.
7. Williams, J. F. 1968. *A Force Analysis of the Hip Joint*. Bio. Med. Eng. Jnl. 3 (8): 365.
8. Rydell, N. W. 1966. *Forces Acting on the Femoral Head Prosthesis*. Acta. Orthop., Scand. Suppl.:88.
9. Frain, Pl. 1985. *Biomécanique de la hanche normale*. Cahiers d'enseignement de la SOFCOT. Expansion Scientifique Française.
10. Dally, J. W.& Riley, W. F. 1978. *Experimental Stress Analysis*. (2nd Edition), McGraw-Hill. Kogakusha.
11. Chousal, J. A. G. 1991. *Processamento de imagem na interpretação de franjas interferométricas em análise experimental de tensões*. (Image processing in the interpretation of interferometric fringes applied to the experimental analysis). M.Sc. Thesis presented to Faculty of Engineering of University of Porto (in portuguese).

Recent Advances in Experimental Mechanics, Silva Gomes et al. (eds) © 1994 Balkema, Rotterdam, ISBN 90 5410 395 7

An attempt at determining the stresses of the human femur with endoprosthesis

R.J.Bedzinski
Technical University of Wroclaw, Poland

ABSTRACT: The paper is concerned with description and results of photoelasticity stress analysis of the human femur fitted with different total hip prostheses. Because of the complex structure of the femur a decision was taken to carry out investigations on three-dimensional models by using the stress frozen photoelasticuty method. The models were made in the 1:1 scale and they retained the differentiated stiffness of the bone and prosthesis.A compostion consisting of epoxy resin was used to make femur bone.

The loading system used made it possible to simulate the loads acting in the joint according to Pauwels model. Investigations of the diffrentiated hip prosthesis were corned out.

A comparison of the stresses in the model of the femur bone without prosthesis with those in the femur with of the differentiated endoprostheses make it possible to estimate influence of the prosthesis stiffness on the distribution of stresses.

1. INTRODUCTION

One of the elements of the human osteoarticular system which is exposed to degenerative - deforming changes is the hip joint. Such factors as transport injuries, a sedentary mode of life or disorders in the metabolism (ecological hazards) contribute to this. The technological progress creates wide possibilities for surgeons and orthopedists to reproduce deformed bone elements. Biomechanical studies, particularly the ones based on experiments, can bring a lot into this field. Clinical trials cannot provide answers to the many questions puzzling orthopedic surgeons because one would have to perform experiments directly on human beings. The application of experimental methods to the investigation of human bone elements can provide us with fresh insights into the very complicated mechanisms functioning in a human being.

Mechanical models quite naturally are not able to render the whole complex structure and functioning of a human being. Therefore there is a need for models which would closely approximate the real conditions.

The present study is devoted to the hip joint, which because of its anatomic structure and functions is most exposed to all kinds of overloads.

Hip alloplasty (Total Hip Arthoplasty) by removing the sick joint and implanting an "artifical joint"in its place terminates coxarthrosis, relieves the pain and reproduces the hip function. The implantation of an artificial joint, particularly the hip joint, has become a routine surgical intervention. Nevertheless, relatively little is known about the distribution of loads and the resulting deformations and the distribution of stresses in bones with an implanted endoprosthesis .

Research on the optimum design and implanting of non-cement endoprostheses has been conducted in many research centres in recent years (Huiskes 1991). Many kinds of endoprostheses differing in their design solutions have been constructed all over the world. Mittelmeier's (OSTEO) and Parthofer-Monch's (AESCULAP) non-cement prostheses are most commonly used (Mittelmeier 1984 ; Parhofer, Monch 1983). Obviously the life of an endoprosthesis is determined by the anatomical and biomechanical conditions, the course of the operation and the construction and the material the endoprosthesis is made of.

The aim of this study was to compare the distribution of stresses in a model of the femoral bone without an endoprosthesis and in models with mounted Mittelmaier (M) and Parhofer-Monch (PM) type prostheses. The test were conducted on three-dimensional models using the photo-elastic method of stress freezing (Wolf 1975) . The results of these test should make it easier for a clinician to select the type of an endoprosthesis depending on the 'clinical' and biomechanical conditions.

2. MODELS CONSTRUCTION AND TEST METHOD

The photo-elastic method of the freezing of stresses utilizes the phenomenon of the transition of some of the model materials from the vitreus state to the high-elastic state when the temperature is increased. The materials behave elastically in both being in the high-elastic state results in temporary double refraction effects which become fixed after the materials have been cooled down below the glassy temperature and unloaded. This process is called the freezing of stresses. When the process is over, the model is divided into thin plates. By examining the photo-elastic effect fixed in the plates one can determine the components of the 'frozen' state of stress. The isochromatic order is proportional to the difference between the extreme stresses lying in the plane perpendicular to the light ray direction. The photo-elastic effect observed in a specimen is associated with quasi-principal stresses whose directions are the projections of the directions of the principal stresses onto the specimen's plane. Hence we have the following realtionships,

$$\sigma'_1 - \sigma'_2 = m \bullet K_f$$

and on the contur when is assumed:

$$\sigma'_1 = \sigma_c = m \bullet K_f$$

where: σ_1' ,σ_2' - the quasi-principal stresses (secondary principle stresses),
    σ_c - the stresses on the contour,
    m   - the isochromatic order,
    K   - the photo-elastic material constant at the temperature of freezing.

The femoral bone models were made on a 1:1 scale by casting in rubber moulds (Fig. 1) (Fessler , Perla 1973)

To match better the properties of the femoral bone model and the endoprosthesis model, the models of these elements were made from materials with differntiated values of the modulus of elasticity. The models of endoprostheses were made of a blend of epoxy resin (Epidian 5) and silica flour added to

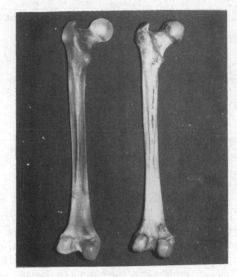

Fig. 1 The original femoral bone (right) and its model (left)

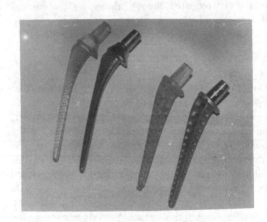

Fig.2 The original endoprothesis PM (left) and M (right) and its model

Table 1. The elastic parameter of the bone , the protheses and the models

	Realy bone			Models	
	cortical bone	cancelous bone	prothesis	Bone	Prothesis
				at the temperature of freezing	
Young's moduls E [MPa]	1,5-2,0 10^4	0,3-1,5 10^3	2,0-2,2 10^5 ass. 2,1 10^7	2,25 10^3	5,85 10^3
Poisson's ratio ν	0,29	0,46	0,3	0,5	0,5

1370

increase the modulus of elasticity of the material at the temperature of frezing. The models of the femoral bone were made of epoxy resin (the Araldit type). The models of the femoral bone with an endoprosthesiis were made by casting in a rubber mould in which an endoprosthesis had been placed. The physical material properties of the bone models and the endoprosthesis models were determined at the temperature of freezing. The determined values of the Young's modulus and the photo-elastic material constant were the following:

- for the bone model material - E = 2.25 GPa,
- for the endoprosthesis model material (epoxy resin + silica flour) - E = 5 . 85 GPa
- the bone model photo-elastic material constant - K_f =1.218   MN/number of isochr. patt.

These properties were determined in a pure bend test. The freezing of both the specimens and the models was conducted for 48 h at the temperature of 105 C.

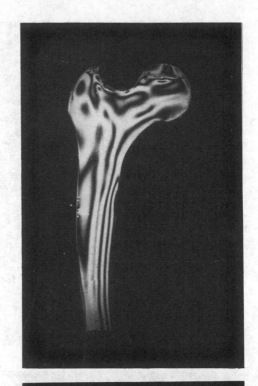

a

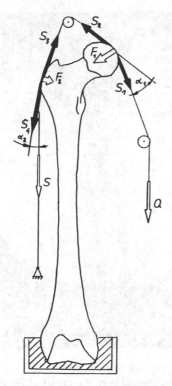

Fig 3. A diagram of the loading system for the testing of the femoral bone model

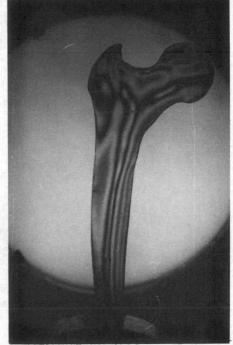

b

Fig. 4. The distribution of full (a) and half (b) isochromatic patterns in the model of the femoral bone.

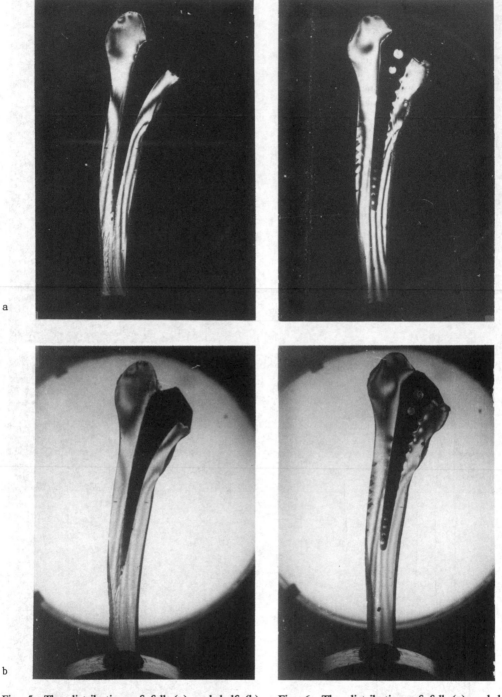

Fig. 5. The distribution of full (a) and half (b) isochromatic patterns in the model of the bone with endoprosthesis of PM type.

Fig. 6. The distribution of full (a) and half (b) isochromatic patterns in the model of the bone with endoprosthesis of M type.

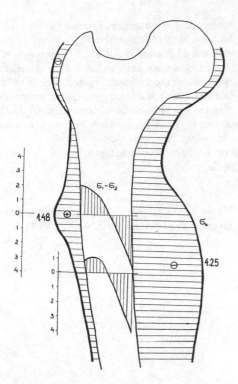

Fig. 7. The distribution of contour stresses σ_c and σ_1-σ_2 in the model of the model of the femoral bone (stresses in number of isochromatic patterns)

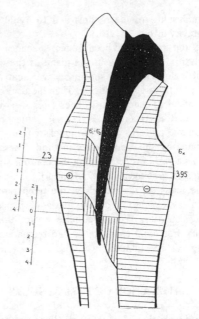

Fig. 8. The distribution of contour stresses σ_c and σ_1 -σ_2 in the model of the bone with on endoprosthesis of the PM type (stresses in number of isochromatic patterns).

3. COURSE OF TESTS AND RESULTS

The material properties of the femoral bone, endoprostheses and the constructed models have been put together in table 1. These data indicate that the correct bone material rigidity ratio has not been fully achieved in the models. We have not quite succeeded in designing and constructing models which would reproduce the structure of the compact part and the spongy part of the bone.

The models of the bone were loaded according to the developed model of (Pauwels 1980) and (Maquet 1985) (fig. 3). Apart from applying a load to the head of the femur, loads induced by the qluteal muscle and the iliotibial band were also applied in the models. The model load conditions were brought as close as possible to the real conditions existing in the hip joint.

The freezing of the stresses in the models was conducted in an oven with programmed temperature control. The models of the bone together with the loading system were immersed in oil to obtain the

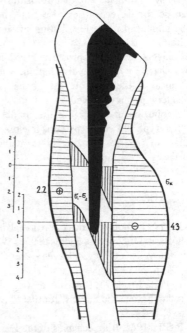

Fig. 9. The distribution of the contour stresses σ_c and σ_1 - σ_2 in the model of the bone with on endoprosthesis of the M type (stresses in number of isochromatic patterns).

uniform heating of the models and to avoid loads from gravity forces.

After the stresses had been frozen, specimens were cut out from the models in the plane of the central femoral bone. The same specimens were cut out in middle part from all the models. Then the distribution of full and half isochromatic lines (figs 4-6) in the specimens was recorded. Fractional values of the isochromatic patterns on the model's contour were determined by the Senarmont's method. These data served as a basis for determining the contour stresses in the tested specimens (figs 7, 8 and 9).

The loads on the models at the temperature of freezing were selected on the basis of papers on models similarity. We were quite aware that the Poisson's fractional condition at the temperature of freezing was not fulfilled.

4. RECAPITULATION OF TEST RESULTS

1. The distribution of isochromatic lines and contour stresses allows one to evaluate the effect of the endoprosthesis construction on the change in the effort of the femoral bone.
2. The distribution of stresses obtained in the model of the bone (without an endoprosthesis) was characteristic for the eccentrics of a bent beam. It should be noticed that the distribution of stresses is quite uniform.
3. The models with endoprostheses exhibited a different character of stress distribution. The effect of the local stiffening of the bone in the subtrochanteric part is observed.
4. The distribution of stresses in the model of the bone with a type PM endoprosthesis is closer to the distribution of stresses in the bone without an endoprosthesis.

REFERENCES

Huiskes R., 1991, Biomechanics of Artificial-Joint Fixation. Basic Orthopaedic Biomechanics. Van C.Mow and Wilson C.Hayes. Raven - Press. New York 1991, p. 375-442.

Fessler H., Perla M., 1973, Precision casting of epoxy-resin photoelastic models. Journal of Strain Analysis, 8, 1: 30-34.

Maquet P.G.J., 1985, Biomechanics of the Hip. Springer - Verlag .

Mittelmeier H., 1984, Total Hip Replacement with Autophor Cement-Free Ceramic Prosthesis

Morscher. The cementless Fixation of Hip Endoprosthesis Springer, Berlin.

Parchofer R., Monch W., 1983, Erfahurungen bei Austanschoperationen von bisher einzementierten Hufttotalenendoprothesen gegen Zementlose Lord und PM - Prothesen. Morscher E. Die zementlose Fixation von Huftendoprothesen. Springer, Berlin.

Pauwels F., 1980, Biomechanic of the Locomotor Apparatus. Springer- Verlag.

Wolf H., 1975, Spannungsoptik, Berlin, Springer-Verlag.

This work was supported by The Scientific Research Communities of Poland, Grant no 4 4248 91 02.

Recent Advances in Experimental Mechanics, Silva Gomes et al. (eds) © 1994 Balkema, Rotterdam, ISBN 90 5410 395 7

Ultrasonic measurement for detecting loosening of a hip prosthesis stem

Yoshihisa Minakuchi & Takaaki Kurachi
Yamanashi University, Japan

Takatoshi Ide & Junji Harada
Yamanashi Medical College, Japan

ABSTRACT: Although it is necessary to find out the slight loosening between hip prosthesis stem and femur, the X-ray and Magnetic Resonance Imaging(MRI) methods look difficult to grasp it. The thickness of the upper side of a normal human femur and the echo waveform reflected from a hip prosthesis stem are investigated via an ultrasonic wave. The ultrasonic wave is emitted toward the femur using a normal probe attached to the upper side of the thigh, the echo waveform reflected from the outside of femur, the inside of femur or the prosthesis stem is measured. The femur thickness using the ultrasonic wave is compared with that of the X-ray CT. Moreover, the echo waveform on the patient is investigated. The proposed ultrasonic method is useful for the measurement of the femur thickness and the loosening between femur and stem.

1 INTRODUCTION

In order to restore the function of a hip joint, the total hip replacement has been widely implanted in recent years. However, relaxation, damage, wear and infection occurring after the total hip replacement necessitate reimplantation. In these cases, the largest problem is the loosening between prosthesis stem and femur. The X-ray and MRI methods are mainly utilized in the diagnosis of the loosening[1,2]. However, these methods are difficult to distinguish a slight loosening. Therefore, the detection of slight loosening after the total hip replacement is required. As one measurement method to overcome this problem, an ultrasonic pulse echo method should reveal the conditions inside the femur. However, only a few studies have been conducted on this issue[3~6].

In this study, the thickness of the upper side of normal human femur and the echo waveform reflected from the hip prosthesis stem are investigated via the ultrasonic wave. First, a simple measurement device is attached to the thigh of a normal femur. The ultrasonic wave is emitted toward the human femur using a normal probe attached to the thigh, the echo waveform reflected from the outside of femur, the inside of femur is measured. This measurement result is compared with that of the X-ray CT. Also, the ultrasonic wave is emitted toward the femur on the patient inserted with the hip prosthesis stem. The echo waveform reflected from the outside of femur, the inside of femur and the stem is measured at the two places on the thigh. Thus, the utility of the proposed ultrasonic method is investigated.

2 MEASUREMENT PRINCIPLE OF ULTRASONIC PULSE ECHO METHOD

When an ultrasonic wave arrives perpendicularly at a boundary surface between dissimilar acoustic impedances in media 1 and 2, which are expressed as the product of density ρ_i (i=1,2) and wave velocity v_i (i=1,2), part of the

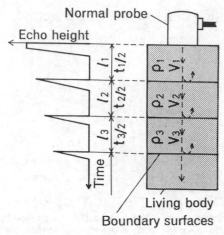

Fig.1 Principle of ultrasonic pulse echo method

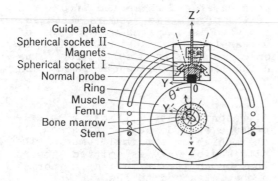

Fig.2 Measuring device

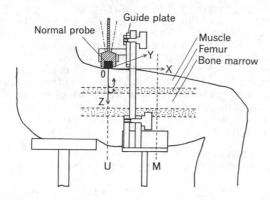

Fig.3 Thigh part equipped with measuring device

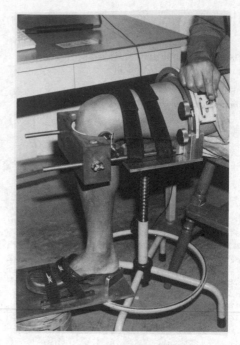

Fig.4 Experimental scene

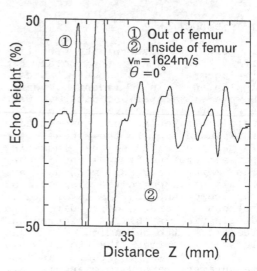

Fig.5 Echo waveform presenting outside and inside of femur ($\theta = 0°$)

wave is reflected and the other part is transmitted.

Figure 1 shows the measurement principle of the ultrasonic pulse echo method, where the ultrasonic wave emitted from the normal probe propagated through the part of the living body. The relationship between echo height and time is shown. Based on Fig.1, the distance l_i between each boundary surface is given as

$$l_i = v_i \times \frac{t_i}{2} \quad (i=1,2,3,\cdots), \quad (1)$$

where v_i is the wave velocity and t_i is the round-trip propagation time.

3 SIMPLE MEASURING DEVICE

In order to measure the upper sectional form of the femur on a normal person and the the thicknesses of femur and bone marrow on the patient inserted with the hip prosthesis stem, Fig.2 shows a section seen from the patient which the simple measuring device is attached to the right thigh. A normal probe (diameter of transducer:13mm, frequency:2.25MHz) is able to move the radius direction (Z-axis) along a guide plate and the circumference direction along the guide of radius 120mm of a ring. Also, the normal probe is fixed to spherical socket I which made of steel. The normal probe can be inclined to the axial direction of femur (X-axis) and the Y-axis direction perpendicular to the X-axis along the spherical socket II. The spherical socket I and the spherical socket II are fixed by utilizing adsorption power of the magnet which is planted to the spherical socket II. Further-more, although a rotation angle of the normal prove in the X-axis direction reads with an indication of the upper part of spherical socket II, a moving magnitude of the Y-axis direction is disregarded because it is small.

4 MEASUREMENT OF FEMUR THICKNESS ON A NORMAL PERSON

4.1 Experimental method
In order to measure the upper sectional form of the femur on a normal person, the simple measuring device is attached to the upper side of the right thigh on two persons as shown in Fig.3.

Namely, the simple measuring device is set to the position of 280mm, 240mm and 265mm, 210mm apart from the top of knee, respectively. An incidence position of the ultrasonic wave is determined the upper position of the ring as $\theta = 0°$. The normal probe moves every $4.1°$ along the guide of radius 120mm of the ring. In each measuring position, the wave velocity of ultrasonic device set up in 1624 mm/s [7] which is the wave velocity through the thigh muscle on the human. After that, the ultrasonic wave is emitted toward the femur. In order to obtain clearly a echo waveform from the outside of femur and the inside of femur, the normal probe is inclined to the axial direction of the femur (X-axis) and the Y-axis direction perpendicular to the X-axis and the radius direction in the Z-axis. Furthermore, the Y' -axis and Z' -axis in this figure are the coordinate that a center of the ring is the origin. As the wave velocity through the femur is 3120m/s[7], the thickness of the femur in each measuring position is calculated from the echo waveform. The sectional form of upper side of the femur is indicated with two dimensional figure. Furthermore, the measurement result is compared with the X-ray CT. Fig.4 shows the practical situation that the simple measurement device is attached to the right thigh.

4.2 Experimental results and considerations
Figure 5 is an echo waveform of the outside of femur ① and the inside of femur ②, that the simple measurement device was attached to the position parted 280mm from the top of knee and the normal probe was located at the top of the ring as $\theta = 0°$. From this figure, the femur thickness is 7.6mm. A concrete method of computation is expressed below.

Fig.5 is the echo waveform that the wave velocity through the thigh muscle is set at $v_m = 1624 m/s$ [7]. As the wave velocity of the femur is $v_b = 3120 m/s$ [7], this differs from the wave velocity v_m. Accordingly,

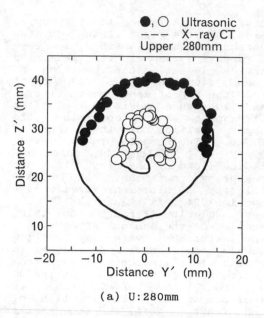

●, ○ Ultrasonic
--- X-ray CT
Upper 280mm

(a) U:280mm

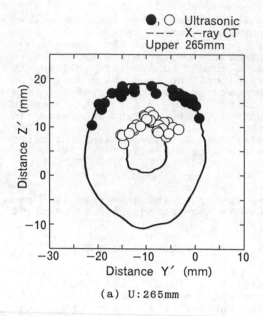

●, ○ Ultrasonic
--- X-ray CT
Upper 265mm

(a) U:265mm

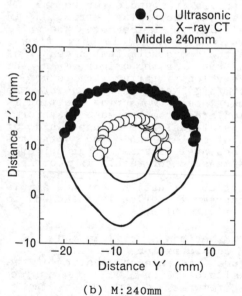

●, ○ Ultrasonic
--- X-ray CT
Middle 240mm

(b) M:240mm

Fig.6 Sections of femur in the
right thigh using ultrasonic
wave and X-ray CT (Person A)

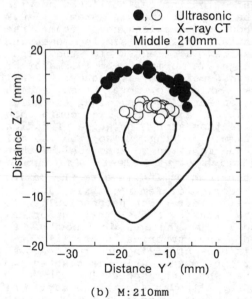

●, ○ Ultrasonic
--- X-ray CT
Middle 210mm

(b) M:210mm

Fig.7 Sections of femur in right
thigh using ultrasonic wave
and X-ray CT (Person B)

the exact distance of the femur l_b is not obtained directly from the distance l_b' indicated on the CRT. Thereupon, l_b is given by the following equation.

$$l_b = l_b' \times \frac{v_b}{v_m} \qquad (2)$$

Furthermore, the incidence direction and the incidence position of the ultrasonic wave are calculated from the radius direction (Z-axis) along the guide plate and the circumference direction (θ) and the rotation angle of the Y-axis direction as shown in Fig.3.

Figures 6 and 7 are the measurement results by the ultrasonic wave and the X-ray CT which were measured at the position 280mm, 240mm and 265mm, 210mm apart from the top of knee, respectively. The position of zero in the Y'-axis and Z'-axis is equal at the center of the ring. The marks ● and ○ in these figures indicate the measuring positions in the outside of femur and the inside of femur by the ultrasonic wave, and solid lines are the sectional forms of femur by the X-ray CT. From these figures, the outside sectional form by the ultrasonic wave agrees approximately with that of the X-ray CT. But, the inside sectional form is disturbed partially. This cause is due to the rough in the inside surface of the femur.

5 MEASUREMENT OF THE FEMUR AND BONE MARROW THICKNESSES ON A PATIENT INSERTED WITH A HIP PROSTHESIS STEM

5.1 Experimental method
In order to measure the thicknesses of femur and bone marrow on the patient inserted with the hip prosthesis stem for 3 months, the simple measurement device is attached to the right groin of the patient. It is considered that the hip prosthesis stem is inserted with the femur on the condition that the major axis of stem section is the direction of $\theta = 60°$ and the minor axis is the direction of $\theta = -30°$ observing from the patient. Thereupon, the ultrasonic wave is emitted toward the femur from the directions of $\theta = 60°$ and $\theta = -30°$.

The echo waveform from the outside of femur, the inside of femur and the stem is measured. The measurement method is carried out in the same manner as described as in section 4.1.

5.2 Measurement results and considerations
Figures 8 and 9 are an echo waveform obtained from the outside of femur ①, the inside of femur ② and stem ③, which the ultrasonic wave is emitted toward the femur from the directions of $\theta = 60°$ and $\theta = -30°$ at the circumference of the thigh. From these waveforms, the thickness of the femur s determined by Eq.(2). If the distance of the bone marrow indicated on the CRT is l_{bm}' and the wave velocity of it is v_{bm}', the exact distance of l_{bm} is given by the following equation.

$$l_{bm} = l_{bm}' \times \frac{v_{bm}}{v_m} \qquad (3)$$

If the wave velocities of the thigh muscle and the bone marrow are $v_m = 1624$ m/s[7], 1500m/s, respectively, the thicknesses of the femur and the bone marrow in Fig.8 are 6.5mm and 2.3mm at $\theta = 60°$, respectively. The thicknesses of the femur and the bone marrow in Fig.9 are 5.0mm and 3.2mm at $\theta = -30°$, respectively. From the echo waveform in Figs.8 and 9, the thicknesses of the femur and the bone marrow on the patient can be measured. The loosening between hip prosthesis stem and femur become a large problem after the total hip replacement. It is clarified that the ultrasonic wave method can be applied to diagnose the loosening.

CONCLUSIONS

The sectional form of the upper side of the normal human femur is measured via an ultrasonic wave, and the obtained result is compared with that of the X-ray CT. Also, the ultrasonic wave is emitted toward the femur on the patient inserted with the hip prosthesis stem. The echo waveform reflected from the femur and the stem is measured, and the following conclusions are obtained.
(1) The sectional form of the

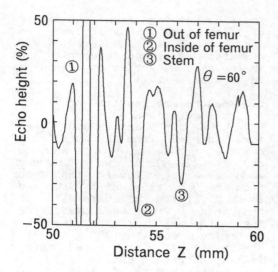

Fig.8 Echo waveform presenting
outside of femur inside of
femur and stem surface.
(θ=60°)

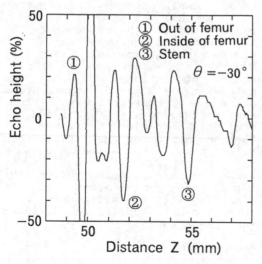

Fig.9 Echo waveform presenting
outside of femur inside of
femur and stem surface.
(θ=-30°)

upper side of the normal human
femur is able to measure
approximately via the ultrasonic
wave using the simple measurement
device.
(2) The thickness of femur agrees
approximately with the measurement
result of the X-ray CT.

(3) The ultrasonic wave method can
be measured the femur thickness on
the patient inserted with the hip
prosthesis stem.

REFERENCES

[1] Meema H. E., Cortical Bone and
Osteoporosis as a Manifestation of
Aging Am. J. Roentgenol., Vol.89,
No.6 (1963), 1287-1295.
[2] Kita H. et al., Magnetic
Resonance Imaging under Existence
of Orthopaedic Implants, Vol.10,
No.5 (1991), 464-470.
[3] Smith, H. W., De Smet, A. A.
and Levine, A., Measurement of
Cortical Thickness in a Human
Cadaver Femur (Conventional
Roentgenography versus Computed
Tomography), Clin. Orthop. Rel.
Res., No.169 (1982), 269-274.
[4] Sigh, S., Ultrasonic Non-
destructive Measurements of
Cortical Bone Thickness in Femur
Cadaver Femur, Ultrasonics,
Vol.27, No.2 (1989), 107-113.
[5] Minakuchi, Y., et al.,
Ultrasonic Measurement of Bone and
Bone Marrow Thicknesses by
Inserting a Hip Prosthesis Stem,
Jpn. Soc. Orthopaedic Biomechanics.
(in Japanese), Vol.14, (1992), 221-
215.
[6] Minakuchi. Y., et al.,
Ultrasonic Measurement of Bone and
Bone Marrow Thicknesses in Femur
Covered with Flesh by Inserting
a Hip Prosthesis Stem, Trans. Jpn.
Soc. Mech. Eng. (in Japanese),
Vol.59, No.561 (1993), 1408-1412.
[7] Minakuchi, Y., et al.,
Ultrasonic Measurement on Sectional
Forms of Human Femur, Prepr. of
Jpn. Soc. Mech. Eng. (in Japanese),
No.940-10 (1994-3).

In-vitro stress analysis of the hip investigating the effects of joint and muscle load simulation

D. Colgan, D. McTague, P. O'Donnell & E. G. Little
Biomechanics/Stress Analysis Group, University of Limerick, Ireland

D. Slemon
Howmedica International Inc., Raheen Industrial Estate, Limerick, Ireland

ABSTRACT: In order to effectively model *in-vitro* the physiological bone strain distribution present at the hip, both the joint and muscle loading must be adequately simulated. The objective of this study is to examine the effects of simulated joint and muscle loading and so enhance the subsequent interpretation of results.

Suitable force magnitudes and directions for the simulated loads on the pelvis and femur were selected on the basis of a thorough review of the relevant literature. Their effects were investigated in a photoelastic coating analysis of a dried hemipelvis and in a three-dimensional embedded strain gauge analysis of cement mantle stresses in a model of an implanted Exeter femoral prosthesis.

The study highlighted the significant effects of the joint and muscle loading combinations and emphasised the need to include realistic load simulation in any *in-vitro* stress analysis of the hip.

1 INTRODUCTION & RELATED STUDIES

In view of the difficulties associated with *in-vitro* full-field stress analysis techniques and the complexities of *in-vivo* muscular and joint loading, considerable simplification is required in order to simulate the biomechanical forces at the joints.

Nevertheless, *in-vitro* stress analyses require satisfactory simulation of both the joint and muscle loads in order to obtain data that is both meaningful and relevant. Accordingly, prior to investigating situations such as the physiological stress levels induced by the introduction of a prosthetic implant etc., it is necessary to validate the loading configuration and to understand the limitations of the *in-vitro* simulation.

This study examines the effects of joint and muscle load simulation on the hip, using two stress analysis techniques, namely; an *in-vitro* analysis of the pelvis using a photoelastic coating technique and a model analysis of cement mantle stresses in the proximal femur employing three-dimensional embedded strain transducers.

Numerous investigations have simulated the joint and muscular forces at the hip in examining biomechanical aspects of the joint. In a review of joint and muscle load simulations relevant to *in-vitro* stress analysis of the hip, Colgan *et al.* (1994) concluded that it is possible, only in a limited manner, to simulate the physiological forces on the hip with some degree of realism, by applying the findings of literature pertaining to joint and muscle forces.

Considering firstly the simulation of the joint load; Crowinshield and Tolbert (1983) simulated only the joint reaction at the hip in an embedded strain gauge analysis of stresses induced by stem/cement interface conditions. However, they justified their selection of the loading condition by claiming that the nature of the study was comparative. Hence, trends were reported as opposed to quantitative data due to the non-physiological load simulation. In contrast, Walker and Robertson (1988) in a photoelastic coating analysis examining the state of strain and load transfer, stated that the advantage of the photoelastic coating method is that it allowed for evaluating quantitative results for surface strains, although such strains weren't actually recorded. This study, however, neglected to consider the effect of the muscular loading; particularly the limitations of using the photoelastic coating technique *in-vitro* given the considerable areas of muscle origin and insertion on the surface of the bone. Other investigations include those of Jacob and Hueggler (1980) and Kareh *et al.* (1988), who both simulated the abductor muscle group along with the joint reaction; however, the effects of these muscle forces on stresses were not considered. All of the quoted studies modelled a 2-dimensional load as opposed to the 3-dimensional loading nature outlined by Paul (1976), with researchers seldom citing sources in justifying the selected loads. Furthermore, the effect of the load simulation on results was rarely investigated.

Tanner *et al.* (1988), in attempting to devise a system for modelling forces on the hip during the single-legged stance, simulated the joint reaction and the abductor muscle group. However, Miles and Dall (1984) in a parametric analysis of femoral prosthesis design, claim that the effect of the joint reaction component in the antero-posterior plane is significant and consequently should be incorporated into simulation techniques. This supports the study of Wroblewski (1980), which stated that the transverse load on the joint is arguably the most important and its effects are significant. Given that considerable forces are induced during walking, the investigation implies that a more dynamic approach to load simulation than the single-legged stance is required.

Secondly, the effect of the muscle forces has been examined in a number of finite element and experimental investigations. In a 3-dimensional model of hip musculature, Dostal and Andrews (1981) found the abductors to have a significant effect on both the joint reaction and the distribution of femoral stresses. This is also evident in the finite element investigations of Rohlmann *et al.* (1983) and Harrigan and Harris (1991). Rohlmann *et al.* (op. cit.) investigated the effects of simulating both the joint reaction and the abductors on stresses in the proximal femur. They reported large tensile stresses on the lateral femur which were only reduced by applying a muscle force resisting in some way the abductors; namely, the ilio-tibial band. This supports the 3-dimensional finite element study of Rybicki (1972) which states that the ilio-tibial band has a significant effect on reducing bending induced in the bone. Furthermore, the strain gauge analysis of Finlay *et al.* (1986) and the study of Rohlmann *et al.* (1983) both concluded that muscles, in acting to resist the action of the joint load, must be included in any simulation of the *in-vivo* hip loading conditions.

In outlining load simulation problems in model testing, O'Connor (1992) states that in order to obtain meaningful data from a stress analysis of the joint, a realistic representation of the *in-vivo* loading configuration must be simulated, due to the effects the forces have on the strain distribution in bone. However, *in-vitro* limitations such as load simplification, photoelastic coating reinforcement, etc. are inevitable and thus the results must be interpreted with caution.

2 METHODS AND PROCEDURES

In investigating the effects of the joint and muscle loads acting on the hip, two separate experimental analyses were carried out; (i) an *in-vitro* photoelastic coating analysis of the pelvis and (ii) an embedded strain gauge analysis of cement mantle stresses in a model of the proximal femur.

2.1 *Pelvis*

The *in-vitro* analysis of the pelvis follows the procedures outlined by Slemon *et al.* (1991). Figure 1 shows the pelvis mounted on the loading bench and the joint and muscle loads.

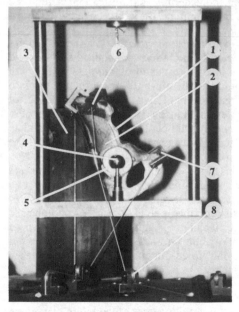

Figure 1. Pelvic loading configuration

A dried hemipelvis (1), coated with a birefringent material (2), was mounted on a specially designed loading rig. The sacroiliac joint, encapsulated in an Araldite resin similar to the methods employed by Petty *et al.* (1980), was rigidly restrained on a mounting block (3). The joint load (4) was applied vertically using a floating frame, thus ensuring unconstrained deflection of the joint. Transmission of the joint reaction to the acetabulum was through an acetabular cup (5). The joint load magnitude, direction and the relative pelvic and femoral orientation shown in Figure 1, simulate the 47% stance phase of gait as defined by Paul (1976), and are rotated to facilitate vertical loading.

The abductor muscle group (6) was simulated acting from the iliac crest utilising the magnitudes presented by Paul (1976). The adductors (7), acting from the pubis, were applied with an assigned value of half the abductor magnitudes. This estimation was necessary due to the lack of adductor force-determination studies in the literature, coupled with the conflicting results regarding muscle activity levels presented by Seirag and Arvikar (1973) and Crowinshield *et al.* (1978). All of the simulated muscle and joint forces were applied using dead weights, via pulleys (8) supported on knife-edge bearings to minimise the adverse effects of friction on the transmission of the applied loads.

The single-legged stance was also simulated, utilising the magnitudes and directions for both the joint reaction and the abductors predicted by McLeish and Charnley (1970). The loading configuration and procedure was subjected to repeatability and reproducibility tests (Slemon, 1991), with the load magnitudes being scaled down to accommodate the *in-vitro* nature of the test hemipelvis.

Testing began by assessing the effects of both the joint load and the muscular loads in isolation. the joint load was then applied with the abductors and adductors added sequentially to the loaded pelvis so as to assess their effects on the strain distribution in the pelvis.

2.2 *Proximal femur*

The joint and muscle forces acting on the proximal femur were simulated in a three-dimensional embedded strain gauge analysis of cement mantle stresses within a model of an Exeter total hip replacement. The model was manufactured three-times full size to facilitate embedding the three-dimensional strain transducers. An actual left femur with an implanted prosthesis was scanned using Computerised Tomography (CT) in order to replicate the geometries of the cortical shell and the cement mantle. Only the proximal half of the femur, however, was modelled due to the size limitations imposed by equipment (ovens etc.) used in the casting procedure (McTague, 1992). Figure 2 shows the model mounted on the test bench and the applied joint and muscle loading.

The cortical shell (1) was simulated using an Araldite epoxy MY750/HY906/Silica flour (identical to that used by Little (1990)), as it has properties similar to cortical bone provided the anisotropic nature of bone is neglected. The implant used in the model was an Exeter type prosthesis (2), while the cement was modelled using CT200/HY907 Araldite epoxy.

The joint load (3) was applied at the head of the prosthesis via a spherical bearing using a floating frame (4). The model was set-up on an angled mounting block (5), simulating the relative joint load directions and femoral orientations outlined by Paul (1976) for the 47%

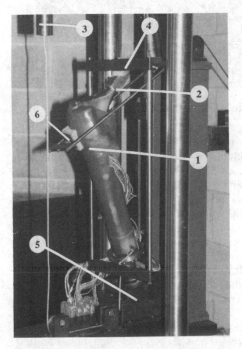

Figure 2. Femoral loading configuration

stance phase of the walking cycle. This represented a 3-dimensional joint reaction and consequently a 2-dimensional load was also simulated to investigate the effect of the torsional component. The magnitude of the joint load was chosen as 2.3 times' body weight based on the *in-vivo* estimates of Davy *et al.* (1988).

The combined muscular force (6) is a resultant of both the abductor muscle group and the ilio-tibial band and was applied acting from the greater trochanter on the femur. The magnitude and direction of the abductors used are based on the estimates of Merchant (1966), which have subsequently been widely supported in the literature. The ilio-tibial band was simulated in accordance with the *in-vivo* magnitudes recorded by Jacob *et al.* (1982), although that investigation disputes the extent to which the muscle is active during late stance.

The cement mantle strains were recorded using three-dimensional embedded strain gauging methods (Rossetto *et al.* 1975, Barbato and Little 1984). Figure 3 illustrates, from left to right, the various stages of transducer manufacture.

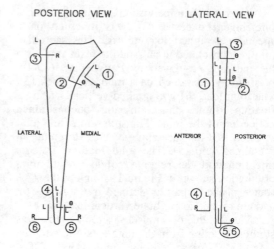

Figure 4. Transducer positions relative to femur

1992), in regions of low strain gradients so as to reduce experimental error.

The effect of the muscular load on the strains was investigated by applying the muscle forces alone. Both the joint and muscle loads were then applied to complete the simulation.

3 RESULTS AND DISCUSSION

3.1 *Pelvis*

The fringes induced on the medial (left) and lateral (right) sides of the pelvis are shown in Figure 5 under no load and at the 47% stance.

The effect of the joint load is to induce bending strains at the upper end of the iliopectineal line near the Sacroiliac Restraint.

The effects of the muscular groups are more difficult to present as their actions applied alone induced bending strains at the fixation that were approximately equal and opposite to the strains induced by the joint load. As a result constructive interference existed with the joint load fringe patterns. This highlighted the considerable influence the muscle forces have on pelvic strain magnitudes. The application of the abductor muscles to the upper ilium resulted in low order strains being induced at the superior iliac fossa, originating from the rigid fixation, and resulting in increased fringe orders. Thus, these muscles increased the magnitude of the joint load, confirming the suggestion that the

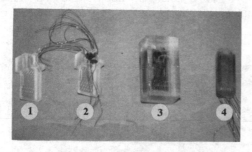

Figure 3. Strain transducer development.

The transducer consists of three miniature rectangular rosettes mounted adjacent to each other on a CT200 carrier (1), on three mutually perpendicular planes. The instrumented carrier (2), is placed in a mould and cast to form a CT200 block (3) containing the embedded gauges, which is then machined (4) and implanted in the model. The rosette orientation is described by Barbato and Little (1984), and yields the nine directional strains required to calculate the strain tensor for a given point.

Transducers were embedded in six sites, three in both the proximal and distal regions relative to the Exeter prosthesis, as shown in Figure 4. The sites were selected, based on finite element and experimental analyses (McTague

abductors act downward to stabilise the pelvis. In contrast, the addition of the adductors induces a three-point bend effect with the joint load and the sacroiliac joint. The adductors and the joint load are largely responsible for the fringes in the acetabular region.

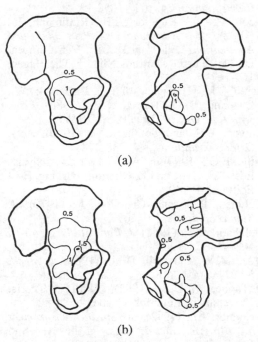

(a)

(b)

Figure 5. Medial and lateral views of hemipelvis with cup inserted; (a) no load, (b) 47% stance.

The simulation of muscles on the pelvis is, however, inhibited by the use of full-field stress analysis techniques such as photoelasticity. The very nature of this surface analysis is such that the vast origins and insertions of the muscles cannot be simulated realistically. The technique is also limited due to its reinforcing effect, which is complex due to the varying thickness of the cortical shell.

Considering the selection of force magnitudes, while there is an abundance of information regarding the reaction between the femur and the acetabulum, it is difficult to select a realistic physiological representation of the muscular loading on the pelvis due to the conflicting information in the literature. Consequently, the effects of the muscle loads applied are difficult to establish. Due to these difficulties in simulating forces acting on the pelvis, it is often necessary to limit *in-vitro*

pelvic studies to parametric investigations where they have restricted application to product design.

3.2 Proximal femur

Relevant experimental data showing the effect of the joint reaction versus the muscular loading for the model of the proximal femur is given in Table 1, showing tensor stresses for sites 1, 5 and 6. The trends evident at site 1 are representative of the three proximal sites, while sites 5 and 6 reflect the distal trends.

Table 1. Tensor stresses produced by individual joint (J) and muscle (M) loads at sites 1,5 and 6.

Stress (MPa)		Proximal medial (1)	Distal medial (5)	Distal lateral (6)
σ_θ	M	.010	-.041	.137
	J	.050	-.005	-.03
σ_L	M	.000	-.694	.950
	J	.011	-.136	-.657
σ_R	M	.011	.162	.020
	J	.051	-.057	-.066

Proximal stresses (Site 1) due to the combined muscle force are very low as the bending moment produced by the muscular load is almost negligible near its point of application. The effect of the joint load alone is to increase the stress magnitudes in the proximal region.

The muscular load dominates the longitudinal stresses in the distal medial (Site 5) and distal lateral (Site 6) regions with some hoop and radial stresses also in evidence, supporting the findings of Rybicki et al. (1972) and Finlay et al. (1986). The joint load produces distal stresses related to bending and axial loading, however, the effect of the torsional component of the joint reaction is small in the distal regions. Proximally, the torsional component increased the compressive hoop stresses in the proximal posterior region, and contributed to the tensile radial stresses evident at site 1 in the proximal medial position. Thus, the 3-dimensional load simulation signals an improvement over the single-legged stance reported in studies to date.

The simulation of the muscles acting on the proximal femur as a single resultant force is, however, a considerable simplification of the physiological loading conditions present *in-vivo,* despite the muscle loads being selected based on numerous force determination studies. While the muscles chosen for this simulation incorporate all the features of proximal femoral loading in the literature, it is nevertheless a considerable simplification of the physiological loading conditions present on the femur. However, the load simulation presented here may represent an instantaneous phase in the walking cycle. The effects of which were demonstrated by Lanyon and Smith (1970) who showed the walking cycle to consist of a number of discrete deformation cycles, each ranging from tensile to compressive. Therefore, quantitative data may possibly be recorded and interpreted in terms of the extent to which the joint and muscular forces have been simulated.

4 CONCLUSIONS

The effect of the joint and muscle forces acting on the hip are significant and, consequently, these forces must be simulated in any *in-vitro* analysis of the joint attempting to predict trends in biomechanical response. However, given the complexities of such simulations, results can only be presented reliably in parametric form and hence application to product design is in its infancy.

5 ACKNOWLEDGEMENT

The authors would like to thank Howmedica Int. Inc., Limerick for supporting this work.

6 REFERENCES

Barbato, G. and Little, E.G. (1984), Tech. note, Institutio Di Metrologia, Turin (Italy)

Colgan, D., Trench, P., Slemon, D., McTague, D., Finlay, J.B., O'Donnell, P. and Little, E.G., (1994), in print *Strain*

Crowinshield, R.D., Johnston, R.C., Andrews, J.G. and Brand, R.A., (1978), *J.Biomechanics*, 11, 75-85

Crowninshield, J. and Tolbert, J., (1983), *J.Biomed. Matls.Res.*, 17,, 819-828

Davy, D.T., Kotzer, G.M., Brown, R.H., Heiple, K.G., Goldberg, V.M., Heiple, K.G.Jr.,

Berilla, J., Burstein, A.H., (1988) *J.Bone Joint Surg.*, 70-A,, 45-50

Dostal, W.F. and Andrews, J.G., (1981), *J.Biomechanics*, 14, 803-812

Finlay, J.B., Bourne, R.B., Landsberg, R.P.D. and Andreae, P., (1986), 19, 723-739

Harrigan, T.P. and Harris, W.H., (1991), *J. Biomechanics*, 24, 1047-1058

Jacob, H.A.C., Huggler, A.H., Ruttimann, B., (1982), In *"Biomechanics: Principles and Applications"*,Eds: Huiskes, R., Van Campen, D., DeWijn, J., Martinus Nijhoff, The Hague, 161-167

Kareh, J., Harrigan, T.P., Burke, D.W., O'Connor, D.O., Harris, W.H., (1988), *Proc. 34th Ann. ORS,* Atlanta (USA), 344

Lanyon, L.E. and Smith, R.N., (1970), *Acta. Orthop. Scandinav.*, 41, 238-248

Little, E.G., (1990) in "Applied Stress Analysis" Eds: T.E. Hyde and E. Ollerton, Barking, First Edition,140-149

McLeish, R.D. and Charnley, J., (1970), *J. Biomechanics*, 3, 191-209

Merchant, A.C., (1966), *J. Bone Jt. Surg.*, 46-A, 462-476

Miles, A.W. and Dall, D.M., (1984), *I. Mech. E.*, 31-36

McTague, E.D., (1992), Ph.D. Thesis, University of Limerick, Chapter 5, 73-82

O'Connor, J.J., (1992), in *"Strain Measurement in Biomechanics"*, Eds: Miles,A. and Tanner,K., Chapman & Hall, 14-38

Paul, J.P., (1976), *Proc.R.Soc.Lond.B.*, 192, 163-172

Petty, W., Miller, G.J. and Piotrowski, (1980), *Bull. Prosth. Res.*, 17, 80-89

Rohlmann, A., Mossner, U., Bergmann, G., Kolbel, R., (1983), *J. Biomechanics*, 16, 727-742

Rossetto, S., Bray, A., Levi, R., (1975), *Exptl. Mech.*, 15, 375-381

Rybicki, E.F., Simonen, F.A., Weiss, E.B., (1972), *J.Biomechanics*, 5, 203-215

Seirag, A. and Arvikar, R.J., (1973), *J.Biomechanics*, 5, 89-102

Slemon, D.P., O'Donnell, P. and Little, E.G., (1991), *Strain*, 11, 143-153

Slemon, D.P., (1991), M.Eng Thesis, University of Limerick, Chapter 5, 34-70

Tanner, K.E., Reed, P.E., Bonfield, G., Rasmussen, L., Freeman, M.A.R., (1988), *J. Biomed. Eng.*, 10, 289-290

Walker, P.S. and Robertson, D.D., (1988), *Clin. Orthop.*, 235, 25-34

Wroblewski, B.M., (1980), *Eng. Med.*, 9, 163-4

Recent Advances in Experimental Mechanics, Silva Gomes et al. (eds) © 1994 Balkema, Rotterdam, ISBN 90 5410 395 7

Linear and Fourier analysis of stress relaxation in the cervical spine

Andrew D. Holmes & Jing Fang
Peking University, Beijing, People's Republic of China

ABSTRACT: Single intervertebral joints (discs) of the cervical spine (neck) have been subject to axial compressive loads over a physiological range of loads and the relaxation recorded. Stress and strain readings obtained at each stage of loading to allow isochronal plots to be drawn for each specimen, which showed a high linear correlation coefficient in all cases, and a suitable linear model was accordingly formulated and fitted to the data.

Fourier transform of the relaxation curve allows the relaxation function to be displayed as a function of frequency, and suggests that the cervical spine has a much wider and smoother distribution of relaxation times, spreading from a few minutes to a few hours.

This work shows that the cervical discs are capable of dissipating a large proportion of the applied strain energy, but can only do this efficiently below about 0.001 Hz, and do not function well as impact load absorbers.

1 INTRODUCTION

The mechanical properties of spinal discs depend on the rate at which they are strained, that is, they show a high degree of viscoelastic behaviour. This has implications in understanding how they function both as isolated elements and synergistically with the whole musculoskeletal system: it has been suggested that the intervertebral discs could act as 'shock absorbers' in the spine as they are much more compliant than the other main phase, the bone of the cervical vertebrae.

The cervical spine is a part of the spine where absorbtion of impact loads (mainly compressive) is likely to be an important feature, whether the source be from external oscillators or from impact loading on the cranium. The possible existence of a non-axial second stiffest axis in the cervical spine (King-Liu and Guo Dai, 1989) also suggests that the compressive forces in the cervical spine may still be the greatest even if the loading is not a true axial compression.

The linear and Fourier analyses of results from relaxation experiments on cervical discs under axial compression are presented in this paper.

2 PREPARATION AND TESTING

All the segments used in this series of experiments were from young males, and showed typically normal anatomy. Two adjacent vertebrae and the intervertebral disc connecting them were used in all cases. The specimen was prepared by cutting through the discs at the adjacent levels and the two end faces were smoothed to flat surfaces using a fine bone file. The two end faces were checked to be mutually parallel to each other and the intervening disc, and the specimen was then attached to a pair of perspex loading plates using short bone screws. The locating holes for the bone screws in the perspex plates were countersunk to allow two flat parallel loading faces aligned with the disc under test. Care was taken at all times during the preparation, mounting, and testing of the specimen to keep it moistened with physiological saline (9.6 g NaCl per litre).

The experiments presented were separated into two parts:

(a) to plot isochronal curves from different stress and strain values for the same segment

(b) to examine the longer term relaxation of segments.

The relaxation curves were recorded over 15 minutes for each point on the isochronal plot, and the specimens rested at zero stress for 45 minutes between successive loadings. An initial test showed that this recovery time was sufficient for the segment to display reproducible relaxation curves. All of the experiments were performed on a standard materials testing machine, and strain readings were taken from a displacement transducer.

3 LINEAR ANALYSIS

A typical isochronal plot for a single specimen at times of 0, 1, and 10 minutes is shown in figure 1.

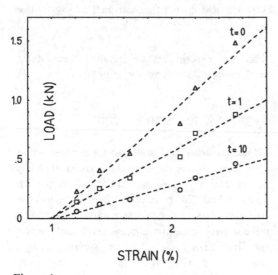

Figure 1

The lowest value obtained for the linear correlation coefficient was found to be greater than 0.9 in all cases for all five specimens tested for linearity. Although the origin of the regression lines from these isochronal plots does not coincide with that of the x-y axis, this is simply a reflection of the quasi-static properties of the segment. A material

that exhibits non-linear stress-strain properties may still exhibit linear stress relaxation, providing that (1) the shape of the isochronal plot for $t=0$ is similar to that seen in quasi-static loading and (2) that the shape of the isochronal plot remains constant with increasing time (Holmes, 1991).

The results of the isochrones justify the modelling of the data by a linear system, by the criteria that the disc is poroelastic and that a single relaxation time is insufficient given the time range over which relaxation is seen (Dorrington, 1980). On the basis of these and the observed experimental phenomena, the simplest model is shown below in figure 2, which contains elastic, plastic, and viscous elements.

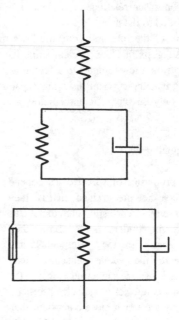

Figure 2

The parameters for the various elements in this model allow a differential equation to be set up, the solution for which gives a form of stress relaxation determined by two simple exponential terms (due to the two independent viscous elements) and a constant stress term (due to the friction plate), such that

$$\sigma(t) = A e^{-\alpha t} + B e^{-\beta t} + C$$

1388

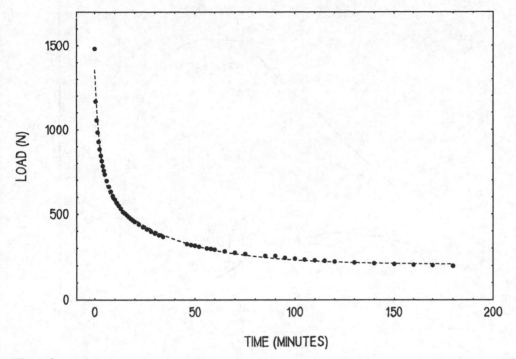

Figure 3

Figure 3 shows a longer stress relaxation curve (3 hrs) with a fitted curve according to the model in figure 2 also shown as a dotted line. Note that no correction from load to stress has been made for any of the results presented here, as the area of loading is a constant, and stress and load are therefore simply related by a constant factor. The fitting routine was performed on a microcomputer and optimised by the downhill simplex method (eg, Press et al, 1986). The values of the relaxation times obtained from this model when fitted to the actual data were approximately 8 minutes and 4 hours, for each of the five segments tested for long term relaxation. The range of these values were about two minutes and half an hour, respectively. There is clearly a reasonably accurate fit between the theoretical model and the actual data, indicating that the relaxation times obtained from this model are representative of the approximate time region over which the disc is undergoing most relaxation and hence dissipating energy at the greatest rate, that is, in the region of minutes to hours.

4 FOURIER ANALYSIS

Linear analysis of stress relaxation curves is disadvantaged in that it typically presupposes the form of the data obtained, by the modelling process, as above. While it is useful for establishing 'ball park' values for relaxation times, the relaxation will be more generally made up of a continuous spectrum of relaxation times. Fourier transform mechanical spectroscopy (FTMS) described by Arridge and Barham (1986) transforms the data into separate spectrums of loss and storage moduli, continuously defined for a frequency range corresponding to end points of the data acquisition time and the duration of the experiment. This makes no assumption on the form of the data (linear or non-linear) and is an accurate and convenient method for examining the stress relaxation data.

Aspden (1991) also included an empirical correction to counter aliasing effects that occur when a discrete Fourier transform (DFT) is performed. A similar method, including the

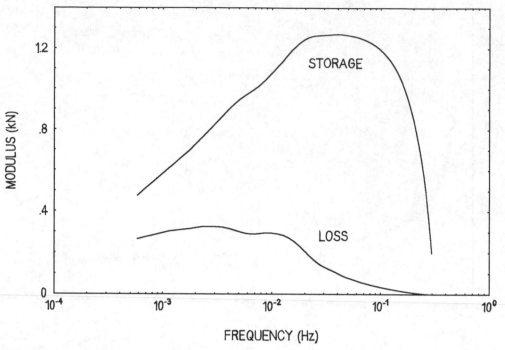

Figure 4

aliasing correction to the loss modulus yields the spectrum shown in figure 4, for the same experimental data as presented in figure 3. As the Fourier transform procedure is extremely sensitive to noise in the signal, the data is smoothed (using cubic B-splines) before passing to the DFT. As the resulting spectrum shows, there is evidence of features in the storage moduli at frequencies indicated by the relaxation times calculated from the linear analysis, but these are not strong, and the spectrum is basically smoothly defined for frequencies corresponding to time ranges from a couple of minutes to a couple of hours (the sharp drop in the storage modulus at higher frequencies is simply an artefact of the aliasing caused by the DFT, and should be ignored).

While the finite loading time of the experiment will cause the high frequency components of the spectrum to be only partially present in the pure relaxation part of the curve (because they will decay substantially during the loading period) there should still be some evidence of any of these components at up to about 10 Hz for this loading

time (0.5 sec). However, separate examination of the high frequency end of the spectrums (using a similar plot to that in figure 4 but for the data collected from the start of relaxation with a higher acquisition rate) revealed a flat response, and the tendency of the loss modulus to zero for frequencies greater than about 0.1 Hz indicated that the response of the disc at these frequencies is almost entirely elastic; that is, the amount of energy dissipated by viscous behavior is minimal.

5 CONCLUSIONS

The viscoelastic properties of the functional unit of the cervical spine (the disc and adjacent vertebrae) have been shown to display linear behavior under axial compression. The simplest appropriate linear model showed a good agreement with the experimental results when fitted to the data. This suggested most substantial relaxation in the period of 8 minutes to 4 hours, and similar values were obtained between specimens. Fourier transform mechanical spectroscopy showed similar results, with a smooth decrease in the loss modulus and no evidence of any particular resonance frequencies. Very little relaxation was seen in the frequency range of Hz and above; the disc does not function as an effective shock absorber of axial loads at these frequencies.

REFERENCES

Arridge, R. G. C., and P. J. Barham 1986. Fourier transform mechanical spectroscopy. *J. Phys. D: App. Phys.* 19: L89-L96.

Aspden, R. M. 1991. Aliasing effects in Fourier transforms of monotonically decaying functions. *J. Phys. D: App. Phys.* 24: 803-808.

Dorrington, K. L. 1980. The theory of viscoelasticity in biomaterials. *Symp. Soc. Exp. Biol.* 34: 289-314.

Holmes, A. D. 1991. Ph.D. Thesis: University of Manchester, Manchester U.K.

King-Liu, Y., and Q. Guo Dai 1989. The second stiffest axis of a beam column: implications for cervical spine trauma. *J. Biomech. Eng.* 111: 122-127.

Press, W. H., B. P. Flannery, S. A. Teukolsky, and W. T. Vetterling 1986. *Numerical recipes: the art of scientific computing.* Cambridge: Cambridge University Press.

Recent Advances in Experimental Mechanics, Silva Gomes et al. (eds) © 1994 Balkema, Rotterdam, ISBN 90 5410 395 7

Experimental equipment for improving the performances of long jump sportsmen

B. Merfea & M. Cibu

Transilvanya University, Braşov, Romania

ABSTRACT: With a view to improve the results of the long jump sportsmen the paper presents a new method which is able to offer to the couple coach-sportsman the possibility to analyse the stage of the beat. The experimental results point out the value and the direction of the forces by which the jumper acts on the beat board and the graphs cocerning the variation of the forces on a vertical and horizontal direction, according to the time.

1 INTRODUCTION

Sportsmen's results in long jump depend, to a great extent, on their assimilation of the jump's technique. We mean by "technique" the whole of the movements an athlete does during his jump.

The long jump consists of two distinct stages, which are respectively subdivide into other two substages:

1. The first stage, which is compose of the spring and the beat.

2. The second stage, which includes the flight and the touch-down.

A charecteristic feature of this event is the feat that the athlete can influence the jump's result, during the flight. We know that, to a great extent, this result depends on the last steps he takes during the spring. The length of the jump depends on the curve described by the centre of gravity, during the filght, and it is influenced by the speed, the take-off angle and the height of the jump. All the mentioned parameters characterize the board beat.

2 THEORETICAL ASPECTS

The beat is the stage during which the jumper, using horizontal speed of the spring, increases it by a set of impulsive actions, whose consequence consists of the take-off from the ground.From a mechanical point of view, the stronger the force developped by the jump is, the more efficient the beat is, for an interval of time, as short as possible.

The stage of the beat consists of three main moments, fig.1:

-the moment when the foot touches the beat board;

-the moment of damping, carried out by a light relaxation of three joints (ankle, knee, hip);

-the moment of the impulse, characterized by an active extention of the above mentioned joints.

Because of the fact that the horizontal running speed must be turned to an obliquely aimed speed, the position of the beat foot on the beat bord is done before the athlete projects his centre of gravity on the board. Setting the foot manner, should not cause an exagerate position backwards of the body. The foot should not remain too mouch on the ground, but it has to act immediatly, pushing obliquely-downwards. It is recommended that, at the touch of the beat board, the foot should be almost stretched 170°. The setting of the foot on the beat board should not be hard and it should be done on the whole sole.

The damping is destined to prepare an advantageous position to obtain a forward-upwards take-off direction. The whole force that appears at the moment when the foot is setting on the beat board, is taken over by the three joints of the inferior limb, and especially by the joint of the knee. In some specialists' opinion, the beat foot is flexed to 145°-150°. In case the angle is larger or smaller than the mentioned values, two big mistakes appear, namely:

1. A too rapid and short push - the knee has been too little flexed - the consequence is that the beat foot is blocked. This makes the jump a high one, but not a long one.

2. A too slow push, too late fulfilled if the foot has been too much flexed. As a cosequence its stretch is done too late, that is why both the height and the length of the jump will be diminished.

At the damping moment, the extensive muscles of the beat foot come to a so-called "arching" stage, creating optimal conditions for the muscular activity that follows: the "explosive" take-off. So, while the foot is passing on the front part of the sole, the component forces that contribute to the take-off should be directed forward and upwards.

When all the movements that take part in the take-off are co-ordinated, and the jumper's centre of gravity is situated over the line of the force applied, the body can be pulsed with the maximum speed upwards and forward.

Fig.1 The stages of the beat

The attack foot exerts a great influence upon the take-off moment. If we decompose the movement, we can notice that there are four points we have to insist upon:

- the attack foot acts from the joint of the hip, and its direction is pointed forward and upwards;

-the spring of the attack foot must be done at a precise moment so that the spring force may be transmitted to the whole body;

-during the take-off, the attack foot will be flexed from the knee, and the thigh, at the end of the take-off, must come to the horizontal line;

-the execution of theese movements has as a consequence the orientation forward of the results of the forces.

The efficiency of the beat technique, followed by the take-off:

-the achievement of an optimal ratio between thee three stages, concerninig their duration;

-the value, the orientation and the variation in time of the forces which appear during each of the three above mentioned moments;

-the position of the body and of its centre of gravity, as well as that of the beat foot.

3 MEASURING TECHNIQUE

If for the determination of the position of the body, of its centre of gravity and the position of the foot we can use the cinematographic technique, for the measurement of the time force variation and for the calculation of an optimum ratio between the three stages, the present paper proposes the usage of an experimental equipment, like the one presented in fig. 2.

The beat board is a plate leaning on four force transducers, for the determination of the action force in a vertical plane Fz, and, leaning on the other four force tranduscers for the action force in a horizontal plane Fx. The measurement set-up includes the eight force transducers, the measur-

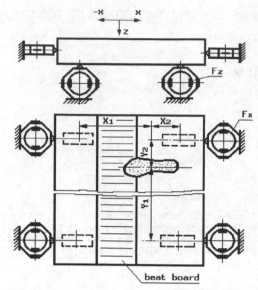

Fig.2 Experimental device for force measurement

ing amplifier, a analogue-to-digital converter, and a IBM computer, fig. 3.

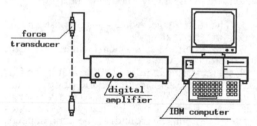

Fig.3 Measurement set-up

The footprint left by the sportsman, permits the determination of the four coordinates x_1, x_2, y_1, y_2, of the pression centre, according to the eight force transducers. The data aquisition concerning the value of the forces exerted, are separately carried out for each force transducer, so that we can determine a force equivalent to Fx and Fz.

4 EXPERIMENTAL RESULTS

The experimental determinations have been done on a device, like the one presented in fig. 2, computer assisted, for the long jumps of an average performance sportsman. The graphs cocerning the variationof the forces on a vertical and horizontal direction, according to the time, presented in the fig.4 and 5, showed the three distinct moments of the sportsman's board beat.

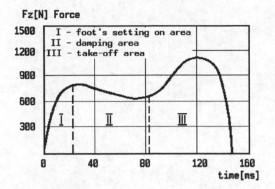

Fig.4 Variation of the force on vertical direction

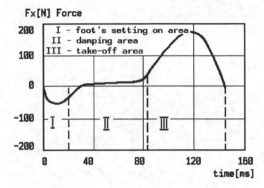

Fig.5 Variation of the force on horizontal direction

So, we can notice that during the first interval of time corresponding to the moment when the foot is setting on the beat board, the force Fz rapidly increases up to a larger value than the static weight of the sportsman. The damping moment is showed by a light diminution of the value of the force Fz, while the force Fx passes from a negative value to a positive one, being kept, however, around the zero value. The take-off starts around the 0.08 [s] value, marked by a sudden increase, both of the force Fz, and of the force Fx. The contact loss between the beating foot and the beating board, which corresponds to the moment when the flight starts, takes place at the 0.12-0.14 [s] moment.

5 CONCLUSIONS

The new conception of the device presented here permits the couple coach-sportsman to get superior performances by the possibility it offers to analyse and determine:
-the value and direction of the forces by which the jumper acts on the beat board.
-the ratio between the duration of the time corresponding to the damping and that of the take-off.
-the total duration of the board beat stage ;
-the value and direction of the take-off speed.

The teaching value of the method was appreciated by the sportsmen, because of the fact that the analyses can be done immediately after the jump, so that , at the next jump the mistakes could be corrected. The experimental method presented here, combined with the high-speed cinematographyc method, lead to the perfecting of the technique that aims to turn to profit the sportsman's physical qualities and, in the end, comes to the improvement of his performances.

REFERENCES

Cibu, M. 1993. *Athelitics*. Brasov: Transilvanya University.
Dumitru, C. 1973. *Athelitics*. Bucuresti: National Sport Institute.
Popescu, I. 1986. *Experimental Research*. Brasov: Transilvanya University.

Recent Advances in Experimental Mechanics, Silva Gomes et al. (eds) © 1994 Balkema, Rotterdam, ISBN 90 5410 395 7

Aeroelastic characteristics of dragonfly wings

S. Sudo
Iwaki Meisei University, Japan

H. Hashimoto & F. Ohta
Tohoku University, Sendai, Japan

ABSTRACT: Experimental studies are presented for the function of flapping insects. Dragonflies are emphasized in this report. The self-excited vibration of dragonfly wings was examined in a uniform smooth wind stream, and the flow characteristics around a dragonfly were investigated with a hot-wire anemometer. Dragonflies of two kinds, *Sympetrum frequens* and *Sympetrum infuscatum*, were tested using a wind tunnel. It was found that the amplitude of the wing vibration increased abruptly when the wind velocity exceeded a certain value. In spite of two pairs of wings, close agreement between flapping frequency and predominant spectrum of the velocity fluctuation behind the dragonfly was obtained.

1 INTRODUCTION

Recently, mechanical engineering has been developing in combination with other fields. Especially, biotechnology and bioengineering have been developing by analyzing and utilizing a variety of functions of organism. Therefore, extensive investigations on the motion of a great many animals have been conducted [R.M.Alexander(1992)]. For example, T.Weis-Fogh(1973) has observed the hovering motions of the chalcid wasp *Encarsia formosa*, and proposed a new mechanism of lift generation. M.J.Lighthill(1973) has devised an ingenious explanation for the fluid-dynamic processes whereby certain insects are able to generate large lift coefficients by use of the so-called' clap and fling' mechanism. C.P.Ellington(1984) re-examined the aerodynamics of hovering insect flight, and presented new morphological and kinematic data for a variety of insects. He also offered an aerodynamic interpretation of the wing kinematics and a discussion on the possible roles of different aerodynamic mechanisms. In spite of many investigations, however, there still remains a wide unexplored domain. Research data on aeroelastic characteristics of insect wings and velocity fluctuations produced behind a insect are scanty, and there are many points which must be clarified.

This paper is concerned with function of flapping insects. Some morphological parameters were measured for a variety of insects : its mass, wing area and flapping frequency. The relationship between total wing area and flapping period

was obtained. Dragonflies were emphasized in this report, because they can fly skillfully in the air. The self-excited vibration of dragonfly wings was examined in a uniform smooth wind stream, and the velocity fluctuations around a dragonfly were measured with a hot-wire anemometer. Dragonflies of two kinds, *Sympetrum frequens* and *Sympetrum infuscatum*, were tested using a wind tunnel. Galloping oscillation of the dragonfly using was measured by use of an optical displacement detector system. It was found that the wave form and the power spectrum of the velocity fluctuation produced around the dragonfly differed in the location.

2 EXPERIMENTAL APPARATUS AND PROCEDURES

2.1 *Morphological parameters*

Morphological parameters were measured for a variety of insects. The parameters provide a description of the morphology of a flying animal : its mass, body length, wing length, wing area and wing mass. First, the mass of the insect bodies was measured with a high analytical electronic balance(A and D HA120M). Secondly, the flapping frequency was measured with an optical displacement detector. This displacement detector was equipped with a laser diode which, via a lens system, emits a nonparallel beam of invisible infrared light(GaAs laser, wavelength:850nm,

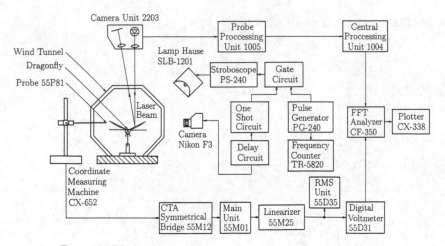

Figure 1. Block diagram of experimental apparatus.

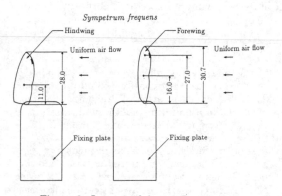

Figure 2. Layout of dragonfly wing.

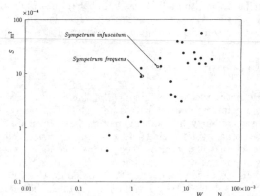

Figure 3. Wing area plotted against total weight for insects.

output power:10mW). When this beam hits a wing surface, a diffused or scattered reflection occurs. The scattered light reflection is then focussed through a lens on a unique semiconductor. The output signal from the photodetector gives the position of the measured wing relative to the gauge probe. The output signal is linearized and can be transmitted in digital form over long distances to the central processing unit. This signal is simultaneously analyzed by a fast Fourier transform analyzer. During measurements, the insect was fastened by a thread.

Detailed measurements of the wing area were made as follows. Immediately after weighing the insects, the all wings were cut from the body. The severed wings were placed between two glass slides and photographed by a camera(Nikon F3). Prints were made at about ×10 magnification. The wing area was measured with a digital planimeter(TAMAYA PLANIX7) on the enlarged photographs.

2.2 *Wind tunnel test*

The experiments on the aeroelastic characteristics of dragonfly wings and the measurements of velocity fluctuation around a dragonfly were conducted in a small low-turbulence wind tunnel [Y.Kohama et al.(1980)], located at the Laboratory for Air-Flow Measurements in the Institute of Fluid Science, Tohoku University. A block diagram of the wind tunnel and measuring devices is shown in figure 1. The test section of the wind tunnel was set in a state of open-jet type, 508mm long. The shape of the contraction exit is a regular octagon of subtense length 293mm. The turbulence intensity at the test section was 0.05–0.15% at the velocity range of 1–14m/s. In the experiments on the aeroelastic characteristics of dragonfly wings, the test wing was mounted parallel to the wind direction as shown in figure 2. The time-dependent amplitudes of the wing vibration was measured with the above optical displacement detector.

In the experiments on the measurements of velocity fluctuation, the live dragonfly was stuck on the wooden needle with the adhesive, and mounted in the test section of the wind tunnel(figure 1). A hot-wire anemometer was used to measure the velocity field. The output signal of the hot-wire anemometer was analyzed by a fast Fourier transform(FFT) analyzer. The results analyzed by FFT were displayed by a plotter. The flapping motion of a dragonfly was recorded by the multi-strobe photograph system.

3 EXPERIMENTAL RESULTS AND DISCUSSION

3.1 *Gross weight and wing area*

Accurate morphological data are a necessary foundation for any aerodynamic study. In figure 3, the gross weight W in insects is plotted against total insect wing area S for a number of species. The experimental data in figure 3 include a wide range of insects, such as, fly, wasp, butterfly, cicada, scarab, longicorn, crane-fly, and so forth. The insects were collected in the fields, Iwaki Japan. Dragonflies of two kinds are specially specified in figure 3, because we mention them in this paper. The gross weight W in insects is equivalent to the lift generated during insect hovering. It can be seen from figure 3 that the more insects have wide wing area, the more insects generate large lift.

3.2 *Flapping motion of dragonfly*

In this paper, dragonflies(Odonata) were emphasized, because they can fly skillfully in the air. Dragonflies are characterized by a large head, a robust thorax, two pairs of wings, and a slender abdomen. Dragonflies of two kinds, *Sympetrum frequens* and *Sympetrum infuscatum*, were used in the experiments. Figure 4 shows the right forewing of a dragonfly *Sympetrum frequens*. It can be seen that the dragonfly wing is composed of much veins and cells. The number of cells of wing is more than 340 for a forewing and approximately 410 for a hindwing. Dragonflies can fly with flapping of the two pairs of wings. Figure 5 shows the output signal from the optical displacement detector. The signal correspond to flapping of the forewing of a dragonfly *Sympetrum infuscatum*. In the figure 5, x_w is one-dimensional coordinates for a dragonfly wing system, and l is the wing length. The wing coordinate system is shown in figure 6. Figure 7 shows the power spectrum for the wing-

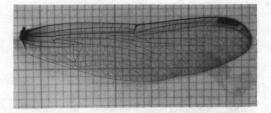
Figure 4. Forewing of dragonfly.

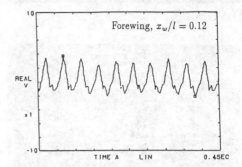
Figure 5. Flapping oscillation of dragonfly.

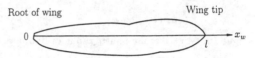
Figure 6. Coordinate system of dragonfly wing.

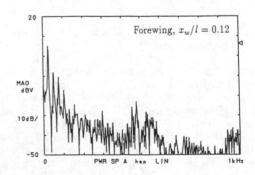
Figure 7. Power spectrum for wingbeat.

beat oscillation in figure 5. The fundamental frequency is in agreement with the dragonfly's flapping frequency. The flapping frequencies of the dragonflies of two kinds were f_d=30–40Hz. Figure 8 shows the multi-strobe photographs of the flapping wings of *Sympetrum frequens*, which was stuck on the wooden needle with the adhesive. It can be seen from figure 8 that the dragonfly twists

(a) Horizontal view

(b) Over view

Figure 8. Multi–strobe photographs of flapping.

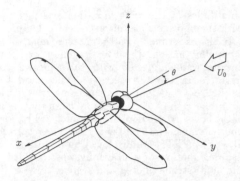

Figure 9. Coordinate system of dragonfly setting.

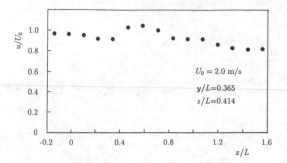

Figure 10. Velocity distribution along x–axis.

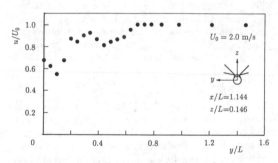

Figure 11. Velocity distribution along y–axis.

its wings and changes the angle of attack time-dependetly during one stroke of flapping.

3.3 Velocity fluctuation and velocity distribution around a dragonfly

Flow characteristics around a dragonfly are investigated in this section. The settled dragonfly *Sympetrum frequens* was mounted in the test section of the wind tunnel. The forewings were fixed at the angle of 60° to the horizontal plane, and the hindwings were fixed at the angle of 0°. The coordinate system to the dragonfly is shown in figure 9. Figure 10 shows the velocity distribution along x axis at z/L=0.414 and y/L=0.365, where L=41.75mm is the body length. In figure 10, u is the local flow velocity, and the velocity of uniform flow is U_0=2.0m/s. Figure 11 shows also the velocity distribution along y axis at x/L=1.144 and z/L=0.146. It can be seen that the velocity distributions are complicated.

In the second stage of the experiments, the live dragonfly was set in the uniform smooth wind stream. Figure 12 shows the change of velocity fluctuations generated by the flapping motion of the dragonfly wings, that is, the dependence on

the position in time series signal of the velocity(at $\theta=\pi/12$). From this figure, it can be seen that the position of maximum amplitude is in the neighbourhood of y/L=0.709. In this experiment, the body length is L=46.52mm, and the wing length is l=39.57mm. The power spectra of the time series signal are shown in figure 13. It can be seen that the peaks at the frequency of f_d are predominant as compared with other peaks. In figure 13, however, the power spectra have the peaks at the frequency of $2f_d$; that is, there is the phase difference between fore and hind pairs of wings.

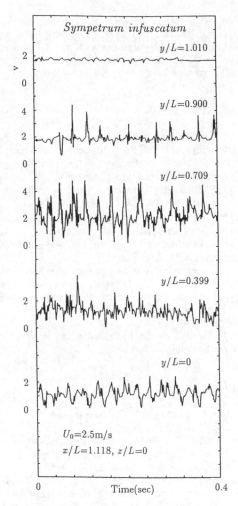

Figure 12. Time series signals of velocities.

3.4 *Aeroelastic characteristics of dragonfly wing*

As was stated previously, the experiments on the aeroelastic characteristics of dragonfly wings were conducted in the wind tunnel (figure 1 and figure 2). When the wind speed U_0 increases slowly from zero, the onset of aeroelastic galloping can be observed at a certain U_0. Figure 14 shows an example of the output signal from the optical displacement detector. These data in figure 14 correspond to the time-varying signal of the forewing oscillation of dragonfly *Sympetrum frequens*. In this experiments, the displacement 1mm was equivalent to the output voltage, 0.309V. In figure 14 the signal shows the flow-induced self-excited oscillation of the dragonfly wing in $U_0 \geq 3.86$m/s. Figure 15 displays the Reynolds number Re vs. the dimensionless maximum total amplitude x_0/l for the

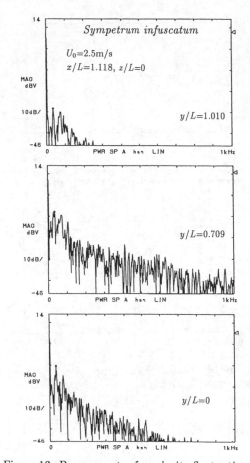

Figure 13. Power spectra for velocity fluctuations.

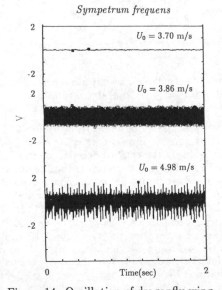

Figure 14. Oscillation of dragonfly wing.

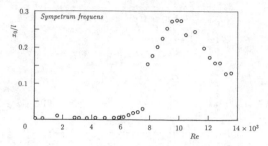

Figure 15. Oscillation amplitude versus Reynolds number for forewing of *Sympetrum frequens*.

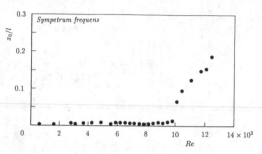

Figure 16. Oscillation amplitude versus Reynolds number for hindwing of *Sympetrum frequens*.

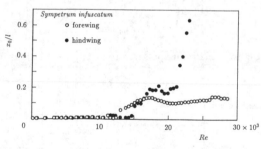

Figure 17. Oscillation amplitude versus Reynolds number for wings of *Sympetrum infuscatum*.

forewing oscillation of *Sympetrum frequens*. The Reynolds number Re was defined as $Re = U_0 l/\nu$, where ν is the kinematic viscosity of the fluid. From figure 15 it can be seen that the flow-induced self-excited oscillation occurs in neighbourhood of $Re \doteqdot 8 \times 10^3$. Figure 16 shows the Reynolds number vs. the dimensionless amplitude for the hindwing oscillation of *Sympetrum frequens*. It is obvious on comparing the two figures 15 and 16 that the hindwing has the higher Reynolds number for the onset of the flow-induced self-excited oscillation. Figure 17 shows the aeroelastic characteristics of the wings of dragonfly *Sympetrum infuscatum*. The tendency is similar in dragon-flies of two kinds *Sympetrum frequens* and *Sympetrum infuscatum*. The fact in figure 15–17 suggest that the forewing acts as a part of sensor of flow speed measurement, and the hindwing acts as the greater part of the lift generation.

4 CONCLUSIONS

The flow-induced self-excited oscillation of dragonfly wings and the flow characteristics around a dragonfly were investigated with the wind tunnel. The results obtained are summarized as follows :

(1) When the dragonfly wing is exposed in the uniform smooth wind, the flow-induced self-excited oscillation of dragonfly wing occurs at the valve above the critical Reynolds number. The critical Reynolds number for hindwing is higher compared with forewing.

(2) The predominant frequency of velocity fluctuations generated by the flapping wings coincides with the flapping frequency of dragonfly in spite of the phase difference flapping between fore and hind pairs of wings.

(3) Heavy insects have wider wing area and longer flapping period.

REFERENCES

Alexander,R.M. 1992. *Exploring biomechanics.* New York: W.H.Freeman and Company.

Ellington,C.P. 1984. The aerodynamics of hovering insect flight; I the quasi-steady analysis. *Phil.Trans.R.Soc. Lond.* B305:1–15.

Kohama,Y., R.Kobayashi & H.Ito 1980. Performance of the small low-turbulence wind tunnel. *Mem.Inst.High Speed Mech., Tohoku Univ.* 48: 119–142.

Lighthill,M.J. 1973. On the Weis-Fogh mechanism of lift generation. *J.Fluid Mech.* 60:1–17.

Weis–Fogh,T. 1973. Quick estimates of flight fitness in hovering animals, including novel mechamisms for lift production. *J.Exp.Biol.* 59:169 –230.

Recent Advances in Experimental Mechanics, Silva Gomes et al. (eds) © 1994 Balkema, Rotterdam, ISBN 90 5410 395 7

Author index